# Springer Series in Surface Sciences

Volume 52

D1743629

This series covers the whole spectrum of surface sciences, including structure and dynamics of clean and adsorbate-covered surfaces, thin films, basic surface effects, analytical methods and also the physics and chemistry of interfaces. Written by leading researchers in the field, the books are intended primarily for researchers in academia and industry and for graduate students.

More information about this series at http://www.springer.com/series/409

Karsten Hinrichs · Klaus-Jochen Eichhorn
Editors

# Ellipsometry of Functional Organic Surfaces and Films

Second Edition

 Springer

*Editors*
Karsten Hinrichs (iD)
Leibniz-Institut für Analytische
  Wissenschaften – ISAS – e.V.
Berlin
Germany

Klaus-Jochen Eichhorn (iD)
Abteilung Analytik
Leibniz-Institut für Polymerforschung
  Dresden e.V.
Dresden, Sachsen
Germany

ISSN 0931-5195          ISSN 2198-4743   (electronic)
Springer Series in Surface Sciences
ISBN 978-3-030-09351-8          ISBN 978-3-319-75895-4   (eBook)
https://doi.org/10.1007/978-3-319-75895-4

This Springer imprint is published by the registered company Springer International Publishing AG
part of Springer Nature
The registered company address is: Gewerbestrasse 11, 6330 Cham, Switzerland

# Foreword

Given the increasingly important role of spectroscopic ellipsometry (SE) as a real-time, non-perturbing, monitoring, and characterization tool in numerous technological and biomedical applications, the editors, authors, and Springer are commended for publishing the 2nd edition of this monograph, which is dedicated to significant applications of SE to functional organic surfaces and thin films. Retained topics from the first edition cover the adsorption of biomolecules at liquid–solid interfaces, smart polymer surfaces and thin films for sensor applications, characterization of nanoparticles and nanostructured surfaces and thin films, thin-film organic semiconductors for photovoltaics and light emitters, and recent developments of SE instrumentation and related techniques over an extended optical bandwidth.

New topics that appear in the 2nd edition include SE studies of bonding of biomolecules on self-assembled monolayers, structure and interactions of hydrated polymer thin films, ellipsometry of solvent-induced swelling at soft polymer interfaces, optical properties of anisotropic thin films of organic dye aggregates, relationship between morphology and optical properties of conjugated polymers, and polarons in conducting polymers.

This updated monograph is a welcome contribution to the expanding ellipsometry literature.

New Orleans, USA
R. M. A. Azzam
University of New Orleans

# Acknowledgements

The editors would like to acknowledge all authors for their contributions and many valuable discussions, and Arnulf Röseler for his lasting support in helping for understanding of the ellipsometric world. In addition, this book would not have been possible without the motivation and support of our families; KJE thanks his wife Sigrun and his children Susanne and Jörg, and KH thanks his wife Claudia and his daughter Daria. For permanent technical support in the preparation of the book, we are indebted to I. Engler and E. Bittrich.

Berlin, Germany                                                    Karsten Hinrichs
Dresden, Germany                                           Klaus-Jochen Eichhorn

# Contents

## Part II  Smart Polymer Surfaces and Films

# Contributors

**Maria Isabel Alonso** Institut de Ciència de Materials de Barcelona (ICMAB-CSIC), Bellaterra, Spain

**Hans Arwin** Department of Physics, Chemistry and Biology, Linköping University, Linköping, Sweden

**Dennis Aulich** Leibniz-Institut für Analytische Wissenschaften – ISAS – e.V., Berlin, Germany

**Damien Aureau** Inst. Lavoisier, CNRS, UMR 8180, University of Versailles St Quentin Yvelines, Versailles, France

**Francesco Bisio** Istituto CNR-SPIN, Genova, Italy

**Eva Bittrich** Leibniz-Institut für Polymerforschung Dresden e.V., Abteilung Analytik, Dresden, Germany

**Lars Bittrich** Leibniz-Institut für Polymerforschung Dresden e.V., Abteilung Verbundwerkstoffe, Dresden, Germany

**Mariano Campoy-Quiles** Institut de Ciència de Materials de Barcelona (ICMAB-CSIC), Bellaterra, Spain

**Maurizio Canepa** OPTMATLAB, Department of Physics, University of Genova, Genova, Italy

**Loredana Casalis** Elettra Sincrotrone Trieste S.C.p.A., Basovizza, Trieste, Italy

**Ornella Cavalleri** OPTMATLAB, Department of Physics, University of Genova, Genova, Italy

**Christoph Cobet** Center of Surface and Nanoanalytics (ZONA), Johannes Kepler University Linz, Linz, Austria

**Klaus-Jochen Eichhorn** Leibniz-Institut für Polymerforschung Dresden e.V., Abteilung Analytik, Dresden, Germany

**Michael Erber** Leibniz-Institut für Polymerforschung Dresden e.V., Abteilung Analytik, Dresden, Germany

**Norbert Esser** Leibniz-Institut für Analytische Wissenschaften – ISAS – e.V., Berlin, Germany

**Dominik Farka** Linz Institute of Organic Solar Cells (LIOS), Johannes Kepler University Linz, Linz, Austria

**Andreas Furchner** Leibniz-Institut für Analytische Wissenschaften – ISAS – e.V., Berlin, Germany

**Michael Gensch** Institut für Strahlenphysik/Institut für Ionenstrahlphysik und Materialforschung, Helmholtz-Zentrum Dresden-Rossendorf, Dresden, Germany

**Jacek Gasiorowski** Center of Surface and Nanoanalytics (ZONA), Johannes Kepler University Linz, Linz, Austria; Linz Institute of Organic Solar Cells (LIOS), Johannes Kepler University Linz, Linz, Austria; EV Group E.Thallner GmbH, St. Florian am Inn, Austria

**Ovidiu D. Gordan** Semiconductor Physics, Technische Universität Chemnitz, Chemnitz, Germany

**Karsten Hinrichs** Leibniz-Institut für Analytische Wissenschaften – ISAS – e.V., Berlin, Germany

**Tino Hofmann** Department of Electrical Engineering, University of Nebraska–Lincoln, Lincoln, NE, USA

**Kenneth Järrendahl** Department of Physics, Chemistry and Biology, Linköping University, Linköping, Sweden

**Meike Koenig** Karlsruhe Institute of Technology (KIT), Institute of Functional Interfaces, Advanced Polymers and Biomaterials, Eggenstein-Leopoldshafen, Germany

**Argiris Laskarakis** Lab for Thin Films-Nanosystems & Nanometrology (LTFN), Department of Physics, Aristotle University of Thessaloniki, Thessaloniki, Greece

**Stergios Logothetidis** Lab for Thin Films-Nanosystems & Nanometrology (LTFN), Department of Physics, Aristotle University of Thessaloniki, Thessaloniki, Greece

**Michele Magnozzi** OPTMATLAB, Department of Physics, University of Genova, Genova, Italy

**Nico B. Eisele** Biosurfaces Unit, CIC BiomaGUNE, Donostia, San Sebastian, Spain; Department of Cellular Logistics, Max Planck Institute of Biophysical Chemistry, Göttingen, Germany

**Thomas W. H. Oates** Leibniz-Institut für Analytische Wissenschaften – ISAS – e.V., Berlin, Germany

**Wojciech Ogieglo** Advanced Membranes and Porous Materials Center, 4700 King Abdullah University of Science and Technology (KAUST), Thuwal, Kingdom of Saudi Arabia

**Pietro Parisse** Elettra Sincrotrone Trieste S.C.p.A., Basovizza, Trieste, Italy

**Simona D. Pop** Leibniz-Institut für Analytische Wissenschaften – ISAS – e.V., Berlin, Germany

**Ralf P. Richter** Biosurfaces Unit, CIC BiomaGUNE, Donostia, San Sebastian, Spain; Department of Molecular Chemistry, J. Fourier University, Grenoble, France; Max Planck Institute for Intelligent Systems, Stuttgart, Germany; School of Biomedical Sciences, University of Leeds, Leeds, UK

**Jörg Rappich** Inst. für Si-Photovoltaik, Helmholtz-Zentrum Berlin für Materialien und Energie GmbH, Berlin, Germany

**Charles Rice** Department of Electrical Engineering, University of Nebraska–Lincoln, Lincoln, NE, USA

**Keith B. Rodenhausen** Department of Chemical and Biomolecular Engineering, University of Nebraska–Lincoln, Lincoln, NE, USA

**Katy Roodenko** Department of Materials Science and Engineering, Laboratory for Surface and Nanostructure Modification, University of Texas at Dallas, Richardson, TX, USA

**Daniel Schmidt** Department of Electrical Engineering, University of Nebraska–Lincoln, Lincoln, NE, USA

**Eva Schubert** Department of Electrical Engineering, University of Nebraska–Lincoln, Lincoln, NE, USA

**Mathias Schubert** Department of Electrical Engineering and Center for Nanohybrid Functional Materials, University of Nebraska-Lincoln, Lincoln, NE, USA

**Ilaria Solano** OPTMATLAB, Department of Physics, University of Genova, Genova, Italy

**Philipp Stadler** Linz Institute of Organic Solar Cells (LIOS), Johannes Kepler University Linz, Linz, Austria

**Guoguang Sun** Leibniz-Institut für Analytische Wissenschaften – ISAS – e.V., Berlin, Germany

**Peter Thissen** Institute of Functional Interfaces (IFG), Karlsruhe Institute of Technology (KIT), Eggenstein-Leopoldshafen, Germany

**Martin Tress** Institut für Experimentelle Physik II, Universität Leipzig, Leipzig, Germany

**Petra Uhlmann** Leibniz-Institut für Polymerforschung Dresden e.V., Abteilung Nanostrukturierte Materialien, Dresden, Germany

**Sylvia Wenmackers** Faculty of Philosophy, University of Groningen, Groningen, The Netherlands

**Florent Yang** Institute for Heterogeneous Material Systems, Helmholtz-Zentrum Berlin für Materialien und Energie GmbH, Berlin, Germany

**Dietrich R. T. Zahn** Semiconductor Physics, Technische Universität Chemnitz, Chemnitz, Germany

**Xin Zhang** Institut für Silizium Photovoltaik, Helmholtz-Zentrum Berlin für Materialien und Energie GmbH, Berlin, Germany

# Introduction to Book Contents

For more than a century, ellipsometry has been utilized by physicists as a nondestructive, absolute, and thin-film-sensitive optical method to determine index of refraction and absorption of solid materials (metals, semiconductors, and oxides). Over the last few decades, user-friendly ready-to-use ellipsometers with application-related accessories have been developed spanning the far-infrared to ultraviolet spectral range. Continuing advances in experimental ellipsometric techniques and theory have enabled researchers to tackle challenges in modern material science such as characterization of superconductors and metamaterials, rough and nanostructured surfaces, complex hybrid films, and plasmonic and magneto-optic samples. [1–3] Nevertheless, if one were to ask a chemist or biologist working with organic surfaces and thin films "What is ellipsometry?", the answer will often be, "It is a nice method to quickly and easily measure film thickness". However, although the technique is considered highly accurate, beyond thickness determination it is often viewed as somewhat exotic and difficult to understand.

This book intends to bridge this gap and aims to overcome certain prejudices ("ellipsometry is a black box…"). It presents ellipsometry to scientists as a versatile method for chemical, biological, and material science applications dealing with small and large organic molecules on surfaces. Prime examples are the study of synthetic polymers with different architectures and functionalities, as well as biomolecules. The analysis of functional surfaces often requires new methods to apply ellipsometry for quantitative, nondestructive, label-free, and contact-less characterization. Most of the authors of this book, as well as we the editors, have been active in the application and development of ellipsometry for many years and were interested in applying ellipsometry to studies of functional organic films. The close cooperation in an interdisciplinary field between chemistry, physics, material sciences, and biotechnology was necessary to tackle such analysis. Nevertheless, it is important to emphasize that there is a broad still developing field of analysis and ellipsometric methods for better understanding of complex and structured samples and materials. In particular, when the probing wavelengths are in the range of structure dimensions, the measured spectra cannot be sufficiently understood within homogeneous layer optical models. For such cases, in particular, promising attempts have been made in combination with numerical simulations (as e.g., rigid coupled wave analysis and finite-element calculations) but also scatterometric as well as Mueller Matrix ellipsometry measurements. Beside these approaches in general, the structural and anisotropic properties over several length scales down to the μm and nm range are of high relevance for thin films and surfaces in many sensoric, opto-electronic, and biomedical applications. With respect to laterally higher resolved measurements, recent methodical developments for ellipsometric- and polarization-dependent measurements in the infrared spectral range are addressing to involve tunable lasers, microfluidic, and near-field measurement concepts.

Of course, a zoo of modern microscopic, spectroscopic, and also physical–chemical methods are known and widely applied for the characterization of organic surfaces and thin films. They play an important role in the "daily life" of the material scientist, for example, XPS, SIMS, AFM, SEM, X-ray and neutron reflection, photoluminescence, fluorescence, FTIR- and Raman spectroscopy, mass spectrometry, inverse GC, contact angle and zetapotential measurements, and many more. Cooperatively applied ellipsometry can determine thicknesses and complementing anisotropic optical and structural properties in a noncontact and nondestructive manner in various environments. The different chapters of this book demonstrate the possibilities, advantages, and problems of application of (mainly spectroscopic) ellipsometry. In comparison to many other methods, ellipsometry as an optical technique is relatively easy to do under normal lab conditions. Using special cells, temperature-dependent in situ experiments in vacuum, gaseous, and liquid ambient are possible. A sometimes more challenging task is the evaluation of the experimental (optical) data to obtain the desired physical and chemical information on the films and surfaces.

In 24 chapters grouped in seven parts, worldwide recognized experts from universities and research institutes give examples and actual results in studies applying ellipsometry to different aspects of functional organic surfaces and thin films.

The first edition of this book having 18 chapters was published as print book and eBook in 2014. It was very successful in the community, e.g., it was one of the top 25% most downloaded eBooks in the relevant SpringerLink Collection. So, we decided in agreement with Springer to publish a revised and extended second edition.

Therefore, we present in the second edition beside some chapters in their original form a lot of updated, improved chapters as well as complete new ones:

As theoretical introduction, C. Cobet gives an overview about the ellipsometric method, including history, basics and principles, experimental techniques, and optical models for data evaluation.

The experimental examples begin with "Biomolecules at surfaces". H. Arwin shows why ellipsometry is an excellent tool to study many aspects of protein adsorption at solid surfaces. DNA structures on silicon and diamond are the focus of the special chapter by S. D. Pop and colleagues.

M. Canepa et al. demonstrated recently that spectroscopic ellipsometry in combination with AFM nanolithography is able to characterize ultrathin self-assembled monolayers and their interaction with biomolecules on surfaces. Analyzing the difference spectra, they describe an in situ hybridization process.

"Smart polymer surfaces and films" are actual materials of interest for applications as organic sensors, actuators, or bioactive/bioinert surfaces. The glass transition in thin polymer films remains a controversial topic; however, M. Erber et al. demonstrate that it may be studied very comfortably by spectroscopic ellipsometry (SE). This chapter is complemented now with a new method for a more precise ellipsometric determination of $T_g$ developed by E. and L. Bittrich.

In situ ellipsometry is necessary to study polymer brushes, hydrogels, and polyelectrolyte multilayers—typical stimuli-responsive systems. E. Bittrich et al. give an overview of recent results of smart polymers and the protein adsorption at these soft organic surfaces. In the actual chapter also, enzymes are included. During the last years, in situ Infrared spectroscopic ellipsometry (IR-SE) comes more and more into play to characterize molecular and supermolecular changes at solid–liquid or solid–gaseous interfaces. A. Furchner shows impressively that in situ IR-SE can provide a lot of information on the structure and interactions in hydrated polymer films with focus on new data on hydrogen-bonding phenomena. The second new contribution in the polymer topic is given by W. Ogieglo. He reviews the important recent developments in the application of ellipsometry in industrial membrane-related studies. Sorption, transport, and penetrant-induced phenomena in membrane-relevant thin films exposed to organic solvents or high-pressure gases are discussed.

In the first chapter of the Part "Nanostructured Surfaces and Organic/inorganic Hybrids", T. Oates demonstrates how systems consisting of nanoparticles and polymers or self-assembled monolayers can be characterized by appropriate ellipsometric methods. In the second part, complicated nanostructured (sculptured) thin films with high anisotropy are presented by K. B. Rodenhausen et al. These highly ordered three-dimensional structures and, moreover, organic attachment onto such surfaces may be characterized by advanced ellipsometric techniques. Similar techniques are necessary to describe polarizing natural nanostructures (e.g., surfaces of beetles) as shown in the last updated part by K. Järrendahl and H. Arwin.

"Thin films of organic semiconductors" play an outstanding role in organic electronics and the development of OPV, OLED, and OTFT. Optical properties from UV to IR range, morphology, and molecular orientation may be excellently characterized by spectroscopic ellipsometry. Large molecules as important polymers, blends, and composites are the focus of the report of S. Logothetidis from Thessaloniki, whereas O. Gordan and D. R. T. Zahn describe ellipsometric measurements on films of small organic molecules. In a new chapter by K. Roodenko and P. Thissen, details on thin films formed by organic dye aggregates are given. They present data on the optical properties obtained by Vis ellipsometry and polarized infrared spectroscopy. New as well is the chapter of M. I. Alonso and M. Campoy-Quiles on conjugated polymers. They discuss the relationship between the morphology and optical properties of these polymers and their blends with fullerenes. The last contribution in this topic given by C. Cobet et al. is concerned with polarons in conducting polymers. The authors use spectroscopic ellipsometry and ATR-IR to study the formation of polarons in chemically and electrochemically doped polymers.

"Developments in Ellipsometric Real-time/in situ Monitoring Techniques" are presented in Part V. A main point here is, again, the solid–liquid interface. It is possible to study the behavior of organic surfaces and thin films in their natural and also (artificial) liquid environment. R. P. Richter et al. show the power of coupled complementary methods, namely QCM-D with spectroscopic ellipsometry (SE). Total Internal Reflection Ellipsometry (TIRE) and SPR-enhanced SE are introduced by H. Arwin as emerging techniques with very high sensitivity and precision for studying adsorption processes. The combination of SE in the mid-infrared spectral range and electrochemistry provides fascinating insights into the chemistry of thin organic films as described by J. Rappich et al. Current results on IR-SE investigation of the maleimidobenzene modified Si surfaces are included now. And last but not least, it is possible to use SE for the inline quality control of organic thin-film fabrication on rigid and flexible substrates. In their part of Chap. 19, S. Logothetidis and A. Laskarakis give an overview on the state-of-the-art in this field.

Infrared ellipsometry (IR-SE) but also other surface-sensitive FTIR spectroscopic methods for the characterization of thin organic films are reviewed by K. Roodenko et al. in Chap. 21. Their focus is on the evaluation of molecular structure and orientation.

Using the brilliant infrared light from a synchrotron, source makes it possible to perform far-field micro-ellipsometric studies with good lateral resolution. In Chap. 22, M. Gensch presents the technical background and interesting applications and outlook.

This book is of great interest to anyone who would like to use spectroscopic ellipsometry to study thin organic films of polymers or small molecules to have an idea on their material optical constants. Thus, the last chapter of the book provides support in this direction: D. Aulich and A. Furchner present a collection of optical constants of organic thin-film materials. Such optical constants are an excellent starting point in the interpretation and optical modeling of spectra of related materials; however, they may vary in details for the specific case.

Karsten Hinrichs
Klaus-Jochen Eichhorn

# References

1. H.G. Tompkins, E.A. Irene (eds.), Handbook of Ellipsometry (Springer, 2005)
2. H. Fujiwara (ed.), Spectroscopic Ellipsometry: Principles and Applications (Wiley, New York, 2007)
3. M. Losurdo, K. Hingerl (eds.), Ellipsometry at the Nanoscale (Springer, Berlin and Heidelberg, 2013)

# Chapter 1
# Ellipsometry: A Survey of Concept

**Christoph Cobet**

**Abstract** Already the first attempts by Paul Drude in the late 19th century demonstrate the abilities of optical polarimetric methods to determine dielectric properties of thin layers. Meanwhile ellipsometry is a well-established method for thin film analysis. It provides material parameters like $n$ and $k$ even for arbitrary anisotropic layers, film thicknesses in the range down to a few Ångström, and ellipsometry is used to analyze the shape of nm-scale surface structures. But, the determination of such manifold information by means of light polarization changing upon reflection at a sample surface requires appropriate optical models. This introductory chapter will provide a general overview and explanation of theoretical and experimental concepts and their limitations. It will introduce the very basic data evaluation steps in a comprehensive manner and will highlight the principal requirements for the characterization of functional organic surfaces and films.

## 1.1 Classification

Ellipsometry and other types of polarimetry are well known optical methods which are used since more then 100 years for analytic purposes. Here, the term ellipsometry is certainly linked to the polarization sensitive optical investigation of planar solid state structures (metals, semiconductors) with polarized light. Optical methods in general benefit from the fact that they are usually non destructive and applicable in various environments. The object under investigation can be stored in vacuum, gas, liquid, and even in solid ambiances as long as the surrounding material is transparent within the spectral range of interest. By taking advantage of the polarizability of light, it is possible to measure for example thin film properties like the refractive index and the thickness with very high accuracy and without the need of a reference. Because of these abilities ellipsometry is meanwhile a very popular method used

C. Cobet (✉)
Center for Surface- and Nanoanalytics, Johannes Kepler
Universität Linz, Altenbergerstrasse 69, 4040 Linz, Austria
e-mail: christoph.cobet@jku.at

© Springer International Publishing AG, part of Springer Nature 2018
K. Hinrichs and K.-J. Eichhorn (eds.), *Ellipsometry of Functional
Organic Surfaces and Films*, Springer Series in Surface Sciences 52,
https://doi.org/10.1007/978-3-319-75895-4_1

in many different application fields. Accordingly, a couple of books, book chapters, and review articles provide already comprehensive information about the method ellipsometry itself and the physical/mathematical background especially for thin film applications [1–7]. Therefore, it is not the intention to repeat here once again all technical details. We would rather like to provide in this chapter an overview about relevant aspects which are needed to empathize the analytical possibilities concerning functional organic surfaces and films. Furthermore, we will address limitations of the method and the underlying physical models.

The common concept behind the methods ellipsometry and polarimetry rests upon the analysis of a polarization change of light which is interacting with the object of interest. Here, we follow one of the definitions given by Azzam [8] in 1976 which was discussed in connection to the 3rd International Conference on Ellipsometry. Accordingly "An ellipsometer (polarimeter) is any instrument in which a TE-EMW—transverse electric electromagnetic wave—generated by a suitable source is polarized in a known state, interacts with a sample under investigation, and the ellipse (the state of polarization) of the radiation leaving the sample is analyzed". This concept implies that both the polarization state of the light before and after interaction with the sample can be modified or determined (Fig. 1.1). Investigations for example of atmospheric and extraterrestrial phenomena where the polarization properties of the light source itself are analyzed or where the light polarization before interacting with the object of interest is not accessible are not considered in this definition [9]. Furthermore, only linear optical effects are considered and phenomena, where the light frequency is changed like in Raman scattering, second harmonic generation and sum frequency processes, are excluded.

With the definition above, ellipsometry can be used to analyze reflected, transmitted, scattered, and diffracted light (Fig. 1.2). Ellipsometric transmission measurements are so far preferentially used to analyze birefringence, optical activity, circular birefringence, and in case of a small absorption also circular dichroism. In this book the discussion is focused on the analysis of organic surfaces and stratified films in reflection type measurements. Thus, the sample is illuminated under an oblique angle of incidence and the specular reflection is analyzed (Fig. 1.2a). Accordingly, all presented theoretical models assume that the analysis takes place in the optical far field where the approximation of plane waves is reasonable i.e. the distance between analyzer/detector and the sample has to be much larger than the wavelength and possible lateral inhomogeneities of the sample.

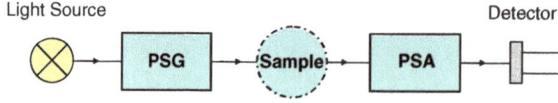

**Fig. 1.1** Principle concept of ellipsometric and polarimetric techniques. The Polarization State Generator (PSG) and Polarization State Analyzer (PSA) may consist of a polarizer or a combination of a polarizer and retardation component

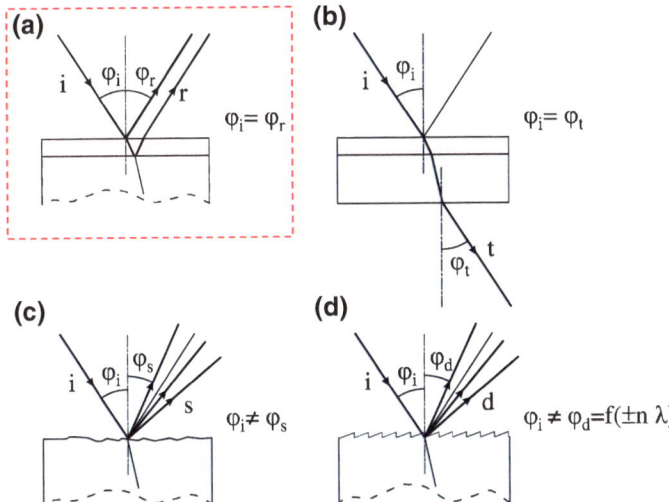

**Fig. 1.2** Fundamental interaction of incident light on different samples under an angle $\varphi_i$: **a** reflection, **b** transmission, **c** scattering, and **d** diffraction. All introduced ellipsometric problems are reflection measurements (*red dashed box*)

The applied optical models assume furthermore monochromatic or quasi-monochromatic electromagnetic waves which are reflected at the sample by retaining total polarization of the incident light. The electromagnetic wave before and after reflection is completely defined by an unique elliptical polarization state which gives the method the name "ellipsometry".

The term "polarimetry", in contrast, is usually used in a more general context including the analysis of non-specular reflected or scattered light from inhomogeneous samples or surfaces (Fig. 1.2b–d). In this context polarimetry is often used as a contact free method in order to determine morphology aspects [10]. Strongly related to scattering processes is a partial depolarization of the light. As we will discuss later, this requires extended optical models. A strict delimitation between ellipsometry and polarimetry, however, is neither possible nor helpful. In reality both terms are used with much overlap and a number of specific approaches are used by related proper names (Sect. 1.5.6).

Bearing in mind that the fundamental electromagnetic theory remains the same for all different regions of the spectrum, it is also not surprising that methods like polarimetry and ellipsometry are applied in much the same way from the region of radio frequencies over the infrared, visible and ultra violet to the X-/$\gamma$-ray spectral range. But due to experimental peculiarities, the knowledge transfer between the communities is unfortunately low. This book will bridge in parts this spacings by including all sections of the "optical" spectral range which includes here the infrared, the visible, and the ultraviolet wavelength/frequency range. Nevertheless, it could be particularly beneficial to consider also applications in the radio, radar, and microwave region. Related to the longer wavelength, the determination of structural and morphological properties in this range is historically stronger in the focus. Respective theoretical models for the data processing are therefore rather sophisticated and can be adapted for the optical spectral range [9, 11].

## 1.2 Historical Context

In a historical review the first observations of the polarization properties of light is directly linked to the discovery that light changes its polarization state after reflection on, for example, glass windows of buildings and is associated with names like Etienne-Louis Malus, David Brewster, and Augustin-Jean Fresnel. In the 1800s the polarization change of reflected light was used in a couple of works to study the optical properties of metals. A first description of elliptically polarized light attributes to Jamin [12–14]. He has observed this polarization after reflection of linearly polarized light on metal surfaces which were decorated with transparent overlayers. It turned out, that the elliptical polarization is the most arbitrary polarization state whose constituting parameters have to be determined when planar homogeneous layers are investigated.[1] For this reason, the name "ellipsometry" was established by Rothen [15] almost 100 years later in 1945 for such kind of measurements. However, a first comprehensive description of the method as a technique to study the optical properties of thin films was given already by Paul Drude in the late 19th century. He was measuring the optical properties of metals under consideration of unintentional and intentional overlayers. Furthermore, he could model the measured polarization changes by an extension/modification of Fresnel's equation, which are originally made for the reflection of light on a single planar interfaces, to the problem of two stacked interfaces [16–18]. With this approach it was possible to determine bulk and film dielectric properties as well as film thicknesses.

70 years later these analytical potentials attract a lot of attention in connection with the invention and development of semiconductor electronics. The investigation of $SiO_2$ films on Si is probably one of the best examples for the abilities of the method until now. On the other hand, the progress in semiconductor electronics considerably accelerates the development of computers and the automation possibilities. With the help of microprocessors it was now possible to build automatic spectroscopic ellipsometers (SE) which made the method much more attractive for a wider community. Large steps forward in development and improvement are associated to the work of Aspnes [19]. This progress also lead to more advanced applications and setups with an appropriate spectral range and a reasonable resolution. In the following different angles of incidence or different polarization states of the incident light were used in order to extract more accurate information from rather complex samples. Meanwhile multi-layer structures, all kinds of optical anisotropy, magneto-optical effects, as well as 3D inhomogeneous structures are accessible. But the final breakthrough for the

---

[1]Possible contributions of unpolarized light are ignored here.

method is definitely linked to the availability of easy-to-use analysis software packages. Hence, it became possible to extract useful information even for complicated sample structures with moderate efforts. In this context it is also apparent why the optical characterization of organic films, which are often anisotropic and inhomogeneous, was mostly restricted to reflection and transmission measurements for a long time. The wide spread developments in the recent years are documented for example in the proceedings of the conference series "International Conferences on Spectroscopic Ellipsometry" [20–24]. Concerning the newer developments we would also refer to a number of publications which provide further details [8, 25–29].

## 1.3 Measurement Principles

### 1.3.1 Data Recording and Evaluation Steps

As it was mentioned, ellipsometry in principal determines polarization changes upon interaction with a sample. Subsequently it is possible to extract, for instance, layer thicknesses or dielectric properties in a "reference free" manner. Thus, two major data evaluation steps are needed in ellipsometry in order to receive information about the sample (Fig. 1.3). In parts they depend on each other. Nevertheless it is helpful to divide the problem in such basic steps and it seems worthwhile to discuss these steps briefly to obtain a general understanding of the method.

All kinds of ellipsometers are primary measuring intensities with light sensitive detectors. These intensities have to be related in the first evaluation step to the polarization change induced by the sample (left hand part of Fig. 1.1). Therefore, each ellipsometer is recording the intensity with different incident light polarizations or analyzer orientations in order to obtain relative intensities. With an appropriate set

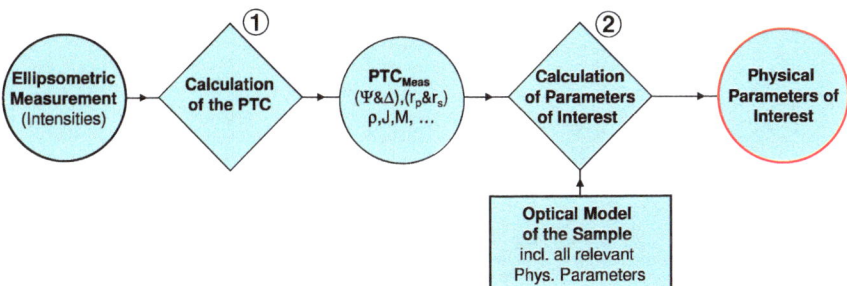

**Fig. 1.3** Basic data evaluation steps in an ellipsometric measurement. The *left hand side* of the flow chart depicts the determination of the polarizing properties of the sample which are represented e.g. by the ellipsometric angles $\Psi$ and $\Delta$ or in more general by polarization transformation coefficients (PTC). In the *right hand part* these polarization parameters are translated in physical sample parameters with the help of a qualitative sample model

of such arrangements one can deduce out of it, how the sample under investigation changes an arbitrary polarization of the incident light. It is evident that a respective theoretical formalism is needed which translates the detector signals to the polarization properties of the sample. As it will be discussed later, the probability of the sample to change the polarization of monochromatic light can be described for an isotropic sample, if no depolarization takes place, by two parameters. Quite often the so-called ellipsometric angles $\Psi$ and $\Delta$ are used. In case of anisotropic structures up to 6 parameters are needed. According to reference [8] these parameters are denoted here as "polarization transformation coefficients" (PTC). If depolarization effects are apparent, the number of parameters increases even further. For the moment it is important to note that the PTC's depend on the angle of incidence, the wavelength, and probably on the sample orientation, too.

A very common and simple ellipsometer is the so-called rotating analyzer ellipsometer. It's principle arrangement and the signal recorded at the detector by rotating the analyzer is shown in Fig. 1.4 (q.v. Sect. 1.5). With at least three different analyzer positions it is possible to assign the sinusoidal dependence of the intensity as a function of the rotation angle $\alpha$ by means of the two $\sin(2\alpha)$ and $\cos(2\alpha)$ Fourier-coefficients $s_2$ and $c_2$, respectively. With the later briefly explained mathematical formalism it is possible to calculate $\Psi$ and $\Delta$ of an isotropic non-depolarizing sample according to

$$\tan\Psi = \sqrt{\frac{1+c_2}{1-c_2}}, \qquad \cos\Delta = \frac{s_2}{\sqrt{1-c_2^2}}. \tag{1.1}$$

In a second step the information about the polarization change by the sample (the PTC's or $\Psi$ and $\Delta$) should be translated in to intrinsic sample properties which are not anymore related to a certain measurement configuration (right hand part in Fig. 1.3). Such intrinsic sample properties are, for instance, layer dielectric functions, layer thicknesses, or volume fractions in inhomogeneous media.

In order to calculate intrinsic properties from the PTC's again, an adequate theoretical description is required. This means that the reflection process depicted in

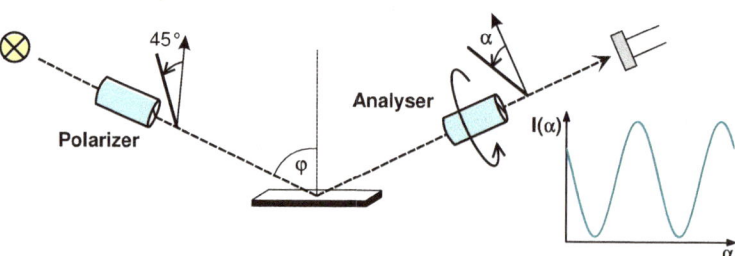

**Fig. 1.4** Principles of a rotating analyzer ellipsometer. Here, the polarizer is fixed with the transmission axis tilted by 45° with respect to the plane of incidence. The intensity signal recorded at the detector for a certain wavelength by rotating the analyzer is of a sinusoidal form with a periodicity of $2\alpha$

Fig. 1.2a has to be specified in more detail. In the very simple and ideal case of a planar abrupt surface of a infinitely thick isotropic sample this connection is given by the well known Fresnel equation. The hereby defined reflection coefficients $r_p$ and $r_s$ for light polarizations parallel and perpendicular to the plane of incidence determine the ellipsometric angles $\Psi$ and $\Delta$:

$$\frac{r_p}{r_s} = \tan \Psi e^{i\Delta}. \tag{1.2}$$

Light reflection from the backside of the sample is neglected in this model. For stratified anisotropic media optical layer models are used in order to calculate the respective PTC's for a given sample structure. In many cases, it is nevertheless possible to define generalized Fresnel equations. The sample parameters are usually obtained by a fit routine comparing the measured PTC's with respective PTC's calculated from the applied optical model.

At this point it is already obvious that the number of parameters which can be deduced is limited. By measuring $\Psi$ and $\Delta$ in a single wavelength measurement at one angle of incidence and sample orientation, ellipsometry can provide two intrinsic sample parameters. Therefore, it has to be ensured that there is a reasonable sensibility to the parameter of interest. In highly absorbing materials it can happen for instance that the penetration depth of light is so small that the electric field in the layer of interest is already damped too strongly. In anisotropic samples the special case might occur in which the electric field vector of the refracted light inside the sample is almost perpendicular to the optical axis of interest. In both examples the sensibility could be low. By using commercially available fit routines, such problems can be tested by means of the so-called standard error. Finally, it has to be ensured that the parameters of interest are not coupled to each other which happens if both of them change the polarization properties of the sample in the same manner. For example, it is sometimes difficult to measure a layer thickness and it's refractive index independently from each other. In a numerical fit, the parameter coupling can be tested by means of the covariance matrix of the standard errors.

The discussion of restrictions in the second evaluation step should emphasize that qualitative information about the sample structure are essential in order to obtain good quantitative results. Indications for deficiencies in the assumed sample structure are for example unexpected interference signatures or an inconsistent dispersion of a deduced dielectric function.

The simple scheme of Fig. 1.3 does not include the important and sometimes demanding step of the definition of appropriate measurement geometries (angle of incidence, sample orientation, etc.) in order to achieve the best possible sensitivity to the sample parameters of interest. It is often not worthwhile to measure just in all possible configurations (e.g. in the whole accessible angle of incidence range). Configurations with low sensitivity to the parameter of interest (e.g. very high or low angles of incidence) may just increase the error in the final result. Appropriate configuration can be chosen by some simple preliminary considerations. If necessary, these can be subsequently modified in an iterative procedure. Thin films are usually

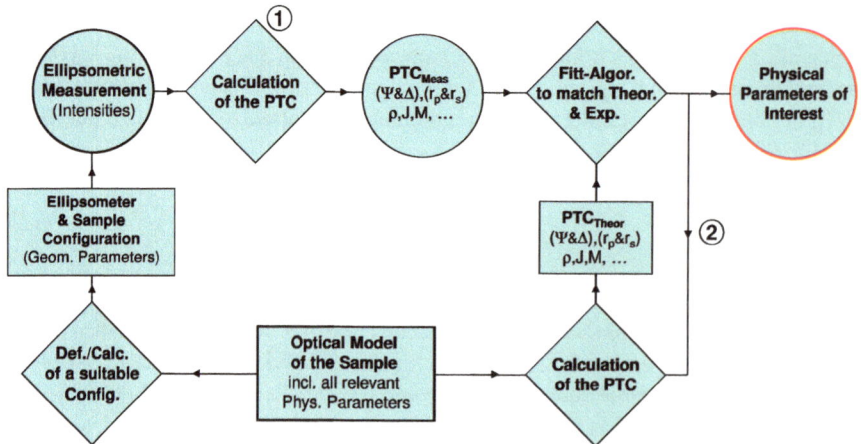

**Fig. 1.5** Extended scheme of the data evaluation in an ellipsometric measurement. The two major calculation steps, which can be found already in the simplified representation of Fig. 1.3 are labeled ① and ②

best measured at incidence angles near the Brewster angle of the respective substrate material. In some cases it is more efficient to calculate the best configuration in a preliminary simulation.

Since all other evaluation steps are based on the chosen measurement configurations a final flow chart of an ellipsometric measurement may appear rather complicated (Fig. 1.5). In this resulting scheme the significance of an appropriate optical model becomes again evident. It is important to remember that ellipsometry is initially reference free measuring how a sample changes the polarization of an incident light beam. All subsequently derived parameters depend on the best possible assumption of the sample structure and the validity of the applied optical models. Following Eugene A. Irene, who has brought this into phrase, this means in turn that if the information about the sample structure is insufficient: "Ellipsometry is perhaps the most surface sensitive technique in the universe. However you often don't know what it is you have measured so sensitively".

### 1.3.2 Determination of Ψ and Δ

The determination of Ψ and Δ or the more generalized PTC's of a sample requires some mathematical tools (calculation step ① in Figs. 1.3 and 1.5). First of all a suitable description of polarized light is needed. Furthermore, all optical components including the sample under investigation have to be represented according to their ability to change the state of polarization. These theoretical tools can be finally used to calculate relative intensities which are measured at a detector or, in turn, to determine the PTC's of the sample from the measured intensities.

### 1.3.2.1  Polarized Light

To simplify the problem as much as possible, assumptions concerning the propagation of light between optical components and the sample are necessary.

- Since linear optical effects are investigated each wavelength $\lambda$ is addressed separately in a quasi-monochromatic approximation.
- The distances between the optical components are much larger than the wavelength and the coherence length of the light. We can assume planar transversal electromagnetic (TEM) waves propagating only in forward direction from the light source to the detector. In other words, all optical components interact independently, one after another, with the light and interferences between them are avoided. This assumption has to be critical reviewed, e.g. for laser light sources where the coherence length is much larger than for conventional light sources and in near field experiments.
- The surrounding medium is assumed isotropic and homogeneous (air, vacuum, water, etc.). Thereby the polarization state does not depend on the propagation direction and can be separately considered.
- The magnetic susceptibility is constantly one in all optical experiments and the polarization of light is fully described by the electric fields.

For the mathematical description of a polarization one can choose without further loss of generality a Cartesian coordinate system so that the light propagates along the positive direction of the $z$-axis. A common description of an arbitrarily planar monochromatic TEM wave is then given by

$$
\begin{aligned}
E_x &= E_{0x} \exp\big[i(kz - \omega t)\big] \exp[i\delta_x], \\
E_y &= E_{0y} \exp\big[i(kz - \omega t)\big] \exp[i\delta_y].
\end{aligned}
\tag{1.3}
$$

Equation (1.3) represents the general case of an elliptical polarization as illustrated in Fig. 1.6. The wave vector $k = 2\pi/\lambda$ and the angular frequency $\omega = 2\pi\nu$ are connected by the known dispersion relation for TEM waves in transparent media. The orientation of the two perpendicular basis vectors in $x$- and $y$-direction is free of choice for the moment. The amplitudes $E_{0x}$ and $E_{0y}$ together with the phases $\delta_x$ and $\delta_y$ for the $x$ and $y$ component, respectively, define the polarization of the light. This set of information can be merged in a so-called Jones vector [30]:

$$
\bar{E}^{Jones} = \begin{pmatrix} E_{0x} \exp[i\delta_x] \\ E_{0y} \exp[i\delta_y] \end{pmatrix}.
\tag{1.4}
$$

However, the absolute phase is not measurable and the absolute intensity ($I \sim E_{0x}^2 + E_{0y}^2$) is an arbitrary value which is not of interest in an ellipsometric measurement. The polarization state is therefore already fully defined by only two parameters: The relative amplitude $\tan \Psi' = E_{0x}/E_{0y}$ and phase $\Delta' = \delta_x - \delta_y$ of the $x$ and $y$ component. Sometimes these two parameters are combined in a single complex number $\chi = 1/\tan \Psi' \exp[i\Delta']$. Another alternative representation refers to the

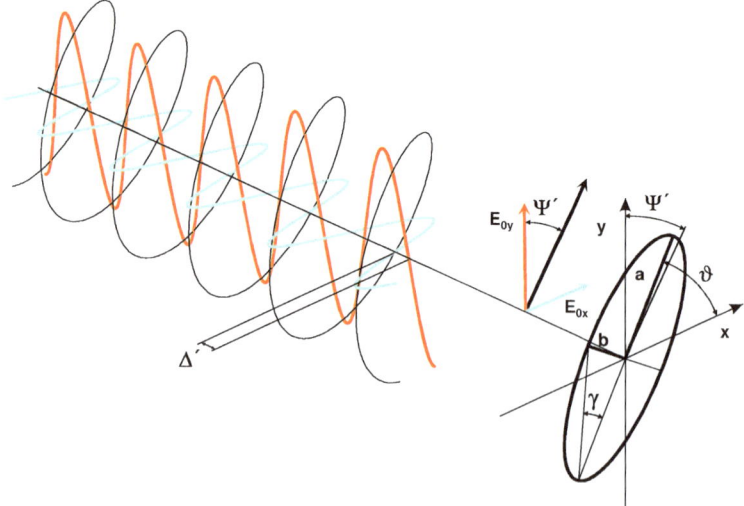

**Fig. 1.6** Elliptically polarized light and the projected polarization ellipse. Mathematically the polarization state is defined by three equivalent parameter sets [$\Psi'/\Delta'$—amplitude quotient and phase difference], [$\vartheta/\gamma$—azimuth angle and ellipticity], and $\chi = 1/\tan\Psi\exp[i\Delta]$

ellipse, which yields from a projection of the electric field vector to the $x$–$y$-plane. The respective parameter pair is given by the azimuth angle $\vartheta$ between the main axis of the ellipse and the $x$-axis and the ellipticity $\gamma$ (Fig. 1.6). Please note that the parameters $\Psi'$ and $\Delta'$, which define here the polarization state of light, are in general not identical with the previously defined $\Psi$ and $\Delta$ values, which describe how the sample changes the polarization. Only in the special historically important case, where the incident light is chosen 45° linearly polarized with respect to the plane of incidence, both parameter sets match. Commercially available ellipsometers determine $\Psi$ and $\Delta$ independent from the selected incident polarization.

So far, all representations of the polarization are only applicable for totally polarized light. Hence, they represent 100% linear, circular or elliptically polarized light and the constituting parameter pairs are of a well defined value. Unfortunately, this is sometimes not sufficient and it is necessary to consider also partial polarizations. Possible sources of partial polarized light are

- non ideal polarizers,
- lateral inhomogeneous samples (e.g. rough surfaces/interfaces or inhomogeneous film thickness),
- a divergent light beam (e.g. a short focal distance results in an uncertain angle of incidence),
- an insufficient spectral resolution and broad spectral line width, respectively.

Independent from the inbound polarization and the source of depolarization, partial polarization means that the well defined polarization is replaced by a statistical mixture of different polarization states. In this view the components of the Jones vector are now time dependent and polarization is measurable just as a time-average. A proper description of the partially polarized light requires therefore a third parameter which for example characterizes the probability $w$ to find a certain polarization state. Related to the fact that effective intensities (time-averaged fields) are finally measured, the three constituting parameters are often expressed in terms of intensities. Complemented by the total intensity they form the 4 Stokes parameters [31]:

$$
\begin{aligned}
S_0 &= I, \\
S_1 &= I_x - I_y, \\
S_2 &= I_{+\pi/4} - I_{-\pi/4}, \\
S_3 &= I_l - I_r.
\end{aligned}
\tag{1.5}
$$

The first Stokes parameter contains simply the total intensity $I$ of the light. $I_x$, $I_y$, $I_{+\pi/4}$, and $I_{-\pi/4}$ are the intensities, which would pass through an ideal linear polarizer oriented with the transmission axis in $x$, $y$, $+\pi/4$, and $-\pi/4$ direction, respectively. $I_l$ and $I_r$ are the intensities, which would pass through ideal left and right circular polarizers. The total intensity resumes to $I = I_x + I_y = I_{+\pi/4} + I_{-\pi/4} = I_l + I_r$. In case of totally polarized light $S_0^2 = S_1^2 + S_2^2 + S_3^2$, while for partially or unpolarized light $S_0^2 \geq S_1^2 + S_2^2 + S_3^2$. The degree of polarization is finally defined by:

$$
P = \sqrt{\frac{S_1^2 + S_2^2 + S_3^2}{S_0^2}} = 2w - 1.
\tag{1.6}
$$

For more comprehensive descriptions of the different representations of polarized light and the relation among these representations we would refer at this point to respective literature about fundamental optics and ellipsometry [2, 4, 9, 32, 33]. Because of similarities in operational concepts to quantum physics, the theory of polarized light is furthermore discussed in a number of rather theoretical publications [11, 32, 34, 35]. An alternative description of partial polarized light by means of coherency-matrices is described for example in references [2, 36].

### 1.3.2.2 Jones and Mueller Matrix Methods

The matrix methods of Jones [30] and Mueller [37–40] are by far the most popular methods in order to describe the linear optical effects of polarizing optical elements. The Jones formalism is based on complex $2 \times 2$ matrices which are applied to the Jones vector as defined in the previous section. The Mueller matrix formalism uses $4 \times 4$ matrices with real elements which are applied to the Stokes parameters arranged now in a column vector $\bar{S} = (S_0, S_1, S_2, S_3)$. The impact of a polarizing element can thus be written as

$$\begin{pmatrix} E'_x \\ E'_y \end{pmatrix} = \hat{J} \begin{pmatrix} E_x \\ E_y \end{pmatrix} = \begin{pmatrix} J_{xx} & J_{xy} \\ J_{yx} & J_{yy} \end{pmatrix} \begin{pmatrix} E_x \\ E_y \end{pmatrix},$$

$$\begin{pmatrix} S'_0 \\ S'_1 \\ S'_2 \\ S'_3 \end{pmatrix} = \hat{M} \begin{pmatrix} S_0 \\ S_1 \\ S_2 \\ S_3 \end{pmatrix} = \begin{pmatrix} M_{00} & M_{01} & M_{02} & M_{03} \\ M_{10} & M_{11} & M_{12} & M_{13} \\ M_{20} & M_{21} & M_{22} & M_{23} \\ M_{30} & M_{31} & M_{32} & M_{33} \end{pmatrix} \begin{pmatrix} S_0 \\ S_1 \\ S_2 \\ S_3 \end{pmatrix}. \tag{1.7}$$

The complex $2 \times 2$ Jones matrix contains 8 parameters including the absolute phase which is not measurable. If the absolute phase and additionally the absolute intensity, which is also not if interest in ellipsometric measurements are ignored, the Jones matrix contains 6 relevant parameters that define the ability to change the polarization. All optical components and any arbitrarily anisotropic samples can be represented by means of these 6 parameters in a Jones matrix as long as no depolarization takes place.

The 16 coefficients of the Mueller matrix contain information about intensities passing through polarizing elements. This includes information about the absolute intensity. Accordingly, the Mueller matrix contains 7 independent parameters, if no depolarization takes place. In turn this means that 9 identities exist among the 16 coefficients of a Mueller matrix in this case [39, 41]. It is thus feasible that the Mueller matrix of a non-depolarizing optical element can be calculated from the respective Jones matrix and vice versa except of the absolute phase [2, 42]. However, in case of depolarization the 16 coefficients of a Mueller matrix become independent and might include manifold orientation depending information about a sample. But notice, the polarization state of the obtained partially depolarized light is always fully characterized by only 3 parameters.

With the help of either the Jones or the Mueller matrix formalism it is now possible to calculate the measurable intensities for different orientations of the polarizing elements in an ellipsometer. Therefore, the matrix representations of all optical elements including the sample under investigation are multiplied in the respective order.

By a comparison of the measured intensities with those calculated, it is finally possible to obtain the unknown sample polarization transformation coefficients (PTC). The latter can be represented for example either by the bare Mueller or Jones matrix coefficients, the ellipsometric angles $\Psi$ and $\Delta$, or the (generalized) complex Fresnel coefficients. The choice of the most convenient representation depends on the sample under investigation and the specific ellipsometer type. Just like the choice which parameters are used, the determination of PTC's in practice also depends strongly on the sample properties and the ellipsometer type. If the intensity is for example continuously measured as a function of the rotation of a polarizing element it is often beneficial to consider the Fourier transformation of the measured sinusoidal signal. In case of an isotropic non-depolarizing sample, the two parameters describing the polarization probability are then encoded in the $\sin(2\alpha)$ and $\cos(2\alpha)$ Fourier coefficients and can be determined thereafter algebraically by a comparison of coefficients.

According to the sampling theorem, it is also sufficient to measure the sinusoidal signal with a minimum of three fixed positions in order to obtain the two Fourier parameters in a fit algorithm. Finally, it is in some cases also possible to measure at four specific positions [4] and to calculate the two PTC's directly from the measured intensities.

## 1.3.3 Fresnel Coefficients

The second crucial step in an ellipsometric measurement now comprises the translation of the obtained PTC's to intrinsic sample properties like the dielectric function or the layer thickness (calculation step ② in Fig. 1.3). This problem of the light matter interaction could be rather complicated in case of increasingly complex sample structures. Therefore, only a few essential conclusions will be introduced which are typically used for analyzing organic film structures.

It is reasonable to consider organic and anorganic materials as homogeneous materials which are well characterized by its macroscopic optical properties i.e. the macroscopic polarizability, dielectric function, or refractive index. This approximation is adequate as long as the wavelength is much larger than the size of the constituting molecules and larger than the unit cell of the crystal. It is often possible to assume a stratified sample structure, which allows a description of the light matter interaction with planar TEM waves. With these two assumptions it is possible to deduce complex reflection and transmission coefficients, which provide a link between the measured $\Psi$ and $\Delta$ or PTC's and the intrinsic sample parameters. The sample parameters are usually determined within a fit procedure. This part of the data evaluation is often called "optical modeling". It is usually the most discriminating step in the data evaluation because it rests on a correctly assigned sample structure which has to be critical reviewed in advance.

### 1.3.3.1 Dielectric Function

The optical properties of a homogeneous material can be encountered in the macroscopic Maxwell equations i.e. the constitutive relations. These relations connect the macroscopic electric field $E$ with the dielectric displacement $D$ as well as the magnetic induction $B$ and magnetic field $H$ according to the polarization and magnetization of the material. Again, three simplifications can be used:

- In the optical frequency range the macroscopic magnetization is always zero.
- The discussed ellipsometric measurements comprise only linear optical effects and higher order contributions can be neglected.
- Spatial dispersion effects are negligible in homogeneous media.

As a result, the response of the material to an electric field is defined by the macro-scopic polarizability $P$.[2] The material equations in a Fourier representation can be written as

$$D[\omega] = \varepsilon_0 E[\omega] + P[E[\omega], \omega]$$
$$= \varepsilon_0 \hat{\varepsilon}[\omega] E[\omega], \tag{1.8}$$
$$B[\omega] = \mu_0 H[\omega].$$

In optical problems the magnetic field strength is connected with the magnetic induc-tion density just by the free space permeability $\mu_0$. The macroscopic electric field strength is connected to the displacement density by the free space permittivity $\varepsilon_0$ and the dielectric tensor $\hat{\varepsilon}$ which depends on angular frequency $\omega$. In absorbing materials $\hat{\varepsilon} = \hat{\varepsilon}_1 + i\hat{\varepsilon}_2$ is a complex tensor and the imaginary part of the tensor components is proportional to energy dissipation and thus to the absorption of the light. In case of isotropic media the dielectric tensor reduces to a scalar dielectric function.

$$\hat{\varepsilon} = \begin{pmatrix} \varepsilon & 0 & 0 \\ 0 & \varepsilon & 0 \\ 0 & 0 & \varepsilon \end{pmatrix} = \varepsilon \begin{pmatrix} 1 & 0 & 0 \\ 0 & 1 & 0 \\ 0 & 0 & 1 \end{pmatrix} \quad \text{(isotropic materials)}. \tag{1.9}$$

As already seen in (1.8) a couple of equivalent quantities can be used in order to describe the linear optical properties of a medium. Most widely used is the complex refractive index $\tilde{n} = n + ik$ where the real part $n$ refers to the refractive index of transparent media. The imaginary part $k$ is the absorption coefficient of a medium. Other common representations of the optical response function and the relations among them are summarized in Table 1.1.

In case of an anisotropic sample with three intrinsic Cartesian optical axes the dielectric tensor can be diagonalized in the form

$$\hat{\varepsilon} = \begin{pmatrix} \varepsilon_x & 0 & 0 \\ 0 & \varepsilon_y & 0 \\ 0 & 0 & \varepsilon_z \end{pmatrix} \quad \text{(anisotropic materials)}. \tag{1.10}$$

The dielectric tensor is defined now by three independent dielectric functions $\varepsilon_x$, $\varepsilon_y$, and $\varepsilon_z$ which determine the different polarizabilities for electric fields in the cor-responding directions. Such a matrix is suitable for biaxial crystals of e.g. triclinic, monoclinic, and orthorhombic symmetry. Uniaxial crystals of e.g. hexagonal, tetrag-onal, trigonal, and rhombohedral symmetry are defined analogue but only with two independent components [43]. As indicated before, in isotropic materials e.g. cubic crystals, all components are equal. If the sample is placed in an arbitrary orientation in the ellipsometer, the matrix (1.10) has to be transposed by a rotation about the Euler angles.

---

[2] $P$ is the spatial average of the induced dipole moments per unit volume.

**Table 1.1** Equivalent quantities for the linear optical properties of homogeneous media

|  | Real part | Imaginary part |
|---|---|---|
| Dielectric function: $\varepsilon$ | $\varepsilon_1 = n^2 - k^2$ | $\varepsilon_2 = 2nk$ |
| Refractive index: $\tilde{n} = \sqrt{\varepsilon}$ | $n = \sqrt{(\varepsilon_1 + \sqrt{\varepsilon_1^2 + \varepsilon_2^2})/2}$ | $k = \sqrt{(-\varepsilon_1 + \sqrt{\varepsilon_1^2 + \varepsilon_2^2})/2}$ |
| Susceptibility: $\chi = \varepsilon - 1$ | $\chi_1 = \varepsilon_1 - 1$ | $\chi_2 = \varepsilon_2$ |
| Optical conductivity: $\sigma$ | $\sigma_1 = \omega\varepsilon_0\varepsilon_2$ | $\sigma_2 = -\omega\varepsilon_0(\varepsilon_1 - 1)$ |
| Loss function: $-\varepsilon^{-1}$ | $\dfrac{-\varepsilon_1}{\varepsilon_1^2 + \varepsilon_2^2}$ | $\dfrac{\varepsilon_2}{\varepsilon_1^2 + \varepsilon_2^2}$ |
| Phase velocity: $v_p = c/\tilde{n}$ | $\dfrac{1}{\sqrt{\varepsilon_0\mu_0}}\dfrac{1}{n}$ | $\dfrac{1}{\sqrt{\varepsilon_0\mu_0}}\dfrac{1}{k}$ |

Organic molecules like sugar, however, have often an intrinsic handedness/chirality. This yields an optical activity and circular dichroism if light is transmitted through a film or liquid solution. The dielectric tensor of such materials is now no longer symmetric. But in case of vanishing absorption (optical activity) the tensor is still Hermitian ($\varepsilon_{ij} = \varepsilon_{ij}*$). The constitutive Maxwell relation is than written as

$$D[\omega] = \varepsilon_a E[\omega] + i\varepsilon_0 G[\omega] \times E[\omega], \qquad (1.11)$$

where $\varepsilon_a$ is the dielectric tensor for vanishing optical activity. The vector $G$ is pointing in the direction of the light propagation and is called the gyration vector [44].

The optical rotation due to the Faraday effect which may emerge in the presents of a magnetic field is defined analogously [43, 44]

$$D[\omega] = \varepsilon_a E[\omega] + i\varepsilon_0 \gamma B[\omega] \times E[\omega]. \qquad (1.12)$$

### 1.3.3.2 Fresnel Equations

As mentioned before the link between sample properties like the dielectric function and its polarization properties given either by $\Psi$ and $\Delta$ or by the PTC's can be obtained by the definition of reflection and transmission coefficients. These complex coefficients determine to which amount the electric field amplitudes of the incident s- and p-polarization component are attributed to the respective fields in the reflected and transmitted beam and determine the relative phase shifts among these components.

For a single interface between two isotropic homogeneous media these relations can be obtained as a result of the boundary conditions committed by the Maxwell equations. They are known as the Fresnel equations of the form

$$r_p = \frac{\tilde{n}_2 \cos \varphi_i - \tilde{n}_1 \cos \varphi_t}{\tilde{n}_2 \cos \varphi_i + \tilde{n}_1 \cos \varphi_t},$$

$$r_s = \frac{\tilde{n}_1 \cos \varphi_i - \tilde{n}_2 \cos \varphi_t}{\tilde{n}_1 \cos \varphi_i + \tilde{n}_2 \cos \varphi_t},$$

$$t_p = \frac{2\tilde{n}_1 \cos \varphi_i}{\tilde{n}_2 \cos \varphi_i + \tilde{n}_1 \cos \varphi_t}, \tag{1.13}$$

$$t_s = \frac{2\tilde{n}_1 \cos \varphi_i}{\tilde{n}_1 \cos \varphi_i + \tilde{n}_2 \cos \varphi_t}.$$

$\tilde{n}_1$ and $\tilde{n}_2$ are the complex refractive indices of the incident and refractive media, respectively, and the angles of incidence $\varphi_i$ and refraction $\varphi_t$ are described by Snell's law ($\tilde{n}_1 \sin \varphi_i = \tilde{n}_2 \sin \varphi_t$).

The assumption of a sample, which consists of only a single perfectly smooth surface, is unfortunately very unrealistic. In practice at least unintentional surface overlayers or a finite surface roughness are not negligible. Nevertheless, the Fresnel equations are often used as a good approximation. The obtained dielectric function is then the so-called pseudo dielectric function [6].

$$\langle \varepsilon \rangle = \sin^2 \phi \left( 1 + \tan^2 \phi \left( \frac{1 - \rho}{1 + \rho} \right)^2 \right), \quad \rho = \frac{r_p}{r_s} = \tan \Psi e^{i\Delta}. \tag{1.14}$$

### 1.3.3.3 Homogeneous Stratified Media

A simple optical layer model of practical importance describes a single isotropic layer ($l$) of the complex refractive index $\tilde{n}_l$ and thickness $d$ on a substrate ($s$) with the complex refractive index $\tilde{n}_s$. Light, which incidences from the ambient ($a$) with the refractive index $n_a$, is reflected on the surface and the interface between layer and substrate. Due to multiple reflections within the layer, the overall reflected electric fields parallel and perpendicular to the plane of incidence add up by a geometric series. The summations gives the Airy formula for the so-called 3-phase model [2, 17, 44–46]:

$$r_p = \frac{r_{al_p} + r_{ls_p} e^{i2\beta}}{1 + r_{al_p} r_{ls_p} e^{i2\beta}},$$

$$r_s = \frac{r_{al_s} + r_{ls_s} e^{i2\beta}}{1 + r_{al_s} r_{ls_s} e^{i2\beta}}, \tag{1.15}$$

where $r_{al_{p/s}}$ and $r_{ls_{p/s}}$ are the Fresnel reflection coefficients on the ambient layer boundary and the layer substrate boundary, respectively (1.13). The phase factor $\beta$ is given by

$$\beta = \frac{2\pi}{\lambda} d \sqrt{n_l^2 - n_a^2 \sin^2 \phi_a} = \frac{2\pi}{\lambda} (d\, n_l) \cos \phi_l. \tag{1.16}$$

$\lambda$ is the wavelength of the light in vacuum and $\phi_a$ the angle of incidence. It will be shown later that generalized complex "Fresnel" coefficients as defined in (1.15) can be obtained also for complex anisotropic structures. Before it should be mentioned that the reflection coefficients of the 3-phase model contain already 5 parameters which determine the optical properties of the sample. Even if the substrate dielectric function is known, we have already one parameter more than a single measurement at a given angle of incidence can deliver. An examination of the phase factor $\beta$ furthermore illustrates the coupling of $n_l$ and $d$. An unambiguous determination of the thickness and the optical properties is possible by means of multiple angle of incidence measurements. A solution for very thin layers with $d \ll 1$ nm will be discussed in Chap. 13.

The problem of a multilayer structure with planar parallel interfaces can be solved with the help of a $2 \times 2$ transfer matrix methods [2, 44, 47–51]. To some extend similar to the Jones matrix formalism, it is possible to connect the electric fields of the forward and backward traveling waves at each interface by a transfer matrix which is defined by the Fresnel coefficients for this specific interface and thus depends on the refractive index on both sides as well as the incident angle. In contrast to the Jones matrix formalism the distance between the interfaces is assumed now to be smaller than the coherence length of the light and interference between the forward and backward traveling waves becomes possible. Consequently, a propagation matrix has to be introduced, which implements a (complex) phase factor to the electric fields crossing a given layer. The reflection and transmission coefficients ($r_p/r_s$ and $t_p/t_s$) of the whole slab are finally obtained by the product of all transfer and propagation matrices in the respective order.

### 1.3.3.4 Anisotropic Media

It is a common property of isotropic media that the reflected and transmitted electric field components parallel and perpendicular to the plane of incidence are independent. Accordingly, the previously defined Fresnel equations and the Airy formulas handle both components independently. Incident light with parallel polarization is not converted by reflection or transmission to perpendicular polarized light and vice versa perpendicular polarized light does not contribute to the parallel polarization. This separation retains also for anisotropic materials as long as the principal optical axes as defined in (1.10) are aligned parallel or perpendicular to the plane of incidence and the sample surface. For such high symmetry configurations it is possible to deduce reflection coefficients analogous to the classical Fresnel equations or the Airy formulas [2, 52]. Solutions for an uniaxial anisotropic bulk sample are summarized in Table 1.2. Solutions for anisotropic layer structures can be found in [2, 52]. The dependency on the sample orientations shows that the measurement of only one $\Psi$ and $\Delta$ pair is not sufficient anymore. Unambiguous results are obtained by measuring in different sample orientations and with different angles of incidence. Thus, sensitivity to an out of plane anisotropy is obtained with a variation of the

**Table 1.2** Reflection coefficients for parallel and perpendicular polarized light reflected at the surface of an uniaxial anisotropic material if the optical axis $c$ coincides with one of the three high symmetry orientations [2, 52]. ($c \parallel x$)—$c$ parallel to the plane of incidence and the surface; ($c \parallel y$)—$c$ perpendicular to the plane of incidence and parallel to the surface; ($c \parallel z$)—$c$ perpendicular to the surface and parallel to the plane of incidence. $\varepsilon_\perp$ and $\varepsilon_\parallel$ correspond to the dielectric tensor components perpendicular and parallel to the optical axis i.e. the ordinary and extraordinary dielectric function

|  | $p$ polarization ($r_{pp}$) | $s$ polarization ($r_{ss}$) |
|---|---|---|
| ($c \parallel x$): | $\dfrac{\sqrt{\varepsilon_\perp \varepsilon_\parallel}\cos\phi_i - \sqrt{\varepsilon_\perp - \sin^2\phi_i}}{\sqrt{\varepsilon_\perp \varepsilon_\parallel}\cos\phi_i + \sqrt{\varepsilon_\perp - \sin^2\phi_i}}$ | $\dfrac{\cos\phi_i + \sqrt{\varepsilon_\perp - \sin^2\phi_i}}{\cos\phi_i - \sqrt{\varepsilon_\perp - \sin^2\phi_i}}$ |
| ($c \parallel y$): | $\dfrac{\varepsilon_\perp \cos\phi_i - \sqrt{\varepsilon_\perp - \sin^2\phi_i}}{\varepsilon_\perp \cos\phi_i + \sqrt{\varepsilon_\perp - \sin^2\phi_i}}$ | $\dfrac{\cos\phi_i + \sqrt{\varepsilon_\parallel - \sin^2\phi_i}}{\cos\phi_i - \sqrt{\varepsilon_\parallel - \sin^2\phi_i}}$ |
| ($c \parallel z$): | $\dfrac{\sqrt{\varepsilon_\parallel \varepsilon_\perp}\cos\phi_i - \sqrt{\varepsilon_\parallel - \sin^2\phi_i}}{\sqrt{\varepsilon_\parallel \varepsilon_\perp}\cos\phi_i + \sqrt{\varepsilon_\parallel - \sin^2\phi_i}}$ | $\dfrac{\cos\phi_i + \sqrt{\varepsilon_\perp - \sin^2\phi_i}}{\cos\phi_i - \sqrt{\varepsilon_\perp - \sin^2\phi_i}}$ |

incidence angle while an in-plane anisotropy requires measurements with different azimuthal sample orientations (Chaps. 10 and 13).

Aspnes [53] has described a solution for the pseudo dielectric function $\langle\varepsilon\rangle$ (1.14) measured on a biaxial crystal. Based on a first-order expansion, which assumes that the anisotropies are small corrections to an isotropic mean value, he obtained

$$\langle\varepsilon\rangle = \varepsilon + \frac{\varepsilon - \sin^2\phi_0}{(\varepsilon - 1)\sin^2\phi_0}\Delta\varepsilon_x - \frac{\varepsilon\cos^2\phi_0 - \sin^2\phi_0}{(\varepsilon - 1)\sin^2\phi_0}\Delta\varepsilon_y - \frac{1}{\varepsilon - 1}\Delta\varepsilon_z, \quad \text{where}$$

$$\varepsilon_x = \varepsilon + \Delta\varepsilon_x, \quad \varepsilon_y = \varepsilon + \Delta\varepsilon_y, \quad \text{and} \quad \varepsilon_z = \varepsilon + \Delta\varepsilon_z.$$

$$(1.17)$$

This relation is not exact, but reveals the very small influence of the $\varepsilon_z$ component normal to the surface, if $|\varepsilon|$ is moderately large. The physical reason is simple. If the material is optically thick ($n_0 \ll n_1$), the incoming light is refracted in the direction of the surface normal and the electric and magnetic field vectors in the material are mostly parallel to the surface. Therefore, it is difficult to measure $\varepsilon_\parallel$ if the optical axis ($c$-axis) is perpendicular to the sample surface and the sample. On the other hand it shows that the anisotropy can often be neglected and the use of an isotropic model yields reasonable results.

In arbitrary anisotropic materials or for arbitrary sample orientation $p$- and $s$-polarizations couple to each other. As a consequence, the previously discussed Fresnel equations as well as the shortly introduced $2 \times 2$ transfer matrix methods are in this case not applicable anymore. Additionally to the mode coupling between the reflected electric field components, the field evolution inside an anisotropic layer depends now on the propagation direction. A solution for this problem was introduced by Teiler, Henvis, and Berreman [54–58] by a $4 \times 4$ transfer matrix formalism. As a result on can obtain generalized reflection coefficients ($r_{pp}$, $r_{ss}$, $r_{ps}$, and $r_{sp}$) [52, 57] which may are used as the polarization transformation coefficients (PTC).

## 1.4 Dielectric Properties

### 1.4.1 Dispersion Models—Lorentz Oscillator

It is by far impossible to provide a common description of the dielectric properties neither for inorganic [33, 59] nor for organic materials [60]. Nevertheless, a relatively good insight could be obtained with some classical considerations. It was mentioned already that the optical properties i.e. the dielectric function rises from the polarizability of the material. Polarizable entities in an organic layer may are the individual molecules were a dipole moment is induced by the electric field of the incident light. In the infrared spectral region this could be obtained by a vibration of ions in the molecule i.e. phonons. In the visible and UV it is mainly the excitation of electrons (excitons) which gives a dipole moment. Both excitations can be described classically by a mechanical harmonic oscillators of a negative and positive charge with the equation of motion. Because of the strong localization of electrons in the individual molecules, organic layers can be treated as a ensemble of uncoupled oscillators. Within the Lorentz oscillator model the time dependent dipole moment of all entities is translated in a polarization density and one can obtain an expression for the dielectric function of a single oscillator (Fig. 1.7)

$$\varepsilon[\omega] = 1 + \frac{f}{\omega_0'^2 - \omega^2 - i\omega\gamma} = 1 + \frac{N}{\varepsilon_0}\alpha[\omega]. \tag{1.18}$$

$\omega_0'$ is the resonance frequency of the oscillator and $\gamma$ the damping factor related to energy dissipation e.g. by scattering precesses. The oscillator strength $f$ is proportional to the number of oscillators per unit volume $N$ while $\alpha$ is the polarizability

**Fig. 1.7** Real and imaginary parts of the dielectric function calculated within the classical Lorentz oscillator model for a single resonance. Such a resonance could be the electronic excitation from the highest occupied molecular orbital (HOMO) to the lowest unoccupied molecular orbital (LUMO) which induces a dipole moment $p$

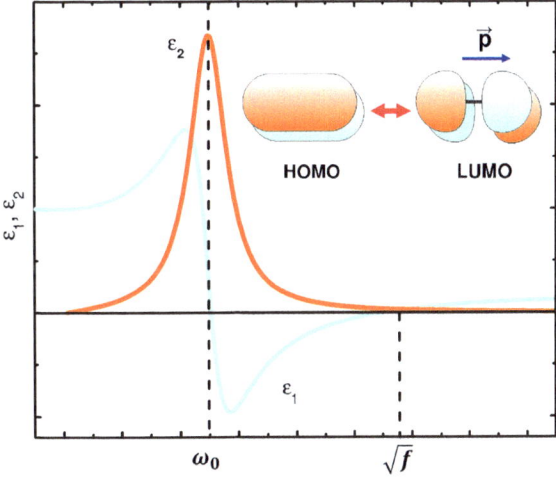

of each entity. The excitation of free electrons in metallic materials is given in this model by an oscillator with a resonance frequency $\omega_0 = 0$ (Chap. 12).

The assumption of totally uncoupled oscillators is of course a very crude approximation. In dense organic layers each molecular dipole is screened at least by the surrounding molecules. Taking this effect into account the dielectric function of isotropic materials is rather given by the Clausius–Mosotti or Lorentz-Lorenz equation [33, 59]

$$\frac{\varepsilon - 1}{\varepsilon + 2} = \frac{N\alpha}{3\varepsilon_0}. \tag{1.19}$$

For a small damping one can find the same expression as in (1.18) but with a slightly shifted eigenfrequency

$$\varepsilon[\omega] = 1 + \frac{f}{\omega_0^2 - \omega^2 - i\omega\gamma} \quad \text{where } \omega_0^2 = \omega_0'^2 - \frac{f}{3}. \tag{1.20}$$

This effect is observed as a red shift of absorption structures while going from gas phase or diluted materials to thin films and finally to bulk materials (q.v. Chaps. 15 and 16). In anisotropic molecular crystals on can observe the so-called Davydov splitting due to different screening components (Chap. 13, O. Gordan et al.).

Organic molecules of course possess not only one oscillator but a couple of different phonon and exciton dipole excitations. Therefore, the whole dielectric function has finally to be approximated by a sum of oscillators:

$$\varepsilon[\omega] = 1 + \sum_n \frac{f_n}{\omega_{0n}^2 - \omega^2 - i\omega\gamma_n}. \tag{1.21}$$

Most of the organic molecules are additionally highly anisotropic. The dipole moments of the different phonon and exciton excitations are usually linked to a certain direction in the molecule and thereby only measurable for respective electric field components. It is of particular importance that different oscillators may emerge in arbitrary orientations (Chap. 3). This anisotropy does not necessarily lead to an optical anisotropy of the material if the molecules are randomly arranged. However, in ordered arrangements, i.e. organic crystals or due to an interface specific bonding, the molecular anisotropy could appear in different aspects. Hereby, the anisotropy of the single molecules is largely conserved. The reason is the strong localization of the electrons within the molecule in contrast to metals and semiconductors. As a consequence it is not always possible to diagonalize the dielectric tensor (1.10) for all wavelengths simultaneously.

Beside the introduced Lorentz oscillator model one can find a huge number of other dispersion relations in literature derived from classical electrodynamic, quantum mechanical, or just empirical consideration. Because of its relevance in the thin film analysis only the model of Cauchy should be briefly mentioned here in addition [61]. The latter describes the refractive index $n$ of a material in the transparent region as a Taylor series in $\omega^2$. The benefit of this model is that it can be used to determine a layer thickness from a spectroscopic ellipsometric measurement. The knowledge about the dispersion of $n$ derived from the Kramers-Kronig relations solves here the problem of parameter coupling between $n$ and the thickness $d$.

### 1.4.2 Inhomogeneous Media and Structured Interfaces

Effective medium approximations (EMA) are used in order to obtain effective dielectric properties for inhomogeneous layers composed of different materials in a certain geometrical arrangement. Such a substitution of heterogeneous media by effective material is possible if

- the constituent particles are smaller than the wavelength (beside the host material),
- but big enough so that dielectric properties of the constituting materials are still the same,
- and if constituent particles are randomly distributed (diffraction and spatial dispersion effects are negligible).

The theory considers local field effects due to the surrounding media and the attendant screening. The effective dielectric function is thus NOT an average of the different constituting material dielectric functions. It can be shown that the effective medium properties are rather than a sum of the respective polarizabilities $\alpha_n$. The effective dielectric function is given by an expression similar to the Clausius–Mosotti relation (1.19) [62]

$$\frac{\varepsilon - \varepsilon_h}{\varepsilon + p\varepsilon_h} = \sum_n f_n \frac{\varepsilon_n - \varepsilon_h}{\varepsilon_n + p_n\varepsilon_h} \qquad (1.22)$$

where $p$ is the so-called depolarization factor with

$$
\begin{aligned}
p &= 0 && \text{no screening,} \\
p &= 1 && \text{2D cylindrical inclusions,} \\
p &= 2 && \text{3D spherical inclusions,} \\
p &\to 1 && \text{maximal screening.}
\end{aligned}
$$

$f_n$ is the volume fraction of the different components. In (1.22) the host material is defined with the dielectric function $\varepsilon_h$ while the Clausius–Mosotti relation (1.19) is using $\varepsilon_h = 1$. The effect of the different depolarization factors is probably best seen by inspecting the extreme cases of none and maximum screening in layered structures, the so-called Wiener bounds (Fig. 1.8).

Table 1.3 summarizes three common EMA solutions for certain configurations. Especially the Bruggeman solution is widely used. Here, we would mention in particular the possibility to mimic a rough surface or interface by means of an effective medium layer [63]. The layer thickness could be determined, if necessary, by atomic

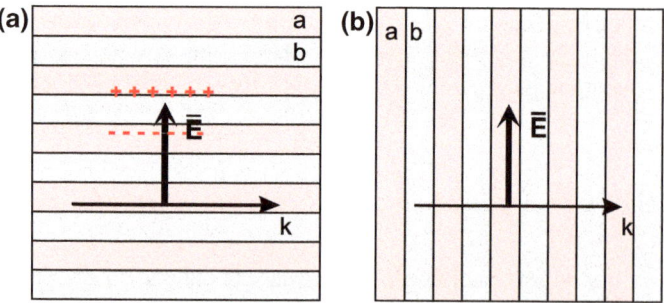

**Fig. 1.8** The effective dielectric function of two media $a$ and $b$ can be calculated within effective medium approximations depending on the topology of mixing. All the different solutions, however, are found between the so-called Wiener bounds (**a**) for a stratified structure parallel to the electric field with maximal screening ($\frac{1}{\langle \varepsilon \rangle} = \frac{f_a}{\varepsilon_a} + \frac{f_b}{\varepsilon_b}$) and (**b**) for a stratified structure perpendicular to the electric field without screening ($\langle \varepsilon \rangle = f_a \varepsilon_a + f_b \varepsilon_b$)

**Table 1.3** Most common effective medium approximations

| Bruggeman [65] | $0 = f \dfrac{\varepsilon_a - \langle \varepsilon \rangle}{\varepsilon_a + 2\langle \varepsilon \rangle} +$ $(1-f)\dfrac{\varepsilon_b - \langle \varepsilon \rangle}{\varepsilon_b + 2\langle \varepsilon \rangle}$ | Randomly mixed particles |
|---|---|---|
| Maxwell-Garnett [66] | $\dfrac{\langle \varepsilon \rangle - \varepsilon_M}{\langle \varepsilon \rangle + 2\varepsilon_M} = f \dfrac{\varepsilon_p - \varepsilon_M}{\varepsilon_p + 2\varepsilon_M}$ | Isolated particles in a matrix |
| Looyenga [67] | $\sqrt[3]{\langle \varepsilon \rangle} = f \sqrt[3]{\varepsilon_a} + (1-f)\sqrt[3]{\varepsilon_b}$ | Heterogeneous mixtures |

force microscopy (AFM). In the visible spectral range the thickness of the effective medium layer corresponds approximately to the root-mean-square (rms) roughness determined in scan range of about $5 \times 5$ µm [64]. Further applications are described in Chap. 9 (T.W.H. Oates).

## 1.5 Ellipsometric Configurations

With increasing amount of different analytical issues and the progress in the technical possibilities, by time also various types of ellipsometric systems have been developed. They differ mainly in the way how the polarization state of the incident light is generated and how the resulting polarization is analyzed (Fig. 1.1). Depending on the analytical requirements, the sensitivity and measurement speed can be optimized by choosing one or the other configuration [68]. Other modifications are made in order to increase the interface sensitivity for instance by attenuated/internal total reflectance (ATR/TIR) and internal total reflection (Chap. 18, H. Arwin).

### 1.5.1  Null-Ellipsometer

A "Null-Ellipsometer" is one of the oldest configurations. It consists of three polarizing elements. Two linear polarizing elements, namely the polarizer ($P$) and the analyzer ($A$) as well as a compensator ($C$). The compensator is placed either between the polarizer and the sample ($PCSA$-configuration) or between the sample and the analyzer ($PSCA$-configuration). As indicated already by the name, sample properties are determined by varying (rotating) two of the three components until the measured intensity is minimized. With this procedure the ellipsometric angles $\Psi$ and $\Delta$ can be directly obtained [69]. The involved nulling procedure is on the other hand a huge disadvantage although the rotation of the polarizers can be meanwhile motorized. Furthermore, all wavelength have to be measured one after another and a parallelization is hardly possible.

### 1.5.2  Rotating Polarizer/Analyzer

The rotating polarizer/analyzer ellipsometer is a photometric configuration, which was used in the first automatic spectroscopic systems [19]. In this $PSA$ configuration either the polarizer ($PSA$-$RPE$) or the analyzer ($PSA$-$RAE$) is rotated. Both the continuous and the so-called "step scan" rotations are used. It was mentioned already in Sect. 1.3.2.2 that $\Psi$ and $\Delta$ are obtained by analyzing the sinusoidal detector signal in terms of the $\sin(2\alpha)$ and $\cos(2\alpha)$ Fourier coefficients. The decision whether the polarizer or the analyzer is rotated depends on the used light source and the position of the monochromator. Hereby, polarization effects of the peripheric components are minimized. The "step scan" mode is usually used in connection with spectrographs and interferometric setups where all wavelengths are recorded in parallel.

With this type of ellipsometer it is possible to determine dielectric functions and layer thicknesses of isotropic or anisotropic absorbing materials with high accuracy. Incident angle scans (variable angle spectroscopic ellipsometry) provide enhanced sensitivity to layer thicknesses or out-of-plane anisotropies. Azimuthal rotation of the sample allows the determination of in-plane anisotropies. A general advantage is the reduced number of optical elements which minimizes alignment errors. Disadvantages arise from sensitivity limitations in case of transparent or metallic samples. Furthermore it is not possible to determine the sign of $\Delta$ or to distinguish between circular and unpolarized light. Thus depolarization effects can not directly measured.

### 1.5.3  Rotating Compensator

In a rotating compensator ellipsometer the linear polarizing elements are fixed and a rotating compensator is added either in $PCSA$ or in the $PSCA$ arrangement. With these modification, errors due to polarization dependent detectors or polarized light sources are avoided. The ellipsometric angles $\Psi$ and $\Delta$ are now decoded in the $2\alpha$ and $4\alpha$ Fourier coefficients of the recorded detector signal [68, 70]. In addition it is possible to determine depolarizations i.e. all four Stokes parameters (1.5). Moreover, it is now possible to determine $\Delta$ with the correct sign and with higher precision in comparison to rotating analyzer systems. Problems mainly arise in connection with the necessary calibration of the compensator. Quarter wave plates are applicable only for one wavelength and even "achromatic" compensator plates feature a distinct wavelength dependence.

Additional information are obtained if also the polarizer is rotated. In these generalized ellipsometric measurements all 6 polarization transformation coefficients (PTC) of a non-depolarizing arbitrary anisotropic sample i.e. the 6 independent Jones matrix coefficients (1.7) can be determined. In case of depolarization one can gain maximal 12 parameters i.e. the first three rows of the Mueller matrix (1.7).

### 1.5.4 Photo-Elastic Modulator Ellipsometer

In these kinds of ellipsometers, a photo-elastic modulator (PEM) is used instead of compensator. The modulation yields here a time dependent change of the retardation. This is achieved technically by a resonant acoustic excitation of an isotropic crystal [71, 72]. The generated modulation is thus typically in the 100 kHz range. By using a PEM instead of the rotating compensator the scan speed is therefore much higher. Also the problem of wobbling light beams is avoided because the system do not contain moving parts. Drawbacks are again the wavelength dependency of the retardation and a fragile calibration of the PEM.

### 1.5.5 Dual Rotating Compensator

The dual rotating compensator configuration (PCSCA) extents the possibilities of the generalized ellipsometry even further. This method is also known as "Mueller matrix" ellipsometry. Quite in general it technically supports the determination of all Mueller matrix coefficients (1.7) [42, 73, 74]. In practice, however, many of them are not independent from each other and the interpretation of the measured Mueller matrix elements in terms of intrinsic physically meaningful sample parameters is a demanding problem. Mueller matrix measurements recorded at different incidence angles and different azimuthal sample orientations may provide perhaps most comprehensive information about the optical response even for complex sample structures [75, 76].

## *1.5.6  Reflection Anisotropy Spectroscopy*

Reflection anisotropy spectroscopy (RAS) is a polarimetric method closely related to ellipsometry with a specific surface and interface sensitivity. It differs only by the angle of incidence which is chosen perpendicular to the sample surface ($\varphi = 90°$). A standard configuration compares, apart from that, to the photo-elastic modulator ellipsometer [77, 78]. In this configuration the polarization of the incidence light does not change upon reflection as long as the sample is isotropic (amorphous glass or cubic crystals like those of Si and Cu). The collected signal at the detector is constant. Any anisotropic surface or self organized anisotropic add-layer of molecules creates a modulation, which can be recorded with very high sensitivity for example with a lock-in amplifier [79–81].

# References

1. H.G. Tompkins, E.A. Irene, *Handbook of Ellipsometry* (William Andrew Publishing, Norwich, 2005)
2. R.M.A. Azzam, N.B. Bashara, *Ellipsometry and Polarized Light*, paperback edn. (North-Holland Personal Library, Amsterdam, 1987)
3. H.G. Tompkins, *A User's Guide to Ellipsometry* (Academic Press, San Diego, 1993)
4. A. Röseler, *Infrared Spectroscopic Ellipsometry* (Akademie-Verlag, Berlin, 1990)
5. U. Rossow, W. Richter, in *Optical Characterization of Epitaxial Semiconductor Layers*, ed. by G. Bauer, W. Richter (Springer, Berlin, 1996), pp. 68–128
6. D.E. Aspnes, in *Handbook of Optical Constants of Solids*, vol. I, ed. by E.D. Palik (Academic Press, Amsterdam, 1985), pp. 89–112
7. D.E. Aspnes, in *Optical Properties of Solids: New Developments*, ed. by B. Seraphin (North-Holland Publishing Company, Amsterdam, 1975)
8. R.M.A. Azzam, Surf. Sci. **56**, 6 (1976)
9. J. Tinbergen, *Astronomical Polarimetry* (Cambridge University Press, Cambridge, 1996)
10. T. Novikova, A. De Martino, S.B. Hatit, B. Drévillon, Appl. Opt. **45**, 3688 (2006)
11. M.C. Britton, Astrophys. J. **532**, 1240 (2008)
12. J.C. Jamin, Ann. Chim. Phys. **19**, 296 (1847)
13. W. Wernicke, Ann. Phys. (Leipz.) **266**, 452 (1887)
14. W. Voigt, Ann. Phys. (Leipz.) **267**, 326 (1887)
15. A. Rothen, Rev. Sci. Instrum. **16**, 26 (1945)
16. P. Drude, Ann. Phys. (Leipz.) **272**, 865 (1889)
17. P. Drude, Ann. Phys. (Leipz.) **272**, 532 (1889)
18. P. Drude, Ann. Phys. **39**, 481 (1890)
19. D.E. Aspnes, A.A. Studna, Appl. Opt. **14**, 220 (1975)
20. J.E. Greene, A.C. Boccara, C. Pickering, J. Rivory (eds.), in *Proceedings of the 1st International Conference on Spectroscopic Ellipsometry*. Thin Solid Films, vol. 234 (Elsevier, Amsterdam, 1993)
21. R.W. Collins, D.E. Aspnes, E.A. Irene (eds.), in *Proceedings of the 2nd International Conference on Spectroscopic Ellipsometry*. Thin Solid Films, vol. 313–314 (Elsevier, Amsterdam, 1998)
22. M. Fried, K. Hingerl, J. Humlíček (eds.), in *Proceedings of the 3rd International Conference on Spectroscopic Ellipsometry*. Thin Solid Films, vol. 455–456 (Elsevier, Amsterdam, 2004)
23. H. Arwin, U. Beck, M. Schubert (eds.), in *Proceedings of the 4th International Conference on Spectroscopic Ellipsometry* (Wiley/VCH, Weinheim, 2008)

24. H.G. Tompkins (ed.), in *Proceedings of the 5th International Conference on Spectroscopic Ellipsometry*. Thin Solid Films, vol. 11 (Elsevier, Amsterdam, 2011)
25. A. Rothen, in *Ellipsometry in the Measurement of Surfaces and Thin Films*, ed. by R.R. Stromberg, J. Kruger, E. Passaglia (Natl. Bur. of Standards, Washington, 1963), pp. 7–24
26. K. Vedam, Thin Solid Films **313–314**, 1 (1998)
27. M. Schubert, Ann. Phys. **15**, 480 (2006)
28. R.M.A. Azzam, Thin Solid Films **519**, 2584 (2011)
29. E.A. Irene, in *Ellipsometry at the Nanoscale*, ed. by M. Losurdo, K. Hingerl (Springer, Heidelberg, 2013), pp. 1–30
30. R.C. Jones, J. Opt. Soc. Am. **31**, 493 (1941)
31. G.G. Stokes, Trans. Camb. Philos. Soc. **9**, 399 (1852)
32. J. Humlíček, in *Handbook of Ellipsometry*, ed. by H.G. Tompkins, E.A. Irene (William Andrew, Norwich, 2005), pp. 3–90
33. M. Born, E. Wolf, *Principles of Optics*, 5th edn. (Pergamon Press, Oxford, 1975)
34. C. Brosseau, *Fundamentals of Polarized Light—A Statistical Optics Approach* (Wiley, New York, 1998)
35. U. Fano, J. Opt. Soc. Am. **39**, 859 (1949)
36. R. Barakat, J. Opt. Soc. Am. **53**, 317 (1963)
37. H. Mueller, J. Opt. Soc. Am. **38**, 661 (1948)
38. P. Soleillet, Ann. Phys. **12**, 23 (1929)
39. R.C. Jones, J. Opt. Soc. Am. **37**, 107 (1947)
40. M.J. Walker, Am. J. Phys. **22**, 170 (1954)
41. G.E. Jellison, in *Handbook of Ellipsometry*, ed. by H.G. Tompkins, E.A. Irene (William Andrew, Norwich, 2005), pp. 237–296
42. F. Le Roy-Brehonnet, B. Le Jeune, Prog. Quantum Electronics **21**, 109 (1997)
43. M. Schubert, in *Handbook of Ellipsometry*, ed. by H.G. Tompkins, E.A. Irene (William Andrew, Norwich, 2005), pp. 637–717
44. P. Yeh, *Optical Waves in Layered Media* (Wiley, New York, 1988)
45. G.B. Airy, Philos. Mag. Ser. 3(2), 20 (1833)
46. H. Hauschild, Ann. Phys. **63**, 816 (1920)
47. F. Abeles, Ann. Phys. Paris **5**, 596 (1950)
48. P.H. Berning, in *Physics of Thin Films*, vol. I, ed. by G. Hass (Academic Press, New York, 1963)
49. J. Humlíček, Opt. Acta **30**, 97 (1983)
50. C.J. Laan, H.J. Frankena, Appl. Opt. **17**, 538 (1978)
51. M.V. Klein, T.E. Furtak, *Optics* (Wiley, New York, 1986)
52. M. Schubert, Phys. Rev. B **53**, 4265 (1996)
53. D.E. Aspnes, J. Opt. Soc. Am. **70**, 1275 (1980)
54. S. Teitler, B.W. Henvis, J. Opt. Soc. Am. **60**, 830 (1970)
55. D.W. Berreman, T.J. Scheffer, Phys. Rev. Lett. **25**, 577 (1970)
56. D.W. Berreman, J. Opt. Soc. Am **62**, 502 (1972)
57. P. Yeh, Surf. Sci. **96**, 41 (1980)
58. H. Wöhler, G. Haas, M. Fritsch, D.A. Mlynski, J. Opt. Soc. Am. A **5**, 1554 (1988)
59. C.F. Klingshirn, *Semiconductor Optics* (Springer, Berlin, 1997)
60. M. Pope, C.E. Swenberg, *Electronic Processes in Organic Crystals and Polymers* (Oxford University Press, New York, 1999)
61. M.A.L. Cauchy, *Memoire sur la dispersion de la lumiere* (Prague, Calve, 1863)
62. D.E. Aspnes, Am. J. Phys. **50**, 704 (1982)
63. D.E. Aspnes, J.B. Theeten, Phys. Rev. B (1979)
64. J. Koh, Y. Lu, C.R. Wronski, Y. Kuang, R.W. Collins, Physics **69**, 1297 (1996)
65. D.A.G. Bruggeman, Ann. Phys. **5**, 636 (1935)
66. J.C. Maxwell-Garnett, Philos. Trans. R. Soc. Lond. Ser. A, Math. Phys. Sci. **203**, 385 (1904)
67. H. Looyenga, Physica **31**, 401 (1965)
68. D.E. Aspnes, Thin Solid Films **455–456**, 3 (2004)

69. F.L. McCrackin, E. Passaglia, R.R. Stromberg, H.L. Steinberg, J. Res. Natl. Bur. Stand. A, Phys. Chem. **67A**, 363 (1963)
70. M. Dressel, B. Gompf, D. Faltermeier, A.K. Tripathi, J. Pflaum, M. Schubert, Opt. Express **16**, 19770 (2008)
71. S.N. Jasperson, Rev. Sci. Instrum. **40**, 761 (1969)
72. O. Acher, E. Bigan, B. Drevilion, Rev. Sci. Instrum. **60**, 65 (1989)
73. R.W. Collins, J. Koh, J. Opt. Soc. Am. A **16**, 1997 (1999)
74. M.H. Smith, Appl. Opt. **41**, 2488 (2002)
75. M. Losurdo, M. Bergmair, G. Bruno, D. Cattelan, C. Cobet, A. de Martino, K. Fleischer, Z. Dohcevic-Mitrovic, N. Esser, M. Galliet, R. Gajic, D. Hemzal, K. Hingerl, J. Humlíček, R. Ossikovski, Z.V. Popovic, O. Saxl, J. Nanopart. Res. **11**, 1521 (2009)
76. B. Gompf, J. Braun, T. Weiss, H. Giessen, M. Dressel, U. Hübner, Phys. Rev. Lett. **106**, 185501 (2011)
77. D.E. Aspnes, A.A. Studna, Phys. Rev. Lett. **54**, 1956 (1985)
78. D.E. Aspnes, J.P. Harbison, A.A. Studna, L.T. Florez, Phys. Rev. Lett. **52**, 957 (1988)
79. P. Weightman, D.S. Martin, R.J. Cole, T. Farrell, Rep. Prog. Phys. **68**, 1251 (2005)
80. T.U. Kampen, U. Kampen, M. Rossow, S. Park Schumann, D.R.T. Zahn, J. Vac. Sci. Technol. B **18**, 2077 (2000)
81. B.S. Mendoza, R. Vázquez-Nava, Phys. Rev. B **72**, 1 (2005)

# Part I
# Biomolecules at Surfaces

# Chapter 2
# Adsorption of Proteins at Solid Surfaces

**Hans Arwin**

**Abstract** Ellipsometry has a very high thin film sensitivity and can resolve sub-nm changes in the thickness of a protein film on a solid substrates. Being a technique based on photons in and photons out it can also be applied at solid-liquid interfaces. Ellipsometry has therefore found many in situ applications on protein layer dynamics but studies of protein layer structure are also frequent. Numerous ex situ applications on detection and quantification of protein layers are found and several biosensing concepts have been proposed. In this chapter, the use of ellipsometry in the above mentioned areas is reviewed and experimental methodology including cell design is briefly discussed. The classical ellipsometric challenge to determine both thickness and refractive index of a thin film is addressed and an overview of strategies to determine surface mass density is given. Included is also a discussion about spectral representations of optical properties of a protein layer in terms of a model dielectric function concept and its use for analysis of protein layer structure.

## 2.1 Introduction

### 2.1.1 Historical Background

Ellipsometry was used already 1932 for studies of organic monolayers by Tronsted et al. [1], but Vroman [2], Rothen and Mathot [3] and Stromberg et al. [4] were most likely among the first to report measurements on protein layers. Azzam et al. [5] demonstrated similar results in studies of immunological reactions and also provided a theoretical framework. Their pioneering work included studies of kinetics of protein adsorption at solid/liquid interfaces. These early investigators found ellipsometry to be a suitable tool for non-destructive analysis of thin films, both ex situ as well as in situ at a solid/liquid interface, but surprisingly the use of the technique

H. Arwin (✉)
Department of Physics, Chemistry and Biology,
Linköping University, 581 83 Linköping, Sweden
e-mail: hans.arwin@liu.se

© Springer International Publishing AG, part of Springer Nature 2018
K. Hinrichs and K.-J. Eichhorn (eds.), *Ellipsometry of Functional Organic Surfaces and Films*, Springer Series in Surface Sciences 52,
https://doi.org/10.1007/978-3-319-75895-4_2

for protein adsorption studies during the last 40 years has been limited to a few laboratories.[1] In addition, with a few exceptions, the types of applications are generally very simple and often limited to single wavelength measurements. This may be due to that most users in the field have a background in biochemistry or medicine and not are sufficiently trained in optics and physics to make full use of the technique. The development in the field during this period has been reviewed several times [6–13].

However, one also finds that about one third of the reports are from the last ten years and new research groups have started to use ellipsometry and more advanced methodology is employed. The applications of the technique on protein adsorption have now expanded to include imaging ellipsometry [14], total internal reflection ellipsometry [15–17], infrared ellipsometry [18, 19], in combination with quartz micro balance [20, 21] and more. Fortunately spectroscopy becomes more and more common and attempts are made to address scientific issues beyond semi-quantitative analysis of layer thickness and simple determination of surface mass density.

## 2.1.2  Opportunities and Challenges

Ellipsometry offers possibilities for true quantitative measurement of thin layers with sub-nm sensitivity and has the in situ advantage to allow monitoring of dynamic processes. It is based on photons-in photons-out and is thus nondestructive and can be applied even at a solid/liquid interface.

The in situ advantage should not be underestimated for protein adsorption studies because: (1) protein layers can to be studied on model surfaces very similar to those in the normal environment for proteins; (2) protein layer surface dynamics can be studied directly; and (3) no labeling of protein molecules is required. These three characteristics facilitate studies of central phenomena in protein adsorption research including competitive adsorption of proteins, protein layer structure and dynamics, protein interaction on surfaces and protein exchange reactions and more. Studies under flow is an example on an additional possibility. It is no doubt that ellipsometry is a convenient and excellent tool to study surface dynamics in biological systems. This major advantage is particular important since living systems are by definition continuously changing and rely on chemical processes, molecular transport, synthesis and degradation.

Which are the challenges then? We may distinguish between *basic research* on protein adsorption performed in research laboratories and *biosensor* applications with goal to get established in clinical laboratories and even in the doctors office. For the research applications, the available ellipsometers on the market have precision, speed and general performance to match the requirements for high-precision bioadsorption studies. It is fair to say that the instrument problem is solved in this

---

[1]A search in Web of Science with topics *ellipsometry* AND *protein* results in more than 1600 hits with one fourth of the hits from a few groups in Sweden. If authors are listed among the 1200 hits one finds that 7 of the 10 with most publications are from Sweden.

context. Of course, depending on the actual scientific question addressed, there are always technical challenges related to how to expose proteins to a surface: in situ or ex situ adsorption; flowing or static adsorption conditions; solution stirring or not; temperature control; etc.

With the instrument problem solved, the two most important scientific challenges are the data evaluation and to select the most appropriate surface for the problem addressed. In in vitro studies of protein adsorption one may always question if the model surface used is relevant. Ellipsometry has this problem in common with many other techniques but a special limitation for ellipsometry is that a relatively flat surface at least a few square mm$^2$ large is required. Adsorption studies on curved surfaces or small particles are not possible. Many ellipsometric protein adsorption studies are traditionally performed on silicon wafers[2] which are extremely flat and lend themselves to surface modification by silanization to change their surface chemistry, physics and energy. Another commonly used material is gold which can be modified using thiol chemistry. From an optical point of view, these two types of surfaces are excellent and provide high optical contrast to protein layers and are readily available. However, the critical question about the biological relevance should always be asked. The second challenge, evaluation of the primary data $\Psi$ and $\Delta$, is a central problem in ellipsometry. For protein layers the main issue is how to simultaneously determine layer thickness and layer refractive index. This will be discussed in some detail in the following sections.

For biosensor applications of ellipsometry, one may say that the challenges are the opposite. The evaluation problem is solved in the sense that sufficient sensitivity is obtainable to detect small amounts of adsorbed protein on a surface. However, ellipsometry is not a method for chemical identification, so chemical and biological specificity must be achieved through biorecognition. This is a central challenge but is rather a biochemical issue than an ellipsometric. The main ellipsometric challenge is on the instrumental side. There are several concepts proposed but systems suitable for clinical tests are not readily available. To be competitive, a system also must be simple to operate and maintain and proof of concept must be well documented. This review is limited to presentation of a few approaches to realize biosensor concepts.

## 2.1.3  Objectives and Outline

The objective of this chapter is to provide an overview of the use of ellipsometry in the life science area with limitation to protein layers. In the methodology sections, hardware configurations will be described very briefly as there are numerous variants found in the literature. Focus will instead be on strategies for evaluation of ellipsometric data. For applications, the more simple applications based on single-wavelength ellipsometry for thickness and surface mass determination will only be summarized

---

[2]A silicon surface always has a thin native oxide or is deliberately oxidized so it would be more correct to say that protein adsorption is done on $SiO_x/SiO_2$ when silicon is used.

as there are several reviews available on this subject. Instead, development including more advanced approaches like imaging and spectroscopy employed for studies of surface dynamics, structural analysis and biosensing will be addressed.

## 2.2   Methodology—Experimental Aspects

Ex situ experiments with ellipsometry on protein layers do not differ from other types of ellipsometric thin film studies and various types of instruments are described elsewhere. Here we discuss some configurations used in in situ applications, i.e. at solid/liquid interfaces.

Null ellipsometry in a PCSA (polarizer-compensator-sample-analyzer) configuration is the traditional ellipsometer used for protein adsorption studies. A PCSA instrument is robust, easy to operate, has high resolution and simple data collection. Among drawbacks are limitations to single-wavelength operation and low speed. If fast dynamics are studied a possibility is to operate a PCSA instrument in off-null mode [22]. This configuration is also convenient for imaging ellipsometry [14]. In more advanced applications involving spectroscopy, photometric instruments like the rotating analyzer configuration are normally employed.

Very important for in situ measurements in liquids is design and features of the cell required to hold the liquid. Most cells are home made and adapted to the particular needs of the experiments conducted. Among the most important factors is how liquids are mixed in the cell. In some designs a flow system is included but it is very hard to avoid dead (unstirred) volumes and there is always an unstirred layer close to the surface under test. Molecules always must diffuse over this layer. Magnetic stirring is often used and has the advantage that the same liquid is in the cell all the time compared to a in flow system. Cells also have windows and proper windows characterization must be performed and included in the data evaluation in a similar manner as for windows in vacuum systems. Additional complications in cell design is if temperature control is required for the surface or molecular interactions studied.

A major difference form measurements performed in air or vacuum is that the ambient, i.e. the liquid, has a refractive index larger than one and, more important, has a dispersion [23]. Furthermore, addition of molecules to the liquid may change its refractive index which in most cases is seen as a change in $\Psi$ which can be mistaken for an adsorption process. However, this change is very fast and occurs as soon as mixing is complete (within seconds), whereas a change due to molecular adsorption on the surface occurs slower (tens of seconds or slower) due to diffusion over the unstirred layer at the interface. To consider in the evaluation of data is also that molecules do not only adsorb on the surface—molecules also desorb at the same time, i.e. there is a replacement. Often the biomolecules replace water molecules and contact adsorbed ions.

## 2.3  Methodology—Modeling Aspects

A data set of $(\Psi, \Delta)$-pairs is obtained in an ellipsometric experiment. In ex situ mode it is possible to perform spectroscopic, variable angle, dynamic and imaging ellipsometry, whereby, wavelength $\lambda$, incidence angle $\theta$, time $t$ and lateral position, respectively, are independent parameters. In in situ mode it may be complicated to vary $\theta$. In a traditional in situ protein adsorption experiment, the time evolution of $\Psi$ and $\Delta$ at single $\lambda$ is monitored which in a modern methodological perspective is very rudimentary as there is much to gain by using spectroscopic ellipsometry.

The basic model for evaluation is that we have a surface with known properties on which there is a layer with thickness $d$ and refractive index $N = n + ik$, where $n$ and $k$ (the extinction coefficient) are the real and imaginary parts of $N$. In addition the fill factor $f$ of the layer may be of interest if density effects are addressed. If spectroscopy is employed, it is also helpful to introduce dispersion models to describe the $\lambda$-variation of $N$. A complementary parameter traditionally used is the surface mass density $\Gamma$ which represents the amount of protein on a surface in units of e.g. ng/mm$^2$. $\Gamma$ is conceptually very easy for a layman to understand but implies a reduction of the information as it combines $d$ and $N$ into one parameter. Figure 2.1 shows a few schematic examples of protein layer structures and also presents the two most common models used for evaluation. The model to the bottom left most closely represents reality in the sense that $d$ is the physical extension of the layer into the ambient and $N$ is its effective refractive index. The model to the bottom right represents a "collapsed" layer with $d_{eq}$ and $N_p$ corresponding to thickness and intrinsic index of a dense protein layer. The latter model is often used when $N$ and $d$ cannot be separated and $N$ is then normally assumed or taken from the literature. One may also consider intermediate models as discussed by Werner and coworkers [24]. They also pointed out that steady-state irreversible adsorption of HSA and fibrinogen on hydrophobic polymers strongly depends on the dynamics.

The examples of layers illustrated in Fig. 2.1 should, to be more precise, be described in an anisotropic model due to form birefringence. However, the out-of-plane (normal to the surface) sensitivity is low in an ellipsometric experiment at oblique incidence on a thin layer so the anisotropy can normally not be resolved. One has to simplify and use isotropic models as shown in Fig. 2.1.

Here we will first discuss strategies to determine $d$ and $N$ followed by presentation of model dispersion functions for protein layers. A short introduction to alternatives to determine $\Gamma$ is also included.

### 2.3.1  Strategies to Determine Both Thickness and Refractive Index

It is often stated that it is impossible, or at least very hard, to resolve both $d$ and $N$ for a thin film in an ellipsometric experiment. The argument is that the product $Nd$ enters into the film phase thickness $\beta$ in the reflection coefficients in the three-phase

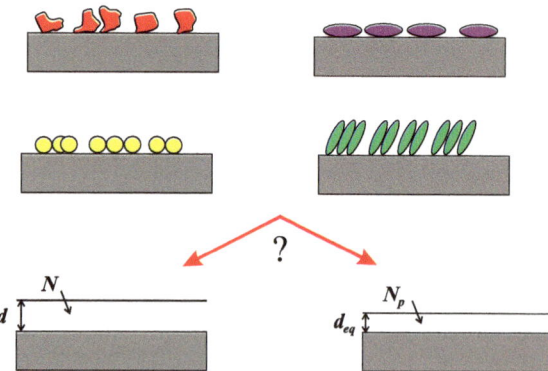

**Fig. 2.1** The schematic sketches *on top* show four different protein configurations including irregular shaped, spherical and ellipsoidal (end-on and side-on) molecules. *Below* is shown two simplified models for evaluation: (*left*) a layer with thickness $d$ and index $N$ representing the layer extension (true thickness) and effective index, respectively; (*right*) a layer with thickness $d_{eq}$ and index $N_p$ representing the equivalent thickness for a dense layer and intrinsic protein index, respectively

model. From this product it is not possible to determine both $d$ and $N$. However, $N$ enters weakly but independent of $d$ into the reflection expressions through the Fresnel interface coefficients which allows a separation of $d$ and $N$ if the accuracy of the data is sufficiently high.

The bottom line is that it all depends on the character of the sample, the measurement conditions and earlier also on the performance of the instrument used. Modern instruments are very precise and are not limiting. The traditionally used silicon surfaces have the disadvantage that mainly $\Delta$ changes upon adsorption of a protein layer due to that silicon is near-dielectric. The change in $\Delta$ can be very large providing very high detection sensitivity but the change in $\Psi$ is often small and in most cases smaller than systematic errors in the system. Effectively there is therefore only one experimental parameter available and two ($d$ and $n$) or three ($d$, $n$ and $k$) model parameters cannot be determined in a single experiment. Usually $N$ is then taken from the literature or assumed.

However, there is no principle hindrance to determine both $d$ and $N$ for a protein layer. This was proven already in the 80s for a 2.4 nm thick layer of bovine serum albumin (BSA) adsorbed on a HgCdTe substrate [25]. Later similar in situ experiments were performed and the spectral dependence of $N$ of a 4.1 nm thick layer of lactoperoxidase on gold was determined as shown in Fig. 2.2 [26].

The strategy used in the two examples above is based on spectroscopic ellipsometric data and uses that $d$ and $n$ can be determined in a spectral region where $k = 0$. Once $d$ is found, $N$ can be determined on a wavelength-by-wavelength basis for all $\lambda$. The thickness determined in this way should be considered as a representation of the extension of the layer and the index is the average layer index as illustrated in Fig. 2.1. The latter would correspond to the intrinsic refractive index of the protein itself only if the layer is 100% dense which rarely is the case for a non-crystalline layer.

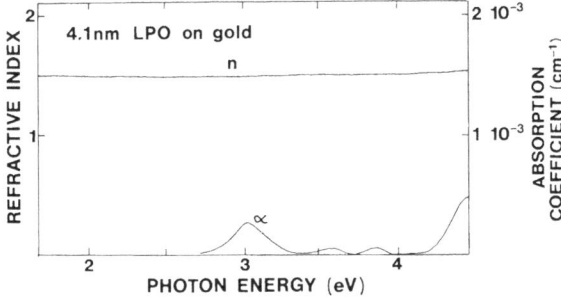

**Fig. 2.2** Real part $n$ of the refractive index of a 4.1 nm layer of lactoperoxidase on gold. Also shown is the absorption coefficient $\alpha = 4\pi k/\lambda$. The Kramers–Kronig counterpart to the $\alpha$-band around 3 eV is not seen in $n$ at the scales used. Reprinted from [26] with permission from Elsevier

Malmsten [27] used interference enhancement to study protein adsorption on silicon with 30 nm thermal oxide and could thereby reduce the disadvantage with silicon. He successfully used accurate single wavelength null ellipsometry data recorded in situ to resolve dynamics in $d$ and $n$ (assuming $k = 0$) for human serum albumin (HSA), immunoglobulin G (IgG), fibrinogen and lysozym. Conclusions about layer structure were possible to draw from these results.

If the interface between the protein layer and the substrate is sufficiently sharp, the Arwin–Aspnes method [28] can be used to find $d$, whereafter the protein layer index can be determined on a wavelength-by-wavelength basis as shown by [29, 30]. The above methods use numerical inversion to find $N$ at fixed $d$. However, if $k = 0$ is assumed, one can use analytic inversion. A fifth degree polynomial equation is then obtained. It can be readily solved and was applied at the air/water interface to layers of arachidic acid and valine gramicidin A with thickness in the range of 2–3 nm [31]. An alternative to wavelength-by-wavelength analysis is to include dispersion models for $N$ as shown by Berlind et al. [32] who used a Cauchy dispersion in the visible part of the spectrum and Arwin et al. [33] who used a more complex model dielectric function in the infrared. The strategies described above are summarized in Table 2.1.

One can also combine ellipsometry with other methods to determine both thickness and porosity of thin organic films. Rodenhausen et al. [20] combined in situ ellipsometry with quartz crystal microbalance measurements to address the ultrathin film limit $2\pi nd/\lambda \ll 1$. This approach is discussed in detail in Chap. 17 in this book.

**Table 2.1** Overview of strategies for determining $N$ and $d$ of thin protein films

| Strategy | Assumption | References |
|---|---|---|
| $\lambda$-by-$\lambda$ | $k = 0$ | [25, 26] |
| Single $\lambda$, interference | $k = 0$ | [27] |
| Arwin–Aspnes method | – | [28] |
| Analytical inversion | $k = 0$ | [31] |
| Dispersion models | Cauchy, Gauss, Lorentz etc. | [32, 33] |

## 2.3.2  Spectral Representations of Protein Layers

As discussed above, a majority of the early ellipsometric protein adsorption studies
were performed with single wavelength methodology and only a single value of $N$
was obtained, often at $\lambda = 546$ nm or $\lambda = 633$ nm. However, spectroscopic ellipsometry comes more an more in use for protein adsorption studies and with spectroscopic
data available it is possible to model $N$ for a protein layer using dispersion models.
Most protein molecules are non-absorbing in the visible spectral range and thus a
protein layer can be assumed to be transparent. A Cauchy model is the basic model
and is defined by

$$n(\lambda) = A + \frac{B}{\lambda^2} + \frac{C}{\lambda^4} \tag{2.1}$$

where $A$, $B$ and $C$ are parameters determined by fitting (2.1) to experimental data.
Sometimes it is sufficient to fit only $A$ and $B$.

Protein molecules have a background adsorption in the ultraviolet (UV) spectral range due to the peptide chain which is seen in the lactoperoxidase-spectrum
in Fig. 2.2. Some proteins may contain functional groups like hemes and exhibit
absorption bands and additional complexity in the modeling of $N$ should be introduced. The heme group absorption around 3 eV in Fig. 2.2 serves as an example. The
peptide backbone resonances and other electronic bands in the UV and visible (VIS)
spectral regions can be modeled with Lorentzian or Gaussian dispersion models.

In the infrared (IR) region, proteins have characteristic but complex absorption
bands which carry information about protein secondary structure and other structural
details. The analysis of these optical features allows for example to determine the
amount of $\alpha$-helix or $\beta$-sheet structure in proteins [34]. These vibrational resonances
can also be modeled with Lorentzian or Gaussian dispersion models. A suitable
overall model dielectric function (MDF) for $\epsilon = N^2$ versus photon energy $E$ is

$$\epsilon(E) = \epsilon_\infty - \sum_j \frac{A_j \Gamma_j E_j}{E^2 - E_j^2 + i\Gamma_j E} - \sum_k \frac{A_k \Gamma_k \bar{\nu}_k}{\bar{\nu}^2 - \bar{\nu}_k^2 + i\Gamma_k \bar{\nu}} \tag{2.2}$$

where $\epsilon_\infty$ is a constant accounting for resonances at energies larger than the spectral
range studied, $A_j$, $\Gamma_j$ and $E_j$ are amplitude, broadening and energy, respectively,
of $j$ UV-VIS resonances, and $A_k$, $\Gamma_k$ and $\bar{\nu}_k$ are amplitude, broadening and energy,
respectively, of $k$ IR resonances. In IR it is customary to express resonance energies
in wavenumbers $\bar{\nu}$ as indicated in the last term in (2.2). $\bar{\nu}$ is related to $E$ by $E = hc_0\bar{\nu}$
where $h$ is Plancks constant and $c_0$ is the speed of light. An example of the use of
(2.2) is given later in this chapter.

## 2.3.3  Determination of Surface Mass Density

If the correlation between $N$ and $d$ cannot be resolved one can present results in
terms of the derived parameter $\Gamma$, the surface mass density. As an example, Cuypers

et al. [35] found that the time evolution of $d$ and $n$ ($k = 0$ was assumed) was very noisy for adsorption of protrombin on chromium, whereas if $\Gamma$ was derived, a considerable reduction in noise occurs. Some advantages with using $\Gamma$ is that it is easy to understand and also directly can be compared with results from radio immunoassays [36, 37] and gravimetric methods. However, a major drawback with $\Gamma$ is that structural information is lost.

The method above is referred to as Cuypers et al. model [35] and is based on that $n$ and $d$ have been determined for a protein layer using ellipsometry. In addition one need the molecular weight, molar refractivity and partial specific volume of the molecules in the layer. The model is derived assuming a Lorentz-Lorenz effective medium for a mixed layer (ambient and biomolecules).

A more frequently used and recommended model for $\Gamma$ was developed by de Feijter et al. [38]. They derived the expression[3]

$$\Gamma = 100\frac{d(n - n_{amb})}{dn/dc} \tag{2.3}$$

where $n_{amb}$ is the refractive index of the ambient (in general a liquid) and $dn/dc$ is the refractive index increment of the protein. The value on $dn/dc$ can be determined with an Abbe refractometer or with prism deviation measurements on protein solutions [23]. Also in de Feijter et al.'s model the noise in $\Gamma$ is strongly reduced compared to in $d$ and $n$. Further possibilities to determine $\Gamma$ and also a comparison among the methods can be found elsewhere [10].

## 2.4 Applications

### 2.4.1 Protein Adsorption and Dynamics on Model Surfaces

Determination of thickness and/or surface mass density is one of the most common applications of ellipsometry in the area of protein adsorption. It is often performed in situ to monitor surface dynamics. These types of applications have been reviewed and further details can be found in [6–13].

### 2.4.2 Studies of Protein Layer Structure

Above we discussed the basic challenge to quantify and monitor the dynamics of protein adsorption. In this section we will address some structural aspects including layer density, layer structure from dynamics of $n$ and $d$, molecular ordering from anisotropy and from analysis of infrared chemical signatures.

---

[3]With a prefactor 100, $\Gamma$ is expressed in ng/cm$^2$ if $d$ is in nm and $dn/dc$ in cm$^3$/g.

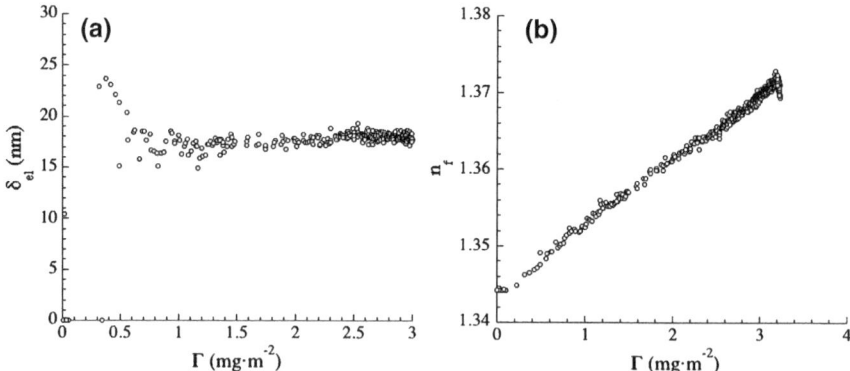

**Fig. 2.3** Thickness (**a**) and refractive index (**b**) versus surface mass density of a layer of IgG evaluated from recordings of in situ ellipsometric data on oxidized silicon. Reprinted from [27] with permission from Elsevier

**Layer Density** A very basic structural parameter of a protein layer is its density. For an adsorbed layer of a rigid protein, there will certainly be density deficiencies. For a more flexible protein, conformation changes may occur upon adsorption but still one cannot expect an ideal homogeneous layer with flat parallel boundaries. Density is strongly related to $n$ and various effective medium approximations (EMA) are used. A very simple density modeling can be done if $n$ for a (more) dense protein layer is known as exemplified by determination of a density deficiency of 30% of a BSA layer on platinum [6] compared to a nominally dense BSA layer on HgCdTe. Ellipsometry can in principle provide $n$ for thin layers as discussed above but it has not been proven yet to have sensitivity to resolve in-depth density profiles for protein monolayers as can be done with neutron reflectometry [39, 40]. However, for thicker protein layers, density depth-profiling is possible as shown by Kozma et al. [41]. They studied several hundred nm thick flagellar filament protein layers on surface activated $Ta_2O_3$. The protein layers were described by five EMA sublayers. By fitting this model to spectroscopic ellipsometry data, they determined the in-depth variation in surface mass density.

**Dynamic Relations Between $n$ and $d$** Malmsten [27] made pioneering work and used single-wavelength ellipsometry to resolve structural details and film formation mechanisms for layers of fibrinogen, $\gamma$-globulin and more proteins on silicon with 30 nm thermal oxide. Figure 2.3 shows that IgG adsorbing on an oxidized silicon surface made hydrophobic by methylation, proceeds with $n$ linearly increasing from the value of the ambient medium to a final value of around 1.37 corresponding to $\Gamma = 3\,mg/m^2$. During the whole adsorption process, $d$ is more or less constant except for low $\Gamma$ where noise is seen. The dimensions of IgG are $23.5 \times 4.5 \times 4.5$ nm and Malmsten concluded that IgG has a near end-on orientation. The relatively low $n$ and $\Gamma$ show in addition that the layer has low packing density. Fibrinogen showed a more complex dynamics with a near-linear change in both $n$ and $d$ [27].

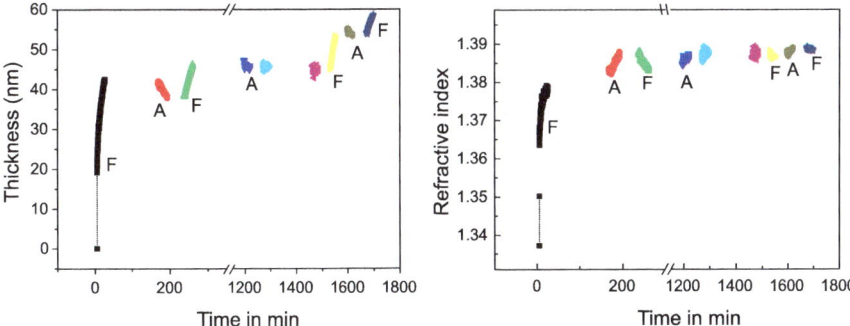

**Fig. 2.4** Time evolution of thickness (*left*) and refractive index (*right*) during formation of a fibrinogen layer matrix on a functionalized silicon substrate. Symbol F indicates fibrinogen adsorption and A layer activation using EDC/NHS. Reprinted from [32] with permission from Elsevier

Further structural studies with ellipsometry along these lines but also including multilayer adsorption were performed by Berlind et al. [32] who studied fibrinogen covalently bonded on functionalized silicon surfaces using affinity ligand coupling chemistry. Adsorption of fibrinogen was monitored in situ with ellipsometry at $\lambda = 500$ nm and ellipsometric spectra were measured at steady-state in the range 350–1050 nm. Figure 2.4 shows $d$ and $n$ during multistep adsorption and in situ chemical activation of fibrinogen layers evaluated on a $\lambda$-by-$\lambda$ basis. Figure 2.5 shows the corresponding change in $\Gamma$ calculated with de Feijter et al.'s formula in (2.3). Fibrinogen matrices with thickness up to 58 nm and with surface mass density $\Gamma$ of 1.6 μg/cm$^2$ were prepared in this way. The first adsorption step results in a fibrinogen layer with $\Gamma = 1$ μg/cm$^2$. A chemical activation using ethyl-3-dimethyl-aminopropyl-carbodiimide and N-hydroxy-succinimide (EDC/NHS) methodology was then performed with intention to promote binding of the next fibrinogen layer. However, only a small thickness increase was observed upon a second fibrinogen adsorption step. Furthermore $\Gamma$ increases monotonically during the whole experiment and when $n$ or $d$ decreases, $\Gamma$ remains constant, i.e. no desorption occurs. This thickness decrease is accompanied with an index increase supporting a densification of the fibrinogen layer. A proposed structural model for the fibrinogen layer is shown in Fig. 2.5 and is further discussed in [32]. Earlier single-$\lambda$ ellipsometry experiments with ex situ activation and incubation showed very different results [42] with a considerable increase in $\Gamma$ after each incubation step more or less proportional to the number of incubation/activation steps performed. These differences are most probably due to the drying steps in the ex situ case.

**Molecular Ordering from Anisotropy** Protein molecules in a layer are normally assumed to be randomly oriented and very few protein layers are crystalline. Even if there is an order, an isotropic layer is normally assumed if the molecules not are uniaxial or biaxial. However, there are some indications that protein molecules can be uniaxial. Sano [43] showed that structural anisotropy in BSA leads to uniaxial molecules with $n$ of 1.744 and 1.563 along the major and minor axes, respectively.

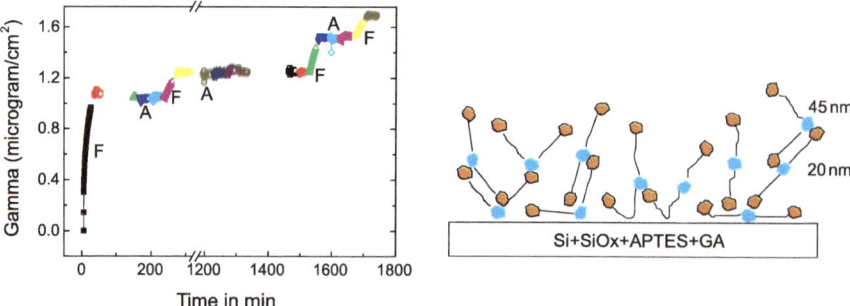

**Fig. 2.5** Time evolution of surface mass density $\Gamma$ (*left*) during formation of a fibrinogen layer matrix on a functionalized silicon substrate. Symbol F indicates fibrinogen adsorption and A layer activation using EDC/NHS. To the *right* is shown a possible structure of a fibrinogen matrix formed by multistep adsorption/activation. Reprinted from [32] with permission from Elsevier

In addition there may be form induced birefringence in protein layers as briefly discussed in the methodology section above. Of course, the resolving power of ellipsometry on protein films may be insufficient to resolve their anisotropy regardless of being intrinsic or form induced. In highly ordered, but thin films of fatty acids prepared by Langmuir-Blodgett techniques, Engelsen [44] demonstrated that anisotropic modeling is relevant. Future refinement of ellipsometric methodology will tell if anisotropy of protein films will add to further understanding of their properties and structure.

**Chemical Structure** Infrared spectroscopy is well established for studies of protein structure. The advantage is that vibrational signatures can be correlated to the secondary structures in protein molecules [34]. However, surprisingly few reports on application of infrared spectroscopic ellipsometry (IRSE) for studies of protein layers are found in the literature in spite of that IRSE has the advantage to provide a quantification of the amount of protein on a surface, i.e. to determine the layer thickness in addition to the IR spectral features. The limited use of IRSE may be due to that the instrumentation is rather expensive and slow. With a pyroelectric DTGS detector, a typical measurement on a protein monolayer may take 12 h or more but using other types of detectors can shorten the measurement time. A rather large sample area of several mm$^2$ is also required. The speed and sample size can be reduced by employing synchrotron infrared spectroscopic ellipsometry as shown by Hinrichs et al. [45]. They studied peptides, proteins and their antibodies and could in particular identify the so called amide bands [19].

IRSE has been applied to determine $N$ of bovine carbonic anhydrase (BCA) adsorbed in a 500 nm thick porous silicon layer [18]. Using Lorentzian resonances, as those in (2.2), five absorption bands, including the amide I, II and A, could be resolved and parameterized. In addition it was found that more BCA per surface area was adsorbed near the surface. Protein monolayers and multilayers have also been studied by IRSE on flat model surfaces [33]. In particular the effect of heating on the secondary structure (the amide bands) were observed. For a multilayer with ten

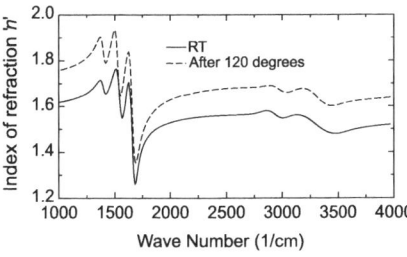
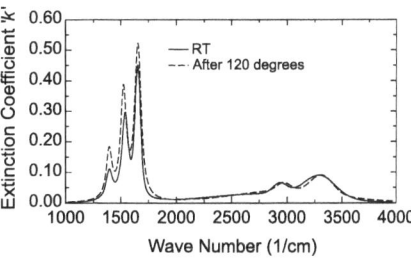

**Fig. 2.6** Real part $n$ (*left*) and extinction coefficient $k$ (*right*) for a fibrinogen monolayer on gold measured at room temperature (RT) before and after heating to 120 °C. Reprinted from [33]

**Table 2.2** Amide band parameters and 90% confidence intervals in a Lorentzian model for $N$ of a 4.1 nm fibrinogen layer on gold

| Resonance | Frequency (cm$^{-1}$) | Broadening (cm$^{-1}$) | Amplitude |
|---|---|---|---|
| – | $1397 \pm 3$ | $40 \pm 8$ | $0.39 \pm 0.09$ |
| Amide II | $1537 \pm 3$ | $40 \pm 8$ | $0.99 \pm 0.09$ |
| Amide I | $1654 \pm 2$ | $44 \pm 4$ | $1.55 \pm 0.09$ |
| Amide A | $2963 \pm 24$ | $141 \pm 77$ | $0.13 \pm 0.04$ |
| Amide A | $3311 \pm 14$ | $260 \pm 45$ | $0.031 \pm 0.03$ |

alternating HSA and anti-HSA layers with a total thickness of 40.2 nm, it was found that heating to 120 °C reduced the thickness around 1 nm and that mainly the amide A band was affected as observed in $N$. Upon heating to 200 °C, major changes in all amide bands occurred.

Also effects of heating a 4.1 nm monolayer of fibrinogen on gold was studied [33]. Figure 2.6 shows that heating of a fibrinogen to 120 °C increases both $n$ and $k$. The thickness decreases to 3.5 nm so effectively a densification of the layer has occurred. A possible explanation is that water has desorbed. The frequencies and corresponding broadenings and amplitudes for the five identified resonance before heating are shown in Table 2.2. Very small changes, except for increase in amplitudes, very found upon heating.

### 2.4.3 Protein Layer Based Biosensing

It was early shown that ellipsometry could be used for detection of biomolecules (see e.g. [3]). The idea of using ellipsometry as a sensor principle is therefore old and was also reviewed several years ago [46]. Several attempts to design systems for end users have been presented but have not been implemented in clinical laboratories so far. The Isoscope or comparison ellipsometer [47] and the fixed polarizer ellipsometer [48] are two examples of suggested point-of-care systems. Also imaging ellipsometry were suggested many years ago as a high through-put system for biosensing [14, 49]. It seems that ellipsometry as a biosensor principle not finds its way from research to clinical laboratories. No user-friendly and cost-efficient instruments suitable for clinical use are commercially available. In spite of this, suggestions for biosensing applications continues to be published which indicates that researches are believers in its potential. The review will here be limited to presentation of some studies addressing two of the most important aspects in this context: (1) imaging ellipsometry readout for achieving high through-put and (2) bioaffinity for achieving specificity.

Imaging ellipsometry provides a means to map the lateral thickness or surface mass density variations on a surface. By using a beam with a large diameter, it is possible to image a large area, e.g. $15 \times 30 \, \text{mm}^2$ [14]. An example of imaging of 4 mm diameter spots of three different proteins is shown in Fig. 2.7. With a focusing lens, small area, e.g. $60 \times 200 \, \mu\text{m}^2$ can be imaged [50]. Recently Gunnarsson et al. [51] demonstrated time-resolved imaging with a sensitivity in surface mass density of $1 \, \text{ng/mm}^2$ and pixel size less than $0.5 \, \mu\text{m}$. Development of affinity biochips with 900 targets and ellipsometric readout has also been reported [52].

The development of imaging ellipsometry for biosensing based on protein layers and as well as other layers has been pioneered by Jin and coworkers [53, 54]. Their applications include detection of monoclonal antibodies from SARS (severe acute respiratory syndrome) using virus, immobilized on silicon substrates, as antigen [55], detection of the protein hormone human somatropin down to $0.0004 \, \text{IU/ml}$ [56], detection of *Riemerella anatipestifer* (bacteria causing septicemia in birds) down to $5.2 \times 10^3 \, \text{CFU/ml}$ using immunoglobulin as sensing layer [57], detection of tumor markers for cancer diagnostics down to $10 \, \text{U/ml}$ [58, 59], biological amplification for detection of alpha-fetoprotein in cancer diagnostics down to $5 \, \text{ng/ml}$ [60], detection of duck hepatitis virus down to $8 \times 10^{-9.5} \, \text{LD}_{50}/\text{ml}$ using polyclonal antibodies on silicon [61], detection of hepatitis B virus markers down to $1 \, \text{mg/ml}$ [62], and more.

Proof of concept of using imaging ellipsometry for immunosensors have also been given by Bae and coworkers. They showed a detection limit of $10 \, \text{ng/ml}$ for insulin [63], detection of the bacteria *Legionella pneumophila* down to $10^3 \, \text{CFU/ml}$ [64] and the bacteria *Yersinia enterocolitica* also with detection limit $10^3 \, \text{CFU/ml}$ [65].

In most cases, the sensitivity of ellipsometry for detection of adsorbing molecules is sufficient for immunoassays. However, if increased sensitivity is required, one can employ SPP-enhancement as discussed in Chap. 18 in this book. An example of such an application, of relevance for diagnosis of Alzheimer's decease, is a label-free direct immunoassay for detection of $\beta$-amyloid peptide (1-16) using monoclonal antibodies immobilized on a gold surface [66].

A complement or alternative to use biochemical specificity is to use optical specificity, e.g. fluorescence or chemiluminescence. Hinrichs et al. [45] suggest the use of infrared spectroscopic mapping ellipsometry. Areas of $6 \times 6 \, \text{mm}^2$ with a resolution of $300 \times 300 \, \mu\text{m}^2$ were mapped and protein amide bands were identified. The drawback is that currently infrared radiation from a synchrotron beam line is required.

In conclusion we find that affinity-based biosensing with ellipsometry can provide sufficiently low detection limits for many important clinical applications and compares well with alternative methods. In addition it is a label-free technique.

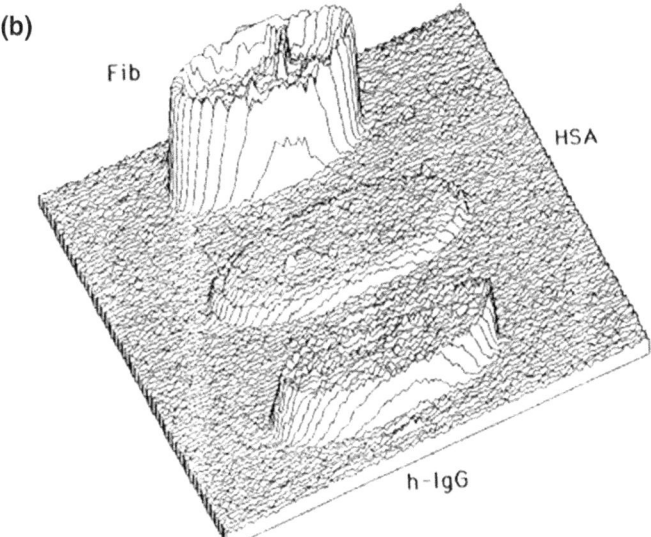

**Fig. 2.7  a** Off-null ellipsometry irradiance measured on a silicon surface patterned with 4 mm diameter protein spots. Fib, HSA and h-IgG corresponds to fibrinogen, human serum albumin and human immunoglobulin G, respectively. **b** Three- dimensional visualization of the irradiance distribution. Reprinted from [14] with permission from American Institute of Physics

### 2.4.4   Other Applications

**Protein Adsorption as a Surface and Thin Film Probe** Protein adsorption has been used as a probe for testing the biocompatibility and other functionalities of surfaces and thin film materials. Gyulai et al. studied biodegradable polyesters for drug carrier systems and used adsorption of BSA to determine how protein repellent the polyester surface is [67]. Mikhaylova et al. used lysozyme and HSA as model proteins with different charge to probe complex adsorption responses on hydroxyl-terminated hyperbranched aromatic polyester thin films [68]. They found that a thicker polyester film resulted in a lower BSA adsorption probably coupled to a higher hydrophilicity. Warenda et al. [69] studied HSA interaction with oligosaccharide-modified hyperbranched poly(ethylene imine) films and found that HSA adsorption was below $50\,ng/cm^2$ under certain conditions. A low protein adsorption is crucial for the use of these films in biosensors. BSA has also been used to test biocompatibility of thin films of tantalum, niobium, zirconium and titanium oxides [70] and HSA to test biocompatibility of thin polymer brushes [71] (see also Chap. 6).

## 2.5   Outlook

It is no doubt that ellipsometry, especially in spectroscopic and imaging modes, is among the most valuable tools for studying protein adsorption on solid surfaces including possibilities to follow dynamics of a thin film structure. Among the new developments now being mature are spatial imaging ellipsometry with potential applications in high-through put screening of bioadsorption but also in surface mapping in general. A representative example is the investigation of light-activated affinity micropatterning of proteins using imaging ellipsometry [72]. Such spatially resolved immobilization of proteins may find applications in surface control of biomaterials and tissue engineering, multi-analyte biosensors, clinical assays and genomic arrays.

Time-resolved spectroscopic ellipsometry and time-resolved imaging ellipsometry have so far only a few applications. Internal reflection ellipsometry is also mature and holds great promises due to its extreme sensitivity if the SPP phenomenon is utilized as is discussed in Chap. 18 of this book. Infrared spectroscopic ellipsometry is a technique with large promises in the life science area in general and in particular for protein adsorption studies. IRSE allows describing composition, structure and layer thickness in the same measurements. Ellipsometry is sometimes combined with other in situ techniques including potentiometry, impedance spectroscopy and quartz-micro balance (see also Chap. 17). The latter has been proven to be a powerful combination in structural analysis of thin protein films [20, 21].

A major problem is technology transfer from the scientific community to colleagues in industrial and clinical laboratories, especially for biosensor applications. Perhaps we have to wait for an ellipsometer on a chip.

# References

1. L. Tronsted, Trans. Faraday Soc. **31**, 1151–1156 (1935)
2. L. Vroman, in *Blood Clotting Enzymology*, ed. by W.H. Seegers (Academic Press, New York, 1967), pp. 279–323
3. A. Rothen, C. Mathot, Helv. Chim. Acta **54**, 1208–1217 (1971)
4. R.R. Stromberg, L.E. Smith, F.L. McCrackin, Symp. Faraday Soc. **4**, 192–200 (1970)
5. R.M.A. Azzam, P.G. Rigbyand, J.A. Krueger, Phys. Med. Biol. **22**, 422–430 (1977)
6. H. Arwin, D.E. Aspnes, Thin Solid Films **138**, 195–207 (1986)
7. H. Arwin, Thin Solid Films **313–314**, 764–774 (1998)
8. H. Arwin, in *Physical Chemistry of Biological Interfaces*, ed. by A. Baszkin, W. Norde (Dekker, New York, 2000), pp. 577–607
9. H. Arwin, Thin Solid Films **377–378**, 48–56 (2000)
10. H. Arwin, in *Handbook of Ellipsometry*, ed. by H.G. Tompkins, E.A. Irene (William Andrew, Norwich, 2005), pp. 799–855
11. H. Arwin, Thin Solid Films **519**, 2589–2592 (2011)
12. H. Elwing, Biomaterials **19**, 397–406 (1998)
13. P. Tengvall, I. Lundström, B. Liedberg, Biomaterials **19**, 407–422 (1998)
14. G. Jin, R. Jansson, H. Arwin, Rev. Sci. Instrum. **67**, 2930–2936 (1996)
15. M. Poksinski, H. Arwin, Thin Solid Films **455–456**, 716–721 (2004)
16. A. Nabok, A. Tsargorodskaya, Thin Solid Films **516**, 8993–9001 (2008)
17. Z. Balevicius, A. Ramanaviciene, I. Baleviciute, A. Makaraviciute, L. Mikoliunaite, A. Ramanavicius, Sens. Actuators B, Chem. 160, 555–562 (2011)
18. H. Arwin, L.M. Karlsson, A. Kozarcanin, D.W. Thompson, T. Tiwald, J.A. Woollam, Phys. Status Solidi A **202**, 1688–1692 (2005)
19. G. Sun, D.M. Rosu, X. Zhang, M. Hovestädt, S. Pop, U. Schade, D. Aulich, M. Gensch, B. Ay, H.-G. Holzhutter, D.R.T. Zahn, N. Esser, R. Volkmer, J. Rappich, K. Hinrichs, Phys. Status Solidi B **247**, 1925–1931 (2010). https://doi.org/10.1002/pssb.200983945
20. K.B. Rodenhausen, T. Kasputis, A.K. Pannier, J.Y. Gerasimov, R.Y. Lai, M. Solinsky, T.E. Tiwald, H. Wang, A. Sarkar, T. Hofmann, N. Ianno, M. Schubert, Rev. Sci. Instrum. **82**, 103111 (2011)
21. E. Bittrich, K.B. Rodenhausen, K.-J. Eichhorn, T. Hofmann, M. Schubert, M. Stamm, P. Uhlmann, Biointerphases **5**, 159–167 (2010)
22. H. Arwin, S. Welin-Klintström, R. Jansson, J. Colloid Interface Sci. **156**, 377–382 (1993)
23. T. Berlind, G. Pribil, D. Thompson, J.A. Woollam, H. Arwin, Phys. Status Solidi C **5**, 1249–1252 (2008)
24. C. Werner, K.-J. Eichhorn, K. Grundke, F. Simon, W. Grählert, H.-J. Jacobasch, Colloids Surf. A Physicochem. Eng. Asp. **156**, 3–17 (1999)
25. H. Arwin, Appl. Spectrosc. **40**, 313–318 (1986)
26. J. Mårtensson, H. Arwin, I. Lundstrom, T. Ericson, J. Colloid Interface Sci. **155**, 30–36 (1993)
27. M. Malmsten, J. Colloid Interface Sci. **166**, 333–342 (1994)
28. H. Arwin, D.E. Aspnes, Thin Solid Films **113**, 101–113 (1984)
29. V. Reipa, A.K. Gaigalas, V.L. Vilker, Langmuir **13**, 3508–3514 (1997)
30. R.G.C. Oudshoorn, R.P.H. Kooyman, J. Greve, Thin Solid Films **284–285**, 836–840 (1996)
31. D. Ducharme, A. Tessier, S.C. Russev, Langmuir **17**, 7529–7534 (2001)
32. T. Berlind, M. Poksinski, P. Tengvall, H. Arwin, Colloids Surf. B Biointerfaces **75**, 410–417 (2010)
33. H. Arwin, A. Askendahl, P. Tengvall, D.W. Thompson, J.A. Woollam, Phys. Status Solidi C **5**, 1438–1441 (2008)
34. J.T. Pelton, L.R. McLean, Anal. Biochem. **277**, 167–176 (2000)
35. P.A. Cuypers, J.W. Corsel, M.P. Janssen, J.M. Kop, W.T. Hermens, H.C. Hemker, J. Biol. Chem. **258**, 2426 (1983)
36. J. Benesch, A. Askendal, P. Tengvall, Colloids Surf. B Biointerfaces **18**, 71–81 (2000)

37. U. Jönsson, M. Malmqvist, I. Ronnberg, J. Colloid Interface Sci. **103**, 360–372 (1985)
38. J.A. de Feijter, J. Benjamins, F.A. Veer, Biopolymers **17**, 1759–1772 (1978)
39. J.R. Lu, X. Zhao, M. Yaseen, Curr. Opin. Colloid Interface Sci. **12**, 9–16 (2007)
40. T.J. Su, J.R. Lu, R.K. Thomas, Z.F. Cui, J. Penfold, J. Colloid Interface Sci. **203**, 419–429 (1998)
41. P. Kozma, D. Kozma, A. Nemeth, H. Jankovics, S. Kurunczi, R. Horvath, F. Vonderviszt, M. Fried, P. Petrik, Appl. Surf. Sci. **257**, 7160–7166 (2011)
42. P. Tengvall, E. Jansson, A. Askendal, P. Thomsen, C. Gretzer, Colloids Surf. B Biointerfaces **28**, 261–272 (2003)
43. Y. Sano, J. Colloid Interface Sci. **124**, 403–406 (1988)
44. D.D. Engelsen, Optical anisotropy in ordered systems of lipids. Surf. Sci. **56**, 272–280 (1976)
45. K. Hinrichs, M. Gensch, N. Esser, U. Schade, J. Rappich, S. Kröning, M. Portwich, R. Volkmer, Anal. Bioanal. Chem. **387**, 1823–1829 (2007)
46. H. Arwin, Sens. Actuators A Phys. **92**, 43–51 (2001)
47. M. Stenberg, T. Sandstrom, L. Stiblert, Mater. Sci. Eng. **42**, 65–69 (1980)
48. R.M. Ostroff, D. Maul, G.R. Bogart, S. Yang, J. Christian, D. Hopkins, D. Clark, B. Trotter, G. Moddel, Clin. Chem. Lab. Med. **44**, 2031–2035 (1998)
49. G. Jin, P. Tengvall, I. Lundström, H. Arwin, Anal. Biochem. **232**, 69–72 (1995)
50. A. Albersdörfer, G. Elender, G. Mathe, K.R. Neumaier, P. Paduschek, E. Sackmann, Appl. Phys. Lett. **72**, 2930–2932 (1998)
51. A. Gunnarsson, M. Bally, P. Jönsson, N. Médard, F. Höök, Anal. Chem. **84**, 6538–6545 (2012)
52. D. van Noort, J. Rumberg, E.W.H. Jager, C.-F. Mandenius, Meas. Sci. Technol. **11**, 801–808 (2000)
53. G. Jin, Y.H. Meng, L. Liu, Y. Niu, S. Chen, Q. Cai, T.J. Jiang, Thin Solid Films **519**, 2750–2757 (2011)
54. G. Jin, Phys. Status Solidi A **205**, 810–816 (2008)
55. C. Qi, J.-Z. Duan, Z.-H. Wang, Y.-Y. Chen, P.-H. Zhang, L. Zhan, X.-Y. Yan, W.-C. Cao, G. Jin, Biomed. Microdevices **8**, 247–253 (2006)
56. Z.-Y. Zhao, G. Jin, Z.-H. Wang, in *Proceedings of the 20th Annual International Conference of the IEEE Engineering in Medicine and Biology Society*, 1998, vol. 2 (1998), pp. 590–593
57. C. Huang, J. Li, Y. Tang, C. Wang, C. Hou, D. Huo, Y. Chen, G. Jin, Mater. Sci. Eng. C Biomim. Mater. Sens. Syst. **31**, 1609–1613 (2011)
58. Y. Zhang, Y. Chen, G. Jin, Sens. Actuators B Chem. **159**, 121–125 (2011)
59. C. Huang, Y. Chen, G. Jin, Ann. Biomed. Eng. **39**, 185–192 (2011)
60. C. Huang, Y. Chen, C. Wang, W. Zhu, H. Ma, G. Jin, Thin Solid Films **519**, 2763–2767 (2011)
61. C. Huang, J. Li, Y. Tang, Y. Chen, G. Jin, Curr. Appl. Phys. **11**, 353–357 (2011)
62. C. Qi, W. Zhu, Y. Niu, H.G. Zhang, G.Y. Zhu, Y.H. Meng, S. Chen, G. Jin, J. Viral Hepat. **16**, 822–832 (2009)
63. Y.M. Bae, B.-K. Oh, W. Lee, W.H. Lee, J.-W. Choi, Biosens. Bioelectron. **20**, 895–902 (2004)
64. Y.M. Bae, B.-K. Oh, W. Lee, W.H. Lee, J.-W. Choi, Mater. Sci. Eng. C Biomim. Mater. Sens. Syst. **24**, 61–64 (2004)
65. Y.M. Bae, B.-K. Oh, W. Lee, W.H. Lee, J.-W. Choi, Anal. Chem. **76**, 1799–1803 (2004)
66. M.K. Mustafa, A. Nabok, D. Parkinson, I.E. Tothill, F. Salam, A. Tsargorodskaya, Biosens. Bioelectron. **26**, 1332–1336 (2010)
67. G. Gyulai, C.B. Pénzes, M. Mohai, T. Lohner, P. Petrik, S. Kurunczi, E. Kiss, J. Colloid Interface Sci. **362**, 600–606 (2011)
68. Y. Mikhaylova, V. Dutschk, M. Müller, K. Grundke, K.-J. Eichhorn, B. Voit, Colloids Surf. A Physicochem. Eng. Asp. **297**, 19–29 (2007)
69. M. Warenda, A. Richter, D. Schmidt, A. Janke, M. Müller, F. Simon, R. Zimmermann, K.-J. Eichhorn, B. Voit, D. Appelhans, Macromol. Rapid Commun. **33**, 1466–1473 (2012)
70. P. Silva-Bermudez, S.E. Rodil, S. Muhl, Appl. Surf. Sci. **258**, 1711–1718 (2011)
71. S. Burkert, E. Bittrich, M. Kuntzsch, M. Müller, K.-J. Eichhorn, C. Bellmann, P. Uhlmann, M. Stamm, Langmuir **26**, 1786–1795 (2010)
72. Z. Yang, W. Frey, T. Oliver, A. Chilkoti, Langmuir **16**, 1751–1758 (2000)

# Chapter 3
# DNA Structures on Silicon and Diamond

**Simona D. Pop, Karsten Hinrichs, Sylvia Wenmackers,
Christoph Cobet, Norbert Esser and Dietrich R. T. Zahn**

**Abstract**  In the design of DNA-based hybrid devices, it is essential to have knowledge of the structural, electronic and optical properties of these biomolecular films. Spectroscopic ellipsometry is a powerful technique to probe and asses these properties. In this chapter, we review its application to biomolecular films of single DNA bases and molecules on silicon and diamond surfaces characterized in the spectral range from the near-infrared (NIR) through the visible (Vis) and toward the vacuum ultraviolet (VUV). The reported optical constants of various DNA structures are of great interest, particularly in the development of biosensors.

## 3.1  Introduction

Nearly two decades ago, Murphy et al. released the first suggestion of long range electron transport in the deoxyribonucleic acid (DNA) helix [1]. Since then, many researchers have aimed to clarify the true electronic behavior of DNA. Most studies report a semiconducting [2–5] to conducting-like [6] character, but also insulating characteristics have been found [7], and later superconductivity if DNA at low temperatures [8]. This lack of consensus is at least partially due to the large variation in the experimental condition, which may influence the molecular conformation,

S. D. Pop (✉) · K. Hinrichs · N. Esser
Leibniz-Institut für Analytische Wissenschaften – ISAS – e.V.,
Schwarzschildstr. 8, 12489 Berlin, Germany
e-mail: simonadorinapop@gmail.com

S. Wenmackers
Faculty of Philosophy, University of Groningen, Oude Boteringestraat 52,
9712 GL Groningen, The Netherlands

C. Cobet
Center for Surface- and Nanoanalytics, Johannes Kepler Universität Linz,
Altenbergerstrasse 69, 4040 Linz, Austria

D. R. T. Zahn
Semiconductor Physics, Technische Universität Chemnitz,
09107 Chemnitz, Germany

© Springer International Publishing AG, part of Springer Nature 2018
K. Hinrichs and K.-J. Eichhorn (eds.), *Ellipsometry of Functional
Organic Surfaces and Films*, Springer Series in Surface Sciences 52,
https://doi.org/10.1007/978-3-319-75895-4_3

the contribution of contact resistance, etc. In any case, the exact nature and mechanism of conduction are still not fully determined and understood [9]. Nevertheless, in the development of the next generation of molecular electronics devices using DNA, which typically involve layers of immobilized DNA molecules, it is essential to gain a better knowledge of the optical and electronic properties of such DNA layers. Hitherto, numerous molecular electronics devices, including nanowires [10–12], transistors [4, 13–15], and magnetic valves [16], have been suggested or some prototypes have been realized. Yoo et al. have demonstrated for the first time the possibility of a DNA-based field effect transistor (FET) operating at room temperature [15], thus opening the research field of DNA-based molecular electronic devices. Charge migration through DNA takes place via the overlap of the $\pi$ orbitals in adjacent base pairs, adenine-thymine (A-T) and guanine-cytosine (G-C). Hence, besides entire DNA molecules, also the single DNA bases are considered to be potential charge transport molecules. Mauricio et al. studied a FET based on a modified single DNA base-guanosine reporting that the prototype bio-transistor gives rise to a better voltage gain compared to carbon nanotubes (CNTs) [17].

A fundamental issue for designing DNA hybrid devices is the determination of structural, electronic and optical properties of such biomolecular films on inorganic substrates. Particularly the knowledge of the dielectric functions and/or of the optical constants—i.e. refractive index—($n$) and absorption coefficient—($k$) becomes essential. The dielectric function is one of the most fundamental magnitudes, which correlates both to electronic and to optical characteristics but also to the structure of the material. Therefore, ellipsometry is a powerful technique that addresses and solves such fundamental aspects.

The purpose of this chapter is to confirm that spectroscopic ellipsometry is a suitable optical method which contributes also to the field of biomolecular films, particularly DNA-based ones. The chapter aims to give a concise survey of the current ellipsometric studies performed so far of DNA structures on technologically relevant substrates such as silicon and diamond.

The majority of currently published ellipsometric studies concentrate on the immobilization and grafting process of DNA onto suitable substrates. These types of DNA experiments are intended to promote biosensor applications: they apply ellipsometry only as a characterization method for the biosensitive layer during the research and development phase. In a minority of the studies, ellipsometry is envisioned as the detection method in the finished biosensor. Recently, Demirel et al. reported such a novel DNA-biosensor based on ellipsometry [18]. In this study the main objective was to use the self-assembled monolayers (SAMs) with functional groups as a platform for oriented (active) immobilization of the probe-ssDNA onto silicon surfaces for better hybridization and detection of the target complementary ssDNA by ellipsometry [18]. Elhadj et al. proved that ellipsometry can discriminate between ssDNA and dsDNA structures, reporting that the refractive index of dsDNA is higher than that of ssDNA by 5% [19]. Other ellipsometric study such as of the covalently attached DNA layers on diamond surfaces were approached by Wenmackers et al. [20]. In this case, the DNA's optical constants derived from ellipsometry data were reported in the UV-VUV spectral range. Previous optical constants of a DNA film on a mesh nickel screen were calculated from standard transmission measurements using the Kramers-Kronig relation [21].

The usage of the DNA structures in the area of FETs often requires the preparation of the DNA films by thermal evaporation under vacuum conditions. The great advantage of such a procedure, known in the literature as organic molecular beam deposition (OMBD) [22], stems from the high degree of molecular ordering onto the substrate, purity, but also film homogeneity. DNA bases were prepared as thin films on hydrogen-terminated H:Si(111) surfaces by OMBD under ultra-high vacuum conditions and their near-IR (NIR)-Vis-VUV dielectric functions have been firstly reported by our group [23]. The anisotropic dielectric functions of the purines (i.e. adenine and guanine) films were derived from a uniaxial optical model, while the dielectric functions of the pyrimidines (i.e. cytosine and thymine) were found to be isotropic. Particularly, the anisotropic dielectric function of guanine was deduced independently in the mid-IR (MIR) and NIR-Vis-VUV spectral ranges. From the analysis of both vibrational and electronic transitions, a planar orientation of the guanine molecule with respect to the silicon surface was inferred [23, 24]. The in situ VUV-SE experiments using synchrotron radiation performed so far either on DNA bases or DNA molecular strands has not revealed any UV-damage, proving that this optical technique is a non-destructive characterization method for this application.

As next, the chapter briefly outlines several aspects regarding the optical modeling used so far in the analysis of the ellipsometry spectra of various DNA structures. Finally, the chapter reviews selective applications of spectroscopic ellipsometry to different DNA structures on inorganic substrates like silicon and diamond.

## 3.2 Dielectric Function

The dielectric constant of a material, $\tilde{\varepsilon}$, at a fixed wavelength of incident light is a complex number that can be expressed as $\tilde{\varepsilon} = \varepsilon_1 + i\varepsilon_2$ or as $\tilde{\varepsilon} = \tilde{n}^2$ with $\tilde{n} = n + ik$, where $n$ is the refractive index and $k$ is the absorption coefficient of the material. The variation of this dielectric constant upon the angular frequency (or photon energy) of the incident light is referred to as dielectric dispersion and can be represented by the dielectric function, $\tilde{\varepsilon}(\omega)$. This function can be described in the linear optics model by the classical oscillator model as

$$\tilde{\varepsilon}(\omega) = 1 + \tilde{\chi}_{el}(\omega) + \tilde{\chi}_{vib}(\omega) + \tilde{\chi}_{fc}(\omega) \quad \text{with } \tilde{\varepsilon}(\omega \to \infty) = \varepsilon_\infty \qquad (3.1)$$

where mainly three frequency-dependent contributions are considered: $\tilde{\chi}_{el}$,—which is the susceptibility of valence electrons involved in the electronic (interband) transitions, $\tilde{\chi}_{vib}$,—which is the susceptibility of the collective vibrations of atoms or intraband transitions, and $\tilde{\chi}_{fc}$, which describes the presence of the free-charge carriers.

**Fig. 3.1** Sketch of an
electronic transition showing
the HOMO (ground state)
and LUMO (first excited
state) levels of a DNA base,
guanine

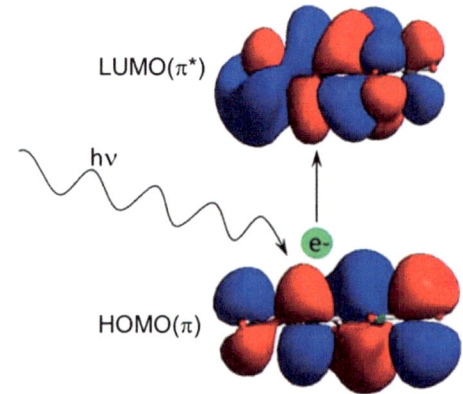

In general case, $\varepsilon$ is a tensor which can be reduced to a complex scalar for isotropic materials or to a direction dependent dielectric function in the situation of anisotropic materials.

In particular, the electronic (interband) transitions are observed from NIR through Vis to VUV spectral photon energy range. Figure 3.1 schematically shows the first electronic transition from the ground state to the first excited state of a DNA base, guanine. The highest occupied molecular orbital (HOMO) to lowest unoccupied molecular orbital (LUMO) electronic transition has a $\pi-\pi^*$ character and the corresponding energy difference is also called the optical band gap, $E_{opt}$.

Now, the measured (effective) dielectric function of an optically isotropic sample (such as DNA film on a substrate) during a spectroscopic ellipsometry measurement is given by the following relation:

$$\langle \varepsilon \rangle = \sin^2 \phi_0 + \sin^2 \phi_0 \tan^2 \phi_0 \left[ \frac{1 - \rho}{1 + \rho} \right] \tag{3.2}$$

where $\phi_0$ is the angle of incidence and $\rho = \tan \Psi e^{i\Delta}$, with $\Psi$ and $\Delta$ as ellipsometric parameters.

The relation (3.2) is not valid in the case of an anisotropic biomolecular film on the substrate, where the optical response of the film should be described by several layers. Further details regarding the optical modeling of the anisotropic samples are given in the introductory chapter on the spectroscopic ellipsometry (Chap. 1) and Chap. 13, which are included in this book.

In the ellipsometric analysis of the optical behavior of the DNA structures two classical oscillator models, namely Lorentz and Gaussian were used so far [18–20, 23, 24]. For the symmetric Lorentz (3) and Gaussian (4) oscillator models, $\varepsilon_2$ is defined by the summation of $n$ Lorentz or Gaussian functions, each being parameterized by the peak energy of each oscillator ($E_{0n}$), its amplitude ($A_n$) and its broadening ($Br_n$) in eV as follows:

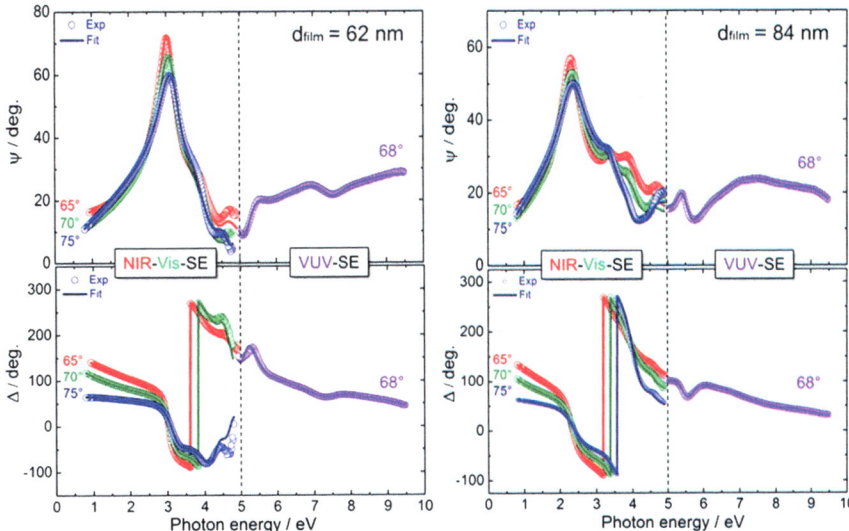

**Fig. 3.2** Two sample analysis in the case of guanine films on H:Si(111) in the NIR-Vis-VUV spectral range. The figure is modified after [23]

$$\varepsilon_2 = \sum_n \frac{A_n Br_n E_{0n}}{E_{0n}^2 - E^2 - i Br_n E} \tag{3.3}$$

and

$$\varepsilon_2 = \sum_n \left[ A_n e^{-((E-E_{0n})/Br_n)^2} + A_n e^{-((E+E_{0n})/Br_n)^2} \right]. \tag{3.4}$$

Figure 3.2 shows a concrete example of two sample analysis of spectroscopic ellipsometry data of DNA base-guanine on silicon. Here, the optical response of both biomolecular films was best described by a uniaxial anisotropic model ($\varepsilon_x = \varepsilon_y \neq \varepsilon_z$) consisting of either two dispersion functions for the refractive index (where $k = 0$) or two sets of Gaussian oscillators (where $k \neq 0$). The optical axes are corresponding to the perpendicular and parallel directions defined with respect to [111] direction of the silicon substrate [23]. The uniaxial optical model was chosen in conjunction with X-ray diffraction measurements. The multi-sample analysis (MSA) procedure has been employed in order to ensure the accuracy of the derived dielectric function, especially for the optical component normal to the surface from the employed uniaxial optical model [25, 26]. In the MSA, several films with different thicknesses, but the same optical constants can be coupled together in the simulation to avoid the correlation between parameters during the simulation procedure. Another option to avoid the correlation between parameters is to vary the angle of incidence [25].

## 3.3  Applications to Thin Biomolecular Films

### 3.3.1  Single DNA Bases

An ellipsometry investigation of the thin films of DNA bases performed by our group revealed that the DNA bases are strongly absorbing in the UV-VUV spectral range, having the absorption onset in the near ultra-violet (UV) region [23]. The DNA base films were grown onto hydrogen-passivated Si(111) surfaces by OMBD under high vacuum conditions. Different film thicknesses, ranging from 40–120 nm, were investigated in order to carry out a MSA with the aim of extracting accurate optical constants using sufficient experimental data. In situ VUV-SE measurements were performed in the energy range between 4 and 9.5 eV under an angle of incidence of 68° by a home-built ellipsometer, using synchrotron radiation as the light source at BESSY II [27, 28]. Afterwards, the biomolecular films were investigated ex situ via spectroscopic ellipsometry (NIR-Vis-SE) in the energy range of 0.8–5 eV under various angles of incidence, using a commercial ellipsometer. Figure 3.3 shows an overview of the measured ellipsometric spectra of a 84 nm thick guanine film and the derived dielectric function in the MIR and NIR to VUV photon energy range, the later from the MSA. The simulated data for the MIR spectral range are thoroughly explained by Hinrichs et al. in the following [24].

In the NIR-VUV spectral range a three-layer model was considered, namely substrate/guanine film/surface roughness. The surface roughness layer described by an EMA approach was neglected in the MIR, since its influence is within the experimental error. On the other hand, in the VUV range, the probing wavelength is much smaller, which would result in surface roughness influences of the ellipsometric spectra above the error of the measurements. The optical response of guanine films was described by a uniaxial model in accordance with X-ray diffraction (XRD) measurements which revealed a planar orientation of the molecules. For the simulation of the NIR-VUV ellipsometric spectra, Gaussian oscillators and for the MIR ellipsometric spectra Lorentzian oscillators have been used. The anisotropic dielectric function of guanine is shown in Fig. 3.3b where the ordinary contribution is larger compared to the extraordinary contribution. This situation points to a mainly planar orientation of the guanine molecules with respect to the silicon substrate in agreement with the XRD. Characteristic vibrational and electronic transitions having their transition dipole moments in the molecular plane are indicated together with the concluded overall molecular arrangement [23, 24]. In the MIR-SE analysis the refractive index $n_{xy\infty} = 1.76$, $n_{z\infty} = 1.45$ and the thickness values determined from visible data have been used as input for determining the dielectric function in this spectral range. These were obtained from the non-absorbing spectral range of (0.8–3 eV) from a two sample analysis, 62 and 84 nm, respectively (see Fig. 3.2).

In the case of the cytosine films, the thickness dependence of the optical properties in the MIR spectral range hindered the usage of a standard MSA. Ellipsometric investigations performed on different film thicknesses of the DNA base-cytosine deposited on H:Si(111) surfaces revealed that IR-SE is additionally sensitive to the

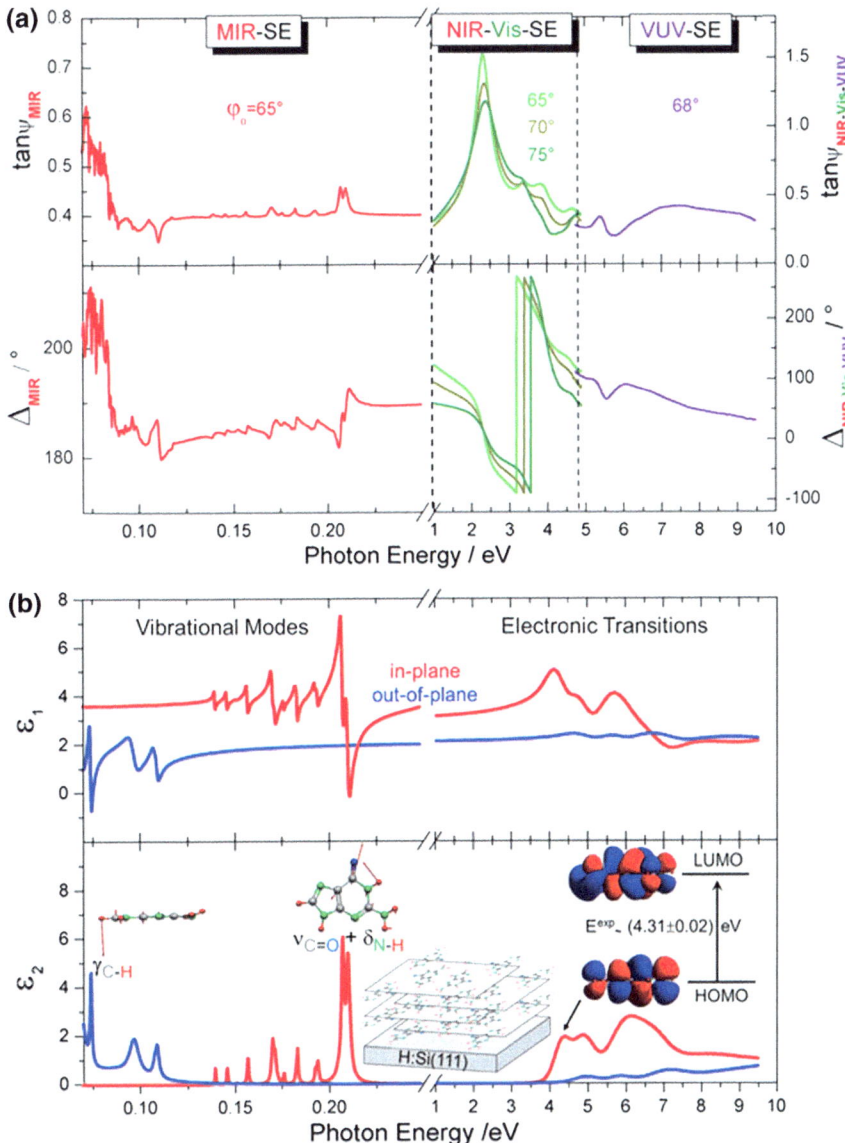

**Fig. 3.3** (**a**) Measured ellipsometric spectra of a 84 nm thick guanine film on H:Si(111); (**b**) Derived dielectric function of highly ordered guanine film from MIR to VUV. The in-plane and out-of-plane contributions are labeled with respect to the substrate's surface. Note the pronounced optical anisotropy related to the dipole orientation of both vibrational and electronic excitations. The overall planar orientation of the guanine molecules onto H:Si(111) surface is schematically shown in the *insets*. The figures are modified after [23, 24]

**Fig. 3.4** Measured MIR ellipsometric spectra of various cytosine films on H:Si(111) taken under an angle of incidence of 60°. In the *inset*, the molecular structure of the cytosine is shown

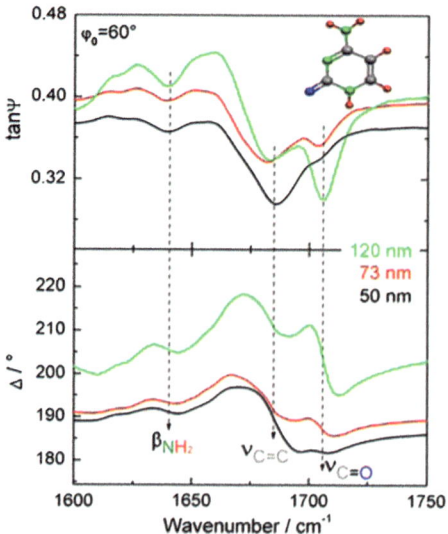

structural changes within the organic films having different thicknesses (see Fig. 3.4). The change in the relative intensities of the in-plane vibrational modes in the frequency range between 1600 and $1750\,cm^{-1}$ with the film thickness denotes a gradual change in the molecular orientation of cytosine from thin to thick layers. In this particular situation, there is no suitable optical model available in the current literature which can describe such complex behavior where the optical modeling is challenged.

### 3.3.2 Single- and Double-Stranded DNA Molecules

As previously mentioned, DNA can be single- or double-stranded (ss-ds); in the latter case, two complementary strands form a double helix. If ssDNA of an unknown sequence binds with ssDNA probes (with a known sequence) thereby forming dsDNA, one can infer the sequence of the unknown DNA. This binding of complementary ssDNA is the working principle of a DNA sensor, in which a layer of probe ssDNA is usually attached to the surface of a suitable material, such as silicon or diamond, using a linker layer.

The characterization of a DNA chip by ellipsometry has been reported by Gray et al. [29]. Figure 3.5a shows the layer structure of the DNA chip in which a 14 base oligonucleotide (ssDNA molecule) is supported by linkers formed on a $SiO_2$/c-Si substrate. The linker used is an organic layer mainly composed of Si–O and O–$(CH_2)_n$ groups and the ssDNA molecule has the following sequence ATCATCTTTGGTGT. The thickness of each layer was determined by a single wavelength ellipsometer ($\lambda = 632.8\,nm$, HeNe laser). From the analysis of the linker/$SiO_2$/c-Si structure, $d = 19.3\,nm$ and $n = 1.460$ were obtained as the thickness and refractive index of

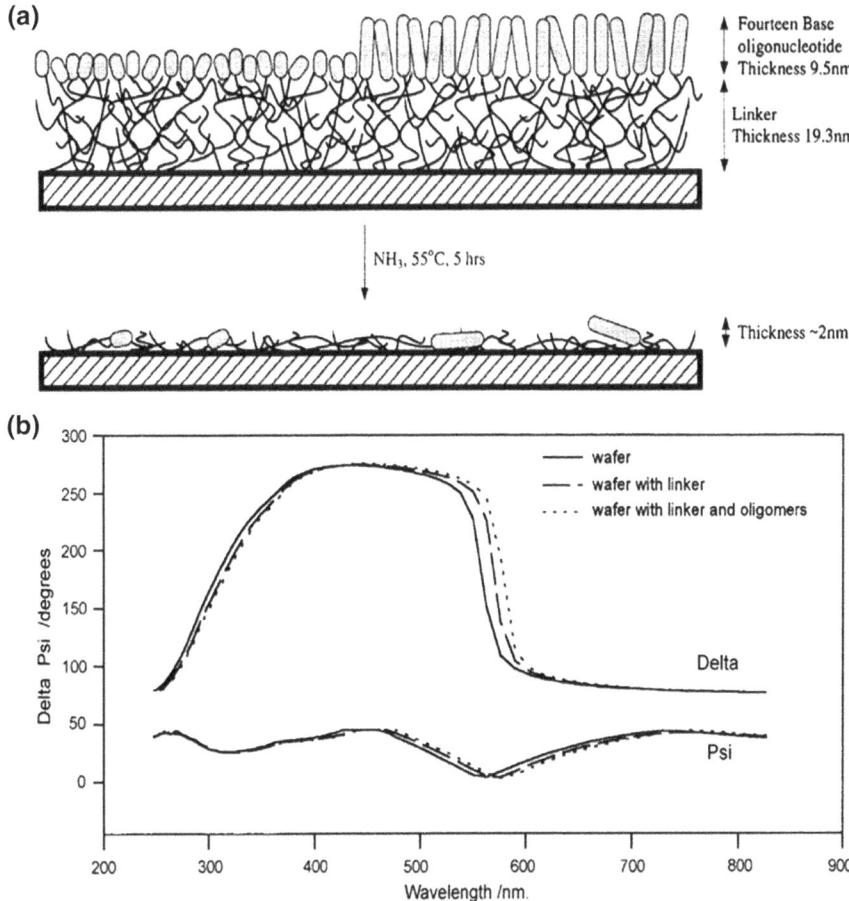

**Fig. 3.5** **a** Layer structure of a ssDNA chip analyzed by spectroscopic ellipsometry. **b** Spectroscopic ellipsometry spectra ($\Psi$, $\Delta$) taken at a point on (i) a wafer, (ii) a linker-covered wafer, and (iii) oligonucleotides (ssDNA) synthesized on the linker-covered wafer. Reprinted with permission from Gray et al., Langmuir **13**, 2833–2842 (1997). Copyright 1997, American Chemical Society

the linker layer, respectively. For $SiO_2$ and c-Si, tabulated optical constants were used. In this case, one can determine $n$ and $d$ of the linker directly from measured ($\Psi$, $\Delta$) values, provided that there are no surface roughness and interface layers. Finally, by analyzing the DNA/linker/$SiO_2$/c-Si structure using $n = 1.460$ (linker), the thickness and refractive index of the ssDNA were estimated to be $d = 9.5\,$nm and $n = 1.462$, respectively. This refractive index is consistent with that reported by S. Elhadj et al. [19]. Often it is required, especially for very thin films, to combine ellipsometry measurements with transmission, reflection data to reduce the strong correlation between $n$ and $d$ [25]. Therefore, in order to confirm the ellipsometry results Gray et al. have additionally used interferometry measurements [29].

**Fig. 3.6** Optical constants ($n, k$) of a DNA monolayer immobilized on a Si/SiO$_2$ substrate. Reprinted with permission from S. Elhadj et al., Langmuir **20**, 5539–5543 (2004). Copyright 2004, American Chemical Society

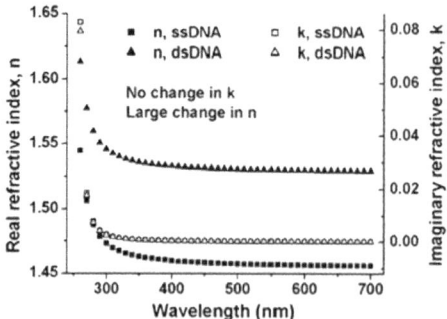

From the study using spectroscopic ellipsometry, it has been reported that the optical constants of DNA can be expressed by the Lorentz model with a single oscillator at 4.87 eV (254 nm) for both ssDNA and dsDNA, and that the refractive index of dsDNA is higher than that of ssDNA by ~5% [19]. S. Elhadj et al. proved that it is basically possible to detect the change from ssDNA to dsDNA by using the ellipsometry technique. The optical constants shown in Fig. 3.6 were obtained from the ellipsometry data using the following four-layer structure: Si/SiO$_2$/SMCC crosslinker and aminosilane/DNA. In order to validate the optical absorption expressed by the Lorentz oscillator at 4.87 eV (254 nm) from the ellipsometry data, UV-Vis absorption spectroscopy was additionally employed. The UV-Vis spectrum reveals a strong absorption at a wavelength of 258 nm (4.81 eV). An excellent consistency within <1.5% deviation between the simulated oscillator energy using the Lorentz function and the measured UV-Vis absorption was an important validation for the model used. The electron delocalization effect in this frequency range is negligible since no change in the absorption coefficient could be observed. The investigated monolayers of ssDNA and dsDNA had the average thicknesses of 6.5 and 4.9 nm, respectively. The difference of 5% in the refractive index $n$ could not be explained by the effect of the increased density due to the increase in the molecular mass from ssDNA to dsDNA, which would give rise to a difference of about 25% [19].

In the context of biosensor research, DNA has also been immobilized on the surface of synthetic diamond. Figure 3.7 depicts the determined optical constants ($n, k$) of a thin ssDNA film on nanocrystalline diamond (NCD) using ellipsometry in the UV-VUV spectral range by Wenmackers et al. [20]. In this work a monolayer of 250 bp (base-pairs) dsDNA was end-grafted on NCD surfaces, using a two-step protocol, involving the photo-attachment of fatty acids to form a linker layer, followed by the covalent DNA attachment using a zero-length cross linker. The resulted dry-dsDNA monolayer had the film thickness of about 9 nm and a refractive index, $n_\infty = 1.51$, as determined from visible energy range which is in agreement with the previous reports [19]. In order to simplify the optical modeling, the measured NCD substrate together with the attached linker has been considered as final substrate for the ds-DNA film. The optical response of the dry-dsDNA film was described by a sum of four Lorentz oscillators and one asymmetric Tauc-Lorentz oscillator at

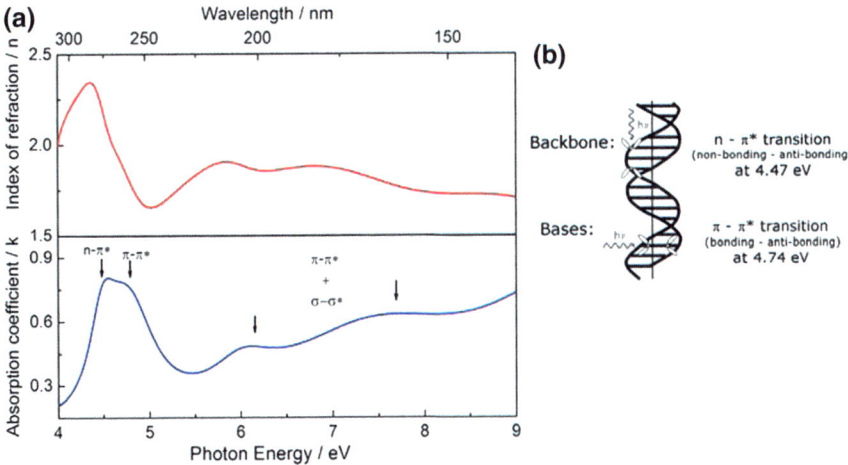

**Fig. 3.7 a** Derived optical constants $(n, k)$ of the covalently attached monolayer of 250 bp dsDNA on NCD (nano-crystalline diamond). **b** The assignment of the first two electronic transitions of DNA molecules is schematically shown. Reprinted with permission from S. Wenmackers et al., Langmuir **24**, 7269–7277 (2008). Copyright 2008, American Chemical Society

**Fig. 3.8** Optical constants, $n$ and $k$, of a 71.5 nm thick dry DNA (sodium salt of calf thymus) film on 750 mesh nickel screens. Reprinted with permission from T. Inagaki et al., J. Chem. Phys. **61**, 4246–4250 (1974). Copyright 1974, American Institute of Physics

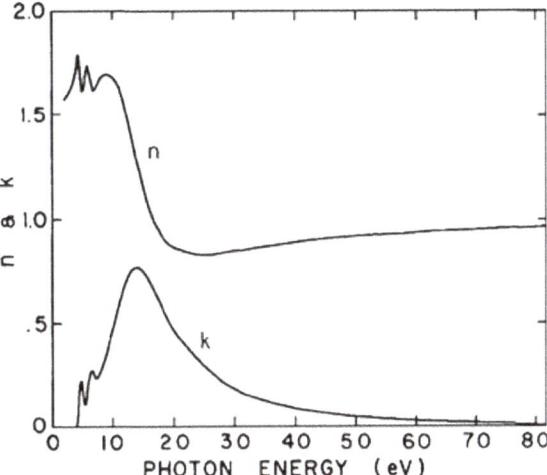

higher photon energies, centered at about 13 eV. The derived line-shape of optical absorption is in perfect agreement with the previous transmission data of a thicker DNA film reported by Inagaki et al. as shown in Fig. 3.8 [21].

The energy positions of the employed Lorentz functions are indicated by arrows. The typical $\pi-\pi^*$ electronic transition of the DNA molecule is observed at 4.74 eV (261 nm) in agreement with the values reported by Inagaki et al. [21]. Although, it is well known that UV-irradiation can damage DNA [30–32], but the presence of this non-shifted $\pi-\pi^*$ electronic transition suggests that UV dose is low enough during the VUV-SE measurement for the DNA layer to remain intact. This absorption band is assigned to the $\pi-\pi^*$ electronic transitions of the single DNA bases [21, 23]. In the higher energy range the absorption structures are dominated by mixed $\pi-\pi^*$ and $\sigma-\sigma^*$ electronic transitions which belong to both single bases and sugar phosphate groups [21].

The steep increase in the absorption above 7 eV is mainly due to $\sigma-\sigma^*$ transitions of the phosphate groups [21]. The lowest electronic transition at 4.47 eV (277 nm) assigned as an $n-\pi^*$ transition is a fingerprint for the molecular orientation, but also discriminates between the ss and ds conformation of DNA molecules [33, 34]. The molecular orientation can be deduced, knowing that the $n-\pi^*$ transition dipole moment is directed along the dsDNA backbone, which is perpendicular to the $\pi-\pi^*$ transition dipole moments of the individual bases. Here, the optical response of the 250 bp dsDNA layer is isotropic, meaning that the DNA molecules are mostly randomly distributed on the NCD substrate. Some general aspects of modeling uncertainty pertaining to this case have been discussed in [35].

Inagaki et al. have calculated the optical constants $(n, k)$ of a 71.5 nm thick dry DNA film on a nickel screen from transmission measurements based on Kramers-Kronig relation (see Fig. 3.8). The optical constants were derived in the photon energy range of 2–82 eV. So far, this is the only experimental evidence revealing a new and strong optical absorption structure of DNA located at about 13.8 eV, in agreement with our previous assumption regarding DNA's absorption in a higher photon energy range above 9 eV. Moreover, this report challenges future ellipsometry investigations of DNA structures using as light source the synchrotron radiation in an extended photon energy range, well-above 9 eV which generally coincides with the limit of the standard laboratory ellipsometers.

## 3.4   Summary

This chapter reviewed the current state-of-art of the applications of ellipsometry to a very important biological system like DNA. From the few currently existing DNA investigations using ellipsometry, it highlighted selective applications to molecular films of single DNA bases as well as of differently configured (ss, ds) DNA molecules attached to technologically relevant substrates, to wit silicon and diamond. During these ellipsometry studies, no UV-damage to the different DNA structures (either as single bases or as molecular strands) has been reported, ascertaining the non-destructiveness of the spectroscopic ellipsometry. The distinction in the refractive index between the different types of molecular DNA strands confirmed the high sensitivity of the ellipsometry technique to structural changes. Some challenges in the optical modeling of the complex ellipsometry data, such as those of the DNA base, cytosine, remain to be solved. Nevertheless, the optical constants (or the dielectric function) of various DNA structures are successfully determined and made accessible for future biotechnological applications.

**Acknowledgements** The financial support by Sächsisches Staatsministerium für Wissenschaft und Kunst, Deutsche Forschungsgesellschaft Graduiertenkolleg 829/1 "Accumulation of Single Molecules to Nanostructures", Bundesministerium für Bildung und Forschung projects 05 622 ESA2, 05 KS4KTB/3, IWT-SBO (project 030219 'CVD Diamond: a novel multifunctional material for high temperature electronics, high power/high frequency electronics and bioelectronics'), FWO-WOG (WO.035.04N 'Hybrid Systems at Nanometer Scale'), the IUAP-P6/42 program 'Quantum Effects in Clusters and Nanowires', the European Community—Research Infrastructure Action under the FP6 "Structuring the European Research Area" Programme (through the Integrated Infrastructure Initiative "Integrating Activity on Synchrotron and Free Electron Laser Science—Contract R II 3-CT-2004-506008"), and the Life Sciences Impulse Program of the transnationale Universiteit Limburg. Financial support by the Ministerium für Innovation, Wissenschaft und Forschung des Landes Nordrhein-Westfalen, the regierende Bürgermeister von Berlin—Senatskanzlei Wissenschaft und Forschung, and the Bundesministerium für Bildung und Forschung is gratefully acknowledged.

# References

1. C.J. Murphy, M.R. Arkin, Y. Jenkins, N.D. Ghatlia, S.H. Bossmann, N.J. Turro, J.K. Barton, Science **262**, 1025 (1993)
2. H.W. Fink, C. Schonenberger, Nature **398**, 407 (1999)
3. D. Porath, A. Bezryadin, S. de Vries, C. Dekker, Nature **403**, 635 (2000)
4. H. Watanabe, C. Manabe, T. Shigematsu, K. Shimotani, M. Shimizu, Appl. Phys. Lett. **79**, 2462 (2001)
5. G. Cuniberti, L. Craco, D. Porath, C. Dekker, Phys. Rev. B **65**, 241314(R) (2002)
6. H.W. Fink, Cell. Mol. Life Sci. **58**, 1 (2001)
7. Y. Zhang, R.H. Austin, J. Kraeft, E.C. Cox, N.P. Ong, Phys. Rev. Lett. **89**(1), 198102 (2002)
8. AYu. Kasumov, M. Kociak, S. Gueron, B. Reulet, V.T. Volkov, D.V. Klinov, H. Bouchiat, Science **291**, 280 (2001)
9. R.G. Endres, D.L. Cox, R.R.P. Singh, Rev. Mod. Phys. **76**, 195 (2004)
10. E. Braun, Y. Eichen, U. Sivan, G. Ben Yoseph, Nature **391**, 775 (1998)
11. O. Harnack, W.E. Ford, A. Yasuda, J.M. Wessels, Nano Lett. **2**, 919 (2002)
12. J. Richter, M. Mertig, W. Pompe, I. Monch, H.K. Schackert, Appl. Phys. Lett. **78**, 536 (2001)
13. E. Ben Jacob, Z. Hermon, S. Caspi, Phys. Lett. A **263**, 199 (1999)
14. Z. Hermon, S. Caspi, E. Ben Jacob, Europhys. Lett. **43**, 482 (1998)
15. K.-H. Yoo, D.H. Ha, J.-O. Lee, J.W. Park, J. Kim, J.J. Kim, H.-Y. Lee, T. Kawai, H.Y. Choi, Phys. Rev. Lett. **87**(1), 198102 (2001)
16. M. Zwolak, M. Di Ventra, Appl. Phys. Lett. **81**, 925 (2002)
17. G. Mauricio, P. Visconti, V. Arima, S. D'Amico, A. Biasco, E. D'Amone, R. Cingolani, R. Rinaldi, Nano Lett. **3**, 479 (2003)
18. G. Demirel, M.O. Caglayan, B. Garipcan, E. Piskin, Surf. Sci. **602**, 952 (2008)
19. S. Elhadj, G. Singh, R.F. Saraf, Langmuir **20**, 5539 (2004)
20. S. Wenmackers, S.D. Pop, K. Roodenko, V. Vermeeren, O.A. Williams, M. Daenen, O. Douhéret, J. D'Haen, A. Hardy, M.K. Van Bael, K. Hinrichs, C. Cobet, M. vandeVen, M. Ameloot, K. Haenen, L. Michiels, N. Esser, P. Wagner, Langmuir **24**, 7269 (2008)
21. T. Inagaki, R.N. Hamm, E.T. Arakawa, L.R. Painter, J. Chem. Phys. **61**, 4246 (1974)
22. S.R. Forrest, Chem. Rev. **97**, 1793 (1997)
23. S.D. Silaghi, M. Friedrich, C. Cobet, N. Esser, W. Braun, D.R.T. Zahn, Phys. Status Solidi (b) **242**, 3047 (2005)
24. K. Hinrichs, S.D. Silaghi, C. Cobet, N. Esser, D.R.T. Zahn, Phys. Status Solidi (b) **242**, 2681 (2005)

25. H. Fujiwara, *Spectroscopic Ellipsometry*, Principles and applications (Wiley, West Sussex, 2007)
26. W.A. McGahn, B. Johs, J.A. Woollam, Thin Solid Films **234**, 443 (1993)
27. R.L. Johnson, J. Barth, M. Cardona, D. Fuchs, A.M. Bradshaw, Nucl. Instrum. Methods Phys. Res. A **290**, 606 (1990)
28. H.G. Tompkins, E.A. Irene (eds.), *Handbook of Ellipsometry* (William Andrew Publishing, Norwich, 2005)
29. D.E. Gray, S.C. Case-Green, T.S. Fell, P.J. Dobson, E.M. Southern, Langmuir **13**, 2833 (1997)
30. C.A. Sprecher, W.A. Baase, W. Curtis, Biopolymers **18**, 1009 (1979)
31. C. Kielbassa, L. Roza, B. Epe, Carcinogenesis **18**, 811 (1997)
32. R.P. Sinha, D.P. Häder, Photochem. Photobiol. Sci. **1**, 225 (2002)
33. A. Rich, M. Kasha, J. Am. Chem. Soc. **82**, 6197 (1960)
34. J.R. Fresco, A.M. Lesk, R. Gorn, P. Doty, J. Am. Chem. Soc. **83**, 3155 (1961)
35. S. Wenmackers, D.E.P. Vanpoucke, Stat. Neerl. **66**, 339 (2012)

# Chapter 4
# Thickness and Beyond. Exploiting Spectroscopic Ellipsometry and Atomic Force Nanolithography for the Investigation of Ultrathin Interfaces of Biologic Interest

Pietro Parisse, Ilaria Solano, Michele Magnozzi, Francesco Bisio, Loredana Casalis, Ornella Cavalleri and Maurizio Canepa

**Abstract** The evaluation of thickness, refractive index, and optical properties of biomolecular films and self-assembled monolayers (SAMs) has a prominent relevance in the development of label-free detection techniques (quartz microbalance, surface plasmon resonance, electrochemical devices) for sensing and diagnostics. In this framework Spectroscopic Ellipsometry (SE) is an important player. In our approach to SE measurements on ultrathin soft matter, we exploit the small changes of the ellipsometry response ($\delta\Delta$ and $\delta\Psi$) following the addition/removal of a layer in a nanolayered structure. So-called $\delta\Delta$ and $\delta\Psi$ difference spectra allow to recognize features related to the molecular film (thickness, absorptions) and to the film-substrate interface thus extending SE to a sensitive surface UV-VIS spectroscopy. The potential of ellipsometry as a surface spectroscopy tool can be boosted when flanked by other characterizations methods. The chapter deals with the combined application of broad-band Spectroscopic Ellipsometry and nanolithography methods to study organic SAMs and multilayers. Nanolithography is achieved by the accurate removal of molecules from regularly shaped areas obtained through the action of shear forces exerted by the AFM tip in programmed scans. Differential height measurements between adjacent depleted and covered areas provide a direct measurement of film thickness, which can be compared with SE results or feed the SE

P. Parisse · L. Casalis
Elettra Sincrotrone Trieste S.C.p.A., s.s. 14 km 163,
5 in Area Science Park, Basovizza, Trieste, Italy

I. Solano · M. Magnozzi · O. Cavalleri · M. Canepa (✉)
OPTMATLAB, Department of Physics, University of Genova,
via Dodecaneso 33, 16146 Genova, Italy
e-mail: canepa@fisica.unige.it

F. Bisio
Istituto CNR-SPIN, C.so Perrone 24, 16152 Genova, Italy

© Springer International Publishing AG, part of Springer Nature 2018
K. Hinrichs and K.-J. Eichhorn (eds.), *Ellipsometry of Functional Organic Surfaces and Films*, Springer Series in Surface Sciences 52,
https://doi.org/10.1007/978-3-319-75895-4_4

analysis. In this chapter we will describe the main concepts behind the SE difference spectra method and AFM nanolithograhy. We will describe how SE and AFM can be combined to strengthen the reliability of the determination of thickness and, as a consequence, of the optical properties of films. Examples will be discussed, taken from recent experiments aimed to integrate SE and AFM nanolithography applied to SAMs and nano layers of biological interest. By analysing in detail the changes of the spectroscopic features of compact versus non-compact layers and correlating such changes with the post-lithography AFM analysis of surface morphology SE unravels the specific versus unspecific adsorption of biomolecules on gold surfaces functionalized with suitable SAMs.

## 4.1  Introduction

The design of bioanalytical devices calls for gentle methodologies dedicated to the accurate surface characterization of sensing elements and sensing processes. Several biosensing strategies rely on the controlled immobilisation of analytes via selective adsorption on specific sites. Self Assembled Monolayers, SAMs, provide a versatile route to achieve the desired surface functionality [1–3]. Functional SAMs conjugate operative simplicity and quality and are integrable in nanopatterning platforms [2, 4, 5].

The immobilisation process can be monitored for example by measuring the induced mass transfer and/or increase of thickness of the surface film. The quantitative analysis of structural and optical properties of SAMs [6], veritable ultrathin films at the nanometre scale, requires specific approaches that ensure the smallest perturbation to the sample; this is a key issue in experiments dealing with delicate biomolecules. Light provides gentle methods of analysis of soft ultrathin films. Optical methods are suitable for the investigation in the aqueous environment of biological samples and are also a simple probe for monitoring relatively slow molecular adsorption/desorption processes [7, 8].

Information about the thickness evolution, and possibly on other morphological/structural parameters of films, has to be "extracted" through the comparison of experimental data with optical models of the layered system under investigation (substrate/film or substrate/interface/film). Within Fresnel optics, the film thickness $d$ is usually a *free* parameter to be determined by fitting [9, 10]. Models are eventually based on some educated guesses about the structure and optical properties of the film. This important aspect, shared by all optical methods, remains sometimes hidden in routine practice of "user friendly" Surface Plasmon Resonance (SPR) commercial instruments, which are very popular in biomedical laboratories as they are able to provide "quick" response. Spectroscopic ellipsometry users should in principle be compelled to confront these issues directly and critically, if they want to derive reliable information from the output of their instruments, that is the somewhat "enigmatic" $\Psi$, $\Delta$ spectra. Since pioneering experiments in the field of SAMs [11] ellipsometry was often confined to the role of ancillary method to obtain estimates of film thickness, which is a key quantity in the science of ultrathin films. It is worth noting that passing from the stage of quick analysis to a more profitable use able to exploit the great potential of SE as an UV-VIS spectroscopic represents a task which requires the development of specific skills. As it is well represented in this book and we hope also in this chapter, the effort is often rewarded with information which go beyond the sole thickness estimate.

Small optical paths inside the film represent an obstacle towards the contemporary determination of thickness and optical properties of ultra thin films [12]. Especially for transparent films the thickness is highly correlated to the refractive index [9, 10, 13]. One can extrapolate *reasonable* values of the film refractive index from bulk references of the same or related substance, and obtain fair thickness estimates. This approach has been often adopted, for example in the popular case of alkanethiols SAMs on gold (see [14] and references therein). Problems may be worse when film and substrate have similar index of refraction in the transparency range. In this case which is encountered for many organic and biologic films on glass is certainly useful to combine SE with an Infrared SE (IRSE) investigation of molecular vibrations [15].

Indeed, in order to exploit the full potential of ellipsometry as an UV-VIS spectroscopy it is to some extents necessary to flank the optical investigation with other thin film/surface characterization methods. Such methods should hopefully exploit the same experimental conditions as to the system preparation and should preferably provide independent information.

An independent determination of thickness, especially for not-too-thin films can be provided, as done e.g. in [16], by X-ray reflectivity measurements but this approach could be perturbative and "complicate" in liquid. Regarding organic SAMs of small molecules in particular, several synchrotron-based methods exploiting aborption or reflectivity of soft X-rays can be profitably employed, with the necessary caution in checking radiation damage effects [17]. Dichroic, resonant soft x-ray reflectivity at the carbon K-edge is a promising method to complement SE regarding anisotropy properties of ultrathin organic films [18].

Electrochemical methods such as impedance spectroscopy (EIS) require SAMs-supporting substrates endowed with specific electron-transfer properties; similarly to ellipsometry, EIS requires a comparison between data and model simulations which can be eventually affected by correlation issues [19].

An interesting option is provided by the combination of Spectroscopic Ellipsometry and Quartz Crystal Microbalance which measures, after calibration, the amount of substance forming the layer, conceptually independent from optical path [20, 21].

We will focus here on another appealing partner of optical spectroscopy methods that is Atomic Force Microscopy (AFM) [22–25]. Usually AFM measurements are used to check the surface morphology of SAMs and are important in the preparation stage to assess the film quality. Naively, one could think to measure the thickness of a film by AFM, using suitable masks for instance. However obtaining reliable and accurate estimates on ultrathin soft matter films by simple imaging is a hard task. Here we will pay attention to so-called AFM nanolithography methods. The most simple and attractive example is perhaps represented by *Shaving* nano-lithography [26] in which the accurate removal of molecules is obtained by exerting a shear force during a programmed areal scan. Once depleted areas with regular shape and sharp contours are achieved, careful differential height measurements with adjacent covered areas provide a direct measurement of film thickness, which can be used to check fitting of SE data or to feed simulation models. Shaving, conceptually simple, is therefore very appealing as it allows to operate on SAMs that have been prepared under conditions that can be exploited also for ellipsometry measurements.

On the side of SE measurements we exploit the detailed analysis of the fine changes of the ellipsometry response (so called $\delta\Psi$, $\delta\Delta$ difference spectra) consequent to the formation of an ultrathin film on a substrate or to the addition/removal of a layer in a nanolayered structure. Our particular approach to difference spectra ($\delta$-spectra) has been refined in a series of experiments [14, 19, 27–29] and aims to extending Ellipsometry to a sensitive surface UV-VIS spectroscopy [30].

The chapter is organized in three sections. In Sect. 4.2 we present and critically discuss the method of difference spectra applied to the analysis of soft matter nanolayers. Several examples are shown, taken from our early works dedicated to organic and biologic Self-assembled monolayers, a few nm thick, with some emphasis on thiolate SAMs on gold. Basic concepts of AFM-based nanolithography are briefly presented in Sect. 4.3. In Sect. 4.4 we address most recent experiments aimed to integrate spectroscopic ellipsometry measurements and AFM nanolithography to obtain detailed information on the thickness and optical properties of nanolayered structures of biological interest. Focus will be devoted to studies about the bilayers formed after the specific immobilization of biomolecules on gold surfaces functionalized with suitable SAMs.

## 4.2 Optical Ellipsometry of Ultrathin Interfaces: Difference Spectra

The amount of change in optical reflectivity and $\Psi$, $\Delta$ parameters after the deposition of ultrathin films depends, through Fresnel relations, on the interplay between the ambient, film and substrate optical properties [31, 32]. Large substrate/film and film/ambient refractive index mismatch as well as the presence of molecular optical transitions in the investigated spectral range increase the sensitivity to the film. The substrate/film optical contrast condition is often fulfilled (with some relevant exception as e.g. glass). More critical is the optical mismatch between ambient and film, condition that is not satisfied for many biologic films in their most natural, aqueous ambient. However, though very small, specially at the monolayer limit, $\delta\Psi$ and $\delta\Delta$ may be in many cases above the limit of sensitivity of good commercial spectroscopic ellipsometers.

Let's take the popular case of SAMs, notably those formed by organosulphur compounds on gold surfaces. Since early developments in the field, optical ellipsometry methods have been considered as routine tools to obtain gentle estimates of the SAM thickness [2, 33]. Approaches polarized on the film thickness determination left aside the actual spectroscopic potential of optical ellipsometry. And indeed, whereas IRSE has been regularly exploited to investigate the adsorption configuration of molecules [34–39], the use of optical ellipsometry to obtain spectroscopic insight on the SAMs has been comparatively much less frequent.

Shi and coworkers were likely the first to propose a SE study dedicated to the dielectric properties of alkanethiols monolayers deposited on polycrystalline Au films [40]. The study, in liquid, considered molecules with different alkyl chain length. Data were presented as difference spectra, as convenient method to emphasize the subtle film-induced changes [9, 41]. The difference spectra were naturally referenced to the bare substrate results, in practice $\delta(\Psi, \Delta) = (\Psi, \Delta)_{film} - (\Psi, \Delta)_{Sub}$ (or $\delta < \epsilon > = < \epsilon >_{film} - < \epsilon >_{substrate}$) [40]. Beyond the analysis of thickness and index of refraction of films, the authors also addressed the SAMs-substrate interface, an important issue to comprehend the electronic transport properties across the interface.

Still considering differential approaches, the so-called *d-parameter* formalism [42] was applied to the analysis of SAMs of aromatic thiols on gold [43]. These molecules, of importance in the field of molecular electronics, differ from simple alkanethiols since they present well-defined optical absorptions. The analysis model allowed to extract, under several assumptions and ignoring the SAM-substrate interface, the polarizability tensor of the adlayer molecules.

The difference-spectra approach was revisited by our group, dealing with alkanethiols and other thiols on gold. High quality data were obtained starting from uniform and *ultraflat* gold substrates. We could distinguish features of difference spectra mainly related to the length of the molecular backbone, substantially determining the film thickness, from those related to the thiol-Au interface layer. $\delta\Psi$ difference spectra revealed a neat transition, at about 500 nm, from positive to negative values in the red-NIR region (see Fig. 4.1). Though rather small (about $-0.1°$), such negative values were reproducible and substantially independent from the molecular chain length. The negative NIR $\delta\Psi$ values could not be reproduced by simulations that assumed a sharp interface [14] between an isotropic, transparent film and the substrate (see also Fig. 4.2). This NIR spectral feature was not present in the case of loosely bound SAMs [19] whereas it has been detected, with minor intensity variability, for all the thiols on Au that have been investigated in several studies, including aminoacids (L-Cysteine) [14] as well as alkane and aromatic dithiols [44–46]. Note that the observation of this feature doesn't depend on the particularly flat nature of gold substrates we used in many works, since negative $\delta\Psi$ NIR values have been observed also in the case of thiols adsorption onto/into nano-granular gold samples [47] formed by the pile-up of nanoparticles [48, 49]. More recently, analogous negative NIR features were also detected for selenolate SAMs on gold [50].

The negative NIR $\delta\Psi$ values could be reproduced by simulations accounting for a so-called *transition* layer which may be related to the formation of the molecule-surface covalent bonds [51]. Effective models (the most practicable consisting of a Bruggeman EMA [51], as it is detailed below and in Fig. 4.2) were used to model the transition layer and allowed to advance an interpretation in terms of a broad-range Drude-like band with spectral weight smoothly increasing toward the IR [14].

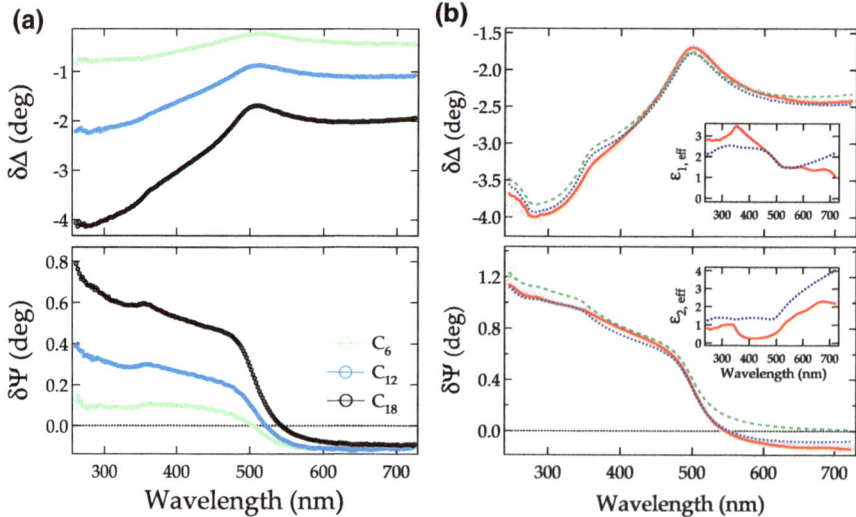

**Fig. 4.1** **a** SE difference spectra (ex-situ) for SAMs of alkanethiols with different alkyl chain length (so-called C6, C12, C18) on gold. Angle of incidence 65°. Note the $\delta\Psi$ transition (at 500–550 nm) to negative values, practically the same for the three molecules. **b** Comparison between experimental spectra (symbols; angle of incidence 70°) and simulations (lines) for C18 SAMs. Dashed lines (green): the model assumed a sharp interface between a transparent layer and the substrate: Negative $\delta\Psi$ values are not reproduced. Dotted (blue) and continuous lines (red) show simulations obtained with effective models for the molecule-Au interface layer whose dielectric functions are shown in the insets (details in [14])

Endorsing an optical reflectivity study [52], this band was assigned to modifications of the scattering of conduction electrons in the near surface region. These modifications could be associated to morphological changes at the nanoscale induced by the formation of the chalcogenide-Au bonds. Several papers, including recent reviews, have discussed the complex nature of the thiol-gold interface. This issue still presents several aspects to be clarified, regarding for instance peculiar dynamical aspects such as the mobility of thiolates, gold adatoms and gold-thiol complexes, to mention a few factors that can affect the electron scattering properties at the interface [53–56].

The transition layer provided an effective rationale of the complex of experiments that we performed on several thiolate SAMs. Other interesting factors may be considered. The most relevant concerns optical anisotropy. Many SAMs, for example well-ordered long-chain alkanethiol or dithiol SAMs on gold films, are uniaxial films. Early works pointed out that in the case of very thin thickness, ellipsometry cannot disentangle uniaxial from isotropic films [32]. Our effective models always considered isotropic layers. Dignam and coworkers, in a seminal paper, showed that if a very thin transparent uniaxial film on an absorbing substrate, such as a metal (in practice our case), is modelled as isotropic, the analysis of data may lead to an effective absorptive behaviour for the film [32]. According to simulations we performed for uniaxial transparent films on Au, with reference to the isotropic film case, anisotropy

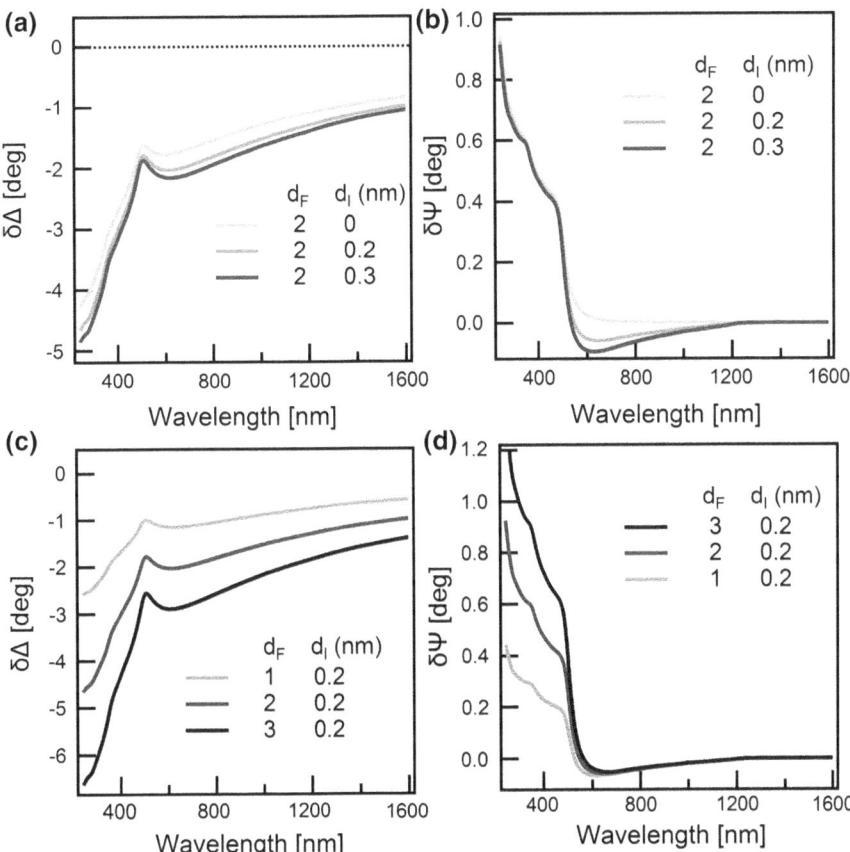

**Fig. 4.2** Simulations of SE data (difference spectra) based on a simple $F|I|S$ model, in which an interface transition layer (BEMA approximation) $I$ is sandwiched between the substrate $S$ and a transparent film $F$. All the simulations were obtained for the indicative value of $n_F = 1.4$. The ambient is vacuum (or dry air). **a–b** Effect of the variation of the interface thickness $d_I$ from 0 to 0.3 nm, at given $d_F$ (2 nm) and BEMA layer fractions ($f_F = 0.5$); note the negative $\delta\Psi$ values for $\lambda > 550$ nm. Instead, the NIR $\delta\Delta$ values depend on the total $d_F + d_I$ thickness. **c–d** Effect of the variation of $d_F$ from 1 to 3 nm, with $d_I = 0.2$ nm and $f_F = 0.5$; note that NIR $\delta\Psi$ values are not affected by $d_F$

is not able to induce the observed NIR negative values and rather, it couples with the film thickness.

In addition we note that sizeable negative $\delta\Psi$ NIR values were also observed in control experiments performed for adsorption of $H_2S$ on gold (unpublished data) and even after the electro-deposition of a copper atomic overlayer on gold [57] where anisotropy effects should be negligible.

The detection of the small SAM-induced spectral changes are fully practicable in in-situ measurements [14, 40, 58], in which the light beam probes the same zone of the sample before and after the film formation. This was also nicely demonstrated on the example of underpotential electro-deposition of an atomic Cu phase on gold films where tiny variations of $\Delta$ and $\Psi$, of the order of $0.3°$ and $0.1°$, respectively, could be monitored cyclically [57].

Regarding ex-situ measurements, which in principle provide a favourable ambient-sample optical mismatch, the macroscopic uniformity of the substrate surface becomes the crucial parameter for the feasibility of the difference spectra approach. In this respect it is worth mentioning that $\delta\Psi$, $\delta\Delta$ values related to low thickness (about 1 nm) SAMs can be of the same size of the sample-to-sample $\Psi$, $\Delta$ variability for common gold substrates and sometimes of the zone-to-zone $\Psi$, $\Delta$ variability observed on individual samples. An inaccurate choice of the substrate reference spectra leads to systematic analysis errors, as it was discussed in other contexts of application of difference spectra [41]. Laterally uniform substrates, protocols ensuring the reproducible SAM formation and systematic averaging over many zones and different samples (before and after the SAM deposition) are usually necessary to obtain meaningful ex-situ results [14, 27].

We note that the "lineshape" and magnitude of $\delta$-spectra depend on all the layers of the multilayer system and basically on the ambient, the film, the substrate [32] and eventually the ambient/film and film/substrate interface layers. The influence of the ambient was discussed e.g. in [14, 28] and will be further discussed in Sect. 4.4. The ambient should be "inert", transparent in the optical range of interest, and provide a good optical mismatch with the film (ex-situ measurements are favoured in this last respect). The ambient can be particularly important in case it can "penetrate" the film, as e.g. in the swelling of polymers films.

The influence of the substrate on difference spectra is particularly evident in spectral regions where films are transparent and it can be visualized by the aid of simulations.

A thorough discussion regarding difference spectra on gold has been proposed e.g. in [29]. Simulations of Fig. 4.2 were calculated for a $F|I|S$ model in which a *transition* layer $I$ is sandwiched between the substrate $S$ and the film $F$, assumed to be transparent. According to early works [51], the dielectric properties of the transition layer were simulated exploiting the Bruggeman Effective Medium Approximation (BEMA):

$$\sum_i f_i \frac{\epsilon_i - \epsilon_e}{\epsilon_i + 2\epsilon_e} = 0 \tag{4.1}$$

where $\epsilon_i$ and $f_i$ are the dielectric functions and volume fractions of the mixing constituents, respectively. The BEMA layer depends on two parameters, the fraction $f_F$ and the interface thickness $d_I$. In the simplest approximation $f_F$ was set to 0.5.

In Fig. 4.2 the effect of the film thickness $d_F$ and of $d_I$ is visualized separately. In panels (a–b), the increase of $d_I$ introduces negative $\delta\Psi$ values in the NIR with the formation of a minimum of $\delta\Psi$ at about 600 nm. The BEMA layer affects to a minor extent also the $\delta\Psi$ UV region. Regarding $\delta\Delta$, note the "parallel" downward shift, related to the $d_F + d_I$ thickness increment. In panels (c–d) the downward shift of

$\delta\Delta$ spectra and the corresponding increase of $\delta\Psi$ in the UV limit are proportional to the increase of $d_F$. $d_F$ doesn't affect the negative $\delta\Psi$ data. Figure 4.2 indicates how (i) positive $\delta\Psi$ values, below 550 nm, and negative $\delta\Delta$ spectra throughout the whole spectral range, are strongly related to the transparent part of the SAM (the molecular backbone) (ii) the negative $\delta\Psi$ values above 550 nm are related only to the interface parameters.

Reflecting our main research interests, we have focused the discussion about films on Au substrates; for gold, other relevant examples will be discussed in Sect. 4.4. Though gold and Si wafers are likely the most popular substrates for SAMs, SE was applied to other substrates such as, for instance Pt, InAs [59–62]; however, only few groups performed the analysis with the aid of difference spectra [46, 63].

In Figs. 4.3 and 4.4 we present difference spectra simulated for a transparent film with refractive index $n_F = 1.45$ on bare Si and $TiO_2$ substrates, respectively. For each substrate, three spectra are shown for film thickness of 1, 2 and 5 nm. The lineshape

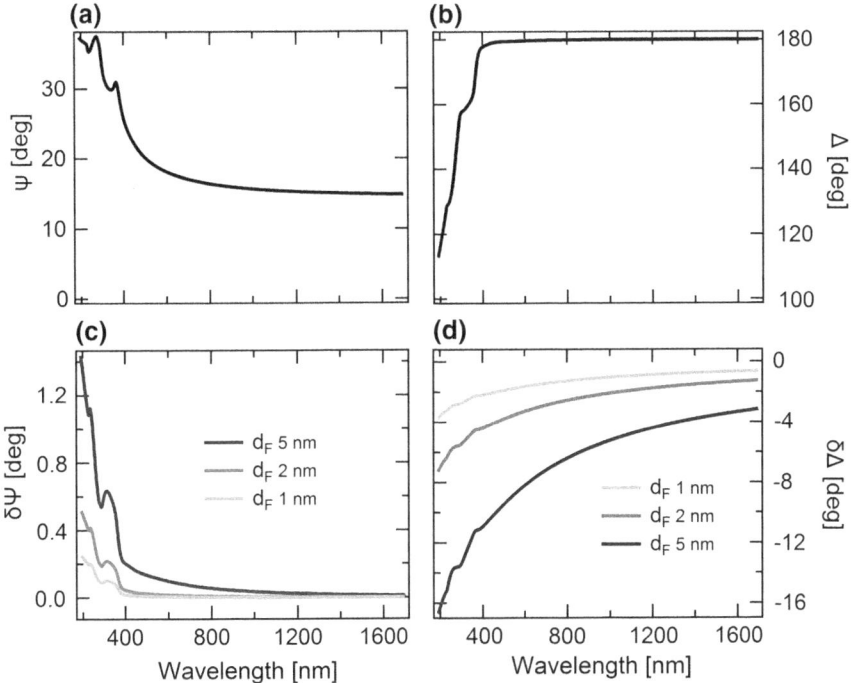

**Fig. 4.3** **a–b** SE spectra calculated at 65° incidence angle for Si. The ambient is vacuum (**c–d**) difference spectra for a transparent film (a Cauchy layer with leading coefficients $A = 1.45$ and $B = 0.01$) and several values of thickness (1, 2, 5 nm) on the Si substrate. $\delta\Delta$ values are strictly negative whereas $\delta\Psi$ are strictly positive and very small, with exception of the spectral region where Si presents an intense absorptive behaviour. Note the UV fine structures which bear memory of the intense absorptions related to Si valence/conduction band parallelism. In the NIR $\delta\Delta$ values are strictly proportional to the film thickness

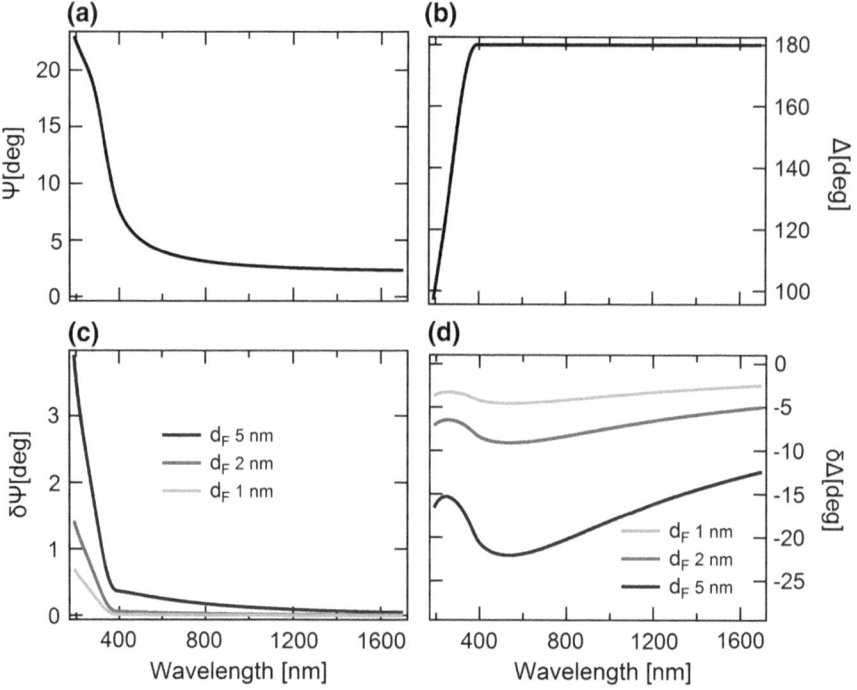

**Fig. 4.4 a–b** SE spectra calculated at 65° incidence angle for $TiO_2$. The ambient is vacuum. **c–d** difference spectra for a transparent film (a Cauchy layer with leading coefficients $A = 1.45$ and $B = 0.01$) and several values of thickness (1, 2, 5 nm) on the oxide substrate. $\delta\Delta$ values are strictly negative whereas $\delta\Psi$ are strictly positive and small, with exception of the UV spectral region after the main absorption threshold (energy gap) of $TiO_2$. In the NIR $\delta\Delta$ values are strictly proportional to the film thickness

of the spectra preserves memory of the substrate optical features. For Si, these are the high energy intense absorptions related to valence/conduction band parallelism and for $TiO_2$ the main absorption threshold (the energy gap) at about 3.2 eV.

Considering the ensemble of Figs. 4.2, 4.3 and 4.4 we can derive some general trends for ultrathin transparent films sharply interfaced to the substrate: $\delta\Delta$ values are strictly negative whereas $\delta\Psi$ are strictly positive and very small, with exception of spectral regions where substrates present an absorptive behaviour [32]. The magnitude of $\delta\Psi$ and $\delta\Delta$ is proportional to the film thickness. $\delta\Delta$ values are strictly linear to the thickness in the NIR range.

The influence of the film optical properties on difference spectra is best appreciated when the molecules present optical transitions. Then difference spectra referenced to the substrate show features representative of UV-VIS absorptions in the form of well-defined anti-resonances (see Fig. 4.5) [27, 28], which can be interpreted by comparison with transmission measurements in solution.

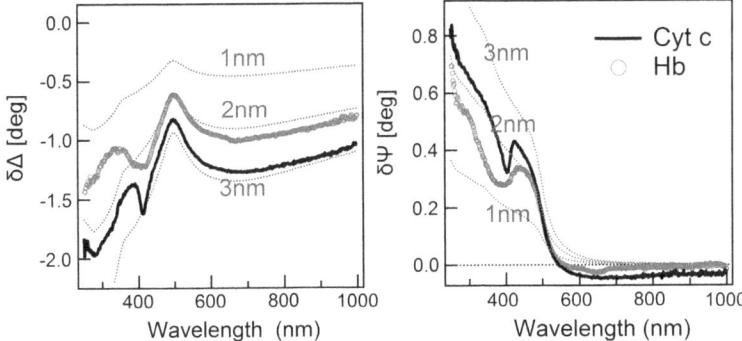

**Fig. 4.5**  In-situ SE difference spectra for Yeast Cytochrome (Cyt c) and hemoglobin (Hb) ultrathin films deposited on gold. Thin lines represent indicative spectra calculated for several values of thickness of a Cauchy transparent layer with $n_F = 1.5$. Simulations of the difference spectra based on a Multi-oscillator model can be found in [28]. The position and shape of the Soret band (the sharp dips at about 410 nm), related to the heme group of Cyt c matched transmission results in solution, suggesting that the adsorbed molecules preserved their native structure [28]. Measurements on Hb show a significant deviation from the native spectral properties

We have neatly detected these extinction features in difference spectra of thiolate Polyacetylenes (PDA) monolayers assembled on ultraflat gold substrates [27] which showed narrow UV molecular absorptions, specific of the carbazolyl-derivatized molecule investigated, and other broader features, in the 500–700 nm wavelength range, which have been interpreted as markers of the $\pi - \pi^*$ transition related to the polymerization state in the so-called red phase [64–66].

Another example regards ultrathin films of metalloproteins. Indeed, the great potential of ellipsometry to investigate biomolecular films was exploited already with Single Wavelength Ellipsometry (SWE) methods; the reader can find examples of application of SWE in [67–71]. Effective interpretative models, condensed in the popular de Feijter's [67] or Cuyper's formulas [67, 68], allow a translation of $\Psi(t)$ and $\Delta(t)$ changes, obtained monitoring the film deposition, into the quantification of the so-called surface mass density [72].

As an evolution of SWE methods, many SE studies focused on the adsorption kinetics of protein films [51, 72–83] and complex multilayers systems [84, 85].

The reader mostly interested in this kind of experiments is addressed to specific review articles [86, 87], to some recent papers [78, 79, 88] and other chapters in this book [89].

Note that in this field, SE is in competition with SPR which is of widespread use in the community of biologists. In the study of ultrathin molecular films, plasmon enhanced ellipsometry (or Total Internal Reflection Ellipsometry, TIRE) combines the advantages of ellipsometry and SPR [9, 90] and ensures a sensitivity which is larger than ordinary SPR [91–94].

Compared to SPR, Spectroscopic ellipsometry allows to look at optical transitions which mark the conformational state of molecules and may provide specific information on the preservation of molecular functionalities, that could be stressed by the interaction with inorganic surfaces [19, 95, 96]. Remarkably, useful insight on these aspects may be inferred even by simple inspection of referenced spectra. Example of application of IRSE in this or related contexts can be found in [72, 97–99]. In [97] the measurements allowed to determine the thickness and index of refraction of a fibrinogen ultrathin (4.5 nm) layer on a Au substrate, and to clearly detect *fingerprints* such as amide I and amide II bands. Reference [99] presents a nice example of use of referenced spectra to characterize the pH-dependent switching of a polymer brush.

The high sensitivity of UV-VIS ellipsometry was recently exploited by our group in the investigation, in buffer ambient and ex-situ, of Yeast Cytochrome c (Cyt c) ultrathin films deposited on ultraflat [28] or nano-grainy [81] gold and on Si wafers [100]. Difference spectra (see Fig. 4.5) exhibit quite sharp absorption features typical of the so-called heme group of this kind of metalloprotein, which can be easily detectable in transmission spectra of molecular solutions of suitable concentration. The observation of heme-related spectral features was exploited to monitor the formation of the biomolecular layer in dynamic measurements performed in a suitable deposition cell [28, 81].

The accurate determination of the position and shape of optical absorptions required a Kramers–Kronig (KK) consistent model of the dielectric function of the protein layer. Similarly to other spectroscopies, a suitable superposition of resonances, usually described by a mix of Lorentzian and Gaussian character, must be devised to reproduce absorption bands; further, the model needs some function representing the background of far resonances. In [28] simulations based on a many-oscillator model allowed an accurate characterization of the absorptions bands of Cyt c adsorbed on Au, and a reliable estimate of film thickness, which turned out of the order of molecular size, testifying the monolayer deposition. The position and shape of the intense B-band (Soret), peaked at about 410 nm and sensitive to the molecule environment, matched transmission results in solution. This was a strong indication that the adsorbed molecules, under proper wet conditions, preserved their native structure [28] after interaction with the substrate. Indeed the fine structure of another absorption band, the so-called Q-band in the 500–600 nm spectral region, would provide even more detailed information on the molecular state, being highly dependent on the oxidation state of the heme group. Q-bands are however far less intense than the main Soret absorption. For the Cyt c/Au system, the observation of Q-bands was practically hindered by substrate-related features, that we briefly discussed above. Q-bands were instead clearly discernible on Si wafers substrates, covered with native oxide, thanks to the smooth character of the contribution of substrate and film thickness to the difference spectra in the spectral region of interest [100]. In this favourable case the $\delta\Psi$ spectra became a fingerprint of the molecular absorption bands, which fine structure could be fully appreciated, practically without the need of simulations. Figure 4.6 thus is well representative of the great potential of SE, through difference spectra analysis, as a sensitive, surface optical spectroscopy of molecular layers endowed with absorptions.

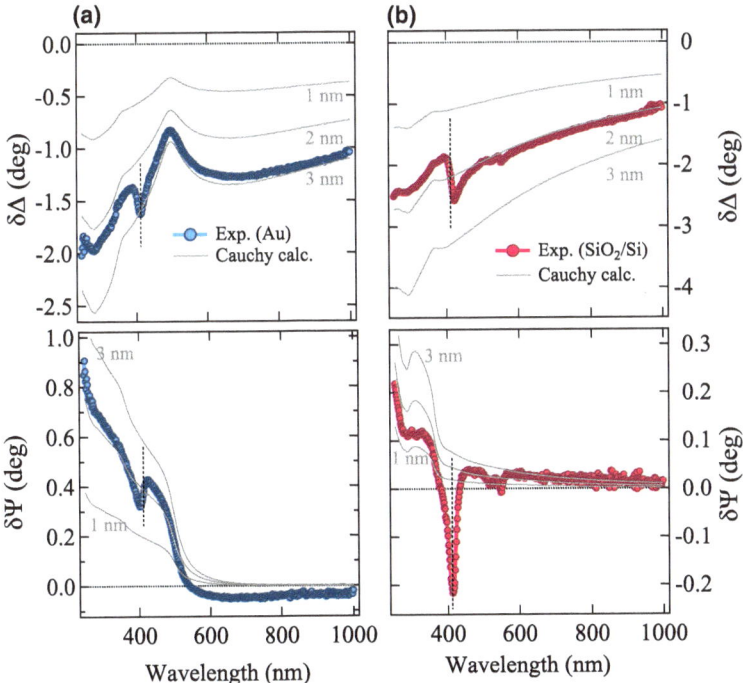

**Fig. 4.6** In-situ SE difference spectra for Yeast Cytochrome deposited on **a** gold and **b** Si-SiO$_2$ substrates. Thin lines represent indicative spectra calculated for several values of thickness of a Cauchy transparent layer ($n_F = 1.5$). At difference with simulated spectra, experimental data show relatively narrow features associated to so-called B-band (Soret), peaked at about 410 nm, and Q-band in the 500–600 nm spectral region, sensitive to the molecule environment. For the Cyt c/Au system the observation of weak Q-bands was practically hindered by substrate-related features. Q-bands are clearly visible on Si wafers (panel b), thanks to the smooth character of the contributions of substrate and film thickness to the difference spectra in the spectral region of interest [100]. In this favourable case the $\delta\Psi$ spectrum is practically a fingerprint of the molecular absorption bands

When thickness increases to a few tens of nm, molecular absorptions may become perceptible directly on $\Psi$, $\Delta$ spectra, as it has been shown on the example of diarylethene photochromic polymers [101] and other chromophores [102]. In this less critical thickness regime, difference spectra are still useful to magnify the effect. Further, difference spectra can emphasize dynamic optical transitions as e.g. photo-induced chromism in polymers [101], as it is exemplified in Fig. 4.7.

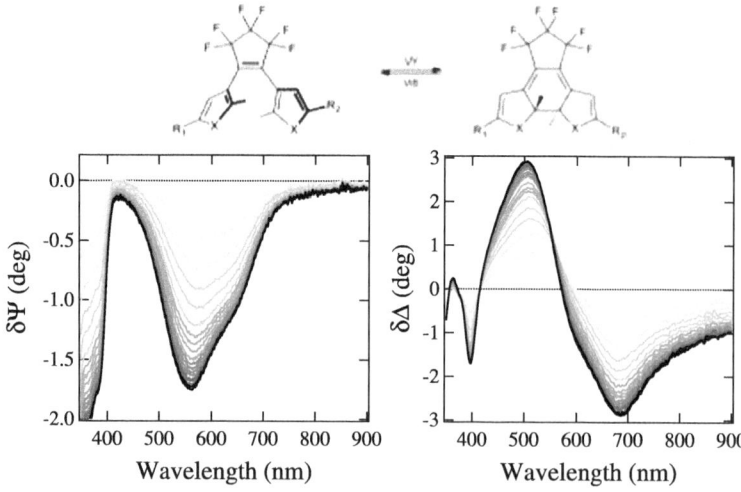

**Fig. 4.7** Dynamic $\delta\Delta(\lambda, t)$, $\delta\Psi(\lambda, t)$ scans observed while monitoring the UV-irradiation induced chromic transition in thin films (about 30 nm) of diarylethene polymers spin cast on SiO$_2$/Si substrates. The transition is reversible, after irradiation with visible light as sketched in the top of the figure. The time interval between lines is 40 s. The variations of the film dielectric function are emphasized in difference spectra formed with respect to the transparent form data [101]. Note that in this case $\delta\Psi$ (left) and $\delta\Delta$ (right) are practically the fingerprint of the variation of extinction and refractive index, respectively, so that the spectra admit a quick interpretation, even without the need of simulations

## 4.3 Atomic Force Nanolithography: Notes on Principles and Application

Atomic Force Microscopy is a powerful technique for the analysis of surface morphology with sub-nanometer resolution. The main difficulty in obtaining reliable measurements of the thickness of ultrathin films with AFM is usually the measurement with respect to the reference substrate. Suitable masks during deposition of the thin film can allow to measure the height of the film across the deposited area and the masked one: this technique is often used in metal deposition or organic thin film deposition in vacuum. Due to the diffusion of atoms/molecules also in the masked region, the evaluation of the height can be hampered by the lack of a sharp interface. This is more critical in the molecular monolayer case, where even small changes can strongly affect the evaluation of the thickness of the deposited film. Another possibility involves the physical scratching of the surface with a sharp blade to create a molecule free region. In this case the residual material is usually deposited at the border of the scratched region, which results in a source of uncertainty in the measurement. Moreover, if the substrate is not sufficiently hard (such as, for example evaporated gold electrodes) the blade can remove also part of the layers underneath the molecular film, thus distorting the results. It is therefore important to remove the molecular layer in a reliable way. AFM-based nanolithography methods are a convenient tool to apply forces sufficient to displace the molecules deposited on the surface [103, 104]. Depending on the type of molecule/substrate bond (chemical bond, electrostatic interaction, etc.) the force can be tuned in order to scratch away the monolayer without damaging the substrate underneath. The nanolithography approach can be declined in two main methods named shaving and grafting, which are sketched in Fig. 4.8. Both methods start with the formation of a precursor SAM (typically thiolated molecules on gold surfaces or silanes on silicon-based surfaces). The sample is placed in a liquid cell for the lithographic action and AFM measurements.

In nanoshaving the liquid cell is usually filled with a solvent in which the molecules anchored on the surface are soluble (typically ethanol for alkanethiols). After calibration for the correct evaluation of the applied forces, the tip is scanned in contact mode on the surface at low load (few nN) to image the surface (Fig. 4.8a). Then a high load (>80 nN) is applied in a selected region (areas with lateral dimension ranging from 20 nm to few microns) (Fig. 4.8b): this force is sufficient to displace from the surface the self-assembled molecules. The high solubility of the molecules in the solvent inhibits re-adsorption of the displaced molecules, thus leaving a hole with the substrate exposed (Fig. 4.8c). After the lithography, the sample can be then carefully imaged to extract the height of the SAM with respect to the substrate. Depending on the solvent used and the quality of the cantilever, very sharp patterns can be obtained with an edge resolution of few nanometers and the height difference between the SAMs and the substrate can be measured with subnanometer precision.

Nanoshaving has been successfully used for the patterning of several SAMs (alkanethiols, DNA oligomers, organic adlayers, etc.) [105–109] as well as for thin films of proteins [110, 111], allowing a precise evaluation of the thickness of the layer, in line with the theoretical predictions and/or the results of complementary spectroscopic techniques.

The ability of AFM tips to locally displace molecules from the surface can be combined with the simultaneous self-assembling of another type of molecules present in the solution: in these conditions one can exploit the Nanografting technique. Firstly introduced in 1997 for the grafting of alkanethiols [112], grafting has been employed for the realization of controlled patterns of molecules, ranging from biorepellent SAMs, to DNA and proteins [29, 113–118]. As mentioned above, the initial conditions for the Nanografting are similar to the ones of Nanoshaving: in Fig. 4.8d–f, we illustrate the steps for the process. The sample is immersed in a solution containing the chosen molecules to be immobilized on the surface in micromolar (1–100 M) concentration (Fig. 4.8d). Imaging of initial monolayer present on the surface is not perturbative; a high load catalyses the exchange of the molecules displaced by the tip with the new ones abundantly present in the solution (Fig. 4.8e). The efficiency of the process depends on several factors such as, for instance, the size of the molecules and their diffusion coefficients, the concentration of the molecules, the number of times that the same area is scanned at high load [119, 120]. The scanned area is finally imaged to observe the effective substitution of the molecules and the creation of the desired pattern on the surface (Fig. 4.8f). In advantage to Nanoshaving, with

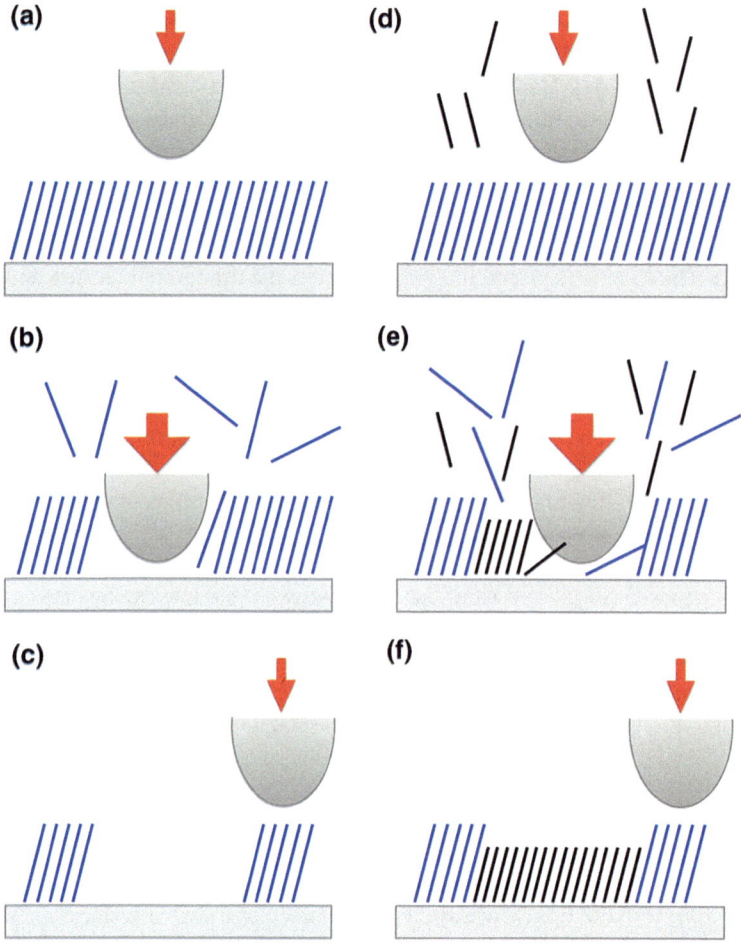

**Fig. 4.8** Schematics of AFM nanoshaving (**a–c**) and nanografting (**d–f**) processes. The microscope tip approaches a surface covered with a self assembled monolayers (**a, d**). Applying a high load (tens of nN), the tip can displace the molecules present on the surface and, either leave an exposed surface in the shaving experiment (**b**) or replace the molecules with others present in solution, creating a confined patch of the new molecules, in a grafting experiment (**e**). The AFM can then used at soft load to measure height differences between different patches (**c–f**)

grafting it is possible to immobilize different molecules in different patches, easing a comparative analysis of the morphological [121, 122], chemical/mechanical [114, 123], electrical [124, 125], functional [115, 126–128] properties of the different grafted molecules. For bio-sensing purposes, the Nanografting of relevant biomolecules (DNA, proteins) in a ethylene glycol-terminated SAM avoids unspecific binding on this bio-repellent surface and eases the observation of specific bio-recognition events on the sensitive areas [117, 129, 130].

Both shaving and grafting methods are prone to uncertainty in the evaluation of thickness, that can be due to an incompleteness of the process (original molecules still present on the surface) or to excessive forces used in the imaging phase. A synergy with spectroscopic techniques and specifically with SE is therefore recommended for a more precise characterization of ultra thin films [22, 29, 121, 131].

## 4.4  Application of SE and Atomic Force Lithography Methods to Ultrathin Soft Matter Films: Case Studies

### 4.4.1  Bio-Inert SAMs: Nanoshaving

SAMs of Oligo(Ethylene Glycol)-terminated alkanethiols (for brevity, T-OEG) deposited on gold substrates have been extensively investigated since early developments of SAM science for their non-fouling properties [1, 11].

Compared to closely packed unsubstituted alkanethiol SAMs, the flexible OEG strands introduce local disorder at the SAM/ambient interface [132–136]. Such an intrinsic local disorder is believed to control the interaction between the OEG moieties and ambient water molecules, effective in preventing the wetting of underlying methylene groups [11]. There is accumulating consensus on the manifold role of the hydration of the OEG part in establishing the bio-inertness of T-OEG SAMs [137–140].

SE and AFM nanoshaving have been recently employed to investigate SAMs of T-OEG molecules on Au. We studied in particular $SH(CH_2)_{11}EG_nOH$ molecules with $n = 3$ and 6, named T-OEG3 and T-OEG6 in the following (see [29] for details about the preparation and other experimental aspects). Figure 4.9 shows representative AFM images and corresponding height profiles of shaved areas in T-OEG3 and T-OEG6 SAMs. Topographic images were obtained in soft contact mode to reduce the effect of the tip load on the SAM height. From the statistical analysis of the height values of the SAM and of the shaved regions, resumed in Fig. 4.10, we were able to evaluate the thickness of the layers, obtaining $2.4 \pm 0.3$ nm and $1.6 \pm 0.4$ for T-OEG6 and T-OEG3, respectively [29].

In-liquid Spectroscopic Ellipsometry data are documented in Fig. 4.11. The data were taken in a proper buffer solution (so-called TE, 10 mM Tris-HCl, 1 mM EDTA, 1 M NaCl, pH $= 7.2$), that it was used in a second stage of the experiment to test the film resistance to protein adhesion. As a consequence of the relatively low ambient/film optical contrast, the values of $\delta\Delta$ and $\delta\Psi$ are rather small and relative uncertainty high, as it is represented by the large error bars. The data would suggest a small thickness difference between the two SAMs; $\delta\Psi$ data would even question any thickness difference. We will return on the interpretation of these data after considering ex-situ data.

Representative results of the ex-situ SE characterization of T-OEG films are shown in Fig. 4.12. The quality of data is neatly increased with respect to the in-liquid

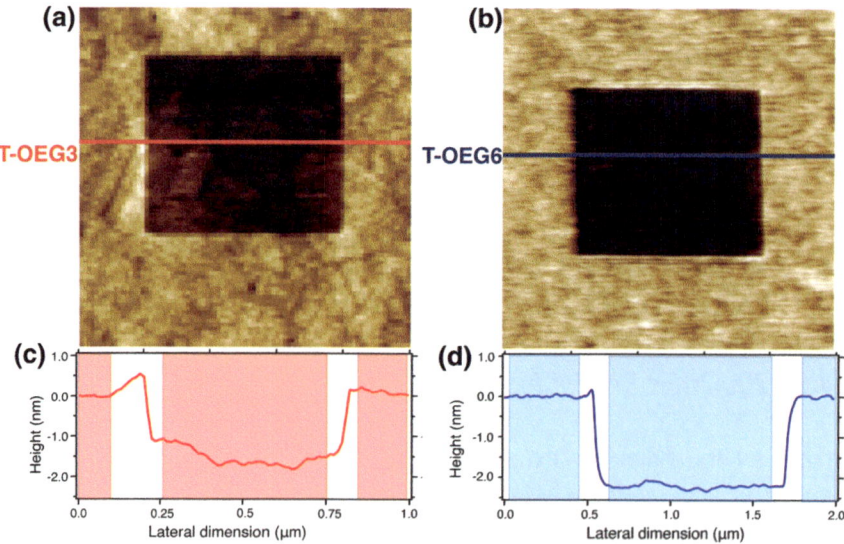

**Fig. 4.9** **a**, **b** AFM micrographs and **c**, **d** representative height profiles across shaved regions in T-OEG SAMs. The T-OEG6 pattern shows a better defined contrast between full and depleted areas

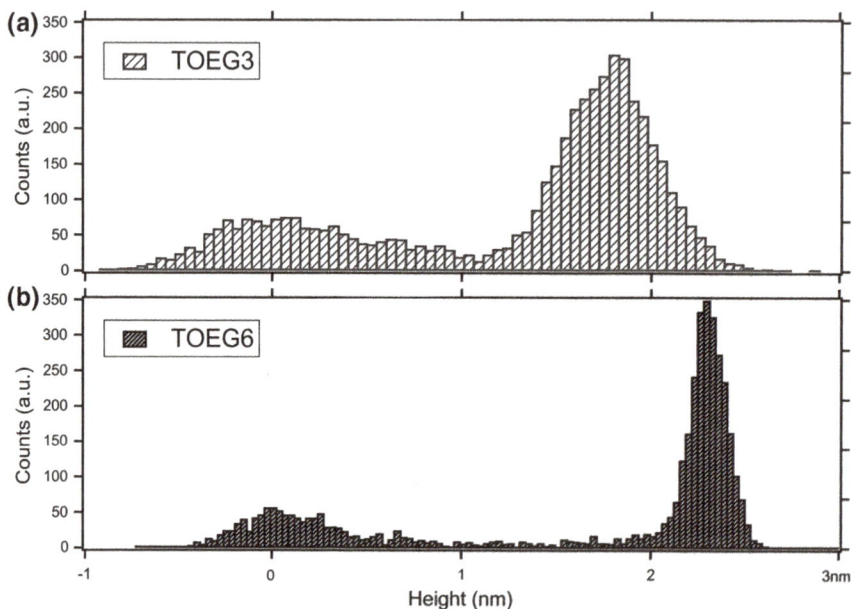

**Fig. 4.10** Determination of SAM thickness based on the statistical analysis of the height values of the SAM and of the shaved regions for several shaved areas. Height histograms for nanoshaved areas in **a** T-OEG3 and **b** T-OEG6. Histograms have been aligned setting arbitrarily the zero at the gold surface. The resulting thickness is $1.6 \pm 0.4$ and $2.4 \pm 0.3$ nm for TOEG-3 and TOEG-6, respectively

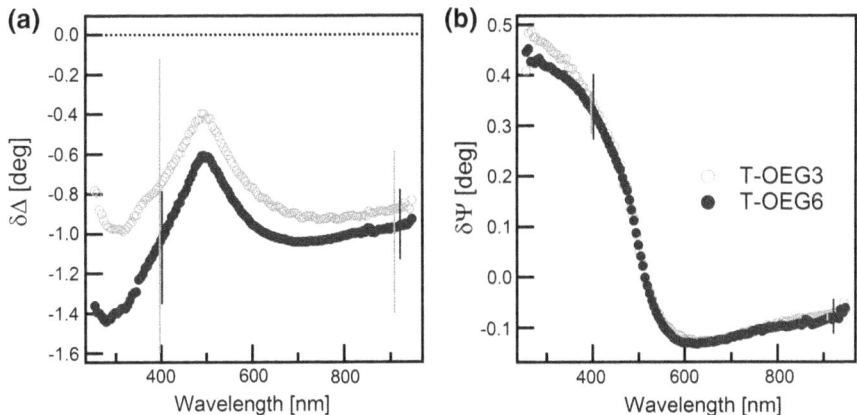

**Fig. 4.11** In-situ difference spectra ($\delta\Delta = \Delta(TOEG + Au) - \Delta(Au), \delta\Psi = \Psi(TOEG + Au) - \Psi(Au)$) for T-OEG SAMS on gold. Measurements were obtained at 65° angle of incidence and are shown in the range of high transparency of water. Thin vertical bars show representative uncertainty in different spectral regions

measurements. In the figure, experimental $\delta\Delta$ and $\delta\Psi$ difference spectra are compared with calculations performed according to a simple $TF|S$ model in which a transparent film ($TF$), modelled with a simple Cauchy formula, is sharply stacked onto the substrate ($S$) [29]. Simulations are presented for three values of the film thickness $d_{TF}$ and for an indicative range of the refractive index $n_{TF}$, which gives rise to the shaded areas in the figure. The shading is intended to provide a representation of the index-thickness correlation. Simulations reproduce the shape of experimental spectra with the "usual" exception of the negative $\delta\Psi$ values in the high reflectivity region of gold, which have been discussed in Sect. 4.2 for other thiolate SAMs.

Within the $TF|S$ model, for a given $n_{TF}$ there is proportionality between $\delta\Delta$ and $d_{TF}$ (Fig. 4.12a); the thickness of T-OEG6 and T-OEG3 SAMs which can be extrapolated from panel A slightly exceeds the value obtained from AFM analysis. Simulated $\delta\Psi$ curves (Fig. 4.12b) in the UV limit, are proportional to the film thickness as well and, as discussed above, are much less affected by interface effects. The shaded areas in $\delta\Psi$ patterns are rather thin. For $\delta\Psi$ the film thickness is less correlated to the value of the $n_{TF}$ parameter. $d_{TF}$ values which can extrapolated from the comparison of $\delta\Psi$ curves with simulations, in the UV range, are in good agreement with AFM measurements.

The apparent discrepancy between the $d_{TF}$ values which would be derived by the disjunct analysis of $\delta\Delta$ and $\delta\Psi$ data, as well as the negative $\delta\Psi$ NIR values, witness the main limitations of the $TF|S$ model. Interface effects can be effectively included in the optical model by using the Bruggeman EMA (so-called Arwin approximation [51]) to model the dielectric function of the interface layer $I$. Fitting the $TF|I|S$ model to the experimental data, a rather well-defined minimum of the MSE was

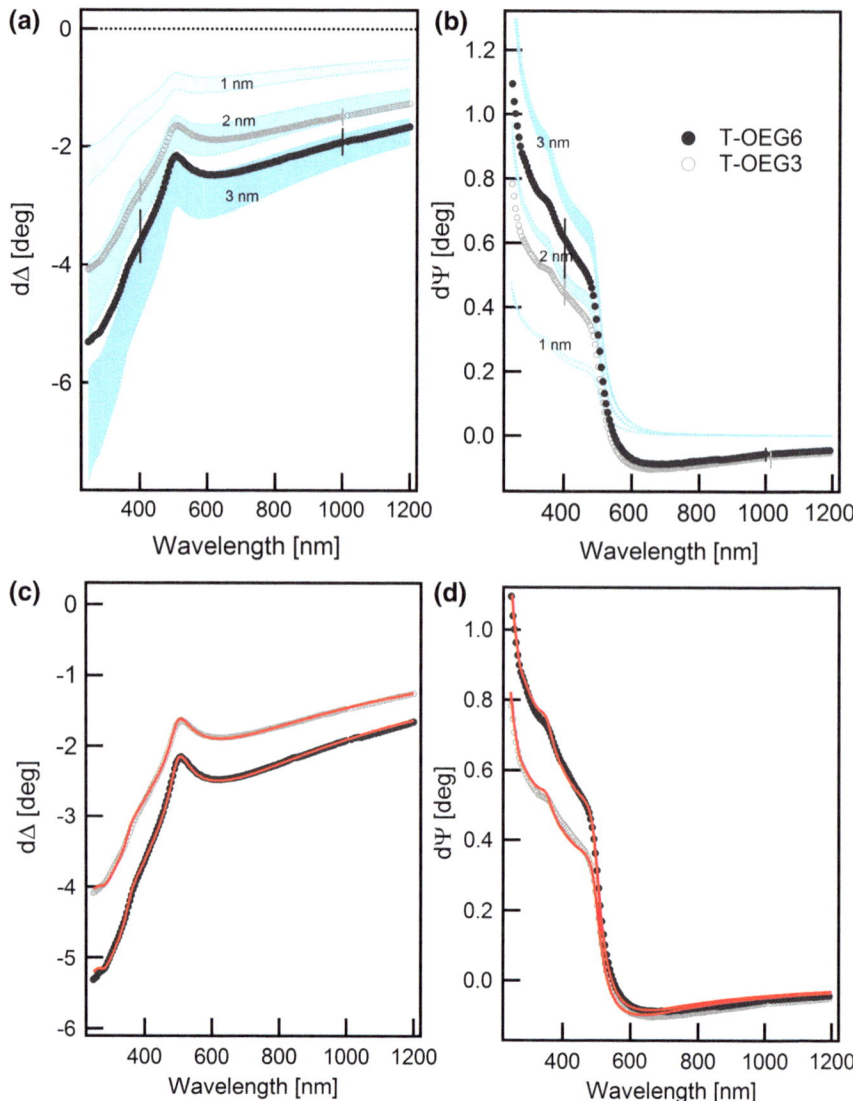

**Fig. 4.12 a–b** Symbols (grey tones): ex situ experimental difference spectra ($\delta\Delta = \Delta(TOEG/Au) - \Delta(Au)$, $\delta\Psi = \Psi(TOEG/Au) - \Psi(Au)$) obtained for the T-OEG SAMs on Au, at 65° angle of incidence. Thin vertical bars show representative uncertainty in different spectral regions. Shading illustrates simulations based on the $TF|S$ model for three values of the film thickness ($d_{TF} = 1, 2$ and $3$ nm). Top and bottom borders of each shaded area in $\delta\Delta$ ($\delta\Psi$) patterns correspond to $n_{TF} = 1.35(1.55)$ and $1.55 (1.35)$, respectively. For a given value of $n_{TF}$, simulated $\delta\Delta$ curves are proportional to $d_{TF}$. **c–d** Comparison between SE data (open circles, grey tones) for T-OEG3 and T-OEG6 films, and best-fit simulations (red lines) at 65 angle of incidence. Simulations have been calculated according to the $TF|I|S$ model. The dielectric function of the interface layer $I$ was modeled by using the Bruggeman EMA (so-called Arwin approximation [51])

found for relatively low values of $n_{TF}$ (1.35–1.40) with $d_{TF} = 2.6$–2.5 (1.9–1.8) nm for T-OEG6 (T-OEG3) [29]. The best-fit curves are reported in Fig. 4.12c–d. For both SAMs the best reproduction of the negative part of $\delta\Psi$ needed an interface thickness $d_I$ of about 0.25 nm, consistent with previous findings on other thiolate compounds [14].

The agreement between the thickness values obtained by ex-situ SE ($TF|I|S$ model) and AFM nanoshaving is satisfactory. Combining information derived from difference ellipsometry spectra with differential AFM measurements considerably strengthens the reliability of the thickness determination and, consequently, the estimate of the refractive index of films [29].

Let's now go back to in-buffer data. Even if the relatively high uncertainty is considered, the model used in the case of dry samples is not able to fit accurately the data in liquid. In order to solve the apparent puzzle one can invoke the hydrophilic nature of the EG units. Hydration of the film induces a smoothing of the optical contrast between the brush-like part of the layer and the aqueous ambient. Such "index matching" effectively removes, from the optical point of view, the outer part of the film. In practice, the measurements become sensitive only to the inner, packed and hydrophobic portion of the molecular film and to the film/substrate interface, as it is witnessed by the observation of negative $\delta\Psi$ values in the NIR, quantitatively similar to those determined ex situ. The similarity between T-OEG3 and T-OEG6 in-situ results then reflects the similar organization and density of the lowest-lying part of the film, especially the $C_{11}$ part, common to the two molecules. It is clear that in-situ SE, for this type of SAMS, prone to partial hydration, is not able to provide reliable values of the film thickness. The advantage, or better, the necessity of comparing SE with AFM measurements in this case is evident. The cross-check between SE and AFM, and the comparison between wet and dry measurements allow obtaining non-perturbative information regarding the vertical density profile of the SAM and the penetration of water into the spatial region of OEG strands. Clearly, the cross-check between SE measurements in air and in-liquid indicates that simulations considering a T-OEG film of homogeneous density provide an average value of the film refractive index which indeed could present a vertical profile.

SE was also exploited to check the resistance of T-OEG SAMs to protein adsorption. The SAMs were exposed to different molecules as for example Cyt c solutions. Representative data, for T-OEG6, are shown in Fig. 4.13. We look again at difference spectra, this time in the form $\delta\Delta, \Psi = \Delta, \Psi(Cyt + TOEG6 + Au) - \Delta, \Psi(TOEG6 + Au)$; in practice the spectra obtained on the T-OEG/Au system are used as the "new" substrate. The data show no significant differences between SE spectra taken after and before exposure to Cyt c: the T-OEG SAM is completely refractory to Cyt c adsorption. Note that this conclusion is drawn from the analysis of broad-band spectra. Instead, repeating the same adsorption experiment on a "simple" octadecanethiol (so-called $C18$) SAM terminated with a $CH_3$ group, well-defined difference spectra are visible, with evident features at about 400 nm,

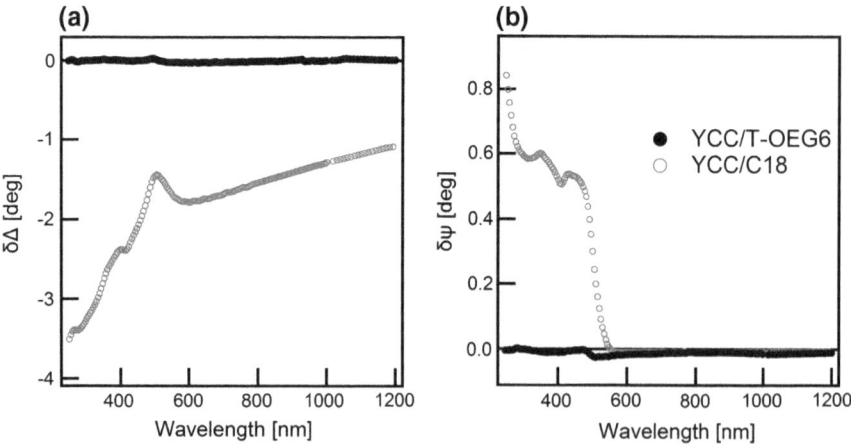

**Fig. 4.13** SE measurements devised to check Yeast Cytochrome C interaction with TOEG-6 and $C18$ alkanethiol SAMs on Au. Dark bullets: Ex-situ difference spectra for Cyt c/TOEG ($\delta\Delta, \Psi = \Delta, \Psi(Cyt + TOEG6 + Au) - \Delta, \Psi(TOEG6 + Au)$). Light grey circles: Ex-situ difference spectra for Cyt/C18 ($\delta\Delta, \Psi = \Delta, \Psi(YCC + C18 + Au) - \Delta, \Psi(C18 + Au)$). Measurements are shown at 65° angle of incidence. The data illustrate, on a broad-band spectroscopic basis, the anti-fouling character of the TOEG SAM

particularly evident in the $\delta\Psi$ spectra. By comparison with our previous studies [28, 100] the spectral features at 400 nm can be safely assigned to the Cyt c Soret band, demonstrating the occurrence of non-specific protein adsorption on the hydrophobic $C18$ SAM. Note that for the Cyt/$C18$ case $\delta\Psi$ values in the NIR are substantially null, as expected for the case of unspecific adsorption.

## 4.4.2 Specific Immobilisation of Proteins on SAMs: Grafting

Suitable terminations can turn a SAM into a specific interface (a "receptor") for binding proteins or enzymes ("analyte"). Going beyond the characterization of a single SAM, we applied AFM nanolithography methods and SE to the analysis of the molecular bilayer formed by the "analyte"/"receptor" interaction.

In this respect, the nitriloacetic acid (NTA for brevity) group, after *loading* with Nickel ions (Ni(II)), becomes able to anchor histidine and his-tagged molecules [141–147]. The resulting NTA+Ni(II)-His coupling scheme is both selective and reversible, following the reaction with competitive ligands [142]. This coupling strategy proved really versatile with a number of bio-oriented applications [148–161]

We studied in particular the adsorption of Small Ubiquitin-like Modifier protein, SUMO for brevity, tagged with six histidine molecules on SAMs of alkanethiols terminated with NTA (AT-NTA) [121, 131]. SUMO is an interesting biomolecule. In eukaryotes, the covalent attachment of SUMO to target proteins can modulate their structure and function, representing a post translational modification involved in several cellular processes. More recently it has been demonstrated that SUMO can help the expression of recombinant proteins when it is used as fusion tag [162, 163]. The structure of SUMO is well-known [164, 165]. The easiness of production make His-tagged SUMO a good candidate for a proof of concept of a SE-AFM combined strategy in the characterization of protein immobilization schemes at surfaces. The use of nanolithograhy tools and of ellipsometry spectra provided us with a solid method to disentangle the effects of Ni(II)-mediated interaction between the AT-NTA layer and the 6His-tagged SUMO and to determine the thickness of the SUMO layer in physiological conditions. We investigated the process in sequential steps: (0) the functionalization of the Au substrate with the receptor NTA-terminated SAM; (1) the Ni(II) loading, positively tested looking at the reaction with "simple" polyhistidine molecules [121]; (2) the Receptor/SUMO interaction [131]; (3) the regeneration of the AT-NTA layer after removal of SUMO by chelating agents [131]. Regarding the AFM side, we preferred a nanografting scheme. T-OEG6 molecules (see Sect. 4.4.1) were grafted in regular micro-holes produced in the AT-NTA layer. The small "carpet" of bio-inert T-OEG molecules acted as reference for the measurement of the height variation of the surrounding AT-NTA-Ni(II) SAM after the interaction with SUMO. The main findings regarding the AFM characterization in liquid are synthesized in Fig. 4.14. Panel (a) shows the 3D topographic image of a T-OEG6 patch grafted inside the AT-NTA layer. The histogram in panel (c) quantifies the well-defined AT-NTA/T-OEG6 height contrast, $1.0 \pm 0.2$ nm. The height difference neatly increased after exposure to His-tagged SUMO, as it is shown in panels (b) and (d). The final value, $3.3 \pm 0.3$ nm, implies a net height of the SUMO layer of $2.3 \pm 0.4$ nm, compatible with the crystallographic dimension of the protein [131].

Ellipsometry measurements were performed on dedicated layers uniformly covering the Au substrates. Ellipsometry was used to disentangle specific versus non specific adsorption of SUMO on the "receptor" layer [131].

Representative SE data (in buffer and ex situ) are shown in Fig. 4.15. The difference spectra are referenced to the AT-NTA film, which plays the role of "substrate" for the SUMO adsorption. $\delta$-spectra measured with or without pre-emptive Ni(II) loading, both in-situ and ex-situ data, are strikingly different as it is shown in Fig. 4.15. In case of Ni(II) pre-treatment, the spectral shape exhibits the KK-related features which we have discussed thoroughly in previous sections for other molecules on gold. These features are markers of the formation of compact SAMs. Without Ni(II) loading $\delta$-spectra are not compatible with the formation of a continuous layer, possibly pointing at a limited unspecific adsorption of SUMO on AT-NTA.

The comparison with models, which leads to the evaluation of the thickness of the His-tagged SUMO layer, reinforced this interpretation. We present diagrams of the same type as in Fig. 4.12, exploiting effective Cauchy models and considering values in the 1.35–1.50 range for the (unknown) index of refraction of the SUMO layer. For in-situ spectra, low experimental values and data dispersion (see error bars

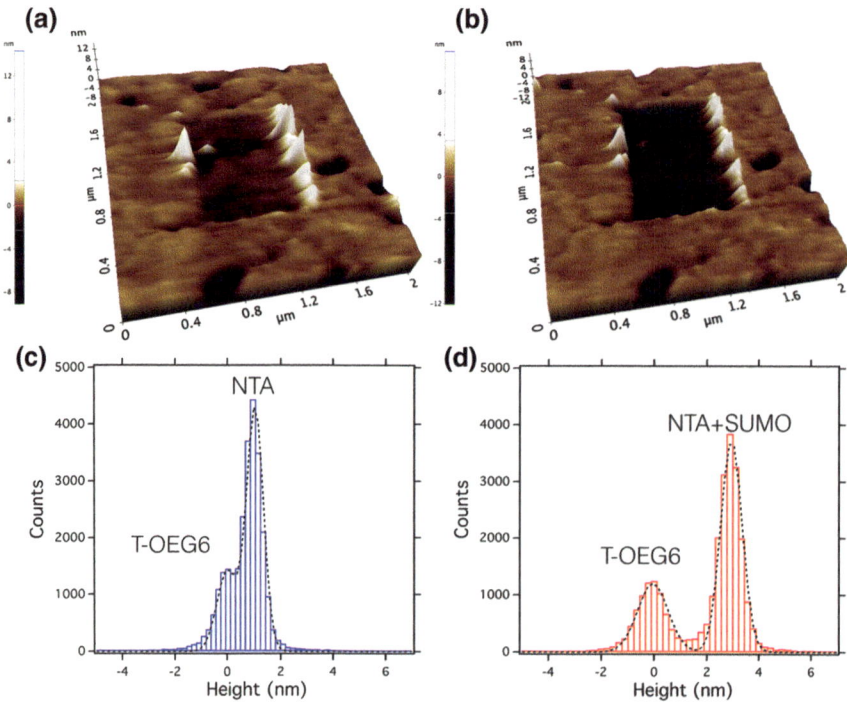

**Fig. 4.14** Three-dimensional AFM images of a nanografted patch of T-OEG6 in a NTA SAM **a** before and **b** after the addition of 6His-tagged SUMO in presence of Ni(II). **c–d** corresponding height histograms. Histograms have been aligned setting arbitrarily the zero at the T-OEG6 grafted surface. Height histograms derived from several grafted areas allow the determination of the NTA and NTA+SUMO film thickness [131]

in Fig. 4.15a) resulted in a large uncertainty in the estimate of the thickness of the SUMO layer, $2 \pm 1$ nm, compatible with AFM findings. The enhanced film/ambient optical contrast in ex situ data allowed a more accurate estimate. Data analysis (see Fig. 4.15c, d) lead to an estimate of $1.8 \pm 0.2$ nm for the effective thickness of the SUMO layer in ex-situ conditions. This value is slightly lower than the AFM estimate. The effect can be attributed to some alteration of the spatial conformation of the protein in dry state with respect to the hydrated conditions [131].

In summary, the combination of AFM and SE provided a rather accurate determination, also in physiological condition, of the thickness value of the SUMO layer and was able to disentangle the effects of Ni(II) in mediating the interaction between the NTA groups and the 6His-tagged SUMO.

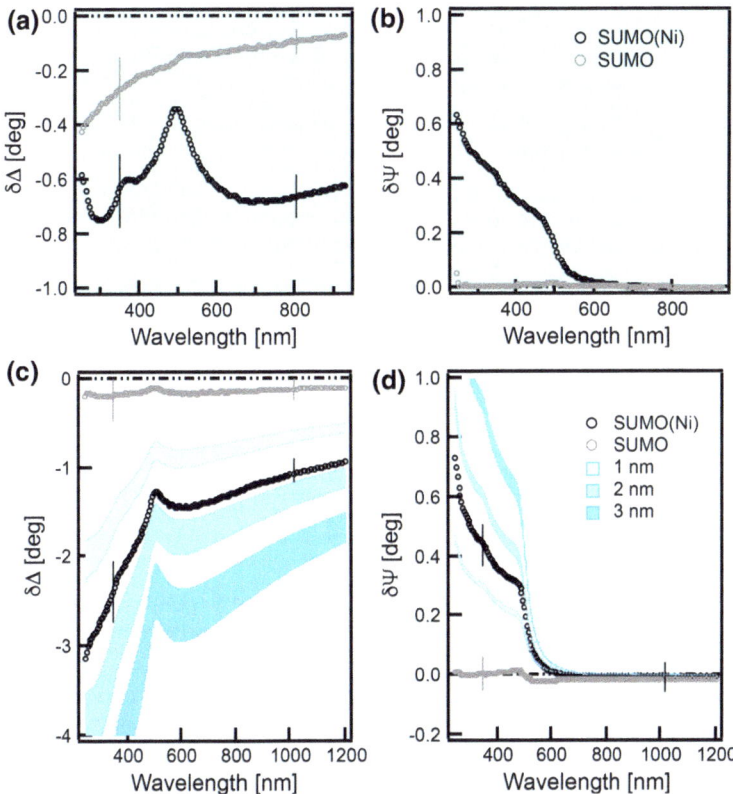

**Fig. 4.15** Exposure to SUMO of AT-NTA SAM. In situ (**a, b**) and ex situ (**c, d**) SE data. Difference spectra ($\delta\Delta$ and $\delta\Psi$, $65°$ angle of incidence) are referenced to the spectra obtained on the precursor AT-NTA SAM prior to the exposure to SUMO. Black symbols: $\delta\Delta$ and $\delta\Psi$ after SUMO adsorption on precursor SAM loaded with Ni(II). Grey symbols: the same, without Ni(II) loading. Thin vertical bars represent the experimental uncertainty related to data dispersion over the samples investigated Shading illustrates simulations based on the $TF|S$ model for three values of the film thickness ($d_{TF} = 1,2$ and $3\,nm$). Top and bottom borders of each shaded area in $\delta\Delta$ ($\delta\Psi$) patterns correspond to $n_{TF} = 1.35 \, (1.50)$ and $1.50 \, (1.35)$, respectively. Shading therefore helps visualizing the thickness/index correlation

## 4.5  Outlook

Spectroscopic ellipsometry is often perceived as a thin film technique where "thin" typically means a thickness ranging from a few tens to hundreds of nanometers. However $\Delta$ and $\Psi$, the output of the ellipsometry experiment, possess an intrinsic sensitivity to films with thickness down to a very small fraction of the wavelength, in the deep realm of nanoscience. As a matter of fact the ellipsometry analysis of ultrathin films (thickness below $10\,nm$) presents many challenging aspects and requires the patient acquisition of specific experimental competences and analytical expertises. Application of SE to ultrathin biologic films presents further specific challenges, as the "average" refractive index of the film is poorly known and often similar to the ambient in case of in-liquid application.

From the analytical point of view, a profitable application of SE to ultrathin films involves the comprehension of the fine changes of the spectral ellipsometry response, $\delta\Psi$, $\delta\Delta$ consequent to the formation of the film on a given substrate or to the addition/removal of a layer in a stack. This is the preferred playground of the method of analysis based on difference spectra, known since early developments of SE [9, 31, 32]. The optical models devised to describe the layered system under scrutiny should be able to provide a clear meaning to the experimental difference spectra. In our attitude, the comparison between experimental and simulated difference spectra represents a key check to assess the quality of analysis. The analysis through difference spectra should be "imperative" in case of films at the monomolecular level and SAMs.

In the chapter we have synthetically presented the evolution of our approach to the comprehension of difference spectra in the study of organic and biologic ultrathin interfaces, which took initial inspiration from seminal works [40, 51]. Our approach is condensed in figures like Figs. 4.5, 4.6, 4.12 and 4.15. Experimental data are initially compared with suitable families of difference spectra calculated for transparent films (in practice Cauchy layers) for indicative ranges of refraction index and thickness. Such a comparison paves the way to estimate the film thickness with an informed view of subtle issues related to index/thickness correlations and to film/substrate and film/ambient optical mismatch. Relatively sharp spectral features related to extinction, often important to assess the functionality of film bio-molecules, cannot be described by Cauchy models, and can be directly identified on spectra, thus demonstrating the potential of optical ellipsometry as a gentle, powerful surface UV-VIS spectroscopy, which can effectively complement SPR experiments.

This first stage of analysis provides in our opinion the correct basis for a second step in which more realistic models are exploited and model parameters are optimized through the fitting routines, traditional of ellipsometry analysis of thin films. To this aim information coming from other surface sensitive diagnostic tools are useful and even necessary. In this perspective we hope to have demonstrated to the reader of this chapter that AFM can be a powerful partner of ellipsometry when it is coupled to a nanolithography method able to deplete regularly shaped areas of the molecular film. Then, differential height measurements between adjacent, empty and pristine areas provide a direct measurement of film thickness, which can be compared with SE results or can feed the analysis of SE data. Remarkably, in case of soft matter films, sharp nanolithography can be achieved in the AFM apparatus itself, by exploiting the atomic force exerted by the tip under high-load regimes, in the so-called shaving and grafting modes.

Improving the accuracy of the thickness determination opens the door to obtaining a more reliable analysis of the dielectric properties of the films, a piece of information that is elusive for such ultrathin, often transparent layers. The SE-AFM combined approach helps defining the sources of systematic uncertainty associated with both kind of measurements, an aspect which is not always addressed in routine applications of the individual methods.

The data we have presented on the His-tag SUMO/AT-NTA system demonstrate the potential of SE for the accurate spectroscopic characterization of molecular recognition processes. This approach paves the way to the application to other ligand/receptor coupling schemes, such as for instance antigen-antibody interactions or DNA hybridisation. Beyond its intrinsic interest, the latter case is particularly attractive as DNA is endowed with intense UV absorptions. These absorptions can be detected in SE difference spectra at the monolayer level, similarly to what we observed on chromophores, and then monitored during the hybridisation process [166].

**Acknowledgements** The authors acknowledge funding from the Italian Ministry of Education (FIRB grant RBAP11ETKA-005). M.C. thanks all the people who collaborated along the years to his SE research on ultrathin organic an biologic films and in particular Mirko Prato and Chiara Toccafondi.

# References

1. K.L. Prime, G.M. Whitesides, Science **252**, 1164 (1991)
2. J.C. Love, L.A. Estroff, J.K. Kriebel, R.G. Nuzzo, G.M. Whitesides, Chem. Rev. **105**, 1103 (2005)
3. A. Hucknall, S. Rangarajan, A. Chilkoti, Adv. Mater. **21**, 2441 (2009)
4. R.K. Smith, P.A. Lewis, P.S. Weiss, Prog. Surf. Sci. **75**, 1 (2004)
5. J.C. Smith, K.-B. Lee, Q. Wang, M.G. Finn, J.E. Johnson, M. Mrksich, C.A. Mirkin, Nano Lett. **3**, 883 (2003)
6. F. Terzi, L. Pasquali, R. Seeber, Anal. Bioanal. Chem. **405**, 1513 (2013)
7. L.S. Jung, C.T. Campbell, T.M. Chinowsky, M.N. Mar, S.S. Yee, Langmuir **14**, 5636 (1998)
8. D.J. Vanderah, R.J. Vierling, M.L. Walker, Langmuir **25**, 5026 (2009)
9. R.M.A. Azzam, N.M. Bashara, *Ellipsometry and Polarized Light* (North-Holland, New York, 1977)
10. H. Fujiwara, *Spectroscopic Ellipsometry: Principles and Applications* (Wiley, New York, 2007)
11. C. Pale-Grosdemange, E.S. Simon, K.L. Prime, G.M. Whitesides, J. Am. Chem. Soc. **113**, 12 (1991)
12. H. Arwin, D.E. Aspnes, Thin Solid Films **113**, 101 (1984)
13. E.A. Irene, Solid State Electron. **45**, 1207 (2001)
14. M. Prato, R. Moroni, F. Bisio, R. Rolandi, L. Mattera, O. Cavalleri, M. Canepa, J. Phys. Chem. C **112**, 3899 (2008)
15. H.G. Tompkins, T. Tiwald, C. Bungay A.E. Hooper, J. Vac. Sci. Technol. A **24**, 1605 (2006)
16. G. Gonella, O. Cavalleri, I. Emilianov, L. Mattera, M. Canepa, R. Rolandi, Mater. Sci. Eng. C **22**, 359 (2002)
17. L. Pasquali, F. Terzi, R. Seeber, S. Nannarone, D. Datta, C. Dablemont, H. Hamoudi, M. Canepa, V.A. Esaulov, Langmuir **27**, 4713 (2011)
18. L. Pasquali, S. Mukherjee, F. Terzi, A. Giglia, N. Mahne, K. Koshmak, V.A. Esaulov, C. Toccafondi, M. Canepa, S. Nannarone, Phys. Rev. B **89**, 045401 (2014)
19. F. Bordi, M. Prato, O. Cavalleri, C. Cametti, M. Canepa, A. Gliozzi, J. Phys. Chem. B **108**, 20263 (2004)

20. K.B. Rodenhausen, T. Kasputis, A.K. Pannier, J.Y. Gerasimov, R.Y. Lai, M. Solinsky, T.E. Tiwald, H. Wang, A. Sarkar, T. Hofmann, N. Ianno, M. Schubert, Review of Scientific Instruments **82**, 103111 (2011)
21. This topic is amply treated in Chapter 10 of this book by K.B. Rodenhausen, D. Schmidt, C. Rice, T. Hofmann, E. Schubert, M. Schubert
22. T. Gesang, D. Fanter, R. Hoper, W. Possart, O.-D. Hennemann, Surf. Interface Anal. **23**, 797 (1995)
23. N.A. Geisse, Mater. Today **12**, 40 (2009)
24. I. Kopf, C. Grunwald, E. Bründermann, L. Casalis, G. Scoles, M. Havenith, J. Phys. Chem. C **114**, 1306 (2010)
25. F. Lu, M. Jin, M.A. Belkin, Nat. Photonics **8**, 307 (2014)
26. G.-Y. Liu, S. Xu, Y. Qian, Acc. Chem. Res. **33**, 457 (2000)
27. M. Prato, M. Alloisio, S.A. Jadhav, A. Chincarini, T. Svaldo-Lanero, F. Bisio, O. Cavalleri, M. Canepa, J. Phys. Chem. C **113**, 20683 (2009)
28. C. Toccafondi, M. Prato, G. Maidecchi, A. Penco, F. Bisio, O. Cavalleri, M. Canepa, J. Colloid Interface Sci. **364**, 125 (2011)
29. I. Solano, P. Parisse, F. Gramazio, O. Cavalleri, G. Bracco, M. Castronovo, L. Casalis, M. Canepa, Phys. Chem. Chem. Phys. **17**, 28774 (2015)
30. M. Canepa, in *Surface Science Techniques*, vol. 51, Springer Series in Surface Sciences, ed. by G. Bracco, B. Holst (Springer, Berlin, 2013), p. 99
31. J.D.E. McIntyre, D.E. Aspnes, Surf. Sci. **24**, 417 (1971)
32. M.J. Dignam, M. Moskovits, R.W. Stobie, Trans. Faraday Soc. **67**, 3306 (1971)
33. A. Ulman, Chem. Rev. **96**, 1533 (1996)
34. C.W. Meuse, Langmuir **16**, 9483 (2000)
35. D. Tsankov, K. Hinrichs, E.H. Korte, R. Dietel, A. Röseler, Langmuir **18**, 6559 (2002)
36. H.G. Tompkins, T. Tiwald, C. Bungay, A.E. Hooper, J. Phys. Chem. B **108**, 3777 (2004)
37. D.C. Bradford, E. Hutter, J.H. Fendler, D. Roy, J. Phys. Chem. B **109**, 20914 (2005)
38. Z.G. Hu, P. Prunici, P. Patzner, P. Hess, J. Phys. Chem. B **110**, 14824 (2006)
39. P.N. Angelova, K. Hinrichs, I.L. Philipova, K.V. Kostova, D.T. Tsankov, J. Phys. Chem. C **114**, 1253 (2010)
40. J. Shi, B. Hong, A.N. Parikh, R.W. Collins, D.L. Allara, Chem. Phys. Lett. **246**, 90 (1995)
41. K.A. Bell, L. Mantese, U. Rossow, D.E. Aspnes, J. Vac. Sci. Technol. B **15**, 1205 (1997)
42. W. Chen, W.L. Schaich, Surf. Sci. **218**, 580 (1989)
43. L.J. Richter, C.S.-C. Yang, P.T. Wilson, C.A. Hacker, R.D. van Zee, J.J. Stapleton, D.L. Allara, Y. Yao, J.M. Tour, J. Phys. Chem. B **108**, 12547 (2004)
44. H. Hamoudi, Z. Guo, M. Prato, C. Dablemont, W.Q. Zheng, B. Bourguignon, M. Canepa, V.A. Esaulov, Phys. Chem. Chem. Phys. **10**, 6836 (2008)
45. H. Hamoudi, M. Prato, C. Dablemont, O. Cavalleri, M. Canepa, V.A. Esaulov, Langmuir **26**, 7242 (2010)
46. H. Hamoudi, K. Uosaki, K. Ariga, V.A. Esaulov, RSC Adv. **4**, 39657 (2014)
47. F. Bisio, M. Prato, E. Barborini, M. Canepa, Langmuir **27**, 8371 (2011)
48. F. Bisio, M. Palombo, M. Prato, O. Cavalleri, E. Barborini, S. Vinati, M. Franchi, L. Mattera, M. Canepa, Phys. Rev. B **80**, 205428 (2009)
49. Regarding the interaction of nanoparticles with SAMs refer to Chapter 9 of this book by T.W.H. Oates
50. M. Canepa, G. Maidecchi, C. Toccafondi, O. Cavalleri, M. Prato, V. Chaudhari, V.A. Esaulov, Phys. Chem. Chem. Phys. **15**, 11559 (2013)
51. J. Mårtensson, H. Arwin, Langmuir **11**, 963 (1995)
52. O. Neuman, R. Naaman, J. Phys. Chem. B **110**, 5163 (2006)
53. R. Mazzarello, A. Cossaro, A. Verdini, R. Rousseau, L. Casalis, M.F. Danisman, L. Floreano, S. Scandolo, A. Morgante, G. Scoles, Phys. Rev. Lett. **98**, 016102 (2007)
54. A. Cossaro, R. Mazzarello, R. Rousseau, L. Casalis, A. Verdini, A. Kohlmeyer, L. Floreano, S. Scandolo, A. Morgante, M.L. Klein, G. Scoles, Science **321**, 943 (2008)

55. E. Pensa, E. Cortes, G. Corthey, P. Carro, C. Vericat, M.H. Fonticelli, G. Benitez, A.A. Rubert, R.C. Salvarezza, Acc. Chem. Res. **45**, 1183 (2012)
56. T. Burgi, Nanoscale **7**, 15553 (2015)
57. M. Prato, A. Gussoni, M. Panizza, O. Cavalleri, L. Mattera, M. Canepa, Phys. Status Solidi C **5**, 1304 (2008)
58. H. Arwin, Thin Solids Films **519**, 2589 (2011)
59. D.Y. Petrovykh, H. Kimura-Suda, A. Opdahl, L.J. Richter, M.J. Tarlov, L.J. Whitman, Langmuir **22**, 2578 (2006)
60. N. Gergel-Hackett, C.D. Zangmeister, C.A. Hacker, L.J. Richter, C.A. Richter, J. Am. Chem. Soc. **130**, 4259 (2008)
61. D.Y. Petrovykh, J.C. Smith, T.D. Clark, R. Stine, L.A. Baker, L.J. Whitman, Langmuir **25**, 12185 (2009)
62. C.A. Hacker, C.A. Richter, N. Gergel-Hackett, L.J. Richter, J. Phys. Chem. C **111**, 9384 (2007)
63. Z. Papa, S.K. Ramakrishnan, M. Martin, T. Cloitre, L. Zimanyi, J. Marquez, J. Budai, Z. Toth, C. Gergely, Langmuir **32**, 7250 (2016)
64. B. Tieke, G. Lieser, G. Wegner, J. Polym. Sci. Polym. Chem. **17**, 1631 (1979)
65. R.W. Carpick, T.M. Mayer, D.Y. Sasaki, A.R. Burns, Langmuir **16**, 4639 (2000)
66. R.W. Carpick, D.Y. Sasaki, M.S. Marcus, M.A. Eriksson, A.R. Burns, J. Phys. Condens. Matter **16**, R679 (2004)
67. J.A. de Feijter, J.A. Benjamins, F.A. Veer, Biopolymers **17**, 1759 (1978)
68. P.A. Cuypers, W.T. Hermens, H.C. Hemker, Anal. Biochem. **84**, 56 (1978)
69. M. Malmsten, J. Colloid Interface Sci. **166**, 333 (1994)
70. H. Elwing, Biomaterials **19**, 397 (1998)
71. P. Tengvall, I. Lundström, B. Liedberg, Biomaterials **19**, 407 (1998)
72. S. Reichelt, K.-J. Eichhorn, D. Aulich, K. Hinrichs, N. Jain, D. Appelhans, B. Voit, Colloids Surf. B **69**, 169 (2009)
73. C. Werner, K.-J. Eichhorn, K. Grundke, F. Simon, W. Grählert, H.J. Jacobasch, Colloids Surf. A **156**, 3 (1999)
74. T. Byrne, L. Lohstreter, M.J. Filiaggi, Z. Bai, J.R. Dahn, Surf. Sci. **602**, 2927 (2008)
75. R.J. Marsh, R.A.L. Jones, M. Sferrazza, Colloids Surf. B **23**, 31 (2002)
76. X.Q. Wang, Y.N. Wang, H. Xu, H.H. Shan, J.R. Lub, J. Colloid Interface Sci. **323**, 18 (2008)
77. S. Lousinian, S. Logothetidis, Thin Solid Films **516**, 8002 (2008)
78. M. Reza Nejadnik, C.D. Garcia, Colloids Surf. B **82**, 253 (2011)
79. D.K. Goyal, A. Subramanian, Thin Solid Films **518**, 2186 (2010)
80. J.L. Wehmeyer, R. Synowicki, R. Bizios, C.D. García, Mater. Sci. Eng. C **30**, 277 (2010)
81. C. Toccafondi, M. Prato, E. Barborini, S. Vinati, G. Maidecchi, A. Penco, O. Cavalleri, F. Bisio, M. Canepa, BioNanoScience **1**, 210 (2011)
82. V. Reipa, A.K. Gaigalas, V.L. Vilker, Langmuir **13**, 3508 (1997)
83. T. Berlind, M. Poksinski, P. Tengvall, H. Arwin, Colloids Surf. B **75**, 410 (2010)
84. K. Spaeth, A. Brecht, G. Gauglitz, J. Colloid Interface Sci. **196**, 128 (1997)
85. A. Nemeth, P. Kozma, T. Hülber, S. Kurunczi, R. Horvath, P. Petrik, A. Muskotal, F. Vonderviszt, C. Hos, M. Fried, J. Gyulai, I. Barsony, Sens. Lett. **8**, 730 (2010)
86. H. Arwin, Thin Solid Films **377–378**, 48 (2000)
87. H. Arwin, in *Handbook of Ellipsometry*, ed. by H.G. Tompkins, E.A. Irene (Andrew, Norwich, 2005), p. 799. (Chap. 12)
88. M.F. Mora, M. Reza Nejadnik, J.L. Baylon-Cardiel, C.E. Giacomelli, C.D. Garcia, J. Colloid Interface Sci. **346**, 208 (2010)
89. In this book, refer in particular to Chapters 1 by C. Cobet, 2 by H. Arwin, 5 by M. Erber et al., 6 by E. Bittrich et al
90. TIRE and SPR-enhanced SE are specifically addressed in Chapter 18 by H. Arwin
91. P. Westphal, A. Bornmann, Sens. Actuators B **84**, 278 (2002)
92. M. Poksinski, H. Arwin, Thin Solid Films **455–456**, 716 (2004)
93. A.V. Nabok, A. Tsargorodskaya, A.K. Hassan, N.F. Starodub, Appl. Surf. Sci. **246**, 381 (2005)

94. A. Nabok, A. Tsargorodskaya, Thin Solid Films **516**, 8993 (2008)
95. J. Mårtensson, H. Arwin, I. Lundström, T. Ericson, J. Colloid Interface Sci. **155**, 30 (1993)
96. J. Mårtensson, H. Arwin, H. Nygren, I. Lundström, J. Colloid Interface Sci. **174**, 79 (1995)
97. H. Arwin, A. Askendahl, P. Tengvall, D.W. Thompson, J.A. Woollam, Phys. Status Solidi (c) **5**, 1438 (2008)
98. G. Sun, D.M. Rosu, X. Zhang, M. Hovestädt, S. Pop, U. Schade, D. Aulich, M. Gensch, B. Ay, H. Holzhütter, D.R.T. Zahn, N. Esser, R. Volkmer, J. Rappich, K. Hinrichs, Phys. Status Solidi (b) **247**, 1925 (2010)
99. D. Aulich, O. Hoy, I. Luzinov, M. Brücher, R. Hergenröder, E. Bittrich, K.-J. Eichhorn, P. Uhlmann, M. Stamm, N. Esser, K. Hinrichs, Langmuir **26**, 12926 (2010)
100. C. Toccafondi, O. Cavalleri, F. Bisio, M. Canepa, Thin Solid Films **543**, 78 (2013)
101. C. Toccafondi, L. Occhi, O. Cavalleri, A. Penco, R. Castagna, A. Bianco, C. Bertarelli, D. Comoretto, M. Canepa, J. Mater. Chem. C **2**, 4692 (2014)
102. C. Akerlind, H. Arwin, F. Jakobsson, H. Kariis, K. Järrendahl, Thin Solid Films **519**, 3582 (2011)
103. L.G. Rosa, J. Liang, J. Phys. Condens. Matter **21**, 483001 (2009)
104. G.-Y. Liu, X. Song, Y. Qian, Acc. Chem. Res. **33**, 457 (2000)
105. L. Verstraete, J. Greenwood, B.E. Hirsch, S. de Feyter, ACS Nano **10**, 10706 (2016)
106. R. Haselberg, F.M. Flesch, A. Boerke, G.W. Somsen, Anal. Chim. Acta **779**, 90 (2013)
107. A. Bonyar, G. Harsanyi, in *Proceedings of the International Spring Seminar on Electronics Technology*, p. 519 (2011)
108. O. El Zubir, I. Barlow, G.J. Leggett, N.H. Williams, Nanoscale **5**, 11125 (2013)
109. C. Lee, E.A. Josephs, J. Shao, T. Ye, J. Phys. Chem. C **116**, 17625 (2012)
110. V. Kolivoska, M. Gal, S. Lachmanova, P. Janda, R. Sokolova, M. Hromadova, Collect. Czechoslov. Chem. Commun. **76**, 1075 (2011)
111. V. Kolivoska, M. Gal, M. Hromadova, S. Lachmanova, H. Tarabkova, P. Janda, L. Pospisil, A.M. Turonova, Colloids Surf. B Biointerfaces **94**, 213 (2012)
112. X. Song, G. Liu, Langmuir **13**, 127 (1997)
113. X. Zhai, H.J. Lee, T. Tian, T. Randall Lee, J.C. Garno, Molecules **19**, 13010 (2014)
114. C. Staii, D.W. Wood, G. Scoles, Nano Lett. **8**, 2503 (2008)
115. C. Rotella, G. Doni, A. Bosco, M. Castronovo, A. De Vita, L. Casalis, G.M. Pavan, P. Parisse, Nanoscale **9**, 6399 (2017)
116. M. Castronovo, D. Scaini, Methods Mol. Biol. (Clifton, N.J.) **749**, 209 (2011)
117. S. Corvaglia, B. Sanavio, R.P. Hong Enriquez, B. Sorce, A. Bosco, D. Scaini, S. Sabella, P.P. Pompa, G. Scoles, L. Casalis, Sci. Rep. **4**, 5366 (2014)
118. E.A. Josephs, T. Ye, J. Am. Chem. Soc. **132**, 10236 (2010)
119. E. Mirmontaz, M. Castronovo, C. Grunwald, F. Bano, D. Scaini, A.A. Ensafi, G. Scoles, L. Casalis, Nanoletters **8**, 4134 (2008)
120. T. Tian, B. Singhana, L.E. Englade-Franklin, X. Zhai, T. Randall Lee, J.C. Garno, Beilstein J. Nanotechnol. **5**, 26 (2014)
121. I. Solano, P. Parisse, O. Cavalleri, F. Gramazio, L. Casalis, M. Canepa, Beilstein J. Nanotechnol. **7**, 544 (2016)
122. J. Liang, M. Castronovo, G. Scoles, J. Am. Chem. Soc. **134**, 39 (2012)
123. J. Te Riet, T. Smit, J.W. Gerritsen, A. Cambi, J.A.A.W. Elemans, C.G. Figdor, S. Speller, Langmuir **26**, 6357 (2010)
124. D. Scaini, M. Castronovo, L. Casalis, G. Scoles, ACS Nano **2**, 507 (2008)
125. J. Liang, G. Scoles, J. Phys. Chem. C **114**, 10836 (2010)
126. P. Parisse, A. Vindigni, G. Scoles, L. Casalis, J. Phys. Chem. Lett. **3**, 3532 (2012)
127. G. Doni, M.D. Nkoua Ngavouka, A. Barducci, P. Parisse, A. De Vita, G. Scoles, L. Casalis, G.M. Pavan, Nanoscale **5**, 9988 (2013)
128. J.N. Ngunjiri, D.J. Stark, T. Tian, K.A. Briggman, J.C. Garno, Anal. Bioanal. Chem. **405**, 1985 (2013)
129. B. Sanavio, D. Scaini, C. Grunwald, G. Legname, G. Scoles, L. Casalis, ACS Nano **4**, 6607 (2010)

130. F. Bano, L. Fruk, B. Sanavio, M. Glettenberg, L. Casalis, C.M. Niemeyer, G. Scoles, Nano Lett. **9**, 2614 (2009)
131. I. Solano, P. Parisse, F. Gramazio, L. Ianeselli, B. Medagli, O. Cavalleri, L. Casalis, M. Canepa, Appl. Surf. Sci. **421**, 722 (2017)
132. A.J. Pertsin, M. Grunze, Langmuir **16**, 8829 (2000)
133. R.Y. Wang, M. Himmelhaus, J. Fick, S. Herrwerth, W. Eck, M. Grunze, J. Chem. Phys. **122**, 164702 (2005)
134. L. Li, S. Chen, J. Zheng, B.D. Ratner, S. Jiang, J. Phys. Chem. B **109**, 2934 (2005)
135. P.S. Johnson, M. Goel, N.L. Abbott, F.J. Himpsel, Langmuir **30**, 10263 (2014)
136. N. Inada, H. Asakawa, Y. Matsumoto, T. Fukuma, Nanotechnology **25**, 305602 (2014)
137. S. Herrwerth, W. Eck, S. Reinhardt, M. Grunze, J. Am. Chem. Soc. **125**, 9359 (2003)
138. H.I. Kim, J.G. Kushmerick, J.E. Houston, B.C. Bunker, Langmuir **19**, 9271 (2003)
139. T. Hayashi, Y. Tanaka, Y. Koide, M. Tanaka, M. Hara, Phys. Chem. Chem. Phys. **14**, 10196 (2012)
140. L.K. Ista, G.P. Lopez, Langmuir **28**, 12844 (2012)
141. E. Hochuli, H. Döbeli, A. Schacher, J. Chromatogr. A **411**, 177 (1987)
142. G.B. Sigal, C. Bamdad, A. Barberis, J. Strominger, G.M. Whitesides, Anal. Chem. **68**, 490 (1996)
143. A. Tinazli, J. Tang, R. Valiokas, S. Picuric, S. Lata, J. Piehler, B. Liedberg, R. Tampé, Chem. Eur. J. **11**, 5249 (2005)
144. L.E. Valenti, C.P. De Pauli, C.E. Giacomelli, J. Inorg. Biochem. **100**, 192 (2006)
145. F. Khan, H. Mingyue, M.J. Taussig, Anal. Chem. **78**, 3072 (2006)
146. F. Cheng, L.J. Gamble, D.G. Castner, Anal. Chem. **80**, 2564 (2008)
147. J.E. Gautrot, W.T.S. Huck, M. Welch, M. Ramstedt, ACS Appl. Mater. Interfaces **2**, 193 (2010)
148. T.T. Le, C.P. Wilde, N. Grossman, A.E.G. Cass, Phys. Chem. Chem. Phys. **13**, 5271 (2011)
149. L. Schmitt, M. Ludwig, H.E. Gaub, R. Tampé, Biophys. J. **78**, 3275 (2000)
150. G.J. Wegner, H.J. Lee, G. Marriott, R.M. Corn, Anal. Chem. **75**, 4740 (2003)
151. V. Gaberc-Porekar, V. Menart, Chem. Eng. Technol. **28**, 1306 (2005)
152. G. Klenkar, R. Valiokas, I. Lundström, A. Tinazli, R. Tampé, J. Piehler, B. Liedberg, Anal. Chem. **78**, 3643 (2006)
153. Y.-C. Li, Y.-S. Lin, P.-J. Tsai, C.-T. Chen, W.-Y. Chen, Y.-C. Chen, Anal. Chem. **79**, 7519 (2007)
154. S.H. Kim, M. Jeyakumar, J.A. Katzenellenbogen, J. Am. Chem. Soc. **129**, 13254 (2007)
155. P. Jain, L. Sun, J. Dai, G.L. Baker, M.L. Bruening, Biomacromolecules **8**, 3102 (2007)
156. I. Nakamura, A. Makino, M. Ohmae, S. Kimura, Macromol. Biosci. **10**, 1265 (2010)
157. W. Shen, H. Zhong, D. Neff, M.L. Norton, J. Am. Chem. Soc. **131**, 6660 (2009)
158. C.-H.K. Wang, S. Jiang, S.H. Pun, Langmuir **26**, 15445 (2010)
159. S. Uchinomiya, H. Nonaka, S. Wakayama, A. Ojida, I. Hamachi, Chem. Commun. **49**, 5022 (2013)
160. M. Sosna, H. Boer, P.N. Bartlett, ChemPhysChem **14**, 2225 (2013)
161. Y.-T. Lai, Y.-Y. Chang, L. Hu, Y. Yang, A. Chao, Z.-Y. Du, J.A. Tanner, M.-L. Chye, C. Qian, K.-M. Ng, H. Li, H. Sun, Proc. Natl. Acad. Sci. **112**, 2948 (2015)
162. T. Panavas, C. Sanders, T.R. Butt, in *SUMO Protocols*, vol. 497, Methods in Molecular Biology, ed. by H.D. Ulrich (Humana Press, New York, 2009), p. 303
163. M.P. Malakhov, M.R. Mattern, O.A. Malakhova, M. Drinker, S.D. Weeks, T.R. Butt, J. Struct. Funct. Genomics **5**, 75 (2004)
164. P. Bayer, A. Arndt, S. Metzger, R. Mahajan, F. Melchior, R. Jaenicke, J. Becker, J. Mol. Biol. **280**, 275 (1998)
165. J. Song, Z. Zhang, H. Weidong, Y. Chen, J. Biol. Chem. **280**, 40122 (2005)
166. I. Solano, Optical spectroscopy methods for the development of biosensors, Ph.D. thesis, University of Genova, 2016

# Part II
# Smart Polymer Surfaces and Films

# Chapter 5
# Glass Transition of Polymers with Different Architectures in the Confinement of Nanoscopic Films

**Michael Erber, Martin Tress, Eva Bittrich, Lars Bittrich and Klaus-Jochen Eichhorn**

**Abstract** The dynamic properties of nanoscopic polymeric films can significantly differ from the well-known bulk properties. In general, with decreasing film thickness the surface to volume ratio increases tremendously and interfacial interactions are expected to dominate the molecular dynamics of geometrically confined polymers. On the one hand, attractive interfacial interactions can inhibit cooperative dynamics and lead to a rise in $T_g$. On the other hand, repulsive interactions may depress $T_g$. However, the order of magnitude of the $T_g$ aberration in nanoscopic films is quite controversially discussed and some scientists even have doubt in the existence of confinement effects for films exceeding 10 nm in thickness. In the last few years, several factors were identified which may mimic confinement effects such as plasticizer effects due to solvent residues, degradation or oxidation processes and crosslinking. In this chapter we try to give a review about the determination and complexity of the glass transition of polymers in nanoscopic films and the unique role of temperature-dependent ellipsometry with its advantages but also methodical challenges therein.

M. Erber · E. Bittrich · K.-J. Eichhorn (✉)
Leibniz-Institut für Polymerforschung Dresden e.V., Abteilung Analytik,
Hohe Str. 6, 01069 Dresden, Germany
e-mail: kjeich@ipfdd.de

M. Erber
e-mail: michael.erber@basf.com

E. Bittrich
e-mail: bittrich-eva@ipfdd.de

L. Bittrich
Leibniz-Institut für Polymerforschung Dresden e.V., Abteilung Verbundwerkstoffe,
Hohe Str. 6, 01069 Dresden, Germany
e-mail: bittrich-lars@ipfdd.de

M. Tress
Institut für Experimentelle Physik II, Universität Leipzig, Linnéstr. 5,
04103 Leipzig, GermanyM. Tressand
e-mail: MartinTress@gmx.de

© Springer International Publishing AG, part of Springer Nature 2018
K. Hinrichs and K.-J. Eichhorn (eds.), *Ellipsometry of Functional Organic Surfaces and Films*, Springer Series in Surface Sciences 52,
https://doi.org/10.1007/978-3-319-75895-4_5

## 5.1 Polymers: A Unique Class of Materials and Their Physical Properties

A polymer (from Greek "poly": many and "meros": part) is a large molecule also called macromolecule characterized by multiple repetitions of constitutive units. Due to its size the addition or removal of one repetitive unit does not substantially change their properties. In fact, the constitutive unit is the smallest periodic unit, which describes the structure of the macromolecular chain completely [1]. Depending on the number of connection points of each constitutive unit, polymers may be terminologically divided into four different classes regarding to their architecture: linear, cross-linked, branched and dendritic polymers as shown in Table 5.1 [2, 3].

Furthermore, dendritic polymers are classified as (a) dendrimers, (b) hyperbranched polymers, (c) dendrigrafts and (d) dendronised polymers.

Besides the architecture, the chemical structure, functional groups and molecular weight are important factors which strongly affect the material properties of a synthetic polymer and provide an extraordinary variety. Because of this diversity of polymeric materials as well as their low cost fabrication, they play an essential and ubiquitous role in everyday life [1, 4]. Typically, the bulk properties are those most often of end-use interest. These are the properties that dictate how the polymeric material actually behaves on a macroscopic scale. Besides the tensile strength, Young's modulus and transport properties, the phase transition temperatures such as the melting point ($T_m$), and the glass transition temperature ($T_g$) are important characteristic quantities of high technological relevance. Among synthetic polymers, melting is generally discussed with regards to semi-crystalline thermoplastics such as PA6, PA66, PET or PE while thermosetting polymers will decompose at high temperatures rather than melt. Due to their statistically molecular arrangement a lot of other polymers show no semi-crystalline behavior at all. They are amorphous.

In case of amorphous polymers, the glass transition temperature is the parameter of particular interest in polymer manufacturing and application. It describes the temperature at which polymers undergo a transition from the rubbery, liquid-like to a glassy solid state and vice versa. In contrast to the melting point the glass transition

**Table 5.1** Terminological classification of polymers into four main categories according to Tomalia and Fréchet: I. linear, II. cross-linked, III. branched, IV. dendritic: (a) dendrimer, (b) hyperbranched, (c) dendrigraft and (d) dendronised polymers. (Modified after [2])

**Fig. 5.1** The specific
volume of a material in
dependence of temperature.
In dependence of the cooling
rate different glassy or
crystalline states may be
achieved which occur at
different temperatures.
(Modified after Donth,
*Glasübergang*,
Akademie-Verlag, Berlin,
1981)

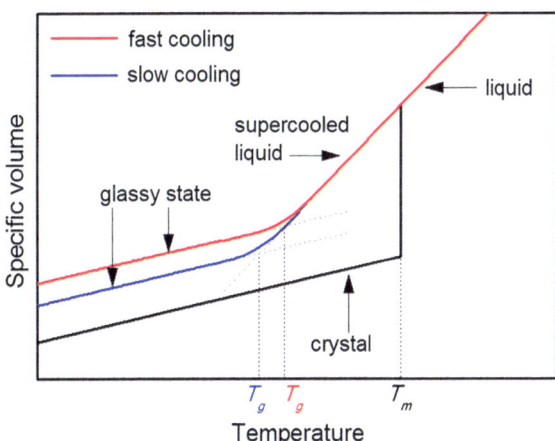

temperature is usually not a clearly defined point, but may extend over a wide temperature range of up to 10 K (Fig. 5.1), depending on time scale of the experiment, molecular weight distribution and thermal history of the polymeric sample.

The melting or crystallization behavior of many materials is a first order phase transition associated with abrupt changes in extensive thermodynamic quantities such as the entropy, enthalpy or volume [4]. In contrast, the glass transition is not a real phase transition! It can be rather seen as solidification without any abrupt change in thermodynamic quantities, and a kinetic origin is nowadays regarded as a plausible explanation [5, 6]. This means, instead of a structural rearrangement (like the development of a highly ordered crystalline state), the mobility of the molecules slows down gradually with decreasing temperature, and as their relaxation time approaches the experimental time scale the material appears solid. Consequently, the detected transition temperature depends on the rate of temperature variation, or the frequency of the oscillating external perturbation applied, to monitor the transition. Therefore, the glass transition is called a dynamic phenomenon, and it is essential to specify such applied rates or timescales (frequencies). Different approaches have been postulated to describe the phenomenon of glassy dynamics in the "bulk" state theory, though neither of them is generally accepted. At this point, the "free-volume theory" [7], "cooperative approaches" [8] or the "Rouse model" [9] should be mentioned.

## 5.2 Nanoscopic Polymer Films versus Bulk Polymers

### 5.2.1 Introduction

In the early 1990s it was reported for the first time that the glass transition temperature $T_g$ of thin polymer films may differ significantly from the bulk value [10–12]. Since

$T_g$ was found to decrease gradually with reduced film thickness below ~100 nm many researchers assumed that the observed $T_g$ shift originated from the influence of geometrical confinement to a very thin layer. To explain this phenomenon manifold effects have been discussed like weak polymer-substrate interactions, sliding modes of chain loops at the free surface [13], or liquid surface layers. However, about 20 years later the debate is still controversial and the phenomenon not yet fully explained [14, 15].

## 5.2.2   True or Mimicked Confinement Effects

Although subsequent investigations found similar trends, the combined results show a strong variation in the thickness dependence of $T_g$ [16]. Even more controversial, some studies did not disclose any change in $T_g$ in films of only a few nanometer thickness [14, 17, 18]. For polystyrene (PS) the huge variety of results is shown in Table 5.2 which evidently cannot be a feature of the material.

Indeed, severe mechanisms affecting $T_g$ and thus mimicking confinement effects were identified and can be summarized as the impact of preparation: plasticizer effects due to remaining solvent [27], agglomeration of solvent at the polymer-solid interface (which explains why thin films keep larger volume fractions of solvent than

**Table 5.2** Some exclusive results reported by different research groups, showing $T_g$ shifts of PS films of thickness $d$ (minimum value under investigation) as measured by ellipsometry. Even for a chemically simple polymer like PS, the variety of Tg aberration is impressive and shows the complexity of the phenomenon

| Author (Year) | $M_w$ (kg/mol) | $d$ (nm) | Shift of $T_g$ (K) | Substrate |
|---|---|---|---|---|
| Keddie [11] (1994) | 501 | 12 | −25 | H-passivated Si |
|  | 2 900 | 12 | −25 | H-passivated Si |
| Forrest [19] (1997) | 767 | 29 | −7 | SiOx |
| Fryer [20] (2000) | 382 | 20 | −20 | HMDS on SiOx |
|  | 382 | 17 | −20 | SiOx |
| Sharp [21] (2003) | 369 | 8 | −15 | SiOx |
| Fakhraai [22] (2005) | 641 | 6 | −32 | Pt |
| Raegen [23] (2008) | 734 | 25 | 0 | SiOx |
|  |  | 6 | −50 | SiOx |
| Kim [24] (2009) | 400 | 23 | −10 | SiOx |
| Mapesa [17] (2010) | 319 | 17 | 0 (±2) | SiOx |
| Tress [25] (2010) | 1 103 | 10 | −2 (±2) | SiOx |
|  | 749 | 12 | −2.5 (±2) | SiOx |
|  | 319 | 19 | −1 (±3) | SiOx |
| Torres [26] (2012) | 575 | 5 | −22 | SiOx |

thick films), thermal history of the sample [28], water sorption [29], and chemical degradation [29].

Nevertheless, real confinement effects are known e.g. from simple molecules in nano-porous host-materials [30]. Hence, it is expected that polymers exhibit similar phenomena - the question is: on which length scale does spatial confinement affect molecular mobility and structural transitions?

### 5.2.3   Make a Long Controversial Story Short: Conclusion

The dynamic glass transition can be assigned to local motions of only a few monomer segments [31]. This corresponds to lengths of less than 1 nm and consequently, alterations of the glassy dynamics of a polymer are expected to appear in a comparable distance from the solid substrate. The direct conclusion from this fact is that $T_g$ shifts of films thicker than $\sim$10 nm are highly likely the result of preparative conditions and after appropriate treatment bulk properties can be recovered [32].

## 5.3   Comparison of Analytical Methods for the Determination of Glass Transition of Polymers in Thin Polymer Films

### 5.3.1   Introduction

Glass transition of polymers can be monitored by means of manifold experimental methods which employ various physical principles (heat capacity, probing the free volume, mechanical stress-strain behaviour, the electrical and optical response of molecular dipoles or optical observation of fluorescence labels, to name a few). Till present, many of these techniques have been developed to enable also thin film measurements as discussed in this section [30, 33–35].

### 5.3.2   Ellipsometry

Ellipsometry, a non-destructive optical technique, is widely applied for the determination of film thickness and optical constants of polymer films [36–39].

By acquiring ellipsometric data in dependence on temperature, extremely small differences in thickness and optical properties can be detected. This unique quality can be used to determine the coefficient of thermal expansion $\alpha$ and from an abrupt change in $\alpha$ the glass transition temperature can be deduced (for details see Sect. 5.4). Besides films supported on a substrate, also free-standing films may be studied with

ellipsometry. Due to the poor sensitivity and several other drawbacks of ellipsometric measurements for free-standing films they are not explained in more detail
[13, 30, 40].

### 5.3.3  Broadband Dielectric Spectroscopy

Broadband dielectric spectroscopy (BDS) uses the fact that many molecules exhibit
dipolar character and, by that, interact with an external electrical field. Therefore,
the sample is connected with two electrodes in a capacitor-like arrangement and an
oscillating electrical field is applied. From the measured (complex) capacitance the
complex dielectric permittivity $\varepsilon^* = \varepsilon' - i\varepsilon''$ can be extracted if the sample geometry is
known. A plot of this quantity in dependence on frequency of the electrical field or
temperature typically exhibits local maxima which are called relaxation processes.
The characteristics of these processes depend on molecular quantities like number
density and mobility of charge carriers or molecular dipoles. One of these molecular
relaxations is called $\alpha$-relaxation which entitles the thermal fluctuation of chain
segments of a polymer and hence, it is assigned to the dynamic glass transition. If
plotted versus temperature, its peak position (the so-called characteristic temperature
$T_\alpha$) at a probe frequency of $f = 2\pi/(100\,\text{s})$ conventionally corresponds to the glass
transition temperature $T_g$ determined by e.g. ellipsometry (or calorimetry) at a heating
rate of 10 K/min. For a comprehensive description of BDS and its applications see
Kremer and Schönhals [41].

Typically, BDS is a volumetric method but the big advantage of a capacitor is the
fact that capacity increases with decreasing thickness - less electrode distance (hence,
sample amount) means larger signal intensity. This enabled a series of studies of thin
polymer films down to $\sim$10 nm [41, 42]. However, at this length scale the probability of electrical short circuits due to surface roughness of the electrodes increases
rapidly and limits the successful investigation of thinner films. Therefore, recently
a novel method was developed which uses ultra-flat highly conductive silicon wafer
dice as electrodes [43]. One of these electrodes is fabricated with highly insulating
silica nano-structures which keep the two electrodes separated from each other. This
sample arrangement solves afore mentioned problems in thin films and enables the
measurement of films with one free interface. The latter makes it an ideal complementary technique for ellipsometry. Investigations which apply this combination of
methods are presented and discussed in detail in Sect. 5.5 [44].

### 5.3.4  AC-calorimetry

Alternating current (AC)-calorimetry is a technique which enables frequency dependent calorimetric measurements. Therefore, instead of a constant heating rate as used
for common DSC experiments the temperature is additionally modulated with a

certain frequency. The resulting signal can be treated as a complex heat capacity and the real part indicates phase transitions by the specific heat, similar to DSC. Furthermore, the dynamic glass transition can be assigned to a local maximum in the imaginary part which coincides with other frequency dependent methods like e.g. BDS. Schick et al. developed an AC-calorimeter capable of measuring ultra-thin polymer films down to 10 nm thickness in a frequency range of 1–10 kHz [35]. Several investigations on PS, PMMA reveal no shift in the dynamic glass transition in films as thin as 10 nm [25, 35].

## 5.3.5  *Other Techniques*

Besides classical DSC experiments, which have been adopted for thin film measurements as well [46, 47], there are several further methods capable of probing $T_g$.

One group are scattering techniques like neutron or X-ray scattering. Thereby, the sample is irradiated with a focussed beam and the spatial distribution of the intensity of the scattered radiation exhibits a pattern which contains information about the structure of the sample. For crystalline materials the geometric structure of the atomic lattice as well as the lattice spacing can be determined, but also for amorphous materials changes in the mean distance between neighbouring molecules are visible. Despite the fact, that these techniques typically require quite large samples, several investigations of thin polymer films were performed by measuring in grazing incidence or stacks of multiple films [48, 49].

Another way to trace especially the dynamic glass transition are fluorescence techniques. Therefore, fluorescent dye molecules are either mixed with the polymer matrix or covalently bonded to the polymer chain [50, 51]. The movements of the dye molecules can be detected with a spatial resolution of up to ~10 nm and are analysed in terms of the auto correlation function [52]. Although these measurements enable direct observation of molecular motion, note that it is the dye molecules which are probed. It is by no means clear that this reflects always the motion of the surrounding polymer matrix or the pure polymer in case of dissolved or covalently bonded dyes, respectively. Hence, this technique requires prudent cross checks with dyes of different size, architecture, chemical behaviour or covalently bonded versus dissolved ones to verify that the result is independent from these parameters and can be assigned specifically to the host polymer matrix.

Furthermore, Fourier-transform infrared spectroscopy (FTIR) was applied to investigate the glass transition of thin polymer films [53]. The intensity of IR absorption bands depends on the amount of material and the concentration of the respective molecular species in the sample. Some polymers exhibit different favorability for possible conformations below and above $T_g$ which can be detected as a change in the ratio of intensities of the corresponding bands.

## 5.4 Concepts of the Determination of Glass Transition Applying Ellipsometry

In 1994, Keddie et al. were one of the first who applied ellipsometry to study the thermal expansion $\alpha$ and the thereof derived $T_g$ of thin polystyrene films [54]. Inspired by their pioneering work and their fascinating results (altered $T_g$ in nanoscopic films) many research groups tried to discover the reasons for the deviation from the well-known bulk properties in either substrate supported films or free-standing films. $T_g$ of a thin polymeric film can be determined applying the ellipsometric quantities in a variety of ways. But in all cases the abrupt change in the thermal expansion is used to extract $T_g$. Depending on the measurement principle of the ellipsometer (null ellipsometry, spectroscopic vis-ellipsometry) different concepts exist which are discussed in more detail in the following.

### 5.4.1 Single Wavelength Ellipsometry

Historically, null ellipsometry as developed by Drude was the basis of the first ellipsometry instrument. The null ellipsometer employs a monochromatic light source usually a laser in the visible wavelength range. The PCSA (polarizer, compensator, sample, analyzer) configuration is applied to determine the ellipsometric parameters at one given wavelength (for more detailed information see Chap. 1). One advantage of null ellipsometry is that the laser beam can be focused on a small spot size, has a higher power than broad band light sources, and, due to its simple measurement principle, it is still the most accurate instrument in the absolute determination of $\Delta$ and $\Psi$. Applying this method, the "raw data" ($\Psi$ and $\Delta$) are monitored at a specific wavelength (preferably at a sensitive one) as a function of temperature. Then, $T_g$ can be determined from the discontinuity ("kink") in $\Psi$ and/or $\Delta$ as a function of temperature (Fig. 5.2) [10]. As explained in Sect. 5.1, $T_g$ is not a precise point at a specific temperature, it is rather a temperature region which makes the evaluation of the position of $T_g$ vague. In order to have definite values, $T_g$ is commonly determined by the intersection point of two straight lines fitted to the temperature dependence of the ellipsometric angles $\Psi$ and $\Delta$ in the rubbery and the glassy regime, respectively.

Despite its advantages of very high accuracy and simplicity of the device, one big drawback is that only one wavelength is used for the measurement. Depending on the film thickness the sensitivity (signal to noise ratio) of $\Delta$ and $\Psi$ changes with wavelength (Fig. 5.3).

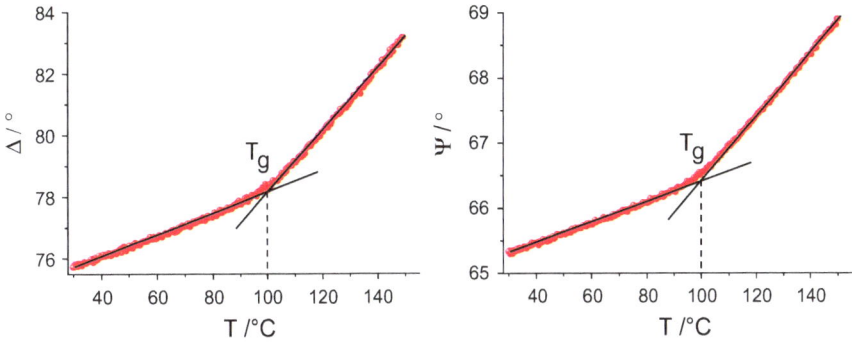

**Fig. 5.2** Ellipsometric angles ($\Psi$ und $\Delta$) as a function of temperature for a 76 nm thin PS film. The ellipsometric angles are determined at a wavelength of $\lambda = 450$ nm which is sensitive for the glass transition. (Modified after [25])

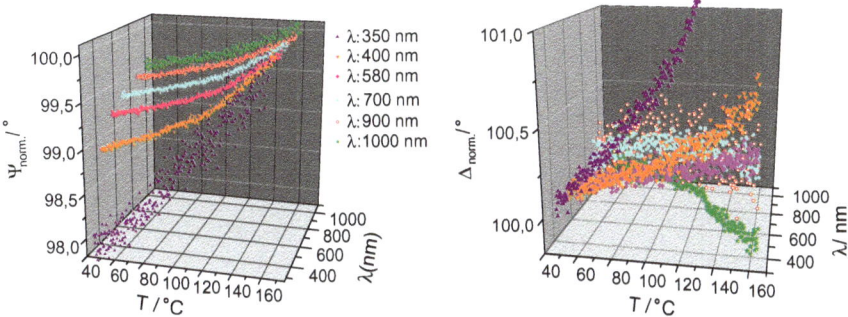

**Fig. 5.3** Ellipsometric angles $\Psi$ and $\Delta$ as a function of temperature and wavelength for a 67 nm thin PMMA film. Depending on the wavelength the sensitivity (signal to noise ratio) and therefore the accuracy of $T_g$ determination is affected

## 5.4.2  Spectroscopic vis-Ellipsometry (SE)

Spectroscopic ellipsometry employs broad band light sources which cover a certain range in the infrared, visible or ultraviolet spectral region. The full spectroscopic data set of $\Psi$ and $\Delta$ can be fitted with an appropriate optical model to yield the thickness $d$ and refractive index $n$ of the polymer film as a function of temperature. Hence, the discontinuity in either $d$ or $n$ as a function of temperature can be used to determine $T_g$. It is important to mention that not the "raw data" $\Delta$ and $\Psi$ at only one individually chosen wavelength are used, in fact the layer parameters $n$ and $d$ include measurements of $\Delta$ and $\Psi$ at many different wavelengths.

The evaluation of $T_g$ can be done (1) by tangent fitting [55], (2) derivation of the thickness data to directly evaluate the thermal expansion [24], but also (3) by using a fit-function for the temperature dependent layer thickness $d(T)$ or the refractive index $n(T)$ [25, 56, 58]. For approach (3) either the fit-parameter $T_g$ or the extremum

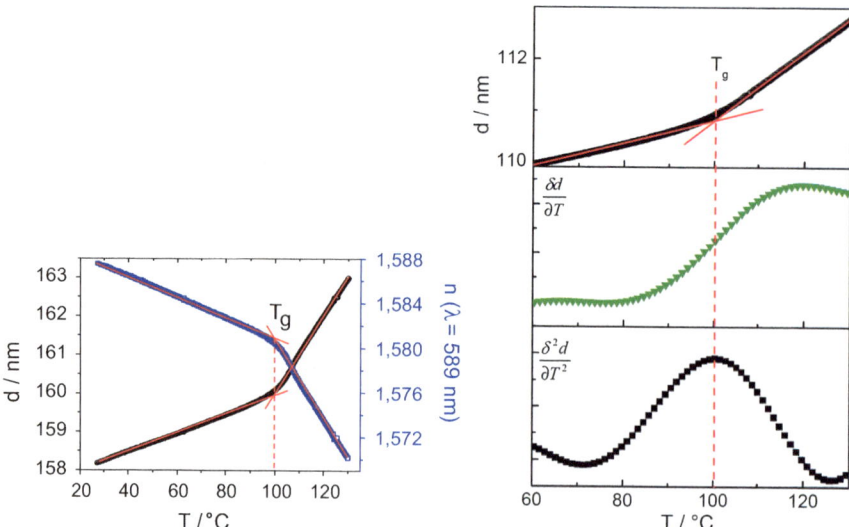

**Fig. 5.4** Film thickness $d$ and refractive index $n$ as a function of temperature *(left)* as well as the first and second derivative *(right)* of a thin PS film. $T_g$ can be determined by the kink in $d$ or $n$ and from the position of the extrema in the second derivatives (shown for film thickness only), respectively

of the 2nd derivative of the fit-function are evaluated. See Fig. 5.4 for examples of tangent fitting and the evaluation of the maximum of the 2nd derivative of a 9th oder polynomial fit-function for $d(T)$.

Ideally, the glass transition process is characterized by a constant coefficient of thermal expansion $\alpha$ in the glassy and the rubbery state and a non-linear change of $\alpha$ in the glass transition region [57]. However, we and others observed that the thermal expansion in the glassy and rubbery state is not necessarily constant but can have a small positive or negative linear slope [24, 25, 48, 58]. This leads to at least quadratic behavior in the thickness data in these temperature regions, and the $T_g$ value obtained by approach (1) becomes sensitive to the temperature ranges fitted by tangents [48]. Non-constant values for $\alpha$ in the glassy and rubbery state are taken into account by approach (2), as well as the use of appropriate fit-functions (3), e.g. a polynomial fit-function as used in [25]. To overcome the drawback of unphysical oscillations introduced by polynomials, we propose a new fit-function for the thermal expansion $\alpha$, based on the formula of Forrest and Dalnoki-Veress [57]:

$$\alpha(T) = \frac{M - G}{2} \frac{\{A(T - T_g) - 1\} e^{\left(-\frac{T - T_g}{w}\right)} + \{B(T - T_g) + 1\} e^{\left(\frac{T - T_g}{w}\right)}}{e^{\left(-\frac{T - T_g}{w}\right)} + e^{\left(\frac{T - T_g}{w}\right)}} + \frac{M + G}{2}$$

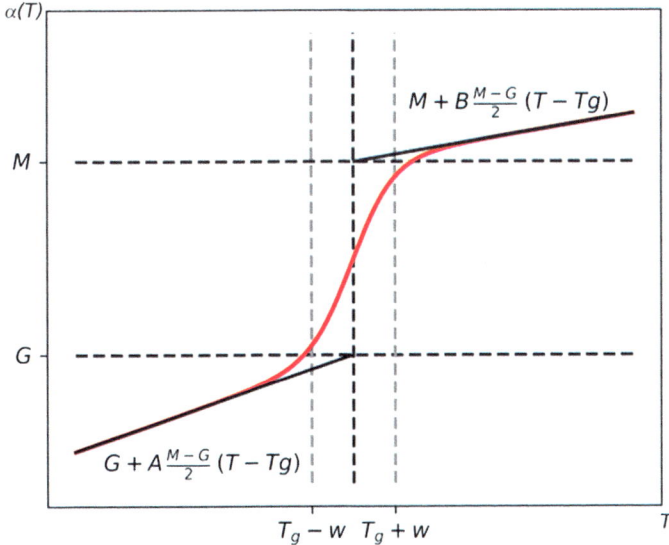

**Fig. 5.5** Sketch to illustrate the trends of the thermal expansion $\alpha$ and the fit parameters G, M, A, B

The fit-function for $\alpha(T)$ additionally reflects a linear behavior of the thermal expansion in the glassy and rubbery state. Parameters are the temperature $T$, $T_w$, the width of transition $w$, as well as the values $G$, $A$ and $M$, $B$ describing the linear behavior in the glassy and the rubbery state, respectively (Fig. 5.5). In the case where $A = 0$ and $B = 0$ (and the value of $\alpha$ is constant in the glassy and rubbery state) this new formula for $\alpha(T)$ matches the formula of Forrest and Dalnoki-Veress exactly. The fit-function for the modelled ellipsometric thickness $d(T)$ is obtained by integration of $\alpha(T)$, while $Li_2$ is the dilogarithm function.

$$d(T) = \frac{M - G}{2}\frac{w^2}{4}\left[(B - A)Li_2\left(-e^{2\left(\frac{T-T_g}{w}\right)}\right)\right.$$

$$+ \frac{2}{w}\left\{(B(T - T_g) - A(T - T_g) + 2)ln\left(e^{2\left(\frac{T-T_g}{w}\right)} + 1\right)\right.$$

$$\left.\left. + \left(\frac{T - T_g}{w}\right)(A(T - T_g) - 2)\right\}\right] + \frac{M + G}{2}(T - T_g) + c$$

With this fit-function for $d(T)$ the glass transition temperature $T_g$ and the width $w$ of the transition can be determined directly from the modeled ellipsometric thickness data, while the signal-to-noise ratio should be improved as compared to the use of direct data for $\alpha(T)$ obtained by derivation of the modeled ellipsometric thickness in approach (2).

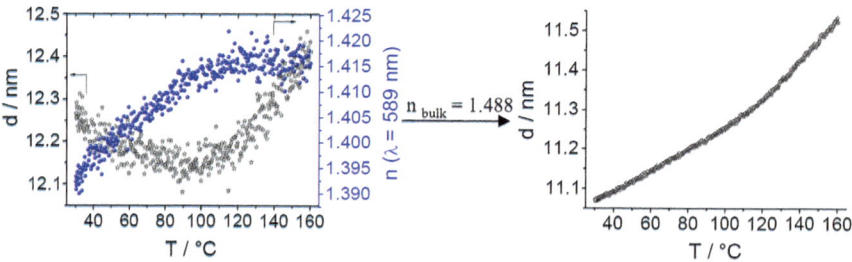

**Fig. 5.6** Film thickness $d$ and refractive index $n$ as a function of temperature for a 12 nm thin PMMA film on Si/SiO$_x$ *(left)* and the corresponding quantities after refractive index fixation to the bulk value ($n_{bulk} = 1.488$ at $\lambda = 589$ nm) *(right)*. The noise in $d(T)$ is minimized, however information in $n(T)$ is not available anymore

At this point it is worth to note, that with decreasing length scale the contrast of the glass transition defined as the ratio of the thermal expansion $\alpha$ of the rubber-like state to that of the glassy state is reduced [59]. As the contrast of the transition decreases, it becomes more and more difficult to identify the transition and by definition it is not possible anymore for a contrast value of 1.

$$Contrast = \frac{\alpha_{rubber}}{\alpha_{glass}} \geq 1 \qquad (5.1)$$

In practice, any data exhibits scatter and noise, and the ellipsometric identification of $T_g$ becomes impossible for very thin films. In our measurements with film thicknesses below 20 nm the overall changes in ellipsometric angles with increasing temperature are so small that the difference in slope above and below $T_g$ is obscured by the noise level in the data. Furthermore, slight non-linearities in the dependence of ellipsometric angles on film thickness are particularly problematic as well as the separate determination of $n$ and $d$ due to parameter correlation for very thin films. Therefore, for films with a thickness below 20 nm the refractive index was fixed to $n_{bulk}$ calculated for thick films ($d > 100$ nm) using a Cauchy dispersion (Fig. 5.6). With this approach the thinnest films ($>10$ nm) were barely analyzable with good accuracy. Below 10 nm film thickness the noise of the data was still adequate but the changes in thickness and refractive index are very small and hence the determination of $T_g$ is afflicted with a distinctive experimental error up to 10 K. Such difficulties lead to an effective lower limit of sensitivity in this technique to determine $T_g$ in extremely thin films.

New methodical developments for the ellipsometric determination of $T_g$ were discussed in recent years to improve the sensitivity of the method, and $T_g$ values are reported for films as thin as 1–2 nm [60, 61]. Additionally, the influence of the free surface roughness on the $T_g$ obtained by ellipsometry was evaluated in detail [62], and the evaluation of the layer fragility from ellipsometric data introduced [63].

## 5.5   Glass Transition in Thin Polymeric Films in Dependence of Film Thickness – Some Exclusive Examples

### 5.5.1   Effect of Polymer Architecture and Functional Groups

#### 5.5.1.1   Linear Polymers

In linear polymers the bulk glass transition is, beyond the chemical composition, mainly determined by the molecular weight, i.e. degree of polymerization, which scales with the chain length. The key quantity thereby is the entanglement molecular weight, which represents the degree of polymerisation above which the chains can form loops to entangle with each other. In this regime $T_g$ is independent of $M_w$, while below the entanglement molecular weight $T_g$ reduces with decreasing chain length.

For free-standing films of PS it has been claimed that even far above the entanglement molecular weight $T_g$ shifts induced by thin film confinement are chain length dependent [40, 57]. Although some publications present coinciding data [55], other reports show conflicting results [35].

Similarly, in the case of PS films supported on a solid substrate a huge variability of results has been published (see Table 5.2). As a matter of fact, some investigations using the complementary techniques SE and BDS on films prepared and measured under identical and appropriate conditions (e.g. controlled annealing, purging with argon) find no effect of the film thickness on $T_g$ down to 5 nm for PS of different molecular weights (Fig. 5.7) [17, 44].

#### 5.5.1.2   Branched and Hyperbranched Polymers

For linear polymers a vast number of controversial experimental results are reported, as discussed in Sect. 5.2.2. To the best of our knowledge, in most cases exclusively linear homopolymers like PMMA or PS were in the focus of such studies [30, 35, 42, 57]. However, other polymer architectures like dendritic macromolecules are likely to exhibit properties very different from their linear counterparts due to their branched, globular structure, their high number and density of functional groups as well as their low viscosity attributable to significantly reduced entanglements [2, 3, 64, 65]. Due to the often tedious and complicated preparative synthesis these materials have rarely been in the focus of confinement investigations.

However, several exclusive polymer systems with special architecture are studied in this context. Studies on PS derivatives ranging from linear to selected star-like and hyperbranched topologies showed that the glass transition is not significantly affected by the polymer topology in films down to 15 nm [66]. In another study Glynos et al. systematically varied the number of arms $f$ and the molecular weight

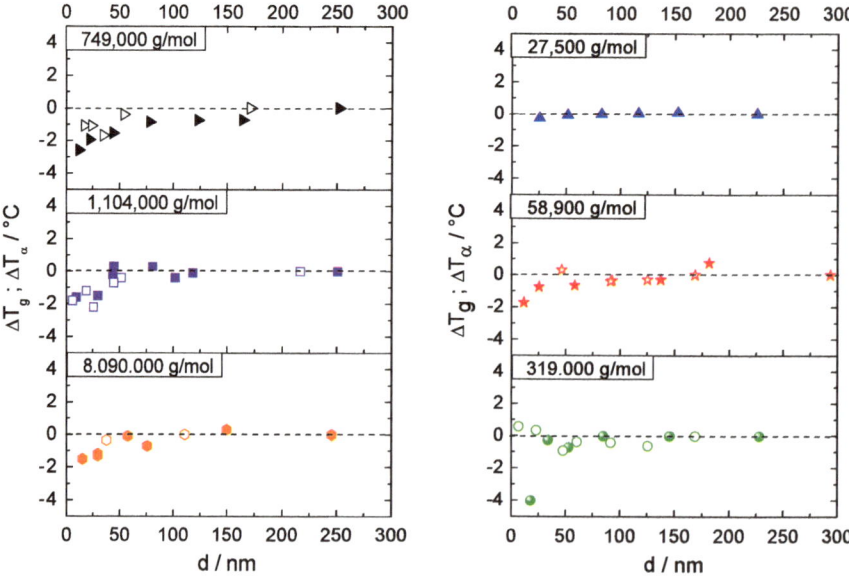

**Fig. 5.7** $T_g$-shift $\Delta T_g$ (measured by SE) of thin films of linear PS with different molecular weight as indicated verses film thickness $d$ (*closed symbols*). Within the experimental accuracy, no deviation from the bulk $T_g$ is found for all investigated molecular weights down to film thicknesses of 10 nm. The study contains complementary measurements by BDS (shift of the alpha relaxation temperature $\Delta T_\alpha$ measured at a frequency of 1 kHz is displayed) down to 5 nm thickness which delivers coinciding results (*open symbols*). (Modified after [25])

per arm of star-shaped PS and observed maximum changes in the $T_g$ of 6 K for 30 nm thin PS films, respectively [67].

Although the topology has no significant influence on the $T_g$ of PS films, varying the interfacial interactions between polyester and the substrate by modification of the functional groups of the polymer induced significant effects. Depending on the nature of interactions, $T_g$ was increased up to 14 K or depressed up to 9 K (Fig. 5.8) [56]. However, especially for hydroxyl-terminated polyester it was found out that these functional groups are temperature sensitive and may undergo crosslinking, although such effects are not known from the corresponding bulk analogues. Of course, such crosslinking may lead to higher glass transition temperatures.

## 5.5.2 Effect of Interfacial Interactions Between Polymer and Substrate

Since a large fraction of molecules in a thin polymer film is in direct contact with the supporting substrate, it is expected that the type of interaction of the polymer with the solid interface is of great importance for the dynamics. It has been proposed that

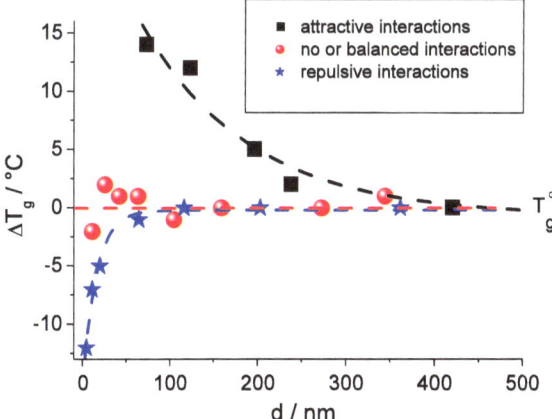

**Fig. 5.8** Impact of interfacial interactions between substrate (Si/SiO$_x$) and systematically function-alized aromatic-aliphatic hyperbranched polyesters on the glass transition temperature in dependence on film thickness $d$ ($T_g$–shifts $\Delta T_g$ measured by SE are displayed). Different functional groups were studied: hydroxyl –OH (*squares*), benzoyl –OBz (*circles*), tert.-butyldimethylsilyl –OSi (*stars*). (Modified after [56])

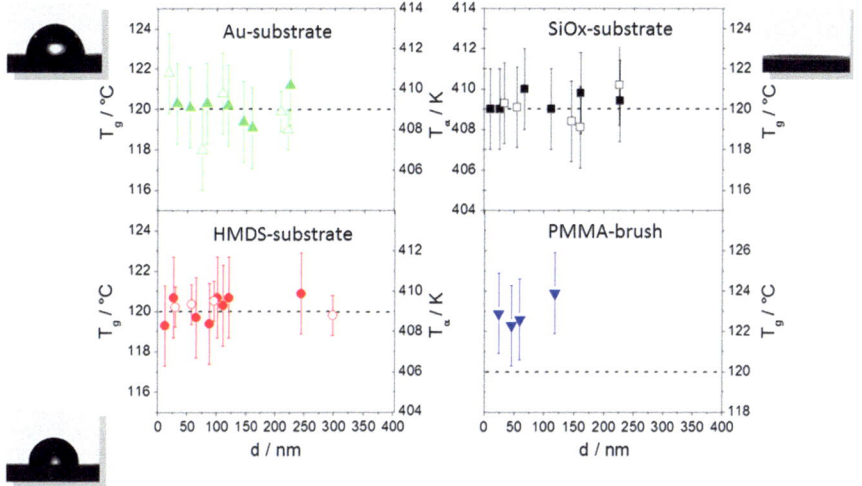

**Fig. 5.9** $T_g$ (*closed symbols*) and $T_\alpha$ (*open symbols*) of atactic PMMA (molecular weight $M_w =$ 319 kg/mol) on different substrates as determined by means of SE and BDS, respectively. The insets show the water contact angle of the differently modified substrates in order to give an impression about the capability and type of interfacial interactions with PMMA on the corresponding substrate. (Modified after [44])

strong interactions result in increasing $T_g$ while weak interactions would cause the opposite [20].

For a systematic investigation of different polymer-substrate interactions PMMA was deposited onto different substrates: (i) hexadimethylsilazane (HDMS) coated silicon, (ii) gold coated silicon, (iii) silicon with a native oxide layer. Additionally, PMMA covalently bond to silicon (PMMA brush) was studied. The interactions are in order of increasing strength; the highly hydrophobic HDMS surface interacts weakly with the polar PMMA, gold is slightly hydrophobic, the native oxide has surface OH-groups which can interact with PMMA, and finally the covalent bond is the strongest type of interaction. Nevertheless, measurements of the glass transition temperature by SE revealed that the type of substrate (HDMS-coated, gold or native oxide) has no impact on $T_g$ in films down to 10 nm thickness (Fig. 5.9). Neither a thickness dependence is detected in any of these substrates nor a difference in $T_g$ even of the thinnest investigated films was found on the different substrates. These results were verified by complementary BDS experiments of identically prepared samples. In the PMMA brushes, $T_g$ was about 3 K larger whereas again no thickness dependence was found in the studied thickness range from ~25 to 120 nm.

## 5.6   Summary and Outlook

The preparation of nanoscopic, well-defined structures as well as the resulting minia-turisation of construction components can be seen as the character of the modern industry. However, due to the dimensional confinement unexpected material proper-ties may occur. In this context, the glassy dynamics in nanoscopic polymer geometries is a highly topical but also controversially discussed subject. Since about 20 years the genesis of confinement effects on $T_g$ is still not completely neither theoretically nor experimentally understood. Certainly, most robust results suggest that $T_g$ devia-tions occur quite below film thicknesses of 15 nm and can be assigned to interfacial interactions. Thereby, strong attractive interfacial interactions will increase $T_g$ in nanoscopic films whereas repulsive interactions will influence $T_g$ vice versa.

Studying films below 15 nm is therefore of special interest. However, investiga-tions require an extremely high sensitivity of the applied analytical method, well-defined polymer systems and corresponding layer stacks as well as clean room conditions to name a few. In this regard, spectroscopic ellipsometry is a quite versatile and unique analytical tool with appropriate sensitivity, simple and robust device setup, fast measurement periods and the possibility to *on-line* monitor films.

Further, this topic is not only in focus of academic science. Especially the glass transition temperature and the closely related stability of polymeric components e.g. flexible solar-cells, e-papers, sensors and actors or computer chips are of particular industrial interest.

# References

1. H.G. Elias, *Von Monomeren und Makromolekülen zu Werkstoffen* (Hüthig & Wepf, Heidelberg, 1996)
2. J.M. Fréchet, D.A. Tomalia, *Dendrimers and Other Dendritic Polymers* (Wiley, New York, 2001)
3. B. Voit, A. Lederer, Chem. Rev. **109**, 5924 (2009)
4. J. Brandrup, E.H. Immergut, E.A. Grulke, *Polymer Handbook* (Wiley, New York, 1999)
5. S.A. Baeurle, A. Hotta, A.A. Gusev, Polymer **47**, 6243 (2006)
6. S. Torquato, Nature **405**, 521 (2000)
7. M.H. Cohen, D. Turnbull, J. Chem. Phys. **31**, 1164 (1959)
8. G. Adam, J. Gibbs, J. Polym. Phys. **46**, 139 (1965)
9. G. Strobl, *The Physics of Polymers* (Springer, Berlin, 2007)
10. J.L. Keddie, R.A.L. Jones, R.A. Cory, Faraday Discuss. **98**, 219 (1994)
11. J.L. Keddie, R.A.L. Jones, R.A. Cory, Eur. Lett. **27**, 59 (1994)
12. G. Reiter, Phys. Rev. Lett. **68**, 75 (1992)
13. K. Dalnoki-Veress, J.A. Forrest, P.G. de Gennes, J.R. Dutcher, J. Phys. IV **10**, 221 (2000)
14. F. Kremer, M. Tress, E.U. Mapesa, J. Non-Cryst. Solids **407**, 277 (2015)
15. M.D. Ediger, J.A. Forrest, Macromolecules **47**, 471 (2014)
16. Y.P. Kalmykov, in *Recent Advances in Dielectric Spectroscopy*, ed. by F. Kremer, E.U. Mapesa, M. Tress, M. Reiche, Molecular Dynamics of Polymers at Nanometric Length Scales: From Thin Layers to Isolated Coils, pp. 163–178
17. E.U. Mapesa, M. Erber, M. Tress, K.J. Eichhorn, A. Serghei, B. Voit, F. Kremer, Eur. Phys. J. Spec. Top. **189**, 173 (2010)
18. M.Y. Efremov, E.A. Olson, M. Zhang, Z.S. Zhang, L.H. Allen, Macromolecules **37**, 4607 (2004)
19. J.A. Forrest, K. DalnokiVeress, J.R. Dutcher, Phys. Rev. E **56**, 5705 (1997)
20. D.S. Fryer, P.F. Nealey, J.J. de Pablo, Macromolecules **33**, 6439 (2000)
21. J.S. Sharp, J.A. Forrest, Phys. Rev. E **67**, 031805 (2003)
22. Z. Fakhraai, J.A. Forrest, Phys. Rev. Lett. **95**, 025701 (2005)
23. A. Raegen, M. Massa, J. Forrest, K. Dalnoki-Veress, Eur. Phys. J. E **27**, 375 (2008)
24. S. Kim, S.A. Hewlett, C.B. Roth, J.M. Torkelson, Eur. Phys. J. E **30**, 83 (2009)
25. M. Tress, M. Erber, E.U. Mapesa, J. Müller, H. Huth, A. Serghei, C. Schick, K.J. Eichhorn, B. Voit, F. Kremer, Macromolecules **43**, 9937 (2010)
26. J.M. Torres, C.M. Stafford, D. Uhrig, B.D. Vogt, J. Poly. Sci. Part B: Poly. Phys. **50**, 370 (2012)
27. J. Perlich, V. Korstgens, E. Metwalli, L. Schulz, R. Georgii, P. Müller-Buschbaum, Macromolecules **42**, 337 (2009)
28. A. Serghei, F. Kremer, Macromol. Chem. Phys. **209**, 810 (2008)
29. A. Serghei, Progr. Colloid Polym. Sci. **132**, 33 (2006)
30. M. Alcoutlabi, G.B. McKenna, J. Phys.-Condens. Matter **17**, R461 (2005)
31. I. Bahar, B. Erman, F. Kremer, E.W. Fischer, Macromolecules **25**, 816 (1992)
32. D. Labahn, R. Mix, A. Schönhals, Phys. Rev. E **79**, 011801 (2009)
33. C.J. Ellison, S.D. Kim, D.B. Hall, J.M. Torkelson, Eur. Phys. J. E **8**, 155 (2002)
34. J.A. Forrest, J. Mattsson, Phys. Rev. E **61**, R53 (2000)
35. G. B. McKenna, Eur. Phys. J.: Special Topics **141**, 291 (2007)
36. R.M.A. Azzam, N.M. Bashara, *Ellipsometry and Polarized Light* (North Holland Publishing Company, Amsterdam, 1977)
37. Guide to using WVASE32, *Spectroscopic Ellipsometry Data Aquisition and Analysis* (J.A. Woollam Co., Inc)
38. H. Fujiwara, *Spectroscopic Ellipsometry: Principles and Applications* (Wiley, Chichester, 2007)
39. H.G. Tompkins, E.A. Irene, *Handbook of Ellipsometry* (Springer, Heidelberg, 2005)
40. K. Dalnoki-Veress, J.A. Forrest, C. Murray, C. Gigault, J.R. Dutcher, Phys. Rev. E **63**, 031801 (2001)

41. F. Kremer, A. Schönhals, *Broadband Dielectric Spectroscopy* (Springer, Heidelberg, 2003)
42. K. Fukao, H. Koizumi, Phys. Rev. E **77**, 021503 (2008)
43. A. Serghei, F. Kremer, Rev. Sci. Instrum. **79**, 026101 (2008)
44. M. Erber, M. Tress, E.U. Mapesa, A. Serghei, K.J. Eichhorn, B. Voit, F. Kremer, Macromolecules **43**, 7729 (2010)
45. H. Huth, A.A. Minakov, A. Serghei, F. Kremer, C. Schick, Eur. Phys. J. Spec. Top. **141**, 153 (2007)
46. Y.P. Koh, G.B. McKenna, S.L. Simon, J. Polym. Sci. Part B: Poly. Phys. **44**, 3518 (2006)
47. V. Lupascu, H. Huth, C. Schick, M. Wubbenhorst, Thermochim. Acta **432**, 222 (2005)
48. S. Kawana, R.A.L. Jones, Phys. Rev. E **6302**, 021501 (2001)
49. T. Miyazaki, R. Inoue, K. Nishida, T. Kanaya, Eur. Phys. J. Spec. Top. **141**, 203 (2007)
50. R.D. Priestley, L.J. Broadbelt, J.M. Torkelson, K. Fukao, Phys. Rev. E **75**, 061806 (2007)
51. C.J. Ellison, J.M. Torkelson, Abstr. Pap. Am. Chem. Soc. **225**, U706 (2003)
52. B.M.I. Flier, M.C. Baier, J. Huber, K. Müllen, S. Mecking, A. Zumbusch, D. Wöll, J. Am. Chem. Soc. **134**, 480 (2012)
53. Y. Zhang, J. Zhang, Y. Lu, S. Yan, D. Shen, Macromolecules **37**, 2532 (2004)
54. G. Beaucage, R. Composto, R.S. Stein, J. Polym. Sci. Part B: Polym. Phys. **31**, 319 (1993)
55. O. Bäumchen, J.D. McGraw, J.A. Forrest, K. Dalnoki-.Veress. Phys. Rev. Lett. **109**, 055701 (2012)
56. M. Erber, A. Khalyavina, K.J. Eichhorn, B. Voit, Polymer **51**, 129 (2010)
57. J.A. Forrest, K. Dalnoki-Veress, Adv. Colloid Interface Sci. **94**, 167 (2001)
58. E. Bittrich, F. Windrich, D. Martens, L. Bittrich, L. Häussler, K.-J. Eichhorn, Polymer Testing **64**, 48 (2017)
59. O. Kahle, U. Wielsch, H. Metzner, J. Bauer, C. Uhlig, C. Zawatski, Thin Solid Films **313**, 803 (1998)
60. M.Y. Efremov, A.V. Kiyanova, J. Last, S.S. Soofi, C. Thode, P.F. Nealey, Phys. Rev. E **86**, 021501 (2012)
61. M.Y. Efremov, C. Thode, P.F. Nealey, Rev. Sci. Instrum. **84**, 023905 (2013)
62. M.Y. Efremov, Rev. Sci. Instrum. **85**, 123901 (2014)
63. T. Lan, J.M. Torkelson, Macromol. **49**, 1231 (2016)
64. K. Inoue, Prog. Poly. Sci. **25**, 453 (2000)
65. E. Malmstrom, A. Hult, J. Macromol. Sci. Rev. Macromol. Chem. Phys. **C37**, 555 (1997)
66. M. Erber, U. Georgi, J. Müller, K.-J. Eichhorn, B. Voit, Eur. Poly. J. **46**, 2240 (2010)
67. E. Glynos, B. Friedberg, A. Chemros, G. Sakellariou, D.W. Gidley, P.F. Green, Macromol. **48**, 2305 (2015)

# Chapter 6
# Polymer Brushes, Hydrogels, Polyelectrolyte Multilayers: Stimuli-Responsivity and Control of Protein Adsorption

Eva Bittrich, Andreas Furchner, Meike Koenig, Dennis Aulich, Petra Uhlmann, Karsten Hinrichs and Klaus-Jochen Eichhorn

**Abstract** The research field of smart polymer surfaces benefits from non-invasive ellipsometric investigations, especially in-situ measurements, monitoring the swelling of polymer films and protein adsorption processes thereon at varying ambient conditions. With ellipsometry in the VIS-range layer thickness and refractive index of the polymer layers can be evaluated. Appropriate models for in-situ measurements will be discussed and results of the influence of solution parameters summarized. In-situ IR-ellipsometry provides information about changes in the vibration band structure for swelling and adsorption processes, where optical modelling in the IR-range yields complementary information about layer thicknesses and structural properties.

E. Bittrich (✉) · K.-J. Eichhorn
Leibniz-Institut für Polymerforschung Dresden e.V., Abteilung Analytik, Hohe Str. 6, 01069 Dresden, Germany
e-mail: bittrich-eva@ipfdd.de

K.-J. Eichhorn
e-mail: kjeich@ipfdd.de

A. Furchner · D. Aulich · K. Hinrichs
Leibniz Institut für Analytische Wissenschaften ISAS e.V., Schwarzschildstr. 8, 12489 Berlin, Germany
e-mail: andreas.furchner@isas.de

D. Aulich
e-mail: dennis.aulich@gmx.de

K. Hinrichs
e-mail: karsten.hinrichs@isas.de

M. Koenig
Karlsruhe Institute of Technology (KIT), Institute of Functional Interfaces, Advanced Polymers and Biomaterials, Hermann-von-Helmholtz-Platz 1, 76344 Eggenstein-Leopoldshafen, Germany
e-mail: meike.koenig@kit.edu

P. Uhlmann
Leibniz-Institut für Polymerforschung Dresden e.V., Abteilung Nanostrukturierte Materialien, Hohe Str. 6, 01069 Dresden, Germany
e-mail: uhlmannp@ipfdd.de

© Springer International Publishing AG, part of Springer Nature 2018
K. Hinrichs and K.-J. Eichhorn (eds.), *Ellipsometry of Functional Organic Surfaces and Films*, Springer Series in Surface Sciences 52,
https://doi.org/10.1007/978-3-319-75895-4_6

115

## 6.1   Introduction

Soft matter films made from special polymer materials are very appealing for the
development of smart surfaces [1–3]. For these polymer films type and rate of
response to environmental stimuli can be regulated by chain length, composition,
architecture, and topology. Different types of film architectures were developed in
the past decades, among them smart polymer brushes [4–7], hydrogels [8], and polyelectrolyte multilayers (PEM) [9–11]. These architectures are sketched in Fig. 6.1.

**Polymer brushes** are formed of polymer chains grafted in close proximity. To
avoid unfavorable interaction the conformational dimension of the polymer chains is
altered as compared to an unperturbed chain in good solvent, and is more extended
away from the substrate [5]. The brushes are usually characterized by the distance
between the grafting sites (the grafting density) and the height of the brush layer.

In the beginning, brushes were investigated because of their positive influence
on the steric stabilization of colloids [12]. But also possible applications like the
formation of chemical gates [13–15], tunable adhesion of biomolecules [16–20],
liquid separation in capillaries [21, 22], mechanosensitive surfaces [23], or sensor
applications with immobilized nanoparticles and enzymes [24–26] were discussed
in recent years for different compositions of polymer brushes. Furthermore, biocompatibility of selected brush architectures could be shown [27–30], and utilized for
the generation of bioactive surfaces [31, 32].

Stimuli-responsive polymer brushes, e.g. pH- or temperature sensitive, offer the
possibility to tune swelling properties dependent on environmental conditions. Temperature sensitive poly(N-isopropylacrylamide)(PNIPAAm) brushes are intensely
investigated because of the lower critical solution temperature (LCST) of PNIPAAm
at 31 °C, close to the human body temperature [33–35]. For pH-sensitive brushes
weak polyelectrolytes like poly(acrylic acid) (PAA) or poly(vinyl pyridine)(PVP)
are interesting because of their specific swelling properties characteristic for weak
polyelectrolyte brushes [16, 36, 37].

**Hydrogels** are three-dimensional cross-linked polymer networks built up of
homopolymers or copolymers. They are hydrophilic and swell in aqueous solutions.
Thus they are able to adsorb large amounts of water or biofluids, while they are
insoluble due to their physical or chemical linking points [38–40]. Physical links

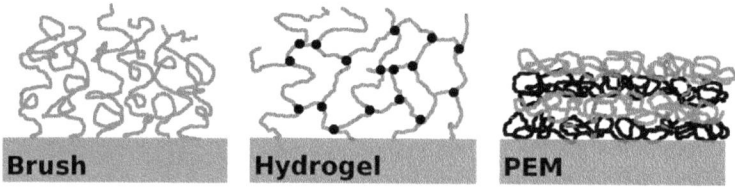

**Fig. 6.1** Schematic drawings of polymer brushes, hydrogels, and polyelectrolyte multilayers
(PEM). For the hydrogel film linking points of the network are marked black. In the PEM polyanionic and polycationic chains are marked in different shades of gray

can be crystallites or entanglements, and chemical linking is achieved by chemical bonds or junctions of branched polymers.

Because of their high degree of hydration, hydrogels are often biocompatible, and have a potential for applications in pharmaceutic or medical devices, for example as contact lens material [41], as material for artificial organs [40], but also as drug release systems [42, 43]. They are most appealing for tissue engineering because their consistency (high degree of hydration, flexibility) is very close to natural tissues [39].

Like polymer brushes, hydrogels can be built up from smart polymers, leading to stimuli-responsive hydrogels, e.g. pH- and temperature sensitive or sensitive to electromagnetic radiation [38]. Common monomers that are used for the preparation are, among others, hydroxyethyl methacrylate (HEMA), N-isopropylacrylamide (NIPAAm) or methacrylic acid (MAA) (see Table 1 in [39]). Especially for drug delivery systems the specific design of polymers with tailored properties is desired.

**Polyelectrolyte multilayers (PEM)** consist of layers of polyions with opposite charges in each layer, and are prepared by layer-by-layer deposition via dip- or spray-coating [9–11]. Alternate adsorption of positively and negatively charged polyions leads to individual layers with control of the layer thickness on the nanoscale. Electrostatic interaction between the polyions and with the charged substrate, but also short range interaction like van der Waals forces, hydrophobic interaction, or hydrogen bonding are the basis for the multilayer built up [10].

PEM are prepared as films but also as microcapsules, where the latter are interesting for applications as microsensors, drug delivery systems or microreactors [44, 45]. The microcapsules are multilayer shells prepared by layer-by-layer deposition on colloidal templates and subsequent dissolving of the core. Biodegradable PEM films are promising as coatings for implant materials and subsequent local drug delivery at specific implantation sites. Here charged and biodegradable biopolymers, e.g. polysaccharides and polypeptides, are used for the film preparation [10, 46].

Stimuli-responsive PEM are also of interest and another approach to applications in medical devices or pharmaceutics, as mentioned for hydrogels and brushes above. For example, for temperature sensitive PEM charged derivatives of PNIPAAm as well as block-copolymers containing PNIPAAm are used. However, a reduced stimuli-responsivity due to a strong interdigitation between adjacent layers in the PEM has to be taken into account [44].

**In-situ ellipsometry with liquid cell** is increasingly used to examine soft matter surfaces in contact to aqueous solutions, to study the swelling of these surfaces, adsorption processes, and the influence of solution parameters. Different setups for the liquid cell were developed, where the most often used cell types have a fixed angle of incidence [35, 47, 48]. The liquid cell is constructed as batch or flow cell. Examples of different cell types are displayed in Fig. 6.2: (a) a simple, heatable batch cell with fixed angle of incidence, we often used for VIS ellipsometry, (b) a flow cell in top view for measurements in the mid-IR spectral range, and (c) sketches of the VIS ellipsometric setup and the flow conditions of a flow cell suitable for kinetic measurements. For possible liquid cell designs of a combined quartz crystal microbalance—ellipsometry setup we refer to Chap. 17. An important aspect of in-situ measurements are the flow conditions in the liquid cell. Flow can have a significant effect on the experimental results especially when monitoring adsorption processes and kinetics. In batch cells concentration gradients can occur, where in common flow cells turbulent flow as well as shearing forces are possible. These concentration inhomogeneities in solution can affect both: kinetics and equilibrium value of the adsorbed amount of molecules.

Within the displayed flow cell (c) in Fig. 6.2, adsorption experiments are performed at stagnation point flow conditions, where the solution impinges the surface perpendicular through a cylindrical channel [49]. For these conditions the hydrodynamics of the mass flux are well defined, and the kinetics of adsorption processes can be addressed. However, setup-specific corrections have to be calculated due to the finite spot size of the probing beam as compared to the infinitesimal stagnation

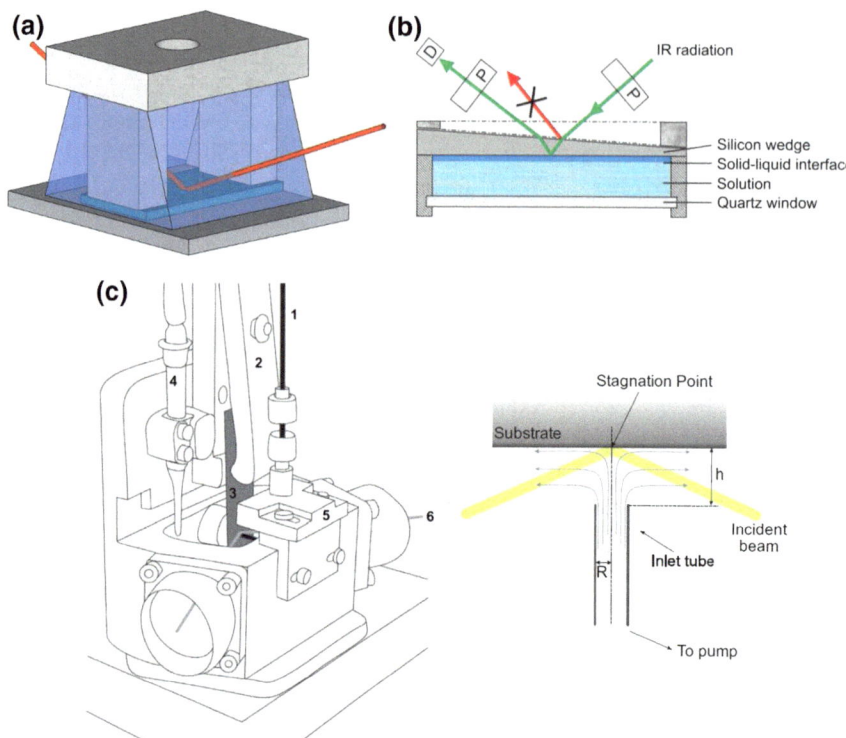

**Fig. 6.2** Sketches of different liquid cell designs: batch cell for VIS-ellipsometry (**a**), top view of an IR flow cell (**b**), and setup as well as schematic flow conditions in top view of a flow cell with stagnation point flow conditions (**c**). The batch cell in (**a**) is equipped with a teflon covering and can be installed on a heating plate. The main components of the flow cell (**c**) are: inlet tube (1), substrate holder (2), substrate (3), drain (4), inlet tube positioner (5), and incident beam (6). Images of the flow cell (**c**) were reprinted with permission from [49] (Copyright 2010 Elsevier), and of the IR-cell (**b**) with permission from [50] (Copyright 2010 American Chemical Society)

flow point, defined as the intersection of the surface and the symmetry axis of the cylindrical channel.

For in-situ ellipsometry in the mid-IR spectral range, it is important that the penetration depth of radiation in water is in the range of several tens of micrometers. Therefore, for in-situ measurements of the solid/liquid interface in aqueous environments a cell with a transparent (silicon) substrate is used as displayed in Fig. 6.2b [37, 50, 51].

Other aspects of liquid cells, that can have an influence on the measurement, are the window properties or the temperature control. These features are discussed in more detail, among others, in the Handbook of Ellipsometry by H. Arwin [48].

## 6.2  Ellipsometry on Swellable, Stimuli-Responsive Surface Layers

The research on swellable stimuli-responsive surface layers benefits from ellipsometric measurements, since on the one hand swollen layer thickness, water content, as well as refractive index of the swollen films can be modeled from VIS ellipsometric data. On the other hand, changes in vibrational bands (e.g. due to dissociation processes) can be correlated to thickness and water content from IR ellipsometric results. The influence of changes in the environmental conditions like temperature, pH, or salt concentration on the layer parameters can be addressed, and stimuli-responsivity of different film architectures investigated.

### 6.2.1  Hydrophilic Polymer Brushes

For the modeling of water swellable polymer brushes, a layer-by-layer built up of a simple box model with sharp interfaces between all individual polymer layers has been considered appropriate for brush layers with wet thicknesses below ca. 100 nm [37, 47, 52, 53]. For brushes with wet thicknesses in the $\mu$m range the brush segment density profile was modeled, using a complementary error function $\phi(z)$, and the wet thickness was derived as twice the first momentum of $\phi(z)$ [54, 55].

Polymer brushes are for example prepared via the "grafting-to" method on silicon substrate with Poly(glycidyl methacrylate) (PGMA) as anchoring layer [56]. For the silicon substrate and the $SiO_2$ layer tabulated refractive indices $n(\lambda)$ can be used [47]. For the very thin PGMA layer the refractive index is fixed to $n_{PGMA} = 1.525$, evaluated from ellipsometric data of a thick layer. For sufficient high layer thicknesses $d$ (depending on the type of polymer) it is possible to model the wavelength dependence of the refractive index of the polymer brush layer according to a Cauchy dispersion (with $k = 0$ for transparent layers):

$$n(\lambda) = A + \frac{B}{\lambda^2} \tag{6.1}$$

Typical layer thicknesses, where the Cauchy dispersion can be applied, start around 10 nm. Below $n$ and $d$ are usually correlated, and the refractive index is fixed to bulk values of the respective polymers.

Furthermore, with the help of an effective medium approach (EMA) the water/buffer content within the swollen brush layer can be determined. The approach according to Bruggeman (6.2) was preferably used [35, 57, 58], which is based on the assumption of a random mixture of polymer and water (or buffer) components with volume fractions of the same magnitude. No host medium can be assigned to one of the components.

$$0 = f_a \frac{\epsilon_a - \epsilon}{\epsilon_a + 2\epsilon} + f_b \frac{\epsilon_b - \epsilon}{\epsilon_b + 2\epsilon} \tag{6.2}$$

Here $f_a$ and $f_b$ are the volume fractions of two components $a$ and $b$ and $\epsilon_{a,b}$ are the corresponding dielectric functions. $\epsilon$ is the dielectric function of the heterogeneous medium. For transparent layers $\epsilon = (n + ik)^2$ equals $n^2$ in the Bruggeman EMA.

*Temperature Sensitive Poly(N-isopropyl acrylamide) (PNIPAAm) Brushes*

PNIPAAm brushes are promising for the formation of responsive surfaces due to their deswelling above a lower critical solution temperature of about 31 °C [34, 35]. Some applications for these smart surfaces, like cell adhesion or adsorption of bacteria, depend critically on changes in short range interactions, that can be introduced by changes in the hydrophobicity of the surface [59, 60]. For other applications the magnitude of conformational change is important, for example in the separation of molecules due to their specific size, or the controlled binding of ligands [61, 62].

The swelling behavior of PNIPAAm brushes was studied with a variety of methods, including neutron reflectometry [63, 64], contact angle measurements [33, 34, 65, 66], surface plasmon resonance [65], atomic force microscopy (AFM) [33, 67, 68], in-situ attenuated total reflection Fourier-transform infrared (ATR-FTIR) spectroscopy [35], quartz crystal microbalance with dissipation monitoring (QCM-D) [66], and ellipsometry [34, 35, 64, 66–69].

VIS- and IR-ellipsometry were applied to study swollen PNIPAAm brushes leading to new insight into the swelling / deswelling process:

First of all the quality of the brush state can be retrieved from the dry brush thicknesses modeled from ellipsometric data, where brush criteria for good and bad solvent conditions exist [34, 35]. Thus the quality of the brush can be estimated, whether a true brush regime or an intermediate regime (mushroom conformation) between stretched chains and unperturbed coil is present.

The deswelling of PNIPAAm brushes was found to depend on the grafting density of the polymer chains. Figure 6.3 shows changes in the buffer content with temperature for several grafting densities at a molecular weight of PNIPAAm of 47,600 g/mol, modeled from VIS ellipsometric results [35]. An increase of the temperature region with maximum deswelling was observed for decreasing grafting densities, pointing at a more heterogeneous deswelling process at low grafting

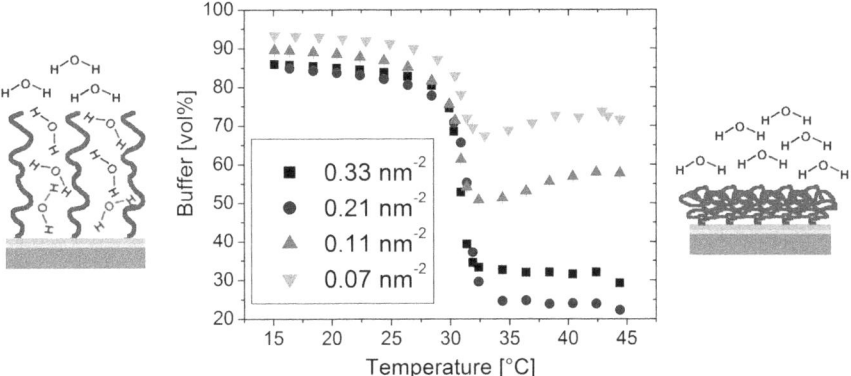

**Fig. 6.3** Percental buffer content in swollen PNIPAAm brushes (47,600 g/mol) on silicon substrate as a function of temperature and grafting density. Measurements were done with a VIS ellipsometer using a batch cell. The data was evaluated with a simple box model. Adapted with permission from [35]. Copyright 2012 American Chemical Society

densities. A transition from the brush to the mushroom regime with increasing temperature possibly occurs at lower grafting densities, affecting the deswelling of the PNIPAAm "grafting-to" brushes.

The buffer content of the swollen brush layers was almost identical below the LCST, but depended on the grafting density above the LCST. For most of the brushes, the buffer content decreased with increasing grafting density, reflecting a higher density of deswollen chains at the surface. An exception was observed for the PN47k brush with the highest grafting density. It is assumed that this points to the fact that the latter films have a brush conformation below and above the LCST, and the deswelling is reduced due to increased interchain interactions.

The decrease in brush quality with increasing temperature is hold responsible for the high "switching amplitude" observed in swollen brush thickness (not shown) and buffer content, which is most desirable for applicational purposes.

Infrared ellipsometry also confirms the deswelling of PNIPAAm brushes with increasing temperature. The infrared spectral range gives access to the molecular vibrations of PNIPAAm and water, which in turn provide complementary insights into structural properties of the brush–water interface. Figure 6.4 shows in-situ tan $\Psi$ spectra of a 132,000 g/mol brush with a grafting density of 0.06 nm$^{-2}$ below and above the LCST [69]. Changes in band amplitude of the water-stretching vibration, $\nu(H_2O)$, in band position of PNIPAAm's CH$_x$ stretching and bending vibrational bands, and in composition of the polymer's amide I and II bands indicate structural changes of the brush that correlate with deswelling effects and changing molecular interactions. This was quantified by fitting the tan $\Psi$ spectra assuming a Bruggeman effective-medium approximation of PNIPAAm and water after (6.2). Specifically, PNIPAAm's dielectric function was modeled by a sum of Voigt oscillators to account for the various molecular vibrations and molecular interactions, such as intra- and intermolecular hydrogen bonding between PNIPAAm's amide groups and

**Fig. 6.4** Measured and fitted tan $\Psi$ spectra of a thin wet PNIPAAm brush ($d_{dry} = 12.6$ nm) below and above the LCST referenced to water spectra at the corresponding temperatures. Adapted with permission from [69]. Copyright 2017 American Chemical Society

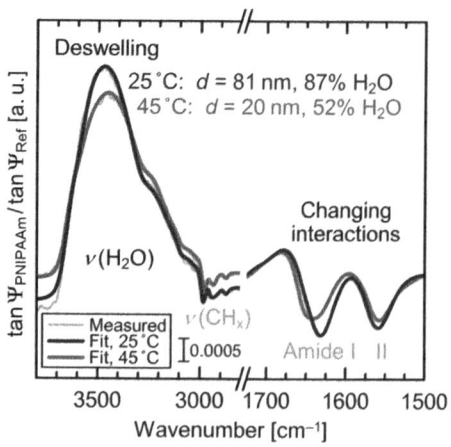

surrounding water molecules. The high spectral contrast in the infrared enabled IR ellipsometry to determine brush thicknesses and hydration states, and to correlate the swelling–deswelling transition with changes in amide–water and amide–amide hydrogen-bond interactions [69].

*pH-Sensitive Brushes*

Polymer brushes, that swell dependent on the concentration of ionic charges in solution, are referred to as sensitive to pH and salt concentration. These brushes are build up from weak or strong polyelectrolytes. Like temperature sensitive films, these brushes are suited for the formation of smart surfaces with possible application in the biotechnological field [70, 71].

Weak polyelectrolyte brushes consisting of poly(acrylic acid) (PAA) swell sensitive to pH and show a characteristical swelling with ionic concentration in solution. Investigations with VIS-ellipsometry provided these characteristic dependency of swollen brush thickness on salt concentration [36, 52, 54], and IR-ellipsometric measurements proved the dissociation process along the tethered polymer chains with changes in the solution pH [50].

In Fig. 6.5 the swollen brush thicknesses dependent on ionic strength for three different pH and grafting densities are displayed [52]. The nonmonotonic characteristic of the thickness, with a maximum of swollen brush thickness at medium salt concentration, agrees well with the theory on the swelling of weak polyelectrolyte brushes [72]. For these weak polyelectrolytes the degree of dissociation depends on the salt concentration in solution, resulting in three different brush regimes with increasing salt concentration: osmotic, salted, and neutral brush. At the ionic strength of maximum swollen brush thickness the osmotic brush at low salt concentrations changes to a salted brush. In the salted regime the degree of dissociation is the same inside the brush and in the bulk solution, while in the osmotic regime the degree of dissociation is reduced in the brush, which leads to a lower swollen brush thickness.

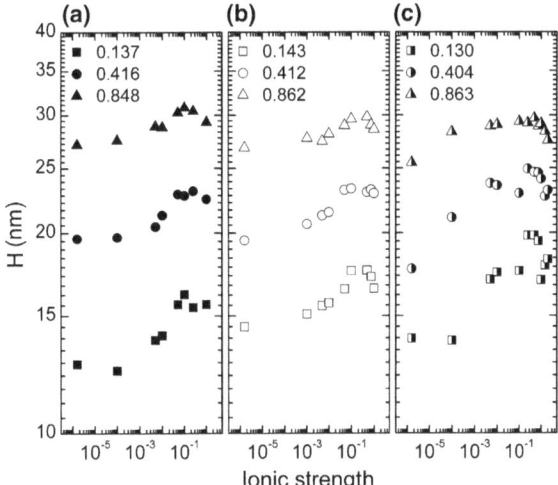

**Fig. 6.5** Swollen brush thickness H of PAA dependent on the ionic strength of solution (in M) for three different pH: pH 4 (**a**), pH 5.8 (**b**), and pH 10 (**c**). The *symbols* (*square, circle* and *triangle*) indicate different grafting densities of PAA (in chains/nm²), as displayed in the *upper corner* of the graphs. Measurements were done with a VIS ellipsometer using a custom-designed solution cell and thicknesses evaluated with a graded EMA model. Reprinted with permission from [52]. Copyright 2007 American Chemical Society

The pH-dependent switching of a PAA mono-brush was also investigated using in-situ Infrared Spectroscopic Ellipsometry (IRSE) [50]. The measured ellipsometric parameters tan $\Psi$ and $\Delta$ can be interpreted by optical simulations giving access to structural parameters as e.g. molecular orientations and optical constants. Depending on amplitude and orientation of the transition dipole moment of a specific molecular vibration, characteristic bands are seen in the ellipsometric spectra. The overall thickness of the PAA brush in the dry state was $\approx 5$ nm. Reversible switching of the polymer brush was studied at titration from pH 2 to pH 10 and back in steps of 1 pH unit. The switching process was observed by monitoring the characteristic vibrational bands of the carboxylic groups of the PAA molecules (Fig. 6.6 left side). Decreasing of the C=O vibrational band at $\approx 1728$ cm$^{-1}$ amplitude and arising of COO$^-$ vibrational bands at $1414$ cm$^{-1}$ and $1560$ cm$^{-1}$ proved the chemical changes in the molecular structure of the brushes due to changes of the pH value in the aqueous solution. Switching the brush in several cycles with increasing and decreasing pH value showed a hysteresis-like behavior (Fig. 6.6 right side).

This hysteresis-like behavior was not reported for polyelectrolyte brushes so far. Lowering the pH value to the starting value of 2 brought the brush back into its initial state. In-situ measurements with spectroscopic ellipsometry in the visible range were used to correlate the observed dissociation with the swelling degree of the brush.

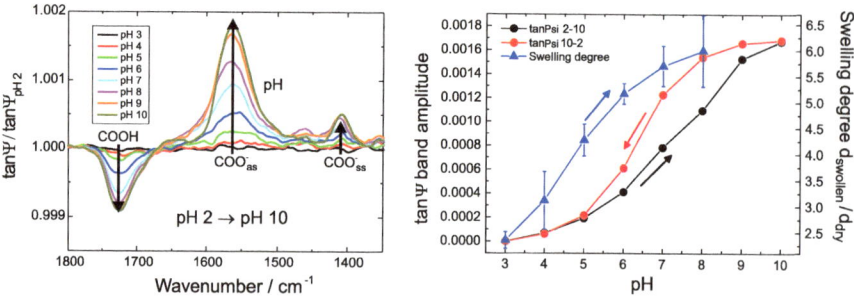

**Fig. 6.6** tan $\Psi$ spectra referenced to the spectrum at pH 2 for the pH-dependent switching of a PAA brush increasing the pH from pH 2 to pH 10 (left side), and the tan $\Psi$ amplitude of the COO⁻ vibration at 1560 cm⁻¹ compared to the swelling degree of the brush (right side). In-situ IRSE was performed with a cell which allows for the penetration of radiation through the backside of a trapezoidal silicon substrate with an angle of incidence of 50°. Reprinted with permission from [50]. Copyright 2010 American Chemical Society

Other methods like streaming potential, contact angle and X-ray standing wave measurements complement the ellipsometric results, proving the pH-sensitive dissociation process of weak polyelectrolyte brushes [50, 73].

pH-sensitive polymers like the polyanionic PAA and the polycationic poly(vinyl pyridine) P2VP can be combined into binary mixed brushes [51, 74]. These films show a very interesting complex swelling behavior as presented in Figs. 6.7 and 6.8. In-situ IR-ellipsometry was used to probe chemical changes in the PAA-P2VP mixed brush [51]. Titration experiments were performed with the mixed brush in the range from pH = 2 to pH = 10. The analysis of the pH-dependent spectral signature, as displayed in Fig. 6.7, provided direct evidence for hypotheses of pH-dependent structural changes in the mixed brush based on previous in-situ VIS-ellipsometric, contact angle and zeta potential results [74]. In contrast to the latter methods, the IR-spectra can be directly associated with structural and chemical properties of the brush. Thus, for the first time, the direct evaluation of the chemical composition and the interaction between two polyelectrolytes in a mixed brush was reported. It was revealed that the mixed brush demonstrates IR spectra which are similar to those of corresponding polymers in aqueous solutions (with the exception of P2VP, which is insoluble at pH > 4; hence, it was not possible to prepare basic solutions of P2VP) at pH = 2 and pH = 10. At around pH = 6.0, the IR-spectrum of the mixed brush resembles that of the PAA-P2VP polyelectrolyte complex prepared by mixing these two polymers in aqueous solution. The observed frequency shifts and changes of band amplitudes are consistent with the model as shown next to the spectra in Fig. 6.7. In basic aqueous solutions, the mixed brush is constituted of swollen and ionized PAA chains, while deprotonated P2VP chains are collapsed. The inverse situation is present in acidic solutions: P2VP chains are protonated, ionized and swollen, while compact coiled PAA chains are not ionized. In the range of pH around the isoelectric point (pH = 4–7), the mixed brush forms a hydrophobic polyelectrolyte complex. In this pH range, both components of the mixed polymer brush are partially ionized,

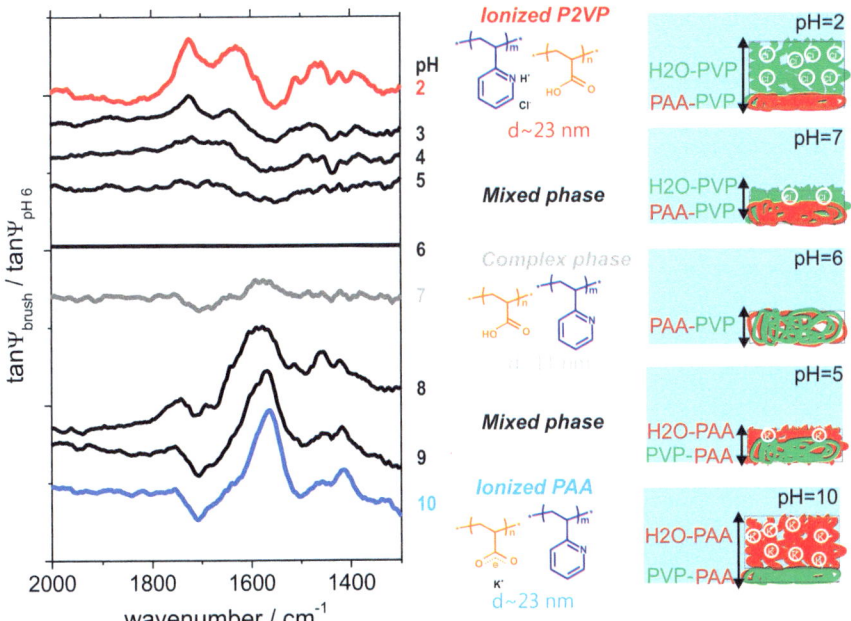

**Fig. 6.7** Referenced IR-ellipsometric tan $\Psi$ spectra of a PAA-P2VP brush for the pH range from pH = 2 to pH = 10 (*left*), schematic drawings of the expected chemical structure of the mixed brush at selected pH values of pH = 2, 7, and 10 (*middle*), and a schematic model for the pH-dependent structures of the mixed brush (*right*). Adapted with permission from [51]. Copyright 2009 American Chemical Society

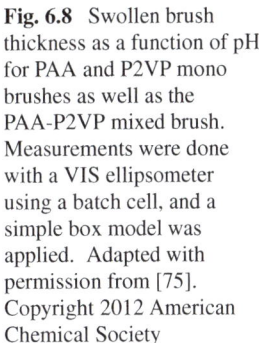

**Fig. 6.8** Swollen brush thickness as a function of pH for PAA and P2VP mono brushes as well as the PAA-P2VP mixed brush. Measurements were done with a VIS ellipsometer using a batch cell, and a simple box model was applied. Adapted with permission from [75]. Copyright 2012 American Chemical Society

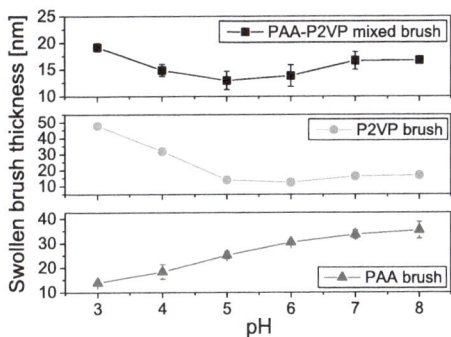

and the polyelectrolyte complex is formed as a result of these ionized functional groups. The pH-dependent swelling behavior due to charge dissociation along the chains is clearly reflected in the VIS-ellipsometric results in Fig. 6.8. The mixed brush is swollen at low and high pH following the characteristics for PAA and P2VP mono brushes, respectively. At intermediate pH the polyelectrolyte complex leads to a deswelling of the mixed brush system.

Polyelectrolytes were also combined with other types of polymers into mixed brushes, e.g. with hydrophobic polystyrene (PS) [76], with PNIPAAm [77], or even in ternary brushes with hydrophilic polyethylene glycol (PEG) and PS [78, 79]. PS leads to a mixed system with sensitivity towards organic solvents. PNIPAAm-PAA brushes were presented as the first pH- and temperature sensitive brush system with unique and complex swelling behavior due to an interaction of the individual chains at certain pH. This system could also be discussed as a thin hydrogel film with physical linkage rather than chemical joints in the network [80]. The swelling behavior of the ternary brush system is similar to the one of pure PAA brushes, for which the water swellable PEG served as a pH-insensitive reference brush [78].

### 6.2.2 Hydrogels

Hydrogels with a wide range of thicknesses can be prepared: relatively thin in the nm range [8, 81] to $\mu$m thick films [82]. Swelling in water and water vapor is investigated, relevant for applications such as contact lens materials [83]. For nm thick hydrogel films a box model with Cauchy layers is successful [81], while for $\mu$m thick films optical gradient profiles are recommended for an appropriate modeling of the swollen hydrogels. Junk et al. developed such gradient profiles for PNIPAAm containing hydrogels on the basis of surface plasmon resonance/optical waveguide mode spectroscopic measurements [82]. In their analysis they divided the hydrogel film into sublayers with uniform refractive index. In Fig. 6.9 the gradient profiles for $n$ ($\lambda = 632.8$ nm) are displayed at different temperatures. The collapse of the hydrogel with increasing temperature is reflected in the decrease of the total layer thickness and in the increase of the refractive index of hydrogel sublayers close to

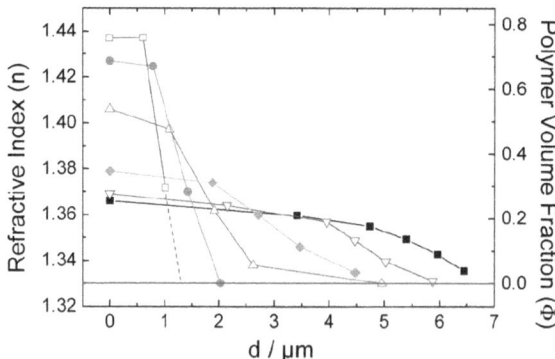

**Fig. 6.9** Profiles of the refractive index and the polymer volume fraction at different temperatures: 42 °C (*empty squares*), 36 °C (*circles*), 32 °C (*upturned triangle*), 28 °C (*diamonds*), 20 °C (*downturned triangle*), and 5 °C (*filled squares*). Measurement data was obtained by combined surface plasmon resonance and optical waveguide mode spectroscopy, and modeling was done by an EMA (Bruggeman) and the Wentzel-Kramers-Brillouin method. Reprinted with permission from [82]. Copyright 2010 American Chemical Society

the substrate surface at higher temperatures. Especially interesting is the refractive index profile close to the transition temperature at 32 °C (upturned triangle). The profile forms an S-shaped curve, where the refractive index is strongly increased close to the substrate, and the decrease is very smooth over $\Delta d \approx 4\,\mu m$. Junk et al. discussed this inhomogeneous profile as characteristic for a collapsed hydrogel state at the film-substrate boundary and a swollen hydrogel state at the film-solution interface.

Toomey et al. found deviations between the measured $\Delta$ and the simulated one at small angles of incidence, when applying the box model for a ca. $1.2\,\mu m$ thick hydrogel film. They improved the model with a three parameter error function for the modeling of roughness at the hydrogel solution interface [84], and obtained a rough interface of ca. $200\,nm$ thickness.

Further PNIPAAm containing hydrogels were examined [85, 86], and n and d of swollen films evaluated dependent on temperature. Schmaljohann et al. investigated micro-patterned films with imaging ellipsometry, where the hydrogel patterns were separated by $60\,\mu m$ wide groves [85]. In Fig. 6.10 3D profiles of $\Delta$ are displayed for the extended hydrogel film at 25 °C (a, $d = 65\,nm$), and two more collapsed states at 30 °C (b, $d = 55\,nm$), and 35 °C (c, $d = 40\,nm$). With this 3D profiles they could show a swelling of the hydrogel layer solely in z-direction, since no change in the resolution of the micro-pattern occurred.

Other temperature sensitive components used are for example PVME [81, 87], PVCL [88], and DMAAm [84]. Hegewald and Gramm et al. focused on the influence of the radiation dose, used for cross-linking of the films by electron beam irradiation, on the temperature sensitive swelling [81, 87]. Toomey et al. concentrated on the swelling of the gel in water at room temperature and compared it to the swelling of a free-standing network [84]. Thus they provided a basis for the discussion of hydrogel properties in confined environments, attached to a surface.

More and more importance is gained by hydrogel films build up from biopolymers like cellulose [89], chitosan [90], or from hyperbranched polymers with oligosaccharide architectures [91]. Cellulose thin films are interesting for example for cell scaffolds, and swelling measurements provided details for the understanding of the impact of chemical modifications on the physical properties of such films. The addition of oligosaccharides in hyperbranched polymer films increases their swelling degree as well as their biocompatibility, and their possible use as drug carrier systems was examined.

**Fig. 6.10** Collapse of a hydrogel pattern monitored by single-wavelength imaging ellipsometry. Displayed is the map of the parameter $\Delta$ for 25 °C (**a**), 30 °C (**b**), and 35 °C (**c**). Reprinted with permission from [85]. Copyright 2005 American Chemical Society

### 6.2.3 Polyelectrolyte Multilayers

For the investigation of polyelectrolyte multilayers (PEM) ellipsometry is one of the most applied techniques, since the measurement of the film thickness (dry, but also in the wet state) during the layer-by-layer build up is a very characteristic parameter of PEM [92, 93]. Starting with the adsorption of polyelectrolytes, ellipsometry helped to verify the theory of adsorption and adsorption kinetics, the influence of flow conditions on the adsorption [94–97], and the structural properties of PEM in the buildup of the multilayers [98]. Influences of the pH [99, 100], the ionic strength [98, 101, 102], thermal treatment [103], as well as the behavior in organic solvents [104] were analyzed.

The influence of the pH of the dipping solution on the deposited layer thicknesses of negatively charged PAA and positively charged poly(allylamine hydrochloride) (PAH) is displayed in Fig. 6.11 [99]. PAA is dissociated about 20% at low pH and reaches virtually full dissociation at pH 6.5. PAH stays completely ionized in the whole low pH regime and starts to loose protons around pH 7. Interestingly, although charge conditions are attractive between polycation and polyanion, the multilayer built up was found to depend considerably on pH, and four different growth regimes were identified by Shiratori et al. The multilayer built up is governed by the charge density of the adsorbing polymer and the surface, as well as the chain conformation, thickness, and free ionic binding sites of the previously adsorbed polymer. Shiratori et al. concluded that these factors have different impact in the individual growth regimes, leading to the observed dependency of layer thickness on pH. They used a single-wavelength ellipsometer and measured the incremental dry thickness of the multilayer after drying at 90 °C for 1 h and storage for one day in air [105].

PEM were also tested for stability and ion permeability in aqueous solution [106], for their sensitivity on shear forces [107], and as new materials for the separation of gases [108]. Often, ellipsometry was complemented by other techniques like e.g. dual polarization interferometry [109], zeta potential [92, 93, 107], X-ray

**Fig. 6.11** Average thickness of individual PAH (dashed line) and PAA (solid line) layers dependent on pH. The dipping solution was at the same pH for both types of polymers. Thicknesses were measured by single-wavelength ellipsometry. Reprinted with permission from [99]. Copyright 2000 American Chemical Society

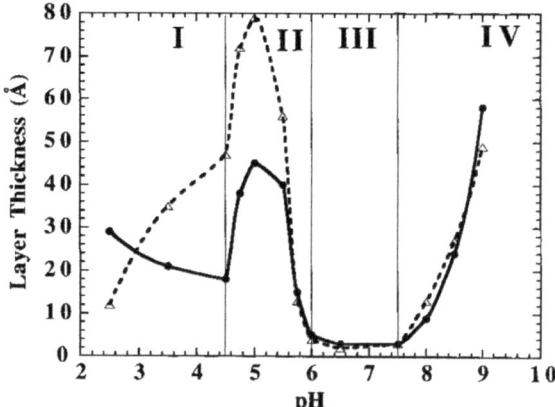

**Fig. 6.12** Index of refraction $n$ and extinction coefficient $k$ of the polycation/dye (PCA$_5$/DR80) multilayer system. The multilayer was modeled by one effective single layer with effective optical constants and total thickness. Reprinted with permission from [112]. Copyright 2002 Elsevier

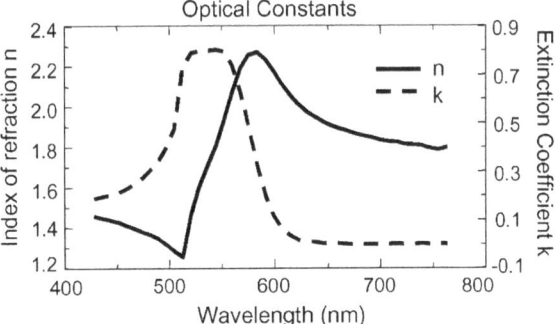

reflectometry [110], UV-VIS spectroscopy [93, 111–113], quartz crystal microbalance with dissipation monitoring (QCM-D) [114], and neutron reflectometry [103, 115].

Polyelectrolytes were further combined with dyes in the multilayer system, e.g. a polycation with a multicharged azo-dye [111–113]. These multilayers are promising for optoelectronic devices or as model for loading of PEM with drugs. The optical constants for the system of the polycation of integral type PCA$_5$ (95 mol.-% of N,N-dimethyl-2-hydroxypropyleneammonium chloride units in the main chain) and the dye Direct Red 80 are displayed in Fig. 6.12. The modeling of the ellipsometric data $\Delta$ and $\tan \Psi$ was done in the following way: The multilayer was modeled as one effective layer, since the sublayers cannot be treated separately. Dye and polycation are expected to have very similar real parts n and the thickness of each sublayer is of the order of Angström, where interfacial roughness is also expected. Firstly, the data above $\lambda = 650$ nm was fitted according to a non-absorbing Cauchy layer.

Secondly, the retrieved multilayer thickness (18.2 nm) was fixed and the optical constants $(n, k)$ fitted in the whole spectral range by a simple point-by-point fit. Above 650 nm the latter fit matched the results obtained with the Cauchy layer. The $k$-spectrum was very similar to UV-VIS spectroscopic results for the same multilayer on glass measured in transmission. Thus the successful modeling of the PEM, absorbing in the VIS-range around 550 nm, could be confirmed. Nevertheless, the fit of $n$ and $k$ could be improved using an oscillator model, which gives a better Kramers-Kronig-consistency. A bilayer thickness of about 3 nm was calculated from these results.

## 6.3 Concepts of Determining the Adsorbed Amount of Protein on Different Polymer Films

Protein adsorption at surfaces can be described quantitatively by the adsorbed amount of protein deposited, as well as protein layer thickness and protein refractive index. The modeling of adsorbed protein films at smooth, rigid surfaces is well described in the literature [48]. A common approach is to model the adsorbed amount with the de Feijter formula (6.3) [116].

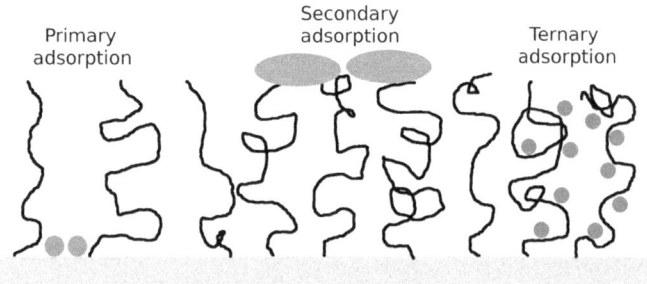

**Fig. 6.13** Modes of adsorption on a grafted polymer layer, displayed for the adsorption on a polymer brush

$$\Gamma = d\frac{n - n_{amb}}{dn/dc} \qquad (6.3)$$

Here $n$ is the refractive index of the protein and $n_{amb}$ of the ambient, $d$ the protein layer thickness, and $dn/dc$ the refractive index increment of the protein. The de Feijter equation is based on the assumption of linearity of the refractive index $n$ with the concentration $c$ of adsorbed molecules, as described in (6.4) [116, 117]. The increase of concentration of adsorbed molecules $\Delta c$ can also be described as $\Gamma/d$.

$$n = n_{amb} + \frac{dn}{dc}\Delta c \qquad (6.4)$$

The in-situ protein refractive index depends on the type of protein, the charge conditions at the adsorbing surface, as well as on pH and ionic concentration of the solution. Refractive indices in-situ were measured for several proteins upon adsorption at hydrophobized silicon with single-wavelength ellipsometry, and are in the range of $n(632\,\text{nm}) \approx 1.37$ [118]. Also a fixed value of $n = 1.46$ as protein refractive index can be found in the literature [119]. For dry albumin layers Arwin et al. examined the dispersion relation with $n = 1.575$ at $\lambda = 632\,\text{nm}$ [120].

The modeling of protein adsorption on swellable polymer films provides new challenges as compared to the adsorption on rigid surfaces. Proteins can penetrate the swellable layers, and three modes of adsorption, primary, secondary and ternary adsorption can occur, as displayed in Fig. 6.13 for the adsorption on polymer brushes [12].

Secondary adsorption is the adsorption of protein on top of the polymer film and can be modeled as described shortly above. Primary adsorption occurs when proteins penetrate the film and adsorb at the substrate, and ternary adsorption describes the direct adherence of proteins at the polymer chains. For the latter two types of adsorption new concepts for the modeling of ellipsometry data are necessary. Since the protein does not form a distinguishable top layer, in these cases a modeling of the protein adsorption with one combined polymer-protein layer is needed.

A modified de Feijter approach is suggested for the calculation of the adsorbed amount of protein from ellipsometric data based on the equation used by Xue et al. for the quantification of very small protein amounts at PNIPAAm brushes [34]. They considered the increase of concentration $\Delta c$ of molecules inside the brush by taking the refractive index of the swollen brush $n_{brush}$ instead of $n_{amb}$ in the de Feijter equation (6.3). When a virtual two layer model for the calculation of $\Gamma$ is applied, this approach can also be used for high adsorbed amounts e.g. at PAA brushes. One layer is given by the polymer brush with protein molecules that have penetrated the brush, and is represented by the swollen brush thickness $d_{brush}$ and the refractive index of the combined polymer-protein layer $n_{comb}$ after adsorption. With the second layer an increase in thickness of the combined layer as compared to the swollen brush thickness is taken into account. Thus the second layer is assumed to be built up solely of (hydrated) protein molecules with the parameters $d_{add} = d_{comb} - d_{brush}$ and $n_{comb}$. Now the change in molecule concentration in layer one is represented by $n_{comb} - n_{brush}$ and in layer two by $n_{comb} - n_{amb}$. With the help of (6.3) a new expression for $\Gamma_{prot}$ can be derived:

$$\Gamma_{prot} = d_{brush} \frac{n_{comb} - n_{brush}}{dn/dc} + d_{add} \frac{n_{comb} - n_{amb}}{dn/dc} \qquad (6.5)$$

$d_{comb}$ and $n_{comb}$ are the parameters modeled for one combined polymer-protein layer with a Cauchy dispersion after the adsorption process. $dn/dc$ is the refractive index increment of the protein. Typically $n(633\,nm)$ is taken for the calculation of $\Gamma$.

For the adsorption of small amounts of protein at PNIPAAm brushes, Xue et al. referenced the ellipsometrically determined amounts with radio assays of [125][I]-labeled protein [34], while for high amounts $(10 - 20\,mg/m^2)$ of RGD-peptides covalently bound to PAA brushes the calculated amount from ellipsometry data was in good agreement to reversephase HPLC-analysis of the extracted amino acids of RGD [121].

The **modeling of the adsorbed protein amount from reflectometric data** is another important approach [117, 122, 123]. With single-wavelength reflectometry the adsorbed amount can be obtained directly from the measurement signal $S$. Kinetics can be addressed due to short measurement times, which are solely determined by the response times of the detecting elements (photo diodes) and the signal processing. The measurement signal $S$ is the ratio of the reflected intensities with polarization parallel and perpendicular to the plane of incidence, which changes upon adsorption of molecules. Dijt et al. showed, that the change in the signal $S$ upon adsorption $\Delta S = S - S_0$ is proportional to the adsorbed amount $\Gamma$ [117].

$$\Gamma = Q \frac{\Delta S}{S_0} \qquad (6.6)$$

For calibration of the system, the baseline Signal $S_0$ is obtained before the measurement by flowing a solution without protein through the cell. $Q$ is a sensitivity factor and depends on the angle of incidence, refractive index, and thickness of the surface layers on the silicon substrate, and the refractive index increment of the adsorbing protein [123].

## 6.4 Protein Adsorption on Different Types of Smart Polymer Brushes

With the help of smart surface layers, the adsorbed amount can be gradually tuned dependent on different parameters like grafting density, or solution pH and temperature. Thus these layers have a high potential for applications in sensoric devices, and as protein release systems, but also as anti-fouling coatings. Two classes of selected brush systems will be discussed: Brushes with very small adsorbed amounts of protein, consisting of poly(ethylene oxide) (PEO) or PNIPAAm, and polyelectrolyte brushes, that allow for rather high adsorbed amounts. These brush layers have a high potential for applications in sensoric devices and as protein release systems. Anti-fouling brushes allow for a low background of non-specific protein adsorption, while functionalization methods (e.g. click-chemistry [124]) can be used to covalently attach functional groups and to investigate specific interaction processes. With polyelectrolyte brushes high amounts of functional proteins (e.g. enzymes) can be immobilized in an active state and with minimal denaturation of the proteins [125, 126]. Additionally, first adsorption results for a ternary brush system including both PEO and PAA will be presented.

### 6.4.1 Polymer Brushes Preventing Protein Adsorption

Protein resistant brushes are of great interest for bio-related applications by preventing non-specific protein adsorption as well as cell/bacteria adhesion. While non-fouling behavior is hard to achieve for complex bio-fluids such as saliva, blood, plasma or cell culture medium, resistance to single-protein solutions is observed for a variety of brushes, among them PEO and PNIPAAm brushes.

*Poly(ethylene oxide) (PEO) Brushes*

It was found that PEO brushes can prevent protein adsorption at artificial surfaces when the brush parameters molecular weight (chain length) and grafting density are selected accordingly [122, 127, 128]. Currie et al. studied the dependence of the adsorbed amount of bovine serum albumin (BSA) on the grafting density for three different chain lengths [122]. They used reflectometry for the determination of the adsorbed amount of protein [117].

An important result of the adsorption of BSA on PEO brushes was the finding of a maximum of the adsorbed amount with grafting density for long PEO chains, where previous experiments only showed the decrease of adsorbed amount with increasing grafting density. Their results are displayed in Fig. 6.14.

**Fig. 6.14** Adsorbed amount $\Gamma$ of bovine serum albumin (BSA) as a function of the grafting density $\sigma$ for three different chain lengths of PEO. Data was obtained by reflectometry using a flow cell. Reprinted with permission from [122]. Copyright 1999 International Union of Pure and Applied Chemistry

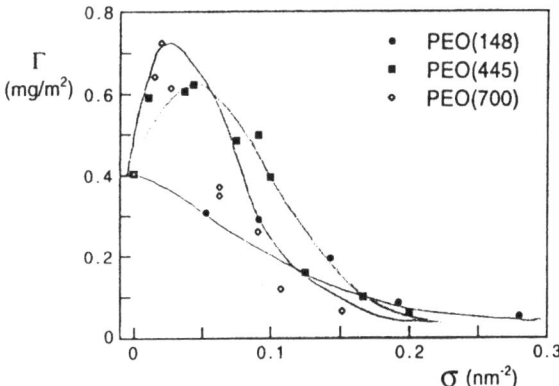

For low as well as high grafting densities adsorption of BSA can be prevented, where for intermediate grafting densities adsorption was evident. Thus they provided important information on a brush design parameter (grafting density $\sigma$) for the built up of soft matter surfaces with resistance to non-specific protein adsorption. This is highly desired, e.g. to prevent biofouling or to specifically bind biomolecules that do not react with the substrate unfavorably.

*Poly(N-isopropyl acrylamide) (PNIPAAm) Brushes*

For PNIPAAm brushes good protein resistance was observed below and above the lower critical solution temperature (LCST) of ca. 31 °C [34, 129]. The protein adsorbed amount was monitored by null-ellipsometry [129] and by spectroscopic ellipsometry [34]. For the discussion of protein resistance the comparison of the ellipsometric parameters $\Delta$ and $\Psi$ before and after the adsorption experiment was better suited than the adsorbed amount $\Gamma$, since $\Gamma$ is more erroneous than the initial parameters. Changes in $\Delta$ and $\Psi$ for the adsorption of albumin at brushes with different molecular weight are displayed in Fig. 6.15.

For the brushes with smallest molecular weight (28,500 g/mol) and thus smallest amount of polymer grafted to the substrate, a marginal protein adsorbed amount of less than 0.1 mg/m² could be detected above the LCST. Increases in $\Delta$ and $\Psi$ for the adsorption at these brushes at 40 °C are visible, while for brushes with higher molecular weight the differences in $\Delta$ and $\Psi$ are close to zero, indicating virtually no changes in the swollen brush parameters (refractive index and thickness) and thus the resistance toward albumin adsorption.

Xue et al. simulated changes in $\Delta$ and $\Psi$ upon protein adsorption and found very low sensitivity of these parameters for adsorbed amounts below 1 mg/m² [34]. They also compared the ellipsometrically determined adsorbed amounts of albumin with results of the adsorption of [125][I]-labeled albumin analyzed with radio assays. They could still detect adsorbed protein, when the ellipsometric measurements showed zero adsorption. Thus a lower sensitivity limit for ellipsometrically measured protein amounts has to be considered. Here null-ellipsometry is better suited than standard

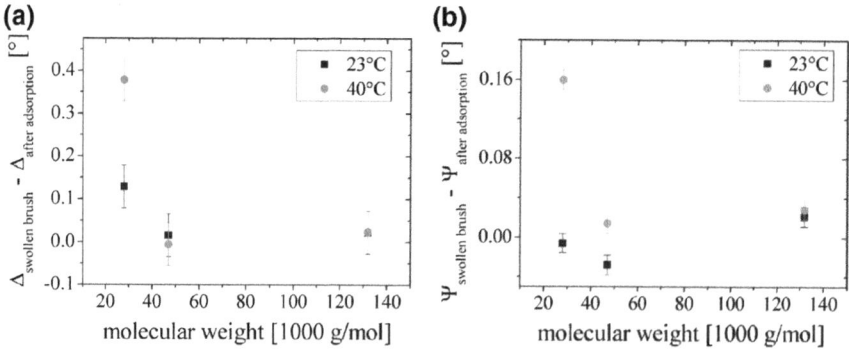

**Fig. 6.15** Changes in the ellipsometric parameters Δ (**a**) and Ψ (**b**) upon adsorption of human serum albumin (HSA) at PNIPAAm brushes with different molecular weight. Data was collected with single wavelength null-ellipsometry at $\lambda = 633$ nm using a batch cell. Reprinted with permission from [129]. Copyright 2010 American Chemical Society

spectroscopic ellipsometry for the investigation of very small adsorbed amounts, because changes in Δ and Ψ as small as 0.01° can be detected. Further increase in sensitivity is obtained with a total internal reflection ellipsometry (TIRE) setup, utilizing the surface plasmon resonance effect of a metal substrate (see Chap. 18 for details).

## 6.4.2 Protein Adsorption at Polyelectrolyte Brushes

Ionic charges along the polymer chains in soft matter surfaces are a characteristic feature of films built of polyelectrolytes (PE), and such films are capable of loading high amounts of protein, as well as releasing them dependent on salt concentration, pH, and the specifics (e.g. isoelectric point) of protein and surface polymer. Thus PE-mediated protein adsorption is investigated very intensely for different surface architectures.

### PAA Brushes

Albumin adsorption at PAA brushes will be presented. These brushes exhibit a characteristic adsorption behavior with high adsorbed amounts around the isoelectric point of the protein and adsorption at the "wrong side" of the IEP, when surface and protein are likewise negatively charged [123, 130, 131]. This phenomenon can be due to charge regulation/ reversal on the proteins, when the charge of weakly charged amino acids due to the local electrostatic potential inside the polyelectrolyte brush is adjusted [132]. But also positive charged patches exist on the protein, that can lead to an entropically favorable replacement of the small counter-ions inside the brush by protein molecules [133]. Penetration of protein molecules inside the brush layer could be demonstrated with small angle X-ray scattering [134].

By use of fixed-angle reflectometry, de Vos et al. characterized the adsorption of albumin at PAA brushes below and above the IEP of the protein dependent on molecular weight and grafting density of the polymers, and on salt concentration [123]. Calculation of the adsorbed amount followed the procedure developed by Dijt et al. [117]. They found an increase of the adsorbed amount with increasing molecular weight and grafting density over the measured pH range. With increasing salt concentration the adsorbed amount is reduced. Here the critical pH for adsorption at the "wrong side" above the IEP of the protein also decreases with increasing salt concentration. The dependency of $\Gamma$ on pH for different salt concentrations is displayed in Fig. 6.16.

The pH dependent adsorption and desorption of HSA on PAA mono brushes was also investigated by IR-ellipsometry [20]. The referenced tan $\Psi$ spectra from pH 5 to pH 7 as shown in Fig. 6.17 are dominated by the amide I and amide II

**Fig. 6.16** Adsorbed amount of bovine serum albumin (BSA) dependent on pH at different ionic strength for the adsorption at PAA brushes (grafting density: $0.1\,\text{nm}^{-2}$, chain length $N = 270$). Measurements were done with fixed angle reflectometry using a stagnation point flow cell. Reprinted with permission from [123]. Copyright 2008 American Chemical Society

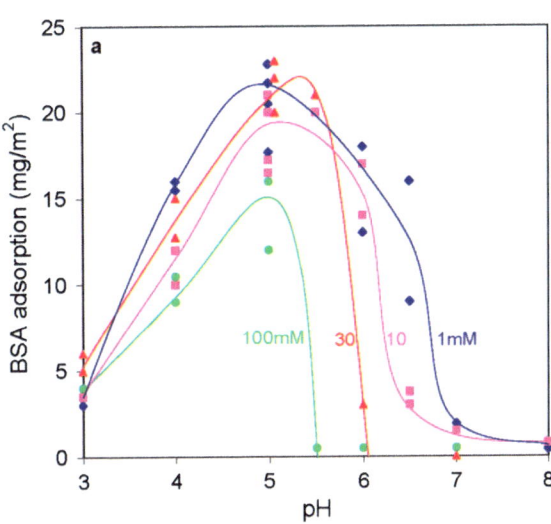

**Fig. 6.17** pH sensitive IR-ellipsometric tan $\Psi$ spectra of HSA desorption from a PAA brush with increasing pH. Spectra are referenced to the buffer solution at pH 5. Reprinted with permission from [20]. Copyright 2015 American Chemical Society

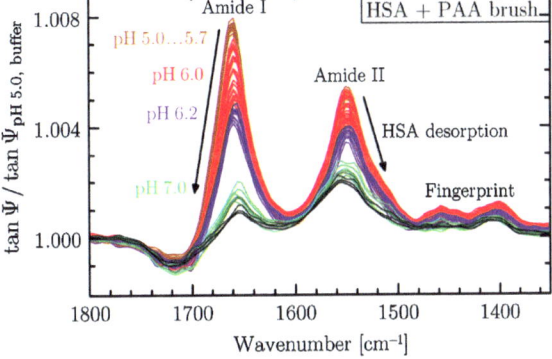

band of adsorbed protein. The height of the amide bands correlate with the amount of adsorbed proteins. From electrostatically attractive conditions at pH 5 towards repulsive conditions at pH 7 subsequent desorption was observed. Changes in the ratio between Amide I and Amide II band can be mainly assigned to changes in the IR signature of the pH-sensitive dissociation of the COOH-groups at the PAA chains (see Fig. 6.6), while the protein structure is preserved. Structural stability for BSA adsorbed at spherical and planar PAA brushes was likewise confirmed by FTIR and ATR-FTIR measurements [135, 136].

Additionally, we used a combination of spectroscopic ellipsometry and quartz crystal microbalance with dissipation (QCM-D) to investigate the adsorption of albumin below and above the IEP of the protein [131] (see also Chap. 17). Quantitative information could be obtained about the amount of buffer components coupled to the PAA brush surface upon swelling and protein adsorption. PAA brushes with more than one anchoring point per single polymer chain were prepared. For the swollen brushes a high amount of buffer was found to be coupled to the brush-solution interface in addition to the content of buffer inside the brush layer. Upon adsorption of bovine serum albumin the further incorporation of buffer molecules into the protein-brush layer was monitored at electrostatic repulsive conditions above the protein IEP. The adsorbed amount of protein $\Gamma_{SE}^{BSA}$, the increase of surface mass determined by QCM-D $\Delta\Gamma_{QCMD}^{ads}$, and the change in amount of buffer at the surface $\Delta\Gamma_{buffer}^{ads}$ for adsorption at the "wrong side" of the IEP at pH 6 is displayed in Fig. 6.18 (central image). The salt concentration was 1 mM. In field I protein is inserted into the measurement cell and adsorption takes place. The cell is filled with buffer to rinse the surface in field II, and in field III protein is desorbed at pH 7.6. An incorporation of excess buffer molecules was observed (Field I from 10 min to ca. 30 min of the ongoing experiment). Thus, viscoelastically coupled buffer molecules are released from the surface in the ongoing adsorption process, indicating an adjustment of

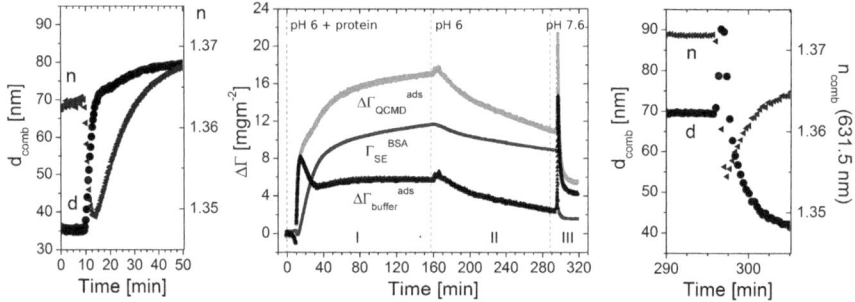

**Fig. 6.18** *Central image*: Changes in the adsorbed amount of molecules (protein and buffer components) at the PAA brush surface as derived from QCMD ($\Delta\Gamma_{QCMD}^{ads}$) and ellipsometry ($\Gamma_{SE}^{BSA}$), as well as the calculated change in the amount of buffer components $\Delta\Gamma_{buffer}^{ads}$. *Alongside* are displayed combined in-situ layer thickness $d_{comb}$ and refractive index $n_{comb}$ for begin (*left*) and end (*right*) of the experiment. A flow cell suitable for combined QCMD-SE measurements was used (see Chap. 17 for details)

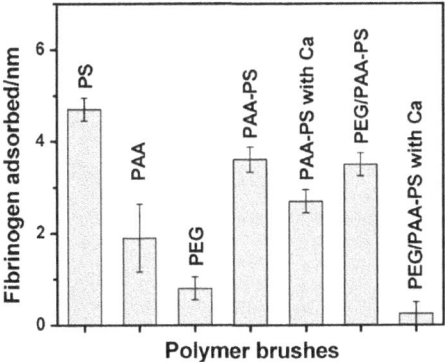

**Fig. 6.19** Dry layer thickness of fibrinogen adsorbed onto different polymer brushes. The protein was adsorbed from PBS buffer solution with 0.6 mg/mL protein concentration. Dry thicknesses were measured by single-wavelength ellipsometry with a fixed $n$ of 1.5 for the protein. Reprinted with permission from [78]. Copyright 2010 John Wiley and Sons

charges and possibly also charge distribution in the combined polymer-protein layer. This peak in the amount of buffer $\Delta \Gamma_{buffer}^{ads}$ is reflected in the in-situ refractive index of the combined polymer-protein layer (left image). Desorption of protein at pH 7.6 (Field III) led to a very high stretching of the polymer-protein layer with peaks in $n$ and $d$ (right image). Additional incorporation of high amounts of buffer could also be observed, indicating the increase of negative charges on the protein molecules at this elevated pH. Thus intermediate states in the adsorption and desorption process can be monitored and interpreted with this combined measurement technique.

*Ternary PEG—PAA-b-PS Brushes*

First results of protein adsorption at ternary brushes are presented. These brushes are built up of poly(ethylene glycol) (PEG) and the block-copolymer PAA-b-poly(styrene) (PS), and were designed to tune hydrophobic interactions of the smart synthetic surface with hydrophobic or amphiphilic molecules in an aqueous environment [78]. These interactions can be altered by changes in pH and salt concentration, which was proven by AFM force-distance measurements. Alongside, swelling measurements with spectroscopic ellipsometry confirmed the pH- and salt sensitivity of the brush surface.

To show the switching of the brush interaction with proteins, fibrinogen was adsorbed at the ternary brush and at the corresponding mono brushes from PBS buffer solution. The thickness of the dry protein layer was evaluated (Fig. 6.19). For the mono brushes adsorption was as expected. Higher fibrinogen thickness was obtained for the hydrophobic PS brush than for the hydrophilic PAA and PEG brushes. Treatment with CaCl$_2$ salt showed no significant change in the fibrinogen thickness for the PAA-b-PS mono brushes. Compared to the adsorption at the latter brushes, a significant effect of the salt treatment could be observed for the adsorption at the

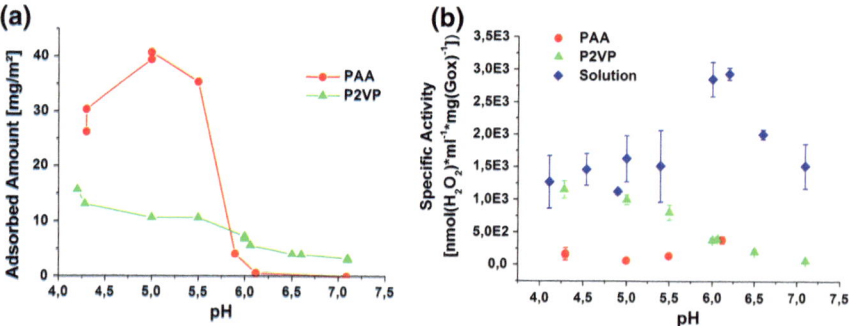

**Fig. 6.20** Adsorbed amount of glucose oxidase at PAA and P2VP brushes dependent on pH as determined from SE measurements (**a**), and comparison of the specific activity of the immobilized enzyme at the brushes and the free enzyme in buffer solution with varying pH (**b**). Error bars represent the standard deviation of at least three measurements. Reprinted with permission from [126]. Copyright 2016 Elsevier

ternary PEG—PAA-b-PS brush, proving the on and off switching of the interaction between surface and protein in the ternary system.

*Activity of Enzymes Adsorbed at Polyelectrolyte Brushes*

Due to the high stability of proteins inside weak polyelectrolyte PAA brushes at low ionic strength a good performance of bioactive molecules like enzymes is expected. This could be confirmed for enzymes such as cellulase [137], $\beta$-glucosidase [125], glucoamylase [138], or glucose oxidase [126].

We investigated the adsorption via ionic interactions of glucose oxidase to anionic PAA- and cationic P2VP-brushes dependent on the pH, and connected adsorbed amounts determined from ellipsometric data with the measurement of the specific activity and the analysis of $\alpha$-helix and $\beta$-sheet content by modeling FTIR-ATR data [126]. High adsorbed amounts were obtained for pH 4.3–5.5 at PAA brushes, while the amount of enzyme immobilized to the P2VP brushes was generally 3–4 times lower in this pH range (Fig. 6.20a). However, the specific activity of the enzyme at P2VP brushes was considerably higher than for immobilized glucose oxidase at PAA brushes (Fig. 6.20b). At pH 4.3 the specific activity at P2VP brush surfaces was even as high as of the free enzyme in solution. At pH 4 secondary structure analysis by ATR-FTIR revealed a decrease of the $\alpha$-helix and an increase of the $\beta$-sheet content of the enzyme at PAA brushes as compared to the free enzyme in solution. This can be interpreted as a loss of the structural ordering of immobilized enzyme, supporting the low values of specific activity observed for these surfaces at comparable pH conditions. For P2VP brushes the $\alpha$-helix and $\beta$-sheet content were the same as for the enzyme in solution, which is in good agreement with the high specific activity at P2VP brushes. The enzyme was also immobilized at mixed brush surfaces of the cationic and anionic polyelectrolytes with temperature-sensitive PNIPAAm and switching of the activity of the enzyme at pH 4 between an active state at 20 °C and a less active state at 40 °C could be achieved.

## 6.5 Conclusion

In this chapter we reviewed the application of ellipsometry for the investigation of soft matter surfaces sensitive to external stimuli. Examples of polymer brushes, hydrogel films, and polyelectrolyte multilayers are presented, and the benefit from ellipsometric measurements is discussed.

In-situ IR and VIS spectroscopic ellipsometry with liquid cell is used to observe swelling of smart surface films, especially polymer brushes and hydrogels. The triggers temperature, pH, and ionic strength of solution are mainly addressed so far by ellipsometric measurements. Prominent polymers for the built up of the smart films are poly(N-isopropylacrylamide) (PNIPAAm) (temperature sensitive) and poly(acrylic acid) (PAA) (pH and salt sensitive). With VIS ellipsometry changes in the swollen film thickness, the in-situ refractive index, and the solution content in the film are analyzed. The simple box model is used for layer thicknesses of the order of 100 nm, where for thicker films in the μm range the density profile of the film has to be considered. IR ellipsometric measurements reveal dissociation processes in the film, e.g. dissociation of COOH groups to $COO^-$ at PAA chains, and swelling processes can be addressed. Due to the high spectral contrast originating from the distinct vibrational bands of water and polymer, IR ellipsometry is very sensitive to brush structure and water content.

The layer-by-layer built up of polyelectrolyte multilayers (PEM) is characterized with VIS ellipsometry, and the optical constants of PEM as well as PEM with dye molecules were determined.

For the calculation of the adsorbed amount of protein from VIS ellipsometric measurements a modified de Feijter approach is presented. Cross-reference with other techniques was done for very low as well as for high adsorbed amounts of proteins/peptides. This approach should in principal also work to describe protein adsorption at hydrogel films.

Single-wavelength ellipsometry and reflectometry are also suitable to monitor the protein adsorption process. Especially reflectometry with a stagnation point flow cell is recommended, when kinetics of adsorption shall be addressed.

Protein adsorption at polymer brushes dependent on grafting density, chain length and solution parameters was investigated with in-situ ellipsometry and in-situ reflectometry. New details of the adsorption process could be provided by these measurements for protein preventing brushes as well as polyelectrolyte mediated protein adsorption. Based on the latter the adsorption of bioactive enzymes to polyanionic and polycationic brushes was connected to the specific activity of the immobilized enzymes.

In conclusion, ellipsometry has a high potential to lead to new insights in the behavior of smart, soft matter surfaces. New directions like light absorbing polymers, the combination of several smart polymers, structured surfaces, or the adsorption of more complex biofluids provide new challenges to adapt the measurement technique and to reference ellipsometry with other methods.

# References

1. T.P. Russel, Science **297**, 964 (2002)
2. I. Luzinov, S. Minko, V. Tsukruk, Progr. Polym. Sci. **29**, 635 (2004)
3. M.A. Cohen Stuart, W.T.S. Huck, J. Genzer, M. Müller, C. Ober, M. Stamm, G.B. Sukhorukov, I. Szleifer, V.V. Tsukruk, M. Urban, F. Winnik, S. Zauscher, I. Luzinov, S. Minko, Nat. Mater. **9**, 101 (2010)
4. S. Minko, J. Macromol. Sci. C: Polym. Rev. **46**, 379 (2006)
5. W.J. Brittain, S. Minko, J. Polym. Sci.: Part A **45**, 3505 (2007)
6. R. Toomey, M. Tirrell, Annu. Rev. Phys. Chem. **59**, 493 (2008)
7. N. Ayres, Polym. Chem. **1**, 769 (2010)
8. I. Tokarev, S. Minko, Soft Matter. **5**, 511 (2009)
9. Y. Lvov, *Protein Architecture - Interfacing Molecular Assemblies and Immobilization Biotechnology* (Marcel Dekker, Inc. New York, 2001), Chap. 6 - Electrostatic Layer-by-Layer Assembly of Proteins and Polyions, pp. 125–167
10. T. Boudou, T. Crouzier, K. Ren, G. Blin, C. Picart, Adv. Mater. **22**, 441 (2010)
11. P. Schaaf, J.C. Voegel, L. Jierry, F. Boulmedais, Adv. Mater. **24**, 1001 (2012)
12. E.P.K. Currie, W. Norde, M.A. Cohen, Stuart. Adv. Colloid Interface Sci. **100**, 205 (2003)
13. M. Motornov, R. Sheparovych, E. Katz, S. Minko, ACS Nano **2**, 41 (2008)
14. A.M. Granville, W.J. Brittain, Macromol. Rapid. Commun. **25**, 1298 (2004)
15. Y.S. Park, Y. Ito, Y. Imanishi, Chem. Mater. **9**, 2755 (1997)
16. P. Uhlmann, N. Houbenov, N. Brenner, K. Grundke, S. Burkert, M. Stamm, Langmuir **23**, 57 (2007)
17. B. Zdyrko, V. Klep, X. Li, Q. Kang, S. Minko, X. Wen, I. Luzinov, Mat. Sci. Eng. C: Mat. Biol. Appl. **29**, 680 (2009)
18. A.L. Becker, K. Henzler, N. Welsch, M. Ballauff, O. Borisov, Curr. Opin. Colloid Interface Sci. **17**, 90 (2012)
19. E. Psarra, U. König, Y. Ueda, C. Bellmann, A. Janke, E. Bittrich, K.J. Eichhorn, P. Uhlmann, A.C.S. Appl, Mater. Interfaces **7**, 12516 (2015)
20. A. Kroning, A. Furchner, D. Aulich, E. Bittrich, S. Rauch, P. Uhlmann, K.J. Eichhorn, M. Seeber, I. Luzinov, S.M. KilbeyII, B.S. Lokitz, S. Minko, K. Hinrichs, A.C.S. Appl, Mater. Interfaces **7**, 12430 (2015)
21. M.D. Miller, G.L. Baker, M.L. Bruening, J. Chromatogr. A **1044**, 323 (2004)
22. A. Feldmann, U. Claußnitzer, M. Otto, J. Chromatogr. B **803**, 149 (2004)
23. J. Bnsow, T.S. Kelby, W.T.S. Huck, Acc. Chem. Res. **43**, 466 (2010)
24. S. Gupta, M. Agrawal, P. Uhlmann, U. Oertel, M. Stamm, Macromol. **41**, 8152 (2008)
25. M. Koenig, F. Simon, P. Formanek, M. Mller, S. Gupta, M. Stamm, P. Uhlmann, Macromol. Chem. Phys. **214**, 2301 (2013)
26. M.E. Welch, T. Doublet, C. Bernard, G.G. Malliaras, C.K. Ober, J. Polym. Sci., Part A: Polym. Chem. **53**, 372 (2015)
27. N. Singh, X. Cui, T. Boland, S.M. Husson, Biomaterials **28**, 763771 (2007)
28. M. Heuberger, T. Drobek, N.D. Spencer, Biophys. J. **88**, 495504 (2005)
29. W.T.E. Bosker, P.A. Iakovlev, W. Norde, M.A. Cohen, Stuart. J. Colloid Interface Sci. **286**, 496 (2005)
30. W.J. Yang, K.G. Neoh, E.T. Kang, S.L.M. Teo, D. Rittschof, Progr. Polym. Sci. **39**, 1017 (2014)
31. E. Psarra, E. Foster, U. König, J. You, Y. Ueda, K.J. Eichhorn, M. Müller, M. Stamm, A. Revzin, P. Uhlmann, Biomacromolecules **16**, 3530 (2015)
32. A. de los Santos Pereira, N.Y. Kostina, M. Bruns, C. Rodriguez-Emmenegger, C. Barner-Kowollik, Langmuir **31**, 5899 (2015)
33. K.N. Plunkett, X. Zhu, J.S. Moore, D.E. Leckband, Langmuir **22**, 4259 (2006)
34. C. Xue, N. Yonet-Tanyeri, N. Brouette, M. Sferrazza, P.V. Braun, D.E. Leckband, Langmuir **27**, 8810 (2011)

35. E. Bittrich, S. Burkert, M. Müller, K.J. Eichhorn, M. Stamm, P. Uhlmann, Langmuir **28**, 3439 (2012)
36. E.P.K. Currie, A.B. Sieval, G.J. Fleer, M.A. Cohen, Stuart. Langmuir **16**, 8324 (2000)
37. Y. Mikhaylova, L. Ionov, J. Rappich, M. Gensch, N. Esser, S. Minko, K.J. Eichhorn, M. Stamm, K. Hinrichs, Anal. Chem. **79**, 7676 (2007)
38. N.A. Peppas, J. Bioact. Compat. Polym. **6**, 241 (1991)
39. N.A. Peppas, P. Bures, W. Leobandung, H. Ichikawa, Eur. J. Pharm. Biopharm. **50**, 27 (2000)
40. S.V. Vlierberghe, P. Dubruel, E. Schacht, Biomacromolecules **12**, 1387 (2011)
41. O. Wichterle, D. Lim, Nature **185**, 117 (1960)
42. N.A. Peppas, A.R. Khare, Adv. Drug Deliv. Rev. **11**, 1 (1993)
43. N.A. Peppas, Curr. Opin. Colloid Interface Sci. **2**, 531 (1997)
44. K. Glinel, C. Dejugnat, M. Prevot, B. Schöler, R.v. Klitzing, M. Schönhoff, Colloids Surf., A **303**, 3 (2007)
45. A.A. Antipov, G.B. Sukhorukov, Adv. Colloid Interface Sci. **111**, 49 (2004)
46. A. Schneider, C. Vodouh, L. Richert, G. Francius, E.L. Guen, P. Schaaf, J.C. Voegel, B. Frisch, C. Picart, Biomacromolecules **8**, 139 (2007)
47. C. Werner, K.J. Eichhorn, K. Grundke, F. Simon, W. Grählert, H.J. Jacobasch, Colloid Surf., A **156**, 3 (1999)
48. H. Arwin, *Handbook of Ellipsometry* (William Andrew Publishing and Springer-Verlag GmbH & Co. KG, 2005), chap. Ellipsometry in Life Sciences, pp. 799–855
49. M.F. Mora, M.R. Nejadnik, J.L. Baylon-Cardiel, C.E. Giacomelli, C.D. Garcia, J. Colloid Interface Sci. **346**, 208 (2010)
50. D. Aulich, O. Hoy, I. Luzinov, M. Brücher, R. Hergenröder, E. Bittrich, K.J. Eichhorn, P. Uhlmann, M. Stamm, N. Esser, K. Hinrichs, Langmuir **26**, 12926 (2010)
51. K. Hinrichs, D. Aulich, L. Ionov, N. Esser, K.J. Eichhorn, M. Motornov, M. Stamm, S. Minko, Langmuir **25**, 10987 (2009)
52. T. Wu, P. Gong, I. Szleifer, P. Vlček, V. Šubr, J. Genzer, Macromolecules **40**, 8756 (2007)
53. S. Reichelt, K.J. Eichhorn, D. Aulich, K. Hinrichs, N. Jain, D. Appelhans, B. Voit, Colloids Surf., B **69**, 169 (2009)
54. M. Biesalski, D. Johannsmann, J. Rühe, J. Chem. Phys. **117**, 4988 (2002)
55. D. Johannsmann, *Functional Polymer Films* (Wiley-VCH Verlag & Co. KGaA, 2011), chap. Investigations of Soft Organic Films with Ellipsometry, pp. 629–647
56. K.S. Iyer, I. Luzinov, Macromolecules **37**, 9538 (2004)
57. D.E. Aspnes, Thin Solid Films **89**, 249 (1982)
58. D.A.G. Bruggeman, Annalen der Physik **24**, 636 (1935)
59. T. Okano, N. Yamada, M. Okuhara, H. Sakai, Y. Sakurai, Biomaterials **16**, 297 (1995)
60. L.K. Ista, V.H. Pérez-Luna, G.P. López, Appl. Environ. Microbiol. **65**, 1603 (1999)
61. P.S. Stayton, T. Shimoboji, C. Long, A. Chilkoti, G. Chen, J.M. Harris, A.S. Hoffman, Nature **378**, 472 (1995)
62. Y.S. Park, Y. Ito, Y. Imanishi, Langmuir **14**, 910 (1998)
63. H. Yim, M.S. Kent, S. Mendez, G.P. Lopez, S. Satija, Y. Seo, Macromolecules **39**, 3420 (2006)
64. S. Micciulla, O. Soltwedel, O. Löhmann, R. von Klitzing, Soft Matter. **12**, 1176 (2016)
65. S. Balamurugan, S. Mendez, S.S. Balamurugan, M.J. O'Brien II, G.P. López, Langmuir **19**, 2545 (2003)
66. P. Zhuang, A. Dirani, K. Glinel, A.M. Jonas, Langmuir **32**, 3433 (2016)
67. Q. Chen, E.S. Kooij, X. Sui, C.J. Padberg, M.A. Hempenius, P.M. Schön, G.J. Vancso, Soft Matter. **10**, 3134 (2014)
68. Y. Yu, M. Cirelli, B.D. Kievit, E.S. Kooij, G.J. Vancso, S. de Beer, Polymer **102**, 372 (2016)
69. A. Furchner, A. Kroning, S. Rauch, P. Uhlmann, K.J. Eichhorn, K. Hinrichs, Anal. Chem. **89**, 3240 (2017)
70. M. Ballauff, O. Borisov, Curr. Opin. Colloid Interface Sci. **11**, 316323 (2006)
71. J.D. Willott, T.J. Murdoch, G.B. Webber, E.J. Wanless, Progr. Polym. Sci. **64**, 52 (2017)
72. P. Gong, T. Wu, J. Genzer, I. Szleifer, Macromolecules **40**, 8765 (2007)
73. R. Zimmermann, W. Norde, M.A. Cohen Stuart, C. Werner, Langmuir **21**, 5108 (2005)

74. N. Houbenov, S. Minko, M. Stamm, Macromolecules **36**, 5897 (2003)
75. E. Bittrich, S. Burkert, K.J. Eichhorn, M. Stamm, P. Uhlmann, *Control of Protein Adsorption and Cell Adhesion by Mixed Polymer Brushes made by the Grafting-to Approach* (American Chemical Society, 2012), chap. 8, pp. 179–193
76. L. Ionov, A. Sidorenko, K.J. Eichhorn, M. Stamm, S. Minko, K. Hinrichs, Langmuir **21**, 8711 (2005)
77. E. Bittrich, M. Kuntzsch, K.J. Eichhorn, P. Uhlmann, J. Polym. Sci., Part B: Polym. Phys. **48**, 1606 (2010)
78. O. Hoy, B. Zdyrko, R. Lupitskyy, R. Sheparovych, D. Aulich, J. Wang, E. Bittrich, K.J. Eichhorn, P. Uhlmann, K. Hinrichs, M. Muller, M. Stamm, S. Minko, I. Luzinov, Adv. Funct. Mater. **20**, 2240 (2010)
79. D. Aulich, O. Hoy, I. Luzinov, K.J. Eichhorn, M. Stamm, M. Gensch, U. Schade, N. Esser, K. Hinrichs, Phys. Status Solidi C **7**, 197 (2010)
80. S.G. Kelmanovich, R. Parke-Houben, C.W. Frank, Soft Matter. **8**, 8137 (2012)
81. S. Gramm, J. Teichmann, M. Nitschke, U. Gohs, K.J. Eichhorn, C. Werner, eXPRESS Polym. Lett. **5**, 970 (2011)
82. M.J.N. Junk, I. Anac, B. Menges, U. Jonas, Langmuir **26**, 12253 (2010)
83. K. Unger, R. Resel, A.M. Coclite, Macromol. Chem. Phys. **217**, 2372 (2016)
84. R. Toomey, D. Freidank, J. Rühe, Macromolecules **37**, 882 (2004)
85. D. Schmaljohann, M. Nitschke, R. Schulze, A. Eing, C. Werner, K.J. Eichhorn, Langmuir **21**, 2317 (2005)
86. Z. Oezyuerek, K. Franke, M. Nitschke, R. Schulze, F. Simon, K.J. Eichhorn, T. Pompe, C. Werner, B. Voit, Biomaterials **30**, 1026 (2009)
87. J. Hegewald, T. Schmidt, K.J. Eichhorn, K. Kretschmer, D. Kuckling, K.F. Arndt, Langmuir **22**, 5152 (2006)
88. W. Higgins, V. Kozlovskaya, A. Alford, J. Ankner, E. Kharlampieva, Macromolecules **49**, 6953 (2016)
89. Y. Müller, I. Tot, A. Potthast, T. Rosenau, R. Zimmermann, K.J. Eichhorn, C. Nitschke, G. Scherr, U. Freudenberg, C. Werner, Soft Matter. **6**, 3680 (2010)
90. M.M.S. Ebrahimi, N. Dohm, M. Müller, B. Jansen, H. Schönherr, Eur. Polym. J. **81**, 257 (2016)
91. A. Richter, A. Janke, S. Zschoche, R. Zimmermann, F. Simon, K.J. Eichhorn, B. Voit, D. Appelhans, New J. Chem. **34**, 2105 (2010)
92. S. Schwarz, K.J. Eichhorn, E. Wischerhoff, A. Laschewsky, Colloids Surf., A **159**, 491 (1999)
93. A. Licea-Claverie, S. Schwarz, F. Simon, O. Urzua-Sanchez, J. Nagel, D. Pleul, K.J. Eichhorn, A. Janke, Colloid Polym. Sci. **283**, 826 (2005)
94. L. Oedberg, S. Sandberg, S. Welin-Klintstroem, H. Arwin, Langmuir **11**, 2621 (1995)
95. A.K. Bajpai, Prog. Polym. Sci. **22**, 523 (1997)
96. N.L. Filippova, Langmuir **14**, 1162 (1998)
97. N.L. Filippova, J. Colloid Interface Sci. **211**, 336 (1999)
98. G. Ladam, P. Schaad, J.C. Voegel, P. Schaaf, G. Decher, F. Cuisinier, Langmuir **16**, 1249 (2000)
99. S.S. Shiratori, M.F. Rubner, Macromolecules **33**, 4213 (2000)
100. X. Gong, Phys. Chem. Chem. Phys. **15**, 10459 (2013)
101. S.L. Clark, M.F. Montague, P.T. Hammond, Macromolecules **30**, 7237 (1997)
102. S. Dodoo, B.N. Balzer, T. Hugel, A. Laschewsky, R. von Klitzing, Soft Mater. **11**, 157 (2013)
103. M. Zerball, A. Laschewsky, R. Köhler, R. von Klitzing, Polymer **8**, 1 (2016)
104. Y. Gu, Y. Ma, B.D. Vogt, N.S. Zacharia, Soft Matter. **12**, 1859 (2016)
105. D. Yoo, S.S. Shiratori, M.F. Rubner, Macromolecules **31**, 4309 (1998)
106. J.J. Harris, M.L. Bruening, Langmuir **16**, 2006 (2000)
107. S. Köstler, V. Ribitsch, K. Stana-Kleinschek, G. Jakopic, S. Strnad, Colloids Surf., A **270–271**, 107 (2005)
108. C. Lin, Q. Chen, S. Yi, M. Wang, S.L. Regen, Langmuir **30**, 687 (2014)
109. S. Edmondson, C.D. Vo, S.P. Armes, G.F. Unali, M.P. Weir, Langmuir **24**, 7208 (2008)

110. A. Laschewsky, B. Mayer, E. Wischerhoff, X. Arys, A. Jonas, Berichte der Bunsengesellschaft für physikalische Chemie **100**, 1033 (1996)
111. S. Dragan, S. Schwarz, K.J. Eichhorn, K. Lunkwitz, Colloids Surf., A **195**, 243 (2001)
112. D. Beyerlein, T. Kratzmüller, K.J. Eichhorn, Vib. Spectrosc. **29**, 223 (2002)
113. E.S. Dragan, S. Schwarz, K.J. Eichhorn, Colloids Surf., A **372**, 210 (2010)
114. T.T.M. Ho, K.E. Bremmell, M. Krasowska, D.N. Stringer, B. Thierry, D.A. Beattie, Soft Matter. **11**, 2110 (2015)
115. A. Zhuk, V. Selin, I. Zhuk, B. Belov, J.F. Ankner, S.A. Sukhishvili, Langmuir **31**, 3889 (2015)
116. J.A. de Feijter, J. Benjamins, F.A. Veer, Biopolymers **17**, 1759 (1978)
117. J.C. Dijt, Kinetics of polymer adsorption, desorption and exchange. Ph.D. thesis, University Wageningen (1993)
118. M. Malmsten, J. Colloid Interface Sci. **168**, 247 (1994)
119. J. Benesch, A. Askendal, P. Tengvall, Colloids. Surf., B **18**, 71 (2000)
120. H. Arwin, Appl. Spectrosc. **40**, 313 (1986)
121. E. Psarra, Biofunctionalization of polymer brush surfaces. Ph.D. thesis, TU Dresden (2015)
122. E.P.K. Currie, J.V. der Gucht, O.V. Borisov, M.A. Cohen, Stuart. Pure Appl. Chem. **71**, 1227 (1999)
123. W.M. de Vos, P.M. Biesheuvel, A. de Keizer, J.M. Kleijn, M.A.C. Stuart, Langmuir **24**, 6575 (2008)
124. S. Rauch, K.J. Eichhorn, D. Kuckling, M. Stamm, P. Uhlmann, Adv. Funct. Mater. **23**, 5675 (2013)
125. K. Henzler, B. Haupt, M. Ballauff, Anal. Biochem. **378**, 184 (2008)
126. M. Koenig, E. Bittrich, U. König, B.L. Rajeev, M. Müller, K.J. Eichhorn, S. Thomas, M. Stamm, P. Uhlmann, Colloids Surf. B: Biointerfaces **146**, 737 (2016)
127. C.G.P.H. Schroën, M.A. Cohen, Stuart, K. van der Voort Maarschalk, A. van der Padt, K. van't Riet. Langmuir **11**, 3068 (1995)
128. W. Taylor, R.A.L. Jones, Langmuir **29**, 6116 (2013)
129. S. Burkert, E. Bittrich, M. Kuntzsch, M. Müller, K.J. Eichhorn, C. Bellmann, P. Uhlmann, M. Stamm, Langmuir **26**, 1786 (2010)
130. A. Wittemann, B. Haupt, M. Ballauff, Prog. Colloid Polym. Sci. **133**, 58 (2006)
131. E. Bittrich, K.B. Rodenhausen, K.J. Eichhorn, T. Hofmann, M. Schubert, M. Stamm, P. Uhlmann, Biointerphases **5**, 159 (2010)
132. W.M. de Vos, F.A.M. Leermakers, A. de Keizer, M.A. Cohen Stuart, J.M. Kleijn, Langmuir **26**, 249 (2010)
133. K. Henzler, B. Haupt, K. Lauterbach, A. Wittemann, O. Borisov, M. Ballauff, J. Am. Chem. Soc. **132**, 3159 (2010)
134. S. Rosenfeldt, A. Wittemann, M. Ballauff, E. Breininger, J. Bolze, M. Dingenouts, Phys. Rev. E **70**, 061403 (2004)
135. A. Wittemann, M. Ballauff, Anal. Chem. **76**, 2813 (2004)
136. C. Reichhart, C. Czeslik, Langmuir **25**, 1047 (2009)
137. O. Kudina, A. Zakharchenko, O. Trotsenko, A. Tokarev, L. Ionov, G. Stoychev, N. Puretskiy, S.W. Pryor, A. Voronov, S. Minko, Angew. Chem. Int. Ed. **53**, 483 (2014)
138. T. Neumann, B. Haupt, M. Ballauff, Macromol. Biosci. **4**, 13 (2004)

# Chapter 7
# Structure and Interactions of Polymer Thin Films from Infrared Ellipsometry

Andreas Furchner

**Abstract** Polymer films play a vital role in technological, industrial, and biomedical applications. Prominent examples are protective anti-fouling coatings, bio/chemical sensors and devices, organic electronics, as well as functional films with tunable surface characteristics. The films' physical and chemical properties strongly correlate with structure and interactions, which also drive function and functionality. Infrared-spectroscopic ellipsometry (IR-SE) enables comprehensive investigations of those properties, as the IR spectral range contains structure-sensitive baselines and material-specific vibrational bands. In situ IR-SE is a powerful monitoring tool for film chemistry in dependence of stimuli like temperature, humidity, solvent type, pH, and solute concentration. Combined with optical modeling, IR-SE can quantitatively probe numerous film properties, such as chemical composition, anisotropy, molecular orientation, thickness, hydration, and molecular interactions. Recent advances of polymer-film characterization are presented in this chapter, showing, among others, various examples from our group *In Situ Spectroscopy* at ISAS Berlin in collaboration with IPF Dresden, including investigations of polymer film composition and orientation, protein adsorption on functional surfaces, swelling behavior of polymer brushes, and molecular interactions in hydrated polymer films.

## 7.1 Introduction

A variety of spectroscopic methods exists for probing the properties of polymers[1–4]. Visible (VIS) ellipsometry, for example, is widely and successfully used to characterize thin polymer films in dry and swollen states (see Chaps. 5, 6, 8, 9, 12, 15, and 16). This chapter deals with infrared ellipsometry, a complementary but powerful polarization-dependent spectroscopic technique for ex situ and in situ investigations of complex polymer thin films. Determining spectroscopically the structural

A. Furchner (✉)
Leibniz-Institut für Analytische Wissenschaften – ISAS – e.V.,
Schwarzschildstraße 8, 12489 Berlin, Germany
e-mail: andreas.furchner@isas.de

© Springer International Publishing AG, part of Springer Nature 2018
K. Hinrichs and K.-J. Eichhorn (eds.), *Ellipsometry of Functional Organic Surfaces and Films*, Springer Series in Surface Sciences 52,
https://doi.org/10.1007/978-3-319-75895-4_7

properties of such films—like blends, multilayers, and mixed brushes—relies on having a significant spectral contrast, that is, a sufficiently large enough difference in the optical constants $n$ (and $k$) of the individual film constituents. Many transparent polymers with Cauchy-like dispersions in the visible spectral range have rather similar refractive indices. The presence of vibrational bands in the infrared often provides substantially higher contrasts, rendering the IR an important region to probe. Moreover, band positions, amplitudes, and shapes are closely linked to numerous physical and chemical film properties. Besides its obvious application to determine a film's optical constants [5–16] (see Chaps. 23 and 24), IR-SE can therefore be employed to access, and consequently better unterstand, such properties including specific chemical bonds, tacticity, molecular orientation, conformation, local chemical environments, and molecular interactions.

Measured IR-SE spectra are quantitatively interpreted via optical multilayer models based on thicknesses and dielectric functions of the individual layers. For many polymer films their IR dielectric function can be described by a sum of vibrational oscillators,

$$\varepsilon(\tilde{\nu}) = \varepsilon_\infty + \sum_i \varepsilon_i^{\text{vib}}(\tilde{\nu}), \tag{7.1}$$

each of which is characterized by position, amplitude, and shape parameters. Here, $n_\infty = \sqrt{\varepsilon_\infty}$ is the high-frequency refractive index. While typical IR bands are of Lorentzian shape, molecular groups involved in interactions like hydrogen bonding tend to be associated with Gaussian-shaped oscillators because of line broadening due to the increased number of local chemical environments. Voigt oscillators [17, 18], with their two line-shape parameters to quantify shapes between the Lorentzian and Gaussian extremes, are suitable for modeling complex band compositions in interacting polymer systems.

Composite and/or hydrated polymer layers can be modeled based on effective-medium theories (see Chap. 1). The effective dielectric function of a hydrated film, for instance, may be calculated according to Bruggeman [19, 20] from the dielectric properties of water and dry polymer,

$$0 = f_{\text{H}_2\text{O}} \frac{\varepsilon_{\text{H}_2\text{O}} - \varepsilon_{\text{eff}}}{\varepsilon_{\text{H}_2\text{O}} + 2\varepsilon_{\text{eff}}} + (1 - f_{\text{H}_2\text{O}}) \frac{\varepsilon_{\text{Polymer}} - \varepsilon_{\text{eff}}}{\varepsilon_{\text{Polymer}} + 2\varepsilon_{\text{eff}}}. \tag{7.2}$$

If the polymer interacts with water, the polymer dielectric function needs to be modified by introducing corresponding additional oscillators, which then allow a simultaneous fit on film thickness, hydration ($f_{\text{H}_2\text{O}}$), and molecular interactions [21]. Film anisotropy, accounted for by replacing the scalar dielectric functions by tensors, can be handled using generalized matrix algorithms (Chap. 1).

To illustrate how IR-SE can probe specific properties of polymer films, this chapter will first give several brief examples concerning composition and molecular orientation of dry films. Afterwards, in situ examples of hydrated polymer films and brushes in aqueous environments will be discussed, focusing on identification, monitoring, and quantification of changes in film structure and molecular interactions.

## 7.2   Composition

Morphology, composition, miscibility, interdiffusion, and interactions at interfaces are important quantities of polymer surfaces, blends, and complex polymer composite films. Because of the high material-specific spectral contrast in the infrared, IR-SE is highly advantageous for the determination of said properties.

Interdiffusion [22] and cross-linking [23], which often occur during aging or thermal annealing of polymers, are particularly important processes to understand. They depend, among others, on structure and interactions of the individual polymers, and are therefore readily detectable via the vibrational fingerprint. Typical infrared marker bands for this purpose are associated with carbonyl, nitrile, amine, ether, ester, or hydroxyl vibrations, but also with ring and backbone vibrations. Monitoring position, amplitude, and shape of these bands provides plenty of qualitative and semiquantitative insights into the properties of the polymer–polymer interface, such as chemical changes during cross-linking. For a detailed and quantitative characterization, optical modeling can be utilized, yielding properties like the interfacial width in miscible and partially miscible polymer systems.

IR-SE was used, for instance, to quantify the annealing-induced mixing in a polymeric layered system of PnBMA [poly($n$-butyl methacrylate)] and PVC [poly(vinyl chloride)] [24]. The authors measured the CO and CC stretching vibrational bands of PnBMA between 1100–1300 cm$^{-1}$, which are sensitive toward molecular interactions of the polymer's ester groups and therefore toward conformational changes upon annealing [25]. Based on optical modeling of individual PnBMA and PVC layers, as well as annealed, completely mixed films, it was possible to distinguish, for partially annealed PnBMA/PVC films, between a bilayer of the two polymers and a three-layer system with an interdiffusion middle layer (see Fig. 7.1).

A similar approach was used by Duckworth et al. [26] to detect whether interdiffusion or complete mixing occured between thin films of PMMA [poly(methyl methacrylate)] and PVDF [poly(vinylidene fluoride)]. Their analysis exploited the dependency of PMMA-specific carbonyl and PVDF-specific CF$_2$ bands on annealing and molecular interactions between the two polymers.

Besides determining the depth composition of a layered system, IR-SE can also be employed for investigating the lateral makeup of complex polymer films. Ionov et al. [27] quantified the chemical composition of an 8 nm thin 1D gradient mixed polymer brush prepared from PS [polystyrene] and PBA [poly(*tert*-butyl acrylate)]. The two polymers were grafted onto a polymer anchoring monolayer by utilizing a heating stage for annealing that produces a temperature gradient across the substrate's surface, thereby gradually varying the grafting properties. The resulting gradient brush was then chemically mapped for PS- and PBA-specific vibrational bands (see Fig. 7.2). PBA's carbonyl stretching band around 1730 cm$^{-1}$ provided a high spectral contrast to the ring vibrations of PS at lower wavenumbers, allowing the determination of the brush composition. A line shape analysis of PBA's carbonyl band in principle also enables one to study how molecular interactions are related to local composition, that is, how PBA–PBA and PBA–PS interactions impact brush

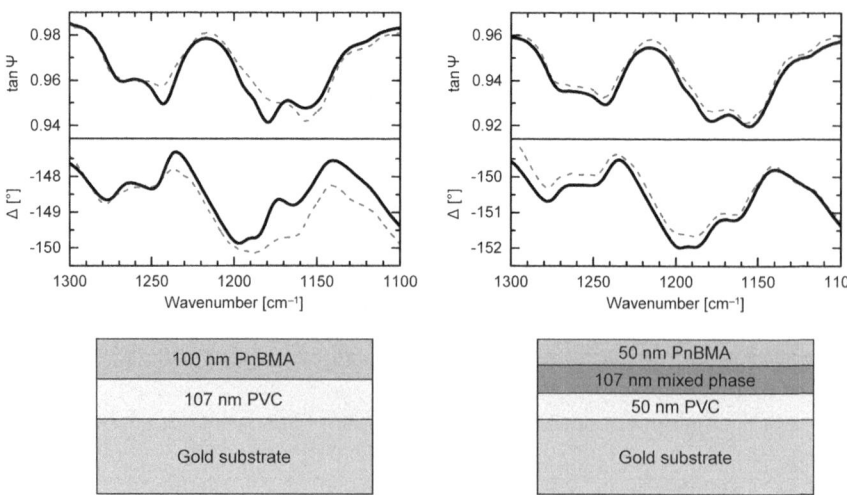

**Fig. 7.1** Measured (dashed) and simulated (solid) IR-SE spectra of partially annealed PnBMA/PVC films (data from [24]). Left: Bilayer model for a 30 min annealed film. Right: Three-layer model accounting for interdiffusion in a 10 min annealed film

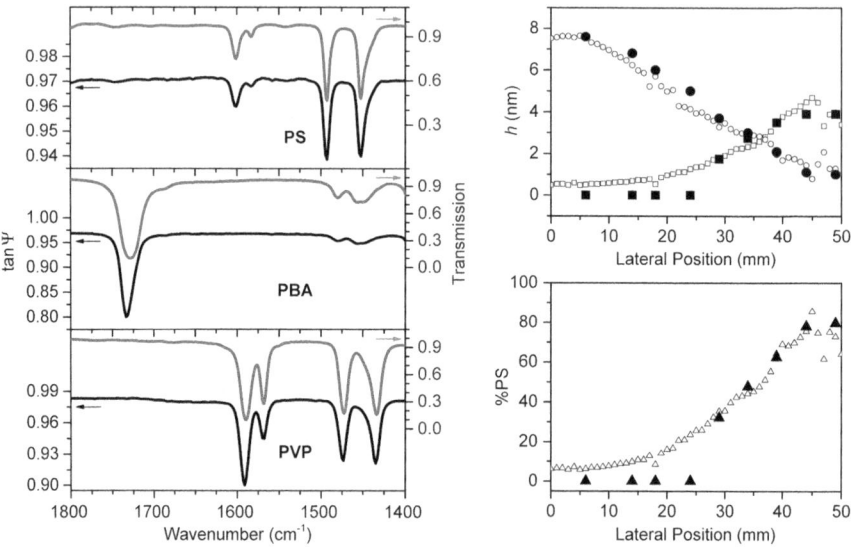

**Fig. 7.2** Left: Reference IR-SE tan$\psi$ (and transmission) spectra of PS, PBA, and PVP [poly(2-vinylpyridine)] for the determination of local brush composition. Right: Laterally resolved relative thickness contributions $h$ of PS (squares) and PBA (circles) in an 8 nm thin gradient PS-*mix*-PBA brush, as well as relative PS volume percentages %PS across the gradient. Closed and open symbols are data from IR-SE and single-wavelength VIS ellipsometry, respectively. Adapted with permission from [27]. Copyright (2005) American Chemical Society

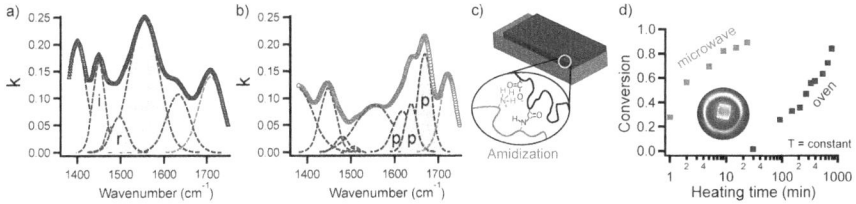

**Fig. 7.3** Amidization reaction in BPEI/PAA multilayers monitored with IR-SE. Absorption coefficient before (**a**) and after (**b**) cross-linking, amidization scheme (**c**) and kinetics (**d**) upon microwave and conventional heating. Adapted from [28]. Copyright (2016), with permission from Elsevier

properties like thickness and grafting density. Such an analysis will be the topic of Sect. 7.4.4.

Concerning cross-linking reactions at polymer–polymer interfaces, IR-SE is a valuable composition monitoring and quantification tool. Lin et al. [28] investigated the cross-linking reaction of BPEI/PAA [branched polyethylenimine/ poly(acrylic acid)] multilayer films. As shown in Fig. 7.3, the authors extracted the absorption index $k$ from IR-SE data between 1350–1750 cm$^{-1}$ using nine vibrational oscillators in the optical model. Besides CH bending modes, which did not change during the reaction, these oscillators are associated with vibrations of antisymmetric (1399 cm$^{-1}$) and symmetric (1552 cm$^{-1}$) COO$^-$ stretching, COOH stretching (1707 cm$^{-1}$), NH$_3^+$ deformation (1496 cm$^{-1}$), NH bending (1632 cm$^{-1}$), as well as amide I related modes around 1670 cm$^{-1}$. The latter become visible as NH$_3^+$ and COO$^-$ groups react to form amide groups (HNCO). This amidization reaction could be induced by conventional heating, or by microwave heating with accelerated reaction kinetics.

Similarly, Simpson et al. [29] followed the formation of covalent Si–O–C bonds upon the reaction between carboxylic acid groups of PAA and silyl (SiH) groups in a vinyl-terminated poly(dimethyl siloxane) cross-linker. They stress that incomplete cross-linking could lead to the continued presence of reactive SiH groups in silicone coatings, causing "lock-up" problems in silicone–acrylic adhesive laminates.

Being able to probe and quantify the individual film constituents is also very valuable for studying effects like film contamination. Bungay et al. [6] investigated silicone films for potential solvent residues. They spincoated 100 nm thick films onto optically thick gold films using CV-1144-O silicone thinned with VM&P Naphtha, a solvent containing different hydrocarbons. Using IR-SE combined with a Bruggeman effective-medium approximation (7.2) to model the film's dielectric function, the residual Naphtha concentration was found to be merely $(0.5 \pm 0.3)$ vol%. As shown in Fig. 7.4, this sensitivity was possible because of the characteristic CH$_x$ bands of Naphtha between 2700 and 3200 cm$^{-1}$.

**Fig. 7.4** Measured (dashed) imaginary pseudo-dielectric function of a 100 nm thick CV-1144-O silicone film compared to calculations (solid) based on different concentrations of residual Naphtha solvent within the film. Reprinted from [6]. Copyright (1998), with permission from Elsevier

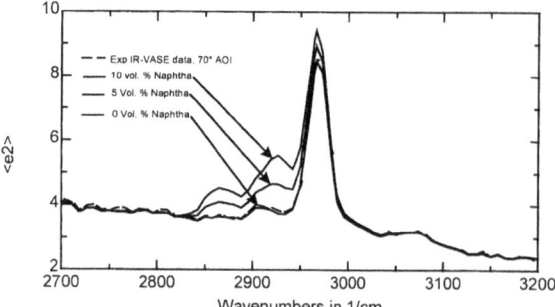

## 7.3 Molecular Orientation

Direct access to the vibrational properties of thin films renders infrared ellipsometry particularly suitable for studying molecular orientation, conformation, packing density, and structure in general. These properties can manifest themselves as anisotropy in the sample's optical response and can therefore be probed with polarized light. While uniaxial film anisotropy can be addressed by ellipsometric measurements at various incidence angles, or by individual p- and s-polarized reflectance measurements, resolving biaxial anisotropy naturally requires additional data at different in-plane sample rotations. However, with regard to molecular orientation, knowledge about film chemistry and direction of transition dipole moments can significantly simplify the infrared analysis, as will be demonstrated later.

Molecular orientation plays an important role in a variety of (polymer) thin-film systems. An instructive example are self-assembled monolayers (SAMs) [30, 31] (see also Chaps. 4 and 9). These ultrathin films usually consist of a head group for anchoring to a substrate, a spacer or tail section, and—if desired for the specific application—a functional end group. It is often the orientation and density of those end groups that are crucial for a successful subsequent functionalization of the monolayer. Typical SAM structural models are depicted in Fig. 7.5. They account for properties like packing density, chain tilt, and chain order, all of which can be explored in the infrared.

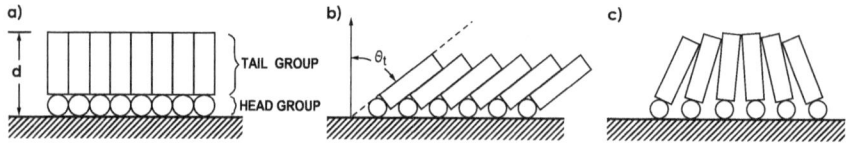

**Fig. 7.5** Structural models of an organized monolayer assembly with closest packed arrangement of head groups and **a** tail groups oriented normal to the substrate surface, **b** tail groups uniformly oriented at an angle $\theta$, and **c** a distribution of tilted tail groups. Adapted with permission from [32]. Copyright (1987) American Chemical Society

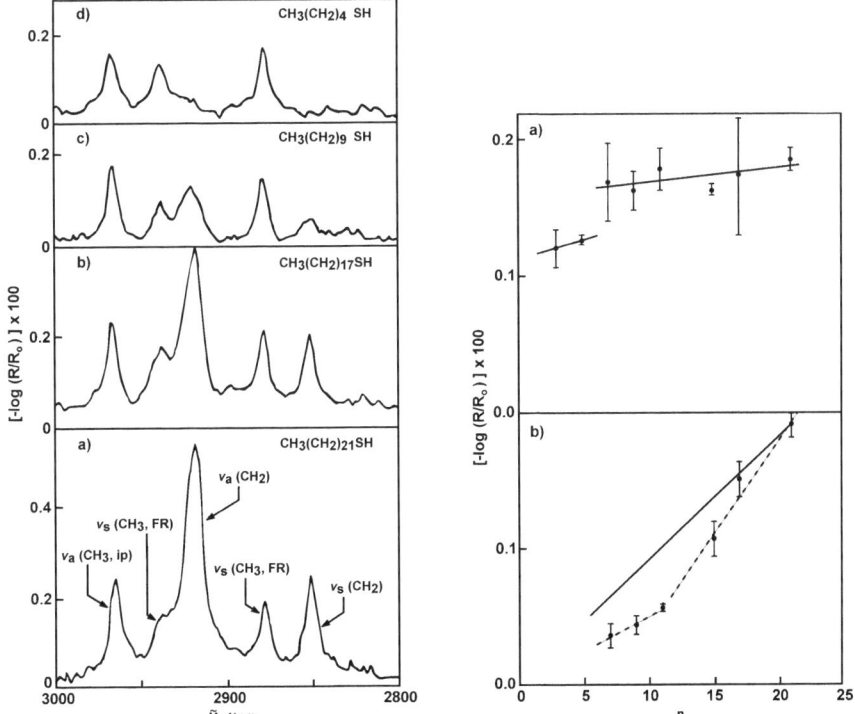

**Fig. 7.6** Left: p-polarized reflection spectra in the methyl and methylene stretching region of self-assembled $n$-alkyl thiol monolayers on gold. Right: Peak heights of $v_a(CH_{3,ip})$ (**a**) and $v_s(CH_2)$ (**b**) bands versus number $n$ of methylene groups. Solid lines in (**a**) are to guide the eye. Solid line in (**b**) is a linear interpolation from the origin. Adapted with permission from [32]. Copyright (1987) American Chemical Society

Infrared spectroscopy was extensively used for studying the structural properties of SAMs [32–35]. Porter et al. [32] used polarized IR spectroscopy to investigate—besides film uniformity, structural integrity, and packing density—the molecular orientation of SAMs, as well as the local chemical environment within the films. They probed the $CH_x$ stretching region of self-assembled $n$-alkyl thiol monolayers [$CH_3(CH_2)_nSH$] on gold, with a number of repeating methylene units between $n = 1$ and 21. Comparing the $CH_x$ peak positions of the monolayers (Fig. 7.6, left) with those of crystalline polymethylene chains (with $n = 21$) and the liquid state (with $n = 7$) provided insights into the local chemical environment of a chain. In monolayers with methylene chain lengths $n > 11$, the peak positions correspond to those of the bulk crystalline phase, whereas for shorter chains with $n < 11$, they resemble those of the liquid phase. Moreover, for alkyl chains with consistent chain orientation, it is expected that the band intensities of the $v(CH_2)$ methylene stretching vibrations increase linearly with $n$, whereas those of the $v(CH_3)$ methyl end group's stretching vibrations remain constant. However, a different situation was found (Fig. 7.6, right), again with a characteristic change in trends around $n = 11$ repeating units (dashed lines). The authors explained the deviations from the linear behavior with differences in average tilt angle and structure. For long-chain monolayers with $n = 15$–21, they found a tilt of about 20–30°, corresponding to a structure similar to that in Fig. 7.5b. Short-chain monolayers with $n < 11$, on the other hand, tend toward slightly higher tilt angles, which they explained by a more disordered chain distribution with decreased packing density and coverage.

A similar study [33] on self-assembled $n$-alkyl thiol monolayers on silver substrates revealed quite different structural properties. For chains longer than about $n = 10$, monolayers were predominantly densely packed and crystalline-like with all-trans conformational sequences that exhibited average tilt angles of about 13°. More tilted structures were hypothesized for shorter-chain monolayers. The smaller tilt angles on silver compared to gold hint at differences in the bonding of the sulfur head group, which was further substantiated by contact-angle measurements.

Deducing the molecular orientation from polarized IR spectra is based on simple trigonometric considerations of the involved transition dipole moments [36, 37]. In all-trans hydrocarbon chains, for instance, the chain axis and the transition dipole moments of the $\nu_{as}(CH_2)$ and $\nu_s(CH_2)$ stretching vibrations are mutually orthogonal. The chain's tilt angle $\theta$ from the surface normal is therefore given by [1]

$$\cos^2 \theta + \cos^2 \theta_{as} + \cos^2 \theta_s = 1. \tag{7.3}$$

The angles $\theta_{as}$ and $\theta_s$ of the two methylene stretching vibrations can be measured via their transition dipole moments

$$
\begin{aligned}
M_{i,x}^2 &= M_{i,\max}^2 \cdot \sin^2 \theta_i \cdot \cos^2 \phi_i, \\
M_{i,y}^2 &= M_{i,\max}^2 \cdot \sin^2 \theta_i \cdot \sin^2 \phi_i, \\
M_{i,z}^2 &= M_{i,\max}^2 \cdot \cos^2 \theta_i,
\end{aligned}
\tag{7.4}
$$

which are proportional to the respective oscillator strengths computable in an optical model. Here, $M_{\max}$ is the principle transition dipole moment, and $\phi$ is the in-plane rotation of a chain. In a uniaxial film, for which $\overline{\cos^2 \phi} = \overline{\sin^2 \phi} = 1/2$, the average tilt angle of both symmetric and antisymmetric methylene stretching vibrations can then be calculated from

$$\frac{M_{i,x}^2}{M_{i,z}^2} = \frac{\sin^2 \theta_i}{2 \cos^2 \theta_i}, \tag{7.5}$$

allowing one to deduce the chain's tilt angle $\theta$ via (7.3).

Using this approach, Rosu et al. [35] determined average chain tilts of 19 and 22° for hexadecanethiol on gallium arsenide and gold, respectively, suggesting a similar organizational structure on both substrates. Monolayers of octanedithiol on GaAs, by contrast, were found to be more disordered with average tilt angles of 30°.

Note that band interpretation is rather straightforward for thin films on metals like silver or gold. Owing to the metal surface selection rule [2], only transition dipole

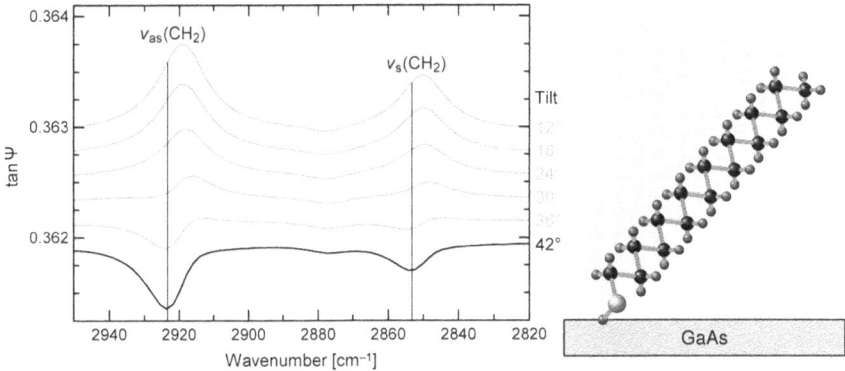

**Fig. 7.7** Calculated tan Ψ spectra of hexadecanethiol self-assembled monolayers on GaAs in dependence of average tilt angle. Data from [35]

components perpendicular to the metal's surface can be observed in a spectrum. The measured band intensities can therefore be directly related with the chain's tilt angle. For arbitrary substrates, though, band positions and line shapes can vary quite dramatically. This is demonstrated in Fig. 7.7 illustrating the sensitivity of IR-SE toward tilt-angle variations. The graph shows calculated tan Ψ spectra of hexadecanethiol on GaAs with heavily tilt-dependent $\nu_{as}(CH_2)$ and $\nu_s(CH_2)$ band amplitudes, positions, and line shapes. The two vertical lines, which indicate the center frequencies of the corresponding oscillators in the optical model, highlight an important circumstance that any IR spectroscopist should internalize: The frequency of a measured peak does not necessarily coincide with the oscillator's resonance frequency!

Another interesting example concerning structure and molecular orientation is that of Langmuir–Blodgett (LB) films. Such films are composed of organic monolayers deposited from the surface of a liquid onto a substrate via immersion or emersion. Tsankov et al. [38] investigated the influence of thermal annealing on the structural properties of LB films consisting of multiple double layers of 2-[4-($N$-dodecanoylamino)phenyl]-5-(4-nitrophenyl)-1,3,4-oxadiazole on gold. The molecule, depicted in Fig. 7.8, has a nonlinear shape with its diphenyl oxadiazole fragment being bent in the molecular plane and twisted with respect to the aliphatic tail. The tails themselves are connected to benzene rings via amide groups, which can facilitate hydrogen bonds between neighboring chains. Strong hydrogen bonds could be inferred from the well-pronounced stretching bands related to bonded NH ($3325$ cm$^{-1}$) and C=O groups ($1680$ cm$^{-1}$). Those bonds were only partly disrupted after film annealing at $130\,°C$. However, the $\nu(CH_2)$ bands diminished in amplitude upon annealing, suggesting molecular rearrangements of the chains. In particular, the observed frequency upshifts by $4$ cm$^{-1}$ could be attributed to an all-trans–gauche conformational transformation. Annealing also resulted in a thickness increase from 60 to 65 nm, hinting at a lowering of the overall molecular tilt angle. This was verified by a tilt-angle determination of the molecule's nitro

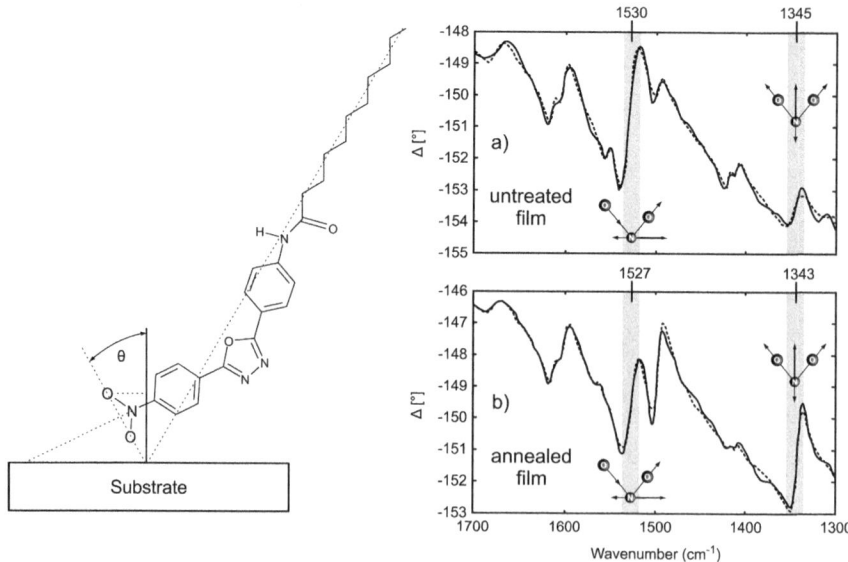

**Fig. 7.8** Left: Molecular orientation of the studied Langmuir–Blodgett film (see text). Right: Measured (solid) and fitted (dashed) $\Delta$ spectra in the $NO_2$ stretching region before (**a**) and after (**b**) annealing at $130\,°C$. Adapted with permission from [38], copyright (2002) American Chemical Society; and from [39], copyright (1998), with permission from Elsevier

head group (angle $\theta$ in Fig. 7.8). Fitting the symmetric and antisymmetric stretching vibrations of $NO_2$ groups at 1345 and 1530 $cm^{-1}$, respectively, the authors found a marked change in tilt angle from 39° for the untreated film to 53° for the annealed film.

One last example of out-of-plane uniaxially anisotropic films are dicyanovinyl-sexithiophene films for photovoltaic applications. A comprehensive IR-SE study revealed that the molecular orientation undergoes changes with increasing film thickness [40]. An isotropic distribution of orientations with an average tilt of 55° (close to the so-called magic angle of 54.7°) was characteristic for 4 nm thin films consisting of about two or three monolayers. Thicker films of 20 nm showed preferential in-plane molecular orientation with average tilt angles of 67°. Such higher tilts are favorable when using sexithiophene as absorber material in organic solar cells, resulting in enhanced photon absorption and thus charge-carrier generation.

Seemingly biaxially anisotropic polymer films are depicted in Fig. 7.9, showing the in-plane anisotropy of stretched PET [poly(ethylene terephthalate)] and PEN [poly(ethylene naphthalate)] films [12]—two materials with relevance for flexible electronic devices. These polymers have several characteristic vibrational modes with transition dipole moments along or normal to the chain axis, hence the observed enhanced or diminished band amplitudes at corresponding sample rotations. While ethylene-glycol wagging (1340 $cm^{-1}$) and the complex ester modes (1260 $cm^{-1}$, 1080–1160 $cm^{-1}$) are parallel to the chain axis, the C=O stretching

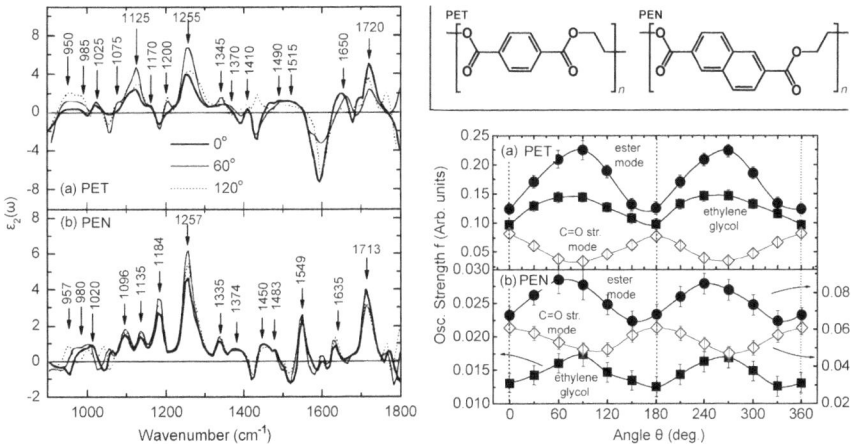

**Fig. 7.9** Left: Imaginary part of the dielectric function of PET and PEN films at 0° (bold line), 60° (solid line), and 120° (dotted line) in-plane sample rotation. Bottom-right: Angular dependence of the oscillator strength of ethylene glycol (squares), ester (circles), and C=O (diamonds) stretching modes. Reprinted from [12], with the permission of AIP Publishing. Top-right: Chemical structures of PET and PEN

mode ($1720 \, cm^{-1}$) is perpendicular to it, as is evident from the projected oscillator strengths (Fig. 7.9, bottom-right). Perhaps unsurprisingly for stretch-oriented films, a refractive-index analysis revealed that both PET and PEN films can be treated as uniaxial materials with their optical axis parallel to their surface.

For further reading on molecular orientation in organic thin films, the reader is referred to Chaps. 3, 13, 20, and 21.

## 7.4  Hydrated Polymer Films

Examples of thin films shown so far dealt with non-hydrated layers under ambient conditions in air. In the following, hydrated films are addressed that are exposed either to air with significant humidity levels or to aqueous environments, including salt and protein solutions. In situ IR-SE on such solid–liquid interfaces is very useful for qualitative and semi-quantitative studies. It was, for instance, successfully applied for monitoring the adsorption of proteins on functional surfaces [41], the pH-responsive dissociation of polyelectrolyte films [42] and mixed brushes [44], the swelling behavior and interactions of end-grafted polyoxazoline brushes [43], the chemical and structural changes during the complex formation of mixed anionic/cationic polyelectrolyte brushes [45], the electrochemical growth of polypyrrole films [46], as well as the hydrophilicity of complex composite polymer surfaces to tune hydrophobic interactions [47]. First quantitative approaches proved promising for modeling the electrochemical grafting of nanometer-thin nitrobenzene [48] and maleimide multilayers [49], the growth process of PSS-doped polyaniline films [50], as well as the swelling–deswelling transition of polyacrylamide brushes [51].

The full potential of in situ IR-SE as an analytical technique became apparent with the use of optical models to gain quantitative information about film hydration, structure, and molecular interactions [21]. This combination of IR-SE and optical modeling enabled detailed quantitative investigations of hydrated polymer films at the solid–liquid interface, and was used, for example, for studying the role of molecular interactions in the stimuli-responsive phase transition of polymer brushes [21].

Before diving into the details of quantitative in situ IR-SE, a typical experimental set-up of ellipsometer plus flow cell will be introduced, followed by examples of polymer films in aqueous environments, with a particular focus on polymer brushes. Their densely grafted polymer chains give rise to fascinating surface properties susceptible to changes induced by external stimuli like temperature or pH changes (see Chap. 6).

### 7.4.1    In Situ Infrared Ellipsometry

In situ IR-SE measurements of solid–vapor or solid–liquid interfaces demand precise control of humidity, temperature, solvent, pH, and other possibly influential environmental parameters. Special in situ cells are thus required. Figure 7.10 shows an example of a temperature-controlled flow cell [41, 44] in an in situ infrared ellipsometer. The cell itself, including in- and outlet tubes, consists of organic polymers like PEEK (polyether ether ketone) or PTFE (polytetrafluorethylen) that exhibit excellent chemical resistance properties needed for measurements in low- or high-pH solutions. A quartz window at the back of the cell allows the possibility to implement additional simultaneous in situ optical techniques like VIS reflection anisotropy spectroscopy [46] (see Chap. 1). Temperature of the cell can be regulated between 15–50 °C with a stability of ±0.05 °C. This precision is imperative because the optical properties of solvents like water are strongly temperature dependent. The whole set-up is purged with dry air to guarantee constant atmospheric conditions and to reduce absorption of IR radiation by atmospheric water vapor.

The polymer film (or any other film for that matter) is prepared on the inner side of the actual window substrate. This might be done either *ex situ* or directly from solution within the cell. The substrate, made from an IR-transparent material like Si or $CaF_2$, has a wedge-shaped form that ensures well-separated inner and outer reflexes preventing interferences from multiple internal reflections. The flow cell can also be modified with electrodes in order to do electrochemistry experiments while performing in situ IR-SE [46].

Note that the configuration in Fig. 7.10 probes the solid–liquid interface under non-ATR conditions. Compared with ATR, the shorter pathlength through the substrate allows access to a wider spectral range (8000–700 cm$^{-1}$ for Si) with retained phase coherence necessary for accurate phase measurements $\Delta$. Depending on external incidence angle at, and refractive index of, the substrate, internal incidence angles

**Fig. 7.10** Schematic of an in situ polarizer/sample/retarder/analyzer IR-SE set-up with temperature-controlled flow cell and infrared-transparent, wedge-shaped window substrate [44]. Depending on film and solvent properties, measurements with and/or without retarder can be performed for sensitive phase determination

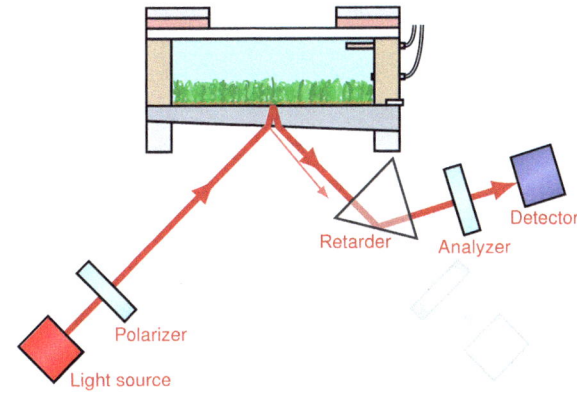

between 5–16° at the substrate/film interface are feasible. This non-zero angle ensures sensitivity toward thin-film thickness and out-of-plane anisotropy [44].

### 7.4.2 Polymer Films and Brushes in Aqueous Environments

Ellipsometric spectra of isotropic polymer films in aqueous environments are dominated by the strong vibrational bands of water that often mask the film signature of interest. This is demonstrated in Fig. 7.11 (left) for thin PGMA [poly(glycidyl methacrylate)] films on Si. While the polymer bands of thicker films are clearly visible, the vibrational features of the 2 nm ultrathin film are buried beneath the water stretching and bending modes. Spectral referencing can circumvent this problem by significantly enhancing the optical contrast. One way of referencing is with respect to spectra of a clean substrate (without polymer film) obtained under similar, or same, well-defined experimental conditions as the film spectra. This approach is often necessary when the solvent's optical properties change during the experiment and again cause too drastic an overlap with the film signature, as is the case, for instance, in temperature-dependent studies. As shown in Fig. 7.11 (right) for the PGMA films, this approach makes visible even the bands of the 2 nm ultrathin film. Another way is self-referencing where ratios $\tan\Psi\,/\,\tan\Psi_0$ and differences $\Delta - \Delta_0$ are recorded with respect to the initial conditions of an in situ experiment, such as the film's state before a protein-deposition study or before exposure to pH variations. Any changes from, respectively, unity and zero are then associated with changes in the optical properties of the solid–liquid interface.

The first straightforward example of an in situ study is presented in Fig. 7.12. In this combined investigation using IR-SE and VIS ellipsometry (see also Chap. 6), the proteins fibrinogen (FIB) and human serum albumin (HSA) were tested for their adsorption behavior on various polymer surfaces, namely a $d_{dry} = 2.5$ nm ultrathin hydrophobic PGMA film, a 14.5 nm thin hydrophilic PNIPAAm brush, a 25.1 nm

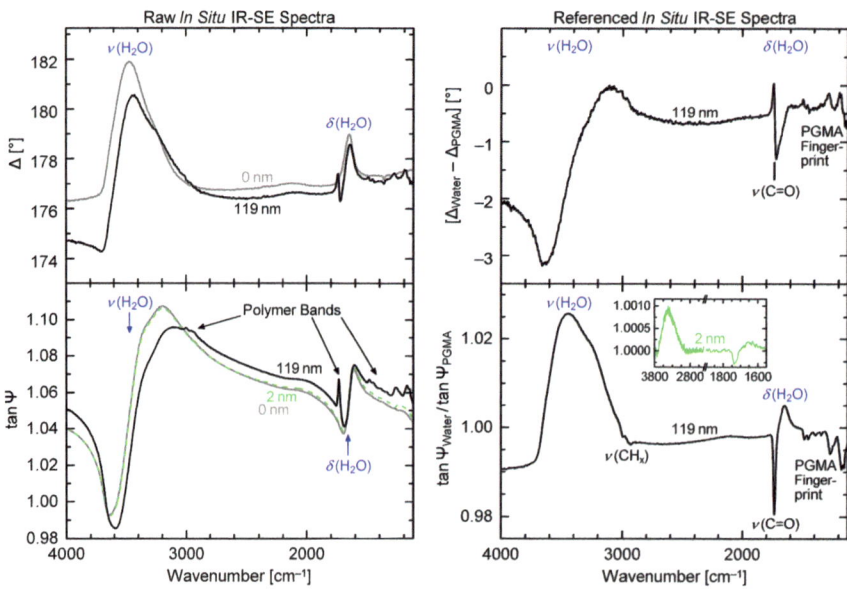

**Fig. 7.11** Left: Raw IR-SE spectra of 0, 2, and 119 nm thick PGMA films in H$_2$O. Right: Same spectra referenced to a clean silicon wedge ("0 nm" on the left)

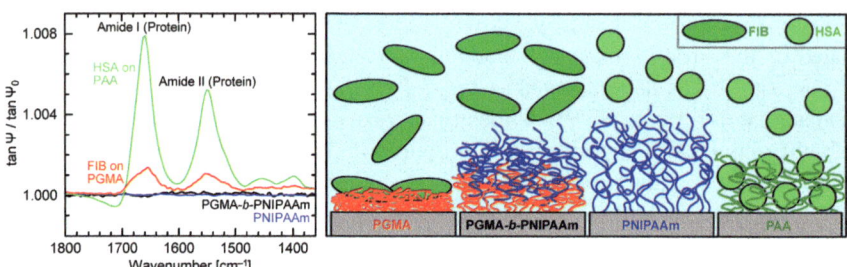

**Fig. 7.12** Self-referenced IR-SE spectra showing protein adsorption and repellent behavior of different polymer (brush) surfaces with respect to fibrinogen (FIB) and human serum albumin (HSA). Initial conditions were protein-free aqueous solutions. The polymer films consist of PGMA [poly(glycidyl methacrylate)], PNIPAAm [poly(N-isopropylacrylamide)], PGMA-b-PNIPAAm, and PAA [poly(acrylic acid)]. Adapted with permission from [41]. Copyright (2015) American Chemical Society

thick PGMA-b-PNIPAAm block-copolymer brush, and a 6.9 nm thin PAA poly-electrolyte brush [41]. The self-referenced spectra exhibit upward-pointing protein-related amide I and amide II bands showing that FIB and HSA adsorbed on PGMA film and PAA brush, respectively. The measured amide band amplitudes correspond to monolayer FIB adsorption on top of PGMA, whereas for PAA they indicate that HSA penetrates into the brush and thereby adsorbs in more copious amounts, in accordance with VIS ellipsometry (26 mg/m$^2$ HSA at pH 5). The shapes of FIB's

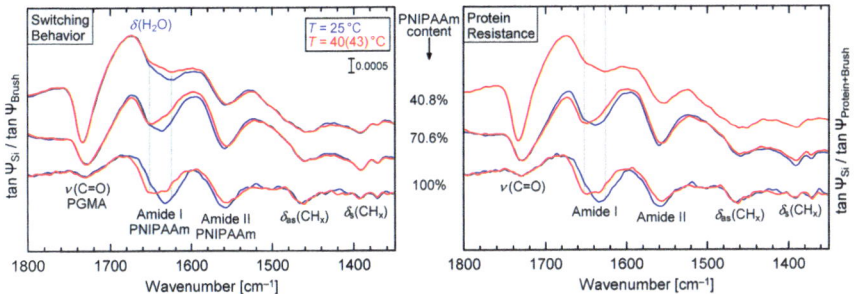

**Fig. 7.13** Left: Cascaded temperature-dependent in situ spectra of a pure PNIPAAm brush (100%, $d = 14.5$ nm dry thickness) and PNIPAAm-*b*-PGMA brushes with 70.6% ($d = 26.8$ nm) and 40.8% ($d = 25.1$ nm) PNIPAAm content measured in protein-free solution. The upward-pointing $\delta(H_2O)$ band overlaps the amide I bands redshifting their indicated band components (vertical lines) in dependence of brush thickness and hydration. Right: Corresponding spectra of pure PNIPAAm brush in HSA solution and PNIPAAm-*b*-PGMA copolymer brushes in FIB solution. Adapted with permission from [41]. Copyright (2015) American Chemical Society

and HSA's amide bands are dictated by the proteins' secondary structures. Different hydrogen-bond patterns, such as alpha-helices or beta-sheets within the proteins, have characteristic impacts on line shape and band composition, often leading to complex amide bands with multiple subbands. A closer analysis of the measured amide bands in Fig. 7.12—in comparison to transmission measurements of protein solutions—revealed that the proteins did not change their structure during adsorption from the solution. This is crucial information for applications that rely on the integrity of the adsorbed molecules.

Being pH-responsive polyelectrolytes [42], PAA brushes can be triggered via changes in solvent pH to ad- and desorb proteins. Such processes were monitored time-dependently with in situ IR-SE, showing good agreement with brush thickness and amount of adsorbed protein determined by VIS ellipsometry [41].

Contrary to PGMA and PAA, both the pure PNIPAAm brush and the PGMA-*b*-PNIPAAm copolymer brush turned out to be resistant toward FIB and HSA adsorption (surface concentration $\ll 0.5$ mg/m$^2$). This repellent behavior was unexpected for the copolymer surface because the idea behind creating those mixed brushes was to obtain thermoresponsive films switchable between a protein repellent and a non-repellent state. The protein-repellent PNIPAAm toplayer was thought to be collapsable via temperature stimulus, thereby surface-exposing the PGMA sublayer that is able to bind proteins. To test whether this switching behavior does occur, several brushes with varying PGMA/PNIPAAm composition were measured below and above PNIPAAm's swelling–deswelling transition temperature [41]. Results are given in Fig. 7.13 (left). Because of referencing with respect to a clean wedge, the data show downward-pointing polymer bands—predominantly the amide I/II and CH$_x$-bending bands of PNIPAAm, and the carbonyl band of PGMA—overlapped by the upward-pointing water bending mode around 1650 cm$^{-1}$. PNIPAAm's amide I band, which is mainly associated with C=O stretching, contains at least two major

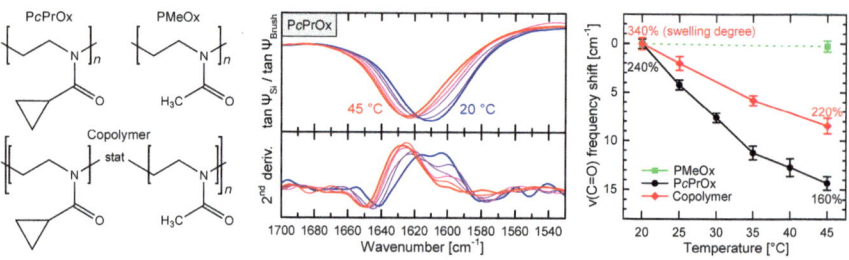

**Fig. 7.14** Left: Chemical structures of PMeOx, P*c*PrOx, and their statistical copolymer. Middle: IR-SE spectra, and their 2nd derivative, of the $\nu$(C=O) band of a P*c*PrOx brush in D$_2$O measured at 20, 25, ..., 45 °C. Right: Relative $\nu$(C=O) upshift (with respect to band maximum at 20 °C) in the three types of polymer brushes compared with swelling degree obtained from VIS ellipsometry. Adapted from [43] in accordance with the Creative Commons Attribution (CC BY) license

components. One, observed at about 1625 cm$^{-1}$, is related to carbonyl moieties fully hydrated by water molecules, whereas the other one at about 1652 cm$^{-1}$ stems in part from C=O groups hydrogen-bonded to H–N moieties of neighboring amide groups. The pure PNIPAAm brush exhibits marked alterations between these two components with increasing temperature, indicating the brush's collapse. This switching behavior is seen to a lesser extent, or not at all, the lower the PNIPAAm content is in the copolymer brushes. These mixed brushes therefore do not collapse like a classical PNIPAAm brush, which can be explained by their structural properties. Cross-linking between different PGMA segments during brush preparation probably causes parts of the PNIPAAm blocks to emerge from within the PGMA sublayer rather than from its surface, resulting in molecular interactions and steric hindrance between PNIPAAm and PGMA in the PGMA/PNIPAAm interpenetration layer, and consequently in a less pronounced swelling–deswelling transition. No differences were observed when comparing the spectra with measurements of the same brushes exposed to protein solutions (Fig. 7.13, right). In other words, no protein adsorption took place on any of the PNIPAAm-containing polymer films because PNIPAAm always dominated their surface properties.

Kroning et al. also investigated molecular interactions in temperature-responsive oxazoline-based polymer brushes between 20–45 °C in H$_2$O and D$_2$O. Specifically, they studied carbonyl–water hydrogen bonding in P*c*PrOx [poly(2-cyclopropyl-2-oxazoline)] and PMeOx [poly(2-methyl-2-oxazoline)] brushes, as well as in copolymer brushes containing 75% *c*PrOx and 25% MeOx. Being ternary amides (see Fig. 7.14), these polymers do not contain N–H groups that could facilitate intra- or intermolecular hydrogen bonds with carbonyl groups. The measured amide I carbonyl bands are therefore comprised of comparitively few subbands related to the differently hydrogen-bonded C=O species. The first subband, associated with free C=O, is seen in the dry but not in the hydrated state in water. The other two subbands are connected with weakly and strongly hydrated C=O groups, implying that all of the brushes' carbonyl groups are at least partially hydrated when exposed to water. For the pure P*c*PrOx brush, the temperature-dependent second derivative of the tan $\Psi$

carbonyl band shows a transition between the two major band components, that is, between stronger hydration at lower temperatures and weaker hydration at higher temperatures. This transition is not as pronounced for the copolymer brush with its lower relative content of temperature-responsive PcPrOx.

As depicted in Fig. 7.14 (right), the measured C=O-stretching transitions compare well with changes in swelling degree determined from VIS ellipsometry. Contrary to the steplike transition of thermoresponsive polymers like PNIPAAm [52], the PcPrOx and copolymer brushes showed continuous $\nu$(C=O) upshifts with increasing temperature, correlating with less polymer–water interactions within the collapsing brushes. Smaller band shifts, but higher swelling degrees, were observed for the copolymer brush, meaning that more water molecules are retained in the copolymer brush because of its higher hydrophilicity. Put another way, the more pronounced the measured C=O transition the stronger the thickness transition, highlighting the close correlation between molecular interactions and brush deswelling. In fact, the nonresponsive and more hydrophilic PMeOx can be used to tune the transition behavior of the copolymer system [53], rendering these types of poly(2-oxazoline) brushes highly interesting for potential applications.

## 7.4.3  Optical Effects and the Role of Water Bands

Nontrivial baselines and spectral overlaps between polymer and water vibrational bands can cause serious problems with regard to data interpretation of transmission and ATR infrared spectra. A common example is the amide I band of peptides and proteins, which is heavily masked by the water bending mode. One way to reveal the true amide band shapes is to use heavy water, the vibrational bands of which occur at much lower wavenumbers compared to normal water. Figure 7.15 illustrates this with IR-SE measurements of a PNIPAAm brush around its swelling–deswelling temperature in $H_2O$ and $D_2O$. As discussed before, during the brush's collapse around $32\,^\circ$C, the amide I band undergoes a transition between two major components associated with hydrated C=O groups at lower temperatures and amide-interacting C=O groups at higher temperatures. Not only are the corresponding vibrational oscillators redshifted in $D_2O$, but the actual transition is less sharp. It extends over a larger temperature range because $D_2O$ forms stronger hydrogen bonds than $H_2O$. Also in more complex polymer systems like proteins, such stronger bonds can potentially influence the film's structural, and thereby functional, properties.

Other strategies to deal with band-overlap and baseline issues are solvent background subtraction and baseline correction methods [2, 54]. The aim is always to suppress, or get rid of, the solvent bands and other unwanted "artifacts" in order to obtain pure polymer spectra suitable for qualititave interpretation, or for quantitative evaluation via oscillator band fits. These strategies, however, can cause serious errors in band interpretations and should thus be avoided whenever possible.

A fundamentally different approach is to use optical models to compare measured with calculated spectra, as is routinely done with ellipsometry. As alluded to in the

**Fig. 7.15** Amide I/II/II'
bands during the
temperature-induced
swelling–deswelling
transition of a PNIPAAm
brush in $H_2O$ and $D_2O$. Note
the absence of an amide II
band for the deuterated
brush. The amide I overlap
with $\delta(H_2O)$ causes smaller
apparent band amplitudes

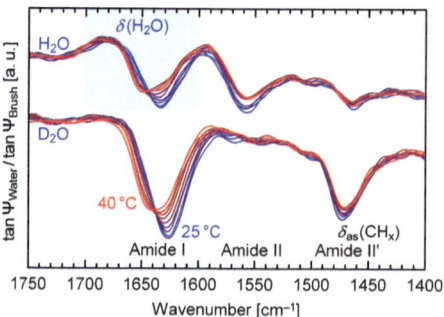

beginning, these models for interpretation of infrared spectra are based on a chemical, physical, and optical description of the hydrated films, automatically accounting for optical effects like baseline drifts due to thin-film interference, band-shape distortions due to $n/k$ mixing, and dielectric effective-medium effects from film hydration [2, 20, 55]. The models do not rely on manipulation procedures like certain ATR corrections or subtraction of solvent background, and therefore allow a more accurate quantitative interpretation of measured spectra.

To illustrate why optical modeling is a powerful tool, it is useful to recall that referenced in situ IR-SE spectra measure the optical contrast of the solid–liquid interface. As shown in Fig. 7.16, derivative-like bands appear in $\Delta$ upon referencing, whereas absorption-band-like downward-pointing film-related bands and upward-pointing solvent-related bands manifest in $\tan \Psi$. Again, it has to be stressed that the observed water bands and baseline drifts are simply a result of the optical contrast between film-on-substrate and no-film-on-substrate data. Any change in the optical properties of the solid–liquid interface will therefore immediately affect bands and baselines [44], rendering them important markers for detecting minute changes of the film's properties. Specifically, amplitude and line shape of both film and solvent bands strongly depend on incidence angle, film swelling, hydration, and complex refractive index. Particularly in the case of water, with its very strong and broad absorption bands, it therefore makes obvious sense not to ignore or remove those bands but to include them in a fit in order to gain a better quantitative understanding of said film properties.

Rappich and Hinrichs were the first to attempt to reproduce the water vibrational bands by optical modeling. They monitored the growth of ultrathin nitrobenzene films on H-passivated Si(111) wedges from a solution of 4-nitrobenzene-diazonium-tetrafluoroborate during the electrochemical processing in diluted sulfuric acid (see Chap. 20 for details). As shown in Fig. 7.17, the presence of symmetric and antisymmetric $NO_2$ stretching bands is clear evidence for successful grafting. Using ellipsometrically determined optical constants of water, they compared the measured nitrobenzene film spectrum to calculated spectra based on a non-hydrated film with varying thickness $d$ and high-frequency refractive index $n_\infty$ (7.1). Although the agreement is not perfect—probably due to small but non-negligible film anisotropy

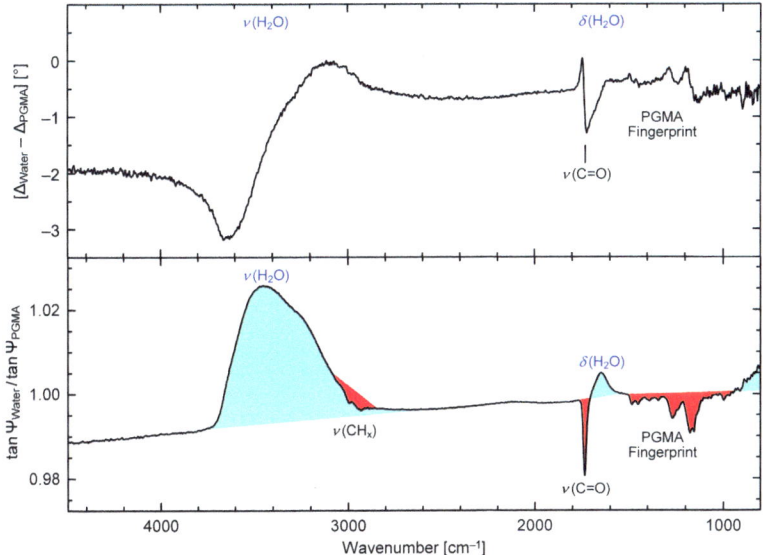

**Fig. 7.16** Ellipsometric spectra of a 119 nm thick PGMA film in water referenced to spectra of a clean silicon wedge

**Fig. 7.17** Measured in situ tan Ψ of a nitrobenzene-modified Si surface referenced to the SiH-covered Si in the same aqueous environment prior to film grafting, compared to calculated spectra showing the sensitivity of $v(H_2O)$ toward changes in film thickness and refractive index. Data from [48]

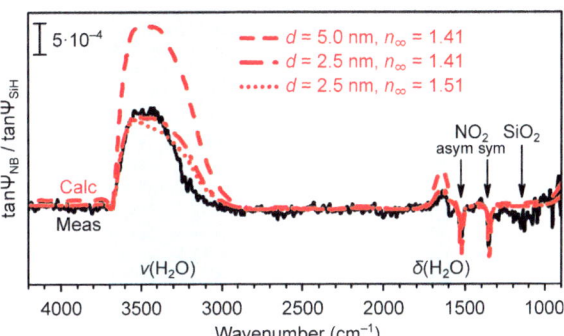

and contributions from overlapping Si–OH vibrations or water molecules aligned at the charged surface—the calculations indicate the usefulness of the solvent-related bands for quantifying, e. g., film thickness.

Better agreement is usually achieved for neutral, non-charged polymer surfaces, allowing one to determine not only the thickness $d_{wet}$ of a potentially swollen film but also to perform a fit on the film's hydration level, i. e., its water content $f_{H_2O}$ via (7.2). This is possible even if both parameters are a priori unknown. As demonstrated in Fig. 7.18, changes in $d_{wet}$ and $f_{H_2O}$ lead to characteristic changes in baseline as well as water band amplitudes and shapes. $v(H_2O)$, for instance, scales almost linearly with $d_{wet}$ but rather non-linearly with $f_{H_2O}$. Moreover, $v(H_2O)$ blueshifts with increasing hydration because the band is comprised of five oscillators with very strong transition

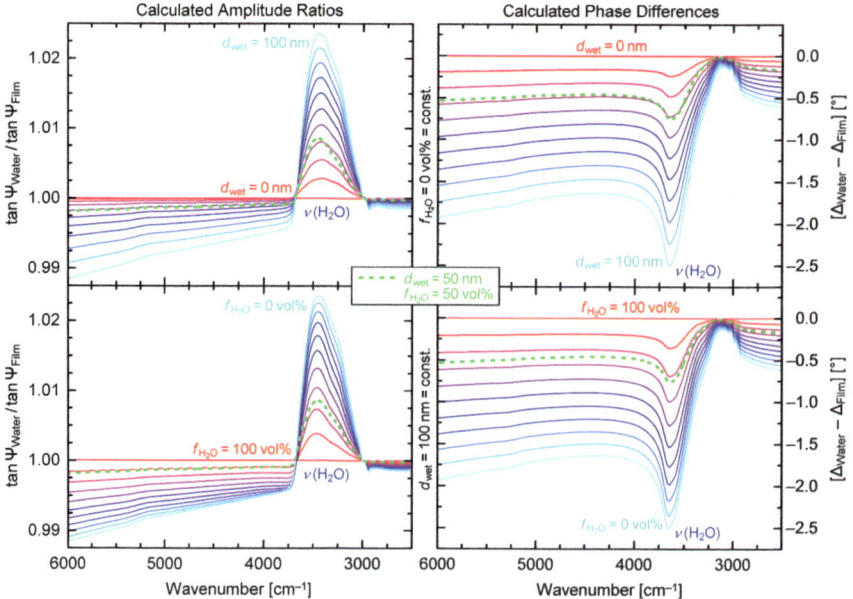

**Fig. 7.18** Effects of varying film thickness $d_{wet}$ and water content $f_{H_2O}$ on calculated ellipsometric spectra in the $\nu(H_2O)$ region of a PGMA film on Si. Upper panels: $d_{wet} = [0, 10, \ldots, 100]$ nm with fixed $f_{H_2O} = 0$ vol%. Lower panels: $f_{H_2O} = [0, 10, \ldots, 100]$ vol% with fixed $d_{wet} = 100$ nm. Dotted lines were calculated with $d_{wet} = 50$ nm and $f_{H_2O} = 50$ vol%

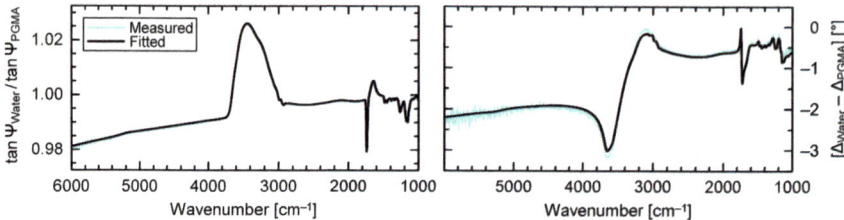

**Fig. 7.19** Measured and fitted in situ IR-SE spectra of a thin PGMA film on Si in water

dipole moments [56]. Also the band shape is undergoing marked alterations upon changes in film thickness. Both effects help decorrelate $d_{wet}$ and $f_{H_2O}$.

Simultaneous fits on thickness and hydration are demonstrated in Fig. 7.19 for the previously introduced 119 nm thick PGMA film. In water the film slightly swells to $d_{wet} = (123 \pm 2)$ nm with a consistently small water content of $f_{H_2O} = (5 \pm 2)\%$. Because of the film's hydrophobic nature, it is not expected that water molecules penetrate through the entirety of the film. This can be reproduced in an expanded model consisting of a non-hydrated, 70–90 nm thick sublayer and a hydrated, 30–50 nm thick toplayer, which is in agreement with AFM surface-inhomogeneity measurements that show film-thickness variations of about $\pm 15$ nm.

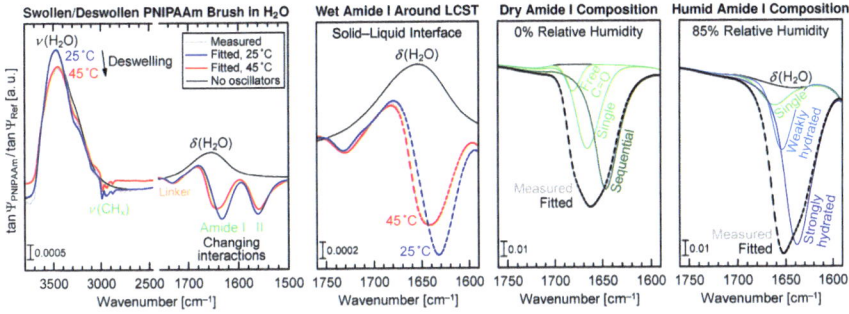

**Fig. 7.20** Measured and fitted in situ IR-SE spectra of swollen and collapsed PNIPAAm films in aqueous and humid environment. Thin solid black lines are calculated spectra without polymer dielectric oscillators. For dry and humid states, individual amide I Voigt oscillator contributions are shown. Adapted with permission from [21]. Copyright (2017) American Chemical Society

### 7.4.4 Structure and Interactions from Quantitative In Situ IR-SE

Of course thickness and hydration are also readily accessible with VIS ellipsometry (see Chaps. 6 and 8), but one of the additional strengths of IR ellipsometry is its ability to study structural changes and molecular interactions. These effects can have considerable and distinct impacts on a vibrational spectrum, most prominently on C=O- and amide-related bands. Nikonenko et al. [57], for example, employed *ex situ* IR-SE to study mucin layers adsorbed on amphiphilic PAA-*b*-PMMA diblock copolymer surfaces. An in-depth analysis of the $\nu$(C=O) region, the protein's amide bands, and bands related to the polymer side-chain could identify adhesive interactions and conformational changes in mucin/polymer double layers. Different components of the C=O stretching bands revealed hydrogen bonds between mucin and polymer without the participation of the mucin's amide groups. From an amide I component analysis, an increase in the proportion of beta-sheets was observed, indicating a more unfolded and aggregated structure of mucin after adsorption.

The obvious aim is to perform such revealing band analyses in situ with the polymer films in contact with aqueous environments. For this purpose, optical modeling of in situ IR-SE spectra is employed based on the Bruggeman effective-medium approximation and a set of vibrational oscillators that describes the differently interacting and non-interacting molecular groups. The detailed example presented is again that of polymer brushes made from PNIPAAm, a temperature-responsive model polymer with regard to structure and interactions of peptides and proteins.

PNIPAAm films measured in aqueous, dry, and humid ambient conditions give rise to rich vibrational spectra [21], as can be seen in Fig. 7.20. Virtually all bands are affected by the presence of water within the film, most noticeably the $\nu$(H$_2$O) region and the polymer-related amide and alkyl-stretching bands. While changes in the $\nu$(H$_2$O) bands are directly related to swelling and hydration effects, band alterations

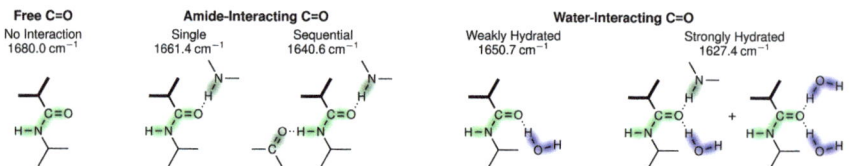

**Fig. 7.21** Carbonyl–amide and carbonyl–water hydrogen-bond interactions in PNIPAAm polymer brushes. Stated wavenumbers refer to fitted amide I oscillator frequencies in the optical model. Adapted with permission from [21]. Copyright (2017) American Chemical Society

in the amide region are evidence for changing molecular interactions associated with PNIPAAm's amide side groups.

Focusing on the tan $\Psi$ amide I band, a second-derivative analysis indicates several subbands, which are related to the various hydrogen-bond interactions depicted in Fig. 7.21. PNIPAAm's C=O moieties can be free (i. e., non-interacting), or involved in intramolecular carbonyl–amide or intermolecular carbonyl–water interactions. Corresponding carbonyl stretching oscillators are progressively redshifted with increasing number and strength of hydrogen bonds [58], and are sufficiently well-separated to allow their accurate determination with IR-SE.

Fitting the data allows a comprehensive analysis of film hydration, interactions, and swelling [21]. Figure 7.22 plots the relative fractions of differently interacting C=O species, the hydration-sensitive oscillator position of antisymmetric methyl stretching, as well as film swelling and number of water molecules per PNIPAAm monomer—all in dependence of the film's water content, as fitted using the Bruggeman equation (7.2). In dry state, the film is dominated by sequentially hydrogen-bonded amide groups, similar to alpha-helices in proteins. Expectedly, amide–amide interactions are replaced by amide–water interactions as the film hydrates in humid air. Approaching 85% relative humidity, all free carbonyl groups become hydrated, while weakly hydrated C=O groups are more and more converted to strongly hydrated ones. At 25 °C in water, the brush is highly swollen and hydrated. In fact, the majority of carbonyl groups is strongly hydrated, with only a few groups remaining involved in pure hydrogen bonds with neighboring amide groups. At 45 °C in water, the brush deswells and dehydrates, with a small percentage of carbonyl groups becoming dehydrated and/or incorporated into sequential amide–amide bonds again. This conversion from amide–water- to amide–amide-interacting carbonyl groups, although expected, leaves most C=O groups unaffected and strongly hydrated, which is in accordance with the considerable amount of water (52 vol%) retained within the collapsed brush. Comparing C=O fractions and swelling behavior, it can be concluded that additional water molecules above 25 vol% water content do not necessarily contribute to specific interactions with PNIPAAm's amide groups, but rather to the overall brush hydration. This hydration is also observed in the $\nu_{as}(CH_3)$ band position corresponding to the polymer's hydrophobic isopropyl side groups.

Interestingly, there was no detectable preferential molecular orientation, neither in the collapsed nor in the stretched state. An isotropic distribution of orientations seems

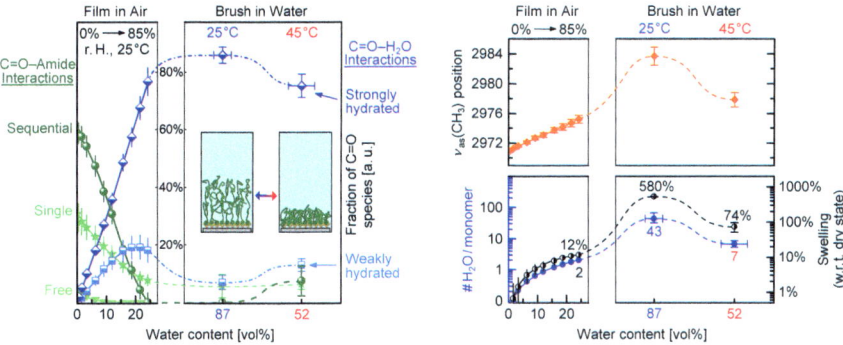

**Fig. 7.22**  Humidity- and temperature-dependent interactions, hydration states, and structure of PNIPAAm films. Left: Amide–amide and amide-water interactions of hydrophilic carbonyl groups. Right-top: Hydration of hydrophobic isopropyl groups. Right-bottom: Number of water molecules per monomer and film swelling with respect to dry-state thickness. Adapted with permission from [21]. Copyright (2017) American Chemical Society

reasonable because the chains are still far away from being completely stretched, even in the brush's highly swollen state.

It has been shown in this section that IR-SE on solid–liquid interfaces can deliver detailed qualitative and quantitative information about the physical and chemical properties of polymer thin films in aqueous environments. Being able to monitor the films while applying external stimuli revealed fascinating surface characteristics. Such investigations are highly interesting for complex polymer film systems like adsorbed peptides, proteins, or membranes.

## 7.5    Future Prospects

This chapter gave an overview of the many uses of infrared ellipsometry for studies of thin polymer films, in particular with regard to investigations of structure and molecular interactions. Current trends in IR-SE are focusing on improving three major aspects of the method: spatial resolution, temporal resolution, and information content. The latter is being addressed, for instance, by infrared Mueller ellipsometry [59, 60]. This generalized version of IR-SE can probe a sample's complete $4 \times 4$ Mueller matrix, thus allowing deeper insights into the properties of depolarizing and anisotropic nanostructured polymer films.

Lateral and temporal resolution are addressed with ellipsometric techniques that make use of brilliant light sources much brighter than standard IR globars (see Chap. 22). Synchrotron IR-SE, for example, was used for mapping chemical composition and dissociation of 3 nm ultrathin pH-responsive PAA brushes with a spatial resolution of $300 \times 800 \, \mu m^2$ [61]. A similar mapping study employed in situ

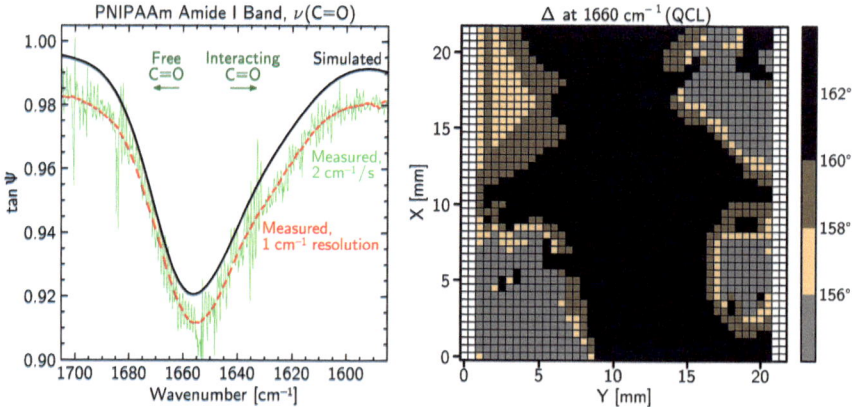

**Fig. 7.23** Inhomogeneous, 90 nm thick PNIPAAm film on gold measured with IR laser ellipsometry. Left: QCL tan $\Psi$ sweeps compared to a simulated spectrum (top) determined from conventional ellipsometry. Right: $\Delta$ homogeneity map of the film's amide I band. Adapted from [65]. Copyright (2017), with permission from Elsevier

synchrotron IR-SE on binary PAA-$b$-PS/PEG brushes in aqueous environments between pH 2 and 10 for determining carboxyl dissociation and film homogeneity [62].

Another interesting approach is IR-ellipsometric microscopy [63, 64] based on commercial reflection/transmission microscopes. Here, the set-up's Cassegrain objective is expanded by two polarizers and a plane-of-incidence defining aperture enabling IR-SE measurements with resolutions down to the diffraction limit. For films of a few 10 nm thickness, measurements with spot sizes of $40 \times 40$ µm$^2$ have been demonstrated, enabling straightforward IR-SE mapping of vibrational bands.

IR laser ellipsometry has recently been introduced and—utilizing tunable quantum cascade lasers (QCLs)—advanced toward spectral studies and mapping characterization of thin polymer films with per-spot resolutions of 60 ms and $120 \times 250$ µm$^2$ [65, 66]. An example is given in Fig. 7.23 showing the amide I region of a 90 nm thick, inhomogeneous PNIPAAm film spectrally measured within seconds and mapped within minutes.

The limits of both lateral and temporal resolution are further pushed by yet another novel technique termed IR nanopolarimetry [67]. Utilizing broad-bandwidths QCLs and a polarization control, this AFM-based method probes a film's frequency-dependent thermal expansion, hence providing direct measurements of IR absorption in domains as small as a few $(10 \text{ nm})^2$. IR nanopolarimetry was successfully used to characterize anisotropic porphyrin supramolecular assemblies, revealing that the aggregate's morphology can be correlated with anisotropic organization and hence different oriented attachment growth mechanisms.

All of the aforementioned techniques hold a bright future for IR-SE characterization of polymer thin films. Temporally improved, nanoscale-resolved measurements will increasingly facilitate kinetic studies of nanostructure and interactions, which is of great importance for film preparation, characterization, process control, and applications.

# References

1. V.P. Tolstoy, I.V. Chernyshova, V.A. Skryshevsky, *Handbook of Infrared Spectroscopy of Ultrathin Films* (Wiley, Chichester, 2003)
2. J.M. Chalmers, P.R. Griffiths, *Handbook of Vibrational Spectroscopy* (Wiley, Chichester, 2006)
3. N. Everall, P.R. Griffiths, J.M. Chalmers, *Vibrational Spectroscopy of Polymers: Principles and Practice* (Wiley, Chichester, 2007)
4. S. Thomas, Y. Grohens, P. Jyotishkumar, *Characterization of Polymer Blends: Miscibility, Morphology and Interfaces* (Wiley-VCH Verlag GmbH & Co, KGaA, 2015)
5. R.T. Graf, F. Eng, J.L. Koenig, H. Ishida, Appl. Spectrosc. **40**, 498–503 (1986)
6. C.L. Bungay, T.E. Tiwald, D.W. Thompson, M.J. DeVries, J.A. Woollam, J.F. Elman, Thin Solid Films **313–314**, 713–717 (1998)
7. T. Heitz, B. Drévillon, J.E. Bourée, C. Godet, J. Non-Cryst, Solids **227–230**, 636–640 (1998)
8. D. Blaudez, F. Boucher, T. Buffeteau, B. Desbat, M. Grandbois, C. Salesse, Appl. Spectrosc. **53**, 1299–1304 (1999)
9. K. Hinrichs, D. Tsankov, E.H. Korte, A. Röseler, K. Sahre, K.-J. Eichhorn, Appl. Spectrosc. **56**, 737–743 (2002)
10. M. Schubert, C. Bundesmann, G. Jakopic, H. Maresch, H. Arwin, N.-C. Persson, F. Zhang, O. Inganäs, Thin Solid Films **455–456**, 295–300 (2004)
11. H.G. Tompkins, T. Tiwald, C. Bungay, A.E. Hooper, J. Phys. Chem. **108**, 3777–3780 (2004)
12. A. Laskarakis, S. Logothetidis, J. Appl. Phys. **99**, 066101 (2006)
13. W.R. Folks, S.K. Pandey, G. Pribil, D. Slafer, M. Manning, G. Boreman, Int. J. Infrared Milli. Waves **27**, 1553–1571 (2006)
14. J.L. Stehle, J.P. Piel, Appl. Surf. Sci. **256S**, S72–S76 (2009)
15. S. Kang, V.M. Prabhu, C.L. Soles, E.K. Lin, W. Wu, Macromolecules **42**, 5296 (2009)
16. H. Arwin, Thin Solid Films **519**, 2589–2592 (2011)
17. D. De Sousa Meneses, G. Gruener, M. Malki, P. Echegut, J. Non-Cryst. Solids **351**, 124–129 (2005)
18. F. Schreier, D. Kohlert, Comput. Phys. Commun. **179**, 457–465 (2008)
19. D.A.G. Bruggeman, Ann. Phys. (Berlin, Ger.) **416**, 636–664 (1935)
20. D.E. Aspnes, Thin Solid Films **89**, 249–262 (1982)
21. A. Furchner, A. Kroning, S. Rauch, P. Uhlmann, K.-J. Eichhorn, K. Hinrichs, Anal. Chem. **89**, 3240–3244 (2017)
22. J. Klein, Science **250**, 640–646 (1990)
23. S. Mavila, O. Eivgi, I. Berkovich, N.G. Lemcoff, Chem. Rev. **116**, 878–961 (2016)
24. K. Hinrichs, M. Gensch, N. Nikonenko, J. Pionteck, K.-J. Eichhorn, Macromol. Symp. **230**, 26–32 (2005)
25. N. Nikonenko, K. Hinrichs, E.H. Korte, J. Pionteck, K.-J. Eichhorn, Macromolecules **37**, 8661–8667 (2004)
26. P. Duckworth, H. Richardson, C. Carelli, J.L. Keddie, Surf. Interface Anal. **36**, 33–41 (2004)
27. L. Ionov, A. Sidorenko, K.-J. Eichhorn, M. Stamm, S. Minko, K. Hinrichs, Langmuir **21**, 8711–8716 (2005)
28. K.L., Y. Gu, H. Zhang, Z. Qiang, B.D. Vogt, N.S. Zacharia, Langmuir **32**, 9118–9125 (2016)
29. T.R.E. Simpson, J.L. Keddie, J. Adhes. **79**, 1207–1218 (2003)
30. A. Ulman, Chem. Rev. **96**, 1533–1554 (1996)
31. J.C. Love, L.A. Estroff, J.K. Kriebel, R.G. Nuzzo, G.M. Whitesides, Chem. Rev. **105**, 1103–1169 (2005)

32. M.D. Porter, T.B. Bright, D.L. Allara, C.E.D. Chidsey, J. Am. Chem. Soc. **109**, 3559–3568 (1987)
33. M.M. Walczak, C. Chung, S.M. Stole, C.A. Widrig, M.D. Porter, J. Am. Chem. Soc. **113**, 2370–2378 (1991)
34. Z.G. Hu, P. Prunici, P. Patzner, P. Hess, J. Phys. Chem. B **110**, 14824–14831 (2006)
35. D.M. Rosu, J.C. Jones, J.W.P. Hsu, K.L. Kavanagh, D. Tsankov, U. Schade, N. Esser, K. Hinrichs, Langmuir **25**, 919–923 (2009)
36. D.L. Allara, R.G. Nuzzo, Langmuir **1**, 52–66 (1985)
37. R. Arnold, A. Terfort, C. Wöll, Langmuir **17**, 4980–4989 (2001)
38. D. Tsankov, K. Hinrichs, E.H. Korte, R. Dietel, A. Röseler, Langmuir **18**, 6559–6564 (2002)
39. E.H. Korte, K. Hinrichs, A. Röseler, Spectrochim. Acta B **57**, 1625–1634 (2002)
40. K. Hinrichs, M. Levichkova, D. Wynands, K. Walzer, K.-J. Eichhorn, P. Bäuerle, K. Leo, M. Riede, Thin Solid Films **525**, 97–105 (2012)
41. A. Kroning, A. Furchner, D. Aulich, E. Bittrich, S. Rauch, P. Uhlmann, K.-J. Eichhorn, M. Seeber, I. Luzinov, S.M. Kilbey II, B.S. Lokitz, S. Minko, K. Hinrichs, A.C.S. Appl, Mater. Interfaces **7**, 12430–12439 (2015)
42. D. Aulich, O. Hoy, I. Luzinov, M. Brücher, R. Hergenröder, E. Bittrich, K.-J. Eichhorn, P. Uhlmann, M. Stamm, N. Esser, K. Hinrichs, Langmuir **26**, 12926–12932 (2010)
43. A. Kroning, A. Furchner, S. Adam, P. Uhlmann, K. Hinrichs, Biointerphases **11**, 019005 (2016)
44. Y. Mikhaylova, L. Ionov, J. Rappich, M. Gensch, N. Esser, S. Minko, K.-J. Eichhorn, M. Stamm, K. Hinrichs, Anal. Chem. **79**, 7676–7682 (2007)
45. K. Hinrichs, D. Aulich, L. Ionov, N. Esser, K.-J. Eichhorn, M. Motornov, M. Stamm, S. Minko, Langmuir **25**, 10987–10991 (2009)
46. G. Sun, X. Zhang, C. Kaspari, K. Haberland, J. Rappich, K. Hinrichs, J. Electrochem. Soc. **159**, H811–H815 (2012)
47. O. Hoy, B. Zdyrko, R. Lupitskyy, R. Sheparovych, D. Aulich, J. Wang, E. Bittrich, K.-J. Eichhorn, P. Uhlmann, K. Hinrichs, M. Müller, M. Stamm, S. Minko, I. Luzinov, Adv. Funct. Mater. **20**, 2240–2247 (2010)
48. J. Rappich, K. Hinrichs, Electrochem. Commun. **11**, 2316–2319 (2009)
49. P. Kanyong, G. Sun, F. Rösicke, V. Syritski, U. Panne, K. Hinrichs, J. Rappich, Electrochem. Commun. **512**, 103–107 (2015)
50. G. Sun, X. Zhang, J. Rappich, K. Hinrichs, Appl. Surf. Sci. **344**, 181–187 (2015)
51. A. Furchner, E. Bittrich, P. Uhlmann, K.-J. Eichhorn, K. Hinrichs, Thin Solid Films **541**, 41–45 (2013)
52. E. Bittrich, S. Burkert, M. Müller, K.-J. Eichhorn, M. Stamm, P. Uhlmann, Langmuir **28**, 3439–3448 (2012)
53. S. Adam, M. Koenig, K.B. Rodenhausen, K.-J. Eichhorn, U. Oertel, M. Schubert, M. Stamm, P. Uhlmann, Appl. Surf. Sci. **421**, 843–851 (2017)
54. P.R. Griffiths, James A. De, J.D.Winefordner Haseth, *Fourier Transform Infrared Spectrometry* (Wiley, Chichester, 2007)
55. M. Miljkovič, B. Bird, M. Diem, Analyst **137**, 3954–3964 (2012)
56. J.-J. Max, C. Chapados, J. Phys. Chem. A **106**, 6452–6461 (2002)
57. N.A. Nikonenko, I.A. Bushnak, J.L. Keddie, Appl. Spectrosc. **63**, 889–898 (2009)
58. A. Barth, Biochim. Biophys. Acta, Bioenerg. **1767**, 1073–1101 (2007)
59. E. Garcia-Caurel, A. Lizana, G. Ndong, B. Al-Bugami, C. Bernon, E. Al-Qahtani, F. Rengnez, A. de Martino. Appl. Opt. **56**, 2776–2785 (2015)
60. G. Ndong, A. Lizana, E. Garcia-Caurel, V. Paret, G. Melizzi, D. Cattelan, B. Pelissier, J.H. Tortai, Appl. Opt. **55**, 3323–3332 (2016)
61. K. Roodenko, Y. Mikhaylova, L. Ionov, M. Gensch, M. Stamm, S. Minko, U. Schade, K.-J. Eichhorn, N. Esser, K. Hinrichs, Appl. Phys. Lett. **92**, 103102 (2008)
62. D. Aulich, O. Hoy, I. Luzinov, K.-J. Eichhorn, M. Stamm, M. Gensch, U. Schade, N. Esser, K. Hinrichs, Phys. Status Solidi C **7**, 197–199 (2010)
63. K. Hinrichs, A. Furchner, J. Rappich, T.W.H. Oates, J. Phys. Chem. C **117**, 13557–13563 (2013)

64. K. Hinrichs, A. Furchner, G. Sun, J. Rappich, M. Gensch, T.W.H. Oates, Thin Solid Films **571**, 648–652 (2014)
65. A. Furchner, C. Kratz, D. Gkogkou, H. Ketelsen, K. Hinrichs, Appl. Surf. Sci. **421**, 440–445 (2017)
66. A. Furchner, G. Sun, H. Ketelsen, J. Rappich, K. Hinrichs, Analyst **140**, 1791–1797 (2015)
67. T. Shaykhutdinov, S.D. Pop, A. Furchner, K. Hinrichs, ACS Macro Lett. **6**, 598–602 (2017)

# Chapter 8
# In Situ Spectroscopic Ellipsometry in the Field of Industrial Membranes

**Wojciech Ogieglo**

**Abstract** Industrial membranes are playing an ever increasing role in the ongoing and necessary transition of our society towards more sustainable growth and development. Already today membranes offer more energy efficient alternatives to the traditional often very energy intensive industrial separation processes such as (cryogenic) distillation or crystallization. For many years reverse osmosis membranes have offered a viable method for the production of potable water via desalination processes and their significance continuously increases. Recently, membrane technology has been demonstrated to play a significant role in potential methods to generate or store energy on an industrial scale. For molecular separations often the key for an efficient membrane operation often lies in the application of an (ultra-) thin organic polymer, inorganic or hybrid selective layer whose interaction with the separated mixture defines the membrane performance. Ellipsometry has started gaining increasing attention in this area due to its large potential to conduct in-situ, non-destructive and very precise analysis of the film-fluid interactions. In this chapter, we aim to review the important recent developments in the application of ellipsometry in industrial membrane-related studies. We briefly introduce the basics of membrane science and discuss the used experimental setups and optical models. Further we focus on fundamental studies of sorption, transport and penetrant-induced phenomena in thin films exposed to organic solvents or high pressure gases. The application of in-situ ellipsometry is discussed for studies of new, promising membrane materials and the use of the technique for emerging direct studies of operating membranes is highlighted.

W. Ogieglo (✉)
Advanced Membranes and Porous Materials Center, 4700 King Abdullah
University of Science and Technology (KAUST), Thuwal 23955-6900,
Kingdom of Saudi Arabia
e-mail: wojciech.ogieglo@kaust.edu.sa

© Springer International Publishing AG, part of Springer Nature 2018
K. Hinrichs and K.-J. Eichhorn (eds.), *Ellipsometry of Functional
Organic Surfaces and Films*, Springer Series in Surface Sciences 52,
https://doi.org/10.1007/978-3-319-75895-4_8

173

## 8.1 Basics of Membrane Science and Technology

Separation of mixtures into streams of pure components is one of the most commonly employed technological unit operations. Traditionally, separations have been achieved with adsorption, centrifugation, crystallization, distillation or extraction [1]. While these frequently employed processes have established themselves over the last century, membranes have gained increasing interest due to their advantages. To familiarize the reader, in the following we attempt to give a very brief, and by no means comprehensive, description of the principles, advantages and challenges, as well as characterization techniques related with membrane technology.

### 8.1.1 Principles of Membrane-Based Separations

A typical membrane module is shown in Fig. 8.1. The feed contains a mixture of at least two components and is separated using a membrane into two streams: retentate and permeate. Depending on the membrane selectivity the two outcoming streams will differ in the composition relative to the feed, thus providing separation. The selectivity is defined as the ability of the membrane to discriminate between the various components of the mixture. It is important to note, that rarely do the membranes fully separate the mixtures. Especially for separations on a molecular level, incomplete separations are common. This needs to be taken into account in the process design. A notable exception is reverse osmosis for potable water production, where salt separations in excess of 99.5% are frequent [2].

Membranes operate on a vast range of size-scales [2]. The conventional filtration separates particles of several tens of microns using porous, for instance, cellulose-based filters. Microfiltration can sieve-out bacteria and viruses. Ultrafiltration deals with larger molecules and proteins, while nanofiltration discriminates between small molecules such as glucose. At the smallest scale, individual ionic species are removed from water in reverse osmosis.

Across this large range of size scales one can observe a trend of gradually moving from porous filters, where the separation is achieved mainly via size-exclusion and described by a pore-flow model to dense membranes where the dominating mechanism is often affinity and molecular diffusion driven (so called solution-diffusion model). In solution-diffusion the components dissolve in the membrane at the side of higher chemical potential (upstream side with for instance higher partial pressure in

**Fig. 8.1** Schematic capturing the essence of a membrane process

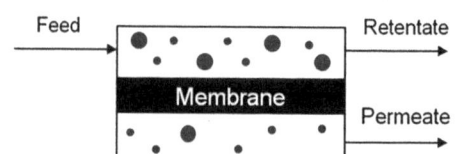

gas separations), diffuse via the dense, usually polymeric network and then desorb at the side of lower chemical potential or the downstream side [3, 4]. The permeability coefficient, $P$, can be related with solubility and diffusivity coefficients, $S$ and $D$, respectively, by a simple relationship: $P = S \cdot D$. For specific penetrant-membrane pairs the $P$ and $S$ can be often viewed as constants and are expressed in a normalized way. For instance in gas separations, $P$ is mass flux normalized for pressure and membrane thickness and $S$ is concentration normalized for pressure at which it is determined.

## 8.1.2   Advantages and Challenges in Membrane Science and Technology

Since the energetically expensive step of phase change (e.g. evaporation in distillation) is avoided membranes are inherently much more efficient than the classical separation unit operations [5]. Membrane technology may also contribute to simplification of the process and its control by limiting the number of moving parts which often require heavy maintenance. The material selection is theoretically unlimited with hundreds of various polymer-support combinations tested over the years.

Among the current challenges in membrane science and technology one should name the propensity to fouling. Fouling is defined as unwanted deposition or growth of a more or less dense film (filtration cake) that first, limits the mass flux over time, second, can irreversibly damage the membrane or, third, may modify the surface properties of the membrane in the unwanted direction (hydrophobicity, zeta potential etc.).

In gas separations an important trade-off effect, manifested as the so called upper-bound, has been identified where the simultaneous achievement of high permeability and selectivity with currently available membrane polymers is impossible despite extensive research efforts [6]. Here, much promise is related with the new developments in materials science, particularly in the rapidly growing metal organic frameworks field [7–9] or graphene-based materials [10].

## 8.1.3   Characterization of Membranes

Membranes are characterized according to their performance, structure and interfacial properties. The performance is often expressed in terms of flux and selectivity. These can be determined by relatively simple methods such as measuring the mass flow over time with gravimetric or volumetric quantification, or more complicated ones like pressure decay in gas separations. In terms of direct characterization of the membrane structure scanning electron microscopy (SEM), optical microscopy or various types of porosimetries have been employed. For the characterization of interfacial properties one could mention the streaming potential or contact angle methods.

Although definitely not among the traditional methods in membrane characterization, spectroscopic ellipsometry has recently gained increasing attention. In particular for separations at a molecular level (nanofiltration, reverse osmosis or gas separation), the preferred membrane structure is a composite consisting of a thin selective film on top of a porous and highly permeable support. While the separation performance is dictated by the dense film, which should be as thin as possible yet defect-free, it is virtually impossible to gain direct information on its properties during operation with the traditional membrane characterization methods. The problem is amplified by the significant controversies regarding the behavior of ultra-thin, particularly glassy films as compared with bulk with some studies indicating significant deviations [11–15] while others arguing against them [16–18]. Spectroscopic ellipsometry with its non-invasive, highly sensitive approach has offered entirely new possibilities to address those issues.

## 8.2 Experimental Configurations and Optical Models for In-Situ Ellipsometry for Thin Films and Membranes

### 8.2.1 Design of Measurement Chambers

Performing in-situ ellipsometry, sometimes referred to as environmental ellipsometry, usually requires a dedicated measurement chamber. The most popular configuration is a trapezoidally-shaped chamber (as viewed from the side) shown in Fig. 8.2. In such a cell the flow of the fluid (which can be liquid, vapor or gas) is gently controlled by external pumps (usually syringe pumps) or gas flow controllers (with possible utilization of a carrier gas in the case of vapor generation) and the cell is dedicated to a single angle of incidence (AOI). This naturally limits the flexibility of the technique so it is preferred to decide on the particular AOI considering the Brewster or pseudo-Brewster angles of the used samples. At Brewster or pseudo-Brewster angles the signal sensitivity is optimal due to full or almost full removal

**Fig. 8.2** Schematics of the most widely used in-situ ellipsometry cell with temperature control and windows perpendicular to AOI = 70° (Reprinted with permission from [20]. Copyright Elsevier 2015)

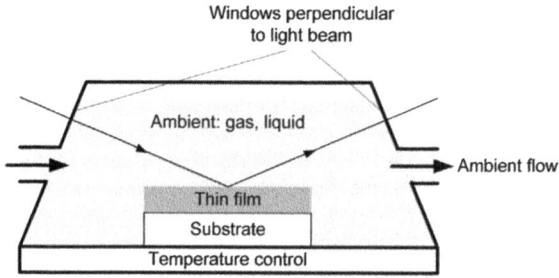

of the p-polarized component [19]. For silicon wafers, chambers with AOI = 70 or 75° are most commonly used. If samples are very weakly reflective (like membrane supports) it is advisable to use as low as possible AOI, typically 85°, to benefit from a gain in reflectance. The use of a focused beam may be very helpful in eliminating the unwanted impact of substrate roughness (particularly for polymer substrates). However in this case, the cell windows' design should take account for the focal length of the optics in the presence of ambients with n > 1.

In most cases the in-situ chambers are made from either stainless steel or glass (quartz). However, with modern rapid fabrication techniques (3D printing) the cell design options have been extended by various thermoplastics or UV-curable polymers. Whatever the material used, the cell construction usually involves the use of windows intersecting the probing light path. As extensively described previously [20, 21] the introduction of windows has several consequences detrimental to the analysis accuracy. First, the presence of the windows at a fixed AOI limits the range of possible angles to one. This itself reduces the measurement accuracy and, to a lesser degree, its precision. Second, windows may be slightly birefringent (anisotropic) which will introduce "delta offsets". These can be calibrated using calibration procedures (like one described in J. A. Woollam Co., Inc., CompleteEASE software manual, [22, 23] or in literature [21, 22]). Third, if the windows are not perfectly symmetrically positioned and the ambient has refractive index, n, significantly larger than 1 (e.g. liquid water), then the cell may become a prism and the various wavelengths will reflect from the sample at slightly different angles. In addition, if focusing optics are used, this configuration may lead to beam defocus and loss of intensity (error increase). All of these reasons point to the large care that needs to be taken when designing and fabricating the in-situ chambers. Because the current 3D printing techniques not always provide sufficient degree of geometrical control we have found, that such made chambers can be used for ambients with n close to 1 but should be avoided for liquids or highly concentrated vapors. The use of condensable organic vapors may also pose a risk of gradual swelling of the measurement chamber which could lead to further distortion of its geometry and accumulation of errors.

High pressure ellipsometry chambers, first introduced by Sirard [22], are in principle following the same guidelines as the chamber shown in Fig. 8.2. They are naturally built in a more robust way from stainless steel and possess thick windows to withstand pressures of several tens of MPa.

A special case of in situ chambers is used in direct studies of operating membranes. Reports of direct studies of membrane structures or, even more so, of membranes under operation are, to date, very rare [24, 25]. We expect, however, that this subfield will grow in the future due to large advantages that ellipsometry can bring to the membrane characterization and fundamental research. One configuration is depicted in Fig. 8.3. The membrane is placed in the plane of beam reflection and sealed from top and bottom with non-swellable O-rings. The feed can be supplied from the top and the permeate withdrawn from the bottom. As such this represents a so called dead-end filtration mode usually suitable for pure components permeation (pure organic solvent permeation) where detrimental effects like concentration polarization can be neglected. The interested reader may find more relevant information on the design

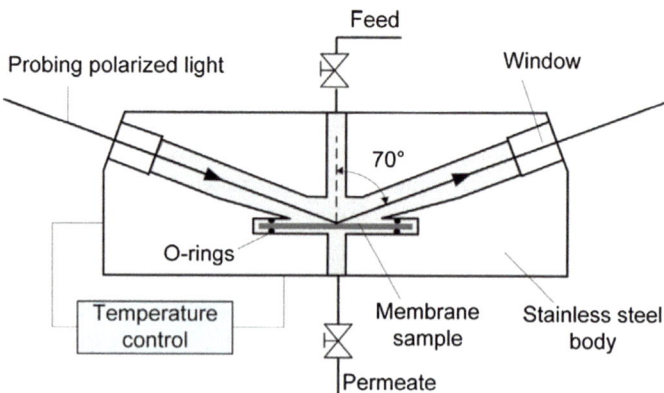

**Fig. 8.3** In-situ ellipsometry high pressure chamber with a possibility of permeation through the membrane (Reprinted with permission from [24]. Copyright Elsevier 2013)

of membrane modules and various non-idealities in multicomponent mixtures in one of the excellent textbooks [2, 26].

## 8.2.2 Optical Modeling for Swelling Films and Membranes

The most dominant way to describe thin membrane films is with uniform, homogenous, fully transparent models (Fig. 8.4a), especially for the determination of equilibrium properties where the possible inhomogeneities or gradients occurring during transient solvent transport have been largely equalized. The large advantage of such description is the usually much smaller number of necessary fit parameters as compared to the number of measured values. This is important as in most in-situ ellipsometry studies only a single AOI is used. With one AOI the detection of gradients (Fig. 8.4b), multilayers (Fig. 8.4c) or anisotropy (Fig. 8.4d) could be hindered, especially for layers below 100–200 nm which are of special interest for membranes. The more complicated optical models shown in Fig. 8.4 have been used only in a very few cases. The graded model was suggestive of the predicted penetrant concentration gradient within a somewhat thicker (much over 1 micron) poly(dimethyl siloxane) (PDMS) membrane [24]. The two-layer model was used to study the anomalous, relaxation-controlled Case II diffusion [23]. The anisotropic model has been found the most appropriate (10 fold reduction of fit error compared to a uniform model) in a study of overshoot swelling behavior in zwitterionic polymers that hold promise in reducing the membrane fouling in nanofiltration [27, 28]. For studies of actual membranes, models similar to the one shown in Fig. 8.4e have been developed. The challenge has been the appropriate description of roughness of porous substrates which produces unwanted light scattering effects [24, 25, 29]. Currently, this challenge is the most critical one to solve in order for ellipsometry

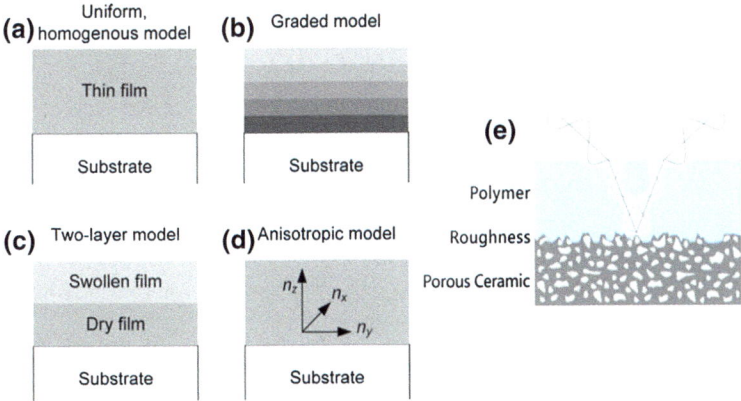

**Fig. 8.4** Optical models used most commonly in membrane-related ellipsometry studies (**a–d**: Reprinted with permission from [20]. Copyright Elsevier 2015; **e**: Reprinted with permission from [29]. Copyright American Chemical Society 2012)

to rapidly progress into the membrane field. One suggestion is to more broadly use relatively smooth substrates like polished $\alpha$-alumina [29], perforated inorganic ones (SmartMembranes GmbH), or polymer-based like polyacrylonitrile or polysulfone. For such substrates the detrimental effects of roughness can be minimized by using focused beams. The polymeric substrates are currently broadly used in membrane research and manufacturing so their use will have an additional benefit of working with realistic systems.

## 8.3 Sorption and Diffusion of Small Molecule Penetrants in Membrane Polymers

### 8.3.1 Swelling and Diffusion Mechanisms in Thin Films

As mentioned before, the starting point for the description of molecular-scale separations is often the solution-diffusion model. Therefore, much attention has been devoted to quantification of the permeability, solubility and diffusivity coefficients. In many cases the knowledge of two of them allows calculation of the third one by solving equation: $P = S \cdot D$. While permeability can only be determined by studying the actual membrane, solubility and diffusivity can be in some cases quantified from characterization of the material alone. The majority of in-situ ellipsometry studies at the edge or directly connected to membrane science have been devoted to determining the solubility, $S$, by investigating the uptake of penetrant molecules in thin films of real or model membrane-related polymer systems. The uptake of penetrant almost universally causes an increase of thickness (swelling) and changes

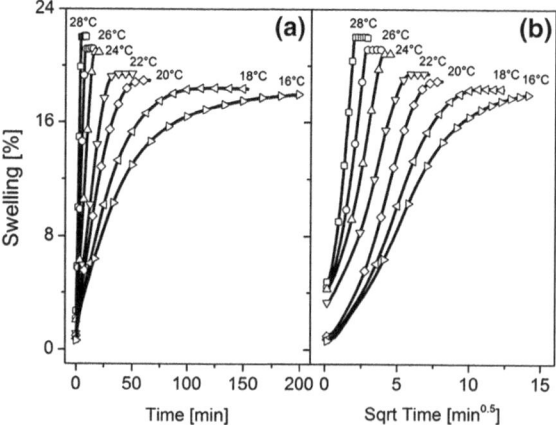

**Fig. 8.5** In-situ ellipsometry used to track changes in swelling magnitude and its mechanism upon traversing glass transition of the polymer-penetrant system. For each curve, the decrease of maximum swelling with reducing temperature is clearly visible together with a change of mechanism from purely relaxation controlled Case II at high temperatures to Fickian relaxation at lower temperatures (Reprinted with permission from [23]. Copyright Elsevier 2013)

of $n$ of the film; either increases or decreases. Both of these effects can be very accurately determined with ellipsometry and then combined with Lorentz-Lorenz, Clausius-Mossotti or effective medium approximation (EMA) relationships [22, 24, 30, 31] to at least approximately calculate the penetrant concentration in films on the order of 100 nm or less. In many studies the determination of equilibrium and dynamics of swelling and $n$ are of sufficient interest and penetrant concentrations are not calculated [23, 27, 28, 32, 33]. One example is shown in Fig. 8.5, where in-situ spectroscopic ellipsometry has been used to study changes in the diffusion mechanism of liquid $n$-hexane through initially glassy polystyrene. In this study both swelling equilibrium and dynamics were studied. The diffusion coefficient (connected with penetrant diffusivity) could be found from the swelling dynamics via semi-empirical Berens-Hopfenberg model [34–36].

Not many other sorption techniques can rival ellipsometry in terms of directly measuring $S$ in thin films with thicknesses in the range of actual membrane selective layers (with quartz crystal microbalance with dissipation, QCM-D as an exception [37]). Studies that consider sorption in bulk samples by gravimetric methods (like magnetic suspension balance or simply by weighing a macroscopic sample after saturation in a solvent) may often be non-representative or rather inaccurate. Even very accurate determination of penetrant concentration inside the bulk polymeric sample does not consider the fact that a thin film due to confinement to the substrate can only swell in one direction perpendicular to the substrate. The 2D confinement in a form of a film has been shown to produce a different Gibbs free energy state as compared with the same concentration during unrestricted three-dimensional swelling [24]. The resulting restriction in polymer chain dilation (additional elastic energy contribution to $\Delta G$) manifests itself at higher solvent uptakes (higher swelling) and reduces the penetrant concentration significantly as compared to bulk swelling. Additionally, weighing samples wet from highly volatile organic solvents has questionable accuracy. The detailed discussion on the nuances of ellipsometry-derived calculations of penetrant concentrations in glassy and rubbery polymer systems has been previously provided [38].

## 8.3.2  Glassy Versus Rubbery Membranes – Polymer Dynamics

In membrane science, key aspects are related with material's ability to discriminate different mixture components. For molecular separations this is usually achieved by trying to design a membrane with maximized permeability differences between the various species. This can be achieved by appropriately manipulating the chemistry of the membrane to affect the solubilities and diffusivities of the respective separated species. The design and characterization of membrane materials has been well described in a book edited by Pinnau et al. [26]. Here, we will only focus on the distinction between the physical state between two important classes of membrane materials: rubbery and glassy polymers. Because the mechanism of small molecule diffusion through a polymer network is directly related with polymer chain mobility rubbers and glasses often vastly differ in terms of diffusivities.

Rubbers used in membrane technology consist of non- or, more often, cross-linked macromolecules whose characteristic chain mobilities are relatively high (relaxation times of much less than 1s). This translates into their glass transition temperatures, $T_g$, being below measurement temperature (usually around room temperature). On the other hand, glassy membrane polymers are frozen-in or kinetically trapped macromolecules, sometimes lightly cross-linked (usually as a result of post-treatment) that show $T_g$-s much above measurement temperatures. Around $T_g$ the characteristic polymer chain dynamics change by many orders of magnitude. In a glassy state the polymer is practically frozen-in with only limited vibrations and rotations of small fragments. In a rubbery state the entire polymer chains remain flexible and able to adjust to external forces almost immediately. The resulting change of slope in the specific volume versus temperature, Fig. 8.6, curve has been extensively studied with ellipsometry [39–43].

The large dynamic difference between glasses and rubbers dictates the way small molecules diffuse through the polymers. While in rubbers the diffusion is usually very efficient and little restricted, like in a liquid phase, in glasses the "hopping" mechanism between neighboring excess free volume elements prevails where the penetrants need to "wait" for the sufficiently large hole to open due to random fluctuations. Glasses, therefore, seem to have certain size-sieving characteristics similar to macroscopically porous filters, although on vastly smaller, molecular scales.

**Fig. 8.6** Various types of volumes distinguished within polymeric materials. The hole free volume is usually thought to be available for host molecules to occupy and diffuse through (Reprinted with permission from [38]. Copyright Elsevier 2014)

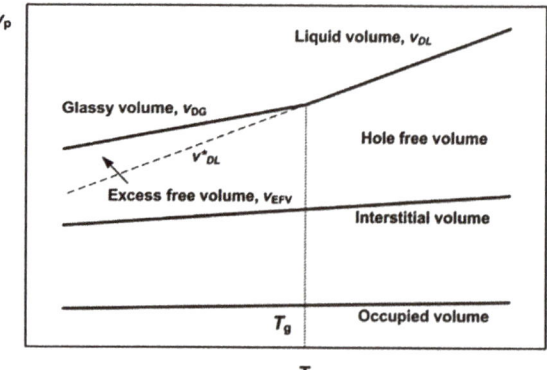

By far the most important rubbery membrane polymer is PDMS, while a range of about ten glassy polymers (mainly polyimides, polyamides, polysulfones or cellulose-variants) have been robustly employed on a large scale. It would be too simple to state that the solubility properties of rubbers and glasses are always different. This heavily depends on the specifics of the given system and the possible non-idealities (like mixed-penetrant phenomena and the possibility of penetrants to severely plasticize the glassy network). However, in general sorption in rubbers is well described by mixing equilibrium liquids (Flory-Huggins, Flory-Rehner theories [44, 45]) and penetrant concentration is linear at low activities (Henry's law) and convex to the X-axis at higher, Fig. 8.7 left. In glasses often the opposite curvature of sorption isotherms is found, concave to the X-axis, Fig. 8.7 right. The sorption in glasses has conceptually been explained by the dual-mode model where the frozen-in

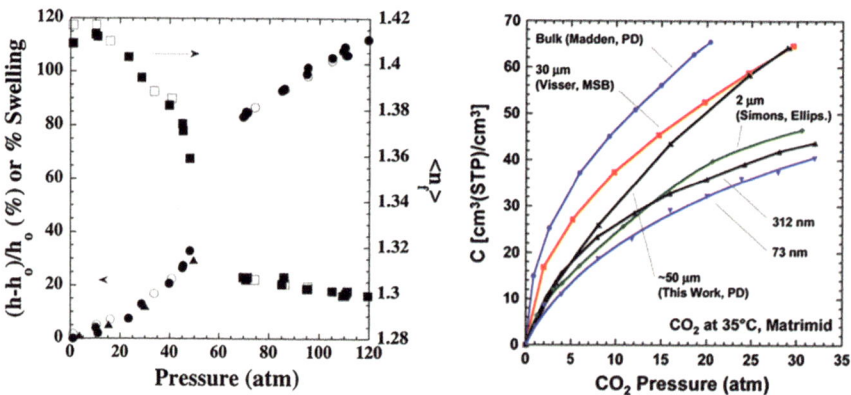

**Fig. 8.7** Various sorption isotherm shapes for sorption in rubbery (left PDMS) and glassy (right Matrimid) polymers in compressed $CO_2$ (left: Reprinted with permission from [14]. Copyright 2001 American Chemical Society, right: Reprinted with permission from [14]. Copyright 2012 American Chemical Society [14, 22])

excess fractional free volume provides sorption sites for incoming molecules. Next to these, the penetrants can also dissolve in the polymer matrix. Although intuitive, the model has several important disadvantages and great progress has recently been achieved in order to augment or replace it [46–49].

### 8.3.3  Membrane Plasticization Studies

The division between rubbery and glassy polymers may become further complicated by the fact that in some cases the initially glassy membrane selective skin may become partly or fully plasticized during operation. An example for the full plasticization are reverse osmosis membranes where the cross-linked, interfacially-polymerized polyamide skin swells significantly in water, fully plasticizes and thus obtains its ability to pass through water while rejecting salts. Ellipsometry has recently been used in one of such studies [50] to quantify the swelling extent as a function of pH.

The issue of plasticization has tremendous significance in membrane technology, particularly in the rapidly growing gas separations (air separation, natural gas production or $CO_2$ capture being the most important branches). Plasticization, defined as a gradual softening and loosening of the rigid glassy network by dissolution of condensable species (mainly hydrocarbons and $CO_2$) progressively reduces membrane selectivities. This in turn leads to problems with maintaining desired stream concentrations. Because the dissolution of penetrants in glasses almost always causes film dilation (swelling) ellipsometry could be used to track it with high precision over long timescales.

One of the first reports on using in-situ ellipsometry to tackle the problem of sorption-induced glassy polymer relaxations, which is closely related to membrane plasticization, has been provided by Wind et al. [51]. The researchers focused on an impact of various annealing procedures on plasticization resistance of polyimide membranes. They also compared thin films (120 nm – measured by ellipsometry) with thick films (40–70 μm – measured by pressure-decay) and were able to show a close connection between plasticization-related increase in permeation and thin film swelling. They were also among the first to suggest that thinner films relax faster than bulk films. This topic has since then been extensively researched by others and ellipsometry has been proven extremely valuable often serving as a method of choice. In fact, the topic of deviations of several important properties of glassy films, such as glass transition temperature and physical aging rates constitutes a separate and growing field of study [12, 14, 16–18]. Some of the recent advances have been described in Chap. 5 of this volume.

Horn et al. have directly focused on the differences in behavior of thin and thick films by utilizing simultaneously the swelling and refractive index information provided by experiments conducted on several glassy membrane materials, including Matrimid, poly(phenylene oxide) and polysulfone [14], Fig. 8.8. Interestingly, the long term high pressure (3.2 MPa) $CO_2$ exposure resulted in first a decrease and then an increase of the refractive index, Fig. 8.8 right. This has led to a conclusion that

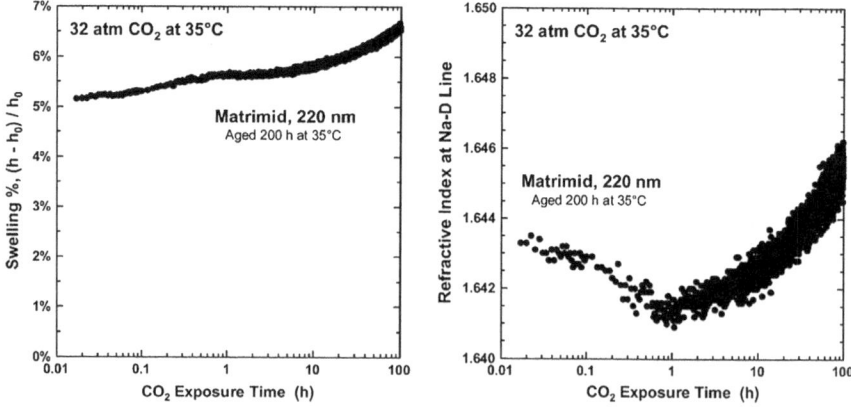

**Fig. 8.8** Swelling (left) and refractive index (right) of a gas separation polymer Matrimid as a function of time under compressed $CO_2$ (Reprinted with permission from [14]. Copyright 2012 American Chemical Society)

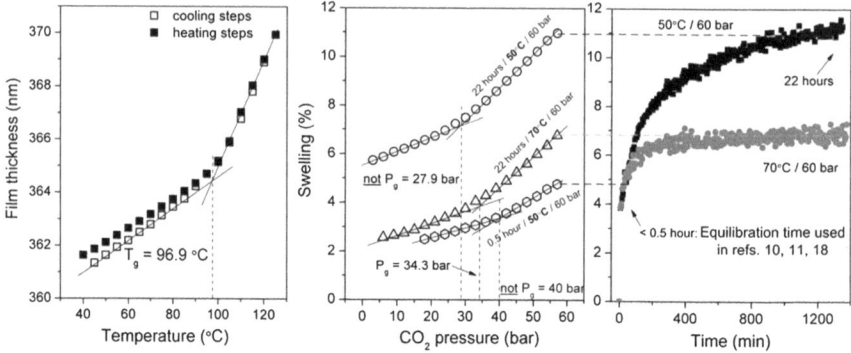

**Fig. 8.9** Comparison of isobaric glass transition ($T_g$) and isothermal, high pressure $CO_2$-induced glass transition ($P_g$) and the impact of history on the apparent position of slope change in the isotherms (Reprinted with permission from [52]. Copyright 2014 American Chemical Society)

plasticization (corresponding to a reduction of ref. index by optically diluting the glassy material) competes with physical aging. The physical aging causes a gradual collapse of the non-equilibrium excess free volume characteristic to a glassy state (corresponding to ref. index increase by matrix densification). Physical aging has been found to dominate over long timescales in thin Matrimid.

Because of its extremely high sensitivity, ellipsometry has recently contributed to observe details of the penetrant-induced glass transition, $P_g$ (analogous to $T_g$), with respect to its dynamics in the vicinity of the transition [52] (Fig. 8.9) and film thickness dependence [53, 54] (Fig. 8.10). As seen in Fig. 8.9, the isobaric (a) and isothermal (b) glass transitions, $T_g$ and $P_g$, respectively, are characterized by a change of the thermal expansion coefficient or in slope of the sorption isotherms.

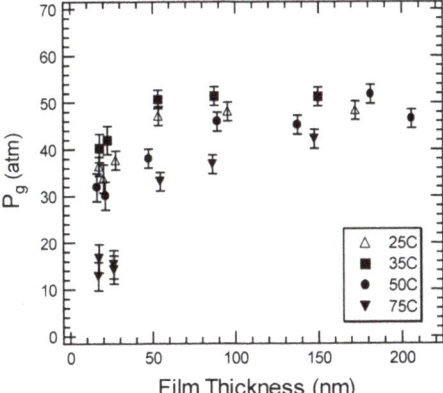

**Fig. 8.10** Impact of film thickness and temperature on the reduction of apparent penetrant induced glass transition in ultra-thin polystyrene exposed to compressed $CO_2$, as measured with high pressure spectroscopic ellipsometry (Reprinted with permission from [54]. Copyright 2004 American Chemical Society)

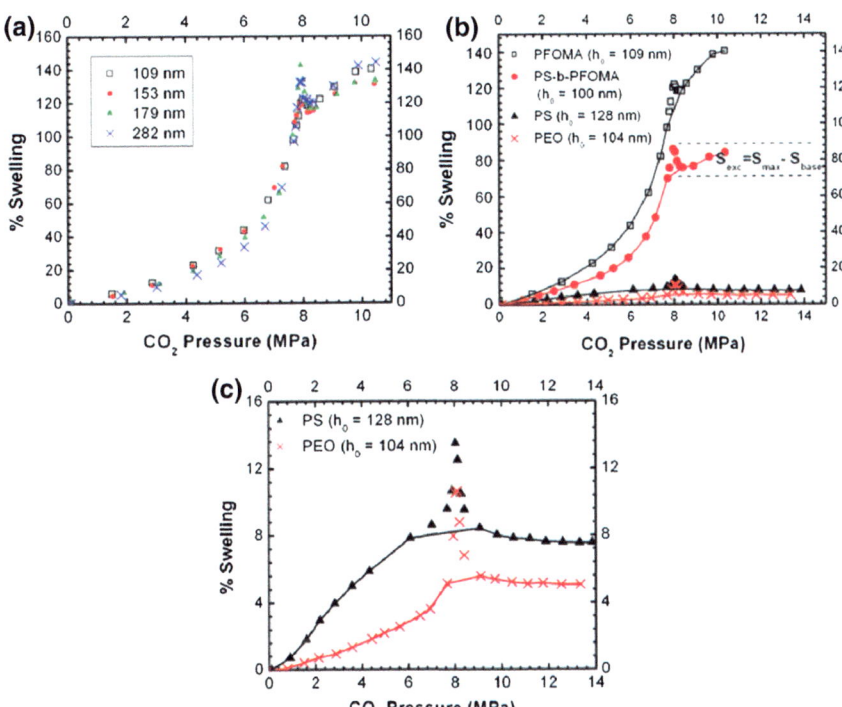

**Fig. 8.11** Anomalous maxima in swelling of several polymers studied across a broad range of pressures in supercritical $CO_2$ (Reprinted with permission from [55]. Copyright Wiley 2007)

However, the vital difference is that $P_g$ shows a much more complex dependence on the total polymer matrix dilation (swelling) as well as time (or sample history). These results have shed new light on the challenge of membrane plasticization and may be used as guidelines for the process design given a particular polymer-penetrant pair. For instance, it might be beneficial to operate membranes much below the pressure range where significant, progressive sorptive relaxations occurs, as seen in Fig. 8.9c. Figure 8.10 shows a result of an important study where the impact of film thickness on the $P_g$ below 100 nm has been detected. This is of paramount importance as one strategy to make membranes more efficient is to reduce skin thicknesses much below 100 nm.

High pressure ellipsometry has also been used to shed light on the peculiar phenomenon of anomalous swelling maxima in supercritical $CO_2$ (Sc-$CO_2$), Fig. 8.11. The high sensitivity of ellipsometry has enabled detection of these maxima in films down to about 100 nm thickness. In their detailed study Li et al. have investigated films of several homopolymers and block co-polymers [55]. They concluded, that the anomalous maxima could not be solely explained by interfacial excess of adsorbed gas. The increased swelling of the whole film must have occurred.

## 8.4   New Generation of Membrane Materials–Metal Organic Frameworks (MOFs) and Polymers of Intrinsic Microporosity (PIMs)

In the last two decades there have been interesting developments in entirely new classes of materials for molecular separations. Two of them are metal organic frameworks (MOFs [56–59]) and polymers of intrinsic microporosity (PIMs [60–65]). MOFs are crystalline materials consisting of coordination bonds between metallic cations (usually transition metals) and multifunctional organic linkers. The almost unlimited possibilities in fine tuning the chemistry and huge opportunities not only in membranes, but also catalysis or gas storage (incl. for $H_2$- or $CH_4$- storage tanks for vehicles) have made the field of MOFs explode in the recent decades. Due to crystalline nature of these materials fabrication of thin, homogenous films is not straightforward. Thus the application of ellipsometry in this field has been limited. One example of a high pressure ellipsometry study of a MOF-like material, based on zeolitic imidazole framework (ZIF-8) has been presented by Cookney et al. [66]. The authors have been able to fabricate continuous ultra-thin films by step-by-step coating with only a single cycle (Fig. 8.12a). The films were shown to produce open, microporous morphologies as evidenced with low refractive index. They could also absorb $CO_2$ under sub-ambient pressures showing almost no swelling (Fig. 8.12b). Contrary to the usually produced large (>micron) crystallites which inherently possess defects or grain boundaries much larger than the separated molecules these nanofilms could potentially be selective when used as membranes.

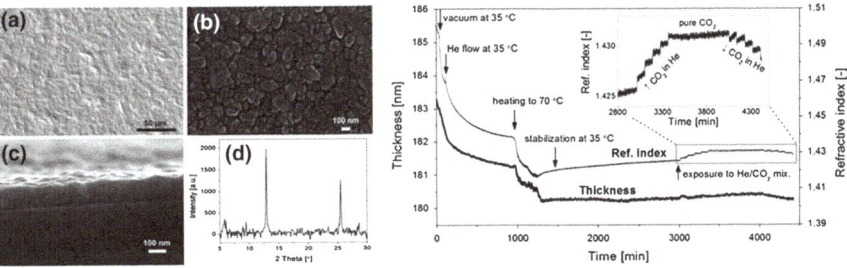

**Fig. 8.12** Left panel: structure of semi-crystalline ZIF-8 nanofilms right panel: in-situ ellipsometry data recorded during first drying with heating and then $CO_2$ sorption and desorption cycles (Reprinted with permission from [66]. Copyright RSC 2014, Published by The Royal Society of Chemistry)

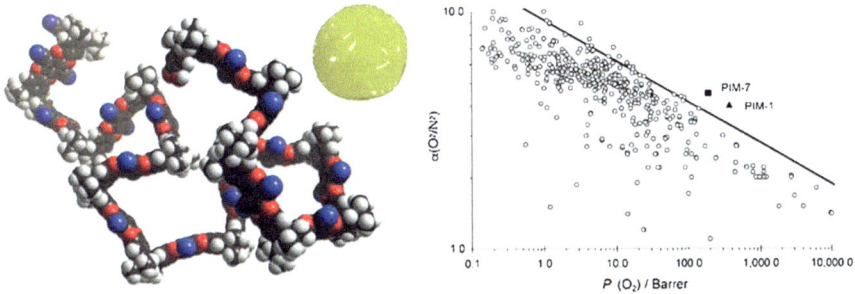

**Fig. 8.13** Left panel: structure of PIM-1 with a picture of a free-standing film, right panel: PIM-1 and PIM-7 show permeabilities and selectivities beyond the upper-bound trade-off relationship (Reprinted with permission from [67]. Copyright RSC 2006, Published by The Royal Society of Chemistry)

The ability to form high quality, homogenous thin films, a prerequisite for high fidelity ellipsometry analysis, is found for the second class of new membrane materials, PIMs. Because of their easy solution-processability and high microporosities these materials hold great promise in large scale membrane applications. A structure of the most studied PIM-1 is shown in Fig. 8.13 left panel. The design principle is introduction of rigid and contorted chains which hinder efficient molecular packing and produce large amount of excess free volumes or microporosities often in the range of 25%. At these high volume fractions the microporosity remains interconnected and produces very efficient channels for molecular diffusion giving rise to extremely large permeabilities. At the same time due to pore size distribution characteristics the PIMs remain selective. A combination of these features allowed this class of materials to break the limits set by the upper-bound trade-off lines, Fig. 8.13 right panel [67].

Swelling of two ultra-thin (50–70 nm), silicon wafer supported PIMs in six organic liquids and water has been studied with in-situ spectroscopic ellipsometry by Ogieglo et al. [31]. The initially ultra-rigid glassy materials have been shown to readily

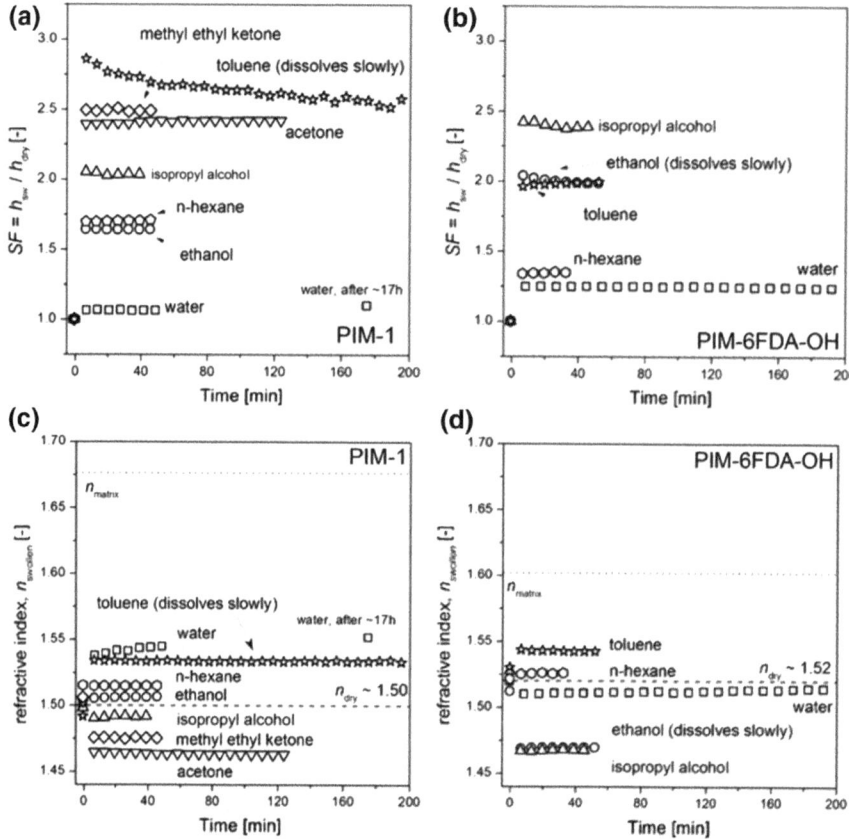

**Fig. 8.14** Swelling and refractive index of two PIMs, PIM-1 and PIM-6FDA-OH, when exposed to six organic solvents and water. All polymers swell significantly in most solvents. The refractive index shows complex behavior with values lower or higher, depending on polymer-solvent pair, than the dry refractive index (Reprinted with permission from [31]. Copyright 2016 American Chemical Society)

swell in organic solvents up to three times and as a result of serious plasticization almost immediately transition into rubbery state. Interplay between sorption into the microporosity and dilation of matrix could be detected in the combined behavior of swelling and refractive index. For the higher swelling solvents (toluene, methyl ethyl ketone) the refractive index has been found to drop below the refractive index of a dry material. This behavior is typical for rubbery polymers where the dissolution of a liquid, low index penetrant inside the liquid-like polymer matrix leads to a drop of the overall optical density. For the lesser swelling solvents (ethanol, *n*-hexane) the refractive index of still largely swollen (70%) films surprisingly increased. This could only be explained by a significant degree of microporosity occupation proving its accessibility to the penetrant molecules. The authors have used a modified effective medium approximation (EMA [38, 68]) to resolve the swelling and microporosity occupation and have provided relationships between penetrant volume fractions and swelling in microporous polymers (Fig. 8.14).

The possibility of PIM-1 microporosity filling by water molecules has also been investigated with ellipsometry by Seok et al. [69]. The authors could exploit this property and have demonstrated outstanding water-vanadium species selectivity in PIM-1 which served as an efficient battery membrane.

## 8.5   Towards Ellipsometry Applied Directly to Membranes

Membranes for molecular-based separations rely on asymmetric geometry where a thin dense film is supported on a porous substrate. The separation performance is mainly dictated by the dense film (also known as membrane skin) and the support needs to provide mechanical and chemical stability as well as low transport resistance. The vast majority of membrane-related ellipsometry studies, including those described in earlier sections, use an approach where the thin film is isolated on a well defined and non-porous substrate (silicon wafer). Such a configuration does not mimic the actual membrane exactly because the perpendicular steady state penetrant transport through the layer is not possible. For several reasons application of ellipsometry to films on porous substrates is challenging. Firstly, for quantification of film's properties ellipsometry requires coherent light reflection from both the free surface and the interface of the film with the underlying substrate. Pores with sizes on the order of light wavelength efficiently scatter light and practically prevent the second reflection. A typical porous substrate may also possess larger than wavelength roughness which scatters light. During membrane fabrication often the not yet solidified skin material flows into the support making the interface ill-defined and producing varying depths of pore intrusion. This further complicates detection of clear optical interfaces. Lastly, the refractive index contrast between typical polymeric skins and polymeric or even inorganic (alumina) substrates is rather poor making the amplitude of psi and delta oscillations vanishingly small.

The first ellipsometry study to directly quantify penetrant sorption in a membrane has been performed by Benes et al. [71]. The authors could determine the porosities and thicknesses of thin silica (15–25%, $\sim$70 nm) and $\gamma$-alumina (51%, $\sim$1.6 $\mu$m) layers, Fig. 8.15 left panel. The thin silica layer has shown a larger heat of sorption for $CO_2$ than bulk silica suggesting that the pores were smaller than in the unsupported material. The authors concluded also that ellipsometry could be a promising tool for in-situ membrane characterization as the technique has been sensitive to penetrant sorption, Fig. 8.15 right panel.

The detailed characterization of a polished $\alpha$-alumina membrane substrate has been performed by Ogieglo et al. [29]. The surface polishing, and the related control of roughness, has been shown critical for the coherent reflection of polarized probing light. The roughness could be correctly described by a graded layer with varying ratio between the dense material and void. Based on the supporting characterization with

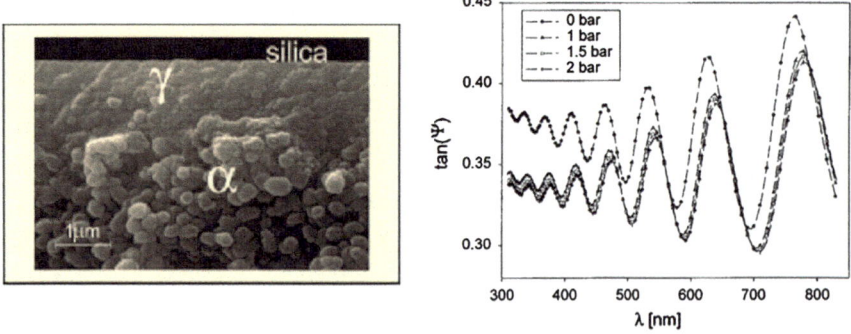

**Fig. 8.15** Left panel: structure of a multilayer inorganic composite membrane analyzed with in-situ spectroscopic ellipsometry, right panel: response of the membrane to adsorbing $CO_2$ (Reprinted with permission from [70]. Copyright Wiley 2001)

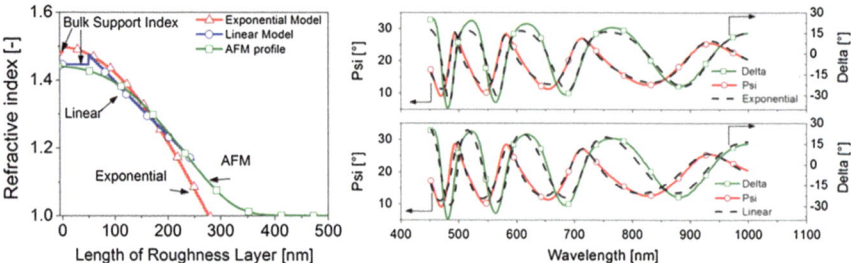

**Fig. 8.16** Left panel: Linear and Exponential optical model-derived refractive index profiles against the AFM-derived profile for $\alpha$-alumina membrane support, right panel: 1.1 μm polysulfone skin modeled with both optical models on top of the porous support (Reprinted with permission from [29]. Copyright 2012 American Chemical Society)

atomic force microscopy (AM) and mercury porosimetry the graded layers have been described with linear or exponential optical models with their own advantages and disadvantages, Fig. 8.16 left panel. The subsequent deposition of a relatively thick, 1.1 μm, polysulfone skin could be well described with multilayer optical models, Fig. 8.16 right panel, thus enabling realistic studies of actual membranes under operation.

The authors have built upon the previous study and applied the developed methodology in the first, to date, simultaneous ellipsometry and permeation fundamental study of a PDMS-based nanofiltration membrane [24]. Swelling of thin (~1 μm) PDMS films has been found to depend on cross-linker content. The more appropriate thermodynamic description has been found with the Flory-Rehner model which includes correction for the elastic energy term in the Gibbs free energy of the swollen films. Confined films dilated significantly more due to only one possible swelling direction (perpendicular to the substrate) as compared with un-confined, bulk films. Under hydrostatic equilibrium (pressure on the feed side equal to pressure on the

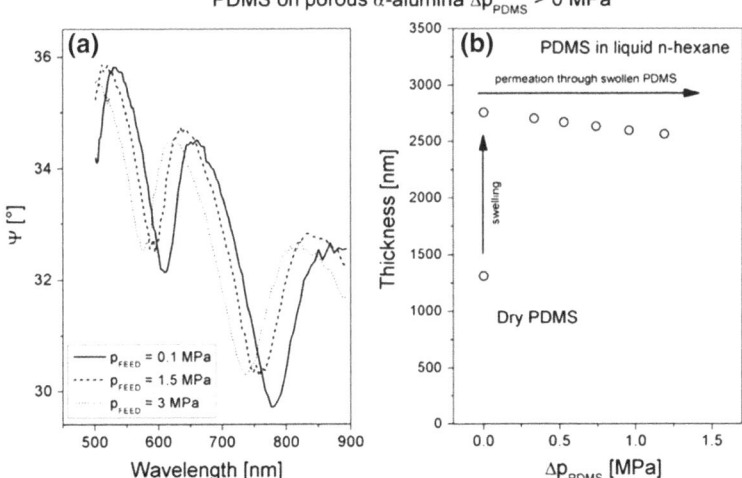

**Fig. 8.17** **a** psi signal from a thin PDMS film supported by $\alpha$-alumina support as a function of increasing driving force **b** the extracted swelling reduction due to rising driving force for solvent permeation (Reprinted with permission from [24]. Copyright Elsevier 2014)

permeate side) no flow of solvent occurred and the swelling of PDMS did not depend on the surrounding pressure. Therefore, the molar volumes of liquid solvent inside the membrane and in the liquid phase have remained equal. When the diving force for the solvent flow was provided (pressure feed > pressure permeate) the swelling has progressively reduced with increasing pressure difference in excellent agreement with the predictions of the solution-diffusion model (Fig. 8.17).

The high pressure spectroscopic ellpsometry has been used by Raaijmakers et al. to study sorption-induced swelling of interfacially-polymerized poly(POSS-imide) membranes synthesized on alumina-based supports [25]. This material represents a hybrid, organic-inorganic giant network consisting of alternating polyhedral oligomeric silsesquioxane and imide groups, Fig. 8.18 left panel. The sorption behavior for the hybrids has been found different from conventional polyimides, Fig. 8.18, right panel. The dilation depended strongly on the linker content and the sorption of both $CO_2$ and $CH_4$ has been found very high due to the large amounts of open space in the networks. Because of the gradual filling of the microporosity at higher pressures the sorbing gases have shown increasing apparent molar volumes that exceeded that of a liquid phase. Further optimization of the materials for $CO_2$-related applications has been indicated possible.

Recently, a mapping ellipsometry method has been presented for non-invasive direct imaging of films deposited on porous substrates immersed in liquids [72], Fig. 8.19. The authors have applied a focused beam (300 $\mu$m short axis) combined with sample chamber translation to map the membrane surface in the presence of a liquid medium. In contrast to the previously described studies, here for the first time a porous polymeric support (polyacrylonitrile, PAN) could be used. The

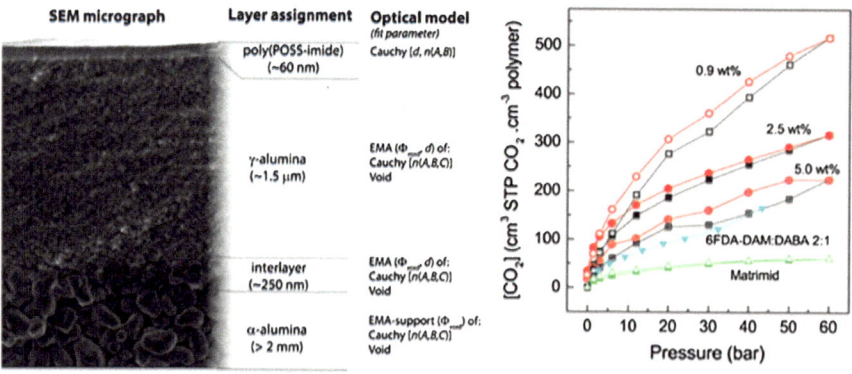

**Fig. 8.18** Left panel: SEM micrograph of the multilayer structure with a thin poly(POSS-imide) separation layer and the parameters of the used optical model, right panel: concentration of the dissolved $CO_2$ in poly(POSS-imide) layers with varying weight fractions of the organic linkers compared with two standard polyimides (Reprinted with permission from [25]. Copyright 2015 American Chemical Society)

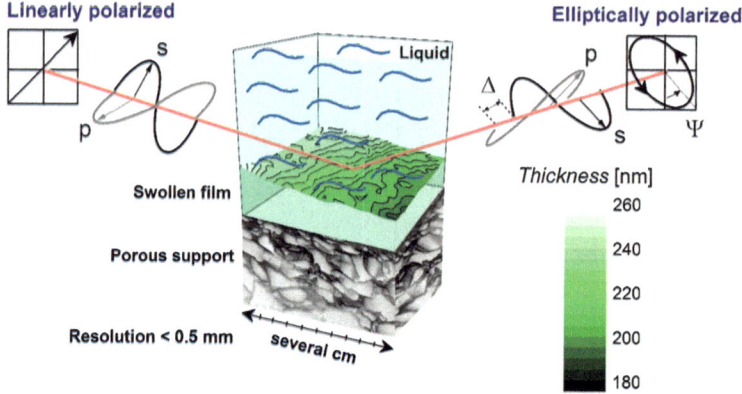

**Fig. 8.19** Ellipsometry imaging method developed to study liquid-immersed thin films on top of porous substrates [72]

characterization of a bare PAN indicated, that its roughness features could be neglected in the optical modeling and a single interface of an EMA-modeled material consisting of dense PAN and void could be reliably used both in dry and wet conditions. The subsequently deposited thin films (200–400 nm) could be well resolved and the coating polymers have been shown to intrude the substrate to various extents depending on the coating method (spin-, dip-coating or floating) or the coating solution concentration. For the first time the pore intrusion within the porous substrate could be measured without the need for membrane destruction. Upon immersion in water the swelling and changes of the refractive index in the system could be determined and laterally resolved over an area of up to $1\,cm^2$. This study has the potential to boost the usefulness of in-situ focused-beam mapping ellipsometry in both the fundamental and application-oriented membrane studies.

## 8.6  Outlook

In-situ spectroscopic ellipsometry entered the field of membrane science in the early 2000s. In the majority of cases the technique has been utilized to study penetrant interactions with thin, dense films supported on dense substrates (silicon wafers, glass slides). In such a way the important membrane properties, such as swelling, diffusion rates and mechanisms or plasticization could be studied in various types of fluids including liquids, vapors or compressed gases. This has generated the much needed understanding on the in-situ performance of ultra-thin films that are pertinent to membrane applications. The future might see a gradual extension of in-situ ellipsometry to study actual membrane systems which involve utilization of porous substrates. Overcoming challenges related with rough, light scattering porous substrates has recently been shown feasible and new possibilities have emerged.

## References

1. W.L. McCabe, J.C. Smith, P. Harriott, *Unit Operations of Chemical Engineering* (McGraw-Hill, New York, 1993)
2. R.W. Baker, *Membrane Technology and Applications* (Wiley, New York, 2004)
3. J.G. Wijmans, R.W. Baker, J. Memb. Sci. **107**, 1–21 (1995)
4. I. Pinnau, J. Memb. Sci. **37**, 81–88 (1988)
5. R. Bounaceur, N. Lape, D. Roizard, C. Vallieres, E. Favre, Energy **31**, 2556–2570 (2006)
6. L.M. Robeson, J. Memb. Sci. **320**, 390–400 (2008)
7. A.O. Yazaydin, R.Q. Snurr, T.-H. Park, K. Koh, J. Liu, M.D. Levan, A.I. Benin, P. Jakubczak, M. Lanuza, D.B. Galloway, J.J. Low, R.R. Willis, J. Am. Chem. Soc. **131**, 18198–9 (2009)
8. J. Yang, 782–835 (2012)
9. R.B. Getman, Y. Bae, C.E. Wilmer, R.Q. Snurr, Chem. Rev. **112**, 703–23 (2012)
10. D.E. Jiang, V.R. Cooper, S. Dai, Nano Lett. **9**, 4019–4024 (2009)
11. J. Spiece, D.E. Martinez-Tong, M. Sferrazza, A. Nogales, S. Napolitano, Soft Matter. **11**, 6179–6186 (2015)
12. J.L. Keddie, R.A.L. Jones, R.A. Cory, Faraday Discuss **98**, 219–230 (1994)
13. R.a L. Jones, Curr. Opin. Colloid Interface Sci. **4**, 153–158 (1999)
14. N.R. Horn, D.R. Paul, Macromolecules **45**, 2820–2834 (2012)
15. N.R. Horn, D.R. Paul, Polymer **52**, 1619–1627 (2011)
16. M. Erber, M. Tress, E.U. Mapesa, A. Serghei, K.-J. Eichhorn, B. Voit, F. Kremer, Macromolecules **43**, 7729–7733 (2010)
17. M. Erber, U. Georgi, J. Müller, K.-J. Eichhorn, B. Voit, Eur. Polym. J. **46**, 2240–2246 (2010)
18. M. Erber, A. Khalyavina, K.-J. Eichhorn, B.I. Voit, Polymer **51**, 129–135 (2010)
19. H. Fujiwara, *Spectroscopic Ellipsometry Principles and Applications* (Wiley, NJ, 2007)
20. W. Ogieglo, H. Wormeester, K.-J. Eichhorn, M. Wessling, N.E. Benes, Prog. Polym. Sci. **42**, 42–78 (2015)

21. K. Tempelman, E.J. Kappert, M.J.T. Raaijmakers, H. Wormeester, N.E. Benes, Surf. Interface Anal. **49**, 538–547 (2017)
22. S.M. Sirard, P.F. Green, K.P. Johnston, J. Phys. Chem. B **105**, 766–772 (2001)
23. W. Ogieglo, H. Wormeester, M. Wessling, N.E. Benes, Polymer **54**, 341–348 (2013)
24. W. Ogieglo, H. van der Werf, K. Tempelman, H. Wormeester, M. Wessling, A. Nijmeijer, N.E. Benes, J. Memb. Sci. **437**, 313–323 (2013)
25. M.J.T. Raaijmakers, W. Ogieglo, M. Wiese, M. Wessling, A. Nijmeijer, N.E. Benes, A.C.S. Appl, Mater. Interfaces **7**, 26977–26988 (2015)
26. Y. Yampolskii, I. Pinnau, B.D. Freeman, *Materials Science of Membranes for Gas and Vapor Separation* (Wiley, New York, 2006)
27. J. de Grooth, W. Ogieglo, W.M. de Vos, M. Gironès, K. Nijmeijer, N.E. Benes, Eur. Polym. J. **55**, 57–65 (2014)
28. W. Ogieglo, J. de Grooth, H. Wormeester, M. Wessling, K. Nijmeijer, N.E. Benes, Thin Solid Films **545**, 320–326 (2013)
29. W. Ogieglo, H. Wormeester, M. Wessling, N.E. Benes, ACS Appl. Mater. Interfaces. **4**, 935–943 (2012)
30. S.M. Sirard, H. Castellanos, P.F. Green, K.P. Johnston, J. Supercrit. Fluids **32**, 265–273 (2004)
31. W. Ogieglo, B. Ghanem, X. Ma, I. Pinnau, M. Wessling, J. Phys. Chem. B (2016)
32. W. Ogieglo, L. Upadhyaya, M. Wessling, A. Nijmeijer, N.E. Benes, J. Memb. Sci. **464**, 80–85 (2014)
33. W. Ogieglo, H. Wormeester, M. Wessling, N.E. Benes, Macromol. Chem. Phys. **214**, 2480–2488 (2013)
34. A.R. Berens, Polymer **18**, 697–704 (1977)
35. A.R. Berens, H.B. Hopfenberg, Polymer **19**, 489–496 (1978)
36. J. Potreck, F. Uyar, H. Sijbesma, K. Nijmeijer, D. Stamatialis, M. Wessling, Phys. Chem. Chem. Phys. **11**, 298–308 (2009)
37. M. Koenig, T. Kasputis, D. Schmidt, K.B. Rodenhausen, K.J. Eichhorn, A.K. Pannier, M. Schubert, M. Stamm, P. Uhlmann, Anal. Bioanal. Chem. (2014)
38. W. Ogieglo, H. Wormeester, M. Wessling, N.E. Benes, Polymer **55**, 1737–1744 (2014)
39. S. Kim, S.a Hewlett, C.B. Roth, J.M. Torkelson, Eur. Phys. J. E **30**, 83–92 (2009)
40. C.B. Roth, J.R. Dutcher, J. Electroanal. Chem. **584**, 13–22 (2005)
41. E.A. Baker, P. Rittigstein, J.M. Torkelson, C.B. Roth, J. Polym. Sci. Part B Polym. Phys. **47**, 2509–2519 (2009)
42. J.L. Keddie, Curr. Opin. Colloid Interface Sci. **6**, 102–110 (2001)
43. H. Richardson, I. López-García, M. Sferrazza, J.L. Keddie, Phys. Rev. E - Stat. Nonlinear, Soft Matter Phys. **70**, 51805 (2004)
44. P.J. Flory, M. Volkenstein, Biopolymers **8**, 699–700 (1969)
45. P.J. Flory, J.J. Rehner, J. Chem. Phys. **11**, 521–526 (1943)
46. I.C. Sanchez, R.H. Lacombe, Macromolecules **11**, 1145–1156 (1978)
47. M. Minelli, S. Campagnoli, M.G. De Angelis, F. Doghieri, G.C. Sarti, Macromolecules **44**, 4852–4862 (2011)
48. M. Galizia, K.A. Stevens, Z.P. Smith, D.R. Paul, B.D. Freeman (2016)
49. M.Z. Hossain, A.S. Teja, J. Supercrit. Fluids **96**, 313–323 (2015)
50. K. Kezia, J. Lee, W. Ogieglo, A. Hill, N.E. Benes, S.E. Kentish, J. Memb. Sci. **459**, 197–206 (2014)
51. J.D. Wind, S.M. Sirard, D.R. Paul, P.F. Green, K.P. Johnston, W.J. Koros, Macromolecules **36**, 6442–6448 (2003)
52. W. Ogieglo, M. Wessling, N.E. Benes, Macromolecules **47**, 3654–3660 (2014)
53. J.Q. Pham, S.M. Sirard, K.P. Johnston, P.F. Green, Phys. Rev. Lett. **91**, 175503 (2003)
54. J.Q. Pham, K.P. Johnston, P.F. Green, J. Phys. Chem. B **108**, 3457–3461 (2004)
55. Y. Li, E.J. Park, K.T. Lim, K.P. Johnston, P.F. Green, J. Polym. Sci. Part B Polym. Phys. **45**, 1313–1324 (2007)
56. N. Nijem, H. Wu, P. Canepa, A. Marti, K.J. Balkus, T. Thonhauser, J. Li, Y.J. Chabal, J. Am. Chem. Soc. **134**, 15201–15204 (2012)

57. G. Lu, O.K. Farha, W. Zhang, F. Huo, J.T. Hupp, Adv. Mater. **24**, 3970–3974 (2012)
58. M.R. Khdhayyer, E. Esposito, A. Fuoco, M. Monteleone, L. Giorno, J.C. Jansen, M.P. Attfield, P.M. Budd, Sep. Purif. Technol. **173**, 304–313 (2017)
59. C. Liu, C. Zeng, T.Y. Luo, A.D. Merg, R. Jin, N.L. Rosi, J. Am. Chem. Soc. **138**, 12045–12048 (2016)
60. S. Harms, K. Rätzke, F. Faupel, N. Chaukura, P.M. Budd, W. Egger, L. Ravelli, J. Adhes. **88**, 608–619 (2012)
61. N.B. McKeown, P.M. Budd, Macromolecules **43**, 5163–5176 (2010)
62. A.F. Bushell, M.P. Attfield, C.R. Mason, P.M. Budd, Y. Yampolskii, L. Starannikova, A. Rebrov, F. Bazzarelli, P. Bernardo, J. Carolus Jansen, M. Lanč, K. Friess, V. Shantarovich, V. Gustov, V. Isaeva, J. Memb. Sci. (2013)
63. T. Emmler, K. Heinrich, D. Fritsch, P.M. Budd, N. Chaukura, D. Ehlers, K. Rätzke, F. Faupel, Macromolecules **43**, 6075–6084 (2010)
64. N. Alaslai, B. Ghanem, F. Alghunaimi, I. Pinnau, Polymer **91**, 128–135 (2016)
65. B.S. Ghanem, R. Swaidan, E. Litwiller, I. Pinnau, Adv. Mater. **26**, 3688–3692 (2014)
66. J. Cookney, W. Ogieglo, P. Hrabanek, I. Vankelecom, V. Fila, N.E. Benes, Chem. Commun. **50**, 11698–11700 (2014)
67. N.B. McKeown, P.M. Budd, Chem. Soc. Rev. **35**, 675–683 (2006)
68. D.A.G. Bruggeman, Ann. Phys. **24**, 636–664 (1935)
69. I.S. Chae, T. Luo, G.H. Moon, W. Ogieglo, Y.S. Kang, M. Wessling, Adv. Energy Mater. **6**, 1600517 (2016)
70. H. Wormeester, N.E. Benes, G.I. Spijksma, H. Verweij, B. Poelsema, Thin Solid Films **455– 456**, 747–751 (2004)
71. N.E. Benes, G. Spijksma, H. Verweij, H. Wormeester, B. Poelsema, AIChE J. **47**, 1212–1218 (2001)
72. W. Ogieglo, I. Pinnau, M. Wessling, J. Membr. Sci. **546**, 206–214 (2018). https://doi.org/10.1016/j.memsci2017.10.027

# Part III
# Nanostructured Surfaces
# and Organic/Inorganic Hybrids

# Chapter 9
# Systems of Nanoparticles with SAMs and Polymers

Thomas W. H. Oates

**Abstract** This chapter reviews the use of spectroscopic ellipsometry (SE) as a characterization tool for nanoparticle-polymer and nanoparticle-SAM hybrids. The development of such materials is based on the drive toward technological applications of new functional organic materials in solar cells, flat screen displays, sensors and organic electronics. For many of these application the optical properties of the materials are of critical importance for the device operation. In this respect, an accurate and complete determination of the frequency-dependent complex dielectric function, $\varepsilon(\omega) = \varepsilon' + i\varepsilon''$, of the materials over a wide spectral range is the primary goal of SE characterization. The major focus of the chapter will be to present optical models that are needed to analyze the data; specifically to develop models that describe the effective dielectric function of a film of NPs supported by, or embedded in, an organic matrix. Starting with the Mie solution to Maxwell's equations, examples of various nanoparticle scattering cross-sections are presented to show the influence of the particle size and material properties. Modeling composites then requires making the step from individual NPs to arrays and composites by using the effective medium approximation. Finally the origin of anisotropy will be described and models for the dielectric tensor elements presented. Examples from the literature will be referred to throughout.

## 9.1 Introduction

Shrinking solid structures to nanometer-scale dimensions results in interesting new physical phenomena. Optical, catalytic and thermal properties are often altered at the nanoscale. These changes are caused by, among other things, increases in the radius-to-wavelength-of-light ratio and the surface-to-volume ratio. A well documented example is the observation of colorful localized plasmon resonances in noble metal

T. W. H. Oates (✉)
Leibniz-Institut für Analytische Wissenschaften – ISAS – e.V., Schwarzschildstr. 8, 12489 Berlin, Germany
e-mail: plaasmatrino@gmail.com

© Springer International Publishing AG, part of Springer Nature 2018    199
K. Hinrichs and K.-J. Eichhorn (eds.), *Ellipsometry of Functional Organic Surfaces and Films*, Springer Series in Surface Sciences 52,
https://doi.org/10.1007/978-3-319-75895-4_9

nanoparticles (NPs) due to confinement of the (usually "free") conduction electrons. Alternatively, the number of atoms in the particle may become small enough that quantum effects may be observed, as in semiconductor "quantum dots" (QDs).

Along with the size, shape and material of which the NP is composed, the dielectric properties of the medium surrounding the NP have a strong influence on the optical response. Increasing the polarizability of the embedding medium will red-shift the plasmon resonance in metal NPs. Organic materials such as polymers and self-assembled monolayers (SAMs), with their wide range of functional properties, make for interesting materials to explore as a supporting matrix. For example, switching the polarizability by temperature-induced phase changes or optical non-linearity allows functional optical switching of organic-NP composites, with great potential in photonics. Alternatively, in the infrared region where organic materials display strong molecular absorption fingerprints, plasmonic NPs are responsible for surface-enhanced infrared absorption (SEIRA), with potential applications in biosensing.

Spectroscopic ellipsometry is ideal for characterizing the linear optical properties of NP/organic systems, in the basic instance providing the effective dielectric function and thickness of thin films and multilayers. Real time ellipsometry measurements are now a standard option on most commercial instruments and provide a wealth of data concerning time-dependent processes down to the millisecond scale, especially for temperature-dependent material properties and reaction kinetics using electrochemistry. There are numerous competing and complementary methods to characterize the linear optical properties of polymer/NP and SAM/NP complexes, the comparative advantages of which were reviewed previously in an article specifically focused on characterizing AuNP/SAM composites [1]. We concentrate here on the ellipsometric techniques, with a special mention of the combination of SEIRA and ellipsometry.

The measurement fundamentals of ellipsometry were presented in the introductory chapter. UV-visible (UV-Vis), Fourier-transform infrared (FTIR), imaging ellipsometry and dynamic in situ ellipsometry are all utilized in the characterization of NP-composites. Of note is the combination of multiple techniques during dynamic measurements. By increasing the information about a sample during processing the uncertainties in the ellipsometric modeling can be substantially reduced. For example, combining microbeam grazing incidence small angle X-ray scattering (μGISAXS) and imaging ellipsometry, Korstgens et al. introduced a new versatile tool for the characterization of nanostructures [2] and demonstrated their technique by characterizing monodisperse colloidal polystyrene nanospheres on a rough solid support which were subsequently coated with a diblock copolymer film, including real time measurements. A similar development is the combination of quartz crystal microbalance measurements with dynamic ellipsometry [3], discussed further in Chap. 17 of this book.

All forms of ellipsometry are a combination of measurement and modeling. The measured ellipsometric angles, $\Psi$ and $\Delta$, do not convey physical information in themselves, although during real-time measurements they do provide feedback with sub-nanometer sensitivity that *something* is changing. What is actually changing, or what the physically relevant material properties are, must be determined by comparing or fitting the measured data with optical models. Only then can information such as the film thickness, effective refractive index and particle fill-factor be determined. The models used in ellipsometry are usually based on the far field reflection from a homogenized material with defined dielectric function or tensor. The local interactions of nanostructures are thus averaged over the entire material. In the following section we will discuss the optical properties of individual nanoparticles in the Mie model, which includes the effect of retardation due to differences in the phase of light at different points on the particle. This will be instructive in considering the applicability of the widely used effective medium models in which the phase retardation is ignored in the long-wavelength limit.

## 9.2 Modeling Nanoparticles: The Mie–Lorenz Solution

Modeling the optical properties of a nanoparticle requires solving Maxwell's equations with appropriate boundary conditions. For the case of an *isolated* nanoparticle this reduces to the problem of determining the absorption and scattering of a sphere by a plane electromagnetic wave. Building on previous work by L. Lorenz, who solved the case for purely dielectric spheres, in 1908 Gustav Mie presented a clear and concise solution for the absorption and scattering cross-sections of spheres with arbitrary dielectric functions [4]. His equations have been applied in the intervening century to numerous physical systems, including rainbows and coronas. In particular the light scattering of suspended nanoparticles is consistently well described by the Mie solution, for which it was originally derived.

In the Mie equations the differential scattering cross sections are defined in terms of the angular intensities. These are determined by expanding a plane wave in spherical vector harmonics, giving a solution which is an infinite sum of harmonics. The harmonics are related to multipoles (dipole, quadrupole, octupole, etc.) of the electric and magnetic fields surrounding the sphere. In the quasistatic (Rayleigh) limit one may ignore the effects of the phase variation of the electric field at different points on the particle (retardation) and the dipolar mode dominates the response. For a spherical particle, surrounded by a medium with dielectric constant $\varepsilon_a$, the polarizability $\alpha$ (defined as $\mathbf{p} = \varepsilon_a \alpha \mathbf{E}_0$, where $\mathbf{p}$ is the dipole moment and $\mathbf{E}_0$ the external field) in the dipolar limit is

$$\alpha = 4\pi \varepsilon_0 a^3 \frac{\varepsilon_m - \varepsilon_a}{\varepsilon_m + 2\varepsilon_a} \tag{9.1}$$

The Mie solution is derived for a spherical particle. It is straightforward to extend the approach to an ellipsoidal geometry in the quasistatic limit. The deviation from spherical symmetry results in splitting of the plasmon resonances in isolated nanoparticles. Noble metal nanorods exhibit two prominent plasmon peaks with frequencies dependent on the aspect ratio (length/width), as well as the influences described above for spherical particles. To account for the shape anisotropy Gans [5] provided an extension to the Mie formulas in the quasistatic limit by splitting the polarizability into Cartesian components. Equation (9.1) then becomes [6]

$$\alpha_i = \frac{4\pi\varepsilon_0 abc(\varepsilon_m - \varepsilon_a)}{3\varepsilon_a + 3L_i(\varepsilon_m - \varepsilon_a)} \qquad (9.2)$$

introducing a depolarization factor, $L_i$, where the subscript $i$ denotes the three principle axes of the ellipse, $x$, $y$, $z$ ($x$ being the long axis), and $a$, $b$ and $c$ are the length of the ellipsoid in those directions, respectively. The depolarization factors are

$$L_x = \frac{1 - e^2}{e^2}\left(-1 + \frac{1}{2e}\ln\frac{1+e}{1-e}\right), \qquad L_{y,z} = \frac{1 - L_x}{2} \qquad (9.3)$$

where $e$ is the rod ellipticity $e^2 = 1 - (b/a)^2$. One should note that since the Gans formula is based on the quasistatic assumption the particle major axis should not be significantly greater than around 20 nm. Larger and more complex shaped particles exhibit multipolar resonance peaks and further lifting of degeneracy due to asymmetrical shapes.

There are numerous computer programs available to calculate the particle scattering cross-sections, and other relevant and derived parameters, using the Mie formulas. The only inputs required to accurately calculate the absorption and scattering cross-sections, apart from the sphere diameter (and ellipticity), are the dielectric functions of the sphere and surrounding material. It is thus important that these are well defined.

## 9.3 Distinguishing NPs Based on Their Dielectric Functions

It is useful to distinguish three types of nanoparticles based on their dielectric functions (equivalently their conductivities). Dielectrics such as glass and polymers have complex dielectric functions characterized by a positive real and small imaginary parts. Below the (screened) plasma frequency, metals have negative real and comparatively large imaginary parts. Semiconductors are similar to dielectrics at frequencies below the band gap, and similar to metals above their band gap; they may be described as metals with very low plasma frequencies, or insulators with small band gaps. These three materials display very different visible and infrared optical properties at the nanoscale. To demonstrate this we need to define dielectric functions that we can use in the Mie solution.

The Lorentz oscillator model presented in the introductory chapter is particularly effective for modeling chemical bonds at infra-red frequencies. If the material is a non-absorbing dielectric, such as a transparent glass or polymer, we may set the broadening term, $\Gamma$, to zero, giving us the Sellmeier model. The commonly used Cauchy model is an approximation of the Sellmeier model in terms of the real refractive index, $n$. Using the Cauchy model in the Mie theory shows the dependence of the scattering cross section on the particle diameter for a dielectric. Figure 9.1 shows the cross section in the UV-Vis-NIR region of dielectric spheres of 100, 200 and 300 nm diameter in vacuum. Particles of the order of 100 nm show extremely small scattering cross-sections. For this reason, dielectric NPs less than 100 nm are not of

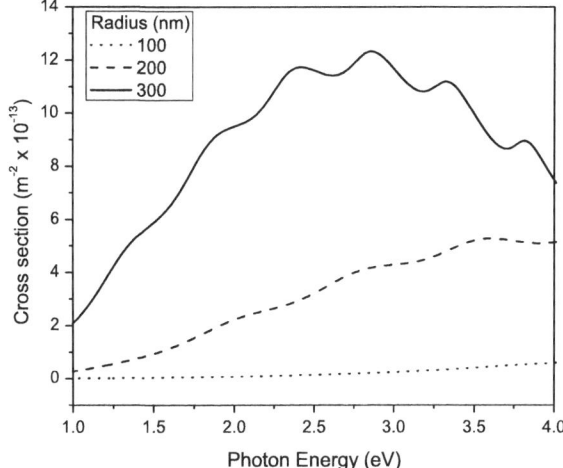

**Fig. 9.1** Mie scattering cross sections for glass spheres in vacuum with radius 100, 200 and 300 nm

great interest for optical composites, except as templates for the fabrication of metal NP arrays. The dielectric function of the embedding medium of NP composites is usually well described by the Cauchy or Sellmeier model.

### 9.3.1   Plasmonic Nanoparticles

In contrast to dielectric NPs, metallic NPs exhibit very strong scattering at the nanoscale, especially noble metals. The optical properties of metals stem predominantly from "free" conduction electrons, resulting in the Drude model presented in the introductory chapter. In real metals the bound electrons also contribute to the dielectric function, especially in the UV region where interband transitions are excited. In the noble metals the lowest energy interband transitions occur at $\hbar\omega = 2.1$, 3.8 and 2.4 eV for Cu, Ag and Au respectively [7]. These correspond to transitions from the $3d$–$4s$, $4d$–$5s$ and $5d$–$6s$ bands, respectively. For energies below the transition edge there is negligible contribution to the imaginary part $\varepsilon''$. The contribution to the real part $\varepsilon'$ may be approximated by a constant offset $\varepsilon_\infty$. The Drude equation thus becomes

$$\varepsilon(\omega) = \varepsilon_\infty - \frac{\omega_p^2}{\omega^2 + i\Gamma\omega} \tag{9.4}$$

The accuracy of this model decreases as the transition energy is approached. The interband transition may be explicitly modeled with multiple Lorentz-oscillators [8] but this does not take into account the actual band-gap. The Tauc–Lorentz (TL) model, on the other hand, explicitly incorporates the band-gap [9] and gives a reasonable representation of the interband transitions in Au [10].

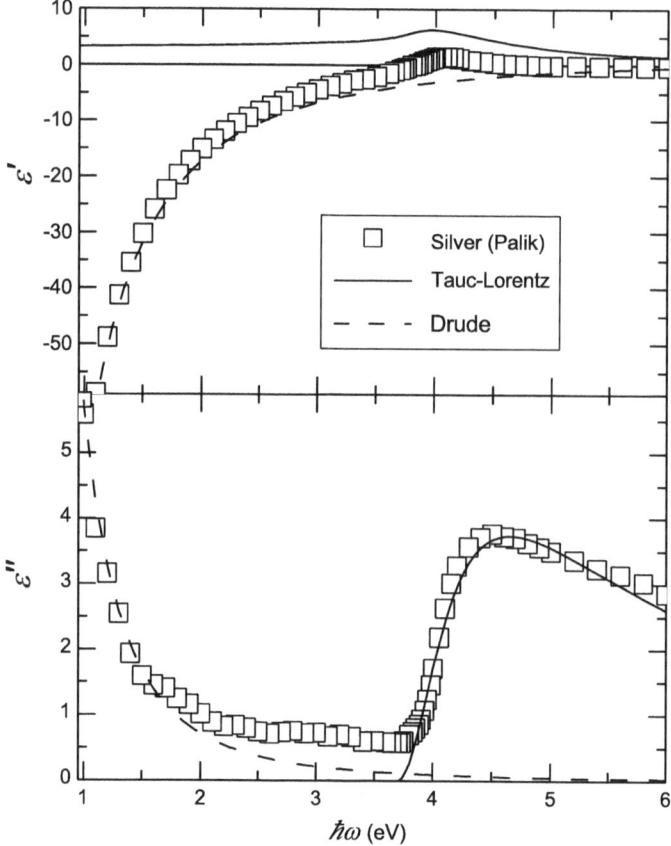

**Fig. 9.2** Dielectric function of silver from the literature [11] fit with a Drude and TL model

The noble metals (or coinage metals) are of great practical interest in nanotechnology for two reasons. Firstly, they are comparatively inert. In particular, gold is widely used in biomedical applications. Secondly, the noble metals have comparatively low damping constants $\Gamma$, especially silver which has a small $\Gamma$ across the entire visible spectrum. Thus the plasmon resonances are quite narrow. Figure 9.2 shows the dielectric function of silver from Palik's Handbook of Optical Constants [11], fit with a Drude to account for the free electrons in the vis-IR and TL model to account for the interband transitions in the UV.

We can now insert the silver dielectric function into the dipolar Mie formula (9.1). Figure 9.3 shows the spectral plane-wave extinction cross-sections of silver particles of 20, 40, 60, 80 and 100 nm diameter in glass, calculated using the MiePlot software [12]. The quasistatic approximation is valid for the 20 nm particle and only a dipolar mode is observed. At larger diameters, the dipolar mode is red-shifted due to retardation. The quadrupolar mode is already clearly visible in the 60 nm particle

**Fig. 9.3** Mie cross section
for silver NPs in air showing
the dipolar resonance
red-shifting with increasing
particle size. Particles of
60 nm radius show an
appreciable electric
quadrupolar resonance, and
from 100 nm octupolar
resonances are clearly
observed

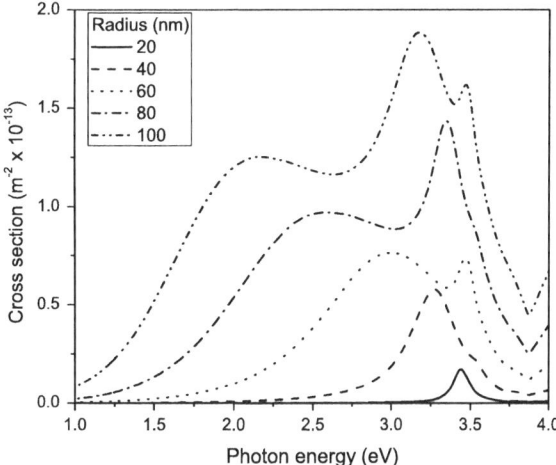

on the high-energy side of the dipolar mode. In the 100 nm particle the quadrupolar
mode is stronger than the dipolar mode and the octupolar mode is also visible.
The broadening of the dipolar oscillation also increases with particle size, both in
Mie theory and experimentally [13]. The broadening occurs due to a dephasing of
the plasmon, arising from both radiative and non-radiative (absorption) relaxation
channels. In the dipolar-dominated small size limit the particle may be considered as
a classical oscillating dipole and the absorbed photons are thus efficiently reradiated.
As the particle size increases the radiative processes are retarded and absorption
processes increase.

For a Drude metal with a dielectric function as given in (9.4), the dipolar particle
plasmon resonance frequency is

$$\omega_{pp}^2 = \frac{\omega_p^2}{\varepsilon_\infty + 2\varepsilon_a} - \Gamma^2 \tag{9.5}$$

For a pure Drude metal with $\varepsilon_\infty = 1$ and $\Gamma \ll \omega_p$ in air ($\varepsilon_a \approx 1$) one obtains
$\omega_{pp} = \omega_p/\sqrt{3}$ [14]. For gold ($\varepsilon_\infty = 10$) and silver ($\varepsilon_\infty = 4$), the actual resonance
frequency is considerably reduced.

### 9.3.2  Quantum Dots

Quantum dots (QDs) are semiconductor NPs which show size-dependent optical
properties. As the dimensions of the particle reduce below the de Broglie wavelength
of the thermal electrons (of the order of 10 nm) the energy levels in the conduction
band become discrete, and quantum effects begin to emerge. Photo-excited electron-
hole pairs become bound (excitons) and limited pathways for de-excitation results in

slow cooling dynamics. The optical properties of QDs are thus in general non-linear and characterization of QD/organic composites by techniques such as spectroscopic ellipsometry is not standard. Nevertheless, reports are emerging of the application of SE to QD composite characterization, although care should be taken to consider non-linear effects. Keita et al. used SE to investigate the optical properties of Si QDs in Si-rich SiN films over the spectral range from 1.5 to 5.9 eV [15]. Morreels et al. used transmission measurements with Kramers–Kronig analysis to determine the dielectric function of PbS QDs in a glass matrix in the 200–1800 nm spectral range [16].

The dispersion of QDs in functional organic materials is already established and the linear optical properties of such composites is of great interest. The use of ellipsometry for optical characterization provides greater accuracy with the additional relative phase information. Aslam et al. used SE to measure the optical response of a photorefractive polymer (poly(N-vinylcarbazole) and the electro-optic chromophore 1-(2-ethylhexyloxy)-2,5-dimethyl-4-(4nitrophenylazo) benzene) composite sensitized by three different types of CdSe/ZnS core/shell nanoparticles [17]. Liang et al. used a layer-by-layer (LbL) technique to fabricate hybrid thin films of conjugated polymers and CdSe NPs, surface passivated with tri-*n* octylphosphine (TOPO). They used SE to monitor the thickness of the layers and explored the photovoltaic properties of the films [18]. Antonello et al. investigated the optical properties of hybrid diphenyldimethoxysilane and ZnS NP using SE and demonstrated tuning the optical density by adjusting the relative concentrations [19].

Apart from the size and shape of the NP, the dielectric function of the NP bulk material defines the optical response of the NP in vacuum. For practical applications the particles must be suspended in a matrix or dispersed on a substrate. The dielectric properties of the embedding medium or substrate greatly influence the optical response. In addition, the structural, biocompatible and hydrophobic properties of the embedding medium also define the potential range of applications for which such a composite might be employed.

## 9.4   Nanoparticles in Polymer Matrices

Polymers may themselves be formed into NPs. Self-assembled monolayers of polymer NPs are routinely used as masks to form periodic arrays of gold and silver NPs; a technique known as "nanosphere lithography" (NSL). Block-co-polymers also form into periodic structures which may be used to define arrays of metallic nanoparticles. Block copolymers also form into NPs known as micelles. The morphology of layers of micelles may be studied by SE. Hong et al. fabricated block copolymer micelle multilayer films by dip-, spin- and spray assisted LbL methods, and studied the porosity by SE [20]. The porosity was readily controlled from 2.6 to 55.9% by manipulating the LbL deposition methods and thickness. Alternatively polymers may be used as stabilizers to prevent agglomeration and degradation of inorganic NPs. Alvarez et al. characterized the long-time and dynamic interfacial tension reduction by polymer-grafted NPs adsorbing from suspension and the corresponding dilation moduli at both xylene-water and xylene-air interfaces [21]. They used ellipsometry to assist with the study of the surface tension. The results suggest that polymer grafted NPs produce significant surface and interfacial tension reductions.

An important concept when considering the physical state of a polymer is the notion of a glass transition temperature ($T_g$). Due to strong long-range intermolecular forces, the transition from the liquid to the solid state occurs over a large temperature range. The glass transition temperature is assigned according to readily observable changes in the thermodynamic properties of the polymer. For example, the heat capacity and expansivity change discontinuously at the glass transition temperature. Ellipsometric investigations of $T_g$ in polymer thin films is reviewed in Chap. 5. We note here a number of such studies that incorporate NPs to influence $T_g$. Herzog et al. investigated the effect of heat treatment on the structure of colloidal thin films of polystyrene NPs in situ using combined GISAXS and ellipsometry [22]. In the vicinity of the glass transition temperature a rapid loss of ordering is observed in spin coated films. Chandran et al. investigated thin films of Au NPs embedded in polystyrene [23]. The temperature dependence of the thickness was monitored by SE and used to determine $T_g$. A strong dependence of $T_g$ on the NP concentration was observed.

Dispersing NPs in polymer significantly affects the optical properties and is strongly dependent on the NP material. Kim et al. determined the complex refractive index of Au-NPs in polystyrene (PS) films using VASE [24]. The composite film were prepared by first grafting thiol terminated polystyrene molecules (PS-SH) onto the surfaces of colloidal gold NPs. These were then dispersed in PS solutions and spin coated onto silicon substrates. The authors observed the LSPR in the spectra, which for homogeneous particle sizes were well described using classical Mie theory and the Drude model. Eita et al. fabricated multilayered thin films composed of nanofibrillated cellulose (NFC), polyvinyl amine (PVAm) and silica nanoparticles on polydimethylsiloxane (PDMS) using a layer-by-layer adsorption technique and measured the thickness using ellipsometry of 2.2 and 3.4 nm per bilayer for the PVAm–NFC system and the PVAm–$SiO_2$–PVAm–NFC system, respectively [25].

Schmidt et al. used SE to demonstrate reversible swelling upon reduction of an electroactive polymer nanocomposite composed of cationic linear poly(ethylene-imine) and 68 vol% anionic Prussian Blue nanoparticles [26]. Using SE they first measured the passive swelling of the composite by monitoring the film thickness as a function of time (Fig. 9.4a). Then they used an applied electric potential to demonstrate active swelling of two (LPEI/PB)30 films subjected to 10 redox cycles. By monitoring the redox cycles using SE the thickness changes for films with different preparation methods could be accurately identified (Fig. 9.4b). This is an excellent example of the highly-sensitive, non-invasive potential of real time ellipsometry.

Noble metal nanoparticles embedded in polymer are also effectively fabricated by co-sputtering, often termed NP/plasma-polymer composites, and recently reviewed by Biederman [27]. Takele et al. studied the LSPR as a function of Ag volume fraction in films of Teflon A, poly(methyl methacrylate) (PMMA) and Nylon 6 using transmission and reflection spectroscopy [28]. They observed that the position,

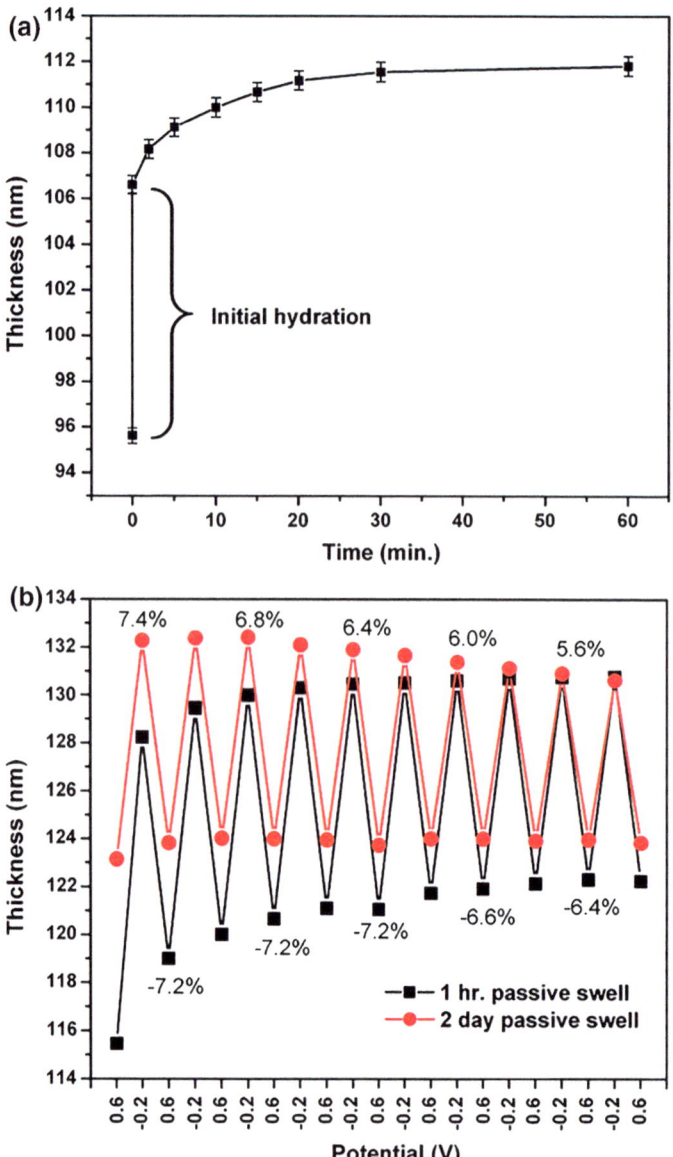

**Fig. 9.4** SE measurement of the time dependent thickness change during swelling of a polymer/NP composite. **a** Initial swelling was observed by immersing the composite in solution. **b** Active swelling was induced by applying repeated redox cycles to the composite film. From [26]. Reprinted with permission from D.J. Schmidt et al., ACS Nano **3**, 2207 (2009). Copyright 2009, American Chemical Society

width and strength of the plasmon resonance depend strongly on the metal filling factor, cluster size and interparticle distance, and also on the embedding polymer matrix. Multilayers of these materials were later demonstrated as effective Bragg reflectors [29].

## 9.5  Nanoparticles and Self-assembled Monolayers

SAMs are molecular assemblies of organic molecules spontaneously adsorbed on inorganic surfaces. They are produced from solution or by gas-phase evaporation, forming ordered domains of three dimensional crystalline or semi-crystalline structures on the substrate surface. The molecules bind, usually by chemisorption, to the substrate, thereby reducing the free energy of the inorganic-ambient interface. The spontaneous ordering of the molecules on the surface further reduces the free energy of the system. The ordered arrangement of molecules changes the interfacial properties of the surface in ways that are well defined and reproducible. Many molecules that form SAMs have a chemically functional "headgroup" with a specific affinity for a substrate. One may also tailor the "tailgroup" to define a chemical functionality for the exposed surface. The most extensively studied class of SAMs is derived from the adsorption of alkanethiols on gold, silver and copper, as well as palladium, platinum and mercury. The high affinity of thiols for the surfaces of noble and coinage metals allows one to generate well-defined organic surfaces with chemical functionalities at the exposed interface.

SE is widely used to investigate SAMs on flat surfaces [30] and forms the basis of other chapters in this book. One of the earliest reports of FTIR ellipsometry of SAMS is by Muese et al. [31]. By comparing models with experimental data, they used FTIR ellipsometry to determine the thickness and molecular structure of a series of alkanethiol monolayers. Gonzalez et al. used in situ SE to control the thickness of evaporated 1,2-dipalmitoyl-sn-3-phosphoglycerocholine membranes deposited onto a silicon substrate [32]. They were able to identify well-defined boundaries for gel, ripple, liquid crystalline and fluid-disordered phases during several heating cycles. Using imaging ellipsometry (IE) they characterized the ripple phase undulations with period of 20.8 nm and height of 19.95 nm along the temperature interval of 34–40 °C. Clusters/agglomerations heights of more than twice the membrane thickness were observed with IE, induced by heating cycles.

Porphyrins are an interesting molecule that form SAMs. Palomaki et al. found good agreement between XRR and spectroscopic ellipsometry thickness measurements of Si(100) supported molecular multilayers fabricated by LbL method utilizing copper(I)-catalyzed azide-alkyne cycloaddition (CuAAC) [33]. Modeling of the ellipsometric data over the full visible region using an oscillator model produces an absorption profile closely resembling that of a multilayer grown on silica glass.

SAMs also form on nanoparticles, most notably gold (Au-NPs), with the circumference of the particle influencing the properties of the tail-group functionality via the curvature of the metal surface. The SAMs also act to stabilize the particles and protect them from chemical interaction and degradation. SAM-stabilized Au-NP films show significant functional properties, such as photoconductance [34]. Malinsky et al. fabricated Ag NP arrays using nanosphere lithography and coated them with alkanethiol SAMs [35]. Using reflection/transmission spectroscopy they made three important conclusions: (1) the resonance frequency of the LSPR linearly shifts to the red 3 nm for every carbon atom in the alkane chain; (2) spectral shifts as large as 40 nm are caused by only 60 000 alkanethiol molecules per nanoparticle, which corresponds to only 100 zmol of adsorbate; and (3) the nanoparticles' sensitivity to bulk external environment is only attenuated by 20% when the nanoparticles are modified with the longest chain alkanethiol (1-hexadecanethiol, 2 nm). Experimental extinction spectra were modeled by using Mie theory for Ag nanospheres with dielectric shells intended to mimic the self-assembled monolayer (SAM) in thickness and refractive index. The authors found that the Mie theory qualitatively predicts the experimentally observed trend that the resonance frequency linearly shifts to the red with respect to shell thickness, or alkanethiol chain length; however, the theory underestimates the sensitivity by approximately a factor of 4.

Numerous ellipsometric studies have been performed to characterize Au-NP/SAM systems. Wang et al. used SE to determine the optical constants of Au NPs linked by dithiols and carboxylic acid functionalized thiols of various lengths [36]. They showed the wavelength of the LSPR to be dependent on the interparticle linker chain length. Jaber et al. used Se to measure the LSPR in 2D arrays of Au–PNIPAM core–shell nanocrystals fabricated using convective deposition and spin-coating [37]. By annealing at 700 °C the polymer shell was removed, while retaining a monolayer of well-separated gold nanoparticles.

Mayya et al. used a host of techniques including ellipsometry to investigate the complexation of cysteine-capped Au NPs with octadecylamine Langmuir monolayers [38]. By controlling the pH during deposition they could control the uniformity of the NPs dispersed on the substrate surface. Combining spectroscopic ellipsometry (SE), Raman spectroscopy, surface potential Kelvin probe microscopy (SP-KPM) and AFM, Giangregorio et al. investigated the interaction of hydrogen with thiol functionalized Au-NPs deposited on silicon substrates by RF sputtering of gold [39]. The kinetics of the thiol self-assembled monolayer (SAM) formation, and on the density of the resulting SAMs were explored. NP films were exposed to a remote $H_2$ plasma and subsequently functionalized by the aromatic (4-methoxyterphenyl-300, 500-dimethanethiol) and aliphatic (dodecanethiol) thiols. They found that remote $H_2$ plasma pretreatments of gold surfaces are effective in improving thiolate adsorption, making SAMs more uniform and densely packed. They also demonstrate that hydrogenation of nanoparticles improves stability of thiol functionalized Au NPs, avoiding their aggregation. The same group prepared SAMs derived of 4-methoxy-terphenyl-300,500-dimethanethiol and 4-methoxyterphenyl-400-methanethiol by chemisorption from solution onto gold thin films and NPs [40]. The SAMs were characterized by spectroscopic ellipsometry to determine their refractive index and extinction coefficient in a spectral range from 0.75 to 6.5 eV. The thermal behavior of SAMs were also monitored using ellipsometry in the temperature range 25–500 °C.

Aureau et al. studied the properties of interfaces of two model surfaces obtained by covalently grafting alkyl chains directly to oxide-free silicon surfaces, either via Si–O–C or Si–C bonds are compared with those currently obtained by attaching organic silane molecules (e.g., (aminopropyl)triethoxysilane) on oxidized silicon surfaces. They used FTIR, Raman spectroscopy and spectroscopic ellipsometry to show that nanopatterned Si–O–C surfaces suffer some oxidation upon attaching nanoparticles, although they remain stable in ambient environments. In contrast, surfaces with Si–C bonds remain oxide free and remarkably stable during and after gold nanoparticle attachment [41].

## 9.6 Effective Medium Approximations

Many of the examples given above present cursory investigations into the material properties using ellipsometry, such as the film thickness or effective optical constants. By monitoring changes in these properties one can infer important information about chemical and physical processes, such as the polymer $T_g$. In the case of NP composites, important questions are the size, shape and fill factor of the NPs in the host matrix. Effective medium models allow one to extract such information from ellipsometric measurements.

When moving from the case of an individual particle to a large ensemble of particles we must take into account the collective behavior of the particles in a medium and their affect on each other. By assuming the material has a spatially-homogeneous *effective* permittivity, the Fresnel equations and those of geometrical optics may still be applied. The effective medium concept reflects the fundamental connection between the macroscopic permittivity of a material and the microscopic polarizabilities of the constituents. The effective medium models are presented in the introductory chapter. Aspnes also gives an excellent contemporary review of effective medium theories [42].

The Maxwell–Garnett theory (MGT) was specifically developed to describe the optical properties of gold nanoparticles in a dielectric. The theory assumes a random ensemble of nanoparticles with a low fill factor, $F$. We should expect that in the limit of very small $F$ the MGT converges to the quasistatic Mie case. For metallic spheres with a Drude dielectric function, the MGT describes a Lorentzian with a resonance frequency of

$$\omega_0^2 = \frac{\omega_p^2(1 - F)}{\varepsilon_\infty + 2\varepsilon_a} \tag{9.6}$$

This formula converges in the limit $F \to 0$ with (9.5) as expected. As $F$ increases the MGT predicts a red-shift of the resonance frequency. This is logical since the resonance frequency of a Drude metal is zero, which would be the case in the limit $F \to 1$.

The MGT was used to model the content of magnetite ($Fe_3O_4$) nanoparticles dispersed in thin films of poly(N-isopropylacrylamide) (PniPAAm) [43]. SE measured were performed for 6 different concentrations of NPs. Using the bulk dielectric functions of magnetite and a Cauchy model for the PniPAAm the effective refractive index and extinction coefficient were determined, as shown in Fig. 9.5. Both the optical parameters are observed to increase markedly with increasing content of nanoparticle. The fill factor values of the NPs in the polymer agreed well with measurements from electron microscopy.

The absence of a size dependence in (9.6) implies that the resonance should not change for particles below the quasistatic limiting diameter of around 20 nm. In practice, for 5 nm particles the resonance is strongly damped and for particles below around 2 nm it completely disappears [44]. This is understood in the context of an

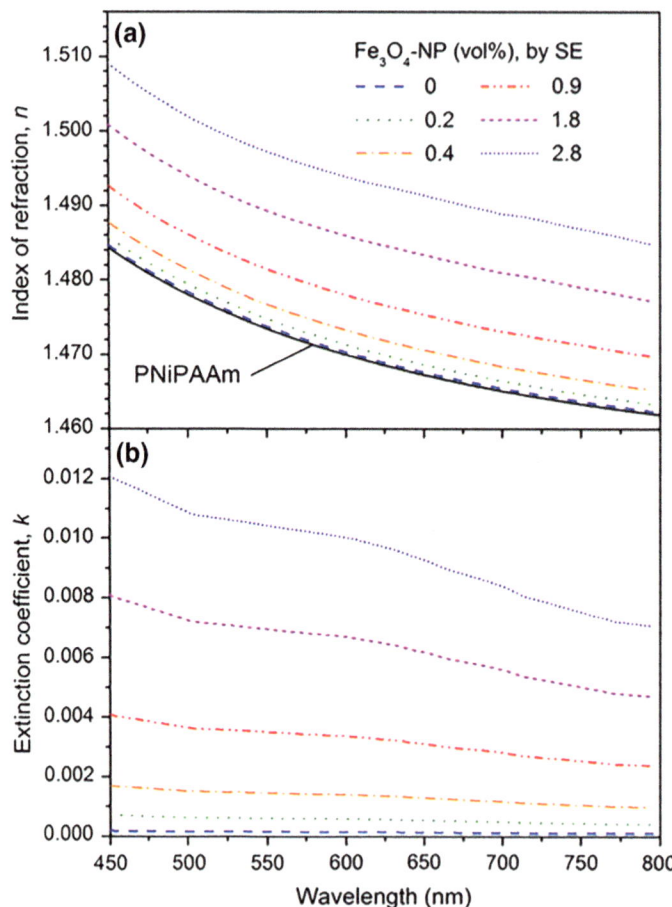

**Fig. 9.5** Effective refractive index and extinction coefficient of magnetite NPs dispersed in Pni-PAAM matrices, determined using the MGT. From [43]. Reprinted with permission from S. Rauch et al., J. Vac. Sci. Technol. A **30**, 041514 (2012). Copyright 2012, American Vacuum Society

increase of the Drude broadening parameter, $\Gamma$, due to effects such as impurities, lattice defects and surface scattering [14]. Note that quantum effects for metal particles are not apparent for particles with more than a few hundred atoms, which corresponds to a diameter below 1 nm [14]. For a bulk crystal, the mean free path $l_\infty$ is determined by electron and phonon interactions. This length scale reflects the distance between memory canceling collisions of the electrons. As long as these collisions are slightly inelastic, the interaction involves electrons close to the Fermi surface and hence:

$$\Gamma_\infty = \frac{v_F}{l_\infty} \tag{9.7}$$

Here $v_F$ is the Fermi velocity of the electrons. For a small particle with radius $a$ the effect is empirically accounted for by introducing an additional term into the broadening [45]:

$$\Gamma(R) = \Gamma_\infty + \frac{Av_F}{a} \tag{9.8}$$

The parameter $A$ accounts for the spherical nature and can as well account for the influence of the chemical environment. For a spherical particle with no chemical effects, $A = 4/3$ [14, 46]. The chemical effect on the $A$ parameter may be determined experimentally by comparing the broadening with microstructural analysis of the particle size. While the broadening can be determined using reflection and transmission measurements [47] it can also be determined using spectroscopic ellipsometry. The method is generally only useful for well-dispersed particles suspended in a film of transparent media such as glass or polymer. Dynamic SE was used to observe the growth of silver nanoparticles in polymer thin films by temperature induced reduction of silver salts [48, 49]. By determining a final $A = 0.41$ using scanning electron microscopy, the data could be modeled to give the particle radius during growth (Fig. 9.6). The plasmon resonance was completely broadened for particles smaller than about 2 nm due to surface damping. This method allows the in situ monitoring of NP growth and can be utilized to stop the growth at any required point by simply cooling the sample.

Schädel et al. used SE to investigate the volume ratios of polymer/NP (CdSe QDs and nanorods with hexanoic acid-treated hexadecylamine or pyridine as the capping ligands) blends by employing effective medium models [50]. They successfully determined the mass ratio of the components, the mass density of the NP including the inorganic crystalline core and the organic ligand layer. A geometrical model for the QDs and nanorods allowed for the estimation of the ligand layer thickness. Warenghem et al. used the MGT to determine the effective optical properties of $BaTiO_3$ NPs embedded in nematic liquid crystals [51]. By combining ellipsometry with refractometry they were able to increase the measured parameters and uniquely determine the ordinary and extraordinary components of the anisotropic dielectric function. Ruiterkamp et al. used effective medium models to fit SE measurements and determine the refractive index of transparent nanocomposites composed of rutile titanium dioxide nanoparticles functionalized with 1-decylphosphonic acid and diethyl

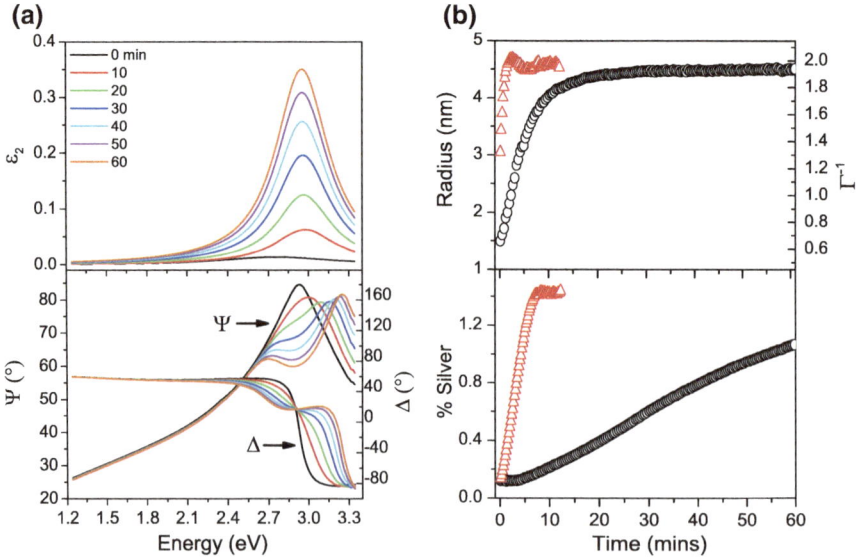

**Fig. 9.6 a** Real time in-situ SE data (*bottom*) and fitted imaginary part of the effective dielectric function (*top*) showing the appearance and growth of a LSPR band due to the nucleation and growth of silver NPs in a PVOH thin film. **b** By scaling the A-parameter in (9.8) from the final size of the NPs, the particle radius (*top*) and silver content (*bottom*) could be inferred from the EMA data for different heating rates (fast: *red triangles*, slow: *black circles*). Adapted from [49]

undec-10-enyl phosphonate in a poly(benzyl acrylate) matrix [52]. They observed an increase from 1.57 for the pure polymer to 1.63 for 14.0 vol.% $TiO_2$ at $\lambda = 586$ nm. Nanocomposite films with particle weight percentages of up to 30% (9.5 vol.%) showed a high light transmittance of around 90% at wavelengths above $\lambda = 400$ nm. Balevicius et al. used TIRE (described in Chap. 18) to investigate the optical response of hybrid multilayer systems of gold nanoparticles dispersed on SAMs [53]. They used the Bruggeman EMA to fit the data and extract the effective complex refractive index of the layers.

## 9.7 Surface Coverage of 2-D Films

The Lorentzian shape of the resonance in the MGT for a Drude metal in a dielectric can be used as a quick approximation to determine the dielectric function from ellipsometric data without explicitly fitting $F$ or other unknowns. Yamaguchi was one of the first to note this [54]. He observed resonance type absorption in discontinuous evaporated noble metal island films (MIFs) and used the MGT to derive a Lorentzian expression for the resonance. This Lorentzian depends on the fill factor of the material, and the energy position of the resonance maximum depends only on $F$, according to (9.6). It can therefore be used to determine the area coverage of the metal film. Doremus [55] exploited this dependence in a similar approach and derived an equation that predicts the resonance maximum in island films of noble metals using the measured bulk dielectric functions of the metal. The maximum is expected to occur when the real part of the dielectric function is related to:

$$\varepsilon'_m = -\frac{(2 + \phi)n_d^2}{1 - \phi} \tag{9.9}$$

where $\phi$ is the projected surface area coverage of the particles on the substrate, and $n_d$ is the refractive index of the substrate. Doremus later empirically demonstrated the applicability of (9.9) using a wide array of published results for various discontinuous metal films [56]. Wormeester et al. showed the effect of $\varepsilon_\infty$ on the resonance position and broadening [57].

Using spectroscopic ellipsometry, a large number of authors have applied the simple Lorentz oscillator approach to determine the dielectric function of gold nanoparticle/organic films [58, 59], electrodeposited gold on anodized and etched Al and Cu films [60], nanoporous silver [61], gold [62], and platinum [63], and hollow gold nanoparticles [64]. The applicability of the Lorentzian assumption and its limitations in the event of larger $\phi$ was demonstrated using in situ real time spectroscopic ellipsometry (RTSE) on films deposited at room temperature and 150 °C [65]. Figure 9.7 shows the effective dielectric functions of the island film deposited at 150 °C as a function of nominal film thickness, determined using both the Arwin–Aspnes method [66] and a Lorentzian expression to approximate the MGT. As more material is deposited and the surface coverage increases, the plasmon resonance red shifts, grows in amplitude and broadens. The Lorentzian fit provides a good approximation of the plasmon frequency but not the broadening or amplitude.

Roth et al. presented a real-time study of the nanostructuring and cluster formation of gold nanoparticles deposited in aqueous solution on top of a pre-structured polystyrene colloidal thin film [67]. Cluster formation takes place at different length scales, from the agglomerations of the gold nanoparticles to domains of polystyrene colloids. By combining in situ imaging ellipsometry and microbeam grazing incidence small-angle X-ray scattering (GISAXS), they were able to identify different stages of nanocomposite formation, namely diffusion, roughness increase, layer build-up and compaction. Roth et al. also used GISAXS to investigate a gradient library of nanoparticles on a polystyrene substrate [68].

Whereas RTSE provides an extended data set in the time domain, intentionally creating films with a thickness gradient provides an extended data set in space. The gradient may be used as a library of particle sizes, shapes and spacings for a combinatorial investigation of the effect of these parameters on the optical properties using multiple analysis techniques. Bhat et al. reported a method to use SE for a quick prediction of the number density of Au NPs bound to surfaces decorated with 3-aminopropyltriethoxysilane (APTES) SAMs or surface-tethered polyacrylamide (PAAm). They used a chemical gradient to create a number density of gold nanoparticles on a substrate. By comparing the results of AFM and SE the authors derived a linear relationship between the number density of NPs and the $\Delta$ value of the SE measurement [69].

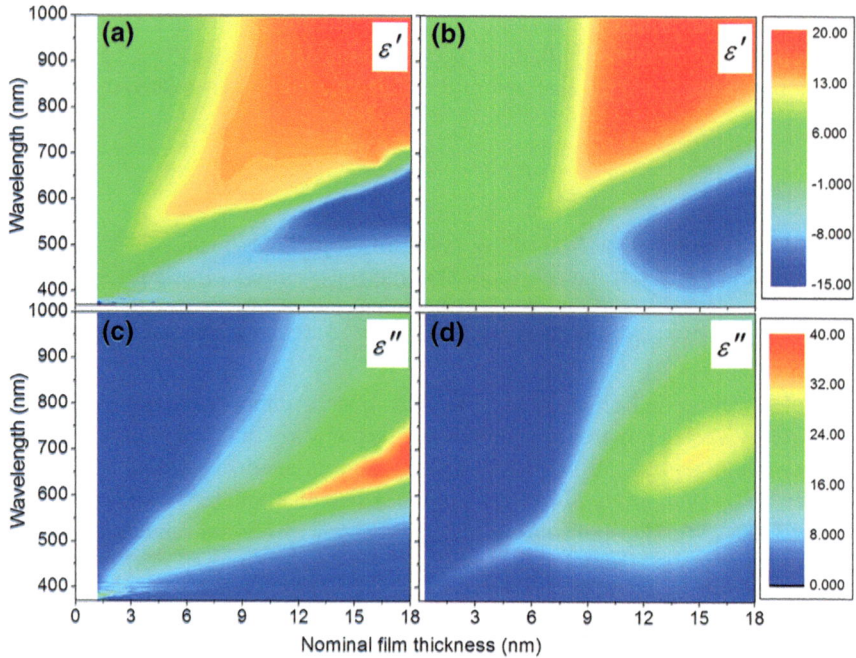

**Fig. 9.7** The dielectric functions of silver films deposited by PVD onto silica. Panels **a**, **c** show the measured complex dielectric function extracted from the in situ data using the inversion method. Panels **b**, **d** show fits of the data using the MGT. Adapted from [65]

Particle gradients are also useful for investigating optimized nanostructured substrates for surface-enhanced Raman (SERS) and infrared (SEIRA) spectroscopies. We have previously compared the Raman enhancement from silver island films with the optical properties determined using SE [70]. Additional information was taken from SEM images to compare the particle dimensions. The spatial resolution of the Raman results was around 1 μm, allowing identification of a dependence in the optimal enhancement morphology on the Raman laser wavelength.

## 9.8 SEIRA

Surface-enhanced spectroscopies exploit the strong electric fields generated at the surfaces of metallic nanostructures. In random and arrayed metallic nanostructures, incident photons excite electronic resonances, resulting in inhomogeneous *local* fields. Since Raman and IR cross-sections are proportional to $E^4$ and $E^2$, respectively, emission from those molecules located within an enhanced-field region are disproportionately enhanced. Thus, compared to a molecular concentration in a uniform field, SERS and SEIRA allow one to detect much lower concentrations of molecules, even down to the single molecule limit.

SEIRA enhancement is significantly weaker than SERS and, since its discovery in the early 1980s [71] SEIRA has received comparatively less attention. However, the normal molecular dipolar resonance cross-section is much greater than the Raman cross-section, and the weak enhancement makes SEIRA comparable in sensitivity to SERS. Recently, renewed interest has been shown in SEIRA and the related technique of surface-enhanced fluorescence (SEF) [72], in part due to the development of fabrication methods for well-defined metallic nano-antennas.

### 9.8.1 SEIRA Substrates

Classic SERS and SEIRA substrates are thin noble-metal island films on glass or silicon, as depicted in Fig. 9.8. The random distribution of nanoparticles results in a number of electric-field-enhanced "hot-spots", which are the source of the scattering enhancement in SERS. In contrast, SEIRA is not a local phenomena but stems from the effective dielectric function of the nanoparticle film mixed with an organic molecule. The first reports of SEIRA demonstrated enhancement factors (EF) of 20, using ATR with silver deposited on 4-nitrobenzoic acid. When coupled with the enhancement from the ATR technique, EFs of $10^4$ were obtained [71]. The technique was studied in depth by the group of Suetaka (later, Osawa group) who showed that SEIRA is also observed by using transmission and external reflection methods [73]. The potential to study SAMs was recognized early [74] and much work was performed during the 90s [75]. More recently, Ataka and Heberle presented interesting results of ATR-SEIRA in differential mode to investigate cyclic voltammetry of lipid membrane proteins [76, 77].

The Osawa group also studied in depth the morphology of the silver island films used as substrates, showing that the maximum enhancements are observed close to the percolation threshold [78]. No advantage was observed for periodic substrates over random arrays [79]. This preliminary work allowed a summary of the SEIRA effect

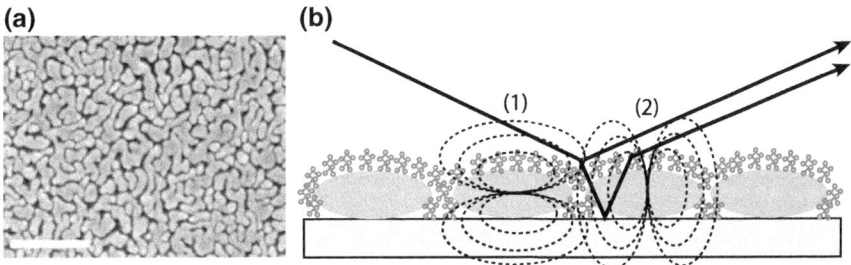

**Fig. 9.8 a** An SEM image of a typical SEIRA substrate composed of a silver island film just before percolation. The scale bar is 500 nm. **b** Schematic of the SEIRA process, showing organic molecules distributed over gold islands. The incident light excites unequal resonances (1) parallel and (2) perpendicular to the substrate surface. Also important in SEIRA is the multiple pathways of light through the thin film

by Osawa in 2001 [80], including; in ATR and external-reflection measurements, the observed band intensities depend on the polarization and angle of incidence of the infrared radiation; vibrational modes that have dipole moment derivative components perpendicular to the surface are preferentially enhanced; metal island films that exhibit SEIRA have a very broad absorption extending from the visible to the mid-infrared, on which the enhanced absorption bands of the adsorbed molecules are superposed.

At our laboratory, Röseler and Korte were the first to apply the newly established technique of FTIR ellipsometry to study the SEIRA effect [81]. Later, Hinrichs et al. showed the ability of IRSE to determine the anisotropy of Langmuir–Blodgett thin films and explored the use of synchrotron radiation for sub-millimeter resolution, with modeling using an EMA [82]. Bradford et al. also demonstrated the advantages of IRSE for SEIRA studies of SAMs on gold substrates, introducing additional layers into the stratified model [83].

## 9.8.2   SEIRA Optical Models

There are two important concepts to consider when modeling the optical properties of SEIRA using EMAs; percolation and anisotropy. Since the films that give maximum SEIRA enhancement are close to the percolation threshold, interactions between particles should be accounted for, which the MGT does not do. The MGT also explicitly distinguishes between the particle and host dielectric functions. In contrast the EMA of Bruggeman makes no such distinction and can therefore be used to predict a percolation event, where the inclusions form a connected network within the host. Another important consideration for a complete optical model of SEIRA is the optical anisotropy of the metal films. In general this has not been taken into account in the past, although it may be of significance for understanding the true nature of SEIRA.

Osawa's group was the first to use the Bruggeman EMA to fit experimental SEIRA data. They showed the basic band shapes and enhancement factors could be well produced. The additional relative phase information contained in the SEIRSE measurements performed by Röseler et al. allowed the direct determination of the real and imaginary parts of the refractive index. Importantly it allowed more precise testing of the Bruggeman EMA used to describe the observed spectra. Caurel et al. used SEIRSE to investigate percolated gold island films with a-Si:H overlayers [84]. They found good agreement with simulations using an anisotropic Bruggeman EMA, but only low enhancement of up to a factor of 5 compared to films without gold. Later the group of Griffiths and Theiss [85] used the Bergman model, discussing for the first time the importance of the percolation threshold on the shape and intensity of the bands.

An example of the Bruggeman EMA to fit the experimental data is shown in Fig. 9.9. The resonance band of the AgCN absorption is clearly observed superposed on a broad resonance due to the silver island film. The superposition of the two

**Fig. 9.9** Comparison of measured and simulated SEIRSE data. The *solid lines* show the measured data of the AgCN band on percolated silver island films. The *dashed line* shows the simulation of using the Bruggeman EMA with a 32% silver content. The silver dielectric function is from the literature [11] and the surrounding dielectric is a Cauchy model with a Lorentzian resonance to account for the AgCN band

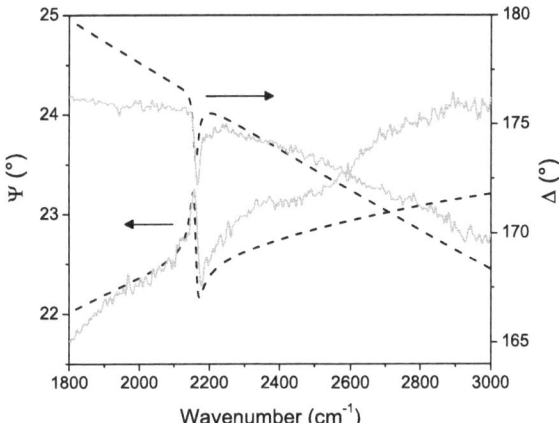

resonances (organic and plasmonic) results in a Fano-type lineshape; a sharp resonance superposed on a broad resonant background [86]. The Bruggeman model with 32% silver content matches reasonably well the measured data.

### 9.8.2.1  Anisotropy

In addition to the anisotropy observed in ellipsoidal nanoparticles, optical anisotropy is also observed due to the coupling of spherical particles with overlapping near-fields. Consider the simple case of a dimer of metallic nanoparticles in the quasistatic limit excited by an electric field parallel to the dimer axis (Fig. 9.10a). The electric fields generated by the plasmonic dipoles will reduce the restoring force on the electrons in the neighboring particle, thereby red-shifting the resonance frequency. Conversely, if the external field is perpendicular to the dipole axis (Fig. 9.10b) the dipole fields will blue-shift the resonance frequency. Since the effect of each particle on the other is reciprocal the plasmons hybridize [87]. In the large-gap limit the effect is akin to the interaction between two isolated dipoles, whilst for small gaps higher order multipoles become important.

Extending the concept to a layer of isolated densely-packed particles on a surface, it is clear that the in-plane resonance (*a*-mode) will be red-shifted, and the out-of-plane resonance (*c*-mode) blue-shifted, with respect to the resonance for an isolated sphere. Thus the dielectric function of the film is uniaxially anisotropic, and the plasmon resonance is split. The two modes are easily observed in the reflection spectrum of p-polarized light at large angles from the surface normal [88], or in ellipsometry measurements [89]. The mode splitting is not predicted by the MGT or Bruggeman formalisms.

However, although the standard MGT does not predict the mode splitting in NP thin films, it does predict two modes when $\varepsilon_{eff}$ is used in thin film models such as the Airy formula. The lower frequency resonance is the *collective* in-plane localized

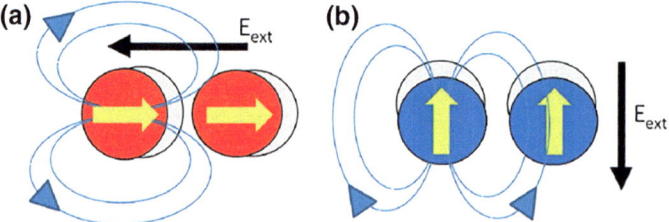

**Fig. 9.10** Dipolar coupling between two adjacent metal NPs. If the external exciting field is **a** parallel to the dimer axis, the induced fields reduce the total field in the particle, thus red-shifting the resonance. In the perpendicular case (**b**), the fields add and blue shift the resonance frequency

plasmon resonance, retarded due to inter-particle coupling and substrate interactions. The higher frequency resonance is the *collective* out-of-plane localized plasmon resonance. Equivalently, it is the plasma oscillation of the entire thin film which occurs very close to $\varepsilon_{eff} = 0$. It is often referred to as the Ferrell mode [90] and is the plasmonic analogue of the phononic Berreman mode [91]. Note that there is no intrinsic resonance in the dispersion of $\varepsilon_{eff}$ at the $c$-mode resonance frequency; the observed resonance arises from defining the boundary conditions of the thin film surfaces. For the resonance to occur in the model the real part of $\varepsilon_z$ is required to transit through zero and the imaginary part must be small. A correct model for $\varepsilon_z$ is important to correctly interpret the physical origins of SEIRA.

### 9.8.3   Recent Advances in SEIRA

With the advent of improved nanofabrication and the rapid interest in plasmonics, there have been a large amount if activity in SEIRA studies in recent years. The Pucci group has recently investigated the use of top-down fabricated gold rods as SEIRA antennas, with organic molecules deposited on the surface. They beautifully show the dependence of the rod length on the localized SPR and how the LSPR is related to the shape and enhancement of the IR absorption spectra [92]. They also presented real time SEIRA measurements during the deposition of gold films in liquid by ATR-SEIRA, followed by in situ SAMs growth [93]. SEIRA has also been demonstrated using novel antenna shapes, include nanoshell aggregates [94], split-ring shaped particles [95], and asymmetric crosses [96]. Recent demonstration of the sensing of Amide bands by top down nanorods has opened the door to SEIRA application in protein array microsensors [97]. The application of FTIR ellipsometry to explore the origins of SEIRA in nanofabricated substrates has great potential.

## 9.9 Conclusions and Outlook

The potential of SE to characterize organic/NP composites has been demonstrated, but by no means is the scientific potential of the field exhausted. The vast range of polymer and SAM materials available for use both as passive and active embedding materials will ensue that a vast range of composites with new material properties are yet to be uncovered. The versatility of SE, for real time measurements with sub-nanometer thickness sensitivity and sub-monolayer surface coverage, makes it an essential method. In addition the ability to monitor processes remotely, non-destructively and in diverse environments all contribute to the attractiveness of SE as a sensing and quality assurance tool.

To date the majority of applications in this field have been simple thickness measurements, with some efforts aimed at parameter retrieval to pinpoint the material effective refractive index. The effective medium models described here are excellent starting points to attempt to quantitatively determine more complex material properties such as the size and fill factor of metal NPs in a polymer matrix. More work is to be done to refine these models for the tasks at hand. In particular the enhancement of IR signals by SEIRA is not completely understood and SE will play a major role in unearthing and exploiting this powerful sensing method.

## References

1. D. Roy, J. Fendler, Adv. Mater. **16**, 479 (2004)
2. V. Korstgens, J. Wiedersich, R. Meier, J. Perlich, S.V. Roth, R. Gehrke, P. Muller-Buschbaum, Anal. Bioanal. Chem. **396**, 139 (2010)
3. K.B. Rodenhausen et al., Rev. Sci. Instrum. **82**, 103111 (2011)
4. G. Mie, Ann. Phys. **25**, 377 (1908)
5. R. Gans, Ann. Phys. **37**, 881 (1912)
6. J. Perez-Juste, I. Pastoriza-Santos, L.M. Liz-Marzan, P. Mulvaney, Coord. Chem. Rev. **249**, 1870 (2005)
7. C. Kittel, *Introduction to Solid State Physics* (Wiley, New York, 1996)
8. A. Vial, A.S. Grimault, D. Macias, D. Barchiesi, M.L. de la Chapelle, Phys. Rev. B **71**, 085416 (2005)
9. G.E. Jellison, F.A. Modine, Appl. Phys. Lett. **69**, 371 (1996)
10. P.G. Etchegoin, E.C. Le Ru, M. Meyer, J. Chem. Phys. **125**, 164705 (2006)
11. D.W. Lynch, W.R. Hunter, *Handbook of Optical Constants of Solids*, ed. by E.D. Palik (Academic Press, New York, 1985)
12. http://www.philiplaven.com/mieplot.htm
13. C. Sonnichsen, T. Franzl, T. Wilk, G. von Plessen, J. Feldmann, New J. Phys. **4**, 93 (2002)
14. U. Kreibig, M. Vollmer, *Optical Properties of Metal Clusters* (Springer, Berlin, 1995)
15. A.S. Keita, A.E. Naciri, F. Delachat, M. Carrada, G. Ferblantier, A. Slaoui, J. Appl. Phys. **107**, 093516 (2010)
16. I. Moreels, D. Kruschke, P. Glas, J.W. Tomm, Opt. Mater. Express **2**, 496 (2012)
17. F. Aslam, J. Stevenson-Hill, D.J. Binks, S. Daniels, N.L. Pickett, P. O'Brien, Chem. Phys. **334**, 45 (2007)
18. Z.Q. Liang, K.L. Dzienis, J. Xu, Q. Wang, Adv. Funct. Mater. **16**, 542 (2006)

19. A. Antonello, G. Brusatin, M. Guglielmi, A. Martucci, V. Bello, G. Mattei, P. Mazzoldi, G. Pellegrini, Thin Solid Films **518**, 6781 (2010)
20. J. Hong, H. Park, Colloids Surf. A Physicochem. Eng. Asp. **381**, 7 (2011)
21. N.J. Alvarez, S.L. Anna, T. Saigal, R.D. Tilton, L.M. Walker, Langmuir **28**, 8052 (2012)
22. G. Herzog et al., Langmuir **28**, 8230 (2012)
23. S. Chandran, J.K. Basu, Eur. Phys. J. E **34**, 99 (2011)
24. J. Kim, H.X. Yang, P.F. Green, Langmuir **28**, 9735 (2012)
25. M. Eita, H. Arwin, H. Granberg, L. Wagberg, J. Colloid Interface Sci. **363**, 566 (2011)
26. D.J. Schmidt, F.C. Cebeci, Z.I. Kalcioglu, S.G. Wyman, C. Ortiz, K.J. Van Vliet, P.T. Hammond, ACS Nano **3**, 2207 (2009)
27. H. Biederman, Surf. Coat. Technol. **205**, S10 (2011)
28. H. Takele, H. Greve, C. Pochstein, V. Zaporojtchenko, F. Faupel, Nanotechnology **17**, 3499 (2006)
29. U. Schurmann, H. Takele, V. Zaporojtchenko, F. Faupel, Thin Solid Films **515**, 801 (2006)
30. M. Prato, M. Alloisio, S.A. Jadhav, A. Chincarini, T. Svaldo-Lanero, F. Bisio, O. Cavalleri, M. Canepa, J. Phys. Chem. C **113**, 20683 (2009)
31. C.W. Meuse, Langmuir **16**, 9483 (2000)
32. H.C. Gonzalez, U.G. Volkmann, M.J. Retamal, M. Cisternas, M.A. Sarabia, K.A. Lopez, J. Chem. Phys. **136**, 134709 (2012)
33. P.K.B. Palomaki, A. Krawicz, P.H. Dinolfo, Langmuir **27**, 4613 (2011)
34. H. Nakanishi et al., Nature **460**, 371 (2009)
35. M.D. Malinsky, K.L. Kelly, G.C. Schatz, R.P. Van Duyne, J. Am. Chem. Soc. **123**, 1471 (2001)
36. L.Y. Wang et al., J. Phys. Chem. C **112**, 2448 (2008)
37. S. Jaber, M. Karg, A. Morfa, P. Mulvaney, Phys. Chem. Chem. Phys. **13**, 5576 (2011)
38. K.M. Mayya, A. Gole, N. Jain, S. Phadtare, D. Langevin, M. Sastry, Langmuir **19**, 9147 (2003)
39. M.M. Giangregorio, M. Losurdo, G.V. Bianco, A. Operamolla, E. Dilonardo, A. Sacchetti, P. Capezzuto, F. Babudri, G. Bruno, J. Phys. Chem. C **115**, 19520 (2011)
40. G. Bruno et al., Langmuir **26**, 8430 (2010)
41. D. Aureau, Y. Varin, K. Roodenko, O. Seitz, O. Pluchery, Y.J. Chabal, J. Phys. Chem. C **114**, 14180 (2010)
42. D.E. Aspnes, Thin Solid Films **519**, 2571 (2011)
43. S. Rauch, K.J. Eichhorn, M. Stamm, P. Uhlmann, J. Vac. Sci. Technol. A Vac. Surf. Films **30**, 041514 (2012)
44. R.H. Doremus, P. Rao, J. Mater. Res. **11**, 2834 (1996)
45. U. Kreibig, C. Vonfrags, Z. Phys. **224**, 307 (1969)
46. U. Kreibig, L. Genzel, Surf. Sci. **156**, 678 (1985)
47. A. Hilger, M. Tenfelde, U. Kreibig, Appl. Phys. B Lasers Opt. **73**, 361 (2001)
48. T.W.H. Oates, E. Christalle, J. Phys. Chem. C **111**, 182 (2007)
49. T.W.H. Oates, Appl. Phys. Lett. **88**, 3 (2006)
50. M. Schadel, K.F. Jeltsch, P. Niyamakom, F. Rauscher, Y.F. Zhou, M. Kruger, K. Meerholz, J. Polym. Sci. Part B Polym. Phys. **50**, 75 (2012)
51. M. Warenghem, J.F. Henninot, J.F. Blach, O. Buchnev, M. Kaczmarek, M. Stchakovsky, Rev. Sci. Instrum. **83**, 035103 (2012)
52. G.J. Ruiterkamp, M.A. Hempenius, H. Wormeester, G.J. Vancso, J. Nanoparticle Res. **13**, 2779 (2011)
53. Z. Balevicius, R. Drevinskas, M. Dapkus, G.J. Babonas, A. Ramanaviciene, A. Ramanavicius, Thin Solid Films **519**, 2959 (2011)
54. S. Yamaguchi, J. Phys. Soc. Jpn. **15**, 1577 (1960)
55. R.H. Doremus, J. Appl. Phys. **37**, 2775 (1966)
56. R. Doremus, Thin Solid Films **326**, 205 (1998)
57. H. Wormeester, E.S. Kooij, B. Poelsema, Phys. Status Solidi A Appl. Res. **205**, 756 (2008)
58. H.L. Zhang, S.D. Evans, J.R. Henderson, Adv. Mater. **15**, 531 (2003)
59. Z.M. Qi, I. Honma, M. Ichihara, H.S. Zhou, Adv. Funct. Mater. **16**, 377 (2006)
60. D.A. Brevnov, C. Bungay, J. Phys. Chem. B **109**, 14529 (2005)

61. H. Pan, S.H. Ko, C.P. Grigoropoulos, Appl. Phys. Lett. **93**, 234104 (2008)
62. M.C. Dixon, T.A. Daniel, M. Hieda, D.M. Smilgies, M.H.W. Chan, D.L. Allara, Langmuir **23**, 2414 (2007)
63. R.A. May, M.N. Patel, K.P. Johnston, K.J. Stevenson, Langmuir **25**, 4498 (2009)
64. D.H. Wan, H.L. Chen, Y.S. Lin, S.Y. Chuang, J. Shieh, S.H. Chen, ACS Nano **3**, 960 (2009)
65. T.W.H. Oates, L. Ryves, M.M.M. Bilek, Opt. Express **16**, 2302 (2008)
66. H. Arwin, D.E. Aspnes, Thin Solid Films **113**, 101 (1984)
67. S.V. Roth et al., J. Phys. Condens. Matter **23**, 254208 (2011)
68. S.V. Roth et al., Appl. Phys. Lett. **88**, 3 (2006)
69. R.R. Bhat, J. Genzer, Surf. Sci. **596**, 187 (2005)
70. T.W.H. Oates, H. Sugime, S. Noda, J. Phys. Chem. C **113**, 4820 (2009)
71. A. Hartstein, J.R. Kirtley, J.C. Tsang, Phys. Rev. Lett. **45**, 201 (1980)
72. S.M. Tabakman et al., Nat. Commun. **2**, 466 (2012)
73. Y. Nishikawa, K. Fujiwara, K. Ataka, M. Osawa, Anal. Chem. **65**, 556 (1993)
74. T. Kamata, A. Kato, J. Umemura, T. Takenaka, Langmuir **3**, 1150 (1987)
75. K. Itoh, K. Hayashi, Y. Hamanaka, M. Yamamoto, T. Araki, K. Iriyama, Langmuir **8**, 140 (1992)
76. K. Ataka, J. Heberle, J. Am. Chem. Soc. **126**, 9445 (2004)
77. X. Jiang, E. Zaitseva, M. Schmidt, F. Siebert, M. Engelhard, R. Schlesinger, K. Ataka, R. Vogel, J. Heberle, Proc. Natl. Acad. Sci. USA **105**, 12113 (2008)
78. Y. Nishikawa, T. Nagasawa, K. Fujiwara, M. Osawa, Vib. Spectrosc. **6**, 43 (1993)
79. T.R. Jensen, R.P. Van Duyne, S.A. Johnson, V.A. Maroni, Appl. Spectrosc. **54**, 371 (2000)
80. M. Osawa, Surface-enhanced infrared absorption (2001)
81. A. Roseler, E.H. Korte, Thin Solid Films **313**, 732 (1998)
82. K. Hinrichs, A. Roseler, K. Roodenko, J. Rappich, Appl. Spectrosc. **62**, 121 (2008)
83. D.C. Bradford, E. Hutter, J.H. Fendler, D. Roy, J. Phys. Chem. B **109**, 20914 (2005)
84. E. Garcia-Caurel, E. Bertran, A. Canillas, Thin Solid Films **398**, 99 (2001)
85. A.E. Bjerke, P.R. Griffiths, W. Theiss, Anal. Chem. **71**, 1967 (1999)
86. A.E. Miroshnichenko, S. Flach, Y.S. Kivshar, Rev. Mod. Phys. **82**, 2257 (2010)
87. E. Prodan, C. Radloff, N.J. Halas, P. Nordlander, Science **302**, 419 (2003)
88. S. Yamaguchi, J. Phys. Soc. Jpn. **17**, 1172 (1962)
89. T.W.H. Oates, M. Ranjan, S. Facsko, H. Arwin, Opt. Express **19**, 2014 (2011)
90. R.A. Ferrell, Phys. Rev. **111**, 1214 (1958)
91. D.W. Berreman, Phys. Rev. **130**, 2193 (1963)
92. F. Neubrech, A. Pucci, T.W. Cornelius, S. Karim, A. Garcia-Etxarri, J. Aizpurua, Phys. Rev. Lett. **101**, 157403 (2008)
93. D. Enders, T. Nagao, A. Pucci, T. Nakayama, M. Aono, Phys. Chem. Chem. Phys. **13**, 4935 (2011)
94. J. Kundu, F. Le, P. Nordlander, N.J. Halas, Chem. Phys. Lett. **452**, 115 (2008)
95. S. Cataldo, J. Zhao, F. Neubrech, B. Frank, C.J. Zhang, P.V. Braun, H. Giessen, ACS Nano **6**, 979 (2012)
96. R. Adato, A.A. Yanik, H. Altug, Nano Lett. **11**, 5219 (2011)
97. R. Adato, A.A. Yanik, J.J. Amsden, D.L. Kaplan, F.G. Omenetto, M.K. Hong, S. Erramilli, H. Altug, Proc. Natl. Acad. Sci. USA **106**, 19227 (2009)

# Chapter 10
# Detection of Organic Attachment onto Highly Ordered Three-Dimensional Nanostructure Thin Films by Generalized Ellipsometry and Quartz Crystal Microbalance with Dissipation Techniques

**Keith B. Rodenhausen, Daniel Schmidt, Charles Rice, Tino Hofmann, Eva Schubert and Mathias Schubert**

**Abstract** Highly ordered three-dimensional nanostructure thin films provide substantially increased surface area for organic attachment and new detection principles due to the new and unique optical and physical properties of the nanostructures. Upon organic material attachment, the optical birefringence of these highly ordered three-dimensional nanostructure thin films changes due to screening of polarization charges. The surfaces of the highly ordered three-dimensional nanostructure thin films are thus suitable candidates for studying organic adsorption for sensing and chromatography applications. We review contemporary research in this area and specifically report the monitoring of organic attachment using the generalized ellipsometry and quartz crystal microbalance with dissipation techniques. Both methods are sensitive to the adsorption of organic layers, on the order of few angstroms to few micrometers in thickness, at the solid-liquid interface. The combinatorial use of both techniques, described in Chap. 17, provides insight toward how organic materials attach within highly ordered three-dimensional nanostructure thin films. We discuss studies of fibronectin protein adsorption, decanethiol chemisorption, and cetyltrimethylammonium bromide adsorption. We also address potential future developments and applications.

K. B. Rodenhausen
Department of Chemical and Biomolecular Engineering, University of Nebraska–Lincoln,
1400 R Street, Lincoln, NE 68508, USA
e-mail: kbrod@engr.unl.edu

D. Schmidt · C. Rice · T. Hofmann · E. Schubert · M. Schubert (✉)
Department of Electrical Engineering, University of Nebraska–Lincoln,
1400 R Street, Lincoln, NE 68508, USA
e-mail: schubert@engr.unl.edu

© Springer International Publishing AG, part of Springer Nature 2018
K. Hinrichs and K.-J. Eichhorn (eds.), *Ellipsometry of Functional Organic Surfaces and Films*, Springer Series in Surface Sciences 52,
https://doi.org/10.1007/978-3-319-75895-4_10

## 10.1 Introduction

Three-dimensional (3D) nanostructure thin films, also known as sculptured thin films (STFs), are a class of materials of great contemporary interest. A STF is a porous layer of nanostructures, which possess distinct shape determined by their growth process. These nanostructures are aligned with respect to a single characteristic axis and may be arranged in a regular pattern, an irregular pattern, or a locally ordered pattern. Nanostructure shapes include columns, chevrons, helices, etc. [8, 19]. All of these geometries imply structural and optical anisotropy because the nanostructures are aligned; more regular ordering of the nanostructures implies a higher anisotropy. Anisotropic STFs exhibit birefringence and dichroism [30–32]. STFs have been actively developed as optical elements, and development of STFs for electronic, thermal, chemical, and biological applications is ongoing or expected in the future [19]. Generalized ellipsometry (GE) is the most appropriate technique for measuring and determining the anisotropic properties of STFs [29–32, 35].

Surface structuring is a common technique for enhancement of signals related to the attachment or presence of chemical or biochemical target molecules in the vicinity of a surface. One such technique, for example, is nanoparticle-based enhancement of electro- or photoluminescence. If an analyte is brought into close proximity of a metallic surface with a spatial dimension much smaller than the wavelength of probing light of a Raman spectroscopy measurement, the signal due to inelastic scattering may be amplified by a factor as high as $10^{14}$; this effect is known as surface-enhanced Raman scattering (SERS) [13]. SERS has been used to detect 4-aminotheophenol that adsorbs onto STFs from ethanol solutions more dilute than $1 \mu g/L$ [1, 37]. Raman and luminescence techniques require large non-linear optical coefficients and luminescent material properties to be detected, respectively, and the nanostructured surface aids in providing local electric excitation field enhancement where target molecules must be present for detection.

Probe molecules that have a binding affinity for an analyte of interest may be attached to the sample surface prior to a measurement. Many of the probes that are ubiquitous and have reached commercial availability require integrated tags for fluorescence, radioactivity, or chemical activity studies, from which binding events may be quantified. These tags have major shortcomings, as they act as intermediaries between the actual binding event and detection, may interfere with adsorption or structural conformation changes, and place requirements on the synthesis of probes to allow the addition of the tag. Surface plasmon resonance (SPR) is a technique currently used for label-free sensing [17]. However, SPR signal responses are non-linear and make comparisons between measurements and references a challenge. The quartz crystal microbalance with dissipation (QCM-D) and GE measurement techniques operate under orthogonal measurement principles and can be used simultaneously on the same sample for cross-validation and determination of porosity parameters for an attached organic layer [25].

Planar chromatography involves separation carried out across a two-dimensional (2D) surface. Planar chromatography is widely used for the analysis of chemicals in fields such as clinical, biochemical, environmental, and pharmaceutical testing. Recent advances in nanostructure preparation have led to the development of highly efficient supports built from monoliths or microfabricated structures that are now being employed in a method known as ultrathin-layer chromatography (UTLC) [3, 15, 20]. While the development of UTLC and its supports has progressed rapidly, the detection mechanism that has been utilized in these methods has been based on ultraviolet/visible absorbance or fluorescence measurements that have been limited in detection levels to approximately $500 \, pg/\mu m^2$ for direct (non-labeled) detection or $5 \, pg/\mu m^2$ for labeled-based detection, respectively, as reported when using capillary flow on a monolithic silica support in UTLC. The past and current optical detection modes for UTLC require analytes that are fluorescent, that use Raman-active labeling (i.e., giving large non-linear optical coefficients), and/or substances with sufficient absorption properties for detection (i.e., linear optical extinction coefficients). In order to yield a sufficient signal intensity change for detection, certain surface geometries are needed for the use of SERS or for accumulating enough analyte within the light beam's path to create sufficient light absorption. As a result, these schemes tend to be limited to highly colored or fluorescent analytes or to labeled substances. Thus, it is expected that these detection modes would be applicable to only a minimal range of substances in current UTLC systems.

In this chapter we describe a new class of surface enhanced detection by the use of a linear optical technique, generalized ellipsometry, and by the use of highly ordered 3D nanostructure thin films with a high-degree of ordered arrangement among the individual, electrically conductive nanostructures. We have recently discovered that such surfaces reveal very strong optical anisotropy due to the ordered arrangement of the nanocolumns [26, 33]. We term this nanostructure-induced anisotropy as surface-enhanced anisotropy here. The surface-enhanced anisotropy is caused by a spatially confined and coherent movement of dielectric charges within highly-ordered 3D nanostructures (Fig. 10.1, left). We have demonstrated that the surface-enhanced anisotropy can be measured accurately and with high precision by ellipsometry from the terahertz to the ultraviolet spectral region [9–11, 26, 33]. Specifically, we employed GE, a method that we developed recently as the appropriate tool for characterizing anisotropic surfaces [26].

The surface-enhanced anisotropy comprises both strong birefringence (i.e., difference in major indices of refraction along the three major axes of the anisotropic materials) and strong dichroism (i.e., difference in major indices of extinction along the three major axes of the anisotropic materials). Here, the "material" is rendered by the average response of a large ensemble of nanostructures because the wavelength of probing light is larger than typical dimensions of one individual nanostructure. The effective major axes **a**, **b**, and **c** of this anisotropy (shown in Fig. 10.6) are found to be oriented along the column, perpendicular to the column and parallel to the slanting plane, and perpendicular to the column and parallel to the substrate surface, respectively. Note, that due to the slanting of the nanocolumns, the average separation between the columns within the slanting plane and between the columns perpendicular to the slanting plane is different. Hence, the two major complex optical constants perpendicular to the columns differ from each other in addition to being generally

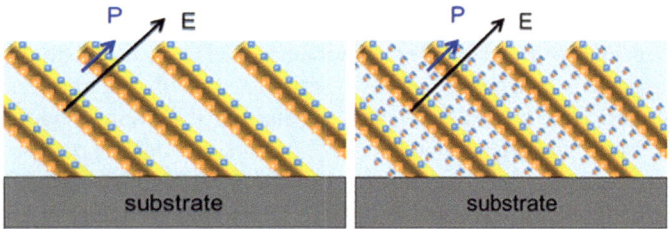

**Fig. 10.1** *Left*: The surface-enhanced anisotropy is caused by anisotropic electric displacement formation within a conductive (metal) columnar thin film of highly-ordered 3D nanostructures on a substrate surface under the influence of an incident electromagnetic wave. $E$ denotes the electrical field component of the electromagnetic wave, and $P$ the induced polarization. If $E$ is oriented parallel to the columns (not shown here), $P$ is much larger since dielectric displacement can form along the column axis. (Note: $P$ is proportional to charge times distance.) The optical axis of this surface-enhanced anisotropy is along the column axis. *Right*: Adsorption of dielectrically polarizable objects (e.g., organic adsorbate molecules) increase the surface-enhanced anisotropy by screening of $P$ perpendicular to the columns, while $P$ parallel remains nearly unaffected. (Typically columnar dimensions: 20–50 nm in diameter and 100–2500 nm in length; lateral separation: 20–150 nm)

different from the complex optical constants for polarization parallel to the column axis. Therefore, these materials respond to electromagnetic radiation like absorbing biaxial "crystals," in general.

The physical origin of the surface-enhanced anisotropy change (Fig. 10.1, right) is a coherent screening effect of the displacement charges within the columnar nanostructures that mostly occurs on the long sides of the columns, while very little screening occurs on the tips of the columns upon attachment of organic molecules. Thereby, the major optical constants for polarization parallel to the columns remain mostly unaffected while the two major optical constants for polarization perpendicular to the column axis change sensitively. It is of great interest here to note that a similar amount of organic attachment in a randomly spatial fashion, for example within a randomly structured nanoporous material, would lead to a fraction of percent change in the material's isotropically averaged optical constants and would hardly be detectable by a linear optical technique. The attachment of the same amount of organic constituent within the highly-ordered 3D nanostructured surface, however, causes a subtle shift of the surface-enhanced anisotropy, and thereby becomes a directly measurable quantity. Such anisotropy change, for example, could be measured by a change in cross-polarized transmittance or reflectance.

The molecular-attachment-induced surface-enhanced anisotropy change can be combined into a highly selective and highly sensitive detection scheme if the nanostructured surface is further functionalized. This functionalization can be done in such a way that only targeted molecules will selectively attach, thereby causing a surface-enhanced anisotropy change. Such surface functionalization can be obtained by using target-selective aptamers that are chemically attached (bonded) to the surface. These aptamers undergo a configuration change when hybridizing with their target

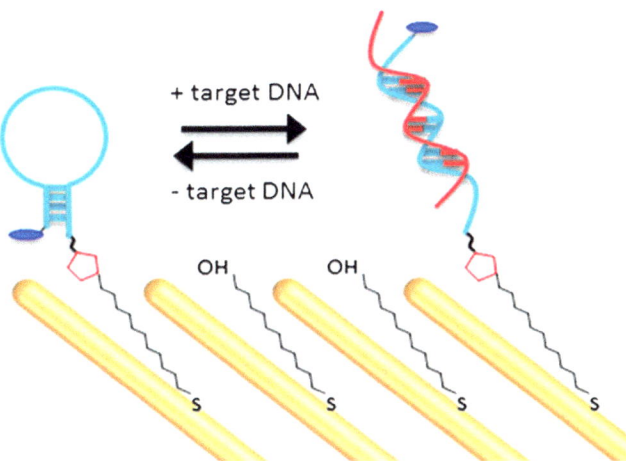

**Fig. 10.2** Principles of target interaction with highly-ordered 3D nanostructures, for example, with aptamer-based DNA. Interaction with target molecules results in attachment of additional organic constituents and surface-enhanced anisotropy change, which can be read-out via ellipsometric methods. Note that the organic molecules are drawn enlarged by about one order of magnitude with respect to the nanostructures

molecule. An example is schematically depicted in Fig. 10.2 for the hybridization reaction with a DNA target molecule. Note that the aptamer image is approximately one order of magnitude enlarged relative to the typical nanostructure dimension.

Measurement and detection of the surface-enhanced anisotropy can be done, for example, either in a source-intensity-modulated, cross-polarized lock-in-based intensity measurement, or in a polarization-modulated, Fourier-analysis-based ellipsometry measurement. The latter provides higher sensitivity to small anisotropy changes with improved accuracy, at the cost of measurement time. The detection of the variation in surface-enhanced anisotropy establishes an intrinsic false-mode suppressing read-out mechanism because only molecules captured (adsorbed) onto the nanostructures will cause a change in surface-enhanced anisotropy.

Due to their porous film characteristics, STFs have higher surface area for chemical adsorption than respective flat surfaces. Thus, STFs can be used as scaffolds for subsequently immobilized probes with surface chemistries that are already understood, such as aptamers or antibodies. The surface concentration of probe relative to that of a flat surface is increased to improve sensitivity to analyte molecules that bind to probes.

The combination of optical methods and the quartz crystal microbalance with dissipation (QCM-D) technique adds further information regarding the porosity of surface layers. Detailed discussion on the theory, implementation, and current progress of combinatorial QCM-D and spectroscopic ellipsometry on flat surfaces is available in Chap. 17. We do point out that future theoretical work is required for application to 3D nanostructured surfaces.

The focus of this chapter is to review contemporary developments to detect and quantify organic adsorption within STFs using the GE and QCM-D techniques. The glancing angle deposition (GLAD) method to fabricate STFs and the atomic layer deposition (ALD) technique to modify STFs are introduced in Sect. 10.2. The GE and QCM-D techniques and analysis methods for chemical detection are discussed in Sect. 10.3. Finally, an overview of ongoing efforts in the field is presented in Sect. 10.4. Discussed are the determination of in-situ monitoring of fibronectin protein adsorption, decanethiol chemisorption, and cetyltrimethylammonium bromide surfactant adsorption, the latter in a transmission setup as a pathway to arrayed detection.

## 10.2 Surface Preparation

The STFs discussed here are grown by electron-beam GLAD. They may be further coated by ALD to fabricate surfaces with specific, further desired chemical and/or material properties.

### 10.2.1 Glancing Angle Deposition

GLAD is a bottom-up fabrication technique to form a layer of nanostructures, also known as a STF, that employs a physical vapor deposition process at oblique angles where the trajectory of the incoming particle flux is not parallel to the substrate normal [8, 16, 19, 24, 35]. The choices of materials and deposition conditions determine the nanostructures' shape, spacing, and morphology [29–32]. Figure 10.3 shows a schematic of the process. Inside an ultra-high vacuum chamber, a target of material is heated by an electron beam. The emitted particles reach the substrate (sample) surface in line-of-sight from the target, where the substrate surface is tilted at an angle with respect to the incident particles. High deposition angles of 80° or greater with respect to the surface normal are commonly used. Discrete nanostructures emerge from the

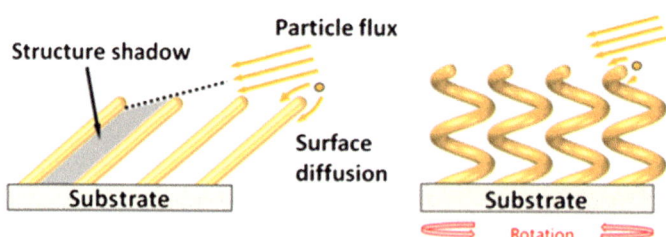

**Fig. 10.3** Glancing angle deposition of columnar thin films with a steady substrate (*left*) and hollow-core helices with a slow and continuous substrate rotation. Reproduced from [28]

initial nucleation sites during the GLAD process and can likewise shadow each other. Thus, an STF comprises individual nanostructures of various lengths on top of a substrate roughened by shadowed nucleation sites. Ideally, the packing fraction of the STF material is uniform with respect to height and column shape across the deposition area. Adatomic diffusion is a dominant mechanism of the GLAD process and affects the nanostructures' morphology [24].

If the sample substrate is rotated continuously during the GLAD process (Fig. 10.3), the direction of the incoming particle flux steadily changes such that spiral- or screw-like nanostructures may be formed. Similarly, if the sample substrate is rotated in discrete increments with intermittent stationary periods, chevron-like nanostructures are yielded.

For applications where control of the spacing between adjacent nanostructures is important, substrates can be seeded by metallic nanoparticles prior to GLAD by a process known as diblock copolymer nanolithography [6, 38]. These nanoparticles may serve as artificially placed initial nucleation sites. Nanoparticles capped by organic materials, such as surfactants or polymers, and dispersed in liquid solution can be close-packed onto a substrate that is dipped in and removed from the solution. The organic capping agent is subsequently removed by plasma etching to yield a pre-patterned substrate for GLAD. The choices of nanoparticle size and organic capping agent determine the spacing distance between nanoparticles and, thus, nanostructures grown by GLAD.

## 10.2.2  Atomic Layer Deposition

ALD is a chemical deposition process that may be used to conformably coat a surface with a thin film whose thickness may be very small, on the order of few Angströms or few nm [18]. The thin film is formed by self-saturating surface reactions. The sample is placed in a vacuum chamber, and an organometallic precursor vapor is introduced that adsorbs to the surface and saturates it over time. If a sample surface comprises 3D structures of high aspect ratio, diffusion may become the dominant mechanism for vapor transport. A purge gas removes excess organometallic precursor. Next, a second precursor vapor (e.g., water) is introduced and chemically reacts with the adsorbed organometallic molecules to form a cohesive layer and eliminate unnecessary precursor ligands. Sometimes this step can be replaced by an alternative treatment, such as plasma or ozone. As illustrated by Fig. 10.4, consecutive intro-duction of precursor vapors and flushing with purge gas allows for the controlled growth of a conformal layer. An ALD coating may provide a desired surface while exploiting favorable structural properties of a different material or an economical use of precious material.

To confer new surface properties to a STF, a homogeneous conformal surface pas-sivation layer can be deposited by ALD following GLAD. Because the nanostructures are 3D, conventional PVD techniques are not useful for post-GLAD modification. Currently available precursors allow, for example, deposition of metal oxide ($SiO_2$,

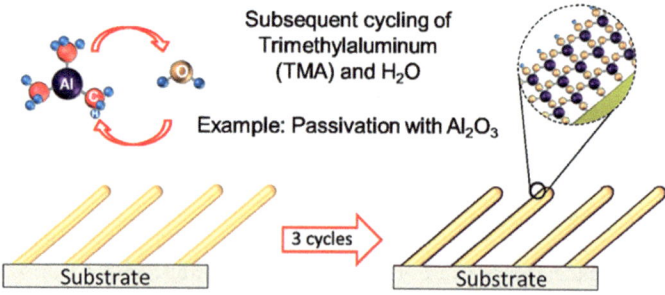

**Fig. 10.4** Schematic of surface passivation for GLAD-prepared highly-ordered 3D nanostructured thin films using ALD. Modified from [33]

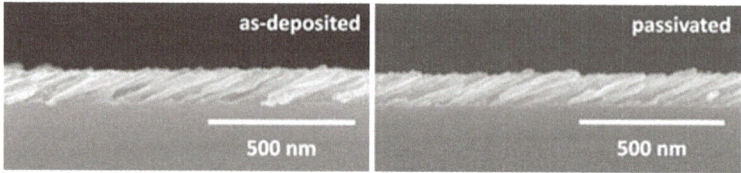

**Fig. 10.5** High-resolution cross-section scanning electron microscope micrographs revealing the structural equivalence before (*left*) and after (*right*) ALD passivation with a 2.4 nm layer of $Al_2O_3$. Reproduced from [33]

$Al_2O_3$, HfAlO, ITO, etc.) and metal (Ru, Pt) conformal overlayers. Such layers may be applied for stabilization against oxidation and as a reproducible, well-defined platform for subsequent linker chemistry for functionalization with biorecognition elements, as shown by Fig. 10.5.

## 10.3 Theory

The GE and QCM-D techniques and the data analysis approaches required to quantitatively determine the amount of organic adsorbate are described here.

### 10.3.1 Generalized Ellipsometry

Generalized ellipsometry is a technique predicated on spectroscopic ellipsometry that draws upon a broader formalism for mathematical treatment of polarized light and that allows the characterization of optically anisotropic materials. Input and output light beams are described by Stokes vectors $\mathbf{S}_{in}$ and $\mathbf{S}_{out}$, respectively, and the optical system (e.g., STF sample) that acts upon the light is described by a $4 \times 4$ element

Mueller matrix **M**, as shown in (10.1),

$$\begin{bmatrix} S_1 \\ S_2 \\ S_3 \\ S_4 \end{bmatrix}_{\text{out}} = \begin{bmatrix} M_{11} & M_{12} & M_{13} & M_{14} \\ M_{21} & M_{22} & M_{23} & M_{24} \\ M_{31} & M_{32} & M_{33} & M_{34} \\ M_{41} & M_{42} & M_{43} & M_{44} \end{bmatrix} \begin{bmatrix} S_1 \\ S_2 \\ S_3 \\ S_4 \end{bmatrix}_{\text{in}} . \tag{10.1}$$

**M** is commonly reported after normalizing all its elements by $M_{11}$. For measurements of a STF consisting of aligned slanted nanocolumns, or a slanted columnar thin film (SCTF), the off-diagonal block elements are zero when the columns align with the ellipsometric plane of incidence (the so-called pseudo-isotropic orientations) [30, 31].

Some or all of the Mueller matrix elements are measurable depending on the ellipsometer's instrumentation configuration. For example, a rotating compensator ellipsometer allows the determination of the first three rows or columns of **M**, depending on the location of the rotating compensator within the experimental setup, while a dual rotating compensator ellipsometer can measure all the elements. Measuring additional Mueller matrix elements allows for a more rigorous data analysis [5, 36].

### 10.3.1.1  Anisotropic Bruggeman Effective Medium Approximation

Effective medium approximations (EMAs) are often used in ellipsometry data analysis schemes for modeling plane-parallel layers that have homogeneously mixed components to determine the volume fractions of constituent materials. The commonly used Bruggeman EMA determines an effective dielectric function for an isotropic mixed layer by rendering the polarizabilities of the constituent materials as equivalent spherical inclusions within a host matrix [4]. For a STF with anisotropic inclusions, the Bruggeman EMA can be modified by introducing depolarization factors $L_{n,j}^{D}$ ($j = a, b, c$) along each of the three orthogonal, major optical polarizability axes (**a**, **b**, and **c**) for the $n$th component, as pictured by Fig. 10.6. The depolarization factors render the now anisotropic polarizability-describing inclusions as ellipsoidal [22]. Thus, three effective dielectric function components, each averaged over the respective polarizability axis, are determined. The anisotropic Bruggeman effective medium approximation (AB-EMA) equations for $m$ constituent materials are

$$\sum_{n=1}^{m} f_n = 1, \tag{10.2}$$

$$\sum_{n=1}^{m} f_n \frac{\varepsilon_n - \varepsilon_{\text{eff},j}}{\varepsilon_{\text{eff},j} + L_{n,j}^{D}(\varepsilon_n - \varepsilon_{\text{eff},j})} = 0, \tag{10.3}$$

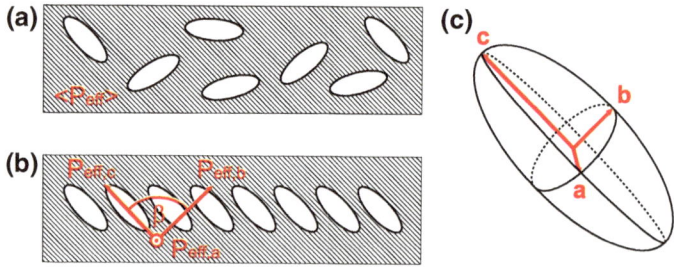

**Fig. 10.6** Effective medium scenarios with mixtures of ellipsoidal inclusions (general case) and a homogeneous host matrix. The mixture with randomly oriented inclusions **a** exhibits an average effective polarizability $\langle \mathbf{P}_{\text{eff}} \rangle$ whereas the mixture with aligned inclusions **b** shows anisotropic properties with three effective polarizability components $\mathbf{P}_{\text{eff},j}$. The major polarizability axes system rendering the biaxial nature of the film is depicted in **c**. Reproduced from [33]

where $\varepsilon_{\text{eff},j}$ is the effective dielectric function along the $j$th axis, $\varepsilon_n$ is the bulk dielectric function of the $n$th constituent material, and $f_n$ is the volume fraction of the $n$th material [7, 11, 33, 34]. For the AB-EMA, one must know or assume the complex index of refraction $N$ ($\varepsilon = N^2 = (n + ik)^2$) of the organic material.

### 10.3.1.2  Optically Determined Surface Parameters

The AB-EMA model yields the volume fraction for each constituent material, and the thickness of the AB-EMA is equivalent to the thickness of the STF. The surface mass density of organic adsorbate on the substrate surface $\Gamma_{\text{GE}}$ can be determined by the following relationship:

$$\Gamma_{\text{GE}} = \rho_{\text{ads}} f_{\text{ads}} d_{\text{STF}}, \tag{10.4}$$

where $\rho_{\text{ads}}$ is the volume mass density of organic adsorbate, $f_{\text{ads}}$ is the organic adsorbate volume fraction parameter, and $d_{\text{STF}}$ is the thickness of the STF. $\Gamma_{\text{GE}}$ is a "pseudo" or effective property parameter because organic material is not actually attaching to the substrate surface but rather to the surface of the nanostructures. The surface mass density of organic adsorbate on the nanostructure surface $\Gamma_{\text{GE}}^{\text{STF}}$ follows by considering the ratio of the STF surface area $A_{\text{STF}}$ and the reference substrate surface area $A_{\text{flat}}$ under the STF, such that

$$\Gamma_{\text{GE}}^{\text{STF}} = \Gamma_{\text{GE}} \frac{A_{\text{flat}}}{A_{\text{STF}}} = \rho_{\text{ads}} f_{\text{ads}} d_{\text{STF}} \frac{A_{\text{flat}}}{A_{\text{STF}}}. \tag{10.5}$$

$\Gamma_{\text{GE}}^{\text{STF}}$ is a real physical property, as it refers to the surface density of mass actually attached onto the available surface of the nanostructures.

## 10.3.2    Quartz Crystal Microbalance with Dissipation

QCM-D is introduced in more detail in Chap. 11. The quartz sensor is used as the experimental substrate, and it is typically coated with a metal or oxide, e.g., Au, Ag, Cr, Ni, $SiO_2$, or Ti. Prior to the in-situ GE and QCM-D measurements described here, the STF is deposited by GLAD onto the coated quartz sensor. ALD, spin-coating, or other optional chemical functionalization may then proceed.

### 10.3.2.1    The Sauerbrey Equation and Nanostructured Surfaces

We recall here the Sauerbrey equation from Chap. 11,

$$\Gamma_{QCM} = -C\frac{\delta\nu_{N_{ov}}}{N_{ov}}, \tag{10.6}$$

where $\Gamma_{QCM}$ is the adsorbate surface mass density, $N_{ov}$ is the harmonic overtone number, $\delta\nu_{N_{ov}}$ is the shift in frequency of a harmonic overtone, and $C$ is $18\,ng/(cm^2\,Hz)$ for a QCM-D sensor with a fundamental overtone frequency of 5 MHz [23, 27]. Equation (10.6) applies to mechanically rigid (non-viscoelastic) experimental systems. STFs under vacuum or air ambient may be considered rigid. But under liquid ambient, the effect of liquid molecules mechanically oscillating with, and thus loading, the QCM-D sensor complicates the interpretation of data. How much liquid is mechanically coupled to the STF? What is the distribution of mechanically coupled liquid molecules in space? To what extent do adsorbate molecules affect the amount or distribution of mechanically coupled liquid molecules? How does one describe the mathematical relationships between (i.e., develop mechanical models for) displacement, restoring, and friction forces associated with the oscillatory shear motion of the 3D nanostructured substrate? These are fundamental questions that current research is seeking to answer.

New modeling approaches for STFs also need to consider the effects of mechanical anisotropy. The QCM-D sensor oscillates along a single axis, and the mechanical response under liquid ambient may be different if a SCTF is aligned parallel or perpendicular with respect to this axis, for example.

Because the 3D surface topography of STFs affects ambient flow patterns during motion of the substrate and the eigenresonances of the substrate's motion may be coupled to the eigenresonances of the STF nanostructures, new mathematical models are necessary for quantifying (or even basic interpretation of) QCM-D data. New modeling approaches are particularly needed for the attachment of large, viscoelastic molecules or macromolecules such as polymers, proteins, or nanoparticles onto STFs under liquid ambient. When air or vacuum is the ambient (negligible friction contributions) and the adsorbate molecules are small and rigid (negligible viscoelastic contributions), deviations from the linear Sauerbrey equation are less severe.

#### 10.3.2.2  Mechano-Acoustically Determined Surface Parameters

The effective mechanically obtained surface mass density parameter $\Gamma_{QCM}$ is directly obtained by (10.6). Like $\Gamma_{GE}$, $\Gamma_{QCM}$ is calculated over the flat reference substrate area. The mechanically obtained surface mass density parameter over the surface of a nanostructure follows the form of (10.5), such that

$$\Gamma_{QCM}^{STF} = \Gamma_{QCM} \frac{A_{flat}}{A_{STF}} = -C \frac{\delta \nu_{N_{ov}}}{N_{ov}} \frac{A_{flat}}{A_{STF}}. \tag{10.7}$$

### 10.3.3  Additional Experimental Considerations

Rotating compensator-based GE and QCM-D allow measurement times on the order of 1 min and several ms, respectively; these periods must be smaller than the time required for the bulk of organic attachment [25]. If instrumentation measurement parameters are unstable with respect to time during the course of an experiment, error is introduced. Such error is particularly relevant as the amount of organic attachment decreases and the duration of attachment increases. Finally, one should consider the effect of adjusting flow parameters during a QCM-D experiment, as variations in pressure within the liquid cell can affect the quartz sensor's mechanical response.

## 10.4  Review of Work in the Field

We now review and discuss work that has been recently accomplished in the field of organic attachment to STFs. Examples include the monitoring of fibronectin protein adsorption [26], the monitoring of decanethiol chemisorption [Rodenhausen et al. unpublished work], and the detection of surfactant by GE in a transmission experimental setup [Rice et al. unpublished work].

### 10.4.1  Fibronectin Protein Adsorption

The AB-EMA described in Sect. 10.3.1.1 was implemented by Rodenhausen et al. to quantitatively determine the amount of fibronectin protein (FN) that adsorbed under a liquid ambient onto a Ti SCTF grown on a Au-coated quartz sensor [26].

Ellipsometry data analysis requires the development of an optical model to yield model-generated data. Relevant parameters of the optical model are varied during a numerical regression-based best-matching procedure of experimental and model-generated Mueller matrix element spectra. A stratified three-layer optical model was used here. The Au-coated quartz substrate was described by a substrate model layer

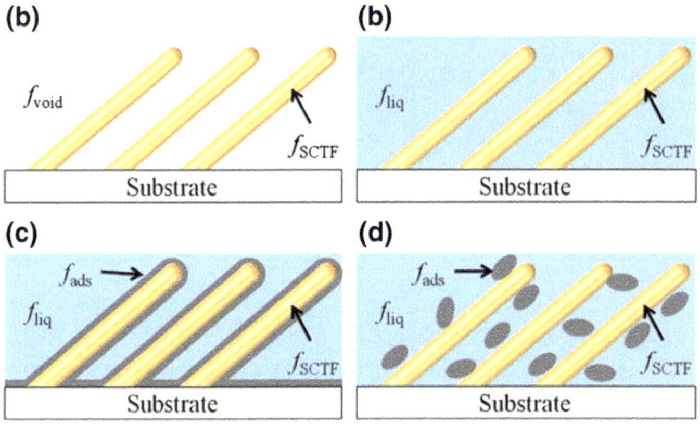

**Fig. 10.7** Illustrations of constituent material fraction regimes. **a** represents a SCTF ($f_{SCTF}$) in air ($f_{void}$) ambient. **b** applies after liquid ($f_{liq}$) replaces air as the ambient, and $f_{liq} = f_{void}$. **c** and **d** represent arrangements of material after an identical amount of analyte ($f_{ads}$) adsorbs and displaces liquid ambient, and $f_{liq} + f_{ads} = f_{void}$. Reproduced from [26]

and characterized prior to SCTF deposition. The SCTF with ambient, organic adsor-bate, and Ti inclusions was described by an AB-EMA layer. An isotropic organic layer was implemented on top of the AB-EMA layer. The two varied parameters during experimental and model-generated data best-matching were the volume fraction of organic adsorbate in the AB-EMA layer ($f_{ads}$) and the thickness of the isotropic organic layer to determine whether the model supported protein adsorbing *within* or forming a separate layer *on top of* the SCTF, respectively. Figure 10.7 shows how fraction parameters represent sample volume spaces. The optical constants of ambient liquid solutions were measured by the minimum deviation technique [2], and the complex index of refraction for FN was assumed to be 1.5.

The Ti SCTF was first characterized by GE measurement of Mueller matrix element spectra at incremental sample orientations comprising an angle of incidence and an in-plane azimuth angle. The sample was then placed in a liquid cell that allowed simultaneous GE and QCM-D measurements [25]. Another GE measure-ment was taken to account for window birefringence effects. The sample in-plane azimuth angle was fixed at 90°, such that the slanting plane of the columns (contain-ing the STF **c**-axis) were oriented normal to the ellipsometry plane of incidence, and the window arrangement of the cell allowed for a 65° angle of incidence.

Repeating in-situ GE and QCM-D measurements were taken of the sample in the liquid cell while standard 1X (10 mM $Na_2HPO_4$, 156 mM NaCl, 2 mM $KH_2PO_4$) phosphate-buffered saline (PBS) at pH 7.4 was pumped through the liquid cell. The only two parameters allowed to vary in the optical model were $f_{ads}$ and the thickness of the isotropic organic adsorbate layer. Throughout the experiment, a constant flow rate of 0.1 mL/min was maintained. During the pumping of blank solution, experimental and model GE and QCM-D parameters were monitored for stability to ensure that no unintended background processes were occurring that

**Fig. 10.8** Select experimental Mueller matrix element spectra from GE measurements just before the FN introduction (*solid lines*) at $t = 145$ min and after FN adsorption and the PBS rinse (*dotted lines*) at $t = 220$ min. Reproduced from [26]

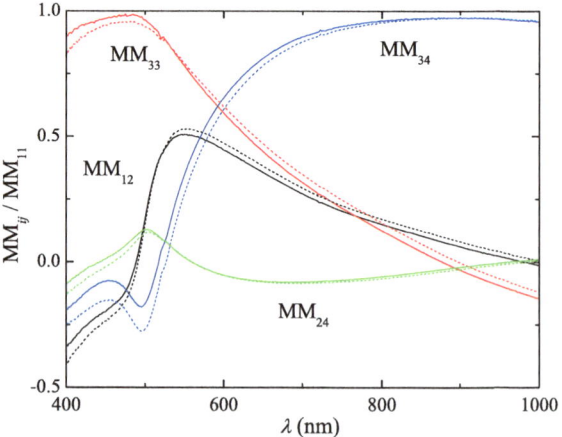

might interfere with the interpretation of results. Once the parameters were deemed sufficiently constant with respect to time ($t = 147$ min), the inlet fluid reservoir was switched to a solution of 10 µg/mL FN prepared in 1X PBS. After 35 min, the inlet reservoir was switched back to the PBS-only solution to rinse passively adsorbed and freely floating FN. GE and QCM-D measurements continued to be taken throughout the adsorption and rinsing processes. The optical constants of the ambient material in the optical model were exchanged between PBS-only solution and FN solution to reflect which ambient was currently flowing into the liquid cell.

The deviations in **M** measured by GE due to FN uptake (Fig. 10.8) were completely accounted for by the optical model with an increase in $f_{\text{ads}}$; the thickness of the isotropic organic adsorbate layer remained negligible, which implied FN adsorbed within the STF (Fig. 10.9). $\Gamma_{\text{GE}}$ and $\Gamma_{\text{QCM}}$ were determined by (10.4) and (10.6), respectively. Figure 10.10 displays these parameters during the FN adsorption and rinsing processes. $\Gamma_{\text{GE}}$ and $\Gamma_{\text{QCM}}$ show striking similarity. As explained in Sect. 10.3.2.1, whether the Sauerbrey equation accurately describes protein adsorption on a STF under liquid ambient is an open question.

Rodenhausen et al. also used cross-sectional scanning electron microscopy (SEM) images to estimate the ratio of $A_{\text{STF}}$ to $A_{\text{flat}}$, which was 4.2. Given an ellipsometrically determined surface mass density of 2 mg/m² for FN on flat Ti [12, 14], the researchers assumed uniform coverage of a nanostructure, such that $\Gamma_{\text{GE}}^{\text{STF}} = 2$ mg/m². From (10.5), the value of $\Gamma_{\text{GE}} = 8.4$ mg/m² was estimated for FN on the Ti SCTF. Figure 10.10 shows peak values of $\Gamma_{\text{GE}} = 14.3$ mg/m² and $\Gamma_{\text{QCM}} = 14.1$ mg/m². The experimental surface mass densities may be larger because the estimation does not consider the surface roughness caused by the GLAD process.

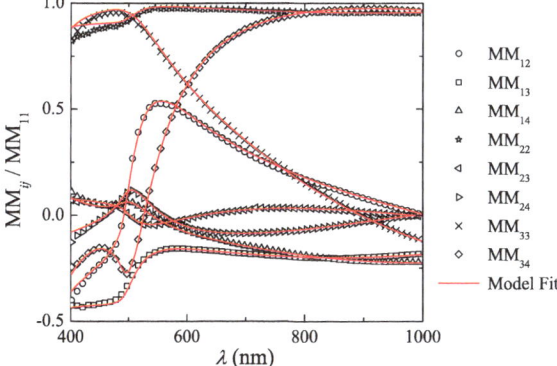

**Fig. 10.9** Model-generated (*lines*) and reduced experimental (*symbols*) Mueller matrix element spectra after FN adsorption and the PBS rinse at $t = 220$ min. Elements $M_{21}$, $M_{31}$, and $M_{32}$ are omitted due to significant overlap with elements $M_{12}$, $M_{13}$, and $M_{23}$, respectively. The thickness of the isotropic adsorbate layer was negligible. Thus, the only model parameter to account for the modulations of the Mueller matrix element spectra was $f_{ads}$. Reproduced from [26]

**Fig. 10.10** Fibronectin adsorption and rinsing processes. $f_{ads}$ on the right axis and (10.4) yield $\Gamma_{GE}$ on the left axis. $\Gamma_{GE}$ shows strong qualitative and quantitative agreement with $\Gamma_{QCM}$, which was calculated via (10.6). Reproduced from [26]

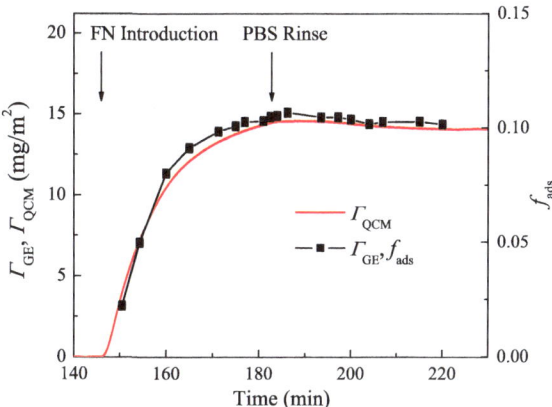

## 10.4.2 Decanethiol Chemisorption

A Ti SCTF coated with Pt via ALD was subsequently chemically functionalized with decanethiol by Rodenhausen et al. to demonstrate the in-situ monitoring of organic chemisorption onto ALD-coated STFs [Rodenhausen et al. unpublished work]. Thiol groups have been used to bind chemical sensor ligands to noble metals, particularly Au [39]. ALD allows the controlled growth of a desired noble metal coating onto a STF made of less expensive material. Although disseminated protocols for ALD deposition of more commonly used Au are not yet available, Pt is an available ALD material. Surface-enhanced birefringence is the approach used here for this first study of organic chemisorption onto an ALD-coated STF.

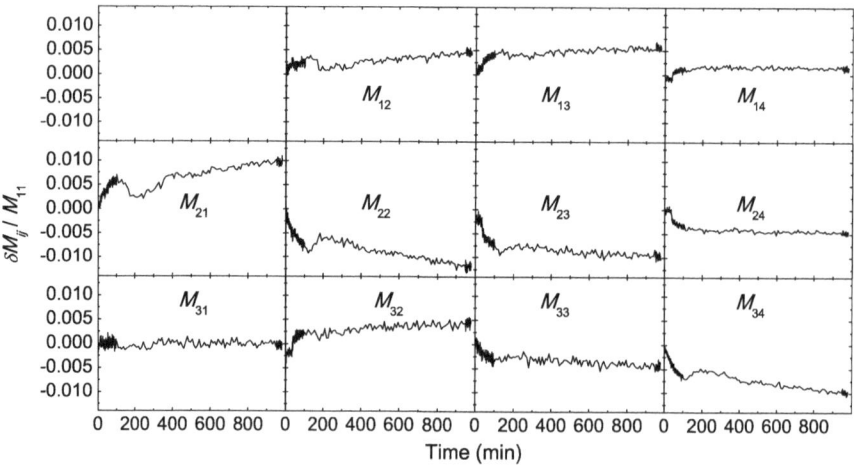

**Fig. 10.11** Mueller matrix element data at $\lambda = 633$ nm measured during chemisorption of decanethiol onto a Pt-coated Ti SCTF. Decanethiol is introduced at $t = 30$ min, pumping is stopped at $t = 100$ min, and pumping resumes with an ethanol rinse at $t = 950$ min

A 70 nm Ti SCTF was deposited on a Au-coated QCM-D sensor. Then, a Pt layer to conformably coat the STF was added via ALD. Ethanol was pumped through the liquid cell at a rate of 0.1 ml/min. The signal stability was monitored for approximately 30 min. Because the oxidation of Pt prevents thiol chemisorption [21], the time the sample was exposed to oxidative environments (e.g., air and ethanol) was minimized after ALD and prior to the introduction of decanethiol solution. At approximately $t = 30$ min, the solution was switched to a 2 mM decanethiol ethanolic solution. Mueller matrix spectra in Fig. 10.11 were acquired approximately once a minute during the first 100 min of the experiment and during a final ethanol rinse step beginning at approximately $t = 950$ min. Between $t = 100$ min and $t = 950$ min, liquid flow was stopped, and an optical measurement was taken once every 10 min. QCM-D data were acquired throughout the optical measurements. The off-diagonal block Mueller matrix elements, which are most sensitive to anisotropy, were used as experimental data for GE data analysis.

A stratified two-layer optical model (similar to that used in Sect. 10.4.1) was used that comprised a Au substrate and a four-component AB-EMA layer. In similar fashion to Sect. 10.4.1, the thickness parameter of an isotropic adsorbate layer on top of the AB-EMA layer was found to converge to zero. The four AB-EMA components were Ti, Pt, ethanol, and decanethiol. The only parameter varied during data analysis was $f_{\text{ads}}$. Because decanethiol is a liquid at standard conditions, its index of refraction and density are known to be 1.458 and 0.824 g/mL, respectively.

$\Gamma_{\text{GE}}$ was determined by applying $f_{\text{ads}}$ and (10.4). The two parameters are shown in Fig. 10.12. Most of the QCM-D response to chemisorption is lost in a baseline drift (not shown). Error from baseline drifts in **M** elements and $\delta\nu$ was made especially

**Fig. 10.12** Decanethiol chemisorption process as monitored by GE. $f_{ads}$ on the right axis and (10.4) yield $\Gamma_{GE}$ on the left axis

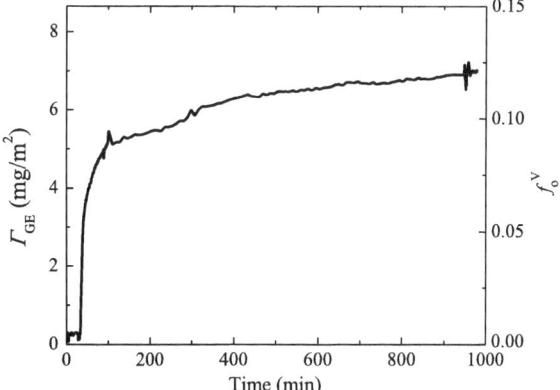

apparent due to the small quantity of chemisorption over a long measurement, nearly 1000 min.

Contact angle measurements of deionized water were made on the SCTF before and after decanethiol chemisorption. The contact angles were 56° and 98°, respectively. The contact angle of 98° seems to follow the trend for alkanethiol self-assembled monolayers (SAM) on flat Pt reported in the literature [21]. However, the ability to evaluate the presence or quality of adsorbate on a STF via contact angle measurement is yet to be determined.

## 10.4.3  Surface-Enhanced Birefringence Chromatography

The surface-enhanced birefringence change upon organic adsorbate attachment can also be detected in a straight-through transmission setup. This concept may lead to applications such as microarrays for detection and chromatography and be applied in conjunction with the ubiquitously referenced "lab-on-a-chip" concept. Additionally, such an experimental setup that provides two-dimensional (2D) spatial mapping would also allow direct side-by-side comparison of STF-coated and flat surfaces [Rice et al. unpublished work].

Rice et al. developed a transmission liquid cell for use with optical measurements. A Ti SCTF was deposited by GLAD onto the interior of one liquid cell window. Cetyltrimethylammonium bromide (CTAB), a cationic surfactant with a hydrophilic ammonium salt "head" group and a hydrophobic hydrocarbon "tail," was used as the organic analyte. At sufficient concentration, surfactants such as CTAB may form micelles, spherical arrangements of individual molecules that are ordered such that the tails are shielded in the interior from water. The critical micelle concentration of CTAB is approximately 1 mM in water. A 2.5 mM aqueous solution of CTAB in aqueous solution with no additives was pumped through the liquid cell, and the adsorption process was monitored with GE.

**Fig. 10.13** Transmission GE data for CTAB adsorption on Ti SCTF. The *top graph* shows experimental (*dotted line*) and model-generated (*solid line*) transmission GE data for normalized Mueller matrix element $M_{24}$. The *bottom graph* shows the CTAB volume fraction $f_{CTAB}$ of the SCTF layer and the thickness parameter $d_{CTAB}$

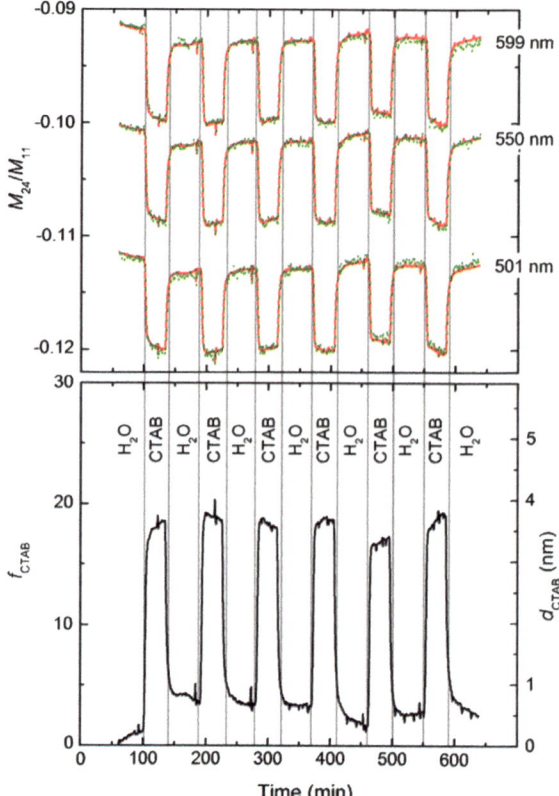

The GE results for CTAB adsorption are shown in Fig. 10.13. Only Mueller matrix element $M_{24}$ was used in this preliminary data analysis because of its high sensitivity to organic adsorption and low drift prior to CTAB exposure relative to the other elements. Experimental and model-generated data for $M_{24}$ are shown in the top graph of Fig. 10.13. The CTAB volume fraction $f_{CTAB}$ of the SCTF layer and the average thickness parameter $d_{CTAB}$ are shown in the bottom graph of Fig. 10.13. $d_{CTAB}$ is calculated from $f_{CTAB}$ using geometry calculations and the dimensions of the SCTF. In similar fashion to the parameter $\Gamma_{GE}^{STF}$, $d_{CTAB}$ is a real parameter calculated over the surface of a given nanostructure. The length of a CTAB molecule is approximately 2.2 nm. The peak value of $d_{CTAB}$ is nearly 4 nm implying that more than a monolayer-amount of CTAB adsorbed, perhaps a bilayer with defects or an undulated layer of micelles.

An intensity-based transmission optical measurement is also possible. The measurement principle here is that formation of a thin film (<10 nm) of adsorbate measured in transmission yields very small intensity changes due to reflection or absorption; on the other hand, if adsorption causes a change in birefringence, a stronger signal may be measured. A pair of crossed polarizers, with the liquid cell between them, allows all source light to be blocked before reaching a charge-coupled device (CCD) array detector. Because the SCTF layers affect the polarization state of light between the crossed polarizers, the azimuth angle difference between the polarizers is not necessarily 90°. Upon the introduction of liquid solution with dissolved organic analyte, the subsequent adsorption of organic material onto the SCTFs modulates the polarization state of light transmitted through the sample and into the second polarizer. Thus, organic adsorption causes light to transmit through the second polarizer. The light intensity is then detected and mapped in two dimensions by the CCD. Measurements after organic adsorption onto windows with SCTF-coated areas and flat areas (for control of the measurement principle) thus provide a proof-of-concept test for optically detecting adsorbate material via birefringence modulation.

It is important to note that because the polarizers are not crossed exactly at 90°, some light will transmit through any flat area of the liquid cell, regardless of the presence of liquid ambient or organic adsorbate. In terms of Mueller matrix elements, the off-diagonal elements are zero for the flat control surface because adsorption of a small organic layer on a flat surface does not cause birefringence; the only Mueller matrix element sensitive to changes of the flat control surface is $M_{11}$.

In the experimental work shown here, the interior of one glass window was coated with a Ti SCTF. The SCTF over part of the window was mechanically removed. Thus, the probing light passed simultaneously through a flat control area and a SCTF-coated area; these areas were differentiated by the CCD and allowed side-by-side comparison. Following 30 min of CTAB exposure, the inlet solution was changed to water to allow rinsing. This cycle was repeated multiple times. 0.1 mM, 1 mM, and 2.5 mM CTAB aqueous solutions were used.

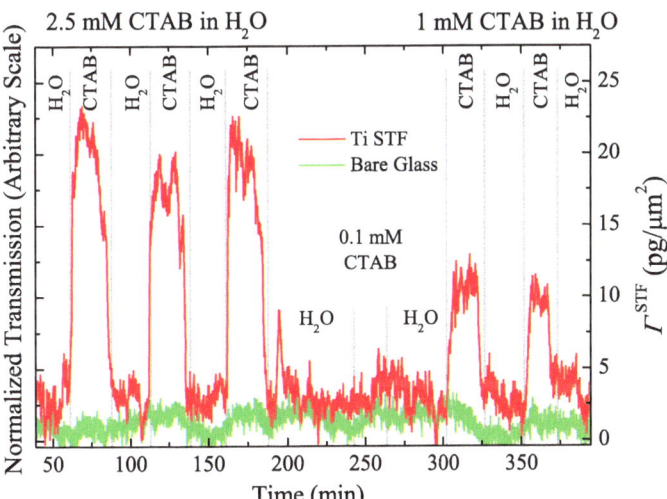

**Fig. 10.14** Transmission data through flat glass and flat glass coated with a Ti STF. The STF allows sensitivity to adsorption of CTAB from 2.5 mM, 0.1 mM, and 1 mM aqueous solutions. The peak surface mass density parameter is inferred from the preceding GE measurement

Figure 10.14 shows changes of measured cross-polarized transmission over two areas of the liquid cell. The peak $\Gamma^{STF}$ value is inferred from the peak $d_{GE}$ value of the preceding GE measurement and an assumed organic volume mass density $\rho_{ads} = 1$ g/mL. The attachment of CTAB only affects the signal of the control very slightly due to light additionally reflected off the organic ultra-thin layer now attached to the glass. This study demonstrates that with STF surfaces and the capability to measure birefringence, one may detect similar amounts of organic adsorbate. These quantities of adsorbate cannot be detected or require much higher instrumental precision to be detected if they attach to a flat surface.

## 10.5  Conclusion

The STF provides an advantageous surface for detecting organic materials because it has a larger surface area for attachment and allows the measurement of optical birefringence modulation. STFs can be made from virtually any material by GLAD and may be functionalized by ALD or chemisorption processes to study various experimental systems. By growing a STF on a quartz crystal sensor, the material may be studied simultaneously by optical, specifically GE, and QCM-D measurements. Recent and ongoing work in this field was reviewed and ultimately points to an expansive future yet to be explored.

## References

1. J.L. Abell, J.M. Garren, J.D. Driskell, R.A. Tripp, Y. Zhao, J. Am. Chem. Soc. **134**, 12889 (2012)
2. T. Berlind, G.K. Pribil, D. Thompson, J.A. Woollam, H. Arwin, Phys. Stat. Solidi C **5**, 1249 (2008)
3. L.W. Bezuidenhout, M.J. Brett, J. Chromatogr. A **1183**, 179 (2008)
4. D.A.G. Bruggeman, Ann. Phys. **416**, 636 (1935)
5. H. Fujiwara, *Spectroscopic Ellipsometry* (Wiley, New York, 2007)
6. R. Glass, M. Arnold, J. Blümmel, A. Küller, M. Möller, J.P. Spatz, Adv. Funct. Mater. **13**, 569 (2003)
7. C.G. Granqvist, D.L. Bellac, G.A. Niklasson, Renew. Energ. **8**, 530 (1996)
8. M.M. Hawkeye, M.J. Brett, J. Vac. Sci. Technol. A **25**, 1317 (2007)
9. T. Hofmann, D. Schmidt, A. Boosalis, P. Kühne, C. Herzinger, J. Woollam, E. Schubert, M. Schubert, Mater. Res. Soc. Symp. Proc. **1409**, CC13 (2012)
10. T. Hofmann, D. Schmidt, A. Boosalis, P. Kühne, R. Skomski, C.M. Herzinger, J.A. Woollam, M. Schubert, E. Schubert, Appl. Phys. Lett. **99**, 081903 (2011)
11. T. Hofmann, D. Schmidt, M. Schubert, in *Ellipsometry at the Nanoscale*, ed. by K. Hingerl, M. Losurdo (Springer, Berlin, 2013)
12. M.B. Hovgaard, K. Rechendorff, J. Chevallier, M. Foss, F. Besenbacher, J. Phys. Chem. B **112**, 8241 (2008)
13. Y.S. Huh, A.J. Chung, D. Erickson, Microfluid. Nanofluid. **6**, 285 (2009)
14. B. Ivarsson, I. Lundström, Crit. Rev. Biocompat. **2**, 1 (1986)

15. S.R. Jim, M.T. Taschuk, G.E. Morlock, L.W. Bezuidenhout, W. Schwack, M.J. Brett, Anal. Chem. **82**, 5349 (2010)
16. T. Karabacak, G.-C. Wang, T.-M. Lu, J. Appl. Phys. **94**, 7723 (2003)
17. R. Konradi, M. Textor, E. Reimhult, Biosensors **2**, 341 (2012)
18. M. Knez, K. Nielsch, L. Niinistö, Adv. Mater. **19**, 3425 (2007)
19. A. Lakhtakia, R. Messier, *Sculptured Thin Flms: Nanoengineered Morphology and Optics* (Wiley, Bellingham, 2005)
20. A.J. Oko, S.R. Jim, M.T. Taschuk, M.J. Brett, J. Chromatogr. A **1218**, 2661 (2011)
21. D.Y. Petrovykh, H. Kimura-Suda, A. Opdahl, L.J. Richter, M.J. Tarlov, L.J. Whitman, Langmuir **22**, 2578 (2006)
22. D. Polder, J.H. van Santen, Physica **12**, 257 (1946)
23. I. Reviakine, D. Johannsmann, R.P. Richter, Anal. Chem. **83**, 8838 (2011)
24. K. Robbie, G. Beydaghyan, T. Brown, C. Dean, J. Adams, C. Buzea, Rev. Sci. Instrum. **75**, 1089 (2004)
25. K.B. Rodenhausen, T. Kasputis, A.K. Pannier, J.Y. Gerasimov, R.Y. Lai, M. Solinsky, T. Tiwald, A. Sarkar, T. Hofmann, N. Ianno, M. Schubert, Rev. Sci. Instrum. **82**, 103111 (2011)
26. K.B. Rodenhausen, D. Schmidt, T. Kasputis, A.K. Pannier, E. Schubert, M. Schubert, Opt. Express **20**, 5419 (2012)
27. G.Z. Sauerbrey, Phys. Hadron. Nucl. **155**, 206 (1959)
28. D. Schmidt, Generalized Ellipsometry on Sculptured Thin Films made by Glancing Angle Deposition, Dissertation, University of Nebraska-Lincoln, (2010)
29. D. Schmidt, E. Schubert, M. Schubert, Phys. Stat. Solidi A **205**, 748 (2008)
30. D. Schmidt, B. Booso, T. Hofmann, E. Schubert, A. Sarangan, M. Schubert, Opt. Lett. **34**, 992 (2009)
31. D. Schmidt, B. Booso, T. Hofmann, E. Schubert, A. Sarangan, M. Schubert, Appl. Phys. Lett. **94**, 011914 (2009)
32. D. Schmidt, A.C. Kjerstad, T. Hofmann, T. Skomski, E. Schubert, M. Schubert, J. Appl. Phys. **105**, 113508 (2009)
33. D. Schmidt, E. Schubert, M. Schubert, Appl. Phys. Lett. **100**, 011912 (2012)
34. D. Schmidt, E. Schubert, M. Schubert, in *Ellipsometry at the Nanoscale*, ed. by K. Hingerl, M. Losurdo (Springer, Berlin, 2013)
35. E. Schubert, F. Frost, H. Neumann, B. Rauschenbach, B. Fuhrmann, F. Heyroth, J. Rivory, E. Charron, B. Gallas, M. Schubert, Adv. Solid State Phys. **46**, 309 (2007)
36. M. Schubert, *Infrared Ellipsometry on Semiconductor Layer Structures: Phonons, Plasmons and Polaritons*, vol. 209, Springer Tracts in Modern Physics (Springer, Berlin, 2004)
37. A. Shalabney, C. Khare, J. Bauer, B. Rauschenbach, I. Abdulhalim, J. Nanophoton. **6**, 061605 (2012)
38. J.P. Spatz, A. Roescher, M. Möller, Adv. Mater. **8**, 337 (1996)
39. W. Yang, J.Y. Gerasimov, R.Y. Lai, Chem. Commun. **20**, 2902 (2009)

# Chapter 11
# Polarizing Natural Nanostructures

**Kenneth Järrendahl and Hans Arwin**

**Abstract**  The first part of this chapter gives a brief description of the polarizing environment we are living in and the possibilities for some animals to detect this polarization. This is followed by a presentation of how animals and plants generate polarized light, usually through reflection from micro- and nanostructures. Special attention is made to scarab beetles reflecting light with a high degree of circular polarization. The use of Mueller matrix spectroscopic ellipsometry to obtain optical and structural properties of the beetle cuticle are demonstrated. Finally some comments on the biological aspects of polarization are made.

## 11.1  Introduction

Polarized light is common in the realm of nature [1–3]. Scattered and reflected light from the sky is linearly polarized to different extent depending on sun elevation and the observed scattering angle. Underwater light is polarized in a similar manner. Another well-known phenomenon is the polarization of light after reflection on water and other horizontal surfaces, predominantly near the Brewster angle. Other types of polarizing surfaces based on layered structures are found on various species among the insects, birds and fishes. The naturally polarized light is dominantly linear but elliptical and circularly polarized light can also be observed [2, 3].

The human eye is, in principle, polarization blind but many living species have polarization vision and use this ability for navigation and possibly also for communication. It is clear that polarization effects play a role in survival on both land and in water.

K. Järrendahl (✉) · H. Arwin
Department of Physics, Chemistry and Biology, Linköping University,
581 83 Linköping, Sweden
e-mail: kenneth.jarrendahl@liu.se

H. Arwin
e-mail: hans.arwin@liu.se

© Springer International Publishing AG, part of Springer Nature 2018
K. Hinrichs and K.-J. Eichhorn (eds.), *Ellipsometry of Functional
Organic Surfaces and Films*, Springer Series in Surface Sciences 52,
https://doi.org/10.1007/978-3-319-75895-4_11

Even if naturally polarized light is abundant and various research studies have been conducted there is a need to learn much more. Biologists continuously develop a comprehensive understanding about how polarization affects animal behaviour and physicists provide descriptions of relations between natural structures and their optical properties. Electron microscopy studies illustrating biological nanostructures are frequently found as well as polarized and unpolarized reflectance studies. Most of these investigations have only provided qualitative information in parallel with computer simulations to verify the observations. However, during the last ten years, optical methodology has developed to a great extent and today quantitative studies are possible. These studies are motivated from pure biological reasons but an increasing interest is also found in the area of biomimetics including efforts to fabricate similar structures for technical applications. The main objective of this chapter is to briefly review polarizing natural nanostructures and report on the rather few studies with polarimetry and ellipsometry that have been conducted.

After some definitions we give a description of the polarizing environment in nature. The ability of some animals to detect polarization is then addressed. This is followed by a presentation of how animals polarize light, usually through reflection from micro- and nanostructures. Special attention is made to scarab beetles reflecting light with a high degree of circular polarization. Finally we conclude with some comments on the biological aspects of polarization and potential applications.

## 11.2 Theoretical Details and Definitions

In this chapter Stokes column matrices $\mathbf{S} = \begin{pmatrix} I & Q & U & V \end{pmatrix}^T$ and Mueller matrices $\mathbf{M} = [m_{ij}]_{4x4}$ will be considered to be normalized ($I = 1$ and $m_{11} = 1$). The degree of polarization $P = \sqrt{P_L^2 + P_C^2}$, the degree of linear polarization $P_L$ and the degree of circular polarization $P_C$ will then be

$$P = \sqrt{Q^2 + U^2 + V^2} = \sqrt{m_{21}^2 + m_{31}^2 + m_{41}^2}, \quad (0 \le P \le 1) \tag{11.1}$$

$$P_L = \sqrt{Q^2 + V^2} = \sqrt{m_{21}^2 + m_{31}^2}, \quad (0 \le P_L \le 1) \tag{11.2}$$

$$P_C = V = m_{41}, \quad (-1 \le P_C \le 1) \tag{11.3}$$

The rightmost expressions in (11.1)–(11.3) are described in terms of the Mueller matrix elements of an interacting material and are true if the incident light is unpolarized, i.e. $\mathbf{S}_{in} = \begin{pmatrix} 1 & 0 & 0 & 0 \end{pmatrix}^T$. The sign of $P_C$ is determined by the handedness of the light with +(-) corresponding to right- (left-) handed polarization. It can also be noted that the ellipticity angle $\varepsilon$ and azimuth angle $\phi$ of the polarization ellipse can be written [4]

$$\varepsilon = \frac{1}{2} \arcsin \left( \frac{V}{\sqrt{Q^2 + U^2 + V^2}} \right) \tag{11.4}$$

$$\phi = \frac{1}{2} \arctan \left( \frac{U}{Q} \right) \tag{11.5}$$

## 11.3 The Polarizing Environment

The most important sources of natural polarized light are presented in this section. We can conclude that polarized light is common in nature and that it is predominantly linear.

**Skylight polarization** One of the main causes of polarized light in nature is scattering of sunlight (and moonlight) in the atmosphere. The solar radiation can generally be treated as unpolarized before it enters the atmosphere and interacts with scattering particles (gases, aerosols, water, ice crystals etc.). The scattering process is rather complex but can, as a first approximation, be treated as Rayleigh scattering assuming that the wavelength of the light is much larger than the size of the scattering particles [1–3].

The principal plane defined by the sun, a scatterer and the observer is presented in Fig. 11.1. The light from the scatterer will be a mixture of unpolarized light and linearly polarized light normal to the principal plane (s-polarized). Thus, the degree of polarization will only have a linear contribution ($P = P_L$) and can be shown [5] to approximately follow the function

$$P_L = \frac{1 - \cos^2(\Theta)}{1 + \cos^2(\Theta)} = \frac{1 - \left(\hat{n}_{sc} \cdot \hat{n}_{su}\right)^2}{1 + \left(\hat{n}_{sc} \cdot \hat{n}_{su}\right)^2} \tag{11.6}$$

Here $\Theta = \theta_{sc} - \theta_{su}$ is the scattering angle defined by the elevation angles to the scatterer $\theta_{sc}$ and the sun $\theta_{su}$. A maximum of $P_L = 1$ occurs when $\Theta = 90°$. In reality $P_L$ will be lower due to factors not included in this approximation. The polarization outside the principal plane can be understood from the rightmost expression in (11.6) with the unit vectors $\hat{n}_{sc}$ pointing from the observer to the scatterer and $\hat{n}_{su}$ pointing from observer to the sun. Accordingly, $P_L$ will be the same in all directions where $\hat{n}_{sc} \cdot \hat{n}_{su}$ is the same.

In some environments, e.g. tropical rain forests, the skylight is strongly reduced. The polarization pattern can still be visible underneath the forest canopy but the degree of polarization will be lower and thus, celestial navigation becomes harder. However, polarization caused by surfaces from vegetation and water then becomes more important and sensing of surface orientation and functions of camouflage may be improved. For many animals polarization vision can therefore be very useful even when brightness is insufficient [6]. Otherwise, very little is known about polarization

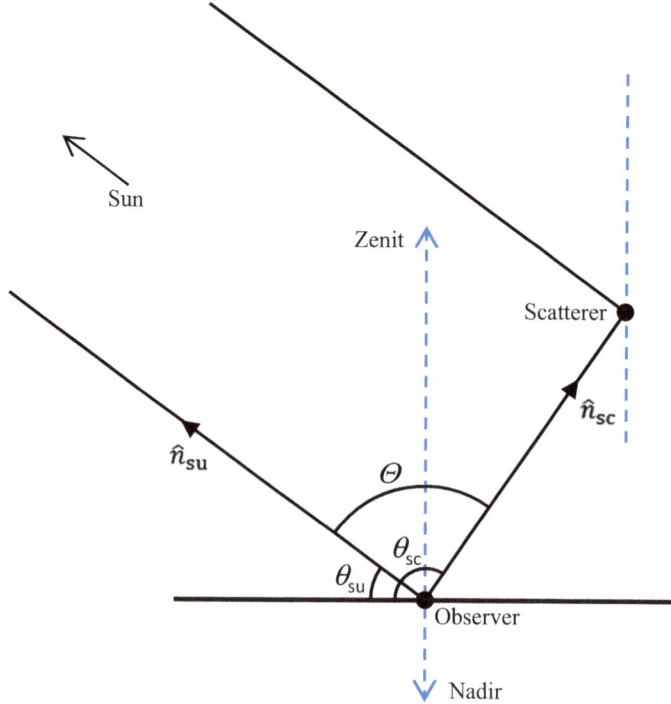

**Fig. 11.1** The principal plane defined by the sun, a scatterer and the observer

of vegetation and its role in nature. Only a few reports are found. Voshchula et al. [7] reports, for example, that linearly polarized light incident on leafs from *Begonia* becomes elliptically polarized upon reflection. The effect on colour perception in insects caused by polarization in light reflected from plants has also been studied but it was found that the effect was very small under natural conditions [8].

Further studies of the sky polarization will reveal more details, for instance on rainbow and cloud phenomena [1, 3] as well as on the four neutral points where $P_L = 0$ [2].

**Polarizing surfaces** A well-known source of polarized light is reflection from surfaces. Fresnel's equations give the reflectance normal to (s) and parallel with (p) the plane of incidence according to

$$R_s = \frac{\sin^2(\theta_i - \theta_t)}{\sin^2(\theta_i + \theta_t)} \text{ and } R_p = \frac{\tan^2(\theta_i - \theta_t)}{\tan^2(\theta_i + \theta_t)} \tag{11.7}$$

where $\theta_i$ is the angle of incidence and $\theta_t$ the angle of refraction. Since the s-reflectance exceeds the p-reflectance for all incident angles, the reflected light will be a mixture of unpolarized light and predominately linearly polarized light normal to the plane of incidence (s-polarized). With incident unpolarized light the degree of linear polarization of the reflected light ($P_L = P$) will be [9]

$$P_L = \frac{\cos^2(\theta_i - \theta_t) - \cos^2(\theta_i + \theta_t)}{\cos^2(\theta_i - \theta_t) - \cos^2(\theta_i + \theta_t)}. \tag{11.8}$$

For an air-dielectric interface $R_p = 0$ and $P_L = 1$ when $\theta_i + \theta_t = 90°$. This is the well-known Brewster's angle ($\theta_i \equiv \theta_B$) which is about 53° for an air/water interface.

**Underwater polarization** The main sources for underwater light are direct sunlight and scattered skylight that reaches an underwater observer through the circular Snell window (having an aperture angle of about 97.5°) as defined by the refraction and total reflection at the water/air surface. Except for elliptically polarized light found near the surface close to the window border, the underwater light will be a mixture of unpolarized light and linearly polarized light. The polarization pattern below the surface of the Snell window can be determined with the same Rayleigh single scattering approximation used in the skylight case. Therefore, (11.6) can be used to estimate the degree of linear polarization also in this case. Here, the reduction of the theoretical $P_L$ values will also depend on the depth [10]. Many other aspects of underwater are addressed by Waterman [11, 12] and others [2, 10].

## 11.4  Polarization Sensitivity and Vision of Animals

The visual systems of many animals are polarization-sensitive [2, 3]. This includes species among the arthropods, chordates and molluscs. In this section we will give some examples from each of these three phyla.

Some animals have an indirect polarization-sensitive vision. They cannot perceive polarization but the photoreceptor response depends on polarization. This type of vision is analogous to a camera with a polarizing filter and may cause false colours. In a more advanced polarization-sensitive vision, the polarization is recognized as a specific property of light similar to wavelength and irradiance. This type of vision could be considered to be an imaging polarimeter. Most studies of polarization-sensitive vision in animals have been focused on possibilities to detect direction and degree of linear polarization in addition to the total irradiance, i.e. the three first Stokes parameters $I$, $Q$ and $U$. Some animals have only a single polarization-sensitive receptor and thus they have to make at least three successive measurements by rotation of the eye, head or body. Other animals have several different types of receptors and can make a simultaneous detection of the direction of polarization and thus perform simultaneous polarimetry.

**Arthropods** Animals in the phylum Arthropoda are invertebrate animals and have exoskeletons. The examples below are from the class Insecta belonging to this phylum or from the subphylum Crustacea.

The honeybee *Apis Mellifera* have compound eyes consisting of many simple eyes called ommatidia [2]. They can detect linear polarization which they use for navigation. They can also communicate direction to food by performing a tail-wagging dance. It has been found, however, that the compound eyes do not serve as polarizers as they are isotropic for on-axis illumination. Instead the polarization sensitivity is located in the photoreceptors which are shaped as narrow tubes called microvilli. The membranes of the microvilli contain the pigment molecule rhodopsin. The microvillar photoreceptors are dichroic and thus polarization sensitive. This is partly caused by form dichroism due to the tubular arrangement leading to a dichroic ratio of up to two but the major effect is due to alignment of the dipoles of the rhodopsin molecules. In vertebrates on the other hand, rhodopsin undergo rotational and translational diffusion which rapidly randomize dichroism. Honeybees have three types of photoreceptors but instead of red, green and blue as in the human eye, they have receptors for green, blue and ultraviolet. Too much polarization sensitivity would affect colour vision and thus mainly the ultraviolet receptors are polarization sensitive and furthermore only those ommatidia directed upwards, i.e. towards the sky, have this property and is used for navigation [2].

Some insects are associated with water and in contrast to honeybees these insects often have enhanced polarization sensitivity downwards. The objective is to resolve water surfaces which in generally has a large horizontal component in the reflected light, especially close to Brewster's angle. For this purpose, the backswimmer *Notonecta glauca* has an orthogonal arrangement of microvilli with horizontal and vertical sensitivity [13]. With these two detector systems, providing the horizontal and vertical irradiances, $S_h$ and $S_v$, respectively, unpolarized light and linearly polarized with azimuth $\phi \pm 45°$ cannot be distinguished. However, $S_h$ and $S_v$ are compared by the insect and if $S_h < S_v$ it holds that $45° < \phi < 135°$ whereas $S_h > S_v$ corresponds to $-45° < \phi < 45°$. The latter would then be recognized as light reflected from a water surface by the backswimmer when flying above a water surface. The photoreceptors are tuned to the UV spectral region as here the water surfaces differs more from the surroundings compared to the visible range and they also have larger polarization contrast. The backswimmer can also hang upside down under the water surface in its resting position but it is not known if the polarization on the ventral side (now pointing upwards) is used for orientation.

Several species of scarab beetles have been found to have polarization sensitivity based on orthogonal microvilli [2] and they use it mainly for navigation. There are even beetles, like *Scarabaeus zambesianus* which uses polarization of the moonlit sky for orientation [14]. Later in this chapter near-circular polarization effects in scarab cuticles will be addressed. It is debated whether beetles really can detect this polarization but it has been reported that *Chrysina gloriosa* has a phototactic response for circularly polarized light and behaved differently depending on polarization [15, 16].

In the subphylum Crustacea, all species have polarization vision and can sense linear polarization [17]. It has been reported that some stomatopods e.g. the mantis shrimp *Gonodactylus smithii* can detect circular polarization [17].

**Chordates** It has been shown that many fish species, e.g. carps [18], besides cones sensitive to blue, green and red light also have ultraviolet sensitive cones in the retina. Furthermore, these ultraviolet cones respond best to vertical polarization whereas the red and green cones respond more to horizontal polarization. The blue cones have very low polarization sensitivity. Several mechanisms for the polarization sensitivity are proposed, e.g. for anchovies it has been suggested that the light passes dichroic filters before being absorbed in the different cones [2]. Some suggestion of the use of polarization vision in fish are navigation during migration, location of plankton, some of which are birefringent and more generally to identify predators, pray and mates [18].

**Molluscs** The subclass Coleoidea belonging to this phylum including octopuses, squids and cuttlefish have in general excellent good vision and other senses. Their single-lens eyes are similar to those of humans but cephalopods are often colour-blind and have maximum sensitivity to wavelengths near 490 nm. However, it has been shown that cephalopods have polarization vision [19] and can be trained by food reward to discriminate between vertical and horizontal polarizations. The polarization detection in cephalopods is done in the retina. In *Octopus vulgaris* as an example, the retina has orthogonally close-packed microvilli coupled to pigments providing polarization sensitivity.

## 11.5   Polarization by Animals

In this section we will give some examples on polarized light originating from animals. As described in Sect. 11.3, scattering and reflection can generate polarized light which in most cases is linearly polarized. This is in principle also true for animal skin or exoskeleton surfaces. We will also discuss the interesting cases where animals generate or reflect light with high degree of circular polarization.

**Arthropods** Among the insects there are several interesting polarization phenomena. Very few animals produce light but the larvae from the fireflies *Photuris lucicrescens* and *Photuris versicolor* have two light-emitting lanterns. Measurements of the luminescence in terms of the left- (right-) handed polarized irradiance $I_L$ ($I_R$), showed that the equation $2\,(I_L - I_R)\,/\,(I_L + I_R) = 2 \cdot P_C$ have opposite signs for the lanterns. That is, the two lanterns emit polarized light with opposite handedness [20].

   Cuticles from many beetles often polarize light due to differences in p- and s-reflectance resulting in a linearly polarized reflection. This can be exemplified with Mueller-matrix data measured on *Coptomia laevis* which at an angle of incidence of $45°$ has $P_C = m_{41}$ close to zero, i.e. no circular polarization capability, but $P_L$ around 0.8 in the spectral range 300 to 900 nm [21]. The rather high $P_L$ values are often attributed to multilayer reflector structures in the cuticle [22, 23]. Even beetles which have pronounced light scattering, may exhibit highly polarized specular reflection. One example is *Cyphochilus insulanus* [24]. Brink et al. made some early studies of directionality, polarization and colour of several moths and butterflies [25]. They also used null ellipsometry to determine the refractive index of the scale material in the moth *Trichoplusia orichalcea* [26]. More recently Brink et al. addressed circular-polarization effects in beetles [27]. Near-circularly polarizing scarab beetles have generated much attention and will be presented in more detail in the next section.

Species of stomatopod crustaceans, also called mantis shrimps, have reflectors that polarize light on their bodies. Two types have been identified and are referred to as red and blue polarization reflectors [28]. The red polarization reflectors are found to be multi-layered structures whereas the blue reflectors contain oval-shaped vesicles scattering the light. Both types of reflectors have a high degree of polarization which is 60–80% in the spectral range 450–550 nm for the shrimp *Hemisquilla californiensis* [28].

**Molluscs** The cuttlefish *Sepia officinalis* exhibit horizontally polarized patterns on the arms and around the eyes [19, 29]. The polarization pattern as well as iridescence is caused by diffraction effects in pigment cells, so called iridophores, containing stacked plates of guanine. The cuttlefish can control the pattern with responses times less than a second. The pattern disappears when camouflaged on the sea bottom as well as during copulation, egg laying and a few other situations.

In some squids, like *Loligo vulgaris*, the largest reflection of iridophores is found in the blue and ultraviolet spectral range with almost 100% degree of polarization [29, 30]. Brink et al. [31, 32] have investigated structural colouration in small spots on the outer surfaces of seashells. In *Helicon pruinosus* [31], a tilted quarter-wave stack is responsible for the iridescence phenomena and the microstructures, composed of aragonite, also exhibit polarization effects. The seashell *Patella granatina* was also investigated and determination of the refractive index of the aragonite containing multilayer stack material was addressed with null ellipsometry [32].

## 11.6  Polarized Light in Reflections from Beetles

In this section we will focus on one of the most interesting polarization effects in nature; beetles which reflect polarized light with a high degree of circular polarization. After a short historical review we will describe ellipsometry studies with emphasis on Mueller matrix spectroscopic ellipsometry (MMSE).

**Background** The polarizing effect found for some beetles is demonstrated in Fig. 11.2. In this basic experiment a collection of beetles are observed through a left-circular polarizer or right-circular polarizer.

The dark appearance of the beetles in Fig. 11.2b shows that the reflected light is to a high extent left-circular. A more careful study reveals that for some wavelength regions and incident angles also right-circular light is reflected. These near-circular polarization effects are otherwise rare in nature but have been found for some beetle species but only in the superfamily Scarabaeoidea. From screening projects [34, 35] and by adopting current taxonomy [36–38], this can be narrowed down to the subfamily Ceratocanthinae in the family Hybosoridae, as well as the

**(a)**                                              **(b)**

**Fig. 11.2**  Images of *Chrysina peruviana*, *Chrysina macropus* and *Chrysophora chrysochlora* taken through **a** a left-polarizing filter and **b** a right-polarizing filter. Modified from [33]

subfamilies Cetoniinae, Dynastinae, Melolonthinae, Phaenomeridinae, Rutelinae and Scarabaeinae in the family Scarabaeidae. However, the beetle taxonomy is continuously evolving, especially after the introduction of molecular phylogenetics, and may change this division in the near future [37, 38]. As a consequence many names are erroneous in previous publications. In this section, the original names from the cited publications are kept but with the present name in parenthesis. Further details can be found in Table 11.1. In terms of different species showing circular polarization effects, the number is close to a thousand [35] but it should be noted that the total number of beetle species is enormous and that most of them have not been investigated or even been discovered yet.

In several works around a century ago the lustrous colours in some butterflies and beetles were examined and discussed [39–43]. In his work from 1911 Michelson [40] reported that the Rutelinae beetle *Plusiotis* (*Chrysina*) *resplendens* reflected light which was "circularly polarized even at normal incidence...". This effect was reported to be greatest in the blue wavelength region. He also mentioned that "...towards the red end of the spectrum traces of circular polarization in the opposite sense appear" and made the conclusion that the observations must "...be due to a 'screw structure' of ultramicroscopic, probably molecular dimensions". Later Gaubert [44] found similar phenomena in the reflection of two other Rutelinae beetles, *Anomala aenea* (*dubia*) and *Chrysina amoena* (*peruviana*), and made comparisons with the optical properties of liquid crystal salts of cholesterol. He also connected the observed polarization properties to lamellar structures of chitin. Mathieu and Farragi [45] investigated several species from the Rutelinae subfamily but also from the Cetoniinae subfamily with special focus on *Calchothea affinis* and *Potosia* (*Proatetia*) *speciossima*.

After this, the polarization studies of beetles where on hiatus for about thirty years until Robinson [34] used polarizers for ocular observations of *Anoplognathus chloropyros* (*chlorpyrifos*), *Cetonia aurata*, *Plusiotis* (*Chrysina*) *resplendens* and

**Table 11.1** A selection of optically studied scarab beetles

| Cetoniinae species | Author, year | Studied in References |
|---|---|---|
| *Calchothea affinis* | van Vollenhoven, 1858 | [45] |
| *Cetonia aurata* | Linneus, 1761 | [21, 34, 45, 46, 56, 61, 65, 66, 70–72, 75–77] |
| *Cotonis mutabilis* | Gory and Percheron 1833 | [45, 67, 68, 73, 75–77] |
| *Ischiopsopha jamesi* | Waterhouse, 1876 | [46] |
| *Ischiopsopha bifasciata* | Quoy and Gaimard, 1824 | [60] |
| *Protatetia cuprea* | Fabricius, 1775 | [75] |
| *Protatetia speciossima* | Scopoli, 1786 | [45, 46] |
| *Pyronota festiva* | Fabricius, 1775 | [57] |
| Rutelinae species | Author, year | Studied in References |
| *Anomala dubia* | Scopoli, 1763 | [44] |
| *Anoplognathus aureus* | Waterhouse, 1889 | [21, 75, 77] |
| *Anoplognathus chlorpyrifos* | Drapiez, 1819 | [34] |
| *Anoplognathus parvulus* | Waterhouse, 1873 | [52, 63] |
| *Anoplognathus viridiaeneus* | Donovan, 1805 | [46] |
| *Calloodes grayanus* | White, 1845 | [52, 63] |
| *Chrysina adelaida* | Hope, 1840 | [62, 82] |
| *Chrysina argenteola* | Bates, 1888 | [21, 33, 61, 70, 71, 75–77, 81] |
| *Chrysina chrysargyrea* | Sallé, 1874 | [33, 81] |
| *Chrysina clypealis* | Rothschild and Jordan 1894 | [62] |
| *Chrysina gloriosa* | LeConte, 1854 | [33, 49, 62, 74, 75, 80, 81] |
| *Chrysina macropus* | Francillon, 1795 | [33, 81] |
| *Chrysina optima* | Bates, 1888 | [49] |
| *Chrysina peruviana* | Kirby, 1828 | [33, 44, 81] |
| *Chrysina resplendens* | Boucard, 1875 | [33, 34, 40, 49, 62, 63, 81] |
| *Chrysophora woodi* | Horn, 1885 | [60, 81] |
| *Chrysophora chrysochlora* | Latreille, 1811 | [60] |
| Scarabaeinae species | Author, year | Studied in References |
| *Phanaeus imperator* | Chevrolat, 1844 | [34] |
| *Scarabaeus festivus* | Harold, 1868 | [46] |

*Phanaeus imperator*. The paper includes a survey conducted at the British museum identifying most of the subfamilies containing circularly polarizing beetles.

Neville and Caveney [46] examined more than 20 species including *Anoplognathus viridiaeneus*, *Lomaptera* (*Ischiopsopha*) *jamesi*, *Phanaeus* (*Scarabaeus*) *festivus* and *Potosia* (*Proatetia*) *speciossima* and made further comparisons with cholesteric crystals. They extended the previous information by presenting optical rotation as a function of wavelength and by showing electron microscopy images of

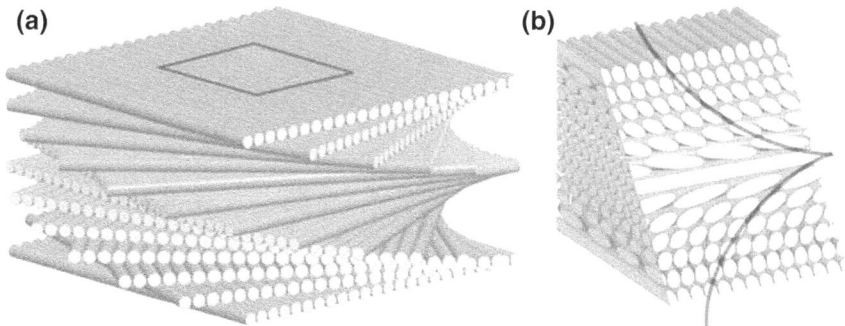

**Fig. 11.3**  A schematic view of the scarab beetle cuticle structure. **a** A half turn of the helicoidal structure. **b** An oblique cut through the structure showing the Bouligand arcs. From [58]

cross-sections of the cuticle structure but incorrectly concluded that the detection of right-handed polarization was a mistake in previous reports.

Studies during the last 40 years have further expanded the knowledge of the cuticle structure [47–53], the surface structure [54–56] and optical properties [35, 46, 49, 52, 56, 57] of scarab beetles. Many of these efforts were summarized in excellent reviews by Leanu and Barfoed [58], Bouligand [59] and Seago et al. [23].

A common conclusion is that the structure in the cuticles of these beetles is a helicoidal arrangement of birefringent chitin layers [58]. A schematic view of the chitin layers twisted one turn is presented in Fig. 11.3a. Figure 11.3b shows an oblique cut through the layer structure and how it reveals the typical arcs defined by Bouligand [47, 50]. The scanning electron microscopy (SEM) image in Fig. 11.4 shows an oblique cross-sectional cut through the exocuticle of *Chrysina resplendens*. The image show parts of two helicoidal structures (h1, h2) separated by a non-helicoidal layer (u). Bouligand arcs are seen in the helicoidal layers.

**Ellipsometry Studies of Scarab Beetles – a Review** Some pioneering work on using spectroscopic ellipsometry to study the polarization properties of light reflected from scarab beetles were performed by Lowrey et al. [60]. They used a spectral ellipsometer set-up to obtain the polarization state of *Chrysophora chrysochlora* and *Chrysina woodi* from the Rutelinae subfamily as well as *Ischiopsopha bifasciata* from the Cetoniinae subfamily. The measurement data were compared with a model where the helicoidal period was chirped with the shortest pitch near the cuticle/ambient surface.

A better understanding of the complex anisotropic structures in beetle cuticle as well as depolarizing properties requires more advanced techniques. Mueller-matrix ellipsometry is then the natural choice as it provides a complete description of specular reflection of a sample. Besides our own early efforts [21, 33, 61] two other groups initiated measurements using Mueller-matrix spectroscopic ellipsometry. Goldstein [62] used a dual-rotating-retarder reflectometer to record Mueller matrices on four *Plusiotis* (*Chrysina*) beetles, *C. adelaida*, *C. clypealis*, *C. gloriosa* and

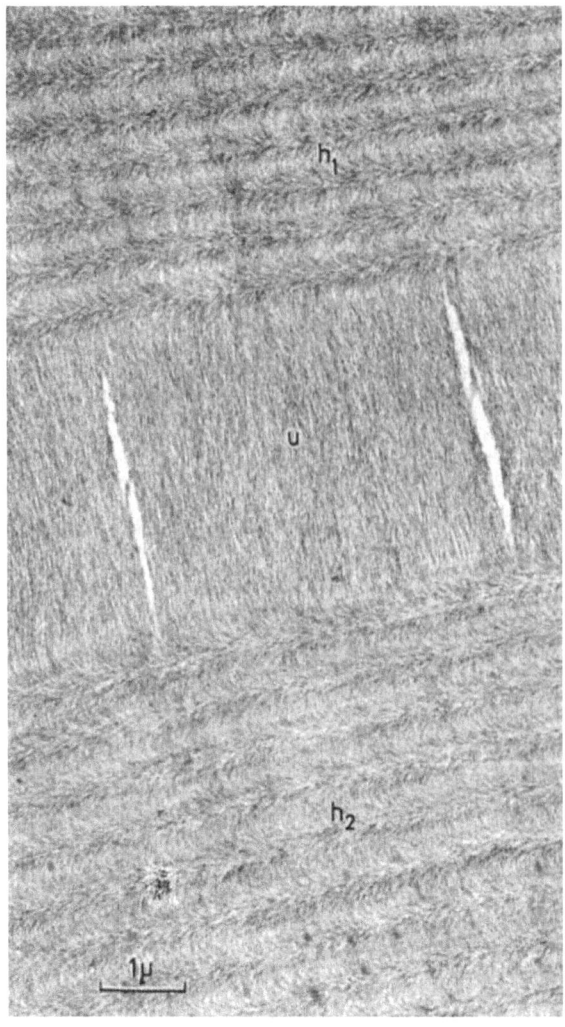

*C. resplendens*, at normal incidence in the spectral range 400–700 nm. From $m_{41}$ being non-zero and negative it was concluded that theses beetles reflected left-handed circularly polarized light, which more correctly should be phrased left-handed elliptically polarized light. In addition *C. resplendens* exhibited $m_{41} > 0$, i.e. right-handed polarization in some parts of the spectrum supporting Michelson observations. The previous studies [57, 60] in the group of Hodgkinson were followed up by employing a low-cost near-normal incidence Mueller-matrix ellipsometer [63]. The full Mueller matrix of four scarab beetles in the spectral range 400–900 nm was recorded. The Cetoniinae beetle *Stephanorrhina guttata*, did not show chirality whereas the Rutelinae beetles *Anoplognathus parvulus* and *Calloodes grayanus* showed left-handed polarization with a high ellipticity ($\varepsilon$ near $-40°$) as well as high degree of polarization. In addition, *C. resplendens* showed right-handed polarization but with lower values of the ellipticity angle and degree of polarization.

In our own studies [21, 33, 61, 64–81] we have access to instrumentation with several important features providing excellent data for understanding of the optical response from beetle cuticles. This includes state of the art optics, wide spectral range, beam focusing down to 50 $\mu$m and a variable angle of incidence. Analysis of off-specular scattered light in the plane of incidence is also feasible. All of these aspects are important to get an extensive view of the polarization properties of the beetles. Furthermore, spectral Mueller-matrix data allow a detailed analysis of optical and structural properties of beetle cuticle.

In the coming sections we present how cuticle polarization features can be determined from Mueller matrices as well as how general electromagnetic modelling is employed to determine cuticle structure. We then describe more advanced Mueller-matrix analysis including determination of allowed optical modes and pitch distribution, sum decomposition to identify basic reflection characteristics and differential decomposition to extract effective linear and circular birefringent and dichroic parameters of cuticle.

**Beetle Cuticle Polarizing Properties Deduced from Mueller Matrices** Below we have chosen to present the polarizance column ($m_{21}$, $m_{31}$, $m_{41}$) of the normalized Mueller matrix in $\lambda - \theta$ contour plots. The polarizance and in particular the $m_{41}$ element, give important information on the elliptical and circular polarization response and many reoccurring patterns are observed. The polarizance $\lambda - \theta$ contour plot for a specimen of *Cetonia aurata* is shown in Fig. 11.5a. At small incident angles $m_{41}$ ($P_C$) is close to $-1$ in a narrow spectral range. An observation of the degree of polarization in the same $\lambda - \theta$ region gives $P$ close to 0.9 [75, 76]. That is, unpolarized light is reflected as near-circular left-handed polarized light in this region. For other wavelengths and at near-normal incidence the degree of polarization is low [77]. For larger incident angles the reflected light is generally linearly polarized and the degree of polarization has a maximum close to 1 at the pseudo-Brewster angle around 55° [77]. This response is also typical for other Cetoniinae beetles but often the left-polarizing region is spectrally broader [75].

The polarizance $\lambda - \theta$ contour plot for a specimen of *Chrysina chrysargyrea* is shown in Fig. 11.5b. Here $m_{41}$ is close to $-1$ in a broad spectral region at small incident angles. With increasing angles this feature becomes spectrally narrower. At large angles, $m_{41}$ instead changes to values close to $+1$ in a narrow spectral range. The degree of polarization is high over most of the $\lambda - \theta$ region. For this beetle, unpolarized light is reflected as near-circular left-handed polarized light for small angles but changes to right-handed polarization at large angles. Similar responses can be seen for other *Chrysina* beetles e.g. *C. argenteola* and golden parts of *C. gloriosa* [75]. Also the well investigated *C. resplendens* has a similar but more complicated response with several shifts between left and right polarization at the same incidence angle. In Fig. 11.6 a few examples of ellipticity angles for a single angle of incidence are shown for a selection of beetles [75] demonstrating the detailed level for evaluation possible from Mueller-matrix data. Further details including $\lambda - \theta$ contour plots

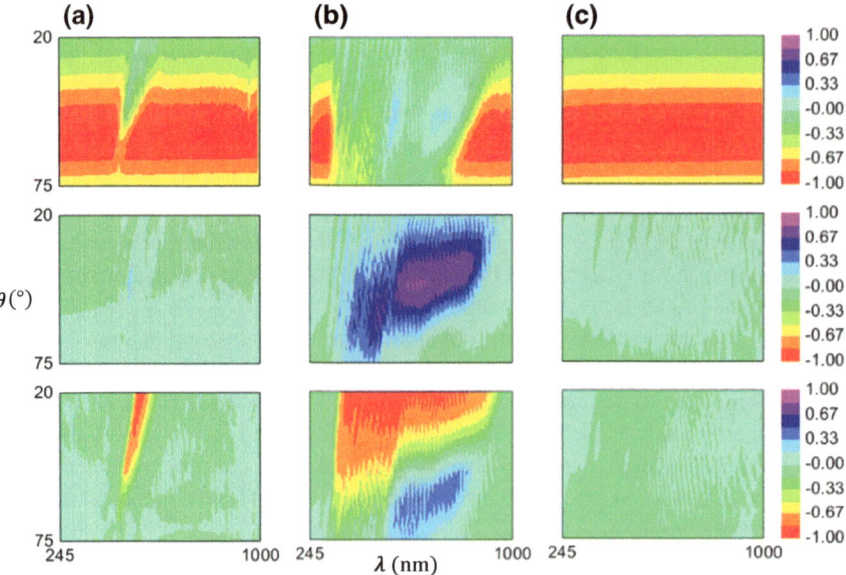

**Fig. 11.5** Polarizance $\lambda - \theta$ contour plots for **a** *Cetonia aurata* **b** *Chrysina chrysargyrea* and **c** *Chrysina gloriosa* (green parts). Data from [21, 33, 80]

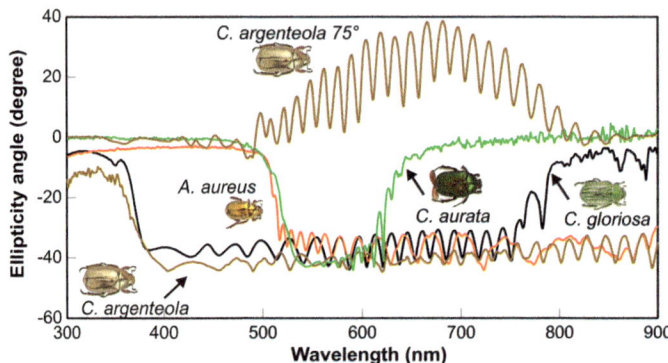

**Fig. 11.6** Ellipticity angle $\varepsilon$ of reflected light for incident unpolarized light as determined from Mueller matrices measured on *C. argenteola*, *C. gloriosa* (golden parts), *C. aurata* and *A. aureus* at an angle of incidence of 20° and for C. argenteola also at 75°

for degree of circular polarization, ellipticity, and azimuth angle for several Chrysina beetles are presented in [81].

It should also be mentioned that some beetles show clear polarization effects in ocular investigations with filters but little response in specular MMSE-measurements. This is exemplified in Fig. 11.5c where a polarizance $\lambda - \theta$ contour plot from a

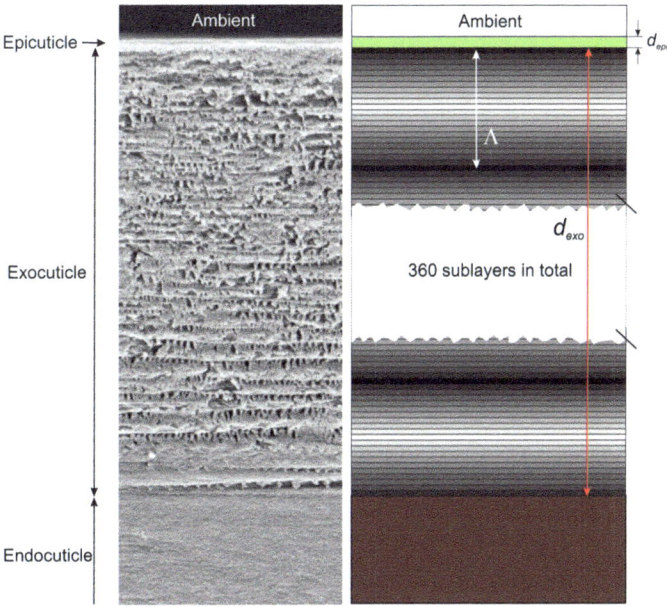

**Fig. 11.7** SEM image of *C. aurata* cuticle (left) and structural model (right) used in regression analysis. The pitch Λ of the twisted structure is indicated as the distance between two planes with the same orientation of the optic axis. From [65]

measurement on green parts of *Chrysina gloriosa* is shown. Other beetles showing this behaviour are, for instance, the highly scattering *Chrysina* beetles *C. macropus* and *C. peruviana* (Fig. 11.2) where the specularly scattered light is linearly polarized. Here reflections with a high degree of circular polarization are instead found when looking off the specular beam [80, 81].

The Mueller-matrix based analysis of polarization properties of reflected light can be generalized from unpolarized incident light to incident light with arbitrary polarization as well as to different types of polarization-state detectors [74]. This is relevant in nature as it is a polarizing environment and in addition beetles and predators may have polarization-sensitive vision.

**Electromagnetic Modelling of Cuticle Structure** Non-linear regression analysis can be employed to model beetle cuticle. The cuticle structure vary in complexity and may include one or more helicoidal structures with different pitch. Other types of layered structures acting as quarter-wave plates and other optical functions are also observed. Here we present a basic model with a stack of twisted biaxial sublayers with a top uniaxial multilayer. Figure 11.7 shows an SEM image of the cuticle of a *C. aurata* specimen and an illustration of the optical model. The stack of 360 consecutively twisted layers mimics a helicoidal structure with thickness $d_{exo}$ and represents the exocuticle and accounts for the circular Bragg reflection effects

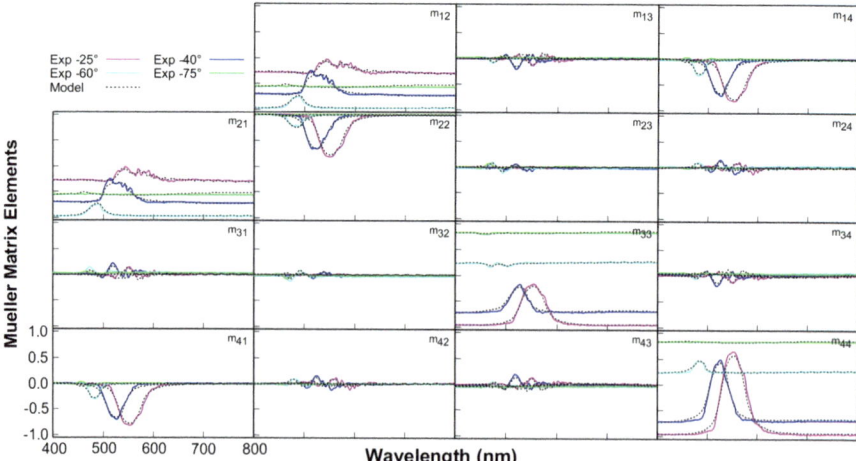

**Fig. 11.8** Mueller-matrix spectra (solid curves) measured at angles of incidence 25°, 40°, 60° and 75° on *C. aurata*. Best fit model-generated spectra using the model in Fig. 11.7 are shown as dashed lines. Only data for 25°, 40° and 60° were used in the regression analysis whereas the model data for 75° are predicted. From [65]

including colour and polarization. An essential layer is the uniaxial top layer with thickness $d_{epi}$ representing the epicuticle which is layered but not chiral.

All layers are assigned optical properties using Cauchy dispersions. The full-turn pitch $\Lambda$ of the helicoidal structure is indicated in Fig. 11.7. In the model $\Lambda$ is allowed to have a small random distribution $\Delta\Lambda$ to account for in depth pitch inhomogeneity. A small absorption is also included which accounts for scattering and implies that the exocuticle effectively becomes semi-infinite. Each of the biaxial sublayers and the uniaxial epicuticle is described with a $4 \times 4$ layer matrix containing the sublayer optical properties [82] and then used in forward calculations of the cuticle Mueller matrix. Non-linear regression is then performed to determine best fit optical and structural parameters including refractive indices, epicuticle thickness, pitch and its distribution. Modelling details for analysis of the cuticle of *C. aurata* of different colours are found in [65, 66].

Figure 11.8 demonstrates an astonishing good match between model data and experimental data for *C. aurata*. Observe that the model can describe the experimental data over the whole visible spectral range and in the angle of incidence range 25° to 75°. For the specimen used to measure the data in Fig. 11.8, the best fit parameters were: $\Lambda = 379$ nm, $\Delta\Lambda = 15.5$ nm and $d_{epi} = 544$ nm. Refractive indices in terms of Cauchy model parameters are found in [65].

**Optical Modes in Cuticle and Modelling of Pitch Grading** Data from many scarab beetles exhibit pronounced interference oscillations overlaid on the surface and Bragg reflection features [21, 73–75]. In the data in Fig. 11.8 these oscillations are suppressed due to absorption in the cuticle of the studied specimen. However, if these

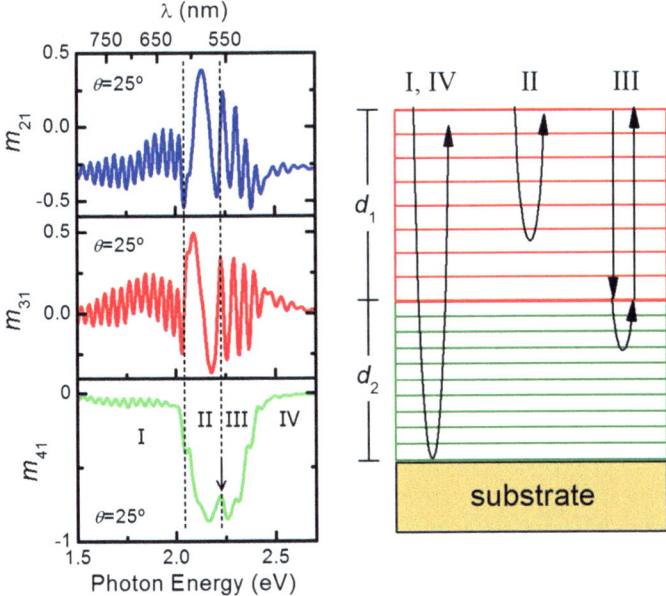

**Fig. 11.9** Spectral dependence of the polarizance vector in the Mueller matrix of a green specimen of *C. mutabilis* at an angle of incidence of 25° (left). Schematics of wave propagation in a structure with two chiral stacks of thicknesses $d_1$ and $d_2$ with pitch $\Lambda_1$ and $\Lambda_2$, respectively. From [67]

oscillations are sufficiently large a phenomenological modelling can be performed by analysing the spectral dependence of their maxima and minima observed in Mueller-matrix elements [67, 68]. These oscillations reveal allowed optical modes in the cuticle. In Fig. 11.9 we show the polarizance vector of the Mueller matrix measured on the beetle *Cotonis mutabilis* and the oscillations are indeed pronounced. Changes with photon energy of these oscillations are clearly visible and after a detailed analysis four spectral regions (I-IV) can be identified as indicated in Fig. 11.9. This allows to determine a structural model with two chiral stacks of thicknesses $d_1$ and $d_2$ comprising the outer part of the exocuticle. Optical modes propagate without attenuation in spectral regions I and IV due to absence of Bragg resonances. The interference oscillations then indicate that the thickness $d_1 + d_2$ of the outer exocuticle is 10.3 μm. In spectral regions II and III, at the top and bottom of outer exocuticle, left-handed polarized light is selectively reflected from chiral stacks due to circular Bragg reflection. In this region the circular Bragg structures effectively act as cavities. The pitch of the Bragg structures can be obtained from an analysis of the oscillations as further detailed in [67]. In some beetle cuticle the pitch distribution is even more complex than a double structure and a general methodology to determine in depth pitch profile using regression analysis is under development [69].

**Sum Decomposition of Mueller Matrices of Beetle Cuticle** When illuminated with natural unpolarized light, we have seen that a beetle with a chiral cuticle can reflect

**Fig. 11.10** The two
non-zero eigenvalues $\lambda_1$ and
$\lambda_2$ obtained by a Cloud
decomposition of Mueller
matrices measured on
*C. aurata* at an angle of
incidence of 20°. The
eigenvalues $\lambda_3$ and $\lambda_4$ are
close to zero. The parameters
$\alpha$ and $\beta$ found in a regression
decomposition fit are also
shown. From [70]

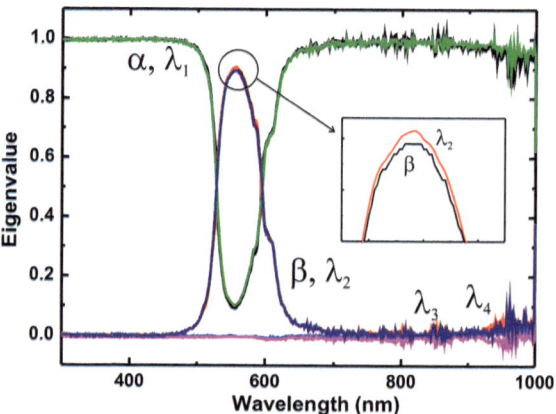

left-handed near-circularly polarized light at near-normal incidence. The observed
Mueller matrix **M** is a non-diagonal depolarizer. Such a depolarizing Mueller matrix
can be decomposed in a sum of four non-depolarizing matrices $\mathbf{M}_i$ in various ways.
We have employed a Cloud decomposition see e.g. [83] which is based on the four
eigenvalues $\lambda_i$ of the covariance matrix of **M**. A Mueller matrix is then expanded as

$$\mathbf{M} = \lambda_1 \mathbf{M}_1 + \lambda_2 \mathbf{M}_2 + \lambda_3 \mathbf{M}_3 + \lambda_4 \mathbf{M}_4$$

The non-depolarizing matrices $\mathbf{M}_i$ are determined from the eigenvalues of the covariance matrix of **M**. Figure 11.10 shows an example of spectral $\lambda_i$ obtained from **M**
measured on *C. aurata*. Only two eigenvalues are non-zero and the cuticle of *C. aurata*
can in a first approximation be described as a sum of two Mueller matrices $\mathbf{M}_1$ and
$\mathbf{M}_2$ with relative weights determined by $\lambda_1$ and $\lambda_2$. These two matrices are identified as those of an ideal mirror and an ideal circular polarizer. As an alternative, a
regression-based decomposition can be performed if the matrices $\mathbf{M}_i$ are known a
priori. The four eigenvalues $\lambda_i$ are then replaced by fit parameters $\alpha$, $\beta$, $\gamma$ and $\delta$. In
Fig. 11.10, a best fit of $\alpha$ and $\beta$ using only a mirror and a circular polarizer is shown.
Further details of the decomposition procedure and applications to more complex
beetle cuticle as well as a generalization to angle-resolved Mueller matrices are found
elsewhere [70, 71].

**Differential Decomposition of Transmission Mueller Matrices of Beetle Cuticle**
For chiral structures it is of interest to quantify its effective circular birefringence and
dichroism. For a homogeneous medium this is readily done by a taking the logarithm
$\mathbf{L} = \ln \mathbf{M}_t$ of a measured transmission Mueller matrix $\mathbf{M}_t$. The elements of the matrix
**L** directly gives all linear and circular birefringent and dichroic properties of the
sample [84] as well as depolarizing properties if any [85]. In Fig. 11.11 we show the
result of a differential decomposition of $\mathbf{M}_t$ measured in transmission on *C. aurata*

**Fig. 11.11** Spectral variation of CB, and CD obtained from a differential decomposition of **M** measured in transmission through the cuticle of a *C. aurata* specimen. From [72]

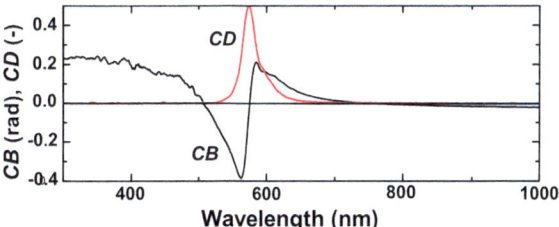

[72]. We can observe that the magnitude of the circular birefringence (CB) in the Bragg resonance region is around 0.4 rad which, with a cuticle thickness of around 20 µm, corresponds to a rotation of polarization of 560°/mm.

## 11.7  Concluding Remarks

Linearly polarized light is very common in nature and many living species can detect this polarization. In many cases the naturally polarized light has a high degree of polarization. Especially the cases with high degree of circular polarization have attracted attention and caused fascination among both physicists and biologists.

The biological significance is an interesting topic and still much is to discover concerning how animals benefit from the polarization phenomena to survive and/or reproduce. It is clear that many species use linearly polarized light for communication and navigation. Polarized light plays an important role in biological signalling but how and to which extent, are questions under debate and areas of active research. There are also possibilities that polarization effects can help to find prey or to avoid predators. A current question is the role of reflections with near-circular polarization. As mentioned above, there are recent reports that *Chrysina gloriosa* can respond on circularly polarized light [15, 16] but it is not clear if this has some biological relevance.

From the research on polarizing natural structures there has been some suggestion on applications. Berthier et al. [86] proposed the possibility to utilize the polarization effects found in some butterflies for anti-counterfeiting of banknotes. Voshchula et al. [7] reported that linearly polarized light incident on leafs from *Begonia* becomes elliptically polarized upon reflection which potentially could be used in agricultural monitoring and early detection of diseases. More examples on structural colour applications and biomimetic design are found in the review by Lenau and Barfoed [58]. A couple of recent examples of biomimetic approaches involving Mueller matrix spectroscopic ellipsometry are found in [87, 88].

We can conclude by stating that the polarizing micro- and nanostructures found in nature are intriguing for both physicists and biologist. They can serve as an inspiration for new applications and also give a deeper understanding of natural systems.

The remarkable optical properties of various species, e.g. some scarab beetles, are yet another reason to maintain biodiversity and stop worldwide habitat destruction.

**Acknowledgements** Our own results presented in this chapter were obtained with the help of Jens Birch, Lia Fernández del Río, Johan Gidholm, Jan Landin, Roger Magnusson, Sergiy Valyukh and Christina Åkerlind from our department. We are also grateful for the fruitful collaboration with Arturo Mendoza-Galván and Eloy Muñoz-Pineda at Cinvestav Querétaro (Mexico), Razvigor Ossikovski and Enric Garcia-Caurel at Ecole Polytechnique Palaiseau (France) and their colleagues. The Swedish Research Council is acknowledged for financial support and Knut and Alice Wallenberg foundation for support to instrumentation. We also want to acknowledge financial support from the Swedish Government Strategic Research Area in Materials Science on Functional Materials at Linköping University (Faculty Grant SFO-Mat-LiU # 2009-00971).

# References

1. G.P. Können, *Polarized Light in Nature* (Cambridge University Press, Cambridge, 1985)
2. G. Horváth, D. Varjú, *Polarized Light in Animal Vision: Polarization Patterns in Nature* (Springer, Heidelberg, 2004)
3. D.H. Goldstein, *Polarized Light* (CRC Press, Boca Raton, 2010)
4. R.M.A. Azzam, N.M. Bashara, *Ellipsometry and Polarized Light* (North-Holland, Amsterdam, 1986)
5. G.S. Smith, Am. J. Phys. **75**, 25 (2007)
6. N. Shashar, T.W. Cronin, L.B. Wolff, M.A. Condon, Biotropica **30**, 275 (1998)
7. V. Voshchula, A.Yu. Zhumar, O.V. Tsaryuk, Biophysics **52**, 418 (2007)
8. G. Horváth, J. Gál, T. Labhart, R. Wehner, J. Exp. Biol. **205**, 3281 (2002)
9. E. Collett, *Polarized Light, Fundamentals and Applications* (Marcel Dekker, New York, 1993)
10. S. Sabbah, A. Lerner, C. Erlick, N. Shashar, Recent Res. Dev. Exp. Theor. Biol. **1**, 1 (2005)
11. T.H. Waterman, Science **120**, 927 (1954)
12. T.H. Waterman, in *Comparative Physiology and Evolution of Vision in Invertebrates B: Invertebrates Visual Centers and Behavior I*, ed. by H. Autrum (Springer, Berlin, 1981)
13. R. Schwind, J. Comp. Physiol. A **154**, 53 (1984)
14. M. Dacke, P. Nordström, C.H. Scholtz, J. Exp. Biol. **206**, 1535 (2003)
15. P. Brady, M. Cummings, Am. Nat. **175**, 614 (2010)
16. E. Warrant, Curr. Biol. **20**, R610 (2010)
17. S. Kleinlogel, A.G. White, PLoS ONE **3**, 1 (2008)
18. C.W. Hawryshyn, Am. Sci. **80**, 164 (1992)
19. N. Shashar, P.S. Rutledge, T.W. Cronin, J. Exp. Biol. **199**, 2077 (1996)
20. H. Wynberg, E.W. Meijer, J.C. Hummelen, H.P.J.M. Dekkers, P.H. Schippers, A.D. Carlson, Nature **286**, 641 (1980)
21. H. Arwin, R. Magnusson, J. Landin, K. Järrendahl, Philos. Mag. **92**, 1583 (2012)
22. J.A. Noyes, P. Vukusic, I.R. Hooper, Opt. Express **15**, 4352 (2007)
23. A.E. Seago, P. Brady, J.-P. Vigneron, T.D. Schultz, J.R. Soc, Interface **6**, S165 (2009)
24. C. Åkerlind, H. Arwin, T. Hallberg, J. Landin, J. Gustafsson, H. Kariis, K. Järrendahl, Appl. Opt. **54**, 6037 (2015)
25. D.J. Brink, J.E. Smit, M.E. Lee, A. Möller, Appl. Opt. **34**, 6049 (1995)
26. D.J. Brink, M.E. Lee, Appl. Opt. **35**, 1950 (1996)
27. D.J. Brink, N.G. van der Berg, L.C. Prinsloo, I.J. Hodgkinson, J. Phys. D. **40**, 2189 (2007)
28. T.-H. Chiou, T.W. Cronin, R.L. Caldwell, J. Marshall, in *Polarization Science and Remote Sensing II*, Proceedings of SPIE, ed. by J.A. Shaw, J.S. Tyo (SPIE, Bellingham, 2005)
29. T.-H. Chiou, L.M. Mäthger, R.T. Hanlon, T.W. Cronin, J. Exp. Biol. **210**, 3624 (2007)

30. L.M. Mäthger, E.J. Denton, J. Exp. Biol. **204**, 2103 (2001)
31. D.J. Brink, N.G. van der Berg, A.J. Botha, Appl. Opt. **41**, 717 (2002)
32. D.J. Brink, N.G. van der Berg, J. Phys. D **38**, 338 (2005)
33. L. Fernández del Río, H. Arwin, J. Landin, R. Magnusson, K. Järrendahl, Image from paper presented at the 7th Workshop Ellipsometry, Leipzig, 5–7 March 2012
34. C. Robinson, Mol. Cryst. **1**, 467 (1966)
35. J.D. Pye, Biol. J. Linn. Soc. **100**, 585 (2010)
36. J. Krikken, Zoologische verhandelingen **210**, 3 (1984)
37. A.B.T. Smith, D.C. Hawkins, J.M. Heraty, Coleopt. Soc. Monogr. **5**, 35 (2006)
38. A.B.T. Smith, Coleopt. Soc. Monogr. **5**, 144 (2006)
39. B. Walter, Die Oberflächen oder Schillerfarben, Braunschweig (1895)
40. A.A. Michelson, Philos. Mag. **21**, 554 (1911)
41. A. Mallock, Proc. R. Soc. Lond. Ser. A **85**, 598 (1911)
42. O.M.F.R.S. Lord Rayleigh, Philos. Mag. Ser. 6 **37:217**, 98 (1919)
43. H. Onslow, Philos. Trans. R. Soc. Lond. Ser. B **211**, 1 (1923)
44. P. Gaubert, C. R. **179**, 1148 (1924)
45. J.-P. Mathieu, N. Faraggi, C. R. **205**, 1378 (1937)
46. A.C. Neville, S. Caveney, Biol. Rev. **44**, 531 (1969)
47. Y. Bouligand, J. Phys. Colloq. **30**, C4 (1969)
48. A.C. Neville, B.M. Luke, Tissue and Cell **1**, 689 (1969)
49. S. Caveney, Proc. R. Soc. Lond. B **178**, 205 (1971)
50. Y. Bouligand, Tissue and Cell **4**, 189 (1972)
51. A.C. Neville, J. Insect Physiol. **23**, 1267 (1977)
52. A. Parker, D.R. Mckenzie, M.C.J. Large, J. Exp. Biol. **201**, 1307 (1998)
53. P. Vukusic, Science **325**, 398 (2009); E.D. Finlayson, L.T. McDonald, P. Vukusic. J. R. Soc. Interface **14**, 20170129 (2017)
54. S.A. Jewell, P. Vukusic, N.W. Roberts, New J. Phys. **9**, 99 (2007)
55. V. Sharma, M. Crne, J.O. Park, M. Srinivasarao, Science **325**, 449 (2009)
56. R. Ossikovski, M. Foldyna, C. Fallet, A. De Martino, Opt. Lett. **34**, 2426 (2009)
57. L. De Silva, I. Hodgkinson, P. Murray, Q.H. Wu, M. Arnold, J. Leader, A. McNaughton, Electromagnetics **25**, 391 (2005)
58. T. Lenau, M. Barfoed, Adv. Eng. Math. **10**, 299 (2008)
59. Y. Bouligand, C. R. Chim. **11**, 281 (2008)
60. S. Lowrey, L. De Silva, I. Hodgkinson, J. Leader, J. Opt. Soc. Am. A **24**, 2418 (2007)
61. K. Järrendahl, J. Landin, H. Arwin, Paper presented at the first NanoCharm Workshop on Advanced Polarimetric Instrumentation, Palaiseau, 7–9 December 2009
62. D.H. Goldstein, Appl. Opt. **45**, 7944 (2006)
63. I. Hodgkinson, S. Lowrey, L. Bourke, A. Parker, M.W. McCall, Appl. Opt. **49**, 4558 (2010)
64. L. Fernández del Río, Optical and Structural Characterization of Natural Nanostructures, Doctoral thesis, Linköping University, 2016
65. H. Arwin, T. Berlind, B. Johs, K. Järrendahl, Opt. Express **21**, 22645 (2013)
66. H. Arwin, L. Fernández del Río, K. Järrendahl, Thin Solid Film. **571**, 739 (2014)
67. A. Mendoza-Galván, E. Muñoz-Pineda, K. Järrendahl, H. Arwin, Opt. Mater. Expr. **4**, 2484 (2014)
68. A. Mendoza, K. Järrendahl, H. Arwin, Materials Today (in press)
69. A. Mendoza-Galván, L. Fernández del Río, K. Järrendahl, H. Arwin In manuscript
70. H. Arwin, R. Magnusson, E. Garcia-Caurel, C. Fallet, K. Järrendahl, M. Foldyna, A. De Martino, R. Ossikovski, Opt. Express **23**, 1951 (2015)
71. R. Magnusson, H. Arwin, E. Garcia-Caurel, K. Järrendahl, R. Ossikovski, Appl. Opt. **55**, (2016)
72. H. Arwin, A. Mendoza-Galvan, R. Magnusson, A. Andersson, J. Landin, K. Järrendahl, E. Garcia-Caurel, R. Ossikovski, Opt. Lett. **41**, 3293 (2016)
73. E. Muñoz-Pineda, K. Järrendahl, H. Arwin, A. Mendoza-Galván, Thin Solid Film. **571**, 660 (2014)
74. L. Fernández del Río, H. Arwin, K. Järrendahl, Mater. Today: Proc. **1S**, 172 (2014)

75. H. Arwin, L. Fernández del Río, C. Åkerlind, S. Valyukh, A. Mendoza-Galván, R. Magnusson, J. Landin, K. Järrendahl, Materials Today (in press)
76. H. Arwin, R. Magnusson, L. Fernández del Río, C. Åkerlind, E. Munoz-Pineda, J. Landin, A. Mendoza-Galván, K. Järrendahl, Mater. Today: Proc. **1S**, 155 (2014)
77. H. Arwin, R. Magnusson, L. Fernández del Río, J. Landin, A. Mendoza-Galván, K. Järrendahl, *Proceedings of SPIE, Bioinspiration, Biomimetics, and Bioreplication, 26 March 2015*, vol. 9429 (2015), p. 942909
78. S. Valyukh, H. Arwin, K. Järrendahl, Opt. Express **24**, 5794 (2016)
79. S. Valyukh, K. Järrendahl, Appl. Opt. **56**, 2510 (2017)
80. L. Fernandez del Rio, H. Arwin, K. Järrendahl, Thin Solid Film. **571**, 410 (2014)
81. L. Fernandez del Rio, H. Arwin, K. Järrendahl, Phys. Rev. E **94**, 012409 (2016)
82. M. Schubert, Phys. Rev. B **53**, 4265 (1996)
83. E. Garcia-Caurel, R. Ossikovski, M. Foldyna, A. Pierangelo, B. Drevillon, A. DeMartino, Advanced Mueller ellipsometry instrumentation and data analysis, in *Ellipsometry at the Nanoscale Chap. 2*, ed. by M. Losurdo, K. Hingerl (Springer, Berlin, 2013), pp. 31–143
84. R.M.A. Azzam, J. Opt. Soc. Am. **68**, 1756 (1978)
85. R. Ossikovski, Opt. Lett. **36**, 2330 (2011)
86. S. Berthier, J. Boulenguez, Z. Bálint, Appl. Phys. A **86**, 123 (2007)
87. R. Magnusson, J. Birch, C-L. Hsiao, P. Sandström, H. Arwin, K. Järrendahl, *Proceedings of SPIE, Bioinspiration, Biomimetics, and Bioreplication, 26 March 2015*, vol. 9429 (2015), p. 94290A
88. A. Mendoza-Galván, E. Muñoz-Pineda, K. Järrendahl, H. Arwin, Opt. Mater. Expr. **6**, 671 (2016)

# Part IV
# Thin Films of Organic Semiconductors for OPV, OLEDs and OTFT

# Chapter 12
# Polymer Blends and Composites

Stergios Logothetidis

**Abstract** The implementation of flexible Organic Electronic (OE) devices in a large variety of consumer applications (generation of electricity, visualization of information, lighting, sensing, etc.) will significantly improve our everyday life. The typical OE device (e.g. a organic photovoltaic cell—OPV) consists of multilayered structures (30–200 nm thick each) fabricated onto rigid (e.g. glass) and/or flexible substrates (as PET, PEN) from transparent and electrically active organic nanolayers. The OE device core is the active or the organic semiconducting (polymer or small molecules) layer, that absorbs photons and generates electric charges, sandwiched between the device electrodes. Finally, the device is encapsulated by barrier layers for the protection of the photoactive layers against degradation and corrosion due to atmospheric gas penetration inside the device. The understanding of the correlation between the polymer blend structure and its optical and electrical properties, and the achievement of the desirable morphology at nanometer scale, is a prerequisite in order to optimize the device performance and stability. Spectroscopic Ellipsometry (SE) from the infrared to the visible and far ultraviolet spectral region has been widely used to provide significant insights on the optical properties, blend morphology, and composition of the polymer blends that are used as active layers in OE devices. In this chapter, we summarize on the latest advances in the implementation of SE from the infrared to the visible and ultraviolet spectral region, for the investigation of the optical and electronic properties, composition profile and structure of polymer nanomaterials that are used as organic semiconductors and transparent electrodes for OE devices, and we discuss the effect of their nanoscale structure on their properties and functionality.

S. Logothetidis (✉)
Lab for Thin Films-Nanosystems & Nanometrology (LTFN),
Department of Physics, Aristotle University of Thessaloniki,
54124 Thessaloniki, Greece
e-mail: logot@auth.gr

© Springer International Publishing AG, part of Springer Nature 2018     271
K. Hinrichs and K.-J. Eichhorn (eds.), *Ellipsometry of Functional
Organic Surfaces and Films*, Springer Series in Surface Sciences 52,
https://doi.org/10.1007/978-3-319-75895-4_12

## 12.1  Introduction

Organic Electronics (OE) is one of the most rapidly emerging sectors of the modern science and technology. This technology is predicted to improve our everyday life in the coming years by the evolution of existing applications, the creation of novel applications and products and by reducing limitations of conventional microelectronics fabrication [1–4]. Among the most exciting applications are the: (a) organic emitting diodes—OLED for displays with superb resolution and conformability, and for lighting in large areas (e.g. roof ceilings and walls), (b) organic photovoltaic cells—OPVs in the form of rolls for generation of electricity everywhere, from tents, walls and portable objects, as bags, (c) organic thin film transistors—OTFTs for performing electronic functions, (d) thin film batteries—TFB for use as portable power sources, and (e) sensors and biosensors [1, 2, 5]. Also, new applications are expected in several sectors, such as in automotive industry (glass with embedded OPVs, interior lighting using OLEDs, etc.), architecture (walls and windows with lighting functionality, rollable television and information systems), healthcare (biosensors for disease detection and vital sign monitoring, etc.) and construction [2, 5]. Among the most intriguing benefits from the use of organic semiconducting materials for deposition onto flexible polymeric substrates include the low cost of the fabrication processes, the mechanical flexibility of the device, the ability to be rolled when the device is not used, the low weight and conformable design [1, 2, 4, 6–8].

The field of flexible organic electronics includes several innovations in materials, such as organic semiconductors (small molecule and polymers) with sufficient electrical conductivity, transparent organic electrodes, hybrid barrier materials for the device encapsulation, and finally, flexible polymeric substrates, as PolyEthylene Terephthalate (PET), and PolyEthylene Naphthalate (PEN) that will replace the Si and glass rigid substrates [5, 9]. The performance, efficiency and lifetime of organic electronic devices are defined by the physical (optical, electronic, electrical, structural, mechanical) properties and the nanoscale morphology of the organic semiconductor, electrode and barrier materials. These should meet specific and advanced requirements, such as high optical transparency, high electrical conductivity, structural stability, ultra low atmospheric gas permeability, film–substrate adhesion, etc. [3–5, 10].

The knowledge of their optical properties in the infrared to ultraviolet spectral region is of considerable importance and it can contribute to the understanding of the bonding and electronic structure and microstructure, their optical transparency as well as the structure-property relationships. In-situ and real-time Spectroscopic Ellipsometry (SE) is a powerful, non-destructive and surface sensitive optical technique that has been used for the investigation of optical, electronic, vibrational, structural and morphological properties, the composition and the growth mechanisms of inorganic and organic bulk materials and thin films [11]. The implementation of real-time SE monitoring and control to large scale production of functional thin films for numerous applications, will lead to the optimization of the materials quality and increase in production yield [11–14].

In this chapter, we will provide an overview of the implementation of SE on the characterization of the optical and electronic properties of polymer:fullerene blends and composites that are used as photoactive nanolayers and transparent electrodes, respectively, for the fabrication of OPVs. This discussion will prove the potentiality of SE for implementation to lab scale processes (as a research tool) and to industrial scale processes (as a powerful quality control tool for the roll-to-roll (r2r) fabrication of flexible OE devices).

## 12.2  Organic Electronics Devices

Among the most important applications of flexible organic electronics are the organic photovoltaic cells (OPVs) and organic light emitting diodes (OLEDs) [2]. The OPVs will be used for solar energy harvesting and conversion towards the generation of electricity for consumer use. There are two different general categories of OPVs; polymer and small molecule OPVs. These two classes of materials are different in terms of their synthesis, and device fabrication processes. The polymer OPVs are processed from solution in organic solvents, whereas small-molecule OPVs are processed mainly using thermal evaporation deposition in a high-vacuum environment. Polymer OPVs are more attractive owing to a number of advantageous features, which include low cost (due to low material consumption and high abundance of organic materials), high optical transparency, mechanical flexibility, and tunable material properties [10, 15–18].

Initially, the planar junction concept (see Fig. 12.1a) has been introduced at the late 1970s, and the first OPVs had achieved power-conversion efficiencies (PCEs) of ~1% [8]. However, this structure has certain limitations, which include a small surface area between the donor and acceptor interfaces and the requirement of long carrier lifetime to ensure that the electrons and holes reach their respective electrodes (metal cathode or transparent anode). These problems can be addressed by introducing a bulk heterojunction (BHJ) structure, which involves mixing donor–acceptor materials in the bulk body of an OPV device (Fig. 12.1b) [19]. Although the BHJ concept can address the exciton dissociation, it has been found that the phase separation between the donor and the acceptor (blend morphology) has a major role in achieving proper charge transport channels for collecting the electrons and holes [20, 21].

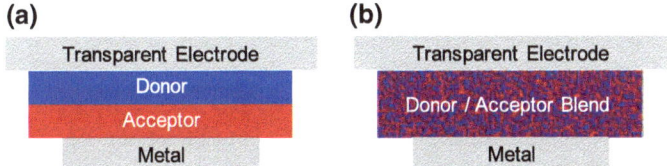

**Fig. 12.1** Typical layer structures of: **a** planar junction OPV, **b** bulk heterojunction OPV

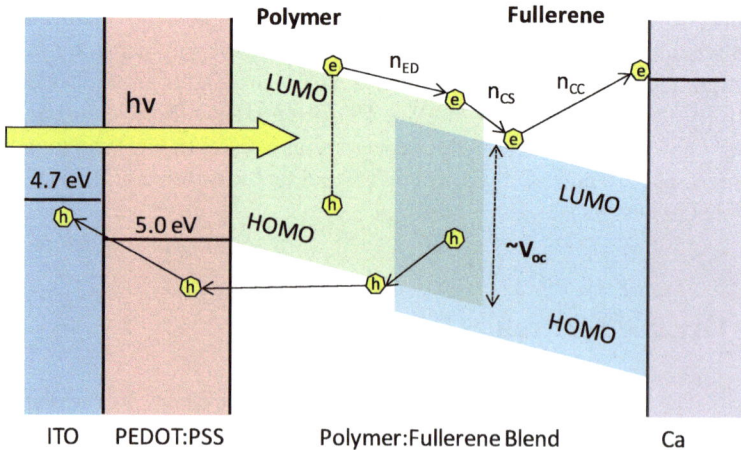

**Fig. 12.2** Principle of operation of a bulk heterojunction OPV

The principle of operation of a polymer:fullerene OPV is shown in Fig. 12.2. The energy difference between the lowest unoccupied molecular orbital (LUMO) of the donor and highest occupied molecular orbital (HOMO) of the acceptor provides the driving force for the dissociation of Frenkel excitons [3, 4, 7, 8, 21, 22]. The two organic semiconductors with electron-acceptor and electron-donor properties are deposited between the anode and cathode layers (with different work functions $\phi$). The bound electron-hole pairs (excitons) are generated within one of the organic layers by light absorption. Although there is a finite probability for exciton dissociation into a positive and a negative polaron within one organic layer (mostly at chemical and/or structural defects of the layer), this process is significantly more efficient directly at a donor-acceptor BHJ with a relative energy-level alignment. The next process is the diffusion of excitons towards the organic-organic interface, where they undergo exciton dissociation that results in electron transfer to the acceptor material and the hole remaining on the electron donor. Finally, the newly generated charge carriers are transported towards the respective electrodes [1, 8, 10, 15–18, 22].

The external quantum efficiency (EQE) of an OPV is calculated as a function of wavelength ($\lambda$) and it is expressed as the ratio between the collected photo-generated charges and the number of incident photons. This is product of four efficiencies ($\eta$): absorption (A), exciton diffusion (ED), charge separation (CS) and charge collection (CC), giving [8]:

$$EQE(\lambda) = \eta_A(\lambda) \times \eta_{ED}(\lambda) \times \eta_{CS}(\lambda) \times \eta_{CC}(\lambda). \quad (12.1)$$

The photo-voltage (or $V_{oc}$) is directly linked to the energy difference between the LUMO level of the acceptor and the HOMO level of the donor, thereby providing the primary driving force for charge separation.

The desired diffusion of excitons towards the donor-acceptor organic interface sets strict requirements on the blend film morphology. The exciton diffusion lengths in polymeric and disordered molecular solids are only a few nanometers, and increases to a few hundred nanometers for highly ordered crystalline materials. Therefore, the average spacing of organic hetero-junctions should be on the length scale of the exciton diffusion lengths. Although, BHJ structures can be fabricated with morphology in the required length scale, however, there are only a few or no continuous percolation paths for charge transport for either electrons or holes within one single phase. Therefore, for the fabrication of efficient OPVs it is crucial to consider and address the materials design (the donor and acceptor must exhibit satisfactory charge mobilities to ensure balanced transport of holes and electrons), the morphology and manipulation, and the interface engineering [6, 8, 20, 23].

Another major application of organic electronic devices is the OLEDs. A schematic energy-level diagram of a multilayer OLED can be seen in Fig. 12.3. The active organic (electroluminescent-EL) layer is formed in the middle of a high work-function ($\phi_1$) anode and a low-work-function ($\phi_2$) cathode. The anode should allow easy hole injection and it consists of a transparent conductive oxide as ITO ($\phi = 4.7\,eV$) or a transparent polymer such as Poly(3,4-ethylenedioxythiophene) poly(styrenesulfonate) (PEDOT:PSS). The cathode should allow easy electron injection and it must have band gap energy higher than the band gap energy of the anode. The cathode consists of metals as Ca, Mg, and Cd, whereas the typical metals, as Ag and Al, have work function values of 5.1 and 4.3 eV, respectively.

By applying an external driving voltage, electrons are injected into the conduction band and holes into the valence band of a semiconducting polymer. The concept of electronic bands in polymers is mainly applicable along the direction of an undisturbed polymer chain, as interchain coupling is rather small. For small-molecular materials, the corresponding energy levels are derived from the LUMO and the HOMO, which are usually confined to one molecule. Upon injection from the electrodes, electrons and holes self-localize to form negative and positive polarons, which

**Fig. 12.3** Energy level diagram of an OLED

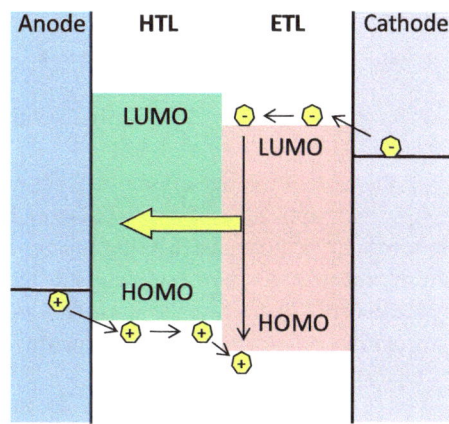

travel under the apparent electric field in opposite directions. When two oppositely charged polarons meet, they can form bound excitons. Exciton formation due to such electrical excitation is governed by spin statistics, which leads to a ratio of 25:75 for singlet excitons (SE)/triplet excitons (TE). In most conjugated organic systems, the lack of heavy atoms in the molecular structure states that only SEs can produce light (radiative decay), while phosphorescence from longer lived triplet states is highly improbable.

## 12.3 Spectroscopic Ellipsometry for the Investigation of Optical and Electronic Properties of OE Nanolayers

Spectroscopic Ellipsometry (SE) is a very powerful technique for the optical characterization of a wide variety of materials (inorganic, organic, composites) in the form of bulk and thin films. Under the appropriate circumstances, SE can determine the thickness, the optical and electronic properties of materials, more accurately than any other technique. Also, SE can provide information on the optical functions, vibrational, electronic, compositional and nanostructural properties, surface roughness, interface layers as well as the volume fraction of multiple phases in composite materials. SE can be implemented for the detailed characterization of all types of materials, such as dielectrics, semiconductors, metals, and organic materials, opaque, semitransparent or even transparent materials [11, 24, 25]. These materials can be deposited either by vacuum and low pressure deposition methods (e.g. evaporation, sputtering, OVPD) or by wet (e.g. spin coating) and printing methods (inkjet, gravure, screen printing, etc.). Its instrumentation relies on the fact that the amplitude and phase of the light polarization changes upon its reflection or transmission through a dielectric interface.

Although the fundamental principles of SE are described in Chap. 1, it can be summarized that SE measures the optical response of bulk materials and thin films as a function of the photon energy $\omega$. The determined properties are the complex refractive index $\tilde{n}(\omega)$ and the complex dielectric function $\tilde{\varepsilon}(\omega)$. The complex refractive index is related to dispersion and absorption of light by the following expression [11]:

$$\tilde{n}(\omega) = n(\omega) + ik(\omega). \tag{12.2}$$

In the case of the reflection of a light beam at the planar interface between two semi-infinite optically isotropic media 0 and 1, when the light beam does not penetrate the medium (1), either due to its high absorption coefficient or its infinite thickness, we are referred to a two-phase (ambient-substrate) system, or a bulk material surrounding by medium (0). In this case the ratio of the $p$-, $s$-Fresnel reflection coefficients, namely the complex reflection ratio is the quantity measured directly by SE and it is given by the expression [11]:

$$\tilde{\rho} = \frac{\tilde{r}_p}{\tilde{r}_s} = \left| \frac{\tilde{r}_p}{\tilde{r}_s} \right| e^{i(\delta_p - \delta_s)} = \tan \Psi \, e^{i\Delta}. \qquad (12.3)$$

In this expression, $\Psi$ and $\Delta$ are the ellipsometric angles, and for a bulk material take values $0° < \Psi < 45°$ and $0° < \Delta < 180°$. From an SE measurement, the complex reflection ratio $\tilde{\rho}$ is estimated, through the calculation of amplitude ratio $\tan \Psi$ and the phase difference $\Delta$. From these two quantities one can extract all the other optical constants of the material. For example, the complex dielectric function of a bulk material with smooth surfaces is directly calculated by the following expression [11, 14]:

$$\tilde{\varepsilon}(\omega) = \varepsilon_1 + i\varepsilon_2 = \tilde{\eta}^2(\omega) = \tilde{\varepsilon}_0 \sin^2 \theta_0 \left\{ 1 + \left[ \frac{1 - \tilde{\rho}(\omega)}{1 + \tilde{\rho}(\omega)} \right]^2 \tan^2 \theta_0 \right\}, \qquad (12.4)$$

where $\theta$ is the angle of incidence of the beam, and $\tilde{\varepsilon}_0$ the dielectric constant of the ambient medium (for the case of air $\tilde{\varepsilon}_0 = 1$).

Although some information on the optical properties can be deduced directly from the SE measurements, quantitative data can be derived from theoretical modeling. This procedure includes the formulation of a theoretical model (which approximates the film architecture and structure of the studied material), and the fitting of the measured $\langle \tilde{\varepsilon}(\omega) \rangle$ to this model by using the desired parameters, as variables in the numerical analysis. The optical and electronic response of the thin films can be deduced by the parameterization of the measured $\langle \varepsilon(\omega) \rangle$ by the use of appropriate theoretical models. These models include, for example, the damped harmonic oscillator (Lorenz model), and the Tauc–Lorentz (TL) model, which calculates the fundamental optical gap $\omega_g$ [26, 27]. These models have been described in detail in Chap. 1.

### 12.3.1   Polymer: Fullerene Blends as Photoactive Layers for OPVs

The most successful OPV active material system that is used mainly as photoactive layers in OPVs is the polymer:fullerene BHJ system. This consists of a blend of a p-type semiconductor (electron donor), such as poly(3-hexylthiophene) (P3HT) with an n-type semiconductor (electron acceptor), such as methanofullerene derivatives (PCBM) [1, 3, 4, 10, 28, 29]. The adoption of $C_{60}$ fullerene and its derivatives to replace the n-type molecules in OPV devices was one of the major breakthroughs in OPV technology. Due to their strong electronegativity and high electron mobility, $C_{60}$ derivatives have become standard n-type molecules in OPV devices [10, 15, 17, 18, 30]. These conjugated molecules are electronically active because of their highly polarizable $\pi$-systems. Efficiencies of more than 5% have been reported for the most studied system consisting of P3HT and PCBM, whereas efficiencies of more than 7% have been realized for low band gap polymers, as benzodithiophene [31].

The high efficiency of OPVs based on P3HT:PCBM can be correlated to the intrinsic properties of the blend components. The regioregular P3HT self-organizes into a microcrystalline structure and, because of efficient interchain transport of charge carriers, its hole mobility is high ($\sim 0.1\,cm^2$/Vs) [15]. Moreover, in thin films interchain interactions cause a red shift of the optical absorption of P3HT, which provides an improved overlap with the solar emission spectrum. PCBM has an electron mobility of $2 \times 10^{-3}\,cm^2$/Vs [15]. Compared to $C_{60}$, the solubility of PCBM in organic solvents is greatly improved, which allows the utilization of deposition techniques requiring highly concentrated solution.

In order to achieve maximum charge generation, a large interfacial area between the polymer and fullerene is required, which can be only achieved by the formation of an optimum distribution of the components. Ideally, more p-type material (polymer) should be located at the interface of the hole collecting anode and more n-type (fullerene) material should be at the electron collecting cathode facilitating collecting of charges from the photoactive layer to the electrodes. Nevertheless, the morphology of BHJ structures cannot be easily controlled. The final blend film structure is affected by the experimental parameters, as blend composition, viscosity, solvent evaporation rate or substrate surface energy. Therefore, there are difficulties to achieve the desired blend morphology for maximum charge generation and transport [5, 23, 32]. Furthermore, the detailed investigation of the BHJ morphology is quite challenging, as few techniques are able to characterize such complex materials with adequate lateral and vertical resolution at the nanoscale.

Several methods have been proposed to control the blend morphology, that include the appropriate solvent choice of spin-coated films [33–36], thermal annealing of blends [37], slow drying, melting of bilayers [38] and vapor annealing [39]. These different methods tend to promote the formation of a phase-separated morphology with crystalline P3HT and PCBM domains and in all cases lead to improved OPV performance [40].

The blend morphology can be determined by the use of SE working in the Vis-farUV spectral region. Prior to the calculation of the optical and electronic properties of the blend, the optical response of the pristine polymer and fullerene materials has to be determined. The calculated bulk dielectric function $\varepsilon(\omega)$ of bulk pristine P3HT is shown in Fig. 12.4. The pristine P3HT was prepared in chlorobenzene (10 mg/ml each) and it was deposited by spin coating at 1000 rpm for 30 s onto glass substrates. Afterwards it was subjected to a thermal annealing process for 5 s on a hot plate at different temperatures in order to study the effect of the annealing on its electronic response. For the analysis of the $\langle \varepsilon(\omega) \rangle$, a theoretical model was used that consists of the layers air/(50%P3HT+50%voids)/P3HT/(glass substrate), whereas the electronic transitions of P3HT were modeled by the use of 3 TL oscillators.

The characteristic electronic transition energies are calculated at 2.04 (singlet excitonic transition), 2.24 and 2.46 eV. From Fig. 12.4 we observe that the increase of the post-deposition thermal annealing is accompanied with an increase of the intensity of the first electronic transition at 2.04 eV. This can be attributed to the increased crystalline ordering of P3HT domains (this peak is assigned to an interchain-delocalized excitation and its intensity has been linked to the degree of P3HT crystallinity [41])

**Fig. 12.4** Calculated bulk
dielectric function $\varepsilon(\omega)$ of
P3HT before and after its
post deposition thermal
annealing to 145 °C and
160 °C

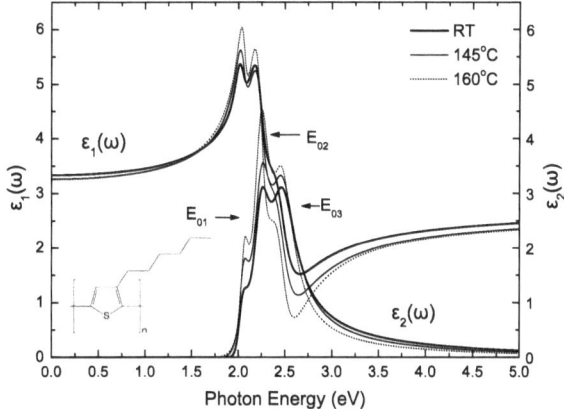

**Fig. 12.4** Calculated bulk dielectric function $\varepsilon(\omega)$ of P3HT before and after its post deposition thermal annealing to 145 °C and 160 °C

**Fig. 12.5** Imaginary part $\varepsilon_2(\omega)$ of the bulk dielectric function of the $PC_{60}BM$ and $PC_{70}BM$ fullerene derivatives

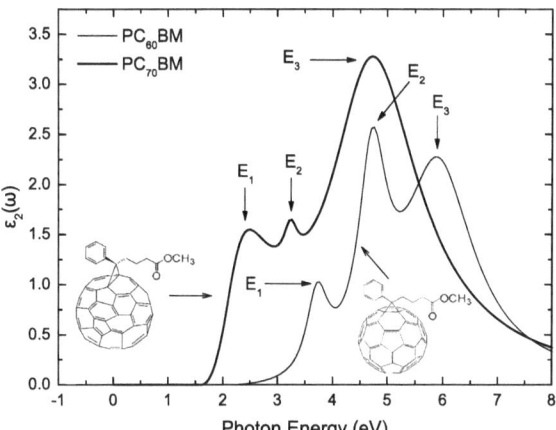

and to the increase of the conjugation length [42]. Although the electronic transition energies were found to be independent on the annealing temperature, the band gap energy reduces from 1.89 (RT) to 1.73 eV ($T = 160$ °C).

Figure 12.5 shows the calculated bulk dielectric function $\varepsilon(\omega)$ of the fullerene derivatives. It is clear that $PC_{60}BM$ (denoted as PCBM) and $PC_{70}BM$ differentiate in terms of their optical band gap and other electronic absorptions. The calculated bulk $\varepsilon(\omega)$ of $PC_{60}BM$ is dominated by three electronic transition energies at 3.7 eV ($S_0 \rightarrow S_{17}$), 4.77 eV ($S_0 \rightarrow S_{37}$) and 5.85 eV ($S_0 \rightarrow S_{56}$), whereas its optical band gap energy is calculated at $\omega_g = 2.09$ eV. Concerning the pristine $PC_{70}BM$, its absorption spectra include the electronic transitions at 2.25, 3.19 eV and a broad absorption at 4.85 eV [40, 43, 44]. From Fig. 12.5 it is clear that $PC_{70}BM$ has a higher absorption in the visible spectral region than $PC_{60}BM$. This is expected to increase the absorption in a fullerene:$PC_{70}BM$ blend and the OPV device efficiency. Moreover, the BHJ morphology is affected by the more hydrophilic behavior of $PC_{70}BM$.

**Fig. 12.6** Measured $\langle \varepsilon(\omega) \rangle$ of two representative P3HT:PCBM thin films; as grown (not annealed) and annealed at $140\,^{\circ}$C for 30 min. Modified after [43]

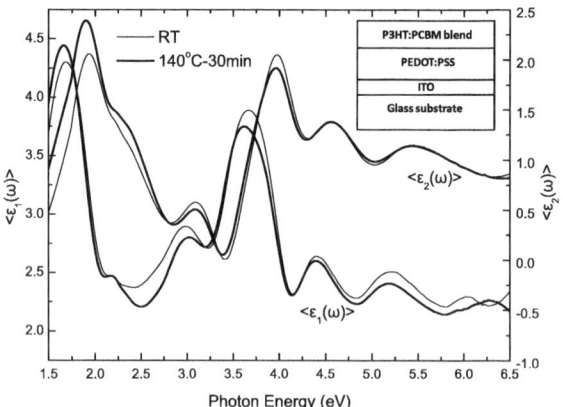

The experimentally measured $\langle \varepsilon(\omega) \rangle$ of P3HT:PCBM blend films is shown in Fig. 12.6, as a function of the photon energy $\omega$ of two representative samples; as grown and annealed at $140\,^{\circ}$C for $t = 30$ min. The SE measurements were performed in the 1.5–6.5 eV energy region with a step of 20 meV and an angle of incidence of $70^{\circ}$. These samples were fabricated onto PEDOT:PSS/ITO/glass structures. The materials used were a regioregular P3HT with purity of 99.995% and mean molecular weight $M_n = 45{,}000$–$65{,}000$ and a PCBM of 99.5% purity. The P3HT:PCBM (1 : 0.8 w/w) solution had a concentration of 18 mg/ml with a chlorobenzene solvent. The P3HT:PCBM films were spin coated onto the PEDOT:PSS/ITO/glass structures with a rotation speed of 1000 rpm for 30 s. After the thin film fabrication, the samples were left to dry for 24 h and afterwards they subjected to thermal annealing treatment at $140\,^{\circ}$C for different durations of time. The changes in the measured $\langle \varepsilon(\omega) \rangle$ are attributed to the different blend morphology and the distribution of the P3HT and PCBM constituents in the blend structure as a result of the annealing process.

In order to obtain quantitative information from the $\langle \varepsilon(\omega) \rangle$, this has been analyzed by the use of a theoretical model consisting of a multilayer stack of air/P3HT:PCBM/PEDOT:PSS/ITO/glass substrate. During the analysis of $\langle \varepsilon(\omega) \rangle$, the P3HT:PCBM film has been approximated as a homogeneous material. Its optical response has been modeled by the TL oscillator model using 5 TL oscillators. Figure 12.7 shows the calculated bulk dielectric function $\varepsilon(\omega)$ as grown and annealed at $140\,^{\circ}$C for 30 min. The optical response of the P3HT:PCBM blend films in the Vis-fUV spectral region includes five optical absorptions that are found at photon energies of 2.05, 2.24, 3.95, 4.65 and 5.89 eV. The first optical absorption at 2.05 eV is attributed to the singlet excitonic transition of the P3HT conjugated polymer whereas the transition at 2.24 eV corresponds to the formation of excitons with phonons [41, 43]. The other three electronic transitions at higher energies are originated from the PCBM and they can be assigned to the electronic transitions $S_0 \rightarrow S_{17}$, $S_0 \rightarrow S_{37}$ and $S_0 \rightarrow S_{56}$, respectively.

Also, significant peak-intensity enhancement is observed at the calculated $\varepsilon(\omega)$ of the blend after the thermal annealing. This can be attributed to the crystallization

**Fig. 12.7** Calculated bulk dielectric function $\varepsilon(\omega)$ of P3HT:PCBM blends, before and after the annealing process based on the best-fit parameters of the analysis of $\langle\varepsilon(\omega)\rangle$. Modified after [43]

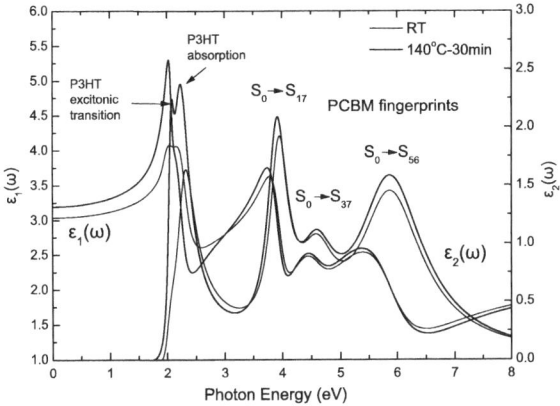

**Fig. 12.8** Evolution of the energy band gap $\omega_g$ of P3HT:PCBM with the annealing time at 140 °C. Modified after [43]

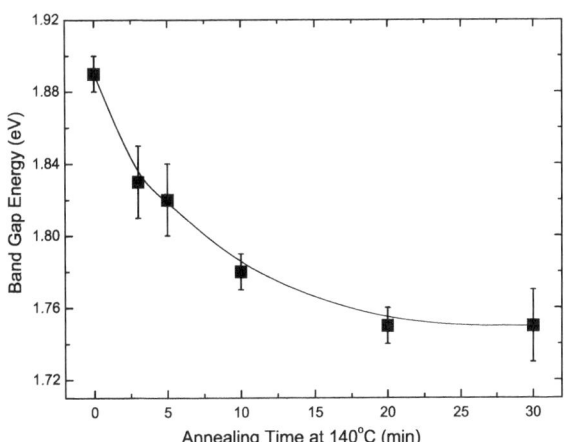

of the P3HT from amorphous to nano-crystalline. The changes in P3HT crystallinity with the annealing temperature can be correlated to the change in the absorbance of the vibronic peak at 2.05 eV. This peak is assigned to an interchain-delocalized excitation and its intensity has been linked to the degree of P3HT crystallinity [41]. From Fig. 12.7 it is also evident that the intensity of this peak is pronounced at the annealed sample leading to the conclusion that P3HT forms a highly ordered structure after the annealing.

The dependence of the fundamental band gap $\omega_g$ on the annealing time, as determined by the analysis of the measured $\langle\varepsilon(\omega)\rangle$ is shown in Fig. 12.8. The calculated band gap $\omega_g$ values of the P3HT:PCBM blends have been found to decrease with the increase of thermal annealing time from 1.89 (not annealed sample) to 1.73 eV (annealed for 30 min).

The kinetics of the molecular rearrangement in the BHJ structure can be understood by the investigation of the vertical distribution of P3HT and PCBM volume fractions in the blend, and its dependence on the experimental conditions (e.g. the

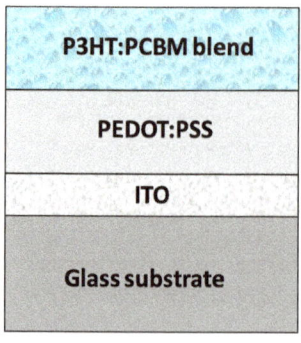

post-deposition thermal annealing) as well as on the device layer structure (e.g. anode or cathode layers). For the study of the effect of the cathode electrode on the P3HT:PCBM blend morphology, we describe below the monitoring of the optical response of P3HT:PCBM blends deposited onto PEDOT:PSS/ITO/glass multilayer structures. These are shown in Fig. 12.9.

For the understanding of the influence of the used substrate, the P3HT:PCBM blend films were deposited onto four different formulations of PEDOT:PSS (Clevios, Heraeus). These are PH1000, PVPAI4083 and PVPCH8000 dispersions (used as received), while a PH1000 solution was modified by the addition of 6% (% v/v) of dimethyl sulfoxide (DMSO) in order to increase its conductivity. These four dispersions were filtered using a 0.45 mm syringe filter (Whatman, PVDF) and were spin-coated onto cleaned and pre-patterned ITO/glass substrates. The PEDOT:PSS/ITO/ glass samples were annealed on a hot plate at 140 °C for 10 min to remove any residual $H_2O$. The blend components include a highly regioregular P3HT material from Rieke Metals (molecular weight $M_n = 10$–30 kDa, poly dispersity index PDI = 1.7–2.2, regioregularity >95%) and PCBM from Solenne BV. A blend solution of P3HT:PCBM (1:0.8 w/w) was prepared with a total concentration of 21.6 mg/ml in chloroform. The solution was stirred for 24 at 55 °C to promote total dissolution and processed without filtering. The P3HT:PCBM blend was spin coated at 700 rpm for 18 s onto the PEDOT:PSS/ITO/glass structures producing thin films of thickness of ∼200 nm. All the above preparation has been conducted in a nitrogen-filled glove box [41, 43]. The optical response of the BHJ was based on the optical response of the polymer and fullerene components. In this model, the information can be provided by the Bruggeman Effective Medium Approximation (BEMA) [11, 45] that can calculate the volume fractions of the polymer and fullerene in different regions in the blend. The volume fractions of P3HT and PCBM have been approximated to change exponentially from the bottom to the top regions of the blend.

Despite the testing of several theoretical models (e.g. homogeneous, linear) for the investigation of the vertical compositional profile of P3HT and PCBM in the BHJ, it was found that this model can provide the most accurate description of the vertical composition profile with the lower value of the minimization function ($\chi^2 = 0.035$). The applicability and reliability of the above model are based on the fact that the penetration depth of the light ensures the extraction of information from the BHJ at all thicknesses. This exponential gradient model can calculate the P3HT and PCBM volume fraction at the bottom layer (layer no. 1) and at the top layer (layer no. 20) by the reference bulk dielectric functions $\varepsilon(\omega)$ of the P3HT and PCBM using the BEMA [11, 12]. These are denoted as %(P3HT)$_i$ and %(PCBM)$_i$ ($i$ = top, bottom). The optical response of the pristine P3HT and PCBM constituents has been previously calculated in order to be used as a reference spectra during the analysis of the $\langle\varepsilon(\omega)\rangle$. Also, the necessary information for the optical response of the glass substrate and the PEDOT:PSS layer are previously determined by the measurement and analysis of plain glass substrate and pristine materials, respectively [41].

Figure 12.10 shows the calculated volume fractions of P3HT and PCBM components as a function of the PEDOT:PSS nano-layer surface energy. The information on Fig. 12.10a corresponds to the volume fractions before the thermal annealing of the device and prior to the cathode metal deposition. On the other side, the volume

**Fig. 12.10** Evolution of the calculated volume fraction of P3HT and PCBM in the blend films as a function of the surface energy for the various PEDOT:PSS substrates: **a** the non-annealed samples and **b** the annealed samples [41]. Reproduced from [41] with permission of The Royal Society of Chemistry

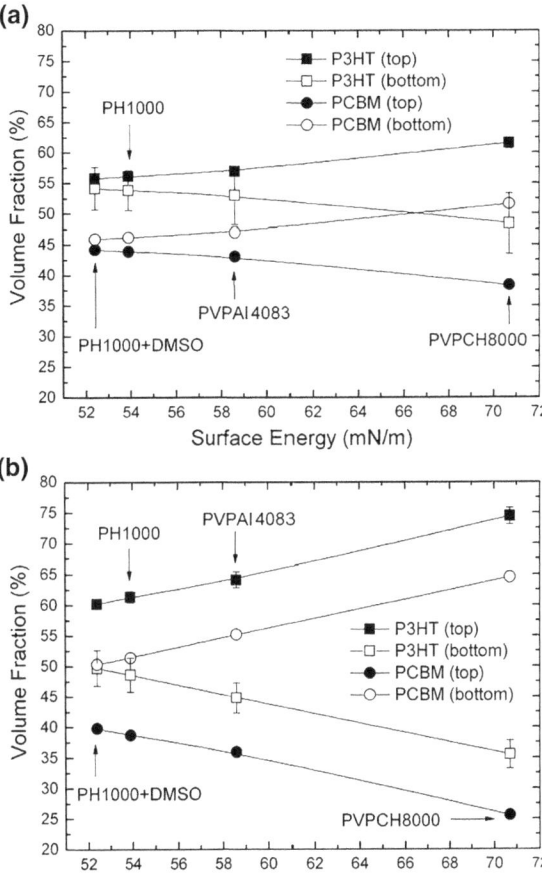

fractions shown in Fig. 12.10b correspond to the annealed samples. In the case of the non-annealed samples, %(P3HT) is found to be 55% and the %(PCBM) is determined at 45% at both the top and bottom regions of the blends deposited on the lower surface energy substrates (PH1000+DMSO, PH1000). This indicates a homogeneous distribution of the P3HT and PCBM in the BHJ layers, which is promoted by the weight ratio of P3HT/PCBM in the blend solution, namely, 1/0.8. On the contrary, the blends deposited onto PEDOT:PSS of higher surface free energy (PVPAI4083 and PVPCH8000), have a non-uniform distribution of the P3HT and PCBM components. The %(P3HT)$_{top}$ increases up to 64% (at the PVPCH8000), whereas an increase of the %(PCBM)$_{bottom}$ at the buried interface close to PEDOT:PSS layer is observed from 46% (at the PH1000+DMSO) to 53% (at the PVPCH8000).

The above behavior is enhanced by the thermal annealing process of the samples. Figure 12.10b shows that the thermal annealing leads to the increase of the %(P3HT)$_{top}$ and %(PCBM)$_{bottom}$ and to the reduction of %(P3HT)$_{bottom}$ and %(PCBM)$_{top}$. This de-mixing process is enhanced when the BHJ layer is deposited onto the higher surface energy buffer layers. As a result, the lowest phase separation occurs in the lower surface energy (most hydrophobic) buffer layers. In these samples we find that %(PCBM)$_{bottom}$ = 50.3% and %(P3HT)$_{bottom}$ = 49.7%, whereas in the top region we find that %(P3HT)$_{top}$ = 60.2% and %(PCBM)$_{top}$ = 39.8%. It is clear from Fig. 12.10b that the largest phase separation takes place at the high surface energy (most hydrophilic) PVPCH8000 substrate [41].

This phenomenon of vertical phase separation, which is well known in polymer blends, has been reported in polymer-fullerene thin films by a number of different groups and is still under discussion [40, 44–46]. That is since, during the spin coating deposition of polymer blends that consist of components with different surface energy, the component with the lower surface tension enriches the film surface. The driving force for this mechanism is the tendency to reduce the free energy of the system [47]. The surface energies of pristine P3HT and PCBM were measured at 24 and 38.2 mN/m, respectively. Consequently, for a typical P3HT:PCBM blend film on a high surface energy substrate, such as PEDOT:PSS, the buried interface is expected to be PCBM-rich while the free surface (exposed to air) is expected to be enriched with the lower surface free energy P3HT. This process is initiated during spin coating and is enhanced during the thermal annealing treatment [41]. The above findings on the lateral phase separation can be supported by the investigation of the surface nano-topography of the BHJ blends by AFM. Figure 12.11 shows the surface nano-topography images of two representative BHJ samples; the non-annealed and the thermally annealed sample to 145 °C for 3 min.

No significant variations were found on the surface nano-topography among the P3HT:PCBM samples deposited onto different PEDOT:PSS substrates. More specifically, the non-annealed samples revealed a very smooth surface with a rms roughness of 0.5 nm and a peak-to-valley distance of 4.3 nm. The low surface roughness suggests that little or no phase separation exist and that the two phases are in a well-mixed state. Indeed, films cast from a high volatile solvent such as chloroform are quenched in a well mixed state, since the solvent evaporates very quickly and freezes the molecular chains in a state far from the thermodynamic equilibrium. However, the

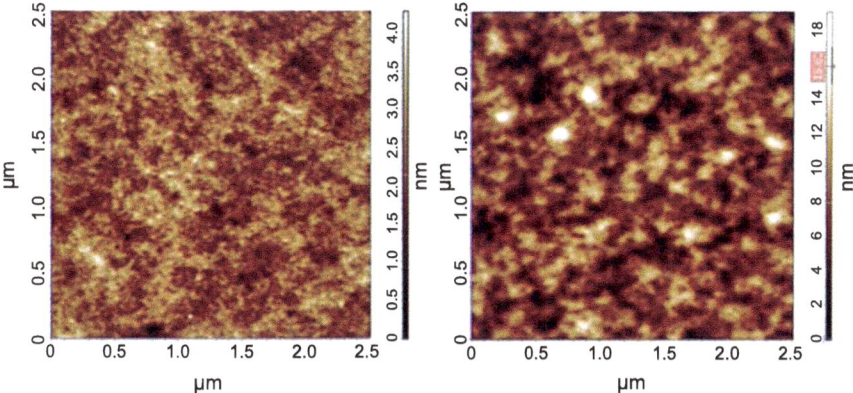

**Fig. 12.11** AFM topography images of the P3HT:PCBM nanolayers. Non-annealed (*left*) and after annealing at 145 °C for 3 min (*right*). Scan size is 2.5 × 2.5 μm [41]. Reproduced from [41] with permission of The Royal Society of Chemistry

thermal annealing above the polymer's glass transition temperature softens the P3HT matrix, and provides the macromolecules the appropriate mobility to crystallize and reach a more thermodynamically favorable state. P3HT forms fibrillar nanocrystals, while PCBM molecules diffuse to form fullerene aggregates or bulky crystals [48, 49]. In the final state, the P3HT crystals, the PCBM molecules or nano-crystals and the P3HT amorphous phase coexist. As a consequence, the rms surface roughness increases to 2.4 nm with a peak-to-valley profile height of 18.9 nm [41].

Finally, the recent developments on the study of the optical properties of BHJ P3HT:PCBM blends includes the investigation of the effect of the optical anisotropy of the blend [50], and the influence of the P3HT regioregularity [51], as well as the use of optical measurements towards the understanding of the vertical phase separation [45, 52–56].

## 12.3.2 PEDOT:PSS for Transparent Electrodes

Despite the enormous effort that is focused on the materials and processes for the fabrication of organic electronic devices, such as OPVs and OLEDs there is still the need for soluble transparent organic electrodes with sufficient optical and electrical response. That is since the inorganic electrodes that are currently used, such as ITO, combine their good electrical response (low resistivity of $\sim 2 \times 10^{-4}\ \Omega$ cm and relatively high work function of $\sim 4.8$ eV) with several disadvantages [5, 9]. These include the high costs for indium, the need for vacuum deposition that is associated with increased temperatures, the oxygen or indium diffusion and the incompatibility with roll-to-roll (r2r) deposition processes on flexible substrates such as PET that have the form of rolls [5, 9].

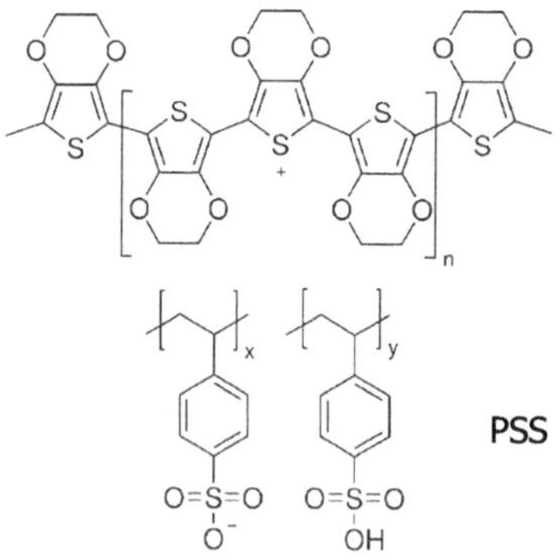

**Fig. 12.12** Chemical structure of the PEDOT:PSS

PEDOT:PSS is a transparent conductive polymer that is used as anode buffer layer in organic electronic devices. It is expected to replace the inorganic, brittle and expensive ITO, as well as other TCO materials, such as zinc oxide. PEDOT:PSS combines metallic-like behavior with solubility and it can be used as hole-injecting material in applications such as sensors, antistatic coatings, solar cells, etc. [5, 7, 9, 58–61]. It consists of a conducting part, PEDOT which is a low molecular weight polymer, insoluble and thus difficult to process and an insulating polymer PSS, which is a high molecular weight polymer which gives the desirable flexibility and also increases the solubility of the system in water, making the whole system easy to process. The oligomer PEDOT segments are electrostatically attached on the PSS polymer chains (see Fig. 12.12) [59–63].

The currently accepted morphology of PEDOT:PSS films (especially films that are fabricated by spin casting or spin coating) is that of a phase segregated material that consists of PEDOT:PSS grains surrounded by a shell formed by excess PSS [49–51]. This structure has been confirmed by several analytical methods [58–60, 62, 65, 66]. More specifically, the material consists of PEDOT-rich spheroid (or pancake-shaped) particles with diameter in the range of 20–30 nm and height of 4–6 nm (see Fig. 12.13). These are actually organized in layers that are separated by quasi-continuous nanometer-thick PSS lamellas. This morphology qualitatively can explain the observed differences in magnitude and temperature dependences of the conductivity in the normal and parallel current directions [64, 67]. That is, the conductivities of spin-coated PEDOT:PSS thin films, measured in lateral and vertical directions with respect to the sample surface, are reported to be highly anisotropic. In the normal direction, the horizontal PSS lamellas were assumed to impose nearest-neighbor hopping between the quasi-metallic PEDOT particles, leading to a strongly reduced conductivity. On the other hand, on the perpendicular direction a higher charge transport is observed between the PEDOT-rich particles that are laterally separated by much thinner barriers, leading to an enhanced conductivity in this direction [64, 67].

The investigation of the optical properties of the PEDOT:PSS layers has been realized also by SE in the infrared energy region by Fourier Transform IR SE (FTIRSE),

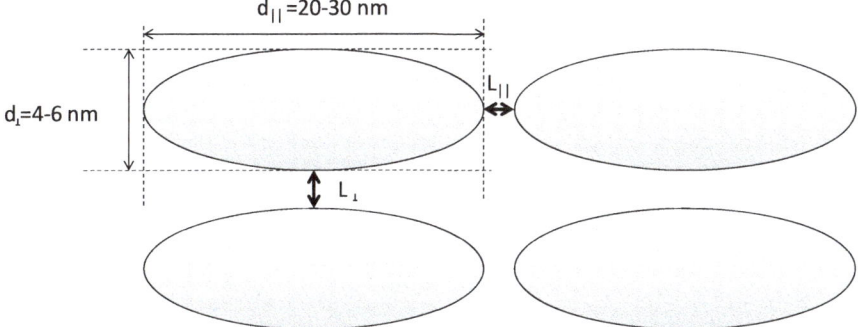

**Fig. 12.13** Schematic representation of the morphological model for PEDOT:PSS. PEDOT-rich spheroid particles organized in layers separated by lamellas of insulating PSS

**Fig. 12.14** Pseudodielectric function $\langle \varepsilon(\omega) \rangle$ of PEDOT:PSS measured in the IR spectral region [9]. Reprinted from S. Logothetidis, A. Laskarakis, Organic against inorganic electrodes grown onto polymer substrates for flexible organic electronics applications, Thin Solid Films **518**, 1245 (2009). Copyright 2013, with permission from Elsevier

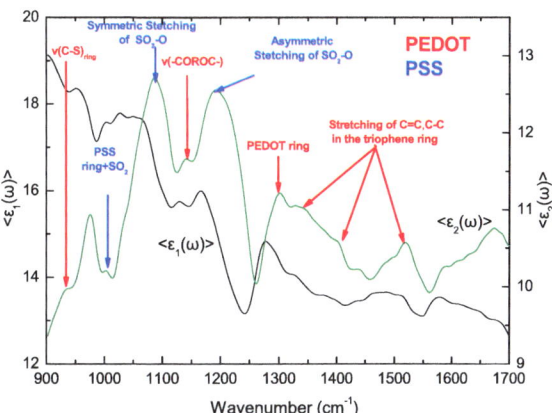

which provides information on the bonding structure, based on the strong absorption bands indicating the contribution of the vibrational modes of the IR active chemical bonds. Figure 12.14 shows the measured $\langle \varepsilon(\omega) \rangle$ of the PEDOT:PSS layer in the IR spectral region by FTIRSE. The main bonding vibrations are found at 1512, 1454, 1394, 1370 and 1168 cm$^{-1}$ from the stretching of C=C and C–C in the thiophene ring. Further vibrations from the C–S bond in the thiophene ring can be seen at 930, 830, 727 and 697 cm$^{-1}$. Vibrations at 1183, 1144–1128, 1093–1076 and 1052–1047 cm$^{-1}$ are assigned to stretching in the alkylenedioxy group. The PSS vibrations appear at 1184 cm$^{-1}$ for the asymmetric stretching of the SO$^{3-}$ group and at 1042 cm$^{-1}$ for the symmetric-stretching of the sulfonic group. At 1130 and 1011 cm$^{-1}$ the in-plane skeleton vibrations and in-plane bending vibrations, respectively, of the benzene ring are shown [9, 68].

The optical properties of PEDOT:PSS films have been extensively investigated also by SE in the Vis-fUV spectral region, for example during its fabrication by electrospray, spin coating and printing techniques [57, 58, 69–72]. The conductivity of PEDOT:PSS depends strongly on its structure and morphology. The conditions for the formation of the desired structure can be achieved either by the synthesis of aqueous solutions of PEDOT:PSS with polar solvents (for example—DMSO), and/or by the implementation of specific post deposition process conditions. Although several explanations have been proposed, it has been mainly reported that the addition of polar solvents produce conformational changes of the PEDOT chains and/or segregation of the excess PSS that affects the film conductivity [61, 64, 66, 73, 74].

In this paragraph, we will present the latest advances on the investigation of the effect of the addition of DMSO polar solvent on the optical and electrical properties of PEDOT:PSS films by the use of Vis-fUV SE. The material that has been used is the formulation PH1000 PEDOT:PSS (1:2.5). The PEDOT:PSS blends were prepared with different concentrations by weight of dimethyl sulfoxide (DMSO by Sigma–Aldrich). The mixture of DMSO and PEDOT:PSS was adjusted, in a ratio by weight of 1:2.5 to give final solutions of 1, 2, 3, 4, 5, 6, 7 and 10% by volume (v/v) of DMSO. Also, the solutions were ultrasonicated for 5 min prior to use. The thin films were deposited by spin-coating on glass substrates. The substrates were cleaned prior to the spin coating first with detergent and then in an ultrasonic bath with deionized water, acetone, isopropanol, methanol and then they were blow dried with nitrogen. The deposition took place inside a nitrogen-filled glove box. The spin coating speed was 2000 rpm for 30 s that resulted in thin films of thickness of ∼85 nm. The spin-coated samples were dried on a hot plate for 10 min at 150 °C inside a $N_2$-filled glove box.

The optical and electronic response of the PEDOT:PSS includes an absorption in the Vis-UV spectral region between 5.3–6.4 eV that can be attributed to the $\pi$–$\pi^*$ transitions of the benzene rings of the PSS part. Also, the absorbance of the more conductive PEDOT part is reported to appear at lower energy values below 1.7 eV [75]. For the determination of the optical response of the PEDOT:PSS with different DMSO ratios, a combination of the Lorentz–Drude model can be used. The Drude model can successfully describe the metallic-like behavior of PEDOT at the low photon energies. The resistivity $\rho$ and the electrical conductivity $\sigma$ of PEDOT:PSS can be calculated by the combination of the Drude model parameters (the plasma energy $\omega_p$ and Drude broadening $\Gamma_D$) that were calculated from the analysis procedure of the measured $\langle \varepsilon(\omega) \rangle$ into the following relation [76]:

$$\sigma = 134.5 \frac{\omega_p^2}{\Gamma_D}. \tag{12.5}$$

The Lorentz oscillator model has been used in order to describe the interband absorption of PEDOT found at energies ∼1 eV. The Tauc–Lorenz (TL) oscillator model can accurately describe the interband transitions of the PSS [60, 66, 77]. We have used 2 TL oscillators to describe the optical absorptions of PSS at energies 5.4 and 6.3 eV. Figure 12.15 shows the evolution of the calculated values of the electronic transition energies of PEDOT and PSS as a function of the DMSO concentration in the samples.

In the case of the electronic transition of the insulating PSS part, we observe that the increase of the DMSO concentration leads to a slight reduction of the $\omega_{01}{}^{PSS}$ from

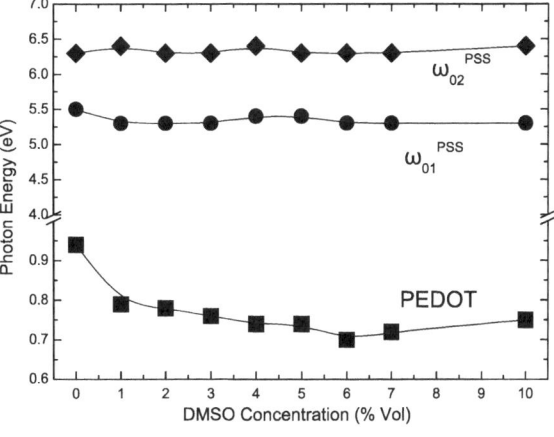

**Fig. 12.15** Dependence of the calculated values of electron transition energies of PEDOT and PSS with the DMSO concentration in the solution. Modified after [78]

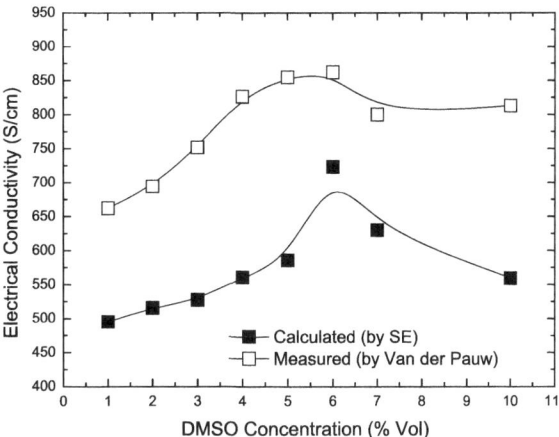

**Fig. 12.16** Calculated (*black squares*) and measured (*white squares*) values of the electrical conductivity of PEDOT:PSS as a function of the DMSO concentration. Modified after [78]

5.5 eV (0% DMSO) to 5.3 eV (10% DMSO). On the contrary, the $\omega_{02}^{PSS}$ remains relatively stable at 6.3 eV. However, in the case of the electronic transition of PEDOT, we observe a reduction from 0.95 eV (pristine PEDOT:PSS) to 0.8 eV, whereas at 6% DMSO concentration it reaches its minimum value to 0.7 eV. At films with higher concentration values the PEDOT electronic transition increases to 0.75 eV. Therefore, at 6% DMSO concentration the lower absorption energy of the PEDOT part, has a higher contribution to the conductivity of the films.

Figure 12.16 shows the evolution of the calculated values of electrical conductivity as a function of the DMSO concentration, together with the measured values by the Van Der Pauw method. Although the measured conductivity of pristine PEDOT:PSS without the addition of DMSO is ∼0.4 S/cm, we observe that the addition of DMSO, significantly improves the electrical conductivity. More specifically, the calculated conductivity by SE increases rapidly from 1 S/cm in the pristine PEDOT:PSS to 495 S/cm in the sample with 1% DMSO, whereas by increasing the % DMSO the calculated conductivity gradually increases up to a maximum value of 723 S/cm at the optimal concentration of 6%. At higher concentration values above 6% the conductivity decreases but it still remains at higher values (560 S/cm) than in the case where the polar solvent concentration is lower than 6%. This behavior is also confirmed by measured conductivity values by the Van Der Pauw method (white squares). These results indicate an improvement of the conductivity to 662 S/cm (1% DMSO) that increases to a maximum value of 862 S/cm at 6% DMSO. After that value, there is a saturation of the conductivity to the range of 813 S/cm at higher solvent concentrations [78].

The increase in the PEDOT:PSS conductivity and therefore in the metallic-like behavior can be attributed to the increase of the average domain size of PEDOT particles in combination to the reduction of the insulating PSS barriers between the PEDOT grains (less energy barriers) as a result of the increasing DMSO concentration [61, 73]. However, after achieving the optimal material morphology (at 6% concentration) the further increase of the solvent concentration leads to further growth of the PEDOT domains to longer distances to each other that results to the reduction of the films conductivity. The differences between the calculated and measured conductivity values can be attributed to the fact that the Drude model can

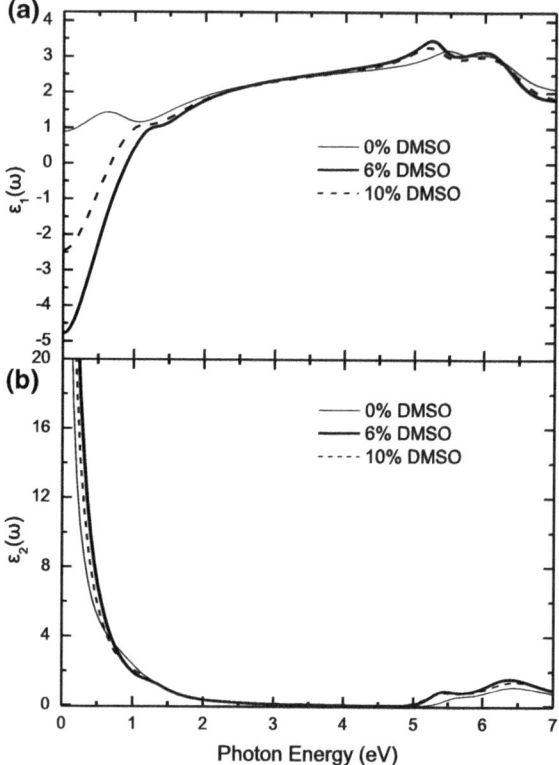

**Fig. 12.17** Calculated real (**a**) and imaginary (**b**) part of the bulk dielectric function $\varepsilon(\omega)$ of the PEDOT:PSS films as a function of the amount of DMSO added in the PEDOT:PSS solution

describe only the local electrical response within the polymer domains. On the other side, the experimentally measured values describe the overall conductivity in the whole area of the sample, which can also involve charge carrier hopping between different polymer grains, and finally to higher conductivity values.

Figure 12.17 shows the calculated (extrapolated) bulk dielectric function $\varepsilon(\omega)$, that does not take into account the contribution of the substrate, based on the best-fit parameters derived from the analysis process for the PEDOT:PSS samples grown with different DMSO concentrations. It is clear that the metallic-like behavior of the PEDOT:PSS is found at the lower energy range below 1 eV where the Drude contribution is shown. The pristine PEDOT:PSS shows a semiconducting behavior whereas the PEDOT:PSS formulations have a more pronounced conductive response. As we can see, the PEDOT:PSS film with 6% DMSO concentration shows the more intense metallic-like behavior with a more profound absorption at low energy values.

## 12.4   Summary and Outlook

In this chapter we provided an overview of the implementation of SE on the optical and electronic properties characterization of polymer:fullerene blends and composites that are used as photoactive nanolayers and transparent electrodes, respectively, for the fabrication of OPVs. Also, it has been demonstrated that SE is a powerful tool to be implemented as a standard method for the determination of the optical properties of state-of-the-art materials used for the fabrication of flexible organic electronic devices. The capability for measurement in an extended spectral region from the IR to the Vis-fUV spectral region can provide significant information on the materials optical and electronic properties, composition, surface and interface roughness, morphology and component distribution (in polymer blends). This will allow the optimization of their functionality and performance that will enable the improvement of the efficiency, performance and long-term stability of a large variety of organic electronic devices (such as OPVs, OLEDs, OTFTs, sensors, etc.), that can be fabricated either by r2r printing processes or by vacuum methods. Moreover, the capability for optical sensing in the nanometer scale combined with fast measurement and analysis times (in the range of ms) will establish SE as a necessary quality control tool for the fabrication of organic electronic devices.

**Acknowledgements**  The authors would like to thank Dr. Despoina Georgiou, Dr. Christos Koidis, Dr. Panagiotis G. Karagiannidis and the other staff of the Lab for Thin Films, Nanosystems and Nanometrology (LTFN) for their contribution. Also, the authors would like to thank Clevios for the supply of the PEDOT:PSS formulations. This work was partially supported by the EC STREP Project OLAtronics, Grand Agreement No. 216211, and by the EC REGPOT Project ROleMak No. 286022.

# References

1. D.M. de Leeuw, E. Cantatore, Mater. Sci. Semicond. Process. **11**, 199 (2008)
2. White Paper OE-A Roadmap (2011)
3. C.J. Brabec, Sol. Energy Mater. Sol. Cells **83**, 273 (2004)
4. P. Kumar, S. Chand, Prog. Photovolt. Res. Appl. **20**(4), 377 (2012)
5. S. Logothetidis, A. Laskarakis, Eur. Phys. J. Appl. Phys. **46**, 12502 (2009)
6. F.C. Krebs, Org. Electron. **10**, 761 (2009)
7. Y. Galagan, I.G. de Vries, A.P. Langen, R. Andriessen, W.J.H. Verhees, S.C. Veenstra, J.M. Kroon, Chem. Eng. Process. **50**, 454 (2011)
8. G. Li, R. Zhu, Y. Yang, Nat. Photonics **6**, 153 (2012)
9. S. Logothetidis, A. Laskarakis, Thin Solid Films **518**, 1245 (2009)
10. H. Hoppe, N.S. Sariciftci, J. Mater. Res. **19**, 1924 (2011)
11. G.E. Irene, H.G. Tompkins (eds.), *Handbook of Ellipsometry* (William Andrew, Norwich, 2005)
12. R.M.A. Azzam, N.M. Bashara (eds.), *Ellipsometry and Polarized Light* (North-Holland, Amsterdam, 1977)
13. A. Laskarakis, S. Kassavetis, C. Gravalidis, S. Logothetidis, Nucl. Instrum. Methods Phys. Res. Sect. B Beam Interact. Mater. Atoms **268**, 460 (2010)
14. S. Logothetidis, in *Thin Films Handbook*, ed. by H.S. Nalwa (Academic Press, San Diego, 2001)
15. X. Yang, J. Loos, S.C. Veenstra, W.J.H. Verhees, M.M. Wienk, J.M. Kroon, M.A.J. Michels, R.A.J. Janssen, Nano Lett. **5**, 579 (2005)
16. P. Peumans, A. Yakimov, S.R. Forrest, J. Appl. Phys. **93**, 3693 (2003)
17. B.C. Thompson, J.M.J. Fréchet, Angew. Chem. Int. Ed. Engl. **47**, 58 (2008)
18. S. Günes, H. Neugebauer, N.S. Sariciftci, Chem. Rev. **107**, 1324 (2007)
19. M. Hiramoto, H. Fujiwara, M. Yokoyama, J. Appl. Phys. **72**, 3781 (1992)
20. W. Ma, C. Yang, X. Gong, K. Lee, A.J. Heeger, Adv. Funct. Mater. **15**, 1617 (2005)
21. G. Li, V. Shrotriya, J. Huang, Y. Yao, T. Moriarty, K. Emery, Y. Yang, Nat. Mater. **4**, 864 (2005)
22. H. Chen, J. Hou, S. Zhang, Y. Liang, G. Yang, Y. Yang, Nat. Photonics **3**, 649–653 (2009)
23. W.-H. Baek, T.-S. Yoon, H.H. Lee, Y.-S. Kim, Org. Electron. **11**, 933–937 (2010)
24. M. Schubert, A. Kasic, T. Hofmann, V. Gottschalch, J. Off, F. Scholz, E. Schubert, H. Neumann, I. Hodgkinson, M. Arnold, W. Dollase, C.M. Herzinger, in *Proceedings of the SPIE*, vol. 4806 (2002), p. 264
25. K. Hinrichs, M. Gensch, N. Nikonenko, J. Pionteck, K.-J. Eichhorn, Macromol. Symp. **230**, 26 (2005)
26. G.E. Jellison, Thin Solid Films **313–314**, 33 (1998)
27. G.E. Jellison, Thin Solid Films **290–291**, 40 (1996)
28. F.C. Krebs, T. Tromholt, M. Jørgensen, Nanoscale **2**, 873 (2010)
29. F.C. Krebs, Sol. Energy Mater. Sol. Cells **93**, 394 (2009)
30. J.C. Hummelen, B.W. Knight, F. Lepeq, F. Wudl, J. Yao, C.L. Wilkins, J. Org. Chem. **60**, 532 (1995)
31. Y. Liang, Z. Xu, J. Xia, S.-T. Tsai, Y. Wu, G. Li, C. Ray, L. Yu, Adv. Mater. **22**, E135 (2010)
32. A. Loiudice, A. Rizzo, G. Latini, C. Nobile, M. de Giorgi, G. Gigli, Sol. Energy Mater. Sol. Cells **100**, 147 (2012)
33. G. Yu, J. Gao, J.C. Hummelen, F. Wudl, A.J. Heeger, Science **270**, 1789 (1995)
34. S.E. Shaheen, C.J. Brabec, N.S. Sariciftci, F. Padinger, T. Fromherz, J.C. Hummelen, Appl. Phys. Lett. **78**, 841 (2001)
35. Y. Kim, S. Cook, S.M. Tuladhar, S.A. Choulis, J. Nelson, J.R. Durrant, D.D.C. Bradley, M. Giles, I. Mcculloch, C. Ha, M. Ree, Nat. Mater. **5**, 197 (2006)
36. V.D. Mihailetchi, H. Xie, B. de Boer, L.M. Popescu, J.C. Hummelen, P.W.M. Blom, L.J.A. Koster, Appl. Phys. Lett. **89**, 012107 (2006)
37. H. Hoppe, N.S. Sariciftci, J. Mater. Chem. **16**, 45 (2006)
38. K. Kim, J. Liu, D.L. Carroll, Appl. Phys. Lett. **88**, 181911 (2006)

39. Y. Zhao, Z. Xie, Y. Qu, Y. Geng, L. Wang, Appl. Phys. Lett. **90**, 043504 (2007)
40. M. Campoy-Quiles, T. Ferenczi, T. Agostinelli, P.G. Etchegoin, Y. Kim, T.D. Anthopoulos, P.N. Stavrinou, D.D.C. Bradley, J. Nelson, Nat. Mater. **7**, 158 (2008)
41. P.G. Karagiannidis, N. Kalfagiannis, D. Georgiou, A. Laskarakis, N.A. Hastas, C. Pitsalidis, S. Logothetidis, J. Mater. Chem. **22**, 14624 (2012)
42. S. Logothetidis, D. Georgiou, Under Preparation (2013)
43. P.G. Karagiannidis, D. Georgiou, C. Pitsalidis, A. Laskarakis, S. Logothetidis, Mater. Chem. Phys. **129**, 1207 (2011)
44. D.S. Germack, C.K. Chan, R.J. Kline, D.A. Fischer, D.J. Gundlach, M.F. Toney, L.J. Richter, D.M. DeLongchamp, Macromolecules **43**, 3828 (2010)
45. Z. Xu, L.-M. Chen, G. Yang, C.-H. Huang, J. Hou, Y. Wu, G. Li, C.-S. Hsu, Y. Yang, Adv. Funct. Mater. **19**, 1227 (2009)
46. T. Agostinelli, T.A.M. Ferenczi, E. Pires, S. Foster, A. Maurano, C. Müller, A. Ballantyne, M. Hampton, S. Lilliu, M. Campoy-Quiles, H. Azimi, M. Morana, D.D.C. Bradley, J. Durrant, J.E. MacDonald, N. Stingelin, J. Nelson, J. Polym. Sci. Part B Polym. Phys. **49**, 717 (2011)
47. Y. Lipatov, Prog. Polym. Sci. **27**, 1721 (2002)
48. C.J. Brabec, S. Gowrisanker, J.J.M. Halls, D. Laird, S. Jia, S.P. Williams, Adv. Mater. **22**, 3839 (2010)
49. S. van Bavel, E. Sourty, G. de With, K. Frolic, J. Loos, Macromolecules **42**, 7396 (2009)
50. S.-Y. Chuang, C.-C. Yu, H.-L. Chen, W.-F. Su, C.-W. Chen, Sol. Energy Mater. Sol. Cells **95**, 2141 (2011)
51. S.-Y. Chuang, H.-L. Chen, W.-H. Lee, Y.-C. Huang, W.-F. Su, W.-M. Jen, C.-W. Chen, J. Mater. Chem. **19**, 5554 (2009)
52. B. Xue, B. Vaughan, C.-H. Poh, K.B. Burke, L. Thomsen, A. Stapleton, X. Zhou, G.W. Bryant, W. Belcher, P.C. Dastoor, J. Phys. Chem. C **114**, 15797 (2010)
53. A. Orimo, K. Masuda, S. Honda, H. Benten, S. Ito, H. Ohkita, H. Tsuji, Appl. Phys. Lett. **96**, 043305 (2010)
54. X. Bulliard, S.-G. Ihn, S. Yun, Y. Kim, D. Choi, J.-Y. Choi, M. Kim, M. Sim, J.-H. Park, W. Choi, K. Cho, Adv. Funct. Mater. **20**, 4381 (2010)
55. J.Y. Oh, W.S. Jang, T. Il Lee, J.-M. Myoung, H.K. Baik, Appl. Phys. Lett. **98**, 023303 (2011)
56. N. Schmerl, G. Andersson, Phys. Chem. Chem. Phys. **13**, 14993 (2011)
57. C. Gravalidis, A. Laskarakis, S. Logothetidis, Eur. Phys. J. Appl. Phys. **46**, 12505 (2009)
58. I. Cruz-Cruz, M. Reyes-Reyes, M.A. Aguilar-Frutis, A.G. Rodriguez, R. López-Sandoval, Synth. Met. **160**, 1501 (2010)
59. B. Friedel, P.E. Keivanidis, T.J.K. Brenner, A. Abrusci, C.R. McNeill, R.H. Friend, N.C. Greenham, Macromolecules **42**, 6741 (2009)
60. A. Nardes, M. Kemerink, R. Janssen, Phys. Rev. B **76**, 1 (2007)
61. J. Ouyang, Q. Xu, C.-W. Chu, Y. Yang, G. Li, J. Shinar, Polymer **45**, 8443 (2004)
62. Z. Xiong, C. Liu, Org. Electron. **13**, 1532 (2012)
63. S.A. Mauger, A.J. Moulé, Org. Electron. **12**, 1948 (2011)
64. A.M. Nardes, R.A.J. Janssen, M. Kemerink, Adv. Funct. Mater. **18**, 865 (2008)
65. F. Herrmann, S. Engmann, M. Presselt, H. Hoppe, S. Shokhovets, G. Gobsch, Appl. Phys. Lett. **100**, 153301 (2012)
66. S. Jonsson, J. Birgerson, X. Crispin, G. Greczynski, W. Osikowicz, A.W.D. van der Gon, W.R. Salaneck, M. Fahlman, Synth. Met. **139**, 1 (2003)
67. A.M. Nardes, M. Kemerink, R.A.J. Janssen, J.A.M. Bastiaansen, N.M.M. Kiggen, B.M.W. Langeveld, A.J.J.M. van Breemen, M.M. de Kok, Adv. Mater. **19**, 1196 (2007)
68. M. Schubert, C. Bundesmann, G. Jakopic, H. Maresch, H. Arwin, F. Zhang, O. Inganas, Thin Solid Films **456**, 295 (2004)
69. T. Ino, T. Hiate, T. Fukuda, K. Ueno, H. Shirai, J. Non-Cryst. Solids **358**, 2520 (2012)
70. M. Garganourakis, S. Logothetidis, C. Pitsalidis, D. Georgiou, S. Kassavetis, A. Laskarakis, Thin Solid Films **517**, 6409 (2009)
71. S.A. Mauger, L. Chang, C.W. Rochester, A.J. Moulé, Org. Electron. **13**, 2747 (2012)

72. M.V. Madsen, K.O. Sylvester-hvid, B. Dastmalchi, K. Hingerl, K. Norrman, T. Tromholt, M. Manceau, D. Angmo, F.C. Krebs, J. Phys. Chem. C **115**, 10817 (2011)
73. S.-I. Na, S.-S. Kim, J. Jo, D.-Y. Kim, Adv. Mater. **20**, 4061 (2008)
74. S.-I. Na, G. Wang, S.-S. Kim, T.-W. Kim, S.-H. Oh, B.-K. Yu, T. Lee, D.-Y. Kim, J. Mater. Chem. **19**, 9045 (2009)
75. C. Koidis, S. Logothetidis, C. Kapnopoulos, P.G. Karagiannidis, A. Laskarakis, N.A. Hastas, Mater. Sci. Eng. B **176**, 1556 (2011)
76. J. Humlicek, A. Nebojsa, J. Hora, M. Stransky, J. Spousta, T. Sikola, Thin Solid Films **332**, 25 (1998)
77. Y. Chen, K.S. Kang, K.J. Han, K.H. Yoo, J. Kim, Synth. Met. **159**, 1701 (2009)
78. A. Laskarakis, P.G. Karagiannidis, D. Georgiou, D.M. Nikolaidou, S. Logothetidis, Thin Solid Films **541**, 102 (2013)

# Chapter 13
# Small Organic Molecules

**Ovidiu D. Gordan and Dietrich R. T. Zahn**

**Abstract** In order to improve devices based on organic thin films like organic light emitting diodes (OLEDs) and organic photo voltaic (OPV) cells, the molecular orientation has to be determined and optimized to increase the carrier mobility and the light emission and absorption within the layers. As many of the organic molecules possess an intrinsic molecular anisotropy, molecular ordering will induce optical and electrical anisotropy in the films. The optical anisotropy can be used to determine the average molecular orientation by modeling the anisotropic dielectric function using ellipsometric measurements. An overview of the procedure, valid for planar molecules, will be given in the first part of this chapter, with the main focus on the Phthalocyanine molecular class. The second part of the chapter focuses on vacuum ultra violet (VUV) ellipsometric measurements and the sensitivity gain at ultra-low coverages. Here the discussion will be restricted to optically isotropic films.

## 13.1 Introduction

Organic molecules, which in solid form exhibit semiconducting properties, were long hailed as candidates for cheap, roll to roll electronics. As of today, organic electronics is not anymore a faraway promise, as the huge success and growth of the smartphones market is powered by innovations like organic light emitting diodes (OLEDs). The recent launch of OLED TVs indicates that the OLED technology already reached a maturation state which ensures the long life time needed in television displays. Even in organic photo voltaic (OPV) applications, higher efficiencies than the limit predicted by the exciton dissociation energy were reached. Plastic electronics using organic field effect transistors (OFETs) is another field where huge progress was made in the past decade, and probably soon we will see commercial organic based

O. D. Gordan · D. R. T. Zahn (✉)
Semiconductor Physics, Technische Universität Chemnitz, 09107 Chemnitz, Germany
e-mail: zahn@physik.tu-chemnitz.de

O. D. Gordan
e-mail: ovidiu.gordan@physik.tu-chemnitz.de

© Springer International Publishing AG, part of Springer Nature 2018
K. Hinrichs and K.-J. Eichhorn (eds.), *Ellipsometry of Functional Organic Surfaces and Films*, Springer Series in Surface Sciences 52,
https://doi.org/10.1007/978-3-319-75895-4_13

radio frequency identification tags (ORFIDs). Compared to OPVs, OFETs, and especially to the already commercial OLEDs, molecular spintronics is still in its infancy. However, what all these applications have in common is the search for the desired high conductivity.

Another advantage using organic molecules is that due to the virtually infinite combination possibilities, new organic molecules can be synthetized with specific properties tailored to the wanted application. Two classes of organic materials are usually mentioned when talking about organic devices: polymers and small molecules. This chapter will focus on the small molecules for which the conductivity is highly dependent on the direction of the overlapping of the $\pi$ orbitals and therefore on the molecular orientation with respect to the electrodes. As most of the organic molecules possess a strong intrinsic molecular anisotropy, maximizing the overlap usually leads to an organic molecular crystal which in turn will possess a large optical and electronic anisotropy. However, even within amorphous layers a preferential molecular orientation can exist, which can be beneficial for device performance not only due to the higher mobility, but also due to an increased light coupling in the active layer. As in both cases, polycrystalline and amorphous layers, the intermolecular forces are mediated via a weak van-der-Waals (VdW) interaction, the molecular order and growth are highly dependent on the deposition conditions as well as on the substrate type. Due to the large parameter space which can influence the molecular growth of the active layer, fast, reliable, and non-destructive characterization techniques have to be employed for understanding the growth process which will lead to device/layer optimization. Optical spectroscopies like ellipsometry, reflection anisotropy spectroscopy (RAS), Raman, and Fourier transform infrared (FTIR) qualify for the above conditions, with ellipsometry and RAS being the fastest in terms of acquisition times as well as being readily adaptable for in situ monitoring. Therefore this chapter will give an overview of ellipsometric applications to thin organic films formed by small molecules. Examples will be given on how the average molecular orientation can be determined from the optical anisotropy as well as examples on the sensitivity of this technique to ultra-low coverage.

## 13.2 Phthalocyanines and Molecular Orientation in Thin Films

An early report on determination of optical properties of organic thin films by ellipsometry by Arwin and Aspnes [1] gives a very good introduction on the effects which surface roughness and density variations of the films have on the measured effective dielectric function, and how this can be treated using effective medium approximations methods, which in turn give microstructural information such as void fraction. The authors note that organic materials often condense in the amorphous state, or as crystals with large unit cells, and their anisotropies can be large, so orientational effects are common [1], and predict that anisotropic modelling of ellipsometric data

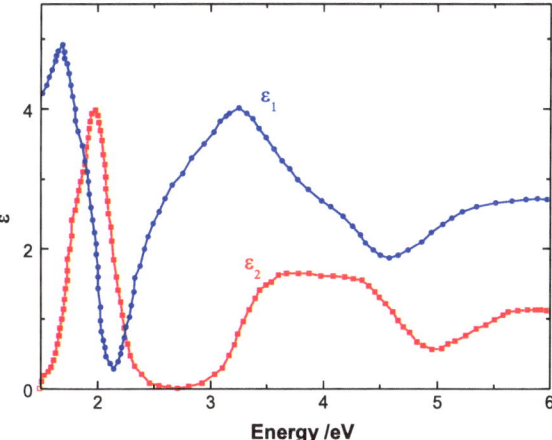

**Fig. 13.1** Dielectric response of an 30 nm $H_2Pc$ layer on a $SiO_2$/Si substrate (after [1]). Reprinted with permission from H. Arwin, D.E. Aspnes, Determination of optical properties of thin organic films by spectroellipsometry, Thin Solid Films **138**, 195 (1986). Copyright 2013, with permission from Elsevier

will be important in the future. They also report, probably for the first time, the dielectric function of a metal free phthalocyanine ($H_2Pc$) film on a silicon (Si) with silicon oxide ($SiO_2$) substrate [1], from an ellipsometric data evaluation, shown in Fig. 13.1. As the Phthalocyanine molecule (Pcs) can be regarded as an 18 or 16 $\pi$-electron system perturbed by the attachment of four benzoid rings, the absorption of Pcs in the UV-visible range is due to the $Q$ band and Soret $B$ band, following the nomenclature of porphyrins [2]. The lowest energy band ($Q$), centred at 2 eV, is due to $\pi-\pi^*$ transitions. The $B$ band is also composed of $\pi-\pi^*$ transitions occurring in the 4 eV region. As both bands are clearly resolved in Fig. 13.1, [1] demonstrated that sophisticated ellipsometric evaluations were not unrealistic.

The results are quite important not only due to the clear indication that ellipsometry can be used to characterize organic thin layers, but also due to the historical importance of the Pcs molecular class. Known for more than one century, since their accidental discovery in 1907 [2], the Pcs were used in many landmark experiments, like being the first molecule imaged with a field emission microscope in 1950 and in 1953 the first organic crystal to yield its structure from X-ray analysis [2]. Being a planar molecule, in most configurations, its central structure is based on the Porphyrin core, and can accommodate a large number of metal ions or oxides in its central ring. A schematic representation of its structure is given in Fig. 13.2, along with an example where the planar structure is distorted due to the vanadyl presence.

Due to their high chemical and thermal stability and their blue or green color, the Pcs were largely used in industry as dyes [2]. It was also quite early noted that the Pc molecules show semiconducting and photoconducting properties [5], which combined with their early discovery made the Pcs being probably the most studied organic molecules [2, 4]. They were proposed for gas sensing applications [6], as wellas for OPVs [7], OFETs [8], and OLEDs [9], the number of publications being so large that is virtually impossible to mention all here.

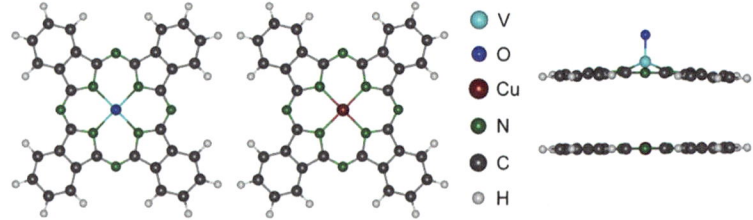

**Fig. 13.2** Schematic representation of VOPc and CuPc (adapted after [3])

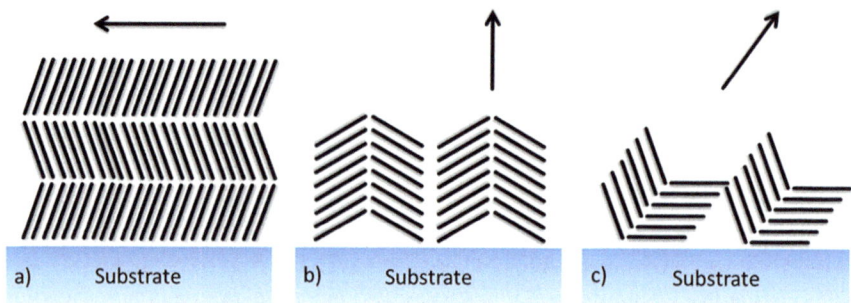

**Fig. 13.3** Types of molecular arrangements of vacuum-sublimed $\alpha$-type films (after [2]). **a** Thin film deposited slowly onto a substrate that does not interact strongly with the first Pc monolayers, **b** molecular stacking in thicker films, **c** thin film deposited onto a substrate that interacts strongly with the initially deposited Pc monolayers. Note the indicated orientation of the stacking axis. Reprinted with permission from H.N.B. McKeown, *Phthalocyanine Materials*, Cambridge University Press (1998). Copyright 2013, Cambridge University Press

Due to their planar structure which allows a fairly large overlap of the $\pi$ orbitals, the Pcs readily form organic crystals. Unlike the covalent bonding in inorganic materials, the VdW interactions are weak. Therefore a number of different forms of crystalline states may be displayed by the same compound. The knowledge of Pc polymorphism is useful for understanding the structure within microcrystalline films prepared by vacuum sublimation [2, 4], as important device properties like conductivity and light absorption/emission are highly dependent on the average molecular orientation.

When deposited by organic molecular beam deposition (OMBD), the Pc molecules usually form polycrystalline films with a typical arrangement of a herringbone like structure, shown in Fig. 13.3 (adapted after [2]). Even if the Pc molecules have a very strong intrinsic optical anisotropy due to its planar structure, which is inherent transmitted to most OMBD films, the dielectric functions reported from ellipsometry before the last decade, consider the layers as being isotropic [1, 10–13]. Here the references refer to $H_2$Pc and CuPc only. One early exception is the work by Debe [14], who tried to approximate the anisotropic dielectric function of CuPc by investigating two different types of oriented films—with the *b*-axis (the molecular stacking axis) perpendicular and respectively parallel to the substrate surface. For each film only the in-plane components of the dielectric function were determined. While as will be shown below, the Debe work [14] was well ahead of its time, the shape of the $Q$ band does not resemble with anything reported in the literature on CuPc. Probably the first reliable report of the Pcs film optical anisotropy was made by Alonso et al. [15], this time for perfluorinated CuPc ($F_{16}$CuPc). Their X-ray investigations indicate that out-of-plane ordering exists and the films have to be uniaxial anisotropic. However, the dielectric function perpendicular to the sample surface could not be unambiguously fitted. Therefore the authors followed a similar procedure like Debe [14], and they used the azimuthal ordering on sapphire ($Al_2O_3$) substrates [15]. Several important observations came out from their ellipsometric work: (i) the film morphologies and structures were found to depend on the substrate used, its temperature during deposition, and also on the layer thickness; (ii) the films on MgO and $SiO_2$/Si are at most uniaxial almost isotropic and biaxial with large in-plane anisotropy on $Al_2O_3$ substrates; and (iii) on cooled substrates it is not possible to obtain ordered films, whereas ordering is achieved for growth on nonintentionally (heated ∼room temperature) or heated substrates [15]. They also define a dichroic ratio $R$ as

$$R = \left| \frac{\varepsilon_{2\perp} - \varepsilon_{2\parallel}}{\varepsilon_{2\perp} + \varepsilon_{2\parallel}} \right|$$

where the subscripts perpendicular and parallel refer to orientation with respect to the sapphire $c$ axis. In this case $R$ takes a value of 0 for isotropic films and approaches 1 for very anisotropic ones.

As visible from Figs. 13.4 and 13.5 the temperature has not only an effect on the magnitude of the dichroic ratio, but also on the shape and magnitude of the $Q$ band. This is associated to a change in the Pc molecules stacking from a face to face to a strongly slipped stacked configuration as reported for crystals of non-planar Pcs [16, 17].

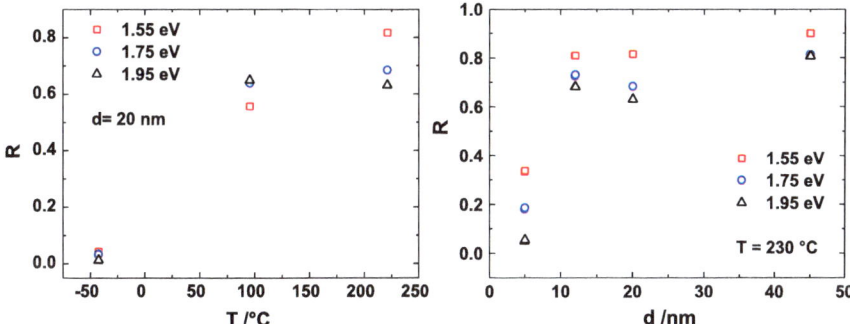

**Fig. 13.4** Calculated dichroic ratios at the energies presented in the legend for $F_{16}$CuPc films on sapphire ($11\bar{2}0$) substrate as a function of substrate temperature (*left*) and as a function of film thickness (*right*) for substrate temperatures of 230 °C (after [15]). Reprinted with permission from M.I. Alonso, M. Garriga, J.O. Ossó, F. Schreiber, E. Barrena, H. Dosh, Strong optical anisotropies of $F_{16}$CuPc thin films studied by spectroscopic ellipsometry, J. Chem. Phys. **119**, 6335 (2003). Copyright 2003, American Institute of Physics

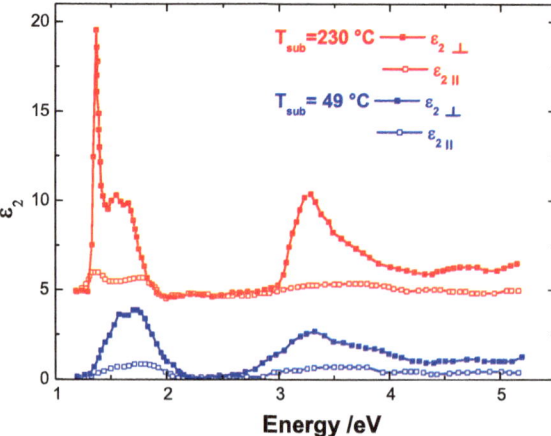

**Fig. 13.5** Comparison of the imaginary part of the dielectric function, $\varepsilon_2$, for $F_{16}CuPc$ films grown at two different substrate temperatures. The *subscripts perpendicular* and *parallel* refer to orientation with respect to the sapphire substrate $c$ axis. For clarity the red spectra are baseline shifted (after [15]). Reprinted with permission from M.I. Alonso, M. Garriga, J.O. Ossó, F. Schreiber, E. Barrena, H. Dosh, Strong optical anisotropies of $F_{16}CuPc$ thin films studied by spectroscopic ellipsometry, J. Chem. Phys. **119**, 6335 (2003). Copyright 2003, American Institute of Physics

Another early report where the authors showed that $H_2Pc$ films have a strong uniaxial anisotropy in the out-of-plane direction can be found in [18]. Here the $H_2Pc$ layers were prepared by OMBD in high vacuum (HV) on different substrates, hydrogen passivated Si(111) and NaCl, which were kept at room temperature during deposition. While ellipsometry measures in a first approximation the dielectric function which lies at the intersection of the incidence plane and sample plane [19], the confidence in the out-of-plane component for uniaxial samples can be increased by using a multi-sample analysis (MSA) procedure [20, 21]. The MSA method was applied in [18] and three different thicknesses ranging from 60 to 155 nm were treated together in a MSA model. The resulting dielectric function of the $H_2Pc$ layers is presented in Fig. 13.6.

Taking into account that the dielectric function for an isotropic $H_2Pc$ layer can be calculated by averaging over twice the in-plane component and once the out-of-plane one, the result from Fig. 13.6 can be directly compared with the one from [1]. As visible there is a remarkable resemblance, not only in the shape and spectral feature positions, but also in the absolute magnitude of $\varepsilon_2$ in the $Q$-band region. Additionally in [18] the extinction coefficient calculated from ellipsometry data was directly compared with the one calculated from the transmission measurements on NaCl substrates. The good agreement between the two different measurements, combined with reflection FTIR studies performed in s- and p-polarizations at 60° angle of incidence, indicates that uniaxial anisotropy has to be considered when modeling the optical response of organic films composed of such planar molecules.

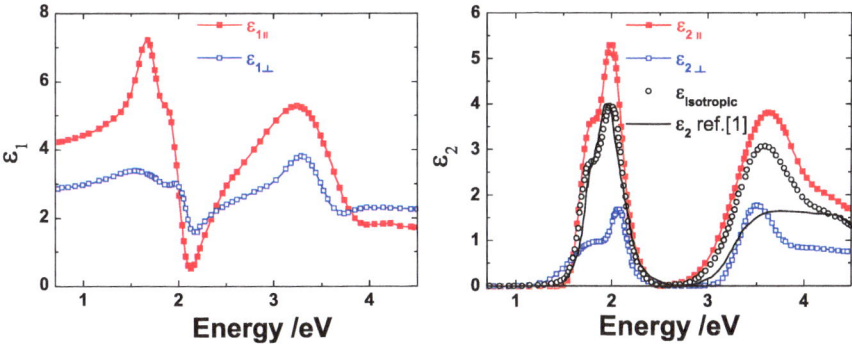

**Fig. 13.6** Anisotropic dielectric function of $H_2Pc$. *Left*—real part, *right*—imaginary part. *Full squares* represent in-plane components, *open squares* out-of-plane components (modified after [18]). With *black open circles* the isotropic average is compared with the result from [1]

In order to model the $Q$ band absorption several approaches were used in literature to calculate the oscillator strength of the optical transitions for various Pc films [22–24]. While this was made by using the second derivative of the dielectric function, calculated from ellipsometric data, for metal tetrasulphonated phthalocyanine films (CuTSPc, NiTSPc and ZnTSPc) [13, 22], the analysis for $H_2Pc$, CuPc, and Fullerene ($C_{60}$) was performed in [24] using a Kramers–Kronig (KK) analysis of the spectrophotometric data (reflection and transmission). Two components were found enough to model the $Q$ band absorption of $H_2Pc$ and CuPc, while three components were used for TSPcs. The $H_2Pc$ $Q$ band components corresponding to $\pi$–$\pi^*$ transitions have the transition dipoles in the molecular plane, as the $\pi$ orbitals are antisymmetric with respect to the Pc plane. For CuPc only one transition is expected due to the higher symmetry induced by the addition of the central Cu atom. As the double structure is visible even in vapor phase (see [24] and references therein), it cannot be attributed to a large Davydov splitting. However, when more components are observed they usually stem from a splitting of the $Q_x$ and $Q_y$ transitions, with dipoles lying in the molecular plane as schematically represented in Fig. 13.7. As the absorption intensity is proportional to the cosine squared of the angle between the electric field vector and the direction of the transition dipole, an average molecular orientation can be determined from the anisotropic dielectric function. A model was proposed by Gordan et al. [18] taking into account the polycrystalline nature of $\alpha$-phase $H_2Pc$ films. An optical anisotropy like the one in Fig. 13.6 thus indicates a molecular growth with the staking axis ($b$-axis) predominantly perpendicular to the substrate surface.

The model described in [18], can be simplified considering a molecule with two equivalently strong dipoles ($\mu$) in the molecular plane, perpendicular on each other. If the molecule has an arbitrary orientation angle $\theta$ with respect to substrate as well as an arbitrary orientation with respect to the $z$ axis perpendicular to the molecular plane, the average orientation can be determined using the schematic representation from Fig. 13.7 as follows.

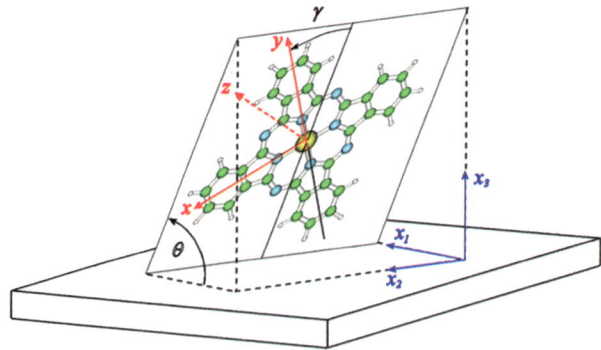

**Fig. 13.7** Schematic representation of a Pc molecule arbitrary oriented on a substrate (modified after [25])

The absorption is proportional to $\sim\mu^2\cos^2(\alpha)$, where $\alpha$ is the angle between the electric field vector and the transition dipole. In the Eigen axis of the sample (see Fig. 13.7) the absorption in $x_3$ direction, perpendicular to the sample surface will be:

$$A_\perp = A_{x_3} \sim \left(\mu^2\cos^2(\gamma) + \mu^2\sin^2(\gamma)\right)\sin^2(\theta) \sim \mu^2\sin^2(\theta).$$

As visible, the angle $\gamma$ smears out, as is expected for two perpendicular, equivalent dipoles. Therefore the rotation of the molecule in the $xy$ plane plays no role in the calculation.

In the sample plane the absorption will be proportional to:

$$A_{x1} \sim \left(\mu^2\cos^2(\gamma) + \mu^2\sin^2(\gamma)\right)\cos^2(\theta) \sim \mu^2\cos^2(\theta),$$
$$A_{x2} \sim \mu^2\cos^2(\gamma) + \mu^2\sin^2(\gamma) \sim \mu^2.$$

If we have a random molecular orientation in the sample plane (i.e. around the $x_3$ direction), the in-plane absorption will average out, and it will be proportional to

$$A_\parallel \sim \frac{1}{2}\mu^2\left(\cos^2(\theta) + 1\right).$$

Therefore the ratio between the area under the in-plane absorption coefficient and the out-of-plane one will be:

$$\frac{A_\parallel}{A_\perp} = \frac{\cos^2(\theta) + 1}{2\sin^2(\theta)}$$

giving an average orientation angle $\theta$:

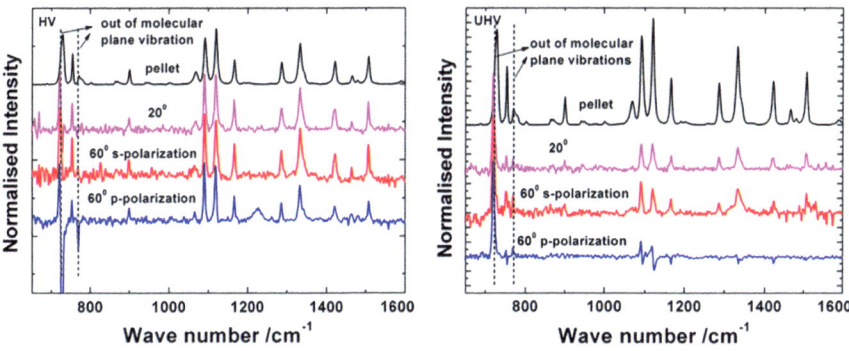

**Fig. 13.8** *Left*—reflection FTIR spectra of a HV CuPc sample and *right* of a UHV CuPc sample. The 722 and 770 cm$^{-1}$ peaks correspond to out-of-plane vibrations of the CuPc molecule while the peak at 753 cm$^{-1}$ and all bands above 800 cm$^{-1}$ are due to molecular in-plane vibrations (modified after [26])

$$\theta = \arccos\sqrt{\frac{2A_{\parallel} - A_{\perp}}{2A_{\parallel} + A_{\perp}}}. \tag{13.1}$$

For a single dipole, e.g. a rod like molecule, the ratio will transform to:

$$\frac{A_{\parallel}}{A_{\perp}} = \frac{\cos^2(\theta)}{2\sin^2(\theta)} \text{ and the average orientation angle to } \theta = \arctan\sqrt{\frac{A_{\perp}}{2A_{\parallel}}}. \tag{13.2}$$

A similar treatment was performed in [26, 27] for CuPc and F$_{16}$PcVO films, respectively. For CuPc grown on hydrogen passivated Si(111) it was found that if the deposition takes place in high vacuum conditions (HV—$8 \times 10^{-7}$ mbar) the molecules adopt an average orientation angle $\theta$ of 40°, while in ultra-high vacuum deposited films (UHV—$6 \times 10^{-10}$) the average orientation angle $\theta$ is around 53°. The authors [26] conclude that the orientation differences may be correlated with the markedly different pressures during growth, but an influence of the purity of the source material cannot entirely be excluded, as 97 and 99% purity materials were used for the HV and UHV deposition, respectively. The anisotropy of the films was also visible in reflection FTIR spectra. As shown in Fig. 13.8 the behaviour of the IR active modes with dipoles perpendicular to the molecular plane have a derivative like shape in p-polarization for the HV films, while this is reversed for the UHV films. A simulation on how anisotropic IR absorption affects the effective optical values is briefly introduced in [28] for some polymer materials.

For F$_{16}$PcVO films from [27] similar results like the ones obtained by Alonso et al. [15] were achieved. The difference is that in [27] the anisotropy refers to in/out-of-sample plane and to a different perfluorinated Pc molecule. From the change in the absorption shape, it was found that the molecules have a co-facial arrangement when grown on fused silica substrates kept at 85 °C, and a head-to-tail (or slipped stacked) configuration on KBr at 42 °C. From the strong in-plane/out-of-plane anisotropy an average orientation angle of 56° was determined for the silica substrates and 3° for the KBr substrates. Here the change in the stacking configuration and orientation is not only due to temperature, but mainly due to the strong molecule–substrate interaction for the KBr case [27]. Also in this case the absorption spectra could be well simulated from ellipsometry data only when considering the films as anisotropic.

A step forward in confirming that the molecular orientation determined from ellipsometry is accurate can be found in [3, 29], where the average orientation angle was also determined from FTIR and MOKE. In [29] the influence of the deposition pressure on the average tilt angle of ZnPc molecules grown on $SiO_2$/Si substrates was investigated with IR spectroscopy and ellipsometry. The average orientation angles calculated from ellipsometry were 61° and 20° for the high pressure $(2.4 \times 10^{-6}$ mbar) and low pressure samples $(3 \times 10^{-8}$ mbar), respectively, while from FTIR reflection data the angles determined were 61° and 19°. The IR analysis was performed comparing the ratios of the molecular in-plane and out-of-plane IR bands, for 15° and 30° angles of incidence (AoIs) for p and s polarized light with respect to KBr pellet transmission spectra, for which the ZnPc molecules are randomly oriented. A formalism like the one presented in [30] was used:

$$\frac{2\sin^2\theta}{\cos^2\theta} = \frac{\text{out of plane ratio}}{\text{in plane ratio}}.$$

Please note that this is equivalent with (13.2) when oriented dipoles are considered. Taking into account the s- and p- reflection FTIR spectra taken at AoIs of 15°, 30°, 45°, and 60°, together with the ellipsometric data, the anisotropic dielectric function of ZnPc was modeled from the IR to the UV-vis region [29] (see Fig. 13.9).

While indisputably the molecular orientation can be more accurately determined form ellipsometry in the IR spectral region than in visible, due to the pure dipole character of the IR absorptions and the typically higher spectral contrast in the infrared range due to spectrally separated vibrational bands, a different approach was used in [3] to have higher determination accuracy in the visible range. Here the

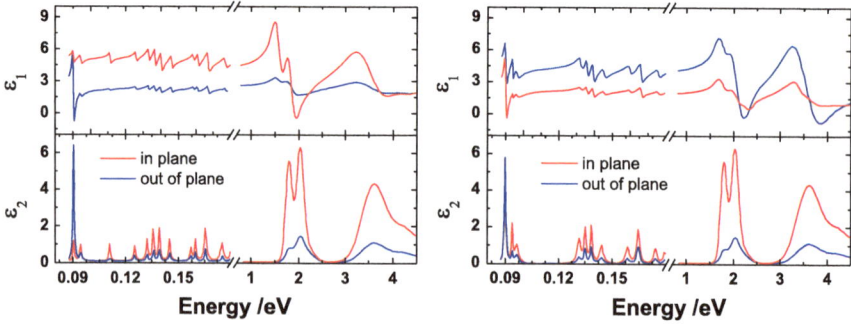

**Fig. 13.9** Complex dielectric function of ZnPc films prepared under high pressure (*left*) and low pressure (*right*) (modified after [29])

magneto-optical Kerr effect (MOKE) was measured in polar configuration for CuPc and VOPc molecules, schematically shown in Fig. 13.2. MOKE spectroscopy is not yet widely applied to organic molecules. Therefore going through the MOKE theory and explaining how the MOKE spectra is analyzed are beyond the scope of this chapter. The influence of the molecular orientation on the magneto-optical response was discussed in [31] and a short description of the procedure applied to the Pc molecular class can be found in [3]. Here three different substrates were used—Si with native oxide and two types of hydrogen passivated substrates treated with 40% hydrofluoric acid (HF) for 2 min and 5% HF for 2 min, respectively. It was found from ellipsometry that for the first two substrates the average CuPc orientation is 85° and 81°, respectively, while for the last substrate type it is 55°. The determination accuracy was estimated to be 4°. The MOKE measurement confirmed the determined values, except for the 85° one. Here the large deviation between the expected and measured ratio of the MOKE $Q$ spectrum could be explained by the fact that the molecules are oriented at 87.5° instead of 85° [3]. The authors conclude that MOKE might be a more accurate method to assess the average molecular orientation in the visible range than ellipsometry, and that the substrate surface roughness induced by different passivation procedures has a large effect on the Pc molecular orientation.

## 13.3 Controlling the Molecular Orientation with a Molecular Template

Besides the Pcs, another class of planar organic molecules intensively studied is the perylene derivatives. An example of a molecular structure from this class is shown below for the 3, 4, 9, 10-perylenetetracarboxylic dianhydride (PTCDA) (Fig. 13.10). Its optical anisotropy in thin films and crystals was thoroughly studied with ellipsometry and RAS [32–34], the mentioned references here being just an early few. While [32, 34] treat uniaxial PTCDA films on Si and GaAs substrates, in [33] the optical anisotropy of a PTCDA single crystal is analyzed in a similar way like described before for $F_{16}$CuPc [15]. One difference worth mentioning is that compared to Pcs the PTCDA grows lying flat on most substrates [32, 34], with two molecules per unit cell rotated by 90° with respect to each other. As the molecular orientation can

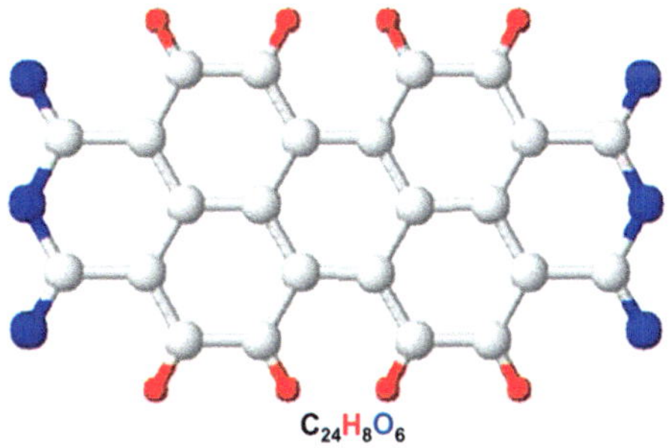

**Fig. 13.10** Schematic representation of the PTCDA molecular structure

$C_{24}H_8O_6$

be determined in a similar way as presented above for Pcs, here we concentrate on the template effect which PTCDA has on $H_2Pc$ [35]. The effect was first studied by Heutz et al. [36] and explained by a theoretical calculation of the VdW interaction [37]. According to the calculations the $H_2Pc$ molecules near PTCDA will adopt a parallel configuration as it is more energetically stable [37]. In [35] several single layers of $H_2Pc$ and PTCDA were prepared on Si with silicon oxide and on glass.

The films showed similar growth modes on both substrates. The MSA ellipsometric analysis indicated that the $H_2Pc$ molecules adopt an average orientation angle of 52° with respect to the substrate while PTCDA lies flat as expected. While the MSA analysis strongly reduces the correlation between the in-plane and out-of-plane components, here the reliability of the model was again increased as the simulated transmittance data using the in-plane optical constants matched perfectly the measured spectra, as shown in Fig. 13.11. Fitting ellipsometry and transmission data together will strongly increase the confidence limit for the out-of-plane components. However, also here the accuracy of the molecular orientation angle was estimated to be ±4°.

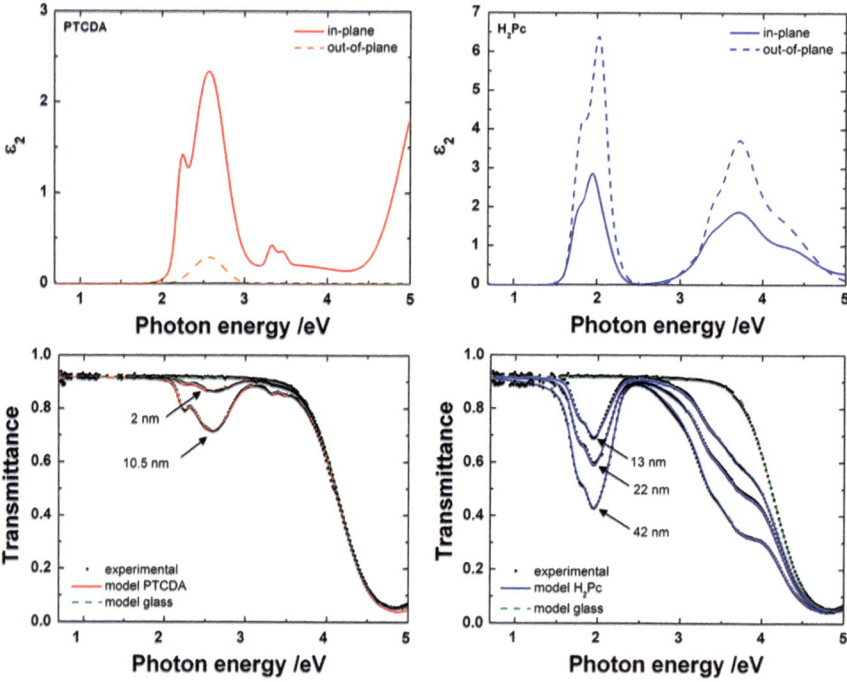

**Fig. 13.11** Imaginary part of the dielectric function of PTCDA and $H_2Pc$ films grown on silicon oxide and glass and below the experimental transmission spectra on glass and the simulated ones using the in-plane components (after [35]). Reprinted from O.D. Gordan, T. Sakurai, M. Friedrich, K. Akimoto, D.R.T. Zahn, Ellipsometric study of an organic template effect: H2Pc/PTCDA, Org. Electron. **7**,521 (2006). Copyright 2013, with permission from Elsevier

When grown on PTCDA the in-plane absorption of $H_2Pc$ becomes stronger than the out-of-plane one, giving an average orientation angle of 25°, which is very close to the value of 27° deduced from X-Ray and reflection IR studies [38]. The strong change in the molecular orientation is visible also when the in-plane components of the films grown on glass and $SiO_2/Si$ are directly compared to the ones grown on PTCDA (see Fig. 13.12).

When looking at the heterostructure transmission spectra from Fig. 13.13, small deviations are visible for the 7 nm $H_2Pc$ film. As this was also visible in the ellipsometric data [35], it indicates that the thin $H_2Pc$ film has slightly different optical

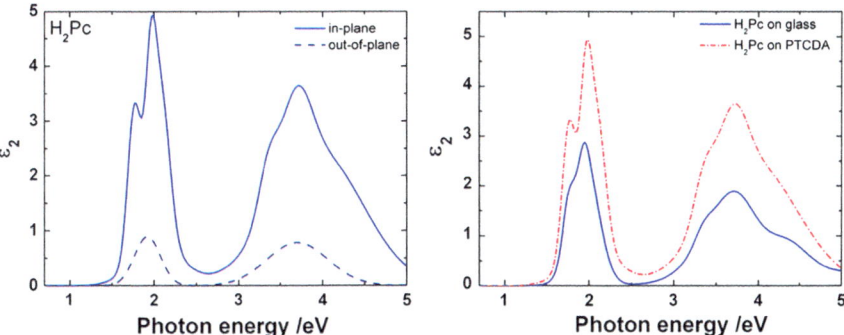

**Fig. 13.12** Imaginary part of the dielectric function of $H_2Pc$ grown on PTCDA (*left*), and comparison between the in-plane imaginary dielectric function of $H_2Pc$ on glass and $SiO_2/Si$ and the in-plane imaginary dielectric function of $H_2Pc$ on PTCDA (after [35]). Reprinted from O.D. Gordan, T. Sakurai, M. Friedrich, K. Akimoto, D.R.T. Zahn, Ellipsometric study of an organic template effect: H2Pc/PTCDA, Org. Electron. **7**,521 (2006). Copyright 2013, with permission from Elsevier

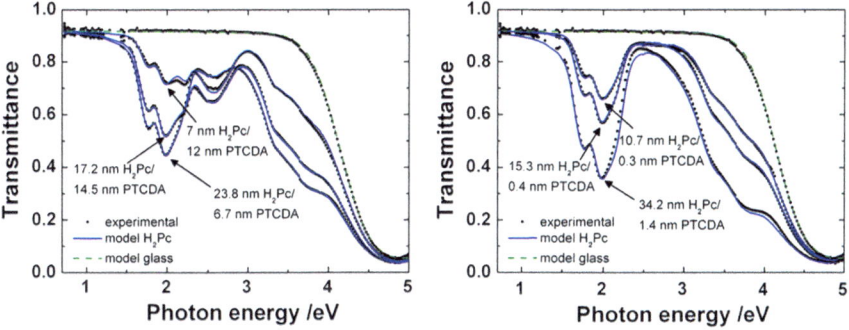

**Fig. 13.13** Transmittance spectra of $H_2Pc$ layers on PTCDA/glass substrates with different PTCDA thicknesses. For comparison the transmittance of the glass substrate is also plotted. The *continuous lines* represent the model simulation of the transmittance using the in-plane components of the dielectric function. Reprinted from O.D. Gordan, T. Sakurai, M. Friedrich, K. Akimoto, D.R.T. Zahn, Ellipsometric study of an organic template effect: H2Pc/PTCDA, Org. Electron. **7**,521 (2006). Copyright 2013, with permission from Elsevier

constants than the thicker films. This can be explained by the packing relaxation of the $H_2Pc$ molecules from a lying flat configuration, close to the PTCDA interface, to the typical $\alpha$-phase herringbone structure [39]. Worth mentioning is also that a PTCDA layer as thin as 3 Å is enough to produce a drastic change in the $H_2Pc$ orientation. Such ways of controlling the molecular orientation can be used to improve device performance for which lying orientation is required.

In a demonstration by Swiggers et al. [40], it was shown that aligning 27% of the pentacene (PEN) grains within a 30° range can lead to an enhancement of the OFET current saturation by a factor of 2.5. The PEN thin films optical anisotropy was studied by Hinderhofer et al. [41] along with a perfluorinated derivative of PEN (PFP). Being one of the most used molecules for OFETs (see [41] and the references therein) due to its high mobility, adding fluor atoms to the PEN structure will change its transport behaviour from $p$ to $n$ type. While both PEN and PFP produce strong anisotropic films, the electronic properties visible from the imaginary part of the dielectric function are quite different for the two molecules. In order to have a more accurate determination of the out-of plane components, besides a MSA analysis, the authors in [41] used the interferences created in a thick $SiO_2$ layer.

## 13.4 Amorphous Organic Thin Films

Even if so far thin film anisotropies were presented only for molecules which easily form organic molecular crystals thus usually giving polycrystalline films, more recently it was shown [42] that even in amorphous films of small organic molecules relevant for OLEDs, a preferential orientation can take place. This can enhance the electric transport [43] and the efficiency in OLEDs [44]. An overview on the results presented in [42] is given in Fig. 13.14. As the extinction coefficients for the materials presented in Fig. 13.14 are higher for the in-plane components (ordinary) than for the out-of-plane ones (extraordinary), the authors conclude that these materials have a preferential horizontal molecular orientation. They define an orientation order parameter $S$ as:

$$s = \frac{1}{2}\langle 3\cos^2\theta - 1\rangle = \frac{k_e - k_o}{k_e + 2k_o}, \qquad (13.3)$$

where $\theta$ is the angle between the molecular long axis and the direction perpendicular to the substrate surface. The indices $o$ and $e$ refer to ordinary and extraordinary, respectively. Taking into account that the angle $\theta$ is complementary to the one defined before, it can be easily shown that (13.3) is a more complicated formulation of (13.2). Here $S$ will be $-0.5$ if the molecules are completely parallel to the surface, 0 for random orientation and 1 if they are perpendicular to the surface [42]. A correlation between $S$ and the OLEDs driving voltage was found [42, 45], indicating again that horizontally oriented molecules provide better electrical characteristics. In [46] the analysis was extended to in situ real time monitoring of amorphous multi-layered structures.

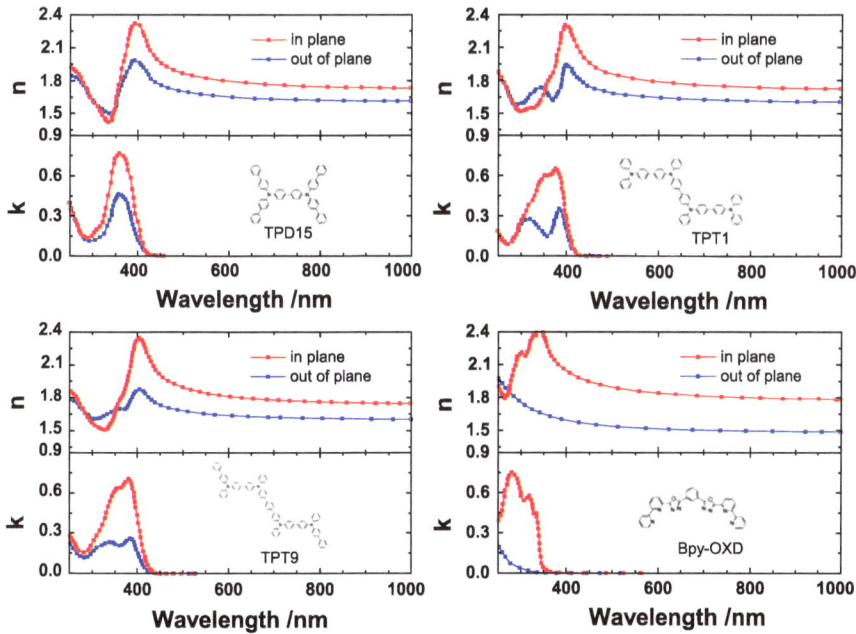

**Fig. 13.14** Selected examples of anisotropic dielectric functions of amorphous organic films with in-plane (*red lines*) and out-of-plane (*blue lines*) for several small organic molecules relevant for OLEDs with their molecular structure schematically shown as *insets* (after [42]). Reprinted with permission from D. Yokoyama, A. Sakaguchi, M. Suzuki and C. Adachi, Horizontal molecular orientation in vacuum-deposited organic amorphous films of hole and electron transport materials, Appl. Phys. Lett. **93**, 173302 (2008). Copyright 2008, American Institute of Physics

## 13.5   Error Sources for Determination of Molecular Orientation

Along this chapter, many examples of modelling optical anisotropies from ellipsometric data were given, and these are just a glimpse of its specific application to small organic molecules. The last decade saw an amazing increase in the number of publications where complex optical anisotropic models were applied to a large number of material systems from IR to VUV. While the $4 \times 4$ matrix algorithm which allows the calculation of the Jones matrices for anisotropic layered structures was presented quite early by Berreman [47], only relatively recently numerical implementations of the formalism [48] became available in commercial software. A very good overview of the different mathematical approaches which can be used to tackle the anisotropic cases is given by Schubert in Chap. 9 of the Handbook of Ellipsometry [49]. Additionally, experimental measurement procedures and approaches for different anisotropic cases are also given in the same chapter. While rigorously correct, the numerical formalisms mentioned before, involve the numerical solution of matrices of nonlinear equations, which are hard to follow from a physical view. Aspnes [19] developed

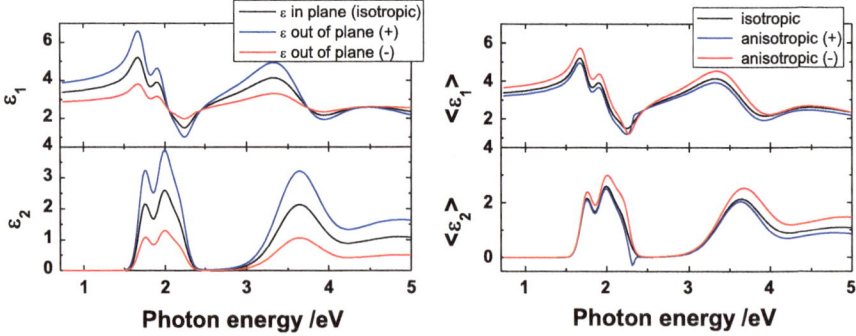

**Fig. 13.15** *Left*: simulated dielectric function of a CuPc: isotropic layer (*black*), and anisotropic with out-of-plane component higher (*blue*) and lower absorption (*red*), respectively. *Right*: the simulated effective response of an isotropic (*black*) and anisotropic half infinite layer using the dielectric function of the isotropic layer for the in-plane component and varying the out-of-plane components like in the left figure

an approximate solution of the ellipsometric equations for a biaxial crystal, based on a first order expansion which assumes that the anisotropies are corrections to an isotropic mean value. This procedure is not exact, but the calculation is performed in the framework of the easily solved isotropic problem and the resulting expansion gives considerable insight in the physics of the ellipsometric measurement [19]. Therefore this work clearly shows in a very intuitive way that ellipsometry measures in a first approximation the dielectric function which lies at the intersection of the incidence plane and sample plane. If the sample has in-plane anisotropy this reduced sensitivity can be overcame by rotating the sample azimuthally. If the sample has uniaxial anisotropy with one Eigen axis parallel to the sample plane, then the Jones matrix becomes diagonal [49]. Therefore unless the sample is thick and transparent, an ellipsometric measurement will not unequivocally yield whether the sample is uniaxial anisotropic or not. On the other side, even for the simple case of an uniaxial substrate, an isotropic treatment could yield a refractive index which is not necessarily between the ordinary and extraordinary one [50]. Therefore, especially for organic crystals which exhibit large optical anisotropies, effective dielectric values measured in different crystallographic direction have to be interpreted with care. An estimation of the differences in measured effective values can be obtained by looking at the simulation[1] presented in Fig. 13.15. Here the in-plane dielectric function of the CuPc layer was artificially modified as if the layer would have $+50\%$ more absorption in the out-of-plane direction $(+)$ or $-50\%$ less absorption in the out-of-plane direction $(-)$. The corresponding real part was generated according to the KK consistency. All simulations were performed without considering surface roughness. For the isotropic case only the in-plane components were used.

---

[1]The simulations were done using the VASE software from J.A. Woollam Co., Inc.

**Fig. 13.16** Simulated effective dielectric function for a 50 nm isotropic and anisotropic film on silicon performed in an analogous way like the simulation in Fig. 13.15. The dielectric functions from the left side of Fig. 13.15 were used for the CuPc layer

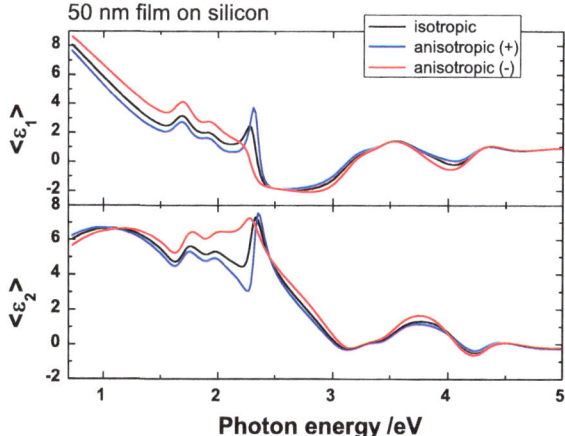

As expected the measured effective dielectric function is equal to the real one only for the isotropic bulk case. Going away from the bulk case, even for a 50 nm film, the measurements will yield a different effective response for an isotropic and anisotropic film with same thickness. However, in both cases, bulk and film, the in-plane dielectric function plays the major role in the sample optical response. Differences of a magnitude like the ones in Fig. 13.16 could be easily introduced by non-idealities like film inhomogeneity and/or surface roughness. Therefore a failure of an isotropic fit is not an unequivocal proof that the sample is uniaxial anisotropic. It is always recommended to have the uniaxial anisotropy confirmed by other investigation techniques (e.g. FTIR, Raman, or MOKE).

Once the presence of the anisotropy is established, several methods can be used to increase the confidence in the out-of-plane component like the MSA analysis. An early estimation of the relative errors for such an analysis, for the case of Poly (3,4-ethylenedioxythiophene) (PEDOT), using the model confidence limits (calculated from correlations between the fit parameters) was given in [51]. It was found that for the in-plane components the relative numerical errors were 1% for the real part and 4% for the imaginary one, while for the out-of-plane absorption the error could be as large as 10%. Another way to increase the confidence limits is to use, whenever possible, transmission data, and/or interference enhancing techniques [52]. Such an approach was used in [53] for anisotropic polyfluorene thin films, and more recently by Wynands et al. [54] for organic films relevant for OPVs. This last reference gives a nice overview about the care one normally should take in evaluating and modelling the data. The error values given above are in no way general, but are highly dependent on the care one takes to reduce parameter correlation.

While so far a typical inaccuracy of $\pm 4°$ was given when determining the average molecular orientation angle from ellipsometry using a 10% determination error for the out-of-plane component, other error sources have to be considered due to the physical simplification of the pure dipole model.

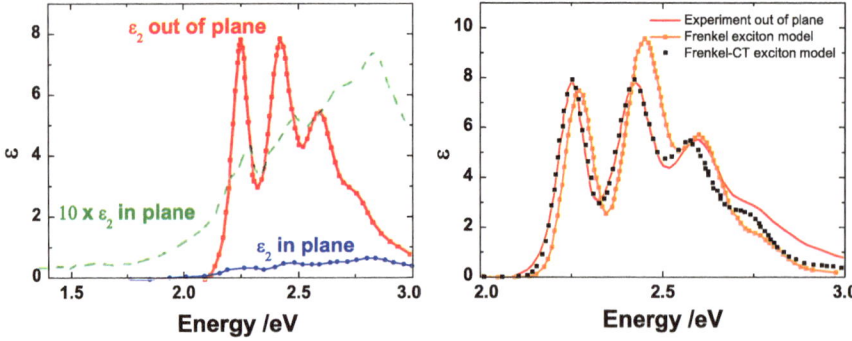

**Fig. 13.17** *Left*: imaginary part of the dielectric function for DIP films and *right*: comparison between Frenkel exciton model and Frenkel-CT exciton model (after [55]). Reprinted with permission from Heinemeyer, R. Scholz, L. Gisslén, M.I. Alonso, J.O. Ossó, M. Garriga, A. Hinderhofer, M. Kytka, S. Kowarik, A. Gerlach, and F. Schreiber, Phys. Rev. B **78**, 085210 (2008). Copyright 2013 by the American Physical Society

A theoretical study [55] applied to diindenoperylene (DIP) thin films on oxidized Si substrates indicated that the observed dielectric function cannot be described by a pure Frenkel exciton. The DIP films were found to be uniaxial anisotropic with an azimuthal random orientation, but highly oriented along the surface normal. The imaginary part of the dielectric function from the MSA modelling procedure is presented in Fig. 13.17. Assuming that the optical transitions arise only from neutral excitations with dipole moments oriented in the molecular plane an average tilt angle of 26° with respect to the surface normal was calculated using a similar formula like the one in (13.2). However, if a combined Frenkel and Charge Transfer (CT) model is used to simulate the optical properties and only the lowest absorption peak at 2.25 eV is used to estimate the molecular orientation, an angle of 17° was found [55], in excellent agreement with the published crystal structure of the thin films [56]. For the Pc class the molecular orientation estimated from the $Q$ band anisotropy was always in very good agreement with other measurement techniques (FTIR, MOKE). Due to the weak VdW interaction and due to the strong exciton binding energy, the assumption that the transitions will keep their Frenkel character, when going to thin films, is usually of common sense for most small organic molecules, however, the spectral range has to be carefully chosen.

## 13.6 Ultra-Thin Films

So far the optical response and the dielectric functions reported here for the organic materials was for relatively thick films (more than 7 nm). The MSA method is valid as long as the optical properties for the films with different thicknesses are the same, regardless of thickness. However, when addressing very thin films in the monolayer and sub-monolayer regime, changes of the intermolecular interaction and other effects are expected, and therefore changes in the optical response. The detailed understanding of nano-scale materials is a key prerequisite for future devices, in the context of the scaling down trend toward the molecular level. In spite of the very high sensitivity of the ellipsometric technique, interpreting the ellipsometry spectra for very low coverages (below few nanometers) down to a monolayer or even sub-monolayer regime remains a challenging task as the optical path of the light through the material is much smaller than the wavelength [57–59]. In this case it is difficult to separate the refractive index of the film and the film thickness. Consequently only the product of these parameters can be uniquely determined [57, 58]. A graphical representation of this experimental limit is presented in Fig. 13.18. In the simulation the Si dielectric function [60] and a constant refractive index were used for the substrates, while a 0.6 nm film with different refractive indexes was considered on top.

The difference $\delta\Delta$ between the $\Delta$ value of the silicon substrate and the $\Delta$ value of the substrate covered with 0.6 nm film with a refractive index from 1.45 to 2.2 is presented in Fig. 13.18. It is clear that up to energies of 3 eV it is very hard to distinguish between the refractive index for films with such a small thickness. Thus, in the higher energy range the refractive index of the film can be determined if the ellipsometer has a measurement accuracy in $\Delta$ better than 0.5°. The changes in the measured $\Delta$ values depend also on the refractive index of the substrate, as can be seen for the simplest case of a substrate with a constant refractive index of 1.5.

In spite of these difficulties McIntyre and Aspnes [61] proved that differential reflection spectroscopy can be a powerful tool in investigating the optical properties of ultra-thin films. Using spectral ellipsometry the optical response of monolayers was successfully investigated as well [62–65]. Rossow et al. [63] presented the effective dielectric function of an As-terminated Si(111) surface compared to the one of a clean Si(111) surface. The dielectric function of the As monolayer was evaluated from a three phase model. In a study by Ritzi [62] the electronic structure of GaP(110) and copper phthalocyanine (CuPc) overlayers was investigated. Esser et al.

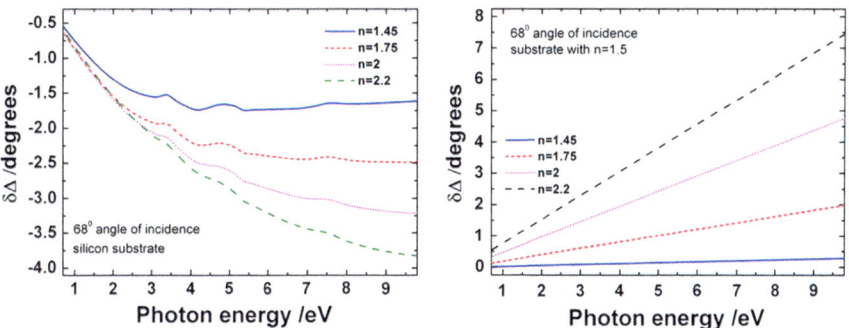

**Fig. 13.18** The difference $\delta\Delta$ between the $\Delta$ value of a bare substrate (*left*—Si substrate, *right*— a substrate with constant refractive index of 1.5) and $\Delta$ values of the covered substrates with a 0.6 nm film with refractive indexes between 1.45 and 2.2

[64, 65] used the combined information from ellipsometry, reflectance anisotropy spectroscopy (RAS), and Raman spectroscopy to characterize the modification of the surface properties induced by hydrogen exposure of GaAs(100) and Sb- terminated GaAs(110). In this study the surface and monolayer electronic properties were evaluated from ellipsometry data. In a more recent work Proehl et al. [66] studied the intermolecular interaction between PTCDA molecules with monolayer resolution by means of differential reflection spectroscopy, revealing shifts in molecular transition energies related to intermolecular coupling. Especially relevant in the context of this chapter are [62, 66] along with another study [67] which showed for both Pc and perylene classes, that shifts and changes of the optical absorption bands are expected as a function of thickness for ultra-low coverages.

When looking at the ellipsometric evaluation of such ultrathin overlayers, a first order approximation for the Fresnel coefficients can be used, if the substrate can be treated as a half infinite media, and the layer thickness is below 1 nm [57, 68]. Then the ellipsometric equation can be simplified to yield a quadratic form. Therefore for a three phase system with $\varepsilon_a$, $\varepsilon_L$, and $\varepsilon_s$ being the complex dielectric functions of ambient, overlayer, and substrate, respectively, the optical response of the system (effective dielectric function) can be written as [69]:

$$\langle \varepsilon \rangle = \varepsilon_s + \frac{4\pi i d}{\lambda} \sqrt{\varepsilon_s - \sin^2 \varphi} \frac{\varepsilon_s (\varepsilon_s - \varepsilon_L)(\varepsilon_L - 1)}{\varepsilon_L (\varepsilon_s - 1)} \qquad (13.4)$$

where $d$ is the layer thickness and $\phi$ the angle of incidence. An experimental demonstration for an organic thin film (Alq$_3$) measured in the VUV range at BESSY II can be found in [70]. The results are presented in Fig. 13.19 for a 0.65 nm film on H–Si(111) [70] along with an additional evaluation obtained for Alq$_3$ films on ZnO substrates. The strong shift of the Alq$_3$ feature on H–Si(111) substrates is probably due to the high polarizability of the silicon. However, besides the high silicon polarizability other effects like inter-molecular interactions, deformation of the Alq$_3$ molecule on the silicon surface and the effective field induced by the arrangement of the Alq$_3$ molecules on the surface should not be neglected.

As the today ellipsometric measurement accuracy requirements are increased by one order of magnitude [71], therefore an evaluation like the one presented before for ultra-thin films could yield realistic results also in the visible range. Along with the increased accuracy, modern ellipsometers have very fast acquisitions times, which make them perfect tools for in situ monitoring. As already mentioned, the substrate type and roughness, pressure conditions, purity of the source materials, and the deposition temperature can affect the growth mode of the organic molecules. Even small substrate off-cut angles could lead to a preferential azimuthal ordering of the molecules [72]. This type of very small azimuthal anisotropies can be easily detected with an ellipsometry related technique like RAS. While the direct investigation of the RAS spectra applied to organic deposition could directly provide useful information [73, 74] related to the growth mode, modelling RAS spectra remains challenging. This difficulty can be overcome by using the approximation formalism demonstrated by Seidel et al. [75] to convert the RAS spectra in pseudo ellipsometric values,

**Fig. 13.19** Comparison between the dielectric functions for a sub-monolayer of Alq3 on H–Si(111) (*blue circles*) and on ZnO (*dashed* and *doted red lines*) obtained using (13.4). The calculated $\varepsilon_2$ from (13.4) is affected by the experimental noise of two different measurements. For comparison a model fit for the Alq3 on H–Si(111) substrate is also shown

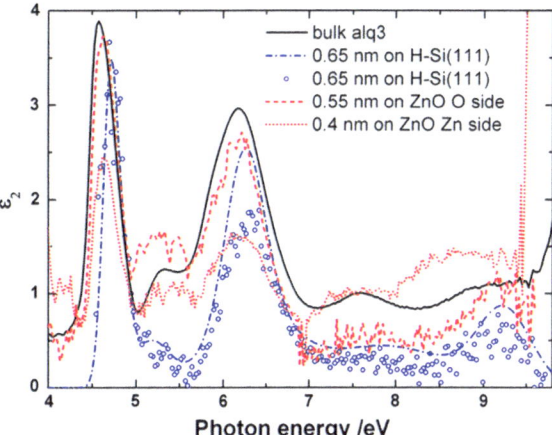

which can then be used to model the data with available ellipsometry software. Therefore, in the context of the vast parameter space which can affect the growth mode and molecular orientation of the organic films, such monitoring techniques combined with appropriate measurements and evaluation procedures could provide the necessary information needed for device optimization.

# References

1. H. Arwin, D.E. Aspnes, Thin Solid Films **138**, 195 (1986)
2. N.B. McKeown, *Phthalocyanine Materials* (Cambrige University Press, Cambrige, 1998)
3. M. Fronk, B. Bräuer, J. Kortus, O.G. Schmidt, D.R.T. Zahn, G. Salvan, Phys. Rev. B **79**, 235305 (2009)
4. C.C. Leznoff, A.B.P. Lever, *Phthalocyanines: Properties and Applications*, vol. 4 (VCH, New York, 1996)
5. R.F. Ziolo, C.H. Griffiths, J. Chem. Soc. Dalton Trans. **11**, 2300 (1980)
6. A.V. Chadwick, P.B.M. Dunning, J.D. Wright, Mol. Cryst. Liq. Cryst. **134**, 137 (1986)
7. H. Yonehara, C. Pac, Thin Solid Films **278**(1–2), 108 (1996)
8. Z. Bao, Adv. Mater. **12**, 227 (2000)
9. D. Hohnholz, S. Steinbrecher, M. Hanack, J. Mol. Struct. **521**, 231 (2000)
10. M.K. Debe, D.R. Field, J. Vac. Sci. Technol. A **9**, 1265 (1991)
11. A.B. Djurisic, C.Y. Kwong, T.W. Lau, W.L. Guo, E.H. Li, Z.T. Liu, H.S. Kwok, L.S.M. Lam, W.K. Chan, Opt. Commun. **205**, 155 (2002)
12. Y. Wu, D. Gu, F. Gan, Opt. Mater. **24**, 477 (2003)
13. E.G. Bortchagovsky, Z.I. Kazantseva, I.A. Koshets, S. Nespurek, L. Jastrabik, Thin Solid Films **460**, 269 (2004)
14. M.K. Debe, J. Vac. Sci. Technol. A **10**, 2816 (1992)
15. M.I. Alonso, M. Garriga, J.O. Ossó, F. Schreiber, E. Barrena, H. Dosh, J. Chem. Phys. **119**, 6335 (2003)
16. G.E. Collins, V.S. Williams, L.-K. Chau, K.W. Nebesny, C. England, P.A. Lee, T. Lowe, Q. Fernando, N.R. Armstrong, Synth. Met. **54**, 351 (1993)
17. R.A. Collins, A. Krier, A.K. Abass, Thin Solid Films **229**, 113 (1993)

18. O.D. Gordan, M. Friedrich, D.R.T. Zahn, Thin Solid Films **455–456**, 551 (2004)
19. D.E. Aspnes, J. Opt. Soc. Am. **70**, 1275 (1980)
20. C.M. Herzinger, B. Johs, W.A. McGahan, J.A. Woollam, W. Paulson, J. Appl. Phys. **83**, 3323 (1998)
21. U. Zhokhavets, R. Goldhahn, G. Gobsch, W. Schliefke, Synth. Met. **138**, 491 (2003)
22. J. Martensson, H. Arwin, Thin Solid Films **188**, 181 (1990)
23. J. Martensson, H. Arwin, Thin Solid Films **205**, 252 (1991)
24. A. Stendal, U. Beckers, S. Wilbrandt, O. Stenzel, C. von Borczyskowski, J. Phys. B At. Mol. Opt. Phys. **29**, 2589 (1996)
25. F. Seidel, O.D. Gordan, D.R.T. Zahn, Ellipsometry of CuPc Films Prepared under Different Vaccum Conditions (in preparation)
26. O.D. Gordan, M. Friedrich, D.R.T. Zahn, Org. Electron. **5**, 291 (2004)
27. O. Gordan, M. Friedrich, W. Michaelis, R. Kröger, T. Kampen, D. Schlettwein, D.R.T. Zahn, J. Mater. Res. **19**(7), 2008 (2004)
28. C. Bungay, T.E. Tiwald, Thin Solid Films **455–456**, 272 (2004)
29. J. Sindu Louis, D. Lehmann, M. Friedrich, D.R.T. Zahn, J. Appl. Phys. **101**, 013503 (2007)
30. R. Scholz, M. Friedrich, G. Salvan, T.U. Kampen, D.R.T. Zahn, T. Frauenheim, J. Phys. Condens. Matter **15**, S2647 (2003)
31. J.L. McInnes, E. Pidcock, V.S. Oganesyan, M.R. Cheesman, A.K. Powell, A.J. Thomson, J. Am. Chem. Soc. **124**, 9219 (2002)
32. T.U. Kampen, A.M. Paraian, U. Rossow, S. Park, G. Salvan, Th. Wagner, M. Friedrich, D.R.T. Zahn, Phys. Stat. Sol. A **188**, 1307 (2001)
33. M.I. Alonso, M. Garriga, N. Karl, J.O. Oss, F. Schreiber, Org. Electron. **3**, 23 (2002)
34. M. Friedrich, Th. Wagner, G. Salvan, S. Park, T.U. Kampen, D.R.T. Zahn, Appl. Phys. A **75**, 501 (2002)
35. O.D. Gordan, T. Sakurai, M. Friedrich, K. Akimoto, D.R.T. Zahn, Org. Electron. **7**, 521 (2006)
36. S. Heutz, R. Cloots, T.S. Jones, Appl. Phys. Lett. **77**, 3938 (2000)
37. S. Yim, S. Heutz, T.S. Jones, Phys. Rev. B **67**, 165308 (2003)
38. T. Sakurai, S. Kawai, J. Shibata, R. Fukasawa, K. Akimoto, Jpn. J. Appl. Phys. **44**, 1982 (2005)
39. T. Sakurai, R. Fukasawa, K. Akimoto, Jpn. J. Appl. Phys. **45**, 255 (2006)
40. M.L. Swiggers, G. Xia, J.D. Slinker, A.A. Gorodetsky, G.G. Malliaras, R.L. Headrick, Brian T. Weslowski, R.N. Shashidhar, C.S. Dulcey, Appl. Phys. Lett. **79**, 1300 (2001)
41. A. Hinderhofer, U. Heinemeyer, A. Gerlach, S. Kowarik, R.M.J. Jacobs, Y. Sakamoto, T. Suzuki, F. Schreiber, J. Chem. Phys. **127**, 194705 (2007)
42. D. Yokoyama, A. Sakaguchi, M. Suzuki, C. Adachi, Appl. Phys. Lett. **93**, 173302 (2008)
43. D. Yokoyama, A. Sakaguchi, M. Suzuki, C. Adachi, Appl. Phys. Lett. **95**, 243303 (2009)
44. J. Frischeisen, D. Yokoyama, A. Endo, C. Adachi, W. Brütting, Org. Electron. **12**, 809 (2011)
45. M. Aonuma, T. Oyamada, H. Sasabe, T. Miki, C. Adachi, Appl. Phys. Lett. **90**, 183503 (2007)
46. D. Yokoyama, C. Adachi, J. Appl. Phys. **107**, 123512 (2010)
47. D.W. Berreman, J. Opt. Soc. Am. **62**, 502 (1972)
48. M. Schubert, Phys. Rev. B **53**, 4265 (1996)
49. H.G. Tompkins, E.A. Irene (eds.), *Handbook of Ellipsometry* (William Andrew/Springer, Berlin, 2005). (Chap. 9)
50. D.J. De Smet, J. Appl. Phys. **76**, 2571 (1994)
51. L.A.A. Pettersson, F. Carlsson, O. Inganas, H. Arwin, Thin Solid Films **313–314**, 361 (1998)
52. W.A. McGahan, B. Johs, J.A. Woollam, Thin Solid Films **234**, 443 (1993)
53. M. Campoy-Quiles, P.G. Etchegoin, D.D.C. Bradley, Synth. Met. **155**, 279 (2005)
54. D. Wynands, M. Erber, R. Rentenberger, M. Levichkova, K. Walzer, K.-J. Eichhorn, M. Stamm, Org. Electron. **13**, 885 (2012)
55. U. Heinemeyer, R. Scholz, L. Gisslén, M.I. Alonso, J.O. Ossó, M. Garriga, A. Hinderhofer, M. Kytka, S. Kowarik, A. Gerlach, F. Schreiber, Phys. Rev. B. **78**, 085210 (2008)
56. M.A. Heinrich, J. Pflaum, A.K. Tripathi, W. Frey, M.L. Steigerwald, T. Siegrist, J. Phys. Chem. **111**, 18878 (2007)

57. D.E. Aspnes, Spectroscopic ellipsometry of solids, in *Optical Properties of Solids-New Developments*, ed. by B. Seraphin (North-Holland, Amsterdam, 1976). Chapter 15
58. R.M.A. Azzam, N.M. Bashara, *Ellipsometry and Polarized Light* (Elsevier, Amsterdam, 1992)
59. U. Rossow, W. Richter, Optical characterization of epitaxial semiconductor layers, *Spectroscopic Ellipsometry* (Springer, Berlin, 1996)
60. B. Johs, J.A. Woollam, C.M. Herzinger, J. Hilfiker, R. Synowicky, C.L. Bungay, SPIE Crit. Rev. Opt. Sci. Technol. **CR72**, 29 (1999)
61. D.E. McIntyre, D.E. Aspnes, Surf. Sci. **24**, 417 (1971)
62. A. Ritz, H. Lüth, Appl. Phys. A **31**, 75 (1983)
63. U. Rossow, U. Frotscher, W. Richter, D.R.T. Zahn, Surf. Sci. **287–288**, 718 (1993)
64. N. Esser, P.V. Santos, M. Kuball, M. Cardona, M. Arens, D. Pahlke, W. Richter, F. Stietz, J.A. Schaefer, B.O. Fimland, J. Vac. Sci. Technol. B **13**, 1666 (1995)
65. P.V. Santos, N. Esser, J. Groenen, M. Cardona, W.G. Schmidt, F. Bechstedt, Phys. Rev. B **52**, 17379 (1995)
66. H. Proehl, T. Dienel, R. Nitsche, T. Fritz, Phys. Rev. Lett. **93**, 097403 (2004)
67. U. Beckers, O. Stenzel, S. Wilbrandt, U. Falke, C. von Borczyskowski, J. Phys. Condens. Matter **10**, 1721 (1998)
68. D.K. Burge, H.E. Bennett, J. Opt. Soc. Am. **54**, 1428 (1964)
69. U. Rossow, W. Richter, Spectroscopic ellipsometry, *Optical Characterization of Epitaxial Semiconductor Layers* (Springer, Berlin, 1996)
70. O.D. Gordan, C. Himcinschi, D.R.T. Zahn, C. Cobet, N. Esser, W. Braun, Appl. Phys. Lett. **88**, 141913 (2006)
71. D.E. Aspnes, Thin Solid Films **455–456**, 3 (2004)
72. M. Nakamura, T. Matsunobe, H. Tokumoto, J. Appl. Phys. **89**, 7860 (2001)
73. T.U. Kampen, A.M. Paraian, U. Rossow, S. Park, G. Salvan, Th. Wagner, M. Friedriech, D.R.T. Zahn, Phys. Stat. Sol. A **188**, 1307 (2001)
74. C. Goletti, G. Bussetti, P. Chiaradia, A. Sassella, A. Borghesi, Org. Electron. **5**, 73 (2004)
75. F. Seidel, L. Ding, O.D. Gordan, D.R.T. Zahn, J. Vac. Sci. Technol. B **30**, 012401–1 (2012)

# Chapter 14
# Optical Dielectric Properties of Thin Films Formed by Organic Dye Aggregates

**Katy Roodenko and Peter Thissen**

**Abstract** Molecular aggregates – the assemblies of dye molecules-possess distinct optical characteristics as compared to their constituent monomeric units. Strong resonant interactions between the molecular electronic transitions of the monomers result in an intense and narrow absorption band at a frequency that depends on the relative orientation of the monomers within the aggregates. The strong absorption bands of these materials makes them attractive for a range of optoelectronic applications. We review the experimental and computational work dedicated to the determination of the optical dielectric functions of molecular aggregates, with an outlook to applications in device engineering.

## 14.1 Introduction

Assemblies of molecules that are held together by non-covalent interactions are known as molecular aggregates. Certain types of molecular aggregates have been known for their superior light absorbing properties through the collective optical response of the constituent monomers [1–5]. Beyond the traditional use of the molecular aggregates in photographic emulsions to increase the spectral sensitivity range [3, 6], molecular aggregates are potentially attractive in chemical sensing [7, 8] and as light harvesting material for application in optical devices [9, 10], such as photodetectors [10] and solar cells [11–13]. The formation of aggregates was independently discovered by Scheibe [14] and Jelley [15] and has remarkable influences on the optical properties of the molecules as compared to those of the monomers. Depending on the molecular order and organization of the monomer units, the

K. Roodenko (✉)
Department of Materials Science and Engineering (MSE), University of Texas at Dallas,
Richardson, TX 75080, USA
e-mail: katy.roodenko@utdallas.edu

P. Thissen
Institute of Functional Interfaces (IFG), Karlsruhe Institute of Technology,
Hermann-von-Helmholtz-Platz 1, 76344 Eggenstein-Leopoldshafen, Germany

© Springer International Publishing AG, part of Springer Nature 2018
K. Hinrichs and K.-J. Eichhorn (eds.), *Ellipsometry of Functional
Organic Surfaces and Films*, Springer Series in Surface Sciences 52,
https://doi.org/10.1007/978-3-319-75895-4_14

aggregation gives rise to a sharp absorption line that is shifted either to the red or the blue side of the spectrum. The aggregates are called J-aggregates (named after Jelley) if they exhibit a bathochromic (shifted to lower wavelength) absorption line. The aggregates that exhibit blue-shifted (hypsochromic) absorption line are called H-aggregates. The sharp absorption peaks of the aggregates are ascribed to the coherent delocalization of the electronic excitation over the constituent monomeric units and are well-described by Frenkel's exciton theory [16, 17]. Reviews of theoretical approaches for modeling of excitations in aggregates and the main experimental techniques for studies of the optical properties of aggregates can be found in [1, 2, 17, 18]. While there is a growing number of the proposed device technologies that utilize the strong, well-defined optical absorption of molecular aggregates, there is only a handful of publications dealing with the determination of the dielectric functions of these materials, the knowledge of which is an important prerequisite for modeling and engineering of the desired device properties. The focus of this chapter is therefore on the experimental and theoretical studies of the dielectric optical functions of thin-film aggregates. The next section (Sect. 14.2) provides a brief overview of the excitonic theories that allow theoretical prediction of the optical and structural properties of molecular aggregates. Review of the computational and experimental work dedicated to the research of the macroscopic optical dielectric response is presented in Sect. 14.3. This section also provides an insight into aggregate-based device design that requires an input of optical constants in the simulations of device properties.

## 14.2   Excitonic Properties and Molecular Structures of Organic Aggregates

Exciton is a bonded electron-hole pair in an insulator or a semiconductor [16]. Excitons play a major role in the absorption of light and the transmission of energy [19–21]. In inorganic semiconductors, an incident photon with an appropriate energy can excite an electron to transition from the valence band into the conduction band. The electron and the oppositely charged hole formed in the valence band attract each other by the Coulomb force. In molecular aggregates, the coupling between electronic transitions of the monomers is through Coulomb dipole-dipole interaction. Whether the electronic excitations is transferred between the monomers depends on the strength of the dipole-dipole interaction as compared to the intra molecular vibronic relaxation. Quantum mechanical classification of the weak and strong coupling regimes was provided by Simpson and Peterson [22], while the coherent delocalization of the electronic excitation across the aggregate is described by Frenkel exciton theory [16].

The absorption band of the molecular aggregates appears energetically blue- or red-shifted as compared to the absorption of the constituent monomers. Kasha et al. [21] had attributed these shifts to the geometric arrangements of molecular monomers

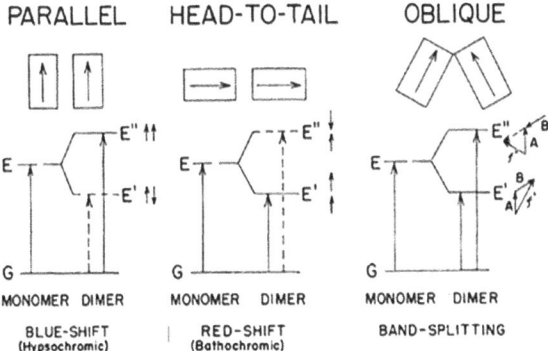

**Fig. 14.1** Energy diagrams for exciton band structures and geometrical arrangements of transition dipoles (small arrows) in molecular dimers. The dashed lines indicate the forbidden exciton state; the solid line represents the allowed exciton state of each dimer [21]. Modified and reprinted with permission from Kasha et al., Radiat. Res. 20, 55–71 (1963). Copyright 1963, Radiation Research Society

in the aggregates. Figure 14.1 illustrates the exciton band structure and the corresponding selection rules for optical transitions for different geometric arrangements of transition dipoles in a molecular dimer.

In the case of a dimer where monomers are stacked in parallel with in-phase transition dipole moments, the excited state energy is increased as compared to the excited state energy of the monomer. When the transition dipoles are out-of-phase, the exciton energy is electrostatically attractive, resulting in a reduction in the excited state energy in the dimer. For head-to-tail aggregates with out-of-phase transition dipole moments, the dipole-dipole interaction is electrostatically repulsive, resulting in increased excited state energy. The in-phase arrangement, on the other hand, produces a reduced excited state energy due to its attractive interaction.

The long-range Coulombic interactions lead to Kasha's exciton model as is summarized in Fig. 14.2. The Coulomb coupling ($J_{Coul}$) between any two molecular units arises from the intermolecular electronic interactions and can be expressed using the point-dipole approximation [23],

$$J_{Coul} = \frac{\mu^2(1 - 3cos^2\theta)}{4\pi\varepsilon R^3} \tag{14.1}$$

where R is the intermolecular distance between mass centers, $\mu$ is the transition dipole moment, $\theta$ is the angle between $\mu$ and R, and $\varepsilon$ is the optical dielectric constant of the medium (see Fig. 14.2a). The angle $\theta$ dictates the "head-to-tail" orientation ($\theta \approx 0$) or the "side-by-side" ("parallel") orientation ($\theta \approx \pi/2$). Inter- molecular coupling leads to delocalized (Frenkel) excitons. When the coupling is limited to nearest neighbors, the energy of the exciton with wave vector k is

$$E_F(k) = E_{s_1} + 2J_{Coul}cos(k) \tag{14.2}$$

**Fig. 14.2** **a** Transition
dipole geometry for
evaluating the Coulombic
coupling under the point
dipole approximation.
Exciton energy bands of
Kasha **b** J-aggregates and
**c** H-aggregates. In J- and
H-aggregates, the optically
allowed exciton (k = 0)
resides at the bottom and top
of the band, respectively. The
packing geometries
associated with each
aggregate type are shown as
insets. **d** Ideal absorption
spectra for J- and
H-aggregates. Reprinted
with permission from
Hestand et al., Acc. Chem.
Res. 50, 341 (2017) [23].
Copyright 2017, American
Chemical Society

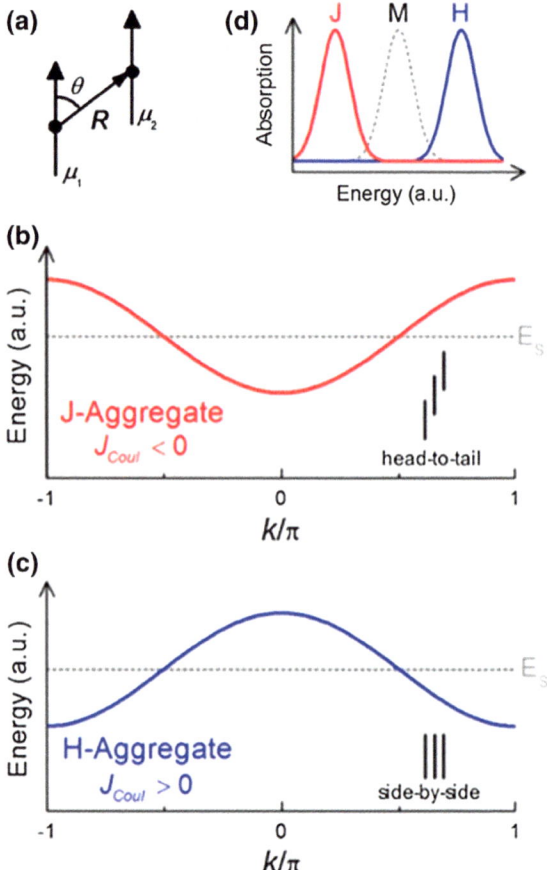

where $Es_1$ is the energy of the HOMO-LUMO electronic transition of the monomer unit. The negative and positive values of $J_{Coul}$ correspond to conventional J-aggregates and H-aggregates, respectively [21], as is schematically represented by the band dispersion $E_F(k)$ in Fig. 14.2a, b [23].

The studies of the structure and morphology of the cyanine aggregates have been a field of active research in recent decades. One of the most extensively studied cyanines is the pseudoisocyanine (PIC, see Fig. 14.3).

In 2002, Guo et al. [24] were able to determine the structure of pseudo-isocyanine interpreting Raman frequencies based on the density functional theory calculations. The molecular structure was calculated to have close C2 symmetry, with a symmetry axis passing through the C-H bond of the central carbon atom. This central carbon atom connects two planar quinolone moieties. These quinolone moieties are twisted by a dihedral angle of about 46°.

The most prominent supramolecular aggregates found in nature are built of the chlorin precursor chlorophyll a [27–29]. Chlorophyll molecules act as light-

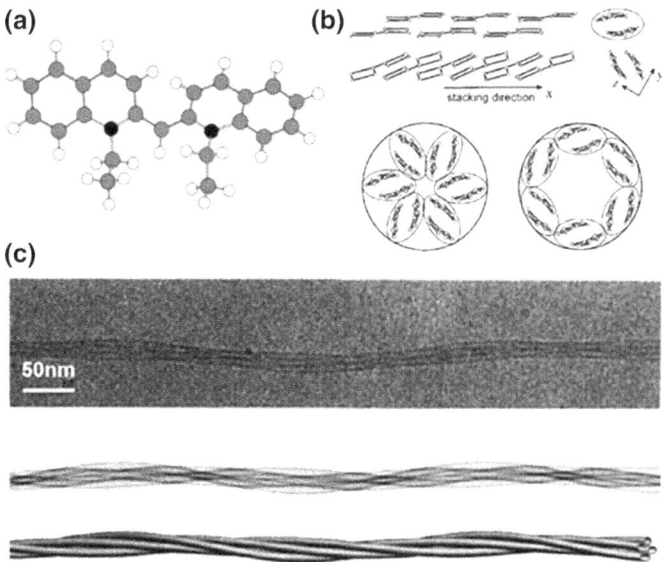

**Fig. 14.3** **a** Optimized structure of pseudoisocyanine (PIC) calculated using B3LYP/6-31G-(dp). White spheres represent hydrogen, gray spheres represent carbon and black spheres represent nitrogen. Modified and reprinted with permission from Guo et al. [24], J. Phys. Chem. B 106, 5447 (2002). Copyright 2002, American Chemical Society. **b** Left side, top: Threadlike arrangement of the PIC molecules derived from the X-ray analysis of PIC chloride single crystals. Both stacks are symmetric to each other, but their inclinations with respect to the image plane differ. The single stack was often used as a structural model for a one-dimensional PIC J-aggregate. The chinoline rings are sketched as rectangles, and the positions of the C-atoms were deduced based on the X-ray analysis. Right side, top: Top view along the x-axis. Taking the hydrogen atoms and the van der Waals radii into account, one stack can be modeled as a thread with an elliptical cross section. Bottom: Two possible models for the PIC J-aggregate: a star-like and a tubular arrangement of six stacks within a cylinder 2.3 nm diameter. Modified and reprinted with permission from Berlepsch et al. [25], J. Phys. Chem. B 104, 8792 (2000). Copyright 2000, American Chemical Society. **c** Top: Cryo-TEM image of a PIC single quadruple helix. Middle: Simulated projection image of this helix. Bottom: Corresponding three-dimensional view. Modified and reprinted with permission from Berlepsch et al. [26], J. Phys. Chem. B 104, 5255 (2000). Copyright 2000, American Chemical Society

harvesting antennas in photosynthesis and have an exceptionally high molar extinction coefficients [29, 30].

The narrow absorption bands of chlorosomal J-aggregates are attractive for implementation in (opto-) electronic devices and had inspired attempts to synthesize artificial nanostructures that can mimic natural light-harvesting systems. Different charge and energy transport properties in ZnChl-based molecular assemblies can be achieved through $\pi - \pi$ stacking, hydrogen bonding, and metal oxygen coordination [27]. For instance, by functionalization of the 31-position of ZnChl with a hydroxyl or methoxy group, tubular or stack assemblies can be induced, respectively (see Fig. 14.4). Detailed reviews of molecular assemblies capable of photo-induced electron transfer and charge transport can be found, for example, in [12, 27, 31, 32].

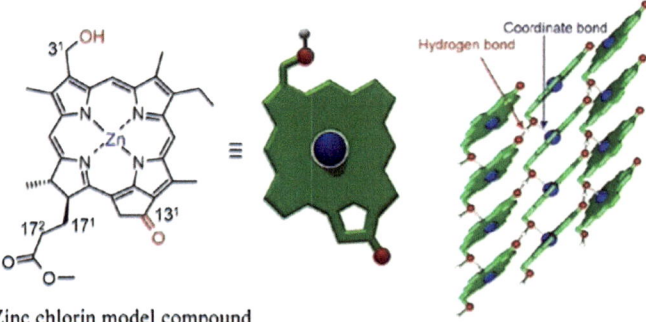

Zinc chlorin model compound

**Fig. 14.4** Chemical structures of the zinc chlorin model compound and schematic representation of a section of tubular assemblies formed by the interplay of hydrogen bonding (red arrow), metal oxygen coordination (blue arrow), and $\pi - \pi$ stacking. Reprinted with permission from Sengupta et al. [27], Acc. Chem. Res. 46, 2498–2512 (2013). Copyright 2013, American Chemical Society

## 14.3 Optical and Structural Properties of Molecular Aggregates

This section reviews efforts dedicated to studies of complex optical dielectric response of organic aggregates in thin films as well as in nanostructures. A great amount of work had been dedicated to the synthesis of aggregates with favorable optical and structural properties, as well as to the development of routes for surface functionalization using molecular aggregates. Much attention had been devoted to the development of the excitonic theories explaining experimentally observed resonances and relating them to the exciton states in aggregates. Optical spectroscopy had been extensively employed to study absorption and emission of the aggregates under different temperature, solvent and monomer concentration conditions. However, there had been relatively few reports on the measurements and calculations of the linear dielectric response of these materials. With the growing availability of surface functionalization techniques applicable to device engineering there is a greater need for tabulated complex dielectric functions of the molecular aggregates. The strong absorption peak of H- and J-aggregates can be generally described by the Lorentz-oscillator model, discussed in the 1st chapter of this book. Numerous studies investigating molecular aggregates and energy transfer in hybrid organic-inorganic systems often disregard the optical anisotropy that molecular aggregates in certain ordered films may possess. The problem of the lack of the published data on optical anisotropy in thin-films of molecular aggregates had been acknowledged, for example, by Cacciola et al. [33]. To justify their calculations, the authors used polarized optical absorption measurements to verify the in-plane isotropic response of the aggregate films that were utilized in their studies. Numerical studies on the influences of disorder and the aggregate chain length on the optical properties, such as the absorption line shape and the superradiant emission rate were performed by Fidder et al. [34], who had expressed average oscillator strength per state at a given energy in terms of the monomer transition dipole moments. Spano et al. [35] performed derivation of nonlinear absorption coefficients for molecular aggregates with site disorder and reviewed formalisms describing absorption and photoluminescence line shapes, based on intra- and intermolecular excitonic coupling, electron-vibrational coupling, and correlated energetic disorder. For practical purposes, however, estimation of the oscillator parameters is frequently performed by directly fitting the absorption and reflection/transmission spectra [36–40] or the ellipsometry data [10, 41–44] obtained from the measurements of thin films of the aggregates. An instructive note on interpretation of absorption spectra in thin films of polymers can be found in [45], where the effects of morphology and interference are discussed in details. Application of classical electromagnetic theory for derivation of the optical parameters of macroscopic samples containing aggregates (i.e. dilute solutions of molecular aggregates or polymer molecules, and molecular crystals) can be found in [46].

Molecular aggregates possess optical anisotropy as dictated by the nature of their ordered stacks of monomers [25, 47, 48]. Pockrand et al. [49, 50] had performed studies of the exciton-plasmon interactions for isotropic and uniaxial anisotropic films on metallic substrates. In this theoretical work, the authors studied interaction of a two-dimensional exciton with a plasmon surface polariton and showed significant differences in the reflectivity spectra that arise due to the interaction with the isotropically oriented transition moments versus anisotropically oriented moments. Another example that addresses anisotropic properties of thin film aggregates is the study of circular and linear birefringence and dichroism of porphyrin J-aggregates by transmission-mode phase modulated ellipsometry in [51]. Recent studies by Shaykhutdinov et al. [52] utilized infrared nanopolarimetry to characterize morphology of zinc tetraphenylporphyrin system that exhibited optical anisotropy induced by H- and J-stacking. There, IR absorption modes that are sensitive to s- and p-polarization were identified using density functional theory (DFT). Experimental observation of these modes was then correlated with the orientation of the monomers within the aggregates and the domain formation during the growth of the supramolecular crystals.

Among the different efforts to study optical dielectric function, the work by Kirstein et al. provides an example of a systematic approach towards characterization of Langmuir–Blodgett (LB) films of cyanine dye molecules on glass substrates [38, 53]. A lipid/dye-aggregate solution was used for molecular deposition by spreading of the positively-charged lipid on an aqueous solution of the anionic dyes. Optical microscopy of LB films revealed the formation of crystallites of regular shapes on the substrate surface. The electron diffraction pattern suggested a herringbone-like arrangement of the dye molecules. The detailed analysis of the optical properties of the LB films showed that the molecules are arranged with their long axes in the crystal plane but with their plane perpendicular to the crystal plane. To determine the oscillator strength values of the absorption peaks that were observed in the optical spectra, the excitonic band structure was calculated based on the extended dipole model [38].

Studies of Langmuir–Blodgett dye aggregate films of merocyanine (MC, a class of polymethyne dyes) and squarylium (SQ, dyes characterized by a four-member ring system derived from squaric acid) using polarized reflection and absorption spectroscopy were reported by Wakamatsu et al. [36]. Numerical simulations of the films were performed based on Fresnel's equations for propagation in stratified medium and the description of the absorption bands based on Lorentz oscillation model. Similar simulations were applied to studies of molecular aggregates measured in attenuated total reflection (ATR) configuration in ~450–600 nm spectral range, retrieving optical properties (the complex refraction index) of SQ H-aggregate films [36, 37].

Dielectric functions of spin-coated cationic pseudoisocyanine (PIC) J-aggregate films were reported in [42, 43]. Using spectroscopic ellipsometry, the complex refractive indices of the PIC films in the monomeric and aggregated forms were measured and the dependence of the complex refractive index on the length of the carbon chain length of N-alkyl substituents was analyzed.

Filho et al. [54] studied luminescent 2,1,3-benzothiadiazole dyes incorporated in liquid crystal matrices. Based on the emission ellipsometry studies, the ordering of the liquid crystal molecules and the aggregation state of the dyes were addressed. Analysis of the degree of polarization pointed to the differences in the dye alignment in the liquid crystal matrix in dependence on the size of the dye molecules.

Studies of nanoparticles that consist of the noble metal (Ag or Au) core and cyanine dye J-aggregate shell had been reported in [39]. The dielectric function of the organic shell aggregate was derived from the experimental data using Kramers-Kronig relations. The shape and position of the peaks in the absorption spectra of the hybrid nanoparticles showed dependence on the composite nanoparticle core/shell size ratio and on the optical properties of the core and shell materials.

Another example of the reported dielectric functions for molecular aggregates is provided in the last chapter of this book ("Optical constants") that lists isotropic optical properties of porphyrins.

Anisotropy of optical dielectric function of spin-coated thin films of thiacarbocyanine (TCC) dye deposited on gold and silicon surfaces using spectroscopic ellipsometry and polarized IR spectroscopy was experimentally studied in [44] and applied to the calculations of energy transfer from a point-like electric-dipole emitter in the vicinity of an absorbing layer (energy acceptor). Atomic-force microscopy (AFM) and optical microscopy images of spin-coated TCC films deposited on a silicon substrate under different conditions are presented in Fig. 14.5.

Dense spin-coated TCC films generally result in a smooth, homogeneous layer as illustrated in Fig. 14.5a. Deposition of the diluted TCC films allows observation of discrete rod-like structures, captured by optical microscopy and AFM in Fig. 14.5b, c.

Figure 14.6 shows ellipsometric data tan $\Psi$ and $\Delta$ obtained from TCC on $SiO_2/Si$ surface (Fig. 14.6a). The surface possessed morphology of a dense-film as presented in Fig. 14.5a. The measured spectra exhibit data characteristic to the films with the uniaxial anisotropy on a semiconducting surface. Specifically, the peak-up features in tan $\Psi$ data on $SiO_2/Si$ surface in Fig. 14.6a (where tan $\Psi$ defined as the ratio between the parallelly and perpendicularly polarized reflection coefficients, tan $\Psi = r_p/r_s$)

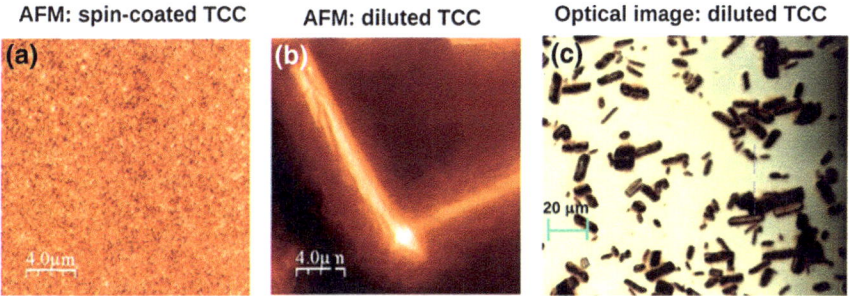

**Fig. 14.5** Images obtained from TCC transferred on $SiO_2$/Si substrates by spin-coating. **a** Dense TCC films, exhibiting smooth, homogeneous films. **b**, **c** AFM data and an optical image obtained from TCC film transferred from diluted solution. Modified and reprinted with permission from K. Roodenko et al. [44], J. Phys. Chem. C. 117, 20186 (2013). Copyright 2013, American Chemical Society

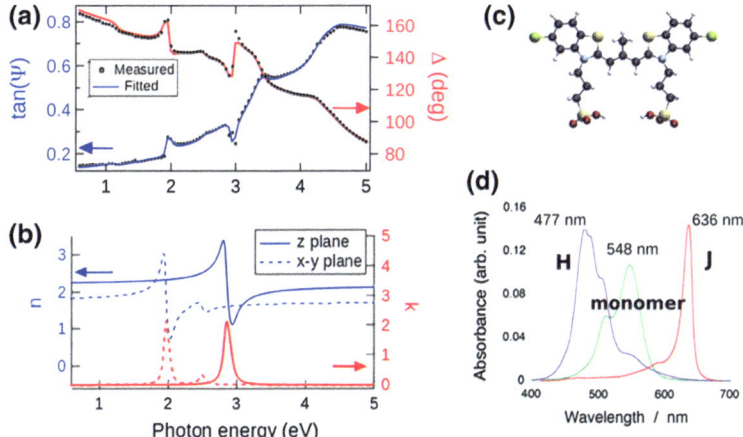

**Fig. 14.6** **a** Ellipsometric parameters tan $\Psi$ (blue, left axis) and $\Delta$ (red, right axis) for TCC deposited on $SiO_2$/Si surface. Dotted lines: data obtained at 70° angle of incidence. Continuous line: fit within the uniaxial anisotropic model. **b** Real and imaginary parts of the complex dielectric function as derived from the ellipsometric measurements; blue: real and red: imaginary parts of the refractive index. The dashed line represents direction parallel to the film, the continuous line represents direction perpendicular to the film plane [44]. **c** Schematic representation of TCC monomer. **d** Spectroscopic data from J-aggregates, H-aggregates, and monomers in solution. Modified and reprinted with permission from Yao et al. [55], J. Phys. Chem. B, 111, 7176 (2007), Copyright 2007, American Chemical Society

are due to the electronic transition dipole moments parallel to the surface and peak-down features are due to those perpendicular to the surface.

Ellipsometric data obtained from TCC on $SiO_2$/Si surface was fitted using standard models of light propagation through stratified media [56]. The absorption peaks were described by two Lorentz oscillators. The resulting real and the imaginary parts of the dielectric function are shown in Fig. 14.6b. Figure 14.6d shows data obtained by Yao et al. [55] from J- and H-aggregates, as well as from the monomers of the TCC in solution. This data suggests that spin-coating of TCC films resulted in a mixed film that exhibits both J- and H-aggregation.

The measured anisotropic dielectric function was further used as an input in model calculations of energy transfer (ET) from a point-like electric-dipole emitter into thin layers of J-aggregate acceptors [44]. The energy donor can be representing, for example, a quantum dot or a molecular fluorophore. The model was applied to compare the effects of a strong anisotropic dielectric resonance with those of an isotropic case on ET. The results illustrated differences in dependence on the layer thickness and distance from the layer. Accurate determination of the anisotropic optical constants can thus become important in engineering of the ET-based hybrid optoelectronic nanostructures.

Isotropic dielectric functions of J-aggregates were implemented in device engineering by several groups [10, 57]. Osedach et al. [10] demonstrated a lab-scale fabrication of near-infrared photodetector that consists of a thin film of the J-aggregating cyanine dye, U3, and transparent metal-oxide charge transport layers. To quantify the exciton diffusion length and the dependence of device performance on layer thicknesses, experimental external quantum efficiency data was modeled using the optical transfer matrix formalism coupled with the one-dimensional diffusion equation. The optical constants for U3 film were determined using ellipsometry and are reprinted from the supplementary information in [10] and shown in Fig. 14.7.

Determination of complex dielectric function of absorbing thin film that consisted of layers of the cationic polyelectrolyte PDAC (poly diallyldimethylammonium chloride) and J-aggregates of the anionic cyanine dye TDBC (the 5,6-dichloro-2-[3-[5,6-dichloro-1-ethyl-3-(3-sulfopropyl)-2(3H)-benzimidazolidene]-1-propenyl]-1-ethyl-3-(3-sulfopropyl) benzimidazolium hydroxide, see Fig. 14.9a) deposited on glass substrate was performed by Bradley et al. [58]. The analysis was performed based on the reflection and transmission data. The thickness measured by atomic AFM was used as the input parameter in the optical models and Kramers-Kronig

**Fig. 14.7** Optical constants for spun-cast U3 films derived from spectroscopic ellipsometry. Reprinted with permission from P. Osedach et al., Appl. Phys. Lett.101, 113303 (2012) (see supporting information therein) [10]. Copyright 2012, AIP Publishing LLC

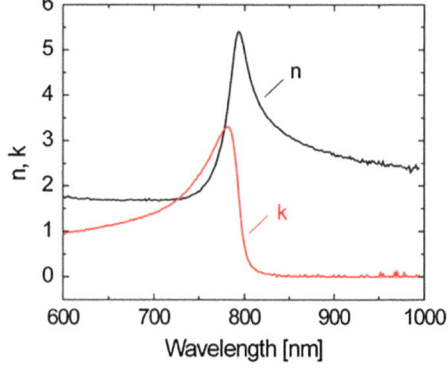

**Fig. 14.8  a** Real and
imaginary components of the
refractive index, (n, k) for a
5.1 ± 0.5 nm thick
PDAC/TDBC film deposited
on a SiO₂ substrate. Inset:
spectral data with fits
calculated from (n, k) values.
Reprinted with permission
from Tischler et al., Optics
Lett. 31, 2045 (2006) [57].
Copyright 2006, Optical
Society of America

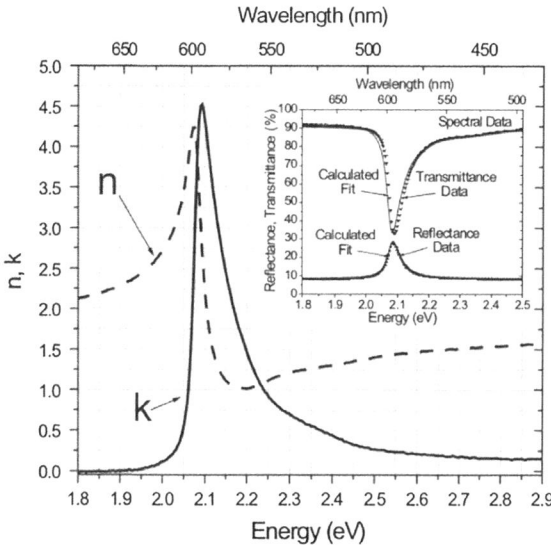

regression was applied for retrieving of the (n, k). This analysis suggested high
absorption constant ($10^6$ cm$^{-1}$) at the peak value of the measured spectrum. The
resulting complex refractive index is shown in Fig. 14.8, while the schematic view
of PDAC/TDBC molecule is shown in Fig. 14.9a.

Based on the measured optical properties of the PDAC/TDBC film, the critically
coupled resonator (CCR) with the PDAC/TDBC thin film as an absorbing layer
[57] was realized experimentally. Here, "coupling" refers to the interaction between
the incident light and the cavity-like system with an absorbing film that acts as the
coupling material. Nearly all of the incident light of a given wavelength is absorbed in
a few-nanometer-thick absorbing film. The structure of the CCR, shown in Fig. 14.9a,
consists of a dielectric Bragg reflector (DBR) as the mirror, and a thin film of organic
material as the absorbing layer. A spacer layer of transparent material that separates
the mirror and absorber layer by the correct distance is required for critical coupling.
The resulting reflectance of the DBR and the spaceral one versus the reflectance of
CCR is shown in Fig. 14.9a, demonstrating absorption of 97% of the incident light
within the 5.1 nm thick absorber layer at the incident wavelength of 591 nm.

We complete this section with a brief note on strong coupling in systems that
consist of emitter (dye molecules) and plasmonic metal structures. Here, "coupling"
refers to the energy exchange between the excitonic and plasmonic systems. Strong
coupling occurs when the rate of coherent energy exchange between the excitonic
and plasmonic systems exceeds the rate of the losses in the system [18, 59, 60].

Generally, the electromagnetic field of the plasmon/exciton systems can be
described as coupled oscillators that act as resonant two-level systems [59, 61].
Alternatively, the system can be described classically, where the optical properties
of the emitter (such as a polymer film coated on a metallic surface) are described

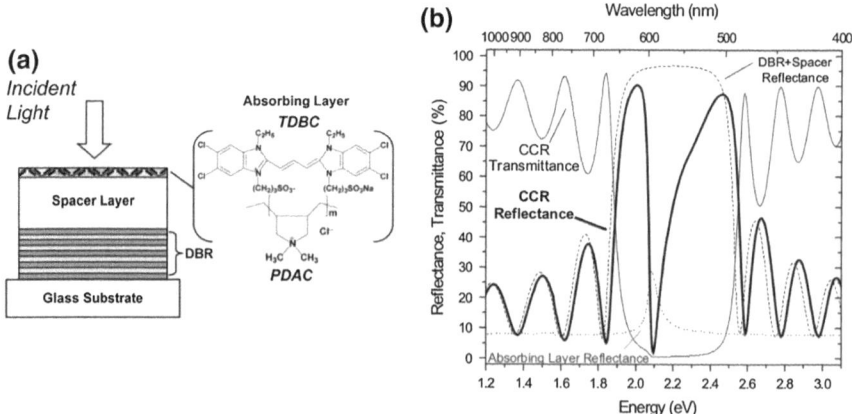

**Fig. 14.9** **a** Critically coupled resonator (CCR) structure and the schematic drawing of the molecule constituting J-aggregate layer that consists of the cationic polyelectrolyte PDAC and the anionic cyanine dye TDBC. **b** Measured reflectance and transmittance for the CCR with spacer thickness of (90 ± 1) nm, along with reflectance data for the neat PDAC/TDBC film and for the dielectric stack consisting of DBR with spacer layer. At wavelength of 591 nm, the CCR absorbs 97% of the incident light. Reprinted with permission from Tischler et al., Optics Lett. 31, 2045 [57]. Copyright 2006, Optical Society of America

by Lorentzian oscillators [59, 62]. In the strong coupling regime, the energy separation between the normal modes of the coupled oscillators is known as vacuum Rabi splitting. Rabi splitting in microcavities and plasmonic nanostructure arrays implementing J-aggregates had been studied, for example, in [60, 63–66] and reviewed in [2, 18, 59]. To illustrate the phenomenon of strong coupling between plasmonic and excitonic resonances, we provide example based on the work by Belessa et al. [60] who studied interactions between surface plasmon (SP) and excitons in J-aggregate emitters. In this example, classical electromagnetic theory was used to describe strong coupling between metallic surface plasmons and excitons, which requires input of the dielectric function of the aggregate layer. A layer containing TDBC dye aggregates was deposited on a thin silver film and reflectometry spectra were obtained in Kretschman configuration [60]. Figure 14.10a shows spectra obtained at varying angles of incident radiation. Figure 14.10b illustrates the dependence of the spectroscopically observed energy positions of the dips as a function of the in-plane wave vector. The two branches in Fig. 14.10b are related to the coupled modes. There is a good agreement between the experimental data (black circles) and the data calculated based on the transfer matrix model (shown by continuous lines). The dispersion curves of each component separately (without coupling) is illustrated by the two dashed lines in Fig. 14.10b. The diagonal dashed line represents the plasmon dispersion of the metallic surface, while the horizontal dashed line at $2100 \, \text{cm}^{-1}$ is the (uncoupled) exciton mode of the organic dye layer. The details of the model that was used to fit the reflectometry spectra can be found in the publication by Bonnard et al. [62]. The dielectric function $\varepsilon_{\text{aggr}}$ was modeled using (14.3):

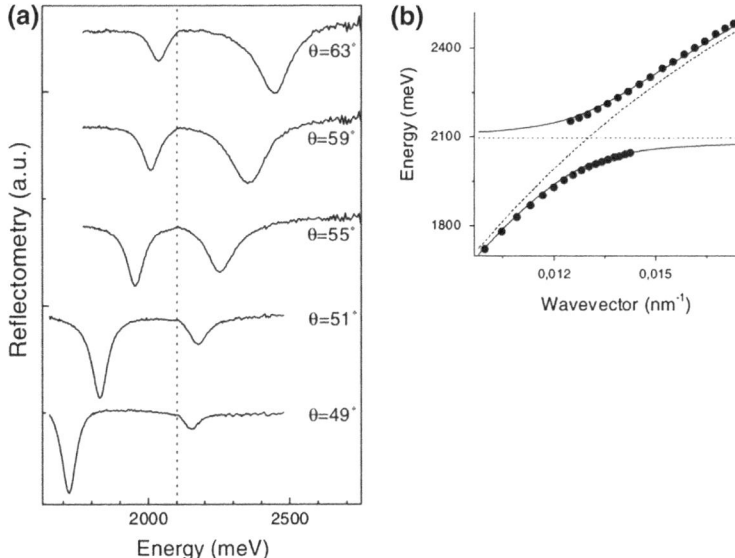

**Fig. 14.10** **a** Reflectometry spectra for different angles as a function of the incident light energy. **b** Position of energy peaks (bold circles) as a function of the wave vector. The dashed line is the dispersion relation of an uncoupled SP calculated with a transfer matrix method. The dotted line is the TDBC exciton energy. The full lines show the calculated polariton energies. Reprinted with permission from Bellessa et al., Phys. Rev. Lett. 93, 036404 (2004) [60]. Copyright 2004, The American Physical Society

$$\varepsilon_{aggr.} = \frac{f q^2 \hbar^2}{m \varepsilon_0 L_z (e_0^2 - e^2 - i \gamma_0 e)} \tag{14.3}$$

where e is the energy, f is the oscillator strength of the TDBC aggregate, $\gamma_0$ is the exciton linewidth, $q$ is the electron charge magnitude, $m$ is the electron mass, $\varepsilon_0$ is the permittivity of the free space, and $L_z$ is the layer thickness. The oscillator strength of the TDBC dye aggregate layer was derived based on the fits of the reflectometry data resulting in $f = 1.8 \times 10^{14} \, cm^{-2}$, while the exciton line width was deduced from absorption measurements from a dye layer supported by a pyrex substrate $\gamma_0 = 49 \, meV$, positioned at the energy $e_0 = 2100 \, meV$.

Other work related to the studies of strongly coupled optical and excitonic states in microcavities containing J-aggregates can be found, for example, in [59, 67–69]. A thorough review by Sukharev and Nitzan [18] provides a detailed discussion on the physics of coupled exciton-plasmon systems, emphasizing the scope and the limits of the macroscopic dielectric response models as well as of those dealing with the microscopic representation of the coupled systems that describe processes that are beyond the simple linear dielectric response (i.e. de-phasing and nonlinear spectroscopy).

## 14.4 Conclusions and Outlook

A growing number of experimental and theoretical research is focused on precise control of the light-matter interaction by utilizing the strong, narrow absorption bands of molecular aggregates. At the same time, reports on the systematic investigations of the optical dielectric functions of these materials are scarce. This chapter highlights research dedicated to studies of the optical parameters of molecular aggregates in relation to applications in device engineering and experimental data analysis. Attention is drawn towards the necessity to expand the experimental and theoretical efforts aimed at the studies of the dielectric functions of the increasing variety of J- and H-aggregates.

## References

1. F. Würthner, T.E. Kaiser, C.R. Saha-Möller, Angew. Chem. Int. Ed. **50**, 3376 (2011)
2. S.K. Saikin, A. Eisfeld, S. Valleau, A. Aspuru-Guzik, Nanophotonics **2**, 21 (2013)
3. R.C. Benson, H.A. Kues, J. Chem. Eng. Data **22**, 379 (1977)
4. A. Mishra, R.K. Behera, P.K. Behera, B.K. Mishra, G.B. Behera, Chem. Rev. **100**, 1973 (2000)
5. T. Kobayashi (ed.), *J-Aggregates* (World Scientific, Singapore, 1996)
6. B.I. Shapiro, Russ. Chem. Rev. **63**, 231 (1994)
7. Q. Wang, Z. Li, D.-D. Tao, Q. Zhang, P. Zhang, D.-P. Guo, Y.-B. Jiang, Chem. Commun. **52**, 12929 (2016)
8. W. Liang, S. He, J. Fang, Langmuir **30**, 805 (2014)
9. C. Curutchet, B. Mennucci, Chem. Rev. **117**, 294 (2017)
10. T.P. Osedach, A. Iacchetti, R.R. Lunt, T.L. Andrew, P.R. Brown, G.M. Akselrod, V. Bulović, Appl. Phys. Lett. **101**, 113303 (2012)
11. E.A. McArthur, J.M. Godbe, D.B. Tice, E.A. Weiss, J. Phys. Chem. C **116**, 6136 (2012)
12. S. Kundu, A. Patra, Chem. Rev. **117**, 712 (2017)
13. J.L. Banal, B. Zhang, D.J. Jones, K.P. Ghiggino, W.W.H. Wong, Acc. Chxem. Res. **50**, 49 (2017)
14. G. Scheibe, Angew. Chem. **50**, 212 (1937)
15. E.E. Jelley, Nature **138**, 1009 (1936)
16. J. Frenkel, Phys. Rev. **37**, 17 (1931)
17. V.V. Egorov, Phys. Procedia **2**, 223 (2009)
18. M. Sukharev, A. Nitzan, Cond-Mat. Phys. (2017), arXiv:170405605
19. V.M. Agranovich, *Excitations in Organic Solids* (OUP Oxford, 2009)
20. T. Förster, Radiat. Res. Suppl. **2**, 326 (1960)
21. M. Kasha, Radiat. Res. **20**, 55 (1963)
22. W.T. Simpson, D.L. Peterson, J. Chem. Phys. **26**, 588 (1957)
23. N.J. Hestand, F.C. Spano, Acc. Chem. Res. **50**, 341 (2017)
24. C. Guo, M. Aydin, H.-R. Zhu, D.L. Akins, J. Phys. Chem. B **106**, 5447 (2002)
25. H. von Berlepsch, C. Böttcher, L. Dähne, J. Phys. Chem. B **104**, 8792 (2000)
26. H. von Berlepsch, C. Böttcher, A. Ouart, C. Burger, S. Dähne, S. Kirstein, J. Phys. Chem. B **104**, 5255 (2000)
27. S. Sengupta, F. Würthner, Acc. Chem. Res. **46**, 2498 (2013)
28. L.P. Vernon, G.R. Seely, *The Chlorophylls* (Academic Press, New York, 1966)
29. G.D. Scholes, G.R. Fleming, A. Olaya-Castro, R. van Grondelle, Nat. Chem. **3**, 763 (2011)
30. W.P. Inskeep, P.R. Bloom, PLANT Physiol. **77**, 483 (1985)

31. Q. Yan, Z. Luo, K. Cai, Y. Ma, D. Zhao, Chem. Soc. Rev. **43**, 4199 (2014)
32. J.L. McHale, J. Phys. Chem. Lett. **3**, 587 (2012)
33. A. Cacciola, C. Triolo, O. Di Stefano, A. Genco, M. Mazzeo, R. Saija, S. Patanè, S. Savasta, ACS Photon. **2**, 971 (2015)
34. H. Fidder, J. Knoester, D.A. Wiersma, J. Chem. Phys. **95**, 7880 (1991)
35. F.C. Spano, Phys. Rev. Lett. **67**, 3424 (1991)
36. T. Wakamatsu, S. Toyoshima, K. Saito, J. Opt. Soc. Am. B **23**, 1859 (2006)
37. T. Wakamatsu, K. Watanabe, K. Saito, Appl. Opt. **44**, 906 (2005)
38. S. Kirstein, H. Möhwald, J. Chem. Phys. **103**, 826 (1995)
39. V.S. Lebedev, A.S. Medvedev, D.N. Vasil'ev, D.A. Chubich, A.G. Vitukhnovsky, Q. Electron. **40**, 246 (2010)
40. G. Wahling, Z. Für Naturforschung A **36**, (1981)
41. S. Pirotta, M. Patrini, M. Liscidini, M. Galli, G. Dacarro, G. Canazza, G. Guizzetti, D. Comoretto, D. Bajoni, Appl. Phys. Lett. **104**, 051111 (2014)
42. V.V. Shelkovnikov, Z.M. Ivanova, A.I. Plekhanov, E.V. Spesivtsev, S.V. Rykhlitsky, J. Appl. Spectrosc. **76**, 66 (2009)
43. A.I. Plekhanov, V.V. Shelkovnikov, Opt. Spectrosc. **104**, 545 (2008)
44. K. Roodenko, H.M. Nguyen, L. Caillard, A. Radja, P. Thissen, J.M. Gordon, Y.N. Gartstein, A.V. Malko, Y.J. Chabal, J. Phys. Chem. C **117**, 20186 (2013)
45. O.P.M. Gaudin, I.D.W. Samuel, S. Amriou, P.L. Burn, Appl. Phys. Lett. **96**, 053305 (2010)
46. H. DeVoe, J. Chem. Phys. **43**, 3199 (1965)
47. D.A. Higgins, P.J. Reid, P.F. Barbara, J. Phys. Chem. **100**, 1174 (1996)
48. H.V. Berlepsch, K. Ludwig, C. Böttcher, Phys. Chem. Chem. Phys. **16**, 10659 (2014)
49. I. Pockrand, A. Brillante, D. Möbius, J. Chem. Phys. **77**, 6289 (1982)
50. I. Pockrand, J.D. Swalen, J.G. Gordon, M.R. Philpott, J. Chem. Phys. **70**, 3401 (1979)
51. O. Arteaga, Z. El-Hachemi, A. Canillas, Phys. Status Solidi A **205**, 797 (2008)
52. T. Shaykhutdinov, S.D. Pop, A. Furchner, K. Hinrichs, ACS Macro Lett. **6**, 598 (2017)
53. S. Kirstein, S. Daehne, Int. J. Photoenergy **2006**, 1 (2006)
54. P. Alliprandini Filho, G.G. Dalkiranis, R.A.S.Z. Armond, E.M. Therézio, I.H. Bechtold, A.A. Vieira, R. Cristiano, H. Gallardo, A. Marletta, O.N. Oliveira, Phys. Chem. Chem. Phys. **16**, 2892 (2014)
55. H. Yao, T. Isohashi, K. Kimura, J. Phys. Chem. B **111**, 7176 (2007)
56. R.M.A. Azzam, N.M. Bashara, *Ellipsometry and Polarized Light* (North-Holland: Sole distributors for the USA and Canada, Elsevier Science Pub. Co., Amsterdam; New York, 1987)
57. J.R. Tischler, M.S. Bradley, V. Bulović, Opt. Lett. **31**, 2045 (2006)
58. M.S. Bradley, J.R. Tischler, V. Bulović, Adv. Mater. **17**, 1881 (2005)
59. P. Törmä, W.L. Barnes, Rep. Prog. Phys. **78**, 013901 (2015)
60. J. Bellessa, C. Bonnand, J.C. Plenet, J. Mugnier, Phys. Rev. Lett. **93**, (2004)
61. R. Houdré, Phys. Status Solidi B **242**, 2167 (2005)
62. C. Bonnand, J. Bellessa, J.C. Plenet, Phys. Rev. B **73**, (2006)
63. A. Salomon, S. Wang, J.A. Hutchison, C. Genet, T.W. Ebbesen, Chem. Phys. Chem. **14**, 1882 (2013)
64. M.M. Dvoynenko, J.-K. Wang, Opt. Lett. **38**, 760 (2013)
65. D. Melnikau, R. Esteban, D. Savateeva, A. Sánchez-Iglesias, M. Grzelczak, M.K. Schmidt, L.M. Liz-Marzán, J. Aizpurua, Y.P. Rakovich, J. Phys. Chem. Lett. **7**, 354 (2016)
66. P.A. Hobson, W.L. Barnes, D.G. Lidzey, G.A. Gehring, D.M. Whittaker, M.S. Skolnick, S. Walker, Appl. Phys. Lett. **81**, 3519 (2002)
67. A. Armitage, D. Lidzey, D.D. Bradley, T. Virgili, M. Skolnick, S. Walker, Synth. Met. **111–112**, 377 (2000)
68. D.G. Lidzey, D.M. Coles, in *Org. Hybrid Photonic Crystal*, ed. by D. Comoretto (Springer International Publishing, Cham, 2015), pp. 243–273
69. J. Wenus, S. Ceccarelli, D.G. Lidzey, A.I. Tolmachev, J.L. Slominskii, J.L. Bricks, Org. Electron. **8**, 120 (2007)

# Chapter 15
# Conjugated Polymers: Relationship Between Morphology and Optical Properties

**Maria Isabel Alonso and Mariano Campoy-Quiles**

**Abstract** In this chapter we will start by briefly summarizing the basic concepts of the electronic structure of conjugated polymers. This will enable the discussion of the relevant descriptions of the dielectric function. We will relate these descriptions to the model parameterizations which are used in advanced ellipsometric analysis of thin films such as those used in devices for organic photovoltaics (OPVs) and light emitting diodes (OLEDs). Amongst other things, such parametric descriptions are useful to deal with structural changes in conjugated polymer thin films. Once the models are presented, we will provide representative examples of the nexus between morphology and optical constants, and how the latter can be employed to infer aspects of the former. First, we will discuss how chain conformation affects the optical properties. Then, we will explain the anisotropic behavior of conjugated polymer films due to their intrinsic molecular anisotropy and review different cases (f. i., oriented films or semicrystalline polymers). We will also describe structural changes that occur upon blending polymers with fullerenes and concomitant variations of the optical properties. Here we will focus on state of the art low band gap polymers mixed with fullerenes. Finally, real-time ellipsometric experiments in which these structure-property relationships can be exploited will be presented.

## 15.1 Introduction

Conjugated polymers are technologically interesting materials sharing the general advantages of polymers and the optical and electrical properties characteristic of semiconductors [1]. This unique combination of features offers substantial benefits for applications in the fields of photovoltaics [2], light emitting diodes [3], and transistors [4], where different molecular structures, morphologies and device architectures provide the required useful functionalities (cf. Chap. 12). Ideally, as

M. I. Alonso (✉) · M. Campoy-Quiles
Institut de Ciència de Materials de Barcelona (ICMAB-CSIC),
Campus UAB, 08193 Bellaterra, Spain
e-mail: isabel.alonso@icmab.es

© Springer International Publishing AG, part of Springer Nature 2018
K. Hinrichs and K.-J. Eichhorn (eds.), *Ellipsometry of Functional
Organic Surfaces and Films*, Springer Series in Surface Sciences 52,
https://doi.org/10.1007/978-3-319-75895-4_15

335

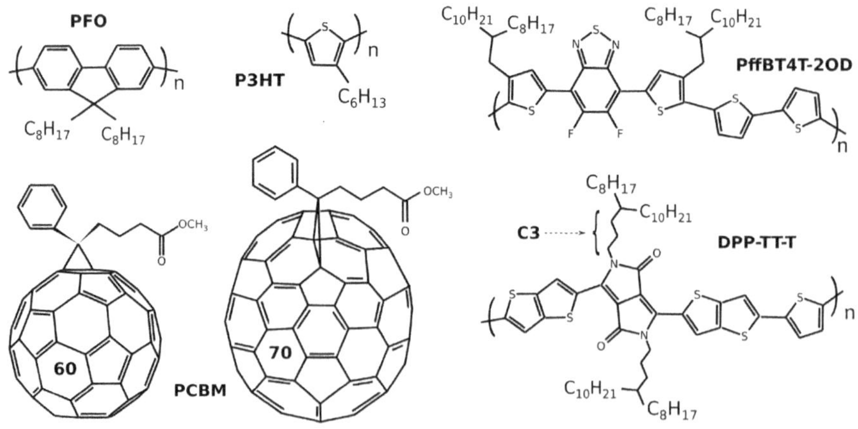

**Fig. 15.1** Chemical structures of conjugated molecules and polymers considered in this chapter

macromolecules, polymers can be designed and synthesized to achieve the desired behavior and can be processed using low-cost solution methods to engineer a wealth of devices. Although solution processing is probably the greatest benefit for polymer electronics, many properties of the polymer electronic devices depend entirely on the processing conditions and with them, their device performance [5]. Some technologically relevant donor polymers and the ubiquitous acceptor fullerene derivative PCBM (Phenyl-C60-butyric acid methyl ester) are depicted in Fig. 15.1. Spectroscopic ellipsometry has emerged as a useful technique for the design of new materials, control of microstructure and characterization and optimization of device-layer films [6]. It helps to evaluate not only structural features such as film thicknesses and composition profiles but it is also the technique of choice to study the basic optical properties of the active organic semiconducting films, which result from the combination of both chemical nature and morphology.

In this chapter, we concentrate on these structure-property relationships, aiming our attention to the UV-VIS spectral range which is relevant for optoelectronic applications. With this focus, we will discuss the dielectric functions of conjugated polymers and how different aspects of their morphology affect them. This will help to explore how the nexus between morphology and optical constants can be exploited to gain information about technologically relevant processes such as blending and monitoring real-time structural evolution during processing.

## 15.2 Basic Concepts

The particular balance of interactions taking place in a conjugated polymer film determines its optical properties. These interactions are strongly affected by morphological variations and therefore there is a strong structure-property nexus. In general, the basic optical behavior will be given by the properties of the polymer backbone. However, the strong differences observed when comparing solution and solid samples indicate that the optical properties are also modified by the final microstructure resulting from the characteristics of the polymer chains (regio regularity, side chains, weight average molar mass and polydispersity) [7] as well as the processing conditions such as the applied coating technique, the choice of solvent, polymer concentration (or viscosity of the polymer solution), drying conditions, and post-deposition treatments [8–10]. The interactions with the substrate may also play a role in determining the final structure, as discussed for some polymers in Chap. 5.

## 15.2.1 Structure and Optical Properties

The structural properties of polymers have been described in Chap. 5 in connection with their physical properties. In the case of conjugated polymers, structural aspects acquire high importance because of their large impact on the electronic structure and concomitantly on their optical properties and suitability for specific applications. The semiconducting character of conjugated monomers and polymers emerges from the bonding via $sp^2$ hybridization among C atoms [1]. The non-hybridized p orbitals from neighboring C atoms overlap forming $\pi$ bonds, which allow for delocalization of electrons leading to conducting or semiconducting electronic structures, depending on the bandwidth that results from the number of interacting C atoms or conjugation length. In common with other organic semiconductors (see Chap. 13), these materials are composed of light atoms and have relatively low packing densities. Consequently, their dielectric constant $\varepsilon_1$ takes low values and the dielectric screening of interactions between electrons is limited. Therefore, the Coulomb interaction and the exchange energy between electron pairs with parallel or anti-parallel spin are much stronger than in traditional inorganic semiconductors [11]. One consequence of the low dielectric screening is that photoexcited electron-hole pairs (aka excitons) are tightly bound and mostly restricted to molecular sites [12, 13]. In addition, inter-molecular interactions are usually of the van-der-Waals type, hence weaker than the intra-molecular covalent bonding. As a result, the optical excitations of the solids have strong localized character and resemble the molecular spectra albeit with some shifts and broadenings that depend on the particular microstructure. Several factors contribute to shift and broaden the spectra and it is difficult to disentangle them.

Considering the polymer as an ensemble of repeating units, the optical spectra will be an inhomogeneously broadened average due to dispersity and different orientations of these polymer molecules. Molecular electronic transitions with vibronic replicas reflect the significant electron-phonon interaction [14]. The frequencies of vibrations that couple to electronic states are usually high, resulting in large total bandwidths that in the polymer frequently give rise to broad electronic transitions. The intensity distribution of the replicas given by the transition probabilities from the 0th vibrational level of the ground state to the mth vibrational level of the excited state, $I_{0-m}$, usually corresponds to a Poisson distribution given by $I_{0-m} = (S^m/m!)e^{-S}$, where S is the Huang-Rhys parameter that indicates the number of quanta involved in the vibrational excitation. The dominant transition is the 0–0 (or $S_{0-1} \leftarrow S_{0-0}$) only when the molecule distortion is small. Variable distances between molecules and conformational disparity are other sources of broadening. These fluctuations have direct influence on the strength of inter- and intra-molecular interactions, thus the conjugation length, and with it also the spectral position of the electronic band [15]. In general, an increased conjugation length leads to a spectral redshift. But this spectral signature can correspond to various structural situations [16]. For example, crystalline polymers are actually composed of both crystalline and amorphous regions. In highly crystalline domains the chains form stacks (lamellas) where interactions are favored and the optical band appears redshifted compared to amorphous domains. Then, the spectral position of the overall band can be correlated with the degree of crystallinity [17]. However, a redshift can also be the result of an increased planarity (larger conjugation length) of the individual chains, especially in the case of short repeat units. In fact, the assignment of spectral signatures observed in a polymer film is quite complex [18]. Many features of the intricate interplay of interactions are captured in the HJ-aggregate model [19] in which the electronically coupled monomers in a single polymer chain behave as a J-aggregate and the π-stacked chains interact as an H-aggregate. The J- or H-character is determined by the sign of the Coulomb coupling which is negative between adjacent repeat units ($J_{intra} < 0$), and positive ($J_{inter} > 0$) between neighboring chains. However, the overall behavior of the HJ-aggregate is not intuitive because it depends on the competition between both mechanisms including the particular configuration of the chain and also the disorder [20]. By the mixing of interactions, relative intensities of the vibronic peaks can also deviate from the Poisson distribution. Finally, structural differences can also affect the observed oscillator strength. These can be associated to anisotropic behavior or to a 3D isotropic increase in absorption due to conformational changes, especially in increased linearity of chains of rigid monomers [21].

Table 15.1 summarizes expected effects of structural variations on the optical properties of organic semiconductors. Usually, combinations of these effects are present in experimental situations. Particular cases of morphological changes will be presented in Sect. 15.3, focusing in different aspects of conformation and crystallinity. We will also deal with anisotropic optical behavior in polymeric films.

## 15.2.2 Ellipsometry and Dielectric Function

The fundamentals of ellipsometric measurements and analysis have been presented in Chap. 1. Particular considerations and precautions to optimize measurements in samples of conjugated polymers and to extract reliable information from them have been recently reviewed [6]. Usually, spectra acquired at multiple angles of incidence (VASE) and on possibly multiple samples (varying film thickness and or substrate), are considered. Then, to extract material properties from VASE data a multilayer optical model must be built describing the sample(s) under consideration [22, 23].

**Table 15.1** Summary of likely effects on the spectral characteristics of optical spectra expected for several structural situations referred to the case of an isolated monomer or repeat unit

| Structural change | Cause | Energy shift | Broadening | Optical density |
|---|---|---|---|---|
| Gas to solid | Polarization | ↓ | ↑ | ≈ |
| J-aggregation | Excitonic coupling (J < 0) | ↓ | ↑ | ↓ |
| H-aggregation | Excitonic coupling (J > 0) | ↑ | ↑ | ↑ |
| Order | Conjugation | ↓ | ↑ | ≈ |
| Curvature | Persistence | ≈ | ≈ | ↑ |

Briefly, the effect of reflection on light polarization is described applying transfer matrix algebra. An overall reflectance matrix is obtained by multiplying individual matrices that describe reflection at each interface and propagation through each individual layer. Therefore, parameters of the model are, besides the number of layers forming the structure, their thicknesses and dielectric functions. These may be taken from available databases or determined from the studied spectra. The dielectric functions of composite materials as well as roughness and intermixing layers are represented by effective-medium models. With these implementations, the reflectance matrix can be evaluated and the parameters of interest fitted by a least-squares method to the measured spectra.

Two main data analysis approaches are commonly used to determine the optical response of the unknown materials. In a simple case of a film on a known substrate, or when only one of the materials in the multilayer is unknown, it is possible to run a point-by-point numerical fit to obtain $n$ and $k$ by fixing the film thickness to that found from the transparency range imposing $k = 0$. Direct point-by-point fits are extremely useful as no assumptions about the spectral dependence of the unknown optical functions are made and arbitrary dispersion relations can be reproduced. An improvement of this numerical procedure is to model $n$ and $k$ by smooth functions such as splines [24] which improves the fit convergence and smoothness of the result. The most common approach, however, is to model the optical response of the unknown materials using parameterized analytical functions of the energy according to the physical properties of the studied materials [23]. The principal disadvantage of parametric models is a loss of flexibility in the description of the dielectric function. Only features assumed to be present by the choice of model will be represented in the fit. They do, however, have advantages such as choosing functions that satisfy Kramers–Kronig consistency and can be appropriately correlated to fundamental electronic properties. Moreover, an analytical expression of the optical constants is then available for use in other models and can be parameterized for descriptions of alloys, for instance.

As already discussed, the optical properties of conjugated polymer films can be quite complex due to the basic molecular nature of the polymer and the fact that the particular conformation is decisive to determine the effective optical properties of the film, both in terms of spectral features (reflecting the electronic density of states) and anisotropy (reflecting orientation). Even in a single-component film, a multi-phase morphology is likely to occur. Because of the abundant morphological variations in polymers, this is a case in which a physical parameterization of the dielectric function that allows describing structural variations and transformations by varying the model parameters is especially useful. As in many other film systems, transparent polymers (or measurements in the transparency region) may be modelled by a Cauchy dispersion [23], which is useful to investigate film thickness or whether there is optical anisotropy, for instance. However, in the majority of cases it is necessary to represent the absorption bands. Several authors performed systematic studies to compare how well different line-shapes described ellipsometry data for thin conjugated polymer and macrocycle films [25–27]. These studies established that it was essential to allow for asymmetric lineshapes in order to reproduce the experimental measurements. Taking into account the physical nature of the electronic excitations in conjugated polymers, the standard critical point model (SCP) provides the most consistent description and it was proven to fit best the experimental data with the minimum number of parameters. It is based on the expression of the dielectric function due to a critical point (CP) [11, 28]:

$$\varepsilon_{CP}(\omega) = C - Ae^{i\phi}(\hbar\omega - E_g + i\Gamma)^n, \tag{15.1}$$

where $n = D/2 - 1$ is related to the dimensionality $D = 3, 1, 0$ of the CP (the dependence for $D=2$ is logarithmic) and the phase angle $\phi$ is a multiple of $\pi/2$ that identifies its kind (maximum, minimum or saddle point) for each $D$. In practice, $\phi$ is allowed to take any value as a phenomenological way of accounting for combined shapes which arise from diverse causes such as many-body interactions or even unresolved bands. The amplitude is given by $A$, the transition energy by $E_g$, and the lifetime broadening by $\Gamma$. The parameter $C$ describes a constant background that takes into account contributions in the UV, beyond the measured energy range. The SCP model considers that the dielectric function can be reproduced by a sum of CPs. Strictly, (15.1) only describes the most resonant term of each CP contribution to the dielectric function and to improve the fitting of CP parameters it is advantageous to consider numerical derivatives of the experimentally obtained dielectric function. Otherwise, additional broad CPs may be added to account for not constant but weakly varying backgrounds. For example, Arwin et al. [25] performed detailed line-shape analyses of second derivative spectra determined point-by-point on polythiophene thin films. They compared symmetric (Gaussian and Lorentzian) and asymmetric (0D and 3D CPs) models and found that only asymmetric lineshapes were satisfactory to explain the experimental measurements. Other authors compared several models to reproduce the dielectric function itself, confirming this conclusion, for instance on films of several metal phthalocyanines [26]. For many conjugated polymers [27], the analysis of the ellipsometric data using the SCP model allowed to distinguish between localized excitons for amorphous polymeric films, and 1D/2D delocalization of the electronic wavefunction for highly crystalline films or containing planar conformations of chains [27]. A description based on a 1D density of states was also employed

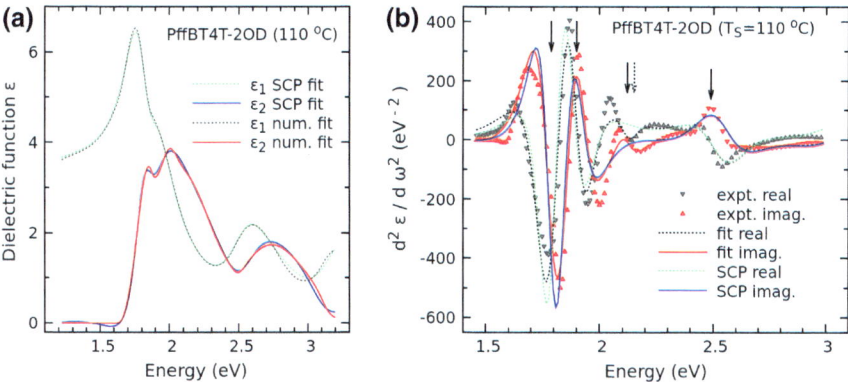

**Fig. 15.2** **a** Dielectric function of a PffBT4T-2OD film blade coated onto a glass slide at a substrate temperature of 110 °C. The displayed curves compare results obtained both numerically (splines) and parametrically (Standard Critical Point model). **b** Numerically built (expt.) and fitted second derivative spectra compared to the derivative calculated from the parameters of the dielectric function fit using the SCP model. Arrows mark the obtained energies in both ways. The shoulder near 2.1 eV is better resolved by the numerical derivative approach and appears much broader and somewhat shifted for the SCP model

by Zhokhavets and co-workers in order to estimate the exciton binding energy ($E_b$) from the dielectric function of polythiophenes [29]. They compared the band edge to a modelled 1D band gap in order to extract $E_b$. The obtained $E_b$ values from this approach were about 0.6 eV for P3OT [29] and also for PFO [30], in reasonable agreement with exciton binding energies obtained with other techniques [31]. Notice that not always parametric models have a clear physical meaning. For instance, fittings with symmetric lines have been used to investigate morphological issues such as anisotropy [32–34] but typically the oscillators included in such models were not related to the electronic structure of the investigated organic semiconductors.

Figure 15.2 shows an example of dielectric function fitting and lineshape analysis on a poly[(5,6-difluoro-2,1,3-benzothiadiazole-4,7-diyl)-alt-(3,3'''-di(2-octyldodecyl)-2,2';5',2'';5'',2'''-quaterthiopene-5,5'''-diyl)] (PffBT4T-2OD) film deposited at 110 °C. This low band gap polymer is one of the record holding materials with over 11% photovoltaic power conversion efficiency. The parametric SCP model dielectric function reproduces quite well the numerical result. In this case, six 0D CPs were needed to obtain a satisfactory fit of the ellipsometry spectra, see Fig. 15.2a, one of them at higher energy and another providing a weakly varying background. The other four CPs (listed in Table 15.2) agree with those present in the numerical second derivative spectra shown in Fig. 15.2b in which the weaker shoulder near 2.1 eV is much better resolved and determined. This comparison illustrates the fact that sharp and isolated transitions can be well determined by fitting the dielectric function but weaker contributions near stronger ones can be distorted by the backgrounds.

**Table 15.2** Comparison of the fitted energies and broadenings of the four peaks plotted in Fig. 15.2b using the SCP model dielectric function or the same model applied to the second numerical derivative

| CP number (eV) | 1 | 2 | 3 | 4 |
|---|---|---|---|---|
| $E_{SCP}$ | 1.80 | 1.89 | 2.15 | 2.50 |
| $E_{2der}$ | 1.79 | 1.90 | 2.12 | 2.49 |
| $\Gamma_{SCP}$ | 0.11 | 0.14 | 0.48 | 0.20 |
| $\Gamma_{2der}$ | 0.14 | 0.12 | 0.10 | 0.20 |

## 15.3 Examples of Structural Variations

Structural variations occurring in conjugated polymer device-layer films are fundamental to tune and optimize the optical properties for a given optoelectronic application. Enormous efforts have thus been dedicated to understand and control the morphologies that result from different processing routes. At the same time, the correlation between those morphologies and the resulting optical behavior is of particular interest. In this context, spectroscopic ellipsometry is most helpful to investigate the link between optical properties and film microstructure.

In general, both thermodynamics and kinetics determine the polymer arrangements in the film and complex structures may form with direct impact on the film's spectral optical behavior. Although different contributions are difficult to disentangle, here we describe three main structural variations that occur in single-phase (one-component) conjugated polymer films and correlate them to observed spectral changes. We distinguish between conformation, crystallinity and anisotropy, which affect in different ways the strengths of the interactions and ultimately the absolute values and/or the energy positions of the optical bands.

### 15.3.1 Conformation

Conformation effects refer to intrinsic molecular properties which result both from steric and electronic interactions. The former derive from the chain regularity, side chain structure, and molecular weight (MW); the latter determine the torsional potentials [35] that regulate the alignment of successive monomers. Various conformation variations and their effect on the optical absorption are schematically depicted in Fig. 15.3. The case of the low-bandgap polymer thieno[3,2-b]thiophene-diketopyrrolopyrrole (DPP-TT-T) is representative and has been well studied [21]. This low-bandgap copolymer shows photostability under prolonged excitation in the low energy absorption band [36] and allows for high field-effect transistor mobilities and good performance as the donor in solar cells [37]. These efficient qualities are correlated to the predominant *trans* conformation of the polymer chains. In this

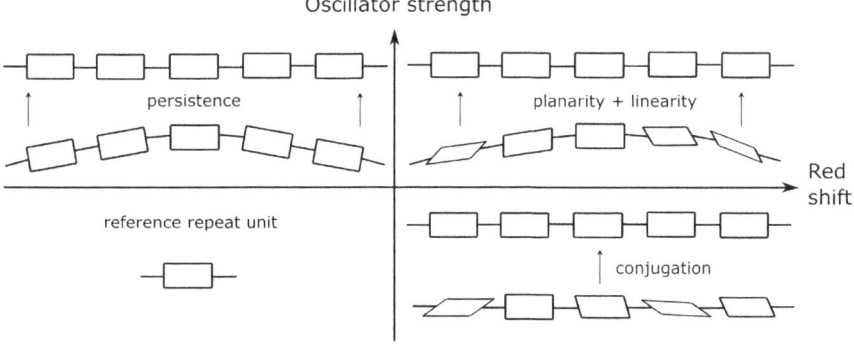

**Fig. 15.3** Schematic representation of the main conformation variations and their effect on the position and strength of the electronic transitions

configuration, the successive monomers are rotated by approximately 180° relative to each other, which leads to more linear oligomer structures than the other limiting conformation case when the successive monomers are equally oriented (*cis*). In the *cis* structure, the backbone is curved within the conjugated plane and results in lower optical absorption [21]. Note that in this case more curved or more linear long oligomers differ in the oscillator strength but not, to a first approximation, in the transition energies. The observed effect is associated to the concept of persistence length, $\lambda_p$, which can be thought as a relative propensity of the polymer to adopt a linear structure. On the other hand, the transition energies are closely related to the conjugation length, which structurally is associated to planarity. It is interesting to think how one chain could be linear but not necessarily planar (if adjacent monomers are rotated with respect to each other). Similarly, one chain could be planar but have more or less curvature, if each repeat unit is bonded in *cis* or *trans* conformations. In addition, the same effect was observed in solution and was insensitive to dilution, confirming the intrinsically high $\lambda_p$ value in DPP-TT-T [21]. Figure 15.4 shows that this effect is enhanced both with the position of the branching point on the polymer side chains [38] and even more with the molecular weight of the polymer, which favors an extended $\lambda_p$.

A different example of conformation variation is the case of the blue-light emitting conjugated polymer poly(9,9-dioctylfluorene) (PFO) [39]. This material appears in a diversity of polymorphic forms [40, 41] with different values of torsion angle between monomers. The most interesting conformer is the β-phase in which the torsion angle between adjacent monomers is 180°. In contrast, most solution processed films contain a broad distribution of torsion angles leading to a wormlike chain conformation or *glassy* structure. In this respect, β-phase represents an example in which planarity is enhanced and thus conjugation length is increased. In fact, the optical features associated to this phase are a new transition altogether. Figure 15.5 shows the refractive indices measured for two spin-coated films of PFO. In the glassy+β sample a fraction of chain segments was driven to adopt the planar conformation,

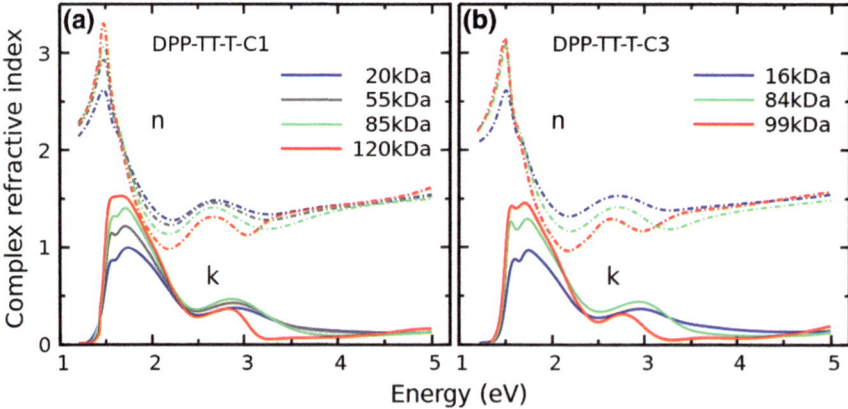

**Fig. 15.4** Optical constants (n, k) for pristine polymer films of DPP-TT-T with **a** C1 and **b** C3 branching points (the synthesis details are given in [38]) and for several number-average molecular weights (adapted from [21])

**Fig. 15.5** Optical constants (n, k) for glassy and glassy +β PFO films (see [27]) as determined by ellipsometry. The low energy absorption feature appears due to the higher conjugation length characteristic of the planar β-phase conformation (see sketch)

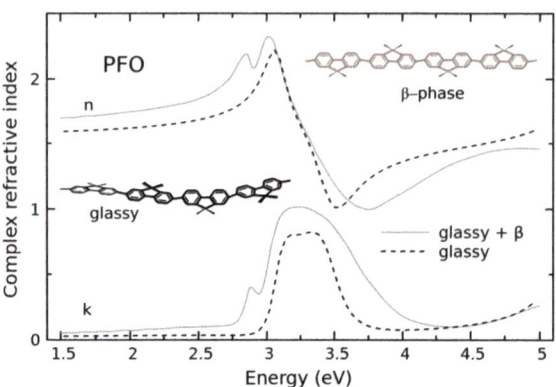

giving rise to a distinct peak at lower energy even if the β-phase fraction is below 10%. In this case, compared to the predominant disordered phase, since the fluorene units are relatively short the well-defined planar chain segments have a higher conjugation length, even if no crystallization takes place [41, 42].

## 15.3.2 Crystallinity

Crystallization refers to the aggregation of the polymer chains in a regular or ordered fashion. Obviously, the degree of stereoregularity and the strength of intermolecular interactions affect the ability of a polymer to form crystalline domains, which are stacks of orderly folded chains (called lamellas). For a material with tendency to

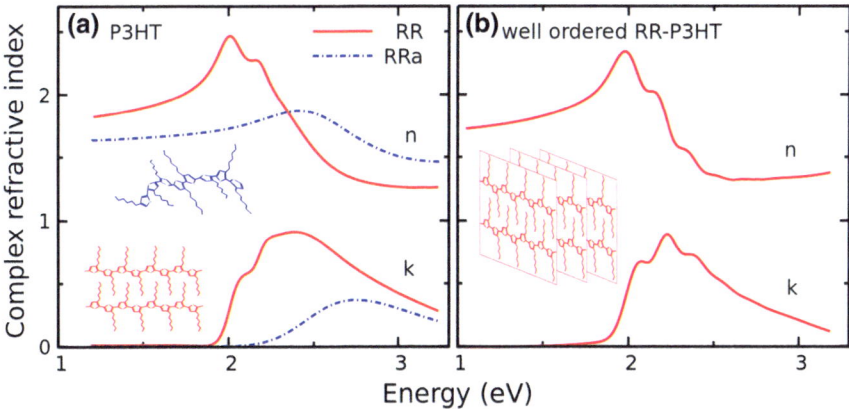

**Fig. 15.6** Optical constants (n, k) for P3HT with different crystallinity. **a** Spectral differences observed in films of amorphous regiorandom (RRa) and device-quality semicrystalline regioregular (RR) material. The sketches illustrate that only the stereoregular chains are able to form crystalline domains. **b** Spectrum of a well-ordered film of crystalline P3HT according to [43]. The sketch indicates that large crystallites are associated to this spectrum

crystallize, the degree of crystallization will strongly depend on molecular weight, as entanglement between chains appears in long molecules. Kinetic factors also have a large influence in crystallization processes: solution processing often results in kinetically-trapped out-of-equilibrium microstructures. However, a highly crystalline polymer is rarely entirely ordered, this means that there are amorphous regions between lamellas and the degree of crystallinity (or paracrystallinity) is a parameter by itself. Frequently, lamellas tend to order forming spherical regions from a nucleation point called spherulites, i.e. spherical crystallites. (In thin films, the equivalent 2D structures with circular symmetry are still called spherulites.) Amorphous domains are still present within the spherulites or between their boundaries.

A well-known example of steric effect occurs in poly (3-hexylthiophene) (P3HT) which contains a mixture of regioisomers. The relative ratio of these regioisomers in the resulting polymer gives the regioregularity. Highly regioregular RR-P3HT tends to crystallize easily and is preferable for applications such as transistors and photovoltaics as both, electrical and optical properties are enhanced (see Chap. 12). On the contrary, the irregular substituent distribution in regiorandom RRa-P3HT forces the thiophene units to twist away from planarity, hindering crystallization [44]. Figure 15.6 compares the refractive indices of regiorandom (amorphous) P3HT and two RR-P3HT samples with different degrees of crystallinity. The spectrum of RRa-P3HT is blue shifted and displays smaller oscillator strength compared to the RR counterparts. The former is due to the reduction in conjugation length associated to the frequent twists forced by steric interactions of the side chains. The latter is a combination of a smaller average density (amorphous chains cannot pack as densely as crystalline ones) and a lower degree of linearity in the chain (as it was discussed in Sect. 15.3.1) The dielectric function is proportional to the density of dipoles, and this is in most cases proportional to the material density (see e.g., [39]). Interestingly, the optical properties are also clearly affected by the degree of crystallinity [20, 45]. As it was discussed in Sect. 15.2.1, the main optical transitions have vibronic replica whose intensity is described by the Huang-Rhys parameter. Samples with higher crystallinity often result in higher intensities for the lowest vibrational levels with respect to higher energy vibronic peaks. Moreover, the broadening of each transition and vibronic sideband increases with disorder. To be able to distinguish vibronic sidebands is, in itself, a sign of chain ordering. For instance, x-ray diffraction shows that both PffBT4T-2OD (Fig. 15.2) and DPP-TT-T (Fig. 15.4) are partially ordered. Increasing ordering in DPP-TT-T by reducing curvature leads not only to higher overall oscillator strength, but also, to a redistribution of peak intensities in favor of the lowest energy transitions, as observed in Fig. 15.4.

## 15.3.3 Anisotropy

Anisotropic optical properties originate both from the intrinsic anisotropy of polymer chains and from the material microstructure. Since most conjugated polymers behave as rigid-rod-like macromolecules, even amorphous solution-processed polymer films may exhibit preferential molecular orientation (in-plane vs. out-of-plane) giving rise to a uniaxial dielectric tensor. Therefore, evaluation of the optical anisotropy is necessary to correctly establish the oscillator strength of the transitions [6]. Higher degree of orientation of the film associated to an increased alignment of the chains tends to enhance the anisotropy, leading to different values of the tensor components as well as varying transition energy positions. This is similar to the case of films made from ordered organic small molecules (see Chap. 13). Certain mixtures of orientations of ordered as well as disordered phases normally coexist in the films giving an effective anisotropic behavior which averages to a higher symmetry [46] than that of a perfectly ordered domain.

Figure 15.7 shows the complex refractive indices determined by ellipsometry on a regioregular (>90%) P3HT film with both random polycrystalline and oriented spherulitic regions [47]. Interestingly, the area where orientation was suppressed was still highly crystalline and its isotropic optical properties showed spectral features typical of crystalline P3HT. An example of a film containing spherulites is shown in the figure. The Maltese cross patterns centered at each spherulite viewed between crossed polarizers evidence their birefringent nature and the radial disposition of polymer chains, which can be parallel or perpendicular to the radial direction. The anisotropic optical response in the measured spherulitic region was well represented, within experimental error, by a uniaxial model with a variable in-plane optic axis aligned with the fibers. In this case, as shown in Fig. 15.7, the index along the fibers is higher than perpendicular to them, indicating that the polymer chains align parallel to the fibers. Although the two reported crystalline polymorphs of regioregular P3HT are monoclinic [48], domain misalignment and amorphous interlamellar material leads to at most orthorhombic-like dielectric tensor, for example in biaxially strained

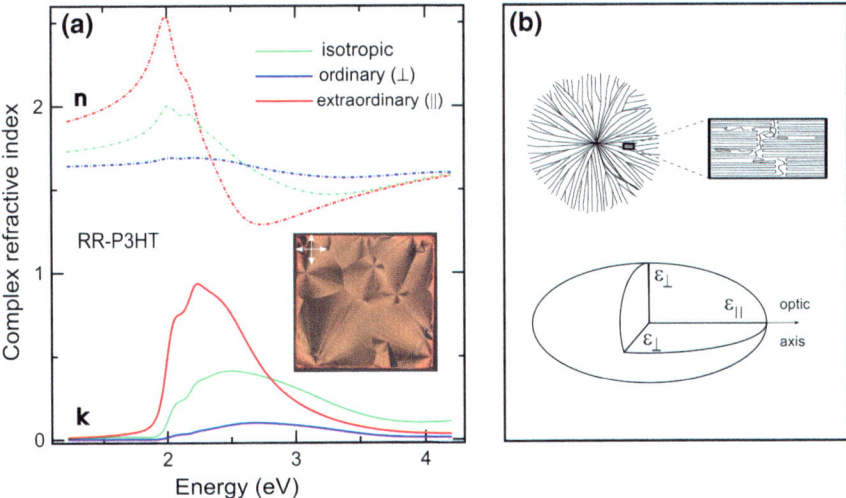

**Fig. 15.7** **a** Complex refractive index (n, k) data in two zones of a RR-P3HT film. The anisotropic components were measured on a spherulite and the isotropic curves correspond to a region where orientation was suppressed. The inset shows a film ($2.5 \times 2.5$ cm$^2$) with spherulites viewed between crossed polarizers. Adapted from [47]. **b** Sketch of a spherulite and preferential orientation of the lamellas within the fibrils according to the deduced uniaxial dielectric tensor represented below

films [44]. Crystallinity can be approximately evaluated by considering that the spectra can be represented by an average of RR and RRa mixtures [49]. For the case of Fig. 15.7, this gives a qualitative estimation of almost complete crystallinity parallel to the fibers whereas the isotropic and the perpendicular spectra contained about 10–20% of RRa spectral contribution [47].

## 15.4   Blending

Mixing different materials into composites is often utilized in the organic semiconductor field. For instance, white lighting can be achieved by blending blue, green and red emitting materials with energy alignment as to produce cascade energy transfer between them. In photovoltaics, blending electron donor and acceptor molecules into the so called bulk heterojunction has emerged as the most effective way of simultaneously obtaining efficient charge separation of the photoinduced excitons while maintaining good light absorption (for which relatively thick films are required). The optical properties of blend films are, in general trends, a combination of the optical properties of the constituents weighted by the corresponding volume fraction. Figure 15.8 shows the complex refractive index of blends of DPP-TT-T with the acceptor PC70BM. The absorption profile includes the transitions of the corresponding components. For instance, the main absorption band of the polymer (Fig. 15.4)

**Fig. 15.8** Optical constants
(n, k) for blend films of
DPP-TT-T-C3 and PC70BM
for several number-average
molecular weights of the
polymer (cf. [21])

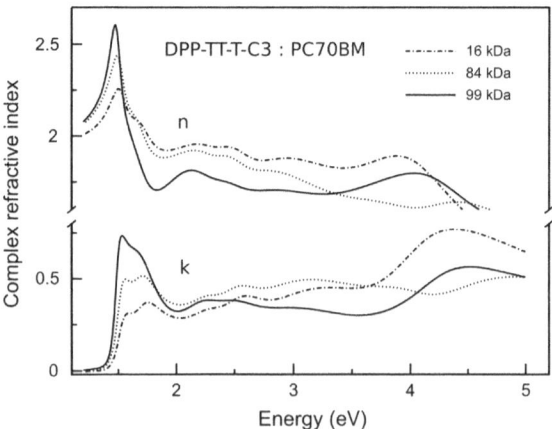

is also easily observed in the blend. Not just that, the increase in absorption obtained with increasing molecular weight is also preserved in the blends.

The new optical transitions that emerge when mixing, such as the charge transfer excitons, have absorption strengths that are several orders of magnitude smaller than the transitions of the individual components [50]. In this respect, one could think of using effective medium approximations (EMAs, see Chap. 1) to describe the optical properties of these blends provided that the two materials are well mixed in comparison to the wavelength of light. In good solar cell systems such as those described in Chap. 12, this is the case as the domain size is limited to a few tenths of nanometers (roughly twice the exciton diffusion length). The major difficulty arises, however, from the fact that during the deposition of the film the two materials interfere in the way each other would solidify when alone. It is known, for instance, that fullerenes tend to partially prevent polymer crystallization. In terms of the optical properties this would mean that the optical properties of each individual component are not the same as the optical properties of that component when mixed. It has been reported, for instance, that the degree of anisotropy decreases with blending [51]. Moreover, the two components may be inhomogeneously distributed in the vertical direction (perpendicular to the plane of the substrate), forming composition profiles of the two species [52]. This will depend on the surface energies as well as the solubilities of the different compounds in the solvent and additives used, and ellipsometry can be employed to determine such segregation.

Since it is difficult to predict a priori the effect of blending on the morphology and subsequent optical properties, blends are often treated as a completely new material, as though it was an alloy. Otherwise, advanced effective medium approximations are required which include the fact that the degree of crystallinity and orientation can vary [53, 54].

When one of the two compounds has a robust morphology, i.e. approximately independent of processing, ellipsometry can be used to evaluate if the optical properties of the other component have changed upon blending. For instance, blends of amorphous polythiophene (RRa-P3HT) with two types of fullerenes revealed that for small amounts of fullerene there is mixing at the molecular level, i.e. the amount of small molecules is below the miscibility limit within the polymer matrix [55]. In this context, ellipsometry was used to deduce the optical properties of the fullerene component in the blend, demonstrating the disappearance of the absorption bump often associated to fullerene aggregation.

## 15.5   Monitoring in Real Time

The strong nexus that exists between the optical and structural properties on conjugated polymer thin films implies that real time ellipsometry can be employed to infer important structural information. Perhaps the simplest case corresponds to following the morphology as a thin film experiences thermally induced phase transitions. In Sect. 15.3.2 we explained that the crystalline and the amorphous versions of a given polymer have substantially different refractive indices. By exposing a thin film to a ramp of temperature, it is possible to actually monitor the different phases directly in a single material film. Figure 15.9. shows the temperature dependent refractive index for regioregular P3HT (after [56]). As the temperature is increased, the film expands reducing like this the density of dipoles and thus the oscillator strength. Thermally activated vibrations statistically decrease the conjugation length, and thus gradual blue shifts are found upon heating. Peaks also broaden with temperature, becoming less well resolved. These general trends can abruptly change when the temperature is raised above a characteristic phase transition temperature. For instance, the speed at which the aforementioned effects happen will accelerate when taking the film above the glass transition temperature. The transition between crystalline and molten film is, perhaps, the one that results in the largest variation in optical properties, as shown in Fig. 15.9. Indeed, the optical properties of the molten film resemble those of the regiorandom P3HT (see Fig. 15.6). It has been shown that the temperature dependence of the ellipsometric raw data can be used to monitor phase transitions [6]. Figure 15.9b, c exemplify this for a RR-P3HT film with a kink observed for the glass transition temperature, while a sudden change is found upon melting (on heating) or crystallization (on cooling).

Since phase transitions can depend on the geometrical confinement imposed by the thin film geometry, their corresponding temperatures often vary with respect to the bulk [57, 58]. The average thermal transition temperature depends thus on film thickness. Recently, ellipsometry has been employed to determine the profile of phase transition temperatures through the depth of thin films. Interestingly, the phase transition temperatures at the free surface have been found to be higher than those at the buried interphase, possibly due to the enhanced polymer packing at the surface granted by the extra free volume [56].

Thermal annealing can also be employed in blend films and monitored upon annealing by means of ellipsometry. Indeed, the phase diagram of polymer and fullerene blends can be determined using this technique [59]. The polymer

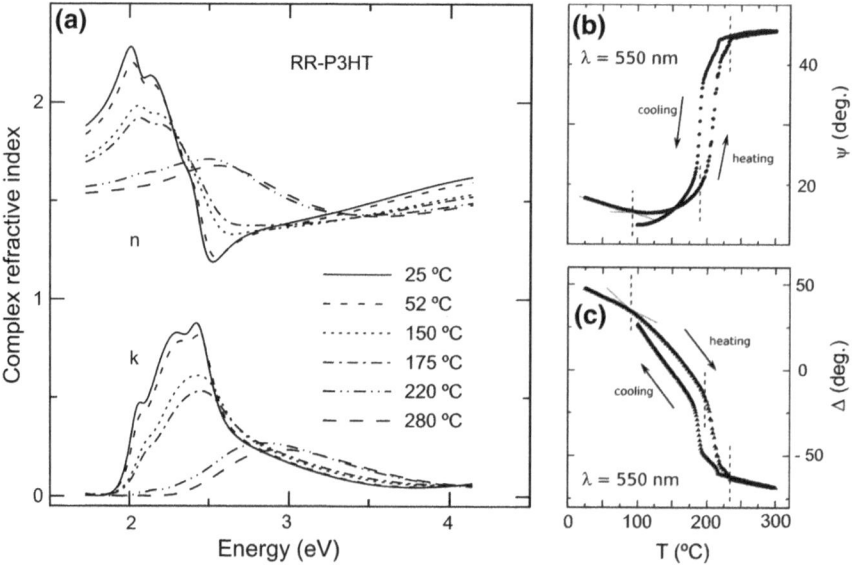

**Fig. 15.9 a** Optical properties of regioregular P3HT as a function of temperature (adapted from reference [56]). The kinks just below 100 °C observed in the measured ellipsometric angles in **b** and **c** are due to the glass transition, and the larger changes near 200 °C correspond to melting or crystallization (see text)

crystallinity, which can be partially frustrated when processing both materials simultaneously, may be restored by thermal or vapour annealing the films [52]. The crystallization process is accompanied by molecular diffusion in the out of equilibrium structure formed from quenching the solution, as ordering could not be achieved without expelling foreign molecules from the inside of a forming crystallite. Correlation between these two processes has been followed using ellipsometry for the case of P3HT blended with PCBM [52]. The complexity of blends makes them, however, a very difficult system to model. An interesting way of investigating diffusion processes isolated from crystallization was proposed by the NIST group [60]. They fabricated bilayer structures of the donor and acceptor types of molecules and ellipsometrically followed the composition of each of the sublayers when applying temperature. They found that the glass transition of the polymer sets the onset for fullerene diffusion. Moreover, they investigated the miscibility limits using this approach.

Film formation has also been investigated using ellipsometry during the deposition itself. The film thinning upon solvent removal can be easily determined spectroscopically. Moreover, the solidification and even the existence of pre-aggregates in solution can be identified with in situ ellipsometry [9, 10]. This technique can also be employed to investigate the role of solvent additives during the film deposition. For instance, ellipsometry collected during drying of a solution of the high performing PffBT4T-ODT (see Fig. 15.1 for chemical structure) blended with PC70BM showed that there are two main drying regimes [61]. First, the wet film thins by evaporating the main carrier solvent, greatly increasing the solid content concentration, and then the solvent additive leaves very slowly the viscous film. During the second regime, the polymer still enhances its crystallization, as the solvent additive mainly dissolves the fullerene.

These very advanced experiments are paving the way for the use of ellipsometry as an in situ quality monitoring tool during the fabrication of roll to roll photovoltaic modules (see Chap. 12) [62, 63].

## 15.6  Summary

In this chapter we have explained the nexus between the ellipsometrically deduced optical properties and the film morphology in polymer semiconductors. We have first given an introduction to the electronic properties of conjugated polymers and how the way in which molecules arrange in the solid state can alter the electronic and optical properties versus isolated polymer chains. We have then correlated these with their dielectric function, describing both the numerical inversion of the ellipsometric data as well as the most common parametric description of the dielectric function. The following sections explained the effect of conformation, crystallinity and molecular orientation on the optical properties. Since blends of different materials are often the technologically relevant architecture, especially in organic photovoltaics through the bulk heterojunction concept, we devoted some attention to the optical properties of blend films focusing on the deviations observed experimentally from the basic mixing theories. In the final section we reviewed some of the latest results on the use of in situ ellipsometry in polymer semiconductors, and how the optical properties can be used as a proxy to address the major morphological changes happening upon phase transition or during film formation.

**Acknowledgements** The authors acknowledge financial support from the Spanish Ministry of Economy and Competitiveness through grant MAT2015-70850-P and the Severo Ochoa Programme for Centres of Excellence in R&D (SEV-2015-0496).

## References

1. A. Köhler, H. Bässler, *Electronic Processes in Organic Semiconductors* (Wiley, New York, 2015)
2. S. Günes, H. Neugebauer, N.S. Sariciftci, Chem. Rev. **107**, 1324–1338 (2007)
3. J.H. Burroughes, D.D.C. Bradley, A.R. Brown, R.N. Marks, K. Mackay, R.H. Friend, P.L. Burn, A.B. Holmes, Nature **347**, 539–541 (1990)
4. K.-J. Baeg, M. Caironi, Y.-Y. Noh, Adv. Mater. **25**, 4210–4244 (2013)
5. J. Liu, Y. Shi, Y. Yang, Adv. Funct. Mater. **11**, 420–424 (2001)
6. M. Campoy-Quiles, M.I. Alonso, D.D.C. Bradley, L.J. Richter, Adv. Funct. Mater. **24**, 2116–2134 (2014)

7. K. Koynov, A. Bahtiar, T. Ahn, R.M. Cordeiro, H.-H. Horhold, C. Bubeck, Macromolecules **39**, 8692–8698 (2006)
8. U. Zhokhavets, G. Gobsch, H. Hoppe, N.S. Sariciftci, Thin Solid Films **451**, 69–73 (2004)
9. M. Campoy-Quiles, M. Schmidt, D. Nassyrov, O. Peña, A.R. Goñi, M.I. Alonso, M. Garriga, Thin Solid Films **519**, 2678–2681 (2011)
10. T. Wang, A.D.F. Dunbar, P.A. Staniec, A.J. Pearson, P.E. Hopkinson, J.E. MacDonald, S. Lilliu, C. Pizzey, N.J. Terrill, A.M. Donald, A.J. Ryan, R.A.L. Jones, D.G. Lidzey, Soft Matter **6**, 4128–4134 (2010)
11. P. Yu, M. Cardona, *Fundamentals of Semiconductors: Physics and Materials Properties* (Springer, Berlin, 2010)
12. U. Rauscher, H. Bässler, D.D.C. Bradley, M. Hennecke, Phys. Rev. B **42**, 9830–9836 (1990)
13. S. Heun, R.F. Mahrt, A. Greiner, U. Lemmer, H. Bassler, D.A. Halliday, D.D.C. Bradley, P.L. Burn, A.B. Holmes, J. Phys. Condens. Matter **5**, 247–260 (1993)
14. J. Cornil, D. Beljonne, Z. Shuia, T.W. Hagler, I. Campbell, D.D.C. Bradley, J.L. Brédas, C.W. Spangler, K. Müllen, Chem. Phys. Lett. **247**, 425–432 (1995)
15. K. Pichler, D.A. Halliday, D.D.C. Bradley, P.L. Burn, R.H. Friend, A.B. Holmes, J. Phys. Condens. Matter **5**, 7155 (1993)
16. S. Giri, C.H. Moore, J.T. Mcleskey, P. Jena, J. Phys. Chem. C **118**, 13444–13450 (2014)
17. M. Campoy-Quiles, V. Randon, M. Mróz, M. Jarzaguet, M. Garriga, J. Cabanillas-González, Org. Photonics Photovolt. **1**, 11–23 (2013)
18. D. Raithel, S. Baderschneider, T.B. de Queiroz, R. Lohwasser, J. Köhler, M. Thelakkat, S. Kümmel, R. Hildner, Macromol. **49**, 9553–9560 (2016)
19. F.C. Spano, C. Silva, Annu. Rev. Phys. Chem. **65**, 477–500 (2014)
20. F.C. Spano, J. Chem. Phys. **122**, 234701 (2005)
21. M.S. Vezie, S. Few, I. Meager, G. Pieridou, B. Dörling, R.S. Ashraf, A.R. Goñi, H. Bronstein, I. McCulloch, S.C. Hayes, M. Campoy-Quiles, J. Nelson, Nat. Mater. **15**, 746–753 (2016)
22. R.M.A. Azzam, N.M. Bashara, *Ellipsometry and Polarized Light* (North-Holland, Amsterdam, 1977)
23. H.G. Tompkins, E.A. Irene, *Handbook of Ellipsometry* (William Andrew Publishing, New York, 2005)
24. M. Garriga, M.I. Alonso, C. Domínguez, Phys. Stat. Sol. B **215**, 247–251 (1999)
25. H. Arwin, R. Jansson, Electrochim. Acta **39**, 211–215 (1994)
26. Z.T. Liu, H.S. Kwok, A.B. Djurisic, J. Phys. D Appl. Phys. **37**, 678 (2004)
27. M. Campoy-Quiles, J. Nelson, D.D.C. Bradley, P.G. Etchegoin, Phys. Rev. B **76**, 235206 (2007)
28. D.E. Aspnes, in *Handbook on Semiconductors chap. 4A*, ed. by T.S. Moss, M. Balkanski (North-Holland, Amsterdam, 1980), pp. 110–154
29. U. Zhokhavets, R. Goldhahn, G. Gobsch, W. Schliefke, Synth. Met. **138**, 491–495 (2003)
30. M. Campoy-Quiles, Ph.D Thesis, Imperial College of London (2005)
31. S.F. Alvarado, P.F. Seidler, D.G. Lidzey, D.D.C. Bradley, Phys. Rev. Lett. **81**, 1082–1085 (1998)
32. M. Tammer, A.P. Monkman, Adv. Mater. **14**, 210–212 (2002)
33. C.M. Ramsdale, N.C. Greenham, Adv. Mater. **14**, 212–215 (2002)
34. M. Losurdo, M.M. Giangregorio, P. Capezzuto, G. Bruno, F. Babudri, D. Colangiuli, G.M. Farinola, F. Naso, Macromol. **36**, 4492–4497 (2003)
35. C. Sutton, T. Körzdörfer, M.T. Gray, M. Brunsfeld, R.M. Parrish, C.D. Sherrill, J.S. Sears, J.-L. Brédas, J. Chem. Phys. **140**, 054310 (2014)
36. S. Wood, J. Wade, M. Shahid, E. Collado-Fregoso, D.D.C. Bradley, J.R. Durrant, M. Heeney, J.-S. Kim, Energy Environ. Sci. **8**, 3222–3232 (2015)
37. H. Bronstein, Z. Chen, R.S. Ashraf, W. Zhang, J. Du, J.R. Durrant, P.S. Tuladhar, K. Song, S.E. Watkins, Y. Geerts, M.M. Wienk, R.A.J. Janssen, T. Anthopoulos, H. Sirringhaus, M. Heeney, I. McCulloch, J. Am. Chem. Soc. **133**, 3272–3275 (2011)
38. I. Meager, R.S. Ashraf, S. Mollinger, B.C. Schroeder, H. Bronstein, D. Beatrup, M.S. Vezie, T. Kirchartz, A. Salleo, J. Nelson, I. McCulloch, J. Am. Chem. Soc. **135**, 11537–11540 (2013)
39. M. Campoy-Quiles, G. Heliotis, R.D. Xia, M. Ariu, M. Pintani, P. Etchegoin, D.D.C. Bradley, Adv. Funct. Mater. **15**, 925–933 (2005)

40. W. Chunwaschirasiri, B. Tanto, D.L. Huber, M.J. Winokur, Phys. Rev. Lett. **94**, 107402 (2005)
41. A. Perevedentsev, N. Chander, J.-S. Kim, D.D.C. Bradley, J. Pol. Sci. Part B Pol. Phys. **54**, 1995–2006 (2016)
42. S.H. Chen, A.C. Su, C.H. Su, S.A. Chen, Macromol. **38**, 379–385 (2005)
43. A.J. Morfa, T.M. Barnes, A.J. Ferguson, D.H. Levi, G. Rumbles, K.L. Rowlen, J. van de Lagemaat, J. Pol. Sci. B Pol. Phys. **49**, 186–194 (2011)
44. M.C. Gurau, D.M. Delongchamp, B.M. Vogel, E.K. Lin, D.A. Fischer, S. Sambasivan, L.J. Richter, Langmuir **23**, 834–842 (2007)
45. J. Clark, J.-F. Chang, F.C. Spano, R.H. Friend, C. Silva, Appl. Phys. Lett. **94**, 163306 (2009)
46. M.I. Alonso, M. Garriga, Thin Solid Films **455–456**, 124–131 (2004)
47. B. Dörling, A. Sánchez-Díaz, O. Arteaga, A. Veciana, M.I. Alonso, M. Campoy-Quiles, Adv. Opt. Mater. **5**, 1700276 (2017). https://doi.org/10.1002/adom.201700276
48. M. Brinkmann, J. Pol. Sci. Part B Pol. Phys. **49**, 1218–1233 (2011)
49. D.M. DeLongchamp, R.J. Kline, E.K. Lin, D.A. Fischer, L.J. Richter, L.A. Lucas, M. Heeney, I. McCulloch, J.E. Northrup, Adv. Mater. **19**, 833 (2007)
50. K. Vandewal, K. Tvingstedt, A. Gadisa, O. Ingänas, J.V. Manca, Nat. Mater. **8**, 904–909 (2009)
51. M. Campoy-Quiles, C. Müller, M. Garriga, E. Wang, O. Inganäs, M.I. Alonso, Thin Solid Films **571**, Part 3, 371–376 (2014)
52. M. Campoy-Quiles, T. Ferenczi, T. Agostinelli, P.G. Etchegoin, Y. Kim, T.D. Anthopoulos, P.N. Stavrinou, D.D.C. Bradley, J. Nelson, Nat. Mater. **7**, 158–164 (2008)
53. S. Engmann, V. Turkovic, P. Denner, H. Hoppe, G. Gobsch, J. Pol. Sci. B Pol. Phys. **50**, 1363–1373 (2012)
54. S. Engmann, V. Turkovic, G. Gobsch, H. Hoppe, Adv. Ener. Mater. **1**, 684–689 (2011)
55. A.A.Y. Guilbert, M. Schmidt, A. Bruno, J. Yao, S. King, S.M. Tuladhar, T. Kirchartz, M.I. Alonso, A.R. Goñi, N. Stingelin, S.A. Haque, M. Campoy-Quiles, J. Nelson, Adv. Funct. Mater. **24**, 6972–6980 (2014)
56. C. Müller, L.M. Andersson, O. Peña-Rodriguez, M. Garriga, O. Inganäs, M. Campoy-Quiles, Macromolecules **46**, 7325–7331 (2013)
57. M. Campoy-Quiles, M. Sims, P.G. Etchegoin, D.D.C. Bradley, Macromolecules **39**, 7673–7680 (2006)
58. A. Roigé, M. Campoy-Quiles, J.O. Ossó, M.I. Alonso, L.F. Vega, M. Garriga, Synth. Met. **161**, 2570–2574 (2012)
59. C. Müller, J. Bergqvist, K. Vandewal, K. Tvingstedt, A.S. Anselmo, R. Magnusson, M.I. Alonso, E. Moons, H. Arwin, M. Campoy-Quiles, O. Inganäs, J. Mater. Chem. **21**, 10676–10684 (2011)
60. D. Leman, M.A. Kelly, S. Ness, S. Engmann, A. Herzing, C. Snyder, H.W. Ro, R.J. Kline, D.M. DeLongchamp, L.J. Richter, Macromolecules **48**, 383–392 (2015)
61. H.W. Ro, J.M. Downing, S. Engmann, A.A. Herzing, D.M. DeLongchamp, L.J. Richter, S. Mukherjee, H. Ade, M. Abdelsamie, L.K. Jagadamma, A. Amassian, Y. Liu, H. Yan, Energy Environ. Sci. **9**, 2835–2846 (2016)
62. S. Logothetidis, Method for in-line determination of film thickness and quality during printing processes for the production of organic electronics (2014), US Patent App. 14/113,125
63. M.V. Madsen, K.O. Sylvester-Hvid, B. Dastmalchi, K. Hingerl, K. Norrman, T. Tromholt, M. Manceau, D. Angmo, F.C. Krebs, J. Phys. Chem. C **115**, 10817–10822 (2011)

# Chapter 16
# Polarons in Conjugated Polymers

**Christoph Cobet, Jacek Gasiorowski, Dominik Farka
and Philipp Stadler**

**Abstract** Conjugated polymers and polymer blends are key components in the development of organic electronics and (photo-) electrocatalysis. In particular, the possibility to produce organic but highly conducting films make these compounds very attractive. Therefore, enormous effort was put in the understanding and improvement of the electrical conductivity of polymer films. Conjugated polymers in their pristine form are mostly insulating or rarely semiconducting. The alternating single and double bonds in each $\pi$-conjugated polymer chain give rise to the formation of a band gap; the HOMO-LUMO gap. Semiconducting or conducting properties are obtained for example by optical, chemical, or electrochemical doping. The doping can be permanent as in the case of the polymer blends like PEDOT:PSS or short term. In both cases, the injected charge carriers commonly self-localize due to the strong electron-phonon interaction which yields in the formation of new quasi-particles called polarons. As a result, characteristic sub-band gap excitations emerge in optical measurements which extend from UV to the medium infrared spectral range. Optical methods in general, and spectroscopic ellipsometry in particular, are thus apparent characterization methods in scientific investigations as well as candidates to solve in-line monitoring and control issues. In the following section, we will briefly review the basic concepts of polymer "doping", the formation of polarons and the origin of sub-band gap excitations. In a survey of methods we will shortly discuss ATR-FTIR and transmission/reflection spectroscopy results. A specific attention will be drawn

C. Cobet (✉) · J. Gasiorowski (✉)
Center of Surface and Nanoanalytics (ZONA), Johannes Kepler
University Linz, Altenbergerstrasse 69, 4040 Linz, Austria
e-mail: christoph.cobet@jku.at

J. Gasiorowski
e-mail: j.gasiorowski@evgroup.com

J. Gasiorowski · D. Farka · P. Stadler
Linz Institute of Organic Solar Cells (LIOS), Johannes Kepler
University Linz, Altenbergerstrasse 69, 4040 Linz, Austria

J. Gasiorowski
EV Group E.Thallner GmbH, DI Erich Thallner Str. 1,
4782 St. Florian am Inn, Austria

© Springer International Publishing AG, part of Springer Nature 2018     355
K. Hinrichs and K.-J. Eichhorn (eds.), *Ellipsometry of Functional
Organic Surfaces and Films*, Springer Series in Surface Sciences 52,
https://doi.org/10.1007/978-3-319-75895-4_16

on the in-situ spectroelectrochemical characterization, since electrochemical doping provides control on the doping level and allows e.g. a quantification of exchanged charges. In-situ ellipsometry could be used to monitor respective changes in the polymer optoelectronic properties. We will not aim for an overview about known types of conducting polymers in general or state of the art developments in organic electronics. The focus is a discussion of the physics of UV-VIS-MIR polaronic and electronic excitations as well as state-of-the-art ellipsometric characterization.

## 16.1   Introduction

In search of cheap and easy to produce materials in semiconductor technology, organic polymers gain a lot of attention due to the enormous flexibility to tailor their physical and chemical properties. On one hand this flexibility is certainly linked to the huge variability in the chemical structure of polymers which can be easily modified and even further extended by producing blends. On the other hand, it is linked to their exceptional electronic properties which combine attributes of localized molecular electron orbitals with electronic band structure properties known from inorganic crystalline semiconductors. Thus it is relatively easy to tune their behavior from insulator to semiconductor and further to a conducting state.

### 16.1.1   π-Conjugation

Organic polymers are basically build up by $sp^2$ hybridized carbon which give rise to the formation of carbon-carbon $\sigma$-bonds (the back bone of the polymer) and double bonds containing $\pi$-electron orbitals. In isolated small molecules or in the monomer units of the polymer, these orbitals are very well described in terms of a localized electron configuration. The asymmetric $\pi$ hybridization of p-orbitals defines for example the highest occupied molecular orbital (HOMO) while the symmetric $\pi^*$ counterpart represents the lowest unoccupied molecular orbital (LUMO). The energetic gap between these two electron levels determines in optical experiments the electronic absorption edge $E_g$. With the formation of polymer chains, which may contain a (periodic) sequence of double bonds, the $\pi$-electronic orbitals could overlap. As a result the localized $\pi$-electrons delocalize in one dimension and form electronic bands along the polymer chains. Polymers with these $\pi$-electron bands are known as $\pi$-conjugated polymers. The strength of the $\pi$-conjugation depend on the specific chemical structure but is also influenced by the conformation of the monomers in the polymer chain. A typical example is the twist of aromatic rings in a polymer chain which reduce the orbital overlap [1, Chap. 20] [2]. A "re-"localization of the $\pi$-electrons in twisted configurations e.g. leads to a measurable higher HOMO-LUMO gap ($E_g$) as a consequence of degenerating electron orbitals and quantum size effect.

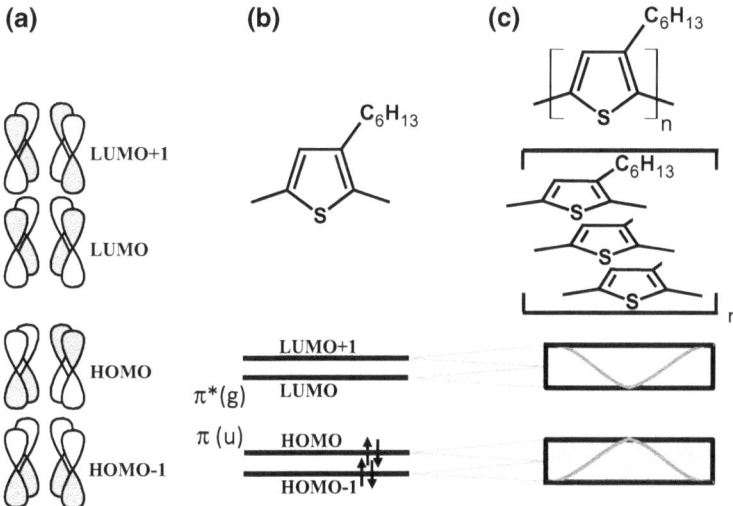

**Fig. 16.1** Schematic representation of filled and empty $\pi$-orbitals in an isolated monomer (**a**), (**b**) and the formation of electronic bands due to $\pi$-conjugation in polymer chains or by intermolecular overlapping orbitals due to a $\pi$-$\pi$-stacking (**c**). The thiophene rings of P3HT are chosen as a representative example. The $\pi$-orbitals are labeled with respect to their energy and occupation highest occupied molecular orbital (HOMO) and lowest unoccupied molecular orbital (LUMO) as well as concerning their symmetry gerade/even (g) and ungerade/odd (u)

The polymer film properties depend on the intra-molecular configuration, but also to a reasonable extent on the aggregation structure. If polymers aggregate in films, Van der Waals forces between the polymer chains could yield locally in a crystal like stacking of the polymer chains. Commonly these structures are discussed in terms of a $\pi$-$\pi$ stacking of the polymers. Independent from the question whether the underlying dipole-dipole interaction is dominated by the $\pi$-electrons, it is evident that the inter-chain overlap of $\pi$-orbitals could also induce an electron delocalization which is now perpendicular to the polymer chains. Different $\pi$-conjugation in combination with the $\pi$-$\pi$ stacking may result locally in a 0D, 1D, 2D, or 3D electronic band structure. The concept of the formation of $\pi$-electron bands is illustrated in Fig. 16.1 by means of poly(3-hexylthiophene) (P3HT) as an example polymer. The symmetry of the system refers to the anisotropy in the electronic and optical properties. The excitation across the HOMO-LUMO transition is only allowed for an electric field parallel to the P3HT polymer chains. However, the final electronic and optical properties of the film depend very much on the ordering of the polymer chains. It should be noted at this point that a typical polymer film contains a domain like structuring and film properties like the conductivity or the effective dielectric function may depend very much on the internal ordering and the domain boundaries.

## 16.1.2  Polarons and Metal-Insulator Transitions

The typical band gap of conjugated polymers is larger than 1.5 eV [3]. The pristine polymers are thus intrinsically insulating and conductivity is usually obtained by charge injection, *it est* by a reduction or oxidation of the polymer chains. The electron orbital/band structure reformation upon charging, however, turned out to be rather complex and the mechanisms assigned to the finally obtained conductivity are in parts still under discussion. A charging of the polymer by adding or removing electrons does not result immediately in an incompletely filled band which could account for a conductivity increase. Adding an excess charge in a polymer chain rather induce locally a new equilibrium geometry which differ from that in the ground state. In inorganic semiconductors a local crystal deformation due to an excess charge is well known and described in terms of polarons - a concept which was first introduced by Landau in 1933 [4]. The formation of these quasi particles, which consist of a charge and a structure deformation, is theoretically described by the electron-phonon interactions. The coupling is generated by the polarization field of the longitudinal optical (LO) phonons. In the "weak-coupling limit" Fröhlich proposed from perturbation theory a Hamiltonian where the interaction strength is linear to the polarization field and expressed by the dimensionless Fröhlich coupling constant $\alpha$ [5]. The formation of polarons and related excitations are thus enhanced in ionic crystals and polar materials. Modifications of the Fröhlich model are, however, required for systems where short-range interaction get essential (small polarons) or in case of reduced dimensionality (1D-polarons). Such systems are better represented in the Holstein, Holstein-Hubbard, or Su-Schrieffer-Heeger model. Concerning a detailed description we would refer at this point to respective review papers [6, 7] and books like those of Alexandrov et al. [8, 9].

The polaron formation in polymers is closely related to the Franck–Condon principle. The latter concept considers conformational changes in molecules due to an electronic excitation by a photon. The time scale of this optical excitation is much smaller than the motion of nuclei and the optical transition thus takes place between the ground state and an electronic plus vibrational excited state which could relax afterwards towards the new equilibrium structure. Polarons in polymers are typically discussed for a persistent charge transfer from an donor/acceptor to the polymer chains or, in the case of the photo induced doping, for screened electron-hole pairs with a long life time. The local relaxation of the polymer chains thus induces new relatively sharp electronic levels. For the discussion of polaronic optical resonances, we would use an empirical description of the new states which refers to an often cited publication of Brédas and Steet [10]. Primarily, it is assumed that adding or removing a charge in $\pi$-conjugated polymers is manly governed by a hole or an excess charge in the $\pi$-electron bands while the $\sigma$ back bone of the polymer remains intact. As depicted in Fig. 16.2 this would yield in first approximation a delocalized hole in the HOMO band or a delocalized electron in the former LUMO band, respectively. However, it was mentioned already that this situation is in most of the polymers energetically not stable. The total energy minimum of the polymer is obtained by a

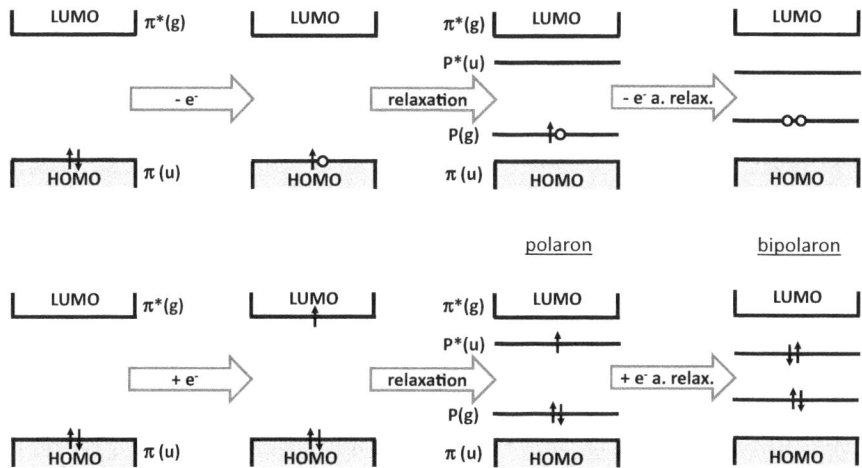

**Fig. 16.2** Illustration of one-electron energy levels in polymer chains and the formation of in-gap polaronic states (P and P*) upon p-type (upper part) and n-type (lower part) doping. The transformation from the pristine, the ionization, the relaxation in a polaronic structure, and further ionization/relaxation in a hypothetical bipolaronic structure is depicted from left to right. The (bi)polaronic in-gap states are localized electronic stats due to the self localization processes in isolated polymer chains

"self localization" of the charge due to a local relaxation of the polymer chain structure (Fig. 16.3). As a consequence of this relaxation the $\pi$-electrons will reorganize locally in a new orbital structure and it is assumed in particular that the unpaired electrons/holes form a new bonding/antibonding orbital pair which is energetically seated in the HOMO-LUMO band gap of the pristine material (Fig. 16.2). Like in the case of inorganic crystals, this new local structure is denoted as a polaron. The spin $\pm 1/2$ character is e.g. detectable in electron spin resonance (ESR) also known as paramagnetic resonance (EPR) measurements [11, Chap. 7] [12]. Furthermore it should be noted that the HOMO band is still completely occupied while the LUMO is still empty. The polaron itself has usually a relatively high effective mass [13] and thus none or only a minor increase of the conductivity is expectable.

In optical spectra the new polaron orbitals, however, should be measurable by some new dipole resonances. Primarily, one could expect a new electronic transition (P1) between the new orbitals formed by the unpaired charge which should occur energetically below the HOMO-LUMO transition (Fig. 16.4). Additional resonances (PE$_g$) may arise if filled double bonds remain in the polaronic section. Such resonances may overlap energetically with the HOMO-LUMO optical transition in the remaining pristine polymer. By symmetry arguments [14] some of the transitions between paired and unpaired $\pi$-electrons at the polaron side are not allowed or suppressed. Transitions between the HOMO and LUMO bands of the remaining pristine polymer and the polaron orbitals are also unlikely because of the small or vanishing spatial overlap. However, as depicted in Fig. 16.4 one could expect a transition (P2)

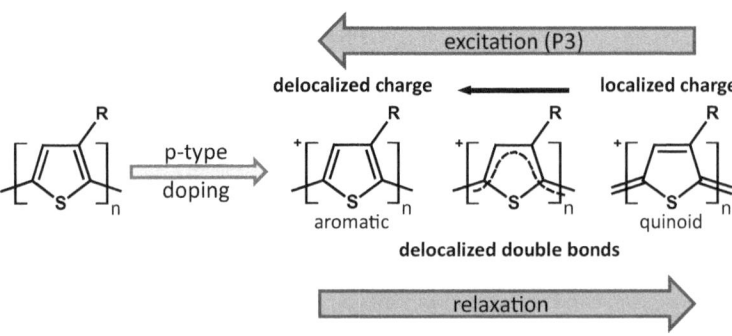

**Fig. 16.3** Schematic representation of the p-type doping induced relaxation of the P3HT polymer structure. Upon the positive charging the polymer chain change locally from aromatic to a quinoid structure. Not shown is the parallel reduction of the inter thiophene bond length and the more planar orientation of the thiophene rings in the quinoid arrangement. Relaxation and excitation processes in the doped state can be understood as (self)localization and delocalization of the charge where a strong delocalization refers to almost homogeneous aromatic structure

between the filled and half filled polaron $\pi$-orbitals [14]. The respective transition energy is typically in the order of 0.3–0.5 eV. In many publications, the authors do not distinguish between the latter transition and an excitation of the polaron charge in its local potential well which we denote in Fig. 16.4 as P3. The P3 transition is explained in the Fröhlich polaron model as a polaron scattering process where the incoming photon is absorbed by the polaron in its ground (initial) state [15]. Upon scattering at least one LO phonon is generated which leaves the polaron in an excited (final) state. The onset of this absorption is thus defined by a "long-wavelength" LO-phonon frequency. But it peaks at somewhat higher energies. The line shape and position of the maximum depend on the Fröhlich coupling constant and the polaron dimensionality [13]. Above the latter absorption peak, additionally LO-phonon side bands may arises. It should be mentioned that probably neither the one-electron model nor the Fröhlich model correctly represents the complexity of the IR excitations. Both contain approximations which are not applicable in the case of strong electron-phonon coupling in polymers.

It was pointed out already by Brédas et al. [10, 16] that the charge induced relaxation of the polymer in a new structure is a "manifestation of a strong electron-phonon coupling". As an example we use in Fig. 16.3 the relaxation of a polythiophene polymer like P3HT after positive charging. In the pristine form the thiophene rings are of an aromatic structure which relaxes in quinoid-like structure upon positive charging. This transformation is accompanied by shortening of the bonds between the thiophene rings [1, Chap. 1] [14, 17]. The before periodically tilted thiophene rings rearrange, furthermore, in a more coplanar structure. A polaron excitation as described before in the Fröhlich model is equivalent to a delocalization of the charge within the polymer chain [18] and the local quinoid-like structure fades out until the polymer chain resembles entirely in an aromatic structure.

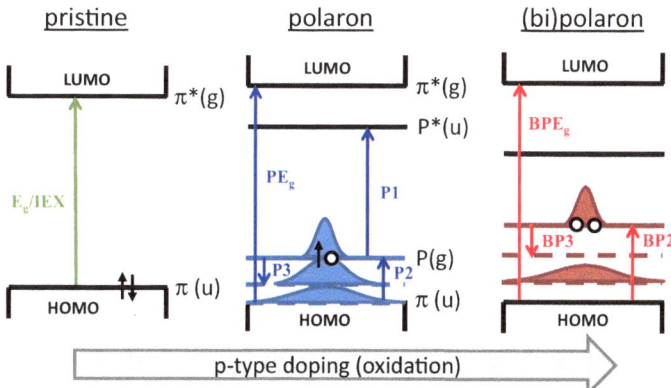

**Fig. 16.4** Schematic illustration of possible optical $\pi$-electron excitations in the undoped (pristine) material and at polaronic or bipolaronic sides in the doped state. We would note that the presented one-electron picture should be used with care. Due to the strong electron-phonon coupling the single partical band theory is not able to describe the full complexity of the proses. The definition of a P2 transition is here motivated from the one-electron view as an excitation from empty to half filled valence states. The definition of P3 excitation is motivated by the Fröhlich polaron model as an excitation of the polaron in its self generated potential well and is probability best described by Fig. 16.3. The dashed lines symbolize the polaron states with increasing charge delocalization. The total energy of the system increase with the increasing charge delocalization which is the polaron binding energy

It should be noted that the discussed new polaron states and the respective electronic transitions are sharp and defined only in the small polaron limit. In the "large polaron" limit intra- and/or inter-chain delocalization due to the $\pi$-$\pi$-stacking yields in a formation of electronic bands as it was described for $\pi$-electron system in the pristine uncharged state [14, 19, 20]. Furthermore intra- and/or inter-chain disorder contribute additionally to a broadening of absorption peaks [21].

A bit more controversially discussed is the electronic structure in case of a "further increased" charging which is frequently described by a situation where a second electron is removed from a polymer chain. If it is energetically favored that the two polarons couple, a localized distortion is obtained which contains two charges of the same type. This quasi particle is called bipolaron and has zero (integer) spin. The polaronic orbitals or bands are now completely empty (p-type doping) or completely filled (n-type doping). The P1 optical absorption between the polaronic orbitals/bands should therefore disappear while the P2/P3 absorption remains or increase at bipolaronic sides (BP2/BP3). Furthermore one could expect that the BP2 bipolaron absorption emerge at somewhat higher photon energies in comparison to the P2 polaron transition due to a further shift of the (bi)polaron in-gap levels. However, neither a vanishing of the NIR-VIS polaron absorption (P1) nor the disappearance of the unpaired spin is an unambiguous indication of a bipolaron formation. A reasonable increase of the polymer conductivity is usually obtained at relatively high charge concentrations. The existence of bipolarons alone, however, cannot explain the increase in the conductivity. The above bipolaron model contains again no half filled electronic bands but locally filled or empty orbitals. The effective mass of a bipolaron should be even larger than this of a single polaron. However, the overlap of (bi)polarons at "higher" charging levels should yield again in a formation of (bi)polaron bands [22, 23]. If the correlation strength exceeds a certain level, these bands split in filled and empty bands which are separated by a small gap (Mott transition). Pure one-dimensional systems are supposed to be always insulating. Likewise metal-insulator transitions are discussed in presents of disorder in the system. Local variation again cause a localization of polaron electron wave functions and in particular one-dimensional systems tend to be insulating (Anderson transition) [24]. Whether the formation of bipolarons or metal like bands could be observed depend very much on the particular polymer and the respective film structure. Some examples will be discussed in the following based on optical/ellipsometric results.

All the distinct electronic orbitals and bands which come along with the formation of polarons, bipolarons, or metallic states produce, as discussed, a number of new dipole allowed optical transitions. These changes of the optical properties in the visible spectral range are that strong that many conjugated polymers were also intensively studied because of their electrochromic properties [25]. The application of conjugated polymers e.g. in electrochromic displays promised higher contrast ratios and faster response times. Most of the conjugated polymers switch between a transmissive and a colored state upon doping. But also switching between different colors is possible [1, Chap. 21].

The strong optical effect of doping is used from the beginning as a probe to study fundamental aspects of the polaron formation. By far most of the optical investigations apply reflection or transmission methods. In the IR spectral range enhanced thin film sensitivity is commonly obtained within attenuated total reflection experiments (ATR). A methodical review of these methods is provide e.g. in Chap. 21 (*Characterization of Thin Organic Films with Surface-Sensitive FTIR Spectroscopy*). But, in particular Fourier-transform infrared (FTIR) results are usually interpreted in a pure chemical manner by means of differential absorption properties of the investigated film without making use of the polarization dependency. Ellipsometric experiments are in comparison still underrepresented although the method could provide quantitative results concerning the complex dielectric function (complex refractive index) and physical parameters like the film thickness. One reason is certainly the relative complexity of the necessary data evaluation (Chap. 1: *Ellipsometry: A Survey of Concepts*). In the following examples, we will discuss the optical response of polarons primarily based on ellipsometric results. But additionally we add a critical review of some basic relationships among the different methods, with the goal to provide a conceptional comparability and to draw the attention on fundamental deficiencies among the optical spectra published in literature.

## 16.1.3   Charge Injection

Before turning to the optical characterization of polarons, we would spend a few words how the charge injection is achieved in the polymers. In analogy to inorganic semiconductor technology, the charge injection in polymers is commonly referred to the synonym "doping". In common, it requires an electron/hole source which could be a chemical element. But in contrast to inorganic semiconductor crystals, this donar/acceptor is typically not inserted in the polymer chain. The finding that a persistent doping is nevertheless feasible was a key step in the discovery of conduction polymers [26, 27] and acknowledged by the Nobel Prize in Chemistry in 2000. In literature one can find a huge number of different "doping" recipes. But in general they can be divided into four principal groups [11]:

(1) **Chemical p- or n-type doping (oxidation or reduction)**: The chemical doping was utilized in the first discovery of polymer conductivity in polyacetylene films. The charge transfer is achieved by exposing and mixing the polymer with oxidizing or reducing reagents which can be considered as acceptor or donor impurities in inorganic semiconductors. In order to obtain p-type doping the polymer could be e.g. exposed to halides like $I_2$ or $Br_2$ which form finally charge transfer complex between the polymer and the $I^-$ or $Br^-$ anions. An closely related method of p-type doping is called "protonation" where polymers containing a nitrogen atom are exposed to acid vapor (e.g. HCl). The strong dipole moment of the acid results in the shift of the lone electron pair at the nitrogen atom generating positive charge. Such doping is applied for example in polyaniline or polyazomethines [28]. A n-type doping can be obtained by exposing polymer thin films to Li [29] or Na atoms which requires a (ulta)high-vacuum chamber.

(2) **Electrochemical p- or n-type doping**: A drawback of the chemical doping is in general the difficulty to control the doping level as well as the problem of inhomogeneity. Electrochemical doping in contrast promise a very good control of the charging level and lateral homogeneous results. The polymer is deposited on a conducting electrode (the working) electrode and attached in an electrochemical cell containing an electrolyte. Common electrolytes used in polymer electrochemistry consist of organic solvents like propylene carbonate, acetone, or acetonitrile and a conducting salts like tetrabutylammonium hexafluorophosphate (TBA-PF6, $C_{16}H_{36}F_6NP$), nitrobenzol, or lithium tetrafluoroborate $LiBF_4$ due to their broad electrochemical window. During the electrochemical doping the working electrode supplies or collects the redox electrons while anions or cations defuse from the electrolyte in the polymer film in order to ensure the charge neutrality. The doping level is controlled by the potential between the polymer and the electrolyte i.e. the counter electrode. Each potential is assigned to a specific doping level defined by the electrochemical equilibrium.

(3) **Field induced "doping" - Charge injection at metal-polymer interfaces**: The concept of a field induced charging refers to the operation principle of polymer based field effect transistors. At interfaces charges can be added or removed from

the $\pi$-orbitals/bands of the polymer due to a contact potential difference which is either externally applied or is a result of a different work function like in a Schottky contact. The main difference in comparison to chemical or electrochemical doping is the absence of counter ions introduced in the polymer. Thus the field induced doping is not persistent if the externally applied field is switched off.

**(4) Photo-induced "doping"**: Photo-induced doping can be obtained only in polymer blends. In such, the film is illuminated with light of a photon energy above the HOMO-LUMO gap. The induced electron-hole pair can be of intra- as well as of inter-chain character. Persistent free charges and the formation of polarons is obtained only if the electron-hole pair separates like in a p-n junction. The life time depend on several factors as the exciton binding energy and the band offset within the polymer blend structure. In specific cases, air can play the role of a n-type semiconductor leading to photooxidation process.

## 16.2   P3HT

Among the (semi-)conducting polymers, poly(3-hexylthiophene)(P3HT) is probably one of the most frequently investigated model systems with applications in organic optoelectronics like solar cells, light-emitting diodes, or transistors [30]. P3HT is a p-type (hole) conducting polymer. In organic photovoltaic devices it is used for example in the active layer as an electron donor material. P3HT is relatively ease to synthesize and shows pronounced optoelectronic properties which make it also an ideal model system in order to study the fundamental properties of doping in polymers. Details concerning the application of P3HT in organic photovoltaic (OPV) and organic light emitting diodes (OLEDs) as well as a discussion of the pristine material optical properties are also presented in Chaps. 12 (*Polymer Blends and Composites*) and 15 (*Conjugated polymers: Relationship between morphology and optical properties*).

### *16.2.1   Dielectric Function*

Figure 16.5 shows the dielectric function (DF) of regioregular (rr-)P3HT as it is known for the pristine undoped state. The presented spectrum was measured by transmission ellipsometry on a 56 nm film in electrolyte at potentials where the P3HT is completely undoped. Unintentional doping by illumination or oxidation from air is thus avoided. The film was spin-casted on glass covered with indium-tin oxide (ITO) [31]. Details about the measurement will be described in the section about the electrochemical doping.

The example P3HT film shows a very strong out-of-plane anisotropy. The oscillator strength of the contributing absorption resonances is strongly reduced in the

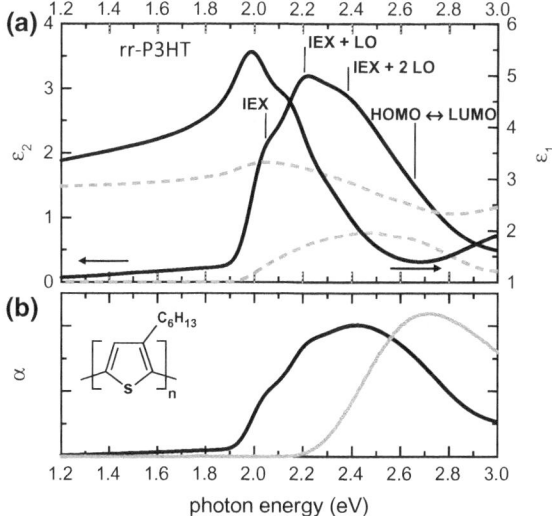

**Fig. 16.5  a** Anisotropic dielectric function of un-doped P3HT measured by transmission ellipsometry. The black line corresponds to the in-plane while the dashed gray line corresponds to the out-of-plane component. The film under investigation was spin casted on a glass substrate covered with ITO and is measured in a 0.1 M Bu$_4$NPF$_6$ electrolyte solution in anhydrous acetonitrile at a potential of $-200$ mV against a Ag/AgCl quasi reference electrode. **b** Absorption spectrum of the un-doped P3HT film calculated from the ellipsometric results (black line) in comparison to an absorption spectrum measured in transmission through a P3HT solution in chlorobenzene (gray line) [31]

DF perpendicular to the surface in the presented spectral range. Such a behavior is typical for P3HT [32] and many other conjugated polymer films [33, 34] where the interaction with the substrate provokes a predominate alignment of the polymer chains parallel to the interface. A detailed ellipsometric study of the anisotropy of P3HT was published in [32]. In this work the authors discuss also the influence of the substrate by comparing spin-coated and drop-cast P3HT films on glass and Si. The orientation selectivity of excitations is explainable by the common nature of the contributing absorption structures in the considered spectral range. All the near band gap optical transitions in P3HT are dipole $\pi$-$\pi^*$ transitions. Their excitation requires an electric field component parallel to the P3HT conjugated polymer chain [35]. An out-of-plane excitation of the $\pi$-$\pi^*$ transitions in the visible and ultra violet spectral range (VIS-UV) is thus dipole forbidden if the polymer chains are all parallel aligned to the surface.

In this connection it should be noted that the measured anisotropies are obtained by assuming a homogeneous film. Such a commonly used assumption actually needs to be confirmed. However, since a strong out-of-plane anisotropy is observed also by "averaging" over the whole film, on can assume that the average volume fraction of the polymer strands with a perpendicular orientation to the surface is generally small. In other words the substrate induced polymer orientation of the polymers retains in the entire film. This behavior can be explained by a relatively strong inter-molecular interaction and is discussed in polymers by means of the $\pi$-$\pi$ stacking [31–34].

The absorption onset of P3HT in the pristine "undoped" form is induced by a distinct resonance at 2.04 eV which is assigned to a delocalized inter-chain singlet exciton (IEX) [36]. Two more resonances follow energetically above the IEX at 2.20 and 2.37 eV almost equidistant in energy. The latter transitions are vibronic sidebands of the IEX with an additional excitation of one or two LO-phonons (Fig. 16.5a). The relatively broad absorption structure between 2.6 and 2.7 eV is finally attributed to the screened intra-chain HOMO-LUMO transition ($E_g$) and thus coincide to $\pi$-$\pi^*$ transition in isolated P3HT polymer chains. The latter association gets evident if one compares the absorption spectrum of the film (calculated from the in-plane DF) with the absorption spectrum measured e.g. in transmission for diluted P3HT. Such a comparison is shown in Fig. 16.5b. The diluted and therefore separated P3HT polymer chains disclose only one relatively broad absorption peak around 2.7 eV which is the just defined intra-chain HOMO-LUMO transition. The distinct film resonances at lower energies are absent. A small red shift of the film HOMO-LUMO transition is explainable by the polarizability of the surrounding P3HT molecules i.e. coupling and screening effects.

The inter-chain character of the IEX was demonstrated based on its dependency on the degree of the $\pi$-$\pi$ stacking and the crystallinity of the film (Fig. 16.1). Generally, it could be shown that the IEX amplitude vanish for totally disordered films or P3HT-blend structures where the individual P3HT chains have no direct contact [37, 38]. Figure 12.4 in Chap. 12 shows for example the effect of thermal annealing on the IEX structures for spin coated P3HT films. Worth mentioning in this connection is the relative large oscillator strength of the HOMO-LUMO transition in the out-of-plane film DF (dashed line in Fig. 16.5a). If we consider that the $\pi$-$\pi^*$ transitions require an electric field component parallel to the polymer chains, we can conclude that a non vanishing absorption in the out-of-plane DF yields from polymers which are upright oriented to the substrate plane. Such a molecular orientation is in particular expected in disordered areas where the IEX contributions are small. A detailed analysis of the optical anisotropy of $\pi$-$\pi^*$ transitions could thus provide already further insights in the inner film structure.

## 16.2.2 Iodine Doping

As mentioned before, chemical doping by exposing conjugated polymers to oxidizing or reducing agents was used already in the first attempts to modify the electronic properties of polymers. Iodine is one of the strongest oxidizing elements and is thus often used in order to obtain p-type doping i.e. a positive charging of the polymer. In general all p-type doping methods should produce very similar changes in the optical absorption spectra. According to the polaron picture described in the introduction, some new transitions in energy below the absorption edge are expected [39, 40].

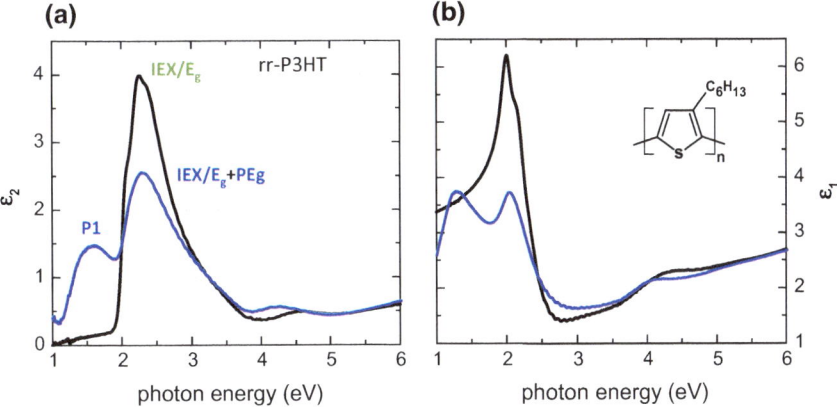

**Fig. 16.6** In-plane dielectric function of P3HT (structure in the inset) in the nominally undoped (black line) and iodine saturated doped state (blue line). The imaginary part of dielectric function is plotted in the left panel (**a**) and the real part in the right panel (**b**) [41]

The iodine doping of P3HT is therefore used as a prototype example in order to demonstrate the doping induced changes in the optical response of polymers.

We use again a rr-P3HT film of 35 nm which was spin casted on an ITO/glass substrate as an example. The particular film was annealed at 150 °C for 3 min in a nitrogen atmosphere. The doping was performed by exposing the sample to iodine vapor for 10 min. A relatively long doping time is used in order to obtain a saturation which allows a subsequent ellipsometric measurement under relatively stable conditions [41]. The DF of the iodine doped P3HT is shown in Fig. 16.6 (blue line) in comparison to the as before measured DF of the pristine film (black line). Within the used measurement geometry the calculated DF represents the in-plane DF component.

After doping the film optical properties disclose two eye-catching modifications. The $\pi$-$\pi^*$ transitions measured before between 2 and 3.5 eV appear now very much reduced in amplitude. At the same time a new peak shows up at 1.6 eV. Both changes resemble the before discussed formation of polaronic states in the band gap of the pristine material (Fig. 16.2). The new absorption feature at 1.6 eV thus refers to the P1 transition as shown in Fig. 16.4. The unpaired spin in the polaron state is proven e.g. in EPR measurements [42].

A closer inspection of the DF changes upon iodine doping discloses further that the remaining, as well as the new appearing absorption features broaden in comparison to the pristine material. This broadening turned out to be not specific for the iodine doping but is rather a result of a more principle property. Among the new polaron transition feature we could expect again a splitting due to the formation of inter chain excitons and respective LO-phonon replica as discussed in the pristine material. The self localization of the positive charge within a polymer chain was already discussed in terms of a local relaxation of the polymer structure. In the case of P3HT, the

$\pi$ electron system change from an aromatic configuration in a quinoid-like structure upon doping (Fig. 16.3) [10, 40]. The relaxation is expected to extend at least over three monomers (Figs. 16.3 and 16.12) and includes a planarization of the monomer units [16, 19, 43]. The polaron electronic states thus broaden due to the spreading of the polarons with in the chain (large polaron approximation) and the interaction of neighboring $\pi$-$\pi$ stacked polymer chains. The latter yields in an additional splitting of polaron states and the formation of bands (2-3D polarons) [14, 22]. Both effects; the delocalization of polaronic states along as well as perpendicular to the polymer chains; were already discussed for the undoped $\pi$ electron orbitals and visualized in Fig. 16.1. Different is now that undoped and doped areas as well as various degrees of localization of the polarons could coexist in the film which leads to an increased complexity and general broadening of the contributing absorption structures.

The quantitative determination of the film DF by means of ellipsometric measurements allows additionally a discussion of the different transition oscillator strength. In the case of the iodine doped P3HT it is already visible by eye that the decrease of the transition strength between 2 and 3.5 eV is not compensated by the new structure around 1.6 eV. At the same time the DF above 3.5 eV does not change on the doping at all. By sum rule arguments, the first moment of the imaginary part of DF is proportional to the density of the contributing ($\pi$) electrons [45, 46]. However, the p-type charging alone could not explain the overall reduction of the transition strength in the spectral range between 1 and 3.5 eV. The missing oscillator strength is thus most likely shifted to transitions in the IR. According to the drawn polaron theory at least one additional electronic transition is expected in the IR - the P2 transition depicted in Fig. 16.4.

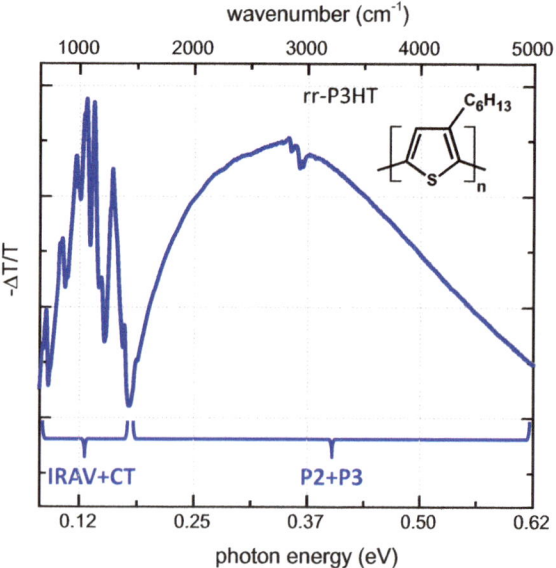

**Fig. 16.7** Iodine doping induced changes in the P3HT optical response measured by ATR-FTIR after 12 s exposure [44]

Typically ATR-FTIR spectroscopy is used to analyze the polaronic absorption of thin films in the IR spectral range. Figure 16.7 shows a differential ATR spectrum of the iodine doped P3HT film [44]. Accordingly, the absorption increases in the whole IR spectral range. But the new absorption emerges clearly in two distinct spectral regions. A broad doping induced absorption band appears above 1800 cm$^{-1}$ which contain various contributions and the line shape differs from the simple Lorentz oscillator model [8, 9]. However, in principle the band is attributed to the $\pi$-electron in-gap transition P2 and the polaronic excitation/relaxation P3 (Fig. 16.4). In the Fröhlich polaron model, the P3 transition is described as a LO-phonon excitation. The onset of the broad P3 absorption band at 1800 cm$^{-1}$ is in this model defined by the long wavelength LO-phonon frequency. In terms of the discussed doping induced structure relaxation in the P3HT polymer chain, the P3 excitation could be interpreted as a delocalization of the polaron as presented in Fig. 16.3.

Below 1800 cm$^{-1}$ new doping induced infrared active vibrations (IRAV) modes appear [47, 48]. Before just Raman active modes are activated by the polaron due to changes in the symmetry relative to the pristine material. The high intensity of IRAV mode absorption is again a result of the electron-phonon coupling. The IRAV-modes may are additionally superimposed by a broad absorption feature which is attributed to a charge transfer transition (CT) between the polymer chains [21].

## 16.3  Electrochemical Doping of P3HT

An obvious problem of the iodine doping and the chemical doping in general is the lack of control concerning the amount of transfered charges or in other words the oxidation/reduction stage of the of the polymers. Closely related to the control problem is also the question of a quantification of the charging. Like in the described iodine doping one can probably find saturation at a certain density but the instability on one hand and the re-crystallization of iodine in the film on the other make it difficult to obtain reproducible results. The electrochemical doping of conjugated polymers in contrast provides full control on the induced charge density by means of the applied electrochemical potential. By cyclic voltammetry (CV) or in chronoamperometry it is furthermore possible to quantify the charge transfer. Electrochemical impedance spectroscopy could provide additionally a determination of the conductivity upon doping variations [49].

Spectroelectrochemical investigations of polymers are therefore already widely used to study doping induced fundamental aspects like a structural relaxation, an insulator to metal transition, changes of the electronic structure and the formation quasi particles like polarons [50–53]. The applied in-situ spectroscopic methods are almost exclusively of absorption and reflection type. An analytical disadvantage of the electrochemical approach could be seen in the fact that the film under investigation is covered with an electrolyte which is usually not transparent over wide areas in the IR spectral range. (FTIR)-ATR spectroscopy has in this connection the advantage that the light transmission through the electrolyte is avoided.

## 16.3.1   (Spectro-)Electrochemistry

For spectroelectrochemical investigations on rr-P3HT, an electrochemical cell like the quartz cuvette as shown in Fig. 16.10b could be used. In such a cell the polymer film is deposited on a conducting substrate (ITO/glass) and is used as the working electrode (WE). The electrochemical potential $E$ of the WE is controlled by a potentiostat with respect to a reference electrode. In the given example a Ag/AgCl wire is used as a quasi reference electrode (QRE) [54]. The potential dependent electrochemical charge transfer due to an oxidation or reduction of the polymer film is measured as a current between the WE and a counter electrode (CE) which is typically made of platinum. A common electrolyte solution is composed of 0.1 M/L $Bu_4NPF_6$ ($Bu_4NPF_6 \rightleftharpoons Bu_4NP^+ + PF_6^-$) in anhydrous acetonitrile.

The electrochemical redox properties of spin casted P3HT films are presented in Fig. 16.8a based on a CV recorded with 10 mV/s [31, 42]. In anodic scan direction, one can identify three major oxidation peaks which contain a couple of sub-structures. In the cathodic scan the P3HT film turns back in the undoped state with respective negative reduction current. The processes are reversible and the cycle can be repeated a couple of times without degradation of the film. Each peak in the CV can be associated with a certain oxidation/doping stage of the film. A precise assignment is difficult. But already the fact that the P3HT film possess more than one or two oxidation levels clearly supports the assumption of inter-chain interactions due to the $\pi$-$\pi$ stacking. The increase of complexity due to the formation of inter-chain structures could increase the number of distinguishable oxidation/doping levels. Beside

**Fig. 16.8** Cyclic voltammogram of P3HT in a $Bu_4NPF_6$-acetonitrile solution recorded with a scan rate of 10 mV/s (**a**) In panel **b** shows the charge density in the film accumulating in the anodic sweep from $-0.2$ to 1.1 V [31]

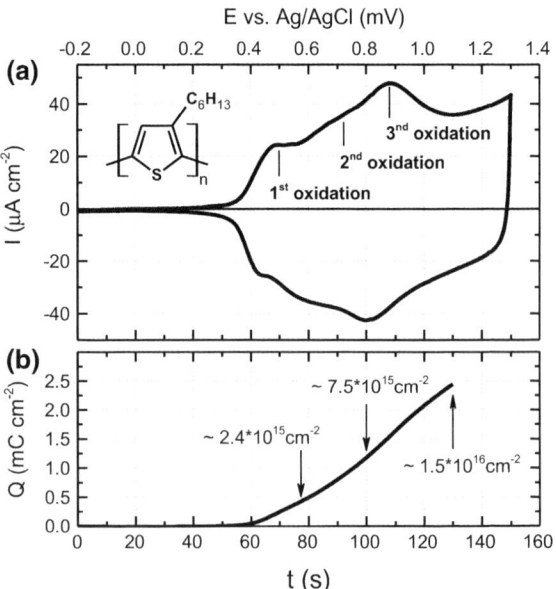

of an identification of oxidation stages, it could be instructive already, to investigate the charge density in the film at different doping levels. The time integral (the first moment) of the anodic current, which is measured in CV, provides this information. The measured current tells us how many electrons are transferred from polymer film to the ITO and leave a positively charge hole in the P3HT polymer chains. The charge neutrality is maintained with the intercalation of an equal number of $PF_6^-$ anions ($A^-$ in Fig. 16.12) in the film. The successive increase of positive charges in the P3HT film upon the electrochemical oxidation is plotted in Fig. 16.8b. With the known film thickness this calculation disclose that the "donor" concentration increase by $3.4 \times 10^{20}$ $cm^{-3}$ upon the first oxidation step. The second and third oxidation steps yield in an overall anodic charge transfer of 1.2 and 2.4 $mC/cm^2$, respectively, which corresponds to a total "donor" concentration in the film of $1 \times 10^{21}$ and finally of about $2 \times 10^{21}$ $cm^{-3}$. An upper limit approximated from the monomer density in the film [31, 55] yields a number of $4$–$5 \times 10^{20}$ thiophene rings per cubic cm. The second oxidation step thus corresponds to a positive charging of about 20% of the thiophene monomers and with the third oxidation this value increase to almost 50%.

In-situ UV-VIS transmission and ATR-FTIR experiments show that the increasing electrochemical charging of the P3HT film leads to a monotonously decrease of all $\pi$-$\pi^*$ transition which have been identified in the pristine polymer (Fig. 16.5) while the NIR and IR absorption structures (P1, P2/P3, IRAV, and CT) monotonously increase at the same time [42]. That the p-type doping induced modifications in the electronic structure are not explainable just by a continuous transformation from the pristine to a oxidized state becomes obvious in the spectral range between 1.2 and 2 eV. The P1 polaron transition structure, which was seen already in the iodine doped films, increase up to 800 mV, but decrease again at higher potentials (Fig. 16.9b). Furthermore, one can observe a splitting of the broad P2/P3 IR absorption in the ATR-FTIR spectra (red arrows Fig. 16.9a). A precise interpretation based on the ATR-results is in fact difficult and the determination of the intrinsic material dielectric properties could facilitate a better identification of transition resonances or their spectral positioning.

## 16.3.2   In-situ Spectroelectrochemical Ellipsometry

The unquestioned advantage of common transmission experiments is on one hand the simplicity of the method. On the other hand it is also of big advantage if the film properties can be investigated in standard glass cuvettes which are typically used in electrochemical experiments. In this connection it seems reasonable to combine the advantages of ellipsometric measurements and the ability to investigate the films in a standard glass cuvette by means of transmission ellipsometric measurements [31]. In such an experiment the light beam is straight transmitted through a glass cuvette, the electrolyte, and the sample which consists of a transparent substrate (e.g. glass), a transparent conducting film (e.g. ITO), and the polymer film under investigation. The sample surface must be tilted against the light beam in order to

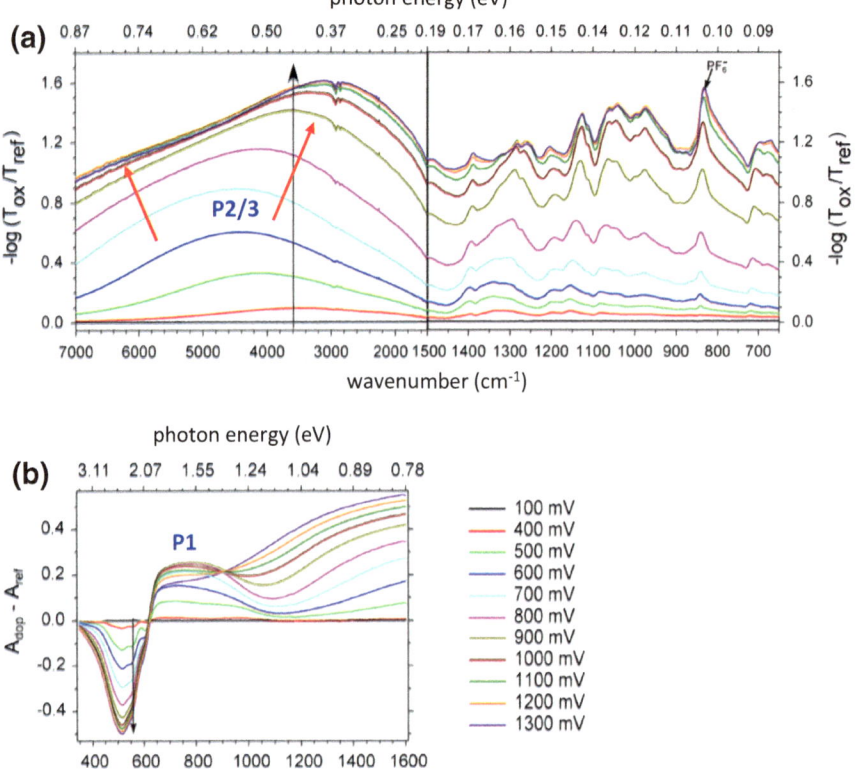

**Fig. 16.9** Electrochemical doping induced changes in the P3HT optical response measured **a** by in-situ ATR-FTIR and **b** by UV-VIS transmission spectroscopy at different potentials between 100 and 1300 mV. The spectra measured at 0 mV are used in this examples as a reference. Reprinted from [42] with permission from WILEY-VCH

allow an ellipsometric analysis by means of the difference in light transmission for parallel and perpendicular polarization components respective the plane of incidence. Polarization effects and intensity losses at the outer cuvette surfaces can be suppressed by two index matching prisms which are attached to the cuvette and thus ensure perpendicular light incidence (lower part Fig. 16.10).

Some attention has to be drawn on the extraction of the film optical properties because several interfaces insight the cuvette contribute to the measure polarization change which is represented by the ellipsometric angles $\Psi$ and $\Delta$. The upper part of Fig. 16.10 shows a schematic example for the light propagation through the internal interfaces. A physical/mathematical description of the light transmission through such an ensemble can be obtained with the known theories of light propagation in media and the reflection/transmission of plan waves at planar interfaces [56–58]. Of major importance in such a structure is the coherence length of the light. Commonly

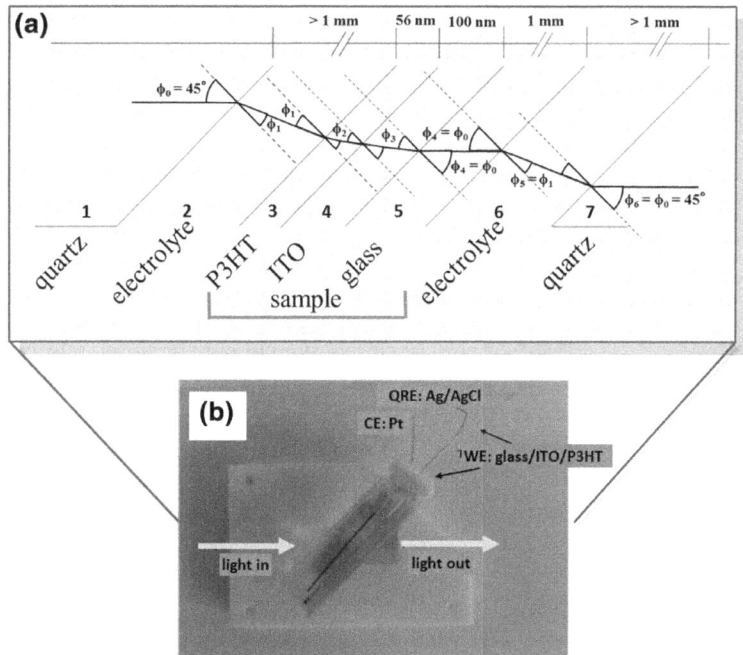

**Fig. 16.10** Schematic diagram of the light propagation through the glass cuvette including a polymer film on an ITO/glass substrate (**a**). Layers represented in red are thinner than the coherence length of conventional light sources and interference has to be taken into account in an ellipsometric layer modeling procedure. In the lower part **b** we show a real experiment image of a filled glass cuvette with connectors for the quasi reference electrode (QRE), the counter electrode (CE) and the ITO/glass working electrode (WE). Reproduced from: [31]

used light sources have a coherence length of a few micrometers. Accordingly, the distance of interfaces like those given by the two sides of the glass substrate is bigger than the coherence length. In our example only the polymer and the ITO films are thinner. Therefore just in these two layers forward and backwards traveling waves have a defined phase relation and interference effects have to be taken into account.

In the case of isotropic media the Abeles matrix method can be applied to solve the problem and to calculate finally the film optical properties [57, 59]. In this formalism the total electric field at a planar boundary between a layer $i$ and $j$ is composed of left and right traveling waves. The respective complex fields are arranged in columnar vectors of the form $E'_i = (E'_{i,l}, E'_{i,r})^T$ and $E_j = (E_{j,l}, E_{j,r})^T$. With the Fresnel transmission and reflection coefficients $t_{ij}(\varepsilon_i, \varepsilon_j, \phi)$ and $r_{ij}(\varepsilon_i, \varepsilon_j, \phi)$ an interface transmission matrix of the form:

$$\hat{H}_{ij} = \frac{1}{t_{ij}} \begin{pmatrix} 1 & r_{ij} \\ r_{ij} & 1 \end{pmatrix}, \tag{16.1}$$

can be defined which connects the fields on both sides of the boundary ($E'_i = H_{ij} E_j$). For a phase correct matching of the fields from the right side of a layer to the left side of the same layer, furthermore, a propagation matrix is defined according to:

$$\hat{L}_i = \begin{pmatrix} e^{-i\beta_i} & 0 \\ 0 & e^{-i\beta_i} \end{pmatrix}, \tag{16.2}$$

where

$$\beta = \frac{2\pi}{\lambda_0} n_i d_i \cos(\phi_i). \tag{16.3}$$

In case of layers, which are much thicker than the coherence length of the light, it is sufficient to use the identity matrix in order to match the electric fields directly to the next boundary. Interference effects are thus exclude. The electric fields on both sides of our glass cuvette with an internal structure as depicted in Fig. 16.10, can be thus calculated with the stack matrix $\hat{S}$ defined by:

$$\begin{pmatrix} S_{11} & S_{12} \\ S_{21} & S_{22} \end{pmatrix} = \hat{H}_{67} \hat{H}_{56} \hat{H}_{45} \hat{L}_4 \hat{H}_{34} \hat{L}_3 \hat{H}_{23} \hat{H}_{12}. \tag{16.4}$$

In transmission experiments with light propagating from the left to the right through the cuvette, we just need the $S_{22}$ component of the stack matrix which defines the overall transmission coefficient

$$t = 1/S_{22}. \tag{16.5}$$

For isotropic media, s- and p-polarized waves are eigenmodes and can be calculated separately with the respective Fresnel equations ((1.13), Chap. 1). The ellipsometric angles $\Psi$ and $\Delta$ can be finally calculated by the known relation:

$$\frac{t_p}{t_s} = \frac{S_{22}^s}{S_{22}^p} = \tan\psi e^{i\Delta}. \tag{16.6}$$

The film optical properties like the dielectric function $\varepsilon$ or the thickness $d$ of the polymer film are determined in a comparison with the experimental results by means of a respective fit algorithm.

The Abeles method is also applicable for anisotropic media, as long as the optical axes are all parallel or perpendicular to the plane of incidence and the surface normal of the interfaces. In such a case the s- and p-polarized waves are still eigenmodes and can be calculated separately. However, as an alternative one can use the Berreman transfer matrix formalism which is implemented nowadays in many ellipsometric software packages. In this formalism forward and backward traveling waves are calculated with differential $4 \times 4$ transfer matrices and an explicit definition of reflection and transmission coefficients at boundaries is avoided. With this approach it is possible to calculate the wave propagation in arbitrary anisotropic media [57, 58, 60, 61]. Interferences in layers thicker than the coherence length of the light can be

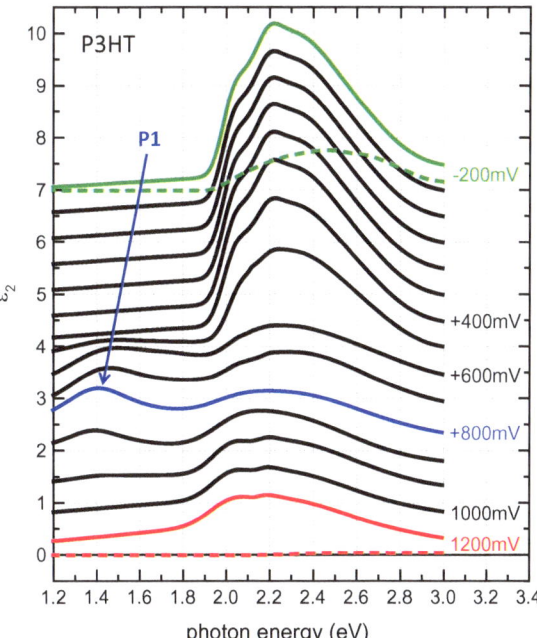

**Fig. 16.11** Imaginary part of the in-plane dielectric function of P3HT at different electrochemical potentials which where measured against a Ag AgCl QRE. The electrochemical doping is connected in this example by an intercalation of $PF_6^-$ anions. An offset of 0.5 is used between the spectra for clarification of the trend and the two dashed line represent the out-of-plane dielectric function at $-200$ and 1200 mV. [31]

suppressed with a simple workaround. By averaging over a number of arbitrarily chosen layer thicknesses (e.g. $\pm10\%$ of the real layer thickness), the interference structures in thicker layers smear out and one can calculated the polarization dependent transmission in terms of $\Psi$ and $\Delta$ also for structures as depicted in Fig. 16.10.

Figure 16.11 shows a serious of dielectric function spectra for electrochemical potentials between $-200$ and $+1200$ mV. The plotted imaginary part of the in-plane DF are obtained by a parametric fit of the measured $\Psi$ and $\Delta$ values. In this fit it was assumed that the dielectric function consist of a number of Gauss-Lorentzian oscillators which differ in the in-plane and out-of-plane DF only concerning their oscillator strength. The parallel received out-of-plane DF's are exemplary shown for the extreme potentials. The undoped film optical properties at $-200$ mV were already used in Sect. 16.2.1 and Fig. 16.5 to discuss the undoped P3HT properties.

In agreement with CV results the DF do not change up to a potential of about 300 mV. With the first and second oxidation step between 350 and 800 mV the oscillator strength of the $\pi$-$\pi^*$ transitions of the pristine material (the IEX, the IEX+LO, and the screened HOMO-LUMO band-to-band transition) continuously decrease by a factor of two. At the same time the polaron related P1 transition appears. The latter P1 resonance reaches a maximal amplitude at 800 mV. At this potential the peak maximum emerge at 1.4 eV. But noticeable is also a significant red shift from the first appearance with increasing potential. Such a behavior is expected in case of an increasing inter- and intra-chain overlap of the polaronic chain sections (Figs. 16.1 and 16.12). The polaronic character of the new absorption structure is additionally confirmed by (EPR) measurements which show an increasing signal and thus prove the unpaired electron spin in localized polaronic states [42]. From CV it is know that about 20% of the thiophene monomers are charge at 800 mV. The high polaron density explains on one hand the strength of the polaron absorption. But it is also apparent that a higher doping could not further increase the number of polarons. If one considers that a polaron in P3HT extend over three or more thiophene units, a further generation of localized polarons is simply not possible.

In connection with the iodine doping of P3HT, we have discussed already a general broadening of the electronic resonances after the iodine exposure. This behavior is clearly traceable in the in-situ ellipsometric spectra. The IEX and the IEX+LO transitions as well as screened HOMO-LUMO band-to-band absorption first decrease in amplitude but broaden in particular upon the second oxidation step. A possible reason for the broadening could be seen in the coexistence of polaronic and pristine sections in the P3HT polymer chains in combination with the already discussed inter and intra-chain overlapping of (bi)polarons orbitals. Regarding transition energies and peak amplitudes/width the electrochemical induced polaronic transition spectrum is truly very similar to those obtained for iodine doped P3HT films (Fig. 16.6) [19, 41, 62]. Best comparable with the saturated iodine doping is probably the electrochemical oxidation at 600 or 700 mV. Both doping types include an incorporation of counter anions, i.e. the $I^-$ and $PF_6^-$, respectively, which may induce additional broadening effects.

Above 800 mV the polaron related P1 transition disappears again although the p-type doping is still increasing. It finally vanishes almost completely at potentials above the third oxidation peak. One might think that the vanishing of the P1 polaron transition, which is accompanied by a disappearing EPR signal, is attributed to the formation of bipolarons as depicted in Fig. 16.4. Such an interpretation would be consistent with the ATR-FTIR results shown in Fig. 16.9a. Aside from the general increase of the IR absorption one can observe a blue and a red shifting feature. The blue shift is in fact expectable for P2 (BP2) transition if one assumes a "further relaxation" due to the pairing of positive charges. On the other hand we have mentioned already the very high doping level above 800 mV. About every second thiophene ring is finally positively charge at 1200 mV and (bi)polarons may strongly overlap. Highly interesting in this connection is the recovery of the IEX features between 2.0 and 2.4 eV in the ellipsometric spectra. The recovery of inter-chain excitonic "fine" structures and the very high doping concentration rather suggest that the P3HT chains entirely relax in a new structure like pictured in the bottom of Fig. 16.12. A formation of bipolarons as a new quasi particle is triggered in a strict definition by an energy gain due to the polymer relaxation which exceeds the Coulomb repulsion between two positive charges. The total Gibbs free energy minimization includes in electrochemical experiments the freely tunable electric potential which could compensate also coulomb repulsion forces. The chemical doping with the strongly oxidizing iodine saturates clearly in a polaron structure and a bipolaron formation seems to be energetically excluded.

Hence, ellipsometric spectra at high electrochemical potentials are explainable in the view of strongly overlapping polarons. They form polaronic bands which

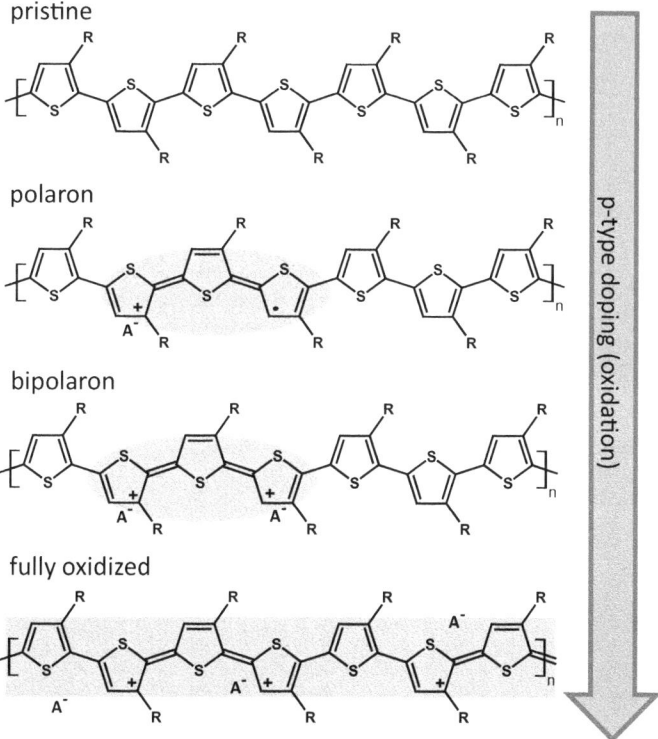

**Fig. 16.12** Illustration of the (bi)polaron bond configuration in P3HT polymer chains at different p-type doping levels

energetically overlap with the HOMO and LUMO bands of the pristine material. This would yield in a band structure where the Fermi level is finally shifted in the HOMO levels of the pristine material (compare Fig. 16.4). Notice, such a model is also totally consistent with EPR measurements.

In electrochemical impedance measurements, it was demonstrated, furthermore, that a first significant conductivity increases is obtained in connection with the third oxidation step [49]. In this potential range the conductivity increases by almost three orders of magnitude up to $\approx 8$ Scm$^{-1}$. Such an increase in the hole-mobility is not consistent with the formation of self localized bipolarons but could be easily explainable if intra- as well as inter-chain polaronic bands merge with the HOMO bands and thus generate a metallic regime due to an incompletely filled valence band [63]. Especially the inter-chain transport is a key feature in the attempt to obtain high film conductivity [64]. It should be noticed that the periodic arrangement of quinoid and aromatic thiophene rings in the polymer chain (Fig. 16.12) could be interpreted as the result of a Peierls's instability [65–68]. At room temperature, the $\pi$-electronic orbitals are equally distributed over the thiophene rings (comp. Fig. 16.3) which correspond to a half filled metallic band. But at low temperature a Peierls's transition is likely in

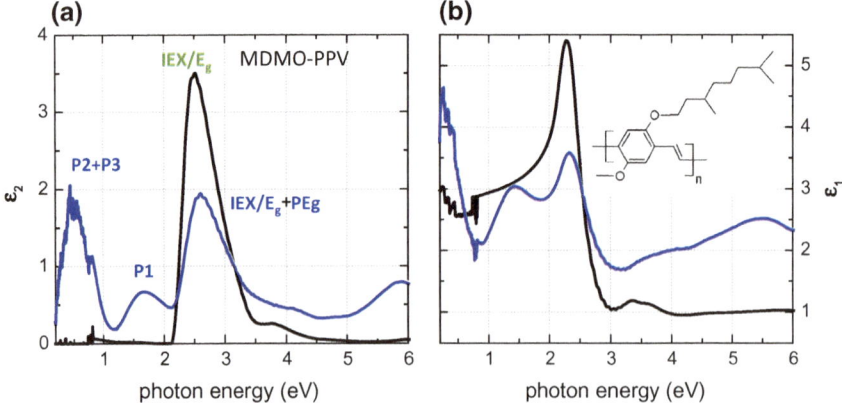

**Fig. 16.13** In-plane dielectric function of MDMO-PPV (structure in the inset) in the nominally undoped (black line) and iodine saturated doped state (blue line). The imaginary part of dielectric function is plotted in the left panel (**a**) and the real part in the right panel (**b**) [72]

particular in 1D and 2D systems [69]. The doubling of the unit cell shown in Fig. 16.12 may yield in such a low temperature state again in an insulating/semiconducting band structure. Worth mentioning in this connection is, additionally, the continuous increase of the out-of-plane anisotropy upon doping (Fig. 16.11). It supposes an increasing ordering within the film. A possible explanation is found in the shorting of bonds between the thiophene rings in the quinoid structure.

## 16.4 MDMO-PPV - Iodine Doping

As a second material example, we show ellipsometric results obtained on poly[2-methoxy-5-(3',7'-dimethyloctyloxy)-1,4-phenylene-vinylene] (MDMO-PPV). Like P3HT it is a semiconducting polymer with application in organic solar cells and OLEDs [70]. Hoppe et al. have determined the DF in the UV-VIS spectral with a combination of reflection and transmission measurement [71]. Here we would like to discuss a comprehensive ellipsometric study where FTIR- and UV-VIS-IR ellipsometry is used [72] to obtain an overview about the polaron optical response. Figure 16.13 shows the DF of MDMO-PPV for the pristine and iodine doped state as well as a graphic formula of the MDMO-PPV polymer. In this experiment the polymer was directly spin casted onto a glass slide from a pyridine solution. Both the UV-VIS as well as the IR measurements were treated in this experiment in a joined point-by-point layer fit in order to obtain a full range film DF. Some artifacts remain at the connection point of the two spectral regions although the same film preparation was used in both experiments. However, the results are, nevertheless, conclusive enough to discuss some general polaron properties in more detail.

**Fig. 16.14** Effective number of electrons ($\pi$-electron density) in MDMO-PPV and its integral evolution in the spectral range of $\pi$-$\pi^*$ transitions. The black line represents the pristine and the blue line iodine doped MDMO-PPV film (comp. Fig. 16.13)

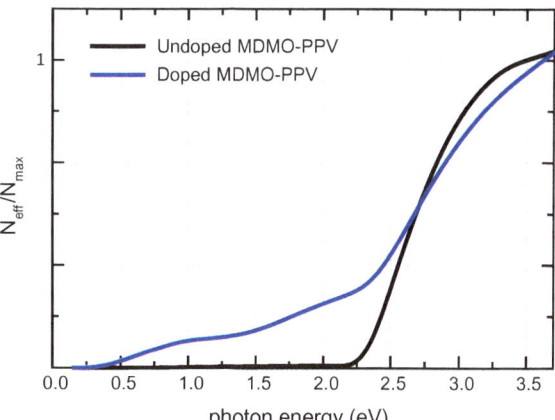

The pristine MDMO-PPV dielectric function (black line in Fig. 16.13) is very much similar to those of P3HT. A single relatively broad absorption peak with maximum at 2.5 eV dominates the UV-VIS spectral range. It is attributed again to $\pi$-$\pi^*$ transitions in the polymer chain. A fine structure due to excitonic or LO-phonon contribution is not visible in the case of MDMO-PPV. One could speculate whether these structures are just not resolved in this experiment or suppressed due to a higher degree of conformational disorder and/or less $\pi$-$\pi$ stacking. The latter assumption is at least consistent with the fact that the film could be unambiguously fitted with an isotropic film DF.

After doping (blue line in Fig. 16.13) two new transition features appear in the DF at $\approx$1.7 eV and $\approx$0.5. Both peaks are described already in terms of the polaron P1 and P2/P3 transitions (Fig. 16.4). In addition a high-energy peak emerge at $\approx$5.9 eV which is probably not relating to the $\pi$ electron system. The broad $\pi$-$\pi^*$ absorption peak around at 2.5 eV decreases, as expected, by about one half. The small but clearly observable blue shift of the peak maximum could be explained by a screening of IEXs and the respective transfer of oscillator strength to higher photon energies like it is seen for P3HT. In the IR below 0.2 eV, new dipole allowed infrared active vibrations (IRAVs) appear [47].

In connection with the iodine doping of P3HT we have discussed already the question whether the oscillator strength of $\pi$-$\pi^*$ transitions of pristine material is entirely shifted to the new $\pi$-electron structures P1-P3 in the NIR and IR. The ellipsometrically determined wide range DF of the MDMO-PPV example allows now a quantitative analysis. According to sum rule arguments the total oscillator or, a bit more precisely formulated, the first moment of the imaginary part of DF is proportional to the effective number of (valence) electrons [45, 46]:

$$N_{eff} \sim \int \omega \varepsilon_2(\omega) d\omega. \tag{16.7}$$

In the present example, we assume that all optical transitions between 0.2 and 3.5 eV are ascribed to the $\pi$ electron system. With the iodine p-type doping, some of the $\pi$-electrons are removed from the polymer chains. However, the comparison of the iodine and electrochemical doping of P3HT has shown that the charging obtained by chemical doping is relatively small in comparison to the total number of $\pi$-electrons in the film. If we thus approximate that the number of $\pi$-electrons is almost constant, the oscillator strength of the attenuate 2.5 eV absorption structure should shift in another spectral range. In order to test this assumption, the accumulative effective number of electrons is plotted for the spectral range between 0.2 and 3.5 eV. The black line in Fig. 16.14 shows the pristine polymer while the blue line is representing the iodine doped state with polarons. In the comparison it becomes evident that the oscillator strength of the $\pi$-$\pi^*$ transitions is solely redistributed among the P1-P3 transitions. A possible doping induced conductivity is ascribe to the formation of metallic $\pi$-bands. These metallic $\pi$-electrons should generate a Drude tail in the DF which may emerge below 0.2 eV. From the sum rule arguments, however, this would add additional contribution to effective number of electrons which would violate the sum rule argumentation. The IR-DF between 0.05 and 0.2 eV, in fact, shows no signature of a Drude like behavior.

## 16.5    Polaron Formation in Push-Pull Polymers

In a last example we discuss a more sophisticated so called push-pull polymer. In such polymers intramolecular dipoles red-shift the band gap to generate a NIR optical activity [73]. A prominent representative of these $\pi$-conjugated polymers is the Poly[(4,8-bis-(2-ethylhexyloxy)-benzo(1,2-b:4,5-b')dithiophene)-2,6-diyl-alt-(4-(2-ethylhexanoyl)-thieno[3,4-b] thiophene-)-2-6-diyl] in short PBDTTT-c [74–76]. In this polymer benzodithiophene and thienothiophene units serve as sequential donors and acceptors [77]. The extraordinary molecular complexity, however, considerably complicates the electronic structure. Simply the number of carbon atoms per monomer increases already by a factor of 4 from P3HT to PBDTTT-c. Additional side-chain are required to obtain solubility. The complexity of the push-pull polymer finally suppresses intermolecular interaction in contrast to the highly-crystalline rr-P3HT.

Figure 16.15 (black line) shows the ellipsometrically determined DF of a pristine PBDTTT-c film [62]. The presented film properties are obtained by a parametric layer fit with an independently determined film thickness and are refined within a point-to-point calculation. Various details in the thereby determined DF differ from the previously discussed polymers although the general line shape remains comparable. The spectral range between 1 and 3 eV is again characterized by the dominating $\pi$-$\pi^*$ absorption structure. But the absorption onset is considerably red shifted due to the orbital interaction of the unified donor (thieno[2,3-b]thiophene) and acceptor (benzodithiophene with alkoxy-side-chains) units. Unlike rr-P3HT, the two sharp peaks at 1.75 and 1.9 eV could not originated from $\pi$-$\pi$ stacking and may show

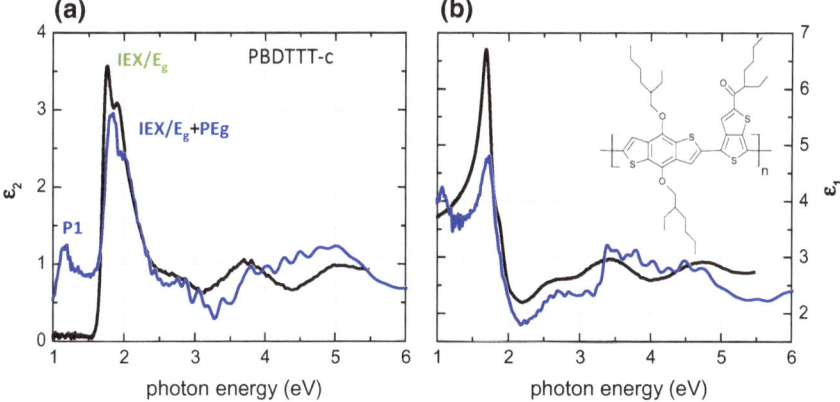

**Fig. 16.15** Dielectric function of PBDTTT-c (structure in the inset) in the nominally undoped (black line) and iodine saturated doped state (blue line). The imaginary part of dielectric function is plotted in the left panel (**a**) and the real part in the right panel (**b**) [62]

rather a fingerprint of the push-pull character. Above 2.5 eV one can observe at least three additional absorption structures. But an identification of the latter UV transition is problematic if one considers the complexity of the structure.

A relatively persistent p-type doping is again achieved by an exposure of the PBDTTT-c film to iodine vapor. Respective ellipsometric spectra are shown in Fig. 16.15 (blue line). The results could not achieve the usual signal to noise level of ellipsometry, due to the limitation in integration time. But the doping effect is clearly visible. For PBDTTT-c one can observe a similar doping impact as it is known for e.g. P3HT. This is a quenching of the main $\pi$-$\pi^*$ absorption structure and in parallel a P1 peak is rising in NIR part of the DF spectrum. The P1 resonance arises at a somewhat smaller photon energy (1.2 eV) consistently to the lower absorption edge of the ground state. But the P1 amplitude as well as the decrease of the original absorption feature is significantly smaller than those of P3HT or MDMO-PPV. The smaller effect upon doping correlates to the higher oxidation potential of PBDTTT-c. Remarkable is the small peak width of the polaronic P1 structure and the absence of the before discussed broadening of all other transition features. As already mentioned, the $\pi$-$\pi$ stacking forces are much smaller in the PBDTTT-c film. Polarons are thus inherently localized in a single polymer chain and thus more close to the 1D polaron picture with self localization in each chain.

Related changes in the IR P2+P3 transition structures are best seen in a direct comparison of the P3HT and PBDTTT-c (ATR) spectra as shown in Fig. 16.16. The broad absorption band of P3HT*, which spreads above the long-wavelength LO-phonon (from $\approx$190 meV) to higher energies, has in PBDTTT-c* a clear peak like maximum at about 380 meV. The latter structure is described as a relaxed excited state in the Fröhlich polaron model [8, Sect. 16.1.2]. The existence of this peak is typical for strong electron-phonon coupling constants ($\alpha > 5$) and the Fröhlich coupling in

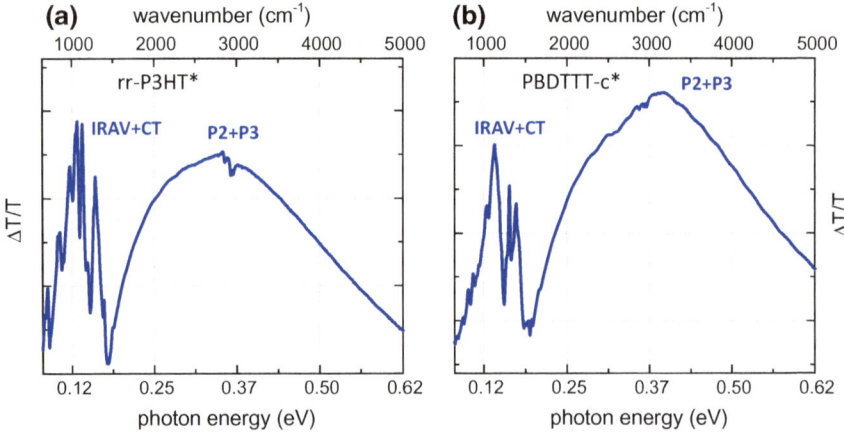

**Fig. 16.16** Comparison of the iodine induced ATR-FTIR modifications in rr-P3HT (left panel **a**) and PBDTTT-c (right panel **b**) after 12 s exposure [62]

fact increases with a lowering of the dimensionality. In conclusion, PBDTTT-c fits much more to the classic 1D Fröhlich polaron model than P3HT or MDMO-PPV although the complexity of the system is much bigger and less polarons are generated by iodine doping.

## 16.6 Comments on Transmission (UV-VIS) and ATR Spectroscopy

Most of the pioneering work [39, 78, 79] and of the following investigations concerning the optical response of polarons in polymers make use of transmission (absorption), ATR, and related spectroscopic techniques. The comparison of the ellipsometrically determined DF of P3HT with the respective film absorption spectrum (black lines in Fig. 16.5) exemplary disclose already some difficulties which has to be considered in the interpretation of the different results. The peak positions and the general line shape differ considerably between the imaginary part of the DF and the absorption spectrum.

The optical properties of a material are typically represented by the complex refractive index ($\tilde{n}$) or the dielectric function ($\varepsilon$):

$$\tilde{n} = n + i\kappa = \sqrt{\varepsilon} = \sqrt{\varepsilon_1 + i\varepsilon_2}. \tag{16.8}$$

In theoretical publications one can find, furthermore, spectra of the real part of the optical conductivity $\sigma_1 = \omega\varepsilon_0\varepsilon_2$, where $\hbar\omega$ is the photon energy and $\varepsilon_0$ is the vacuum permittivity. All three quantities are discussed under assumption that the film is optically isotropic. This is a reasonable approximation if we consider that transmission experiments are usually assembled in a normal incident geometry and that the polymer films are in-plane isotropic.

The primarily determined quantity in transmission experiments is a relative intensity i.e. the transmittance $T$. Disregarding reflections at the film boundaries, the transmittance is a function of the film thickness $d$ and the wavelength of the light:

$$T = \frac{I}{I_{ref}} = e^{-\frac{2\omega}{c}\kappa d} = e^{-\alpha d}. \tag{16.9}$$

The herein used parameter $\alpha$ is the absorption coefficient (Fig. 16.5b). The absorbance of a material is calculated from the transmittance with the negative decadic logarithm (Lambert–Beer law):

$$A = -log(T/T_{ref}) = \Delta\alpha\, d\, log(e) \tag{16.10}$$

The **absorbance $A$ is** thus **proportional to the absorption coefficient** $\alpha$ and the film thickness $d$ and is therefore commonly used in order to discuss the material optical properties based on transmission experiments. The herein used approximations are valid as long as the absorption of light insight in the entire film is much bigger than the reflected intensity (e.g. in thick films with small $\alpha$). The decadic logarithm is used by historical reasons.

In Fig. 16.17 we use the results of a simulated transmission experiment in order to illustrate peculiarities which may occur in transmission experiments based on a single Lorentz oscillator. The comparison discloses two major problems. First, the energy position of the absorbance maximum does not coincide with the resonance energy of the electronic transition. Second, the line shape and amplitude of the absorbance structure is altered significantly by the multiple light reflection at boundaries and a conceivable determination of the dielectric function by a Kramers-Kronig-analysis would be falsified.

ATR experiments are analyzed under the assumption that a light beam, which is n-times reflected at the boundary between the film and the ATR-crystal under total reflection conditions, loses intensity due to the absorption in the evanescent field in the film. The intensity reduction upon each reflection can be calculated for the two principle polarizations by means of the absolute values of the respective reflection coefficients [80, 81]:

$$R_s = |r_s|^2 = 1 - \frac{n_p cos(\theta)}{(n_p^2 - n^2)\sqrt{n_p^2 sin^2(\theta) - n^2}} n\kappa \quad \text{and} \tag{16.11}$$

$$R_p = |r_p|^2 = 1 - \frac{n_p n\kappa(2n_p^2 sin^2\theta - n^2)cos(\theta)}{(n_p^2 - n^2)((n_p^2 - n^2)sin^2\theta - n^2)\sqrt{n_p^2 sin^2(\theta) - n^2}} n\kappa, \tag{16.12}$$

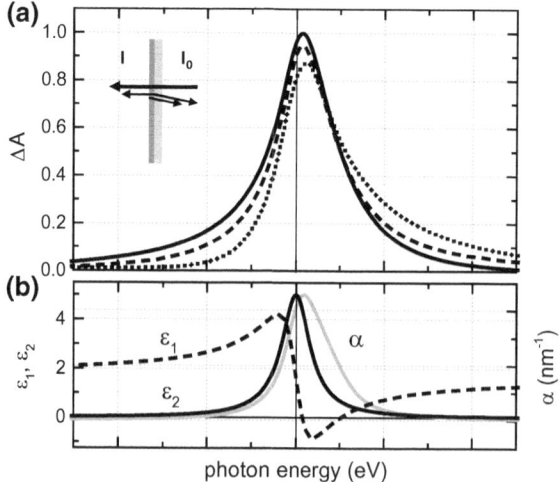

**Fig. 16.17** Simulated differential absorbance spectra **a** for a thin film with a single Lorentz oscillator absorption. The respective dielectric function is plotted in the lower panel (**b**). The differential absorbance is calculated for a normal incidence experiment where the amplitude of the oscillator is set to zero in the reference spectra. Three different sample scenarios are simulated: a free standing 50 nm film (solid line), a film on a glass substrate (dotted line), a film on a glass substrate with a 50 nm ITO coating (dashed line). The absorption coefficient $\alpha$ (gray line) is shown in **b** for comparison

where $n_p$ is the refractive index of the ATR-prism and $\theta$ is the angle of incidence (typically values are between 40 and 60°). These relations are derived with the assumption of weak absorption in the film [81] and assume that the total reflection condition is maintained for each wavelength. Under the latter condition, any reflectivity different from one is due to absorption (energy dissipation) in the film. The second therm in (16.11) and (16.12) can be thus associated to the right hand side of (16.9). The total transmittance $T$ through the ATR prism is:

$$T = R^n \qquad (16.13)$$

ATR results are often presented in two different forms; either in terms of $\Delta T / T_{ref}$ or by means of $-log(T / T_{ref})$. In case of weak absorption the second term in (16.11) and (16.12) is small and one can use the approximation $-ln(1 - x) = x$. Under consideration of the principle relation $log(y^n) = n * ln(y)/log(e)$ it is thus evident that both representations of the ATR results are equivalent. In particular it is worth mentioning that the **ATR results are proportional to $n\kappa$ and thus to $\varepsilon_2$**.

A comparison of ATR-results and the respective film DF is again graphically demonstrated with simulated spectra containing a single Lorentz-oscillator (Fig. 16.18). For weakly absorbing films (oscillator amplitude 10 times smaller than in the lower panel) the ATR-peak amplitude and line shape matches very good with the DF. But just for slightly higher oscillator amplitudes the total reflection condition

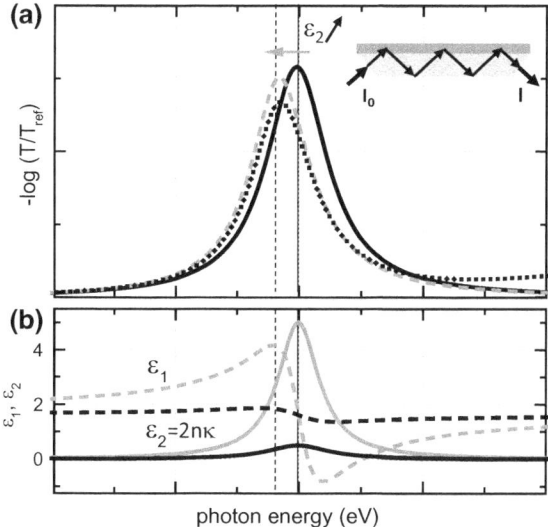

**Fig. 16.18** Simulation of relative ATR transmission spectra **a** for a thin film with a single Lorentz oscillator absorption. The respective dielectric function is plotted in the lower panel (**b**). The relative ATR transmission is calculated for an angle of incidence of 50° and a film of 5 nm where the amplitude of the oscillator is set to zero in the reference spectra. Three different sample scenarios are simulated: a weakly absorbing film (oscillator amplitude 10 times smaller than in **b**) on a high refractive index ATR-prism ($n = 3$) (solid line), with a 5 nm ITO coating (dotted line), and without ITO but with the higher oscillator amplitude (dashed line)

is already falsified and the obtained peak position shifts to lower photon energies towards the maximum in the real part of the film dielectric function (dashed line in Fig. 16.18). In real experiments and in particular in the spectral range of strong and narrow IR phonon absorption lines it could be problematic to ensure whether the weak absorption limit (the ATR-conditions) is valid. The situation gets even more complicated if further layers like a conducting ITO film are introduced (dotted line). The dispersion of the real and imaginary part of the dielectric function of those films superimpose with the film response. It should be noted that these effects are not eliminated with the reference measurement and further increase if interferences effects come into play. It is generally also difficult to deduce correct absolute values for the absorption coefficient. The latter depends on the number of reflections, the propagation length in the film and is finally influenced by multiple reflections at the entrance and exit surface of the ATR-prism. However, with the optical models used in common ellipsometric software packages a complete description is possible. An ellipsometric determination of film dielectric properties ($\varepsilon_1$ and $\varepsilon_2$) with a single total internal reflection is for example presented in [82].

**Acknowledgements** The authors would like to thank Kurt Hingerl, Niyazi S. Sariciftci, Helmut Neugebauer, and Reghu Menon for their valuable comments and enlightening discussions. Furthermore, we acknowledge manifold contributions of Achim W. Hassel, Günther Knör, Jan Philipp

Kollender, Andrei I. Mardare, Kerstin Oppelt, Thomas Plach, Stefanie Schlager, Matthew S. White, Karin Wiesauer, and Cigdem Yumusak for the results presented here.

# References

1. *Handbook of Conducting Polymers, Conjugated Polymers: Theory, Synthesis, Properties, and Characterization*, vol. I–II, 3rd edn. ed. by T.A. Skotheim, J.R. Reynolds (CRC Press, Boca Raton, London, New York, 2006)
2. J.L. Brédas, G.B. Street, B. Thémans, J.M. André, J. Chem. Phys. **83**, 1323 (1985)
3. E. Bundgaard, F.C. Krebs, Sol. Energy Mater. Sol. Cells **91**, 954 (2007)
4. D.T. Haar, *Collected Papers of L.D. Landau* (Gordon and Breach, Science Publishers, New York, London, Paris, 1965)
5. H. Fröhlich, Adv. Phys. **3**, 325 (1954)
6. J.T. Devreese, J. Phys. Condens. Matter **19**, 26 (2006)
7. J.T. Devreese, A.S. Alexandrov, Rep. Prog. Phys. **72**, 066501 (2009)
8. *Polarons in Advanced Materials*, ed. by A.S. Alexandrov (Springer, Dordrecht, 2007)
9. D. Emin, *Polarons* (Cambridge University Press, Cambridge, New York, Melbourne, 2013)
10. J.L. Brédas, G.B. Street, Acc. Chem. Res. **18**, 309 (1985)
11. A.J. Heeger, N.S. Sariciftci, E.B. Namdas, *Semiconducting and Metallic Polymers* (Oxford University Press, Oxford and New York, 2010)
12. S.-I. Kuroda, Int. J. Mod. Phys. B **9**, 221 (1995)
13. F.M. Peeters, J.T. Devreese, Phys. Rev. B **36**, 4442 (1987)
14. D. Beljonne et al., Adv. Funct. Mater. **11**, 229 (2001)
15. S.N. Klimin, J. Tempere, J.T. Devreese, Phys. Rev. B **94**, 1 (2016)
16. J.L. Brédas, R.R. Chance, R. Silbey, Phys. Rev. B **26**, 5843 (1982)
17. D. Bertho, C. Jouanin, Phys. Rev. B **35**, 626 (1987)
18. A.A. Bakulin et al., Science (80-. ) **335**, 1340 (2012)
19. R. Österbacka, C. An, X.M. Jiang, Z. Vardeny, Science (80-. ) **287**, 839 (2000)
20. O. Bubnova, X. Crispin, Energy Environ. Sci. **5**, 9345 (2012)
21. C.M. Pochas, F.C. Spano, J. Chem. Phys. **140**, 244902 (2014)
22. S. Stafström et al., Phys. Rev. Lett. **59**, 1464 (1987)
23. D.J. Thouless, Phys. Rev. Lett. **39**, 1167 (1977)
24. B.I. Shklovskii, A.L. Efros, in *Electronic Properties of Doped Semiconductors*, vol. 45, Springer Series in Solid-State Sciences, ed. by M. Cardona (Springer, Berlin, Heidelberg, 1984)
25. P.R. Somani, S. Radhakrishnan, Mater. Chem. Phys. **77**, 117 (2002)
26. H. Shirakawa et al., J. Chem. Soc. Chem. Commun. **578** (1977)
27. A.J. Heeger, Angew. Chemie **40**, 2591 (2001)
28. J. Gasiorowski et al., J. Phys. Chem. C **117**, 2584 (2013)
29. Y. Taguchi et al., J. Am. Chem. Soc. **128**, 3313 (2006)
30. R. Ludwig, Angew. Chemie **115**, 3580 (2003)
31. C. Cobet et al., Submitt. to Adv. Mater. Interfaces (2017)
32. U. Zhokhavets, G. Gobsch, H. Hoppe, N.S. Sariciftci, Thin Solid Films **451–452**, 69 (2004)
33. U. Zhokhavets et al., Chem. Phys. Lett. **418**, 347 (2006)
34. M. Campoy-Quiles, P.G. Etchegoin, D.D.C. Bradley, Phys. Rev. B **72**, 045209 (2005)
35. T. Tsumuraya, J.-H. Song, A. Freeman, Phys. Rev. B **86**, 075114 (2012)
36. E. Lioudakis, A. Othonos, I. Alexandrou, Y. Hayashi, Appl. Phys. Lett. **91**, 111117 (2007)
37. Y. Kim et al., Nat. Mater. **5**, 197 (2006)
38. P.G. Karagiannidis et al., Mater. Chem. Phys. **129**, 1207 (2011)
39. Z. Vardeny et al., Phys. Rev. Lett. **56**, 671 (1986)
40. A.J. Heeger, S. Kivelson, J.R. Schrieffer, W.P. Su, Rev. Mod. Phys. **60**, 781 (1988)
41. J. Gasiorowski et al., J. Phys. Chem. C **118**, 16919 (2014)

42. C. Enengl et al., ChemPhysChem **17**, 3836 (2016)
43. Y. Furukawa, J. Phys. Chem. **100**, 15644 (1996)
44. J. Gasiorowski, Dissertation, Johannes Kepler University Linz (2013)
45. *Handbook of Optical Constants of Solids*, vol. 111, ed. by E.D. Palik (Academic Press, San Diego, Chestnut Hill, 1998)
46. P.Y. Yu, M. Cardona, *Fundamentals of Semiconductors*, 3rd edn. (Springer, Berlin, Heidelberg, New York, 1996)
47. E. Ehrenfreund, Z. Vardeny, O. Brafman, B. Horovitz, Phys. Rev. B **36**, 1535 (1987)
48. A. Girlando, A. Painelli, Z.G. Soos, J. Chem. Phys. **98**, 7459 (1993)
49. J. Gasiorowski, A.I. Mardare, N.S. Sariciftci, A.W. Hassel, J. Electroanal. Chem. **691**, 77 (2013)
50. N.S. Sariciftci et al., J. Chem. Phys. **96**, 7164 (1992)
51. A.K. Agrawal, S.A. Jenekhe, Chem. Mater. **8**, 579 (1996)
52. H. Neugebauer et al., J. Chem. Phys. **110**, 12108 (1999)
53. C. Kvarnström et al., Synth. Met. **101**, 66 (1999)
54. A.W. Hassel, K. Fushimi, M. Seo, Electrochem. Commun. **1**, 180 (1999)
55. T. Erb et al., Adv. Funct. Mater. **15**, 1193 (2005)
56. H.G. Tompkins, *A User's Guide to Ellipsometry* (Academic Press Inc, San Diego, 1993)
57. R.M.A. Azzam, N.M. Bashara, *Ellipsometry and Polarized Light* (North-Holland Publishing Company, Amsterdam, New York, Oxford, 1987)
58. P. Yeh, *Optical Waves in Layered Media* (Wiley, New Yorke, Chichester, Weinheim, Brisbane, Singapore, Toronto, 1988)
59. M.V. Klein, T.E. Furtak, *Optik* (Springer, Berlin, Heidelberg, 1988)
60. M. Schubert, Phys. Rev. B **53**, 4265 (1996)
61. H. Tompkins, E.A. Irene, *Handbook of Ellipsometry* (Springer, Heidelberg, 2005)
62. C. Cobet et al., Sci. Rep. **6**, 35096 (2016)
63. S. Panero, S. Passerini, B. Scrosati, Mol. Cryst. Liq. Cryst. **229**, 97 (1993)
64. P. Kar, *Doping Conjugated Polymers* (Wiley, Hoboken, 2013)
65. B. Horovitz, Solid State Commun. **88**, 983 (1993)
66. J. Wosnitza, *Fermi Surfaces of Low-Dimensional Organic Metals and Superconductors* (Springer, Berlin, Heidelberg, 1996)
67. D. Jérome, H.J. Schulz, Adv. Phys. **31**, 299 (2006)
68. C. Cobet, E. Speiser, in *Defin. Anal. Opt. Prop. Mater. Nanoscale A Collect. Thoughts, Opin. Ideas Data that Have Matur. Over Years Exploit. Ellipsom. a Range Characterisation Needs*, ed. by M. Losurdo (Ges. für Mikro- und Nanoelektronik, Wien, 2010)
69. D.K. Campbell, A.R. Bishop, K. Fesser, Phys. Rev. B **26**, 6862 (1982)
70. S.E. Shaheen et al., Appl. Phys. Lett. **78**, 841 (2001)
71. H. Hoppe, N.S. Sariciftci, D. Meissner, Mol. Cryst. Liq. Cryst. **385**, 113 (2002)
72. J. Gasiorowski et al., J. Phys. Chem. C **117**, 22010 (2013)
73. C. Duan, F. Huang, Y. Cao, J. Mater. Chem. **22**, 10416 (2012)
74. H.-Y. Chen et al., Nat. Photonics **3**, 649 (2009)
75. J. Hou et al., J. Am. Chem. Soc. **131**, 15586 (2009)
76. J. Gasiorowski, A.I. Mardare, N.S. Sariciftci, A.W. Hassel, Electrochim. Acta **113**, 834 (2013)
77. K.G. Jespersen et al., J. Chem. Phys. **121**, 12613 (2004)
78. J.L. Brédas, J.C. Scott, K. Yakushi, G.B. Street, Phys. Rev. B **30**, 1023 (1984)
79. M.J. Nowak, S.D.D.V. Rughooputh, S. Hotta, A.J. Heeger, Macromolecules **20**, 965 (1987)
80. N.J. Harrick, *Internal Reflection Spectroscopy* (Wiley, New York, 1967)
81. M. Milosevic, in *Internal Reflection and ATR Spectroscopy*, vol. 176, Chemical Analysis, ed. by M.E. Vitha (Wiley, Hoboken, New Jersey, 2012)
82. M. Poksinski, H. Arwin, *Proteins Solid-Liquid Interfaces*, Principles and Practice (Springer, Berlin, Heidelberg, 2006), pp. 105–118

# Part V
# Developments in Ellipsometric Real-Time/In-situ Monitoring Techniques

# Chapter 17
# Coupling Spectroscopic Ellipsometry and Quartz Crystal Microbalance to Study Organic Films at the Solid–Liquid Interface

**Ralf P. Richter, Keith B. Rodenhausen, Nico B. Eisele and Mathias Schubert**

**Abstract** Spectroscopic ellipsometry (SE) and quartz crystal microbalance with dissipation monitoring (QCM-D) have become popular tools for the analysis of organic films, from a few Angstroms to a few micrometers in thickness, at the solid–liquid interface. Because of their different working principles, both techniques are highly complementary, providing insight into optical and mechanical properties, respectively. The combination of SE and QCM-D in one setup is not only attractive because this information becomes available at the same time on the same sample, but also because the correlation of SE and QCM-D responses can provide novel insight that is not accessible with either technique alone. Here, we discuss how the combined setup is implemented in practice and review current data analysis approaches that are useful with regard to the correlation of both methods. Particular attention is given to the novel insight that can be obtained by the combination of both techniques, such as the solvation, density and lateral organization of organic films.

R. P. Richter (✉) · N. B. Eisele
Biosurfaces Unit, CIC BiomaGUNE, Donostia, San Sebastian, Spain
e-mail: r.richter@leeds.ac.uk

R. P. Richter
Department of Molecular Chemistry, J. Fourier University, Grenoble, France

R. P. Richter
Max Planck Institute for Intelligent Systems, Stuttgart, Germany

R. P. Richter
School of Biomedical Sciences, University of Leeds, Leeds, UK

K. B. Rodenhausen
Department of Chemical and Biomolecular Engineering, University of Nebraska-Lincoln, Lincoln, NE, USA

N. B. Eisele
Department of Cellular Logistics, Max Planck Institute of Biophysical Chemistry, Göttingen, Germany

M. Schubert
Department of Electrical Engineering and Center for Nanohybrid Functional Materials, University of Nebraska-Lincoln, Lincoln, NE, USA

© Springer International Publishing AG, part of Springer Nature 2018                                391
K. Hinrichs and K.-J. Eichhorn (eds.), *Ellipsometry of Functional Organic Surfaces and Films*, Springer Series in Surface Sciences 52,
https://doi.org/10.1007/978-3-319-75895-4_17

## 17.1   Introduction

Thin organic films are important for chemical and biological detection, biomaterials, detergent, and surface property tuning applications. Continuous progress in methods for the immobilization of organic molecules, and in the patterning of surfaces at the nano, micro and macro scale, has entailed improved control on the confinement of molecules and enables the creation of films of increasing complexity and functionality. As the level of sophistication of the films advances, methods are required to characterize their formation and their physico-chemical properties in detail.

Optical and acoustical methods are widely employed for non-invasive, label free, in-situ monitoring of organic film formation. Among them, spectroscopic ellipsometry (SE) and quartz crystal microbalance with dissipation monitoring (QCM-D) stand out in several ways. First, both techniques can be used on a variety of substrates, including metal, metal oxide and polymer coatings. Second, thanks to a continuous development over more than a century, the mechanism of signal formation is rather well understood. As a consequence, material parameters such as adsorbed amounts, and film thickness, optical or viscoelastic properties can be extracted from the measured parameters in a quantitative manner. Third, both techniques can in principle probe films over a large thickness range, from Angstroms up to micrometers. These features render SE and QCM-D particularly versatile with regard to the quantitative analysis of organic films at the solid–liquid interface. In other popular optical mass-sensitive methods (e.g., surface plasmon resonance), specialized surface coatings are required and/or the sensitivity decays rapidly with increasing distance from the substrate. These methods, as well as other acoustical methods (e.g. shear acoustic wave devices), appear to be more promising for standardized sensing applications, because of their ease of miniaturization and/or parallelization, yet they tend to be limited when it comes to fully quantitative data analysis.

Thanks to their different transducer principles, QCM-D and SE provide different yet highly complementary information: QCM-D is sensitive to mechanical properties, and SE is sensitive to optical properties. Moreover, both techniques provide information about the amount of deposited material. Importantly, solvent contributes differently to SE and QCM-D responses. In the case of QCM-D, the shear oscillation of the sensor surface entails the motion of solvent that is trapped within the organic thin film. The adsorbed mass measured by QCM-D therefore always contains a contribution of hydrodynamically coupled solvent. SE, on the other hand, is sensitive to changes in the refractive index of the film as compared to the ambient liquid medium, and adsorbate areal mass densities can be determined. By exploiting this difference in mass sensitivity, information about the solvent content (e.g. porosity) can be extracted from combined SE/QCM-D measurements, a quantity that cannot generically be obtained with either technique alone.

In this chapter, we review current data analysis approaches that are useful with regard to the correlation of SE and QCM-D and describe how combined in-situ measurements can be implemented in practice. Particular attention is given to the novel insight that can be obtained by the combination of both techniques, such as the solvation, density and lateral organization of organic films.

## 17.2 Theory

Here, we will review a selection of approaches for the quantitative analysis of SE and QCM-D data. Focus is put on methods for which a correlation of both data sets is particularly useful, providing novel information that cannot be obtained with either technique alone.

### 17.2.1 Approaches to Analyze SE Data

The analysis of ellipsometry data requires an appropriate layer model that accounts for the properties of the substrate, the overlayer(s) and the ambient medium, and possibly for interface roughness. Each layer in this approach requires an appropriate description of its dielectric function $\varepsilon$, which is the square of the complex index of refraction $n + i\kappa$, where $n$ is the refractive index and $\kappa$ the extinction coefficient.

Ellipsometry data is typically represented by the two ellipsometric angles $\Psi$ and $\Delta$. In in-situ spectroscopic ellipsometry, these are monitored as a function of wavelength $\lambda$, typically at a single angle of incidence. Because the equations relating $\Psi(\lambda)$ and $\Delta(\lambda)$ to the layers' thickness and dielectric functions are nonlinear, numerical nonlinear regression methods ("fitting") are required to match measured and model-calculated ellipsometry spectra. The details of this analysis are beyond the scope of this work, and we direct interested readers to more thorough discussions in the literature [1–3].

We restrict ourselves to considering an optically transparent and isotropic, smooth film that is attached to a planar substrate and immersed in a transparent ambient solvent. The optical properties of such a film are fully determined by two parameters: the film thickness $d$ and the wavelength-dependent refractive index $n(\lambda)$. Assuming non-absorbing films ($\kappa = 0$) does not impair the generality of the considerations; films with a complex index of refraction ($\kappa \neq 0$) can be included as well. The film thickness can range from an Angstrom to a few micrometers (and even beyond).

Even though very simple, this scenario represents a reasonable first approximation for many organic thin films. Typically, the organic film will contain both adsorbate and solvent and the characteristic length scale of the internal structure is much smaller than the wavelength of the probing light. In this case, the film's optical properties can be adequately described through effective medium approximations. In this context, the index of refraction becomes an effective film property, $n_{\text{eff}}$. Moreover, the thickness should be considered an effective and method-specific property $d_{\text{eff}}^{\text{SE}}$, because the interface between film and ambient is in practice not ideally sharp and the interfacial contrast is sensed differently by different techniques.

The effective medium approximation can be used to relate $n_{\text{eff}}$ to the refractive indices of the organic adsorbate $n_{\text{o}}$ and the ambient solvent $n_{\text{a}}$ and to quantify the constituents' volume fractions $f_{\text{o}}^{\text{V}}$ and $f_{\text{a}}^{\text{V}}$. Another parameter of interest is the areal adsorbate mass density. This quantity corresponds to the mass of organic adsorbate per unit area of the substrate on which the film is deposited. It is often sloppily called adsorbed mass, hence denoted here as $m_o$, and given by

$$m_o = d_{eff}^{SE} \rho_o f_o^V, \tag{17.1}$$

where $\rho_o$ is the volume density of the organic adsorbate when immersed in the ambient solvent (i.e. the inverse of the partial specific volume). To describe, how $f_o^V$, $f_a^V$ and $m_o$ are determined, we will distinguish two different approaches. The first approach is restricted to what we call ultra-thin films, whereas the second approach can also be employed for thicker films. We will see that the two approaches relate to two distinct effective medium approximations. From the comparison of both approaches, a third method to determine $m_o$ emerges that is simpler yet more approximative.

### 17.2.1.1 Approach 1: Virtual Separation

A virtual separation approach for arbitrarily segregating a heterogeneous ultra-thin film into separate homogeneous sublayers from an ellipsometric modeling point of view has been previously described [4]. The main points are summarized here. Through a $4 \times 4$ matrix modeling approach, also known as the Berreman-formalism [2, 5], a transfer matrix $\mathbf{T}$ can completely describe a linear optical system. $\mathbf{T}$ consists of an ordered product of partial transfer matrices ($\mathbf{T}_p$), which describe homogeneous constituent sublayers, and matrices that describe incident (ambient; $\mathbf{L}_a$) and exit (substrate; $\mathbf{L}_f$) mediums, such that

$$\mathbf{T} = \mathbf{L}_a^{-1} \mathbf{T}_{p,1}^{-1} \cdots \mathbf{T}_{p,N}^{-1} \mathbf{L}_f. \tag{17.2}$$

The partial transfer and medium matrices typically do not multiplicatively commute. However, if the ultra-thin film limit is met, i.e., when $d \ll \lambda/(2\pi n)$, the partial transfer matrix product of (17.2) can be approximated as a matrix sum. Thus, the partial transfer matrices become interchangeable. A porous ultra-thin film that has an effective thickness $d_{eff}^{SE}$ and an effective index of refraction $n_{eff}$ can therefore be virtually separated into two layers. The partial transfer matrix describing constituent ambient inclusions ($d_a$, $n_a$) may be moved via additive commutation to the top of the layer stack, forming an ambient-ambient interface and leaving a layer of pure adsorbate material ($d_o$, $n_o$, with $d_{eff}^{SE} = d_o + d_a$).

Virtual separation implies that the ellipsometric response in the ultra-thin film limit depends on the product $n_o d_o$, and that it is not possible to resolve $n_{eff}$ and $d_{eff}^{SE}$. In practice, one can therefore model ultra-thin films as a (virtual) layer with refractive index $n_o$, which needs to be known, and thickness $d_o$, which can be determined through ellipsometry. The areal adsorbate mass density can then be calculated by

$$m_o = d_o \rho_o. \tag{17.3}$$

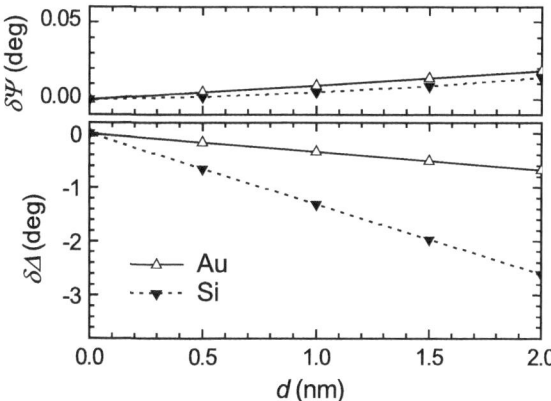

**Fig. 17.1** Changes of $\Psi$ and $\Delta$, $\delta\Psi$ and $\delta\Delta$, with thickness for a transparent layer ($n_o = 1.50$) covering ideally flat, optically opaque surfaces of Au and Si, respectively, at a wavelength of 633 nm, in ambient water ($n_a = 1.333$) and at an angle of incidence of 65°

In contrast, it is not possible to determine $f_o^V$ or $f_a^V$ of ultra-thin films from ellipsometry alone, because $n_{eff}$ and $d_{eff}^{SE}$ cannot be resolved. We will see later that QCM can provide the additional information required for this separation.

**A Stringent Definition of the Ultra-thin Film Limit** Variations in $\Psi$ and $\Delta$ with thickness for an assumed layer on top of ideally flat, optically opaque surfaces of a metal (Au) and a dielectric (Si), respectively, within ambient water are depicted in Fig. 17.1. The calculated data were obtained for a wavelength of 633 nm, yet they are representative over the entire near infrared to visible spectral range for most organic layers. For the small thickness values that are considered here, the changes in both SE parameters are almost linear with thickness. The slope for $\Psi$ is about 2 orders of magnitude smaller than for $\Delta$. Instrumentation with high precision in $\Psi$ and $\Delta$ is required to follow changes in the formation of organic layers with small thickness on metal and other surfaces in liquid ambient. In order to exploit both $\Psi$ and $\Delta$, the error bars for $\Psi$ must be two orders of magnitude better than for $\Delta$. In most current instrumentation, however, the error bar (precision) for $\Psi$ is only one order of magnitude better, and therefore many authors have reported the inability to differentiate between $n_{eff}$ and $d_{eff}^{SE}$ of the organic layer if the layer is very thin.

State-of-the-art ellipsometry instrumentation may reach a precision of a few thousandths of degrees in $\Psi$ and a few hundredths of degrees in $\Delta$, and better, but these values crucially depend on the time period over which multiple measurements are averaged while keeping experimental conditions constant. In fact, the latter may dramatically deteriorate the error bars when monitoring organic layer formation under fast changing kinetic conditions. Because no standardized criteria are available under which manufacturers are specifying their instrument limits, a hard number of current limits is difficult to obtain. Substantial improvement in obtaining better precision can be made when $\Psi$ and $\Delta$ are measured over multiple wavelengths simultaneously, for example with multiple-channel instrumentation capable of reading the ellipsometric values at multiple wavelengths at the same time. Such instrumentation is conveniently available nowadays and in widespread use for characterization of physical and chemical thin film deposition processes. Because the parameter $d_{\text{eff}}^{\text{SE}}$ does not depend on wavelength, the $n_{\text{eff}}d_{\text{eff}}^{\text{SE}}$ product correlation is substantially relaxed when data are analyzed over a wide spectral range.

A criterion for selecting the ultra-thin film approximation as a valid approach for data analysis is suggested here by whether or not variations in $\Psi$ are known upon the organic layer formation, assuming that variations in $\Delta$ are detected. Changes in $\Psi$ being known or unknown (the same being true for $\Delta$) means that a measurement of $\Psi$ reports the previous value of $\Psi$ to have moved outside the range or not, respectively, of the typical precision window of the current instrument. Then one may "switch" between analysis of $n_{\text{eff}}$ and $d_{\text{eff}}^{\text{SE}}$ using data points of $\delta\Psi$ and $\delta\Delta$, or $n_{\text{eff}}$ or $d_{\text{eff}}^{\text{SE}}$ using data points of $\delta\Delta$ only, i.e., the ultra-thin film situation, respectively. The precision is here defined as the "narrowness" of data points between repeated measurements under constant experimental conditions, i.e., the standard deviation in a standard distribution. Note that the difference between the reported $\Psi$ and the true $\Psi$, while the latter may never be known, defines the accuracy of the instrumentation. While the accuracy of the instrument matters for calibration purposes, however, it is less relevant here since we are interested in monitoring small changes due to the organic layer formation.

The situation when the wavelength equivalent $\lambda/(2\pi n_{\text{eff}})$ becomes much larger than the film thickness, for organic transparent layers, and which we refer to here also as the ultra-thin film situation, can be tied to the error bars (or windows of precision) within which one can determine $\Psi$ and $\Delta$. Because changes in $\Psi$ are two orders of magnitude smaller than in $\Delta$ with $d_{\text{eff}}^{\text{SE}}$ and/or $n_{\text{eff}}$ (Fig. 17.1), the not knowing of $\Psi$ triggers the change to the ultra-thin film situation for organic films at approximately 2 nm assuming error bars of $\pm 0.01°$ in $\Psi$. Additional error sources such as angle of incidence variations may increase this thickness further. This criterion also places perspectives on future needs. If one is interested in measuring $n_{\text{eff}}$ and $d_{\text{eff}}^{\text{SE}}$ for smaller films one should improve the precision of $\Psi$ first, as in general the precision of $\Psi$ must be two orders of magnitude better than for $\Delta$.

**Linear Averaging** The effective medium approximation for ultra-thin layers that emerges from this virtual separation approach corresponds to the linear averaging of the dielectric constants $\varepsilon = n^2$ of the embedded materials [4]

$$n_{\text{eff}}^2 = f_o^V n_o^2 + f_a^V n_a^2. \tag{17.4}$$

The number of constituents is not limited to two and can be expanded by adding multiple constituents' parameters to the right side of (17.4).

### 17.2.1.2 Approach 2: Lorentz

For thicker films, $n_{\text{eff}}$ and $d_{\text{eff}}^{\text{SE}}$ can be resolved. The Lorentz equation [6] is based on an effective medium approximation that considers the mixing of two components, here organic solute and ambient solvent

$$\frac{n_{\text{eff}}^2 - 1}{n_{\text{eff}}^2 + 2} = f_{\text{o}}^V \frac{n_{\text{o}}^2 - 1}{n_{\text{o}}^2 + 2} + f_{\text{a}}^V \frac{n_{\text{a}}^2 - 1}{n_{\text{a}}^2 + 2}. \tag{17.5}$$

Again, the number of components is not limited to two and can be expanded. Since $f_{\text{o}}^V + f_{\text{a}}^V = 1$, this equation can be solved for the organic adsorbate fraction

$$f_{\text{o}}^V = \left(\frac{n_{\text{eff}}^2 - n_{\text{a}}^2}{n_{\text{eff}}^2 + 2}\right)\left(\frac{n_{\text{o}}^2 + 2}{n_{\text{o}}^2 - n_{\text{a}}^2}\right). \tag{17.6}$$

Furthermore, the areal adsorbate mass density can be calculated from (17.1) and (17.6) as

$$m_{\text{o}} = d_{\text{eff}}^{\text{SE}} \rho_{\text{o}} \left(\frac{n_{\text{eff}}^2 - n_{\text{a}}^2}{n_{\text{eff}}^2 + 2}\right)\left(\frac{n_{\text{o}}^2 + 2}{n_{\text{o}}^2 - n_{\text{a}}^2}\right). \tag{17.7}$$

Provided that $\rho_{\text{o}}$ and $n_{\text{o}}$ are known, $f_{\text{a}}^V$, $f_{\text{o}}^V$ and $m_{\text{o}}$ can hence be determined from SE data alone.

### 17.2.1.3  Approach 3: De Feijter

To understand how the linear averaging and the Lorentz effective medium approximations relate to each other, it is instructive to recast (17.3) and (17.7) into a different form. We make the ansatz

$$m_{\text{o}} = \frac{d_{\text{eff}}^{\text{SE}} (n_{\text{eff}} - n_{\text{a}})}{\text{d}n/\text{d}c}, \tag{17.8}$$

where we introduce the refractive index increment $\text{d}n/\text{d}c$. With the linear averaging approach (17.3), we find

$$\text{d}n/\text{d}c \equiv \frac{n_{\text{o}}^2 - n_{\text{a}}^2}{\rho_{\text{o}}(n_{\text{eff}} + n_{\text{a}})}, \tag{17.9}$$

whereas with the Lorentz approach (17.7),

$$\text{d}n/\text{d}c \equiv \frac{(n_{\text{eff}}^2 + 2)(n_{\text{o}}^2 - n_{\text{a}}^2)}{\rho_{\text{o}}(n_{\text{eff}} + n_{\text{a}})(n_{\text{o}}^2 + 2)}. \tag{17.10}$$

It can be seen that the two refractive index increments are not the same: they adopt the same form in the limit of very dense films ($n_{\text{eff}} \to n_{\text{o}}$) but differ for dilute films ($n_{\text{eff}} \to n_{\text{a}}$). Figure 17.2 provides an example of the variations of $\text{d}n/\text{d}c$ as a function of adsorbate volume fraction. $n_{\text{a}} = 1.333$, $n_{\text{o}} = 1.6$ and $\rho_{\text{o}} = 1.36$ g/cm$^3$ were chosen to represent a film of proteins in aqueous solution. The data shows that the refractive index increment varies by less than 10% with adsorbate volume fraction for both effective medium approximations, albeit in different ways.

More generally, the variations of $\text{d}n/\text{d}c$ with $f_{\text{o}}^V$ are small as long as the refractive index difference $n_{\text{o}} - n_{\text{a}}$ is small ($|n_{\text{o}} - n_{\text{a}}| \ll n_{\text{a}}$). Fortuitously, this condition

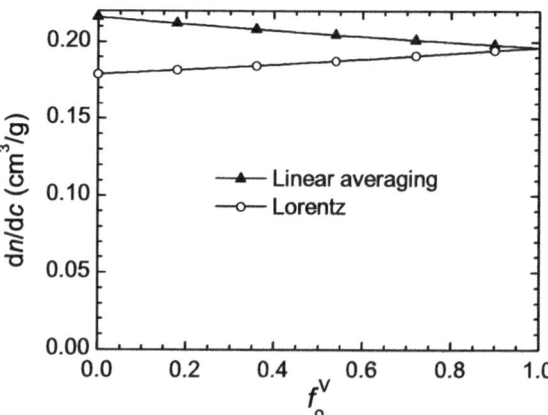

**Fig. 17.2** Refractive index increment *versus* adsorbate volume fraction for the linear averaging ((17.9); *closed triangles*) and Lorentz ((17.10); *open circles*) effective medium approximations, respectively, using $n_o = 1.6$, $n_a = 1.333$ and $\rho_o = 1.36$ g/cm$^3$

is fulfilled for many organic adsorbate/solvent mixtures of practical relevance. In these cases, $dn/dc$ can be treated as a material constant that characterizes a given adsorbate/solvent mixture. The assumption of a concentration-invariant $dn/dc$ is convenient by its simplicity and indeed frequently used with other optical methods. Equation (17.8), when employed together with a concentration-invariant $dn/dc$, is known as De Feijter's equation [7] (see also Chaps. 2 and 6).

#### 17.2.1.4 Determination of Material Constants

In order to determine $m_o$, $f_o^V$ or $f_a^V$ with the above-described methods, either $\rho_o$ and $n_o$ (approaches 1 and 2) or $dn/dc$ (approach 3) need to be known. Knowledge of $n_a$ is also necessary, yet is usually accessible with good accuracy, through the literature, bulk refractometry measurements or by ellipsometry.

The direct determination of $n_o$ may prove difficult, for example, if the organic material in its pure form is either not available or forms a powder. If enough material is available, then $dn/dc$ can be measured directly, e.g. through refractometry on reference solutions of defined solute concentration in the solvent of interest [7, 8]. The facile experimental access to $dn/dc$ constitutes another practical advantage of the De Feijter approach.

If the amount of available material is limiting the direct measurement of $dn/dc$, $\rho_o$ and $n_o$ may be estimated from the properties of the constituent chemical groups [8, 9]. For example, Zhao et al. [8] found that the molar refractivity $R$ (a measure of a molecule's polarizability) and the partial specific volume $\rho_o^{-1}$ of proteins can be determined with good accuracy from weighted averages of tabulated values for the amino acids from which they are made. Only little spread between proteins was observed, with mean values of $R = 0.253$ cm$^3$/g and $\rho_o = 1.36$ g/cm$^3$. The refractive index can then be calculated through the Lorentz-Lorenz equation [10, 11]

$$R\rho_0 = \frac{n_0^2 - 1}{n_0^2 + 2}. \tag{17.11}$$

For proteins, for example, we find $n_0 = 1.6$. In the limit of low coverage, this corresponds to $dn/dc = 0.18 \, \text{cm}^3/\text{g}$ (Fig. 17.2), in good agreement with experimentally determined values.

One should keep in mind that the above-described procedures are based on measurements in solutions which are isotropic, because the solute molecules can adopt random orientations. In contrast, organic thin films may be optically anisotropic, since the adsorbate adopts a preferred orientation on the surface (e.g. in self-assembled monolayers or supported lipid bilayers). In this case, $n_0$ or $dn/dc$ determined in solution may not adequately describe the optical properties of the adsorbates on the surface. More complex models that account for birefringence would be required for a rigorous treatment. While these will not be treated here, we note that a suitably adjusted $dn/dc$ can also provide reasonable solutions.

### 17.2.2  Approaches to Analyze QCM-D Data

The quartz crystal microbalance with dissipation monitoring measures changes in the resonance frequency $\Delta f$ and energy dissipation $\Delta D$ of a quartz sensor crystal at several overtones $i$ upon interaction of soft matter with its surface (Fig. 17.3). The sensor crystal is typically coated with metal films on the top and bottom faces to act as electrodes. The top face, bare or coated with additional layers (e.g. silica), is used as the experimental substrate. The QCM-D response is sensitive to the areal mass density and the mechanical properties of the surface-bound film. To a first approximation, a decrease in $f$ indicates a mass increase, while high (low) values of $\Delta D$ indicate a soft (rigid) film.

With regard to data analysis, it is useful to consider two different types of films, namely, laterally homogeneous films and films made of a monolayer of discrete particles [13]. Films in which the length scale of the sample's internal structure is smaller than the film thickness can be treated as laterally homogeneous films through an effective medium approximation. In the case of QCM-D, and in contrast to ellipsometry, such an approach is not appropriate for films consisting of discrete nanoscale objects (proteins, liposomes, viruses, nanoparticles) adsorbed on the surface, where the film thickness is about the same as the particle diameter. Principal analysis approaches are outlined in the following. For an in depth discussion, the reader is referred to the literature [13, 14].

**Coupled Solvent** A primary film parameter that is in many cases quantitatively accessible through QCM-D is the areal mass density. A feature that distinguishes QCM-D from ellipsometry, and other optical mass-sensitive methods, is that hydrodynamically coupled solvent contributes to the measured areal mass density

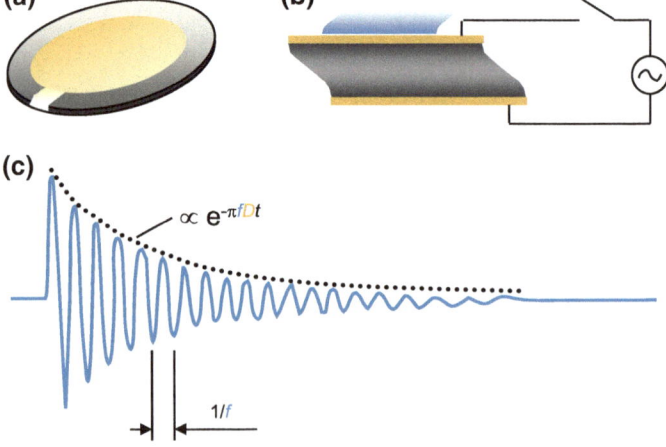

**Fig. 17.3** QCM-D sensor and sensing principle. **a** Commonly used QCM-D sensors have a diameter of 14 mm, a thickness of ~300 μm, and a fundamental resonance frequency of ~5 MHz. The active sensing area (not shown) is confined to a central, circular spot of ~5 mm diameter (the size decreases somewhat with the overtone number $i$ [12]). **b** Application of oscillatory voltage results in a cyclical shear deformation of the sensor. In the ring-down method, which is used by QCM-D, the driving voltage is intermittently switched off and the decay in time of the oscillation is monitored. **c** From the decay curve, the resonance frequency $f$ and the energy dissipation $D$ are extracted. Depending on the driving frequency, different overtones $i = 3, 5 \ldots, 13$ are excited. $f$ and $D$ (or equivalently, the bandwidth $\Gamma$) can also be obtained by another QCM method, based on (acoustic) impedance analysis (not shown) [13]

$$m^{\mathrm{QCM}} = m_{\mathrm{o}} + m_{\mathrm{a}}. \tag{17.12}$$

From the QCM data alone, the mass fractions of organic adsorbate $f_{\mathrm{o}}^{\mathrm{m}} = m_{\mathrm{o}}/m^{\mathrm{QCM}}$ and coupled ambient solvent $f_{\mathrm{a}}^{\mathrm{m}} = m_{\mathrm{a}}/m^{\mathrm{QCM}}$ to the mass response cannot be resolved. In contrast, if $m_{\mathrm{o}}$ is available from a parallel measurement by ellipsometry, $m_{\mathrm{a}}$ can be readily quantified. Hence, the different sensitivity to solvent is a key complementary feature of the two techniques.

#### 17.2.2.1 Homogeneous Films

**Sauerbrey Equation** For films that are sufficiently rigid, the areal mass density can be determined through the Sauerbrey equation [15]

$$m^{\mathrm{QCM}} = \rho_{\mathrm{eff}} d_{\mathrm{eff}}^{\mathrm{QCM}} = -C \frac{\Delta f_i}{i}, \tag{17.13}$$

where $i$ is the overtone order ($i = 1, 3, 5, \ldots$). The mass sensitivity constant $C$ is determined by the material properties of the quartz sensor. For the most common sensors with a fundamental resonance frequency of about 5 MHz, $C = 18$ ng/cm$^2$/Hz. A sufficient criterion for the applicability of the Sauerbrey equation is that the variation in $\Delta f_i$ as a function of $i$ (the overtone dispersion) is small. Equivalently, for 5 MHz sensors, $\Delta D/(-\Delta f_i/i) \ll 4 \times 10^{-7}$ Hz$^{-1}$ [13].

The film thickness can be readily estimated through (17.13), if the densities of the organic adsorbate $\rho_o$ and the ambient solvent $\rho_a$ are similar ($\rho_{\text{eff}} \approx \rho_o \approx \rho_a$). If the densities are different, it can be quantified with the help of the ellipsometrically determined $m_o$

$$d_{\text{eff}}^{\text{QCM}} = \frac{1}{\rho_a}\left[m^{\text{QCM}} - m_o\left(1 - \frac{\rho_a}{\rho_o}\right)\right]. \tag{17.14}$$

For the same reasons as outlined in the case of ellipsometry, the thickness should be considered an effective and method-specific property.

**Viscoelastic Modeling** If the film is laterally homogeneous and gives rise to significant dissipation ($\Delta D_i > 0$) then the viscoelastic model can be applied. The QCM-D response is treated by a continuum model based on the analysis of shear wave propagation in viscoelastic media [14, 16]. The linear viscoelastic properties of the film are determined by the frequency-dependent storage and loss moduli. The storage modulus $G'(f)$ describes material elasticity and the loss modulus $G''(f) = 2\pi f \eta$ is a measure of the film's viscosity $\eta$. Within a limited frequency range, such as it is accessible by QCM-D, the frequency dependencies are often well-approximated by power laws: $G'(f) = G'_o(f/f_o)^{\alpha'}$ and $G''(f) = G''_o(f/f_o)^{\alpha''}$, where $f_o$ is an arbitrarily chosen reference frequency and the exponents $\alpha'$ and $\alpha''$ are constant. Additional parameters required to fully determine the properties of the homogeneous film are the density $\rho_{\text{eff}}$ and the thickness $d_{\text{eff}}^{\text{QCM}}$.

The products $\rho_{\text{eff}} G'(f)$, $\rho_{\text{eff}} G''(f)$ and $\rho_{\text{eff}} d_{\text{eff}}^{\text{QCM}} = m^{\text{QCM}}$ as well as $\alpha'$ and $\alpha''$ can be extracted through numerical fitting of QCM-D data at several overtones. Knowledge of the film density is required to obtain the actual values of $d_{\text{eff}}^{\text{QCM}}$, $G'(f)$ and $G''(f)$ from the fitting. This is trivial if $\rho_o$ and $\rho_a$ are similar. If they are different, $\rho_{\text{eff}}$ can be quantified with the help of the ellipsometrically determined $m_o$ through

$$\rho_{\text{eff}} = \frac{\rho_a}{1 - \frac{m_o}{m^{\text{QCM}}}(1 - \frac{\rho_a}{\rho_o})}, \tag{17.15}$$

as can be seen from (17.14). Independent knowledge of some of the film properties is also helpful for a model with so many fitting parameters to obtain meaningful results. If the film is sufficiently thick to determine $d_{\text{eff}}^{\text{SE}}$ with good accuracy, and if one assumes that $d_{\text{eff}}^{\text{QCM}} = d_{\text{eff}}^{\text{SE}}$, then the thickness (or the total areal mass density $\rho_{\text{eff}} d_{\text{eff}}^{\text{SE}}$ of the solvated film) can actually be provided by ellipsometry. The comparison of $d_{\text{eff}}^{\text{QCM}}$ and $d_{\text{eff}}^{\text{SE}}$, however, is not trivial and should be considered with care. In the practically relevant case of a polymer brush, for example, the two thickness parameters have been found to differ by several fold [16].

Dedicated software for the numerical fitting of QCM data in terms of film properties is available.[1] Reference [17] gives a detailed description of the viscoelastic modeling procedure, including the determination of joint confidence regions.

### 17.2.2.2 Monolayers of Discrete Nanoscale Objects

In contrast to effective medium approximations in ellipsometry, the effective medium approximation for mechanical properties is not appropriate for films consisting of discrete nanoscale objects (e.g. globular proteins, liposomes, viruses, nanoparticles), where the film thickness is about the same as the particle diameter. In such films, an additional energy dissipation mechanism operates that is not captured by the continuum model and that is intimately related to the mode of attachment of the particles to the substrate. To correctly reproduce the full behavior of such systems, finite element modeling is required that considers explicitly the mode of attachment of the particles to the substrate. This approach has only recently emerged and its full potential remains to be explored. The reader is referred to the original literature [13, 18, 19] for information about how this approach can be used to probe mechanical properties. For our purposes, we consider only the information that can be obtained through the quantification of hydrodynamically trapped solvent.

Provided that the films are sufficiently rigid, the areal mass density $m^{QCM}$ can still be extracted through (17.13). Correlation with $m_o$, determined by ellipsometry, provides the hydrodynamically trapped solvent $m_a$ through (17.12). In the limit of high coverage, almost all solvent in the interstitial space between particles contributes to the QCM frequency shift, i.e. the areal mass density corresponds to the mass of particles and liquid within a surface adlayer that is as thick as the particles are high [18, 20, 21]. The film thickness can in this case be estimated through (17.13) or (17.14).

**Trapped-Solvent Coat Model** At lower coverage, the situation can be rationalized in terms of coats of hydrodynamically trapped liquid that surround each adsorbed particle. In reality, there is no sharp line separating trapped solvent from free solvent, but the coats are a useful approximation to the complex hydrodynamic effects that come into play in these films. An empirical model that assumes that the shape of the liquid coat around each adsorbed particle is fixed, and that particles are randomly distributed on the surface, has proven remarkably successful in quantitatively reproducing the coverage-dependent contribution of trapped solvent to $m^{QCM}$ (Fig. 17.4) [22, 23]. If the contribution of hydrodynamically trapped solvent, $f_a^m = 1 - m_o/m^{QCM}$, is determined in a coverage dependent manner, this model can be used to extract quantitative information about the organization of particles on the surface. If particle weight and lateral distribution are known, particle size and height-to-width ratio can

---

[1]The software QTM is freely available for download (http://www2.pc.tu-clausthal.de/dj/software_en.shtml), and features an automated analysis of joint confidence regions for the obtained results. The software QTools is available from Biolin Scientific (http://www.q-sense.com).

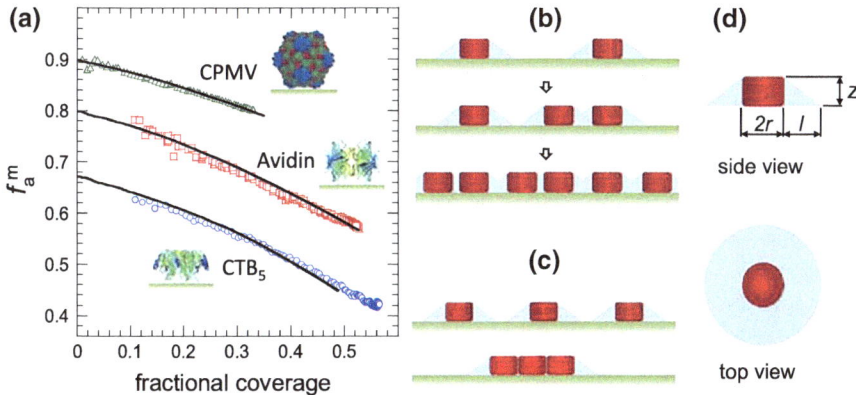

**Fig. 17.4** Empirical approach to describe the contribution of trapped solvent to $m^{QCM}$ for monolayers of discrete nanosized objects. **a** Plots of $f_a^m$ against fractional surface coverage, determined from combinations of optical mass-sensitive techniques (reflectometry or ellipsometry) [22, 23] with QCM-D for the adsorption of particles of various sizes and height-to-width ratios: cow-pea mosaic virus (CPMV; *green triangles*), cholera toxin B5 subunit (CTB$_5$; *blue circles*) and avidin (*red squares*). The particles' structure and orientation on the surface are also shown (*insets*; CPMV has a diameter of 28 nm, sizes of CTB$_5$ and avidin are enlarged 4-fold with respect to CPMV). Generally, $f_a^m$ decreases with increasing coverage. The magnitude of $f_a^m$ and the rate of its decrease with coverage are sensitive to the particles' height-to-width ratio and internal liquid content. *Black lines* are fits to the data with the trapped-liquid coat model, described in detail in **b–d**. **b** Phenomenologically, the trapped liquid can be rationalized as a coat (*light blue*) that surrounds each adsorbed molecule (*red*). With increasing coverage, these coats increasingly overlap, leading to a decrease in $f_a^m$. **c** The amount of trapped liquid also depends on the lateral organization of surface-bound material. A compact cluster traps less liquid than if the particles were dispersed homogeneously across the surface. In this way, the measured areal mass density becomes sensitive to the lateral organization of surface-bound material [23]. **d** To fit the data shown in **a**, particles were modeled as upright cylinders carrying liquid coats with the shape of truncated cones. It was further assumed that the particles were randomly distributed on the surface. Fitting the experimental data in **a** with $z$ and $l$ as free parameters yields $l/z = 1.35 \pm 0.15$. The steepness $l/z$ of the coat is independent of the particle size (sizes up to 30 nm were tested) and height-to-width ratio (ratios between 0.3 and 1 were tested) [23]. To what extent these simple geometrical assumptions work for particles with arbitrary shape is subject to discussion and needs further research. Adapted from [13]

be determined. If the particle size and orientation on the surface are known, information about the lateral distribution of particles, such as the degree of clustering (e.g., protein oligomerization) can be obtained [23]. Dedicated software to simulate the coverage-dependent $f_a^m$ is available.[2]

---

[2]A MATLAB routine to predict the coverage-dependent $f_a^m$ as a function of the particle size, mass and aspect ratio, and for a selected set of lateral distribution scenarios is freely available for download (http://www.rrichter.net).

## 17.3    Practice

### 17.3.1    *Experimental Setup*

Figures 17.5 and 17.6 illustrate the experimental setup for combined SE/QCM-D measurements. The SE and QCM-D are computer-controlled via (separate) user interfaces, and the temperature of the fluid cell is controlled through the QCM-D interface. SE and QCM-D data are acquired simultaneously, and typically continuously, with a time resolution on the order of one data set per second. A proper design of the fluid cell and the fluid handling system is critical for the quality of the obtained data.

#### 17.3.1.1    Fluid Cells

**Closed Fluid Cell** A closed fluid cell for combined SE/QCM-D measurements (Fig. 17.7a) is commercially available. The closed design facilitates sequential exposure of different sample and rinsing solutions to the sensor surface, through application of a pump-driven flow at defined rates to the outlet tubes. Typically, a minimum of ∼0.3 ml of each sample solution is required. Sample handling can be implemented manually (e.g., [23]) or through an automated fluid handling system (e.g., [24]).

The reaction chamber is formed by the circular sensor surface and an opposite wall of equal shape, separated by ∼1 mm using an O-ring that also defines the chamber diameter of ∼12 mm. An inlet and an outlet for the exchange of liquid

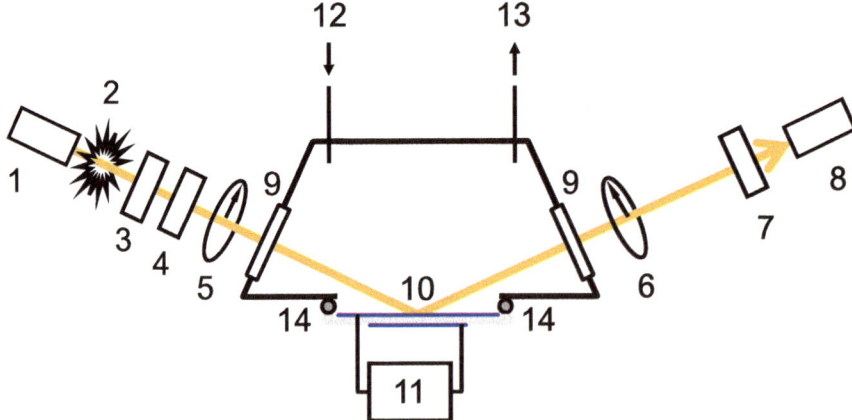

**Fig. 17.5** Schematic of the combined SE/QCM-D setup. *1*—light source, *2*—unpolarized light, *3*—polarizer, *4*—rotating compensator, *5*—polarized light, *6*—polarized light altered by sample surface, *7*—analyzer, *8*—detector, *9*—optical windows, *10*—QCM sensor surface, *11*—QCM sensor control, *12*—fluid inlet, *13*—fluid outlet(s), *14*—O-ring for sealing

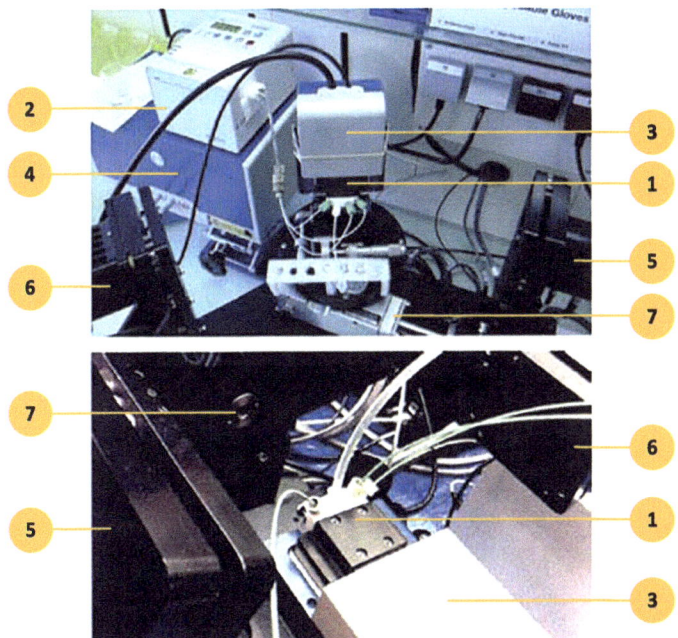

**Fig. 17.6**  Photos of combined SE/QCM-D setups, built from a Q-Sense E1 QCM-D system (Biolin Scientific) and M2000 SE systems (J.A. Woollam Co.) with a horizontal (*top*) and a vertical (*bottom*) plane of incidence, respectively. *1*—closed SE/QCM-D fluid cell with tubing for fluid flow, *2*—peristaltic pump, *3*—E1 QCM-D module, *4*—QCM-D electronics unit, *5*—SE light source unit, *6*—SE detector unit, *7*—SE goniometer

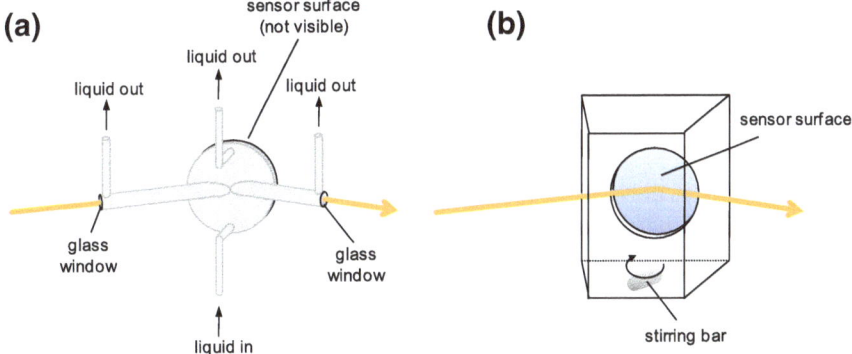

**Fig. 17.7**  Schematic of closed (**a**) and open (**b**) fluid cells. Reproduced with permission from [23]. Copyright 2010 American Chemical Society

are positioned opposite one another close to the O-ring on the non-sensor wall. In addition, two liquid-filled cylindrical channels of ∼2 mm diameter are connected to the reaction chamber with an angle of 65° relative to the sensor surface normal. At its extremity, each light channel is equipped with an additional liquid outlet and a BK7 glass window. The probing light beam enters and leaves the chamber through the glass windows. The windows are mounted at perpendicular incidence and with minimal stress, to minimize any effect on the polarization of the probing beam.

**Open Fluid Cell** One of the authors has recently reported an open fluid cell for combined SE/QCM-D measurements (Fig. 17.7b) [23]. The open design allows for a small volume of concentrated sample solution to be injected with the aid of a pipette, and a stirrer ensures rapid homogenization (i.e. within a few seconds) of the ∼2 ml cell content. This design is particularly useful to realize conditions of uniformly accessible surfaces in still solution.

Based on a traditional cuvette design [25], the cell has a trapezoid-shaped bottom and an open top. The sensor surface is integrated into the cuvette's back wall. The light beam enters and leaves at perpendicular incidence through the side walls, which are made from BK7 glass. A magnetic stirring bar is placed on the bottom of the cuvette and driven by an external motor at speeds of typically ∼200 rpm. The open fluid cell requires the plane of incidence of the probing light beam to be horizontal with an angle of incidence of 70°.

### 17.3.1.2 Uniformly Accessible Surfaces

Frequently, the rate of binding at surfaces is limited by mass transport—through diffusion or convection—of the molecules in solution [26]. In these cases, it is of particular importance that the transport conditions are designed such that the sensor surface is homogeneously accessible. While QCM-D and SE are sensing on the same surface, the size and shape of the sensing regions do not match exactly. To properly correlate the two techniques, one has to ensure that deposition occurs homogeneously across the sensor surface over a sufficiently large area. This has implications for the fluidic conditions that should be used if adsorption is mass-transport limited. A simple and reliable way to achieve the condition of a uniformly accessible surface with the available setups is to work in still solution, i.e. to rapidly inject and homogenize the sample and then to track the binding in the absence of any flow or stirring. Reference [23] provides a detailed description of how this can be implemented.

## 17.3.2 Data Acquisition Procedure

Figure 17.8 illustrates the combined SE/QCM-D data acquisition procedure. Before the dynamic and combined measurement, three separate SE-only data sets are taken. First, an SE measurement, denoted here as *SE-NoCell*, is taken on the appropriately prepared QCM-D sensor surface in the absence of the fluid cell. Next, the sensor is installed in the fluid cell, and a second SE measurement *SE-Cell* is acquired. In parallel, QCM-D measurement *QCM-Cell* is started. The QCM-D measurement

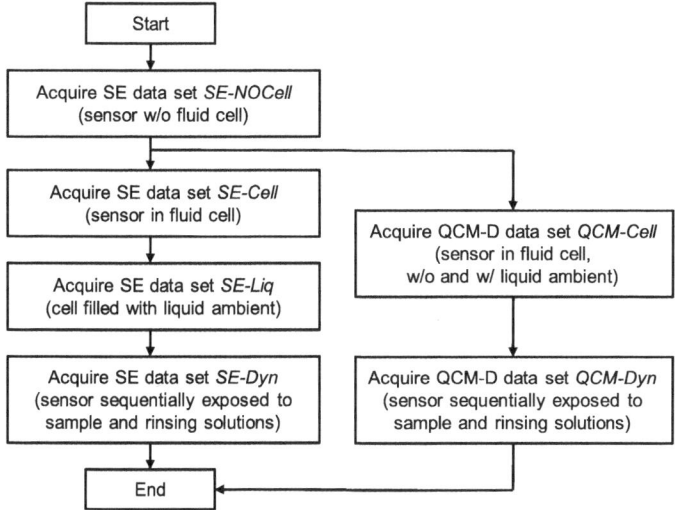

**Fig. 17.8** Combined SE/QCM-D data acquisition flowchart

is useful to ascertain that the QCM-D sensor works with appropriate stability and noise levels. Next, the liquid cell is filled with the ambient liquid of choice, and SE measurement *SE-Liq* is then taken. At this step, it is important to ascertain that the liquid cell is free of air bubbles. A deformed shape and/or heterogeneous brightness of the light beam after passage through the fluid cell can provide indications of bubbles in the light path. Alternatively, the beam path through the fluid cell can also be directly examined with the aid of dental mirrors (one over each window to prevent being temporarily blinded by bright light from the SE). Too low changes in the QCM-D responses upon switching from air to liquid and/or unstable responses in liquid (in particular for $i = 1$ or 3) are a good indicator for the presence of bubbles on the sensor surface. Throughout all measurements, the SE angle of incidence is set to the liquid cell's optical port angle (e.g., 65° for the closed fluid cell).

At this point, the dynamic and combined measurement is ready to begin. Acquisition of time-resolved SE (*SE-Dyn*) and QCM-D (*QCM-Dyn*) data is ideally started simultaneously to enable proper correlation of data at a given time point. In practice, a small delay (typically <1 s) cannot be avoided, because acquisition is started manually through a mouse click.

### 17.3.3  Data Analysis Procedure

The raw SE (i.e. $\Psi(\lambda, t)$ and $\Delta(\lambda, t)$) and QCM-D (i.e. $\Delta f(i, t)$ and $\Delta D(i, t)$) data are analyzed through the protocol summarized by Fig. 17.9.

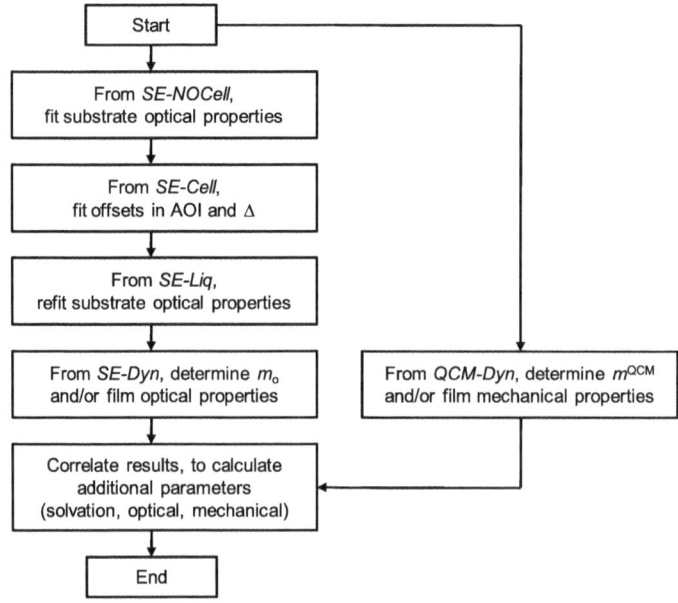

**Fig. 17.9** Combined SE/QCM-D data analysis flowchart

### 17.3.3.1    Areal Mass Densities

The optical model described in Sect. 17.2.1 can now be built by incorporating the three separate SE data sets. Substrate parameters, offsets due to window and angle-of-incidence (AOI) effects, and substrate modification due to liquid exposure are accounted for by best-matching *SE-NoCell*, *SE-Cell*, and *SE-Liq* data, respectively, to data generated by the optical model. A Cauchy layer with $d_{\mathrm{eff}}^{\mathrm{SE}}$ and $n_{\mathrm{eff}}(\lambda)$ is then added to the optical model to represent the organic film. Depending on the approach of choice, either $d_{\mathrm{eff}}^{\mathrm{SE}}$ and $n_{\mathrm{eff}}(\lambda)$ (for Lorentz and de Feijter) or $d_{\mathrm{o}}$ (for virtual separation) are treated as variable fit parameters for each individual data set (i.e. at a given time) in the dynamic SE data set *SE-Dyn*. Through (17.8), (17.7) or (17.3), $m_{\mathrm{o}}$ is then determined. For organic films made up from several discrete layers, the optical model can be expanded by additional layers.

The QCM-D data can be treated using Sauerbrey's equation or viscoelastic modeling to extract $m^{\mathrm{QCM}}(t)$, and with the viscoelastic model the (time-dependent) film mechanical properties can also be quantified. Unlike for the optical data, the mechanical properties of the substrate do usually not need to be precisely known to determine the properties of the organic overlayer(s), because most substrates—including typical metal or metal oxide layers—are well-described as rigid and non-dissipative layers that obey Sauerbrey's equation.

**Table 17.1** Determination of film solvation parameters from $m_o$ and $m^{QCM}$. $\rho_a$ and $\rho_o$ are assumed to be known. Note that some of the equations simplify considerably, if $\rho_o \cong \rho_a$

| Parameter | Determined through formula |
|---|---|
| Trapped solvent areal mass density | $m_a = m^{QCM} - m_o$ |
| Film density | $\rho_{eff} = \dfrac{\rho_a}{1 - \frac{m_o}{m^{QCM}}\left(1 - \frac{\rho_a}{\rho_o}\right)}$ |
| Adsorbate mass fraction | $f_o^m = \dfrac{m_o}{m^{QCM}}$ |
| Trapped solvent mass fraction | $f_a^m = 1 - \dfrac{m_o}{m^{QCM}}$ |
| Adsorbate volume fraction | $f_o^V = \dfrac{\rho_{eff}}{\rho_o}\dfrac{m_o}{m^{QCM}}$ |
| Trapped solvent volume fraction | $f_a^V = 1 - \dfrac{\rho_{eff}}{\rho_o}\dfrac{m_o}{m^{QCM}}$ |
| Film thickness | $d_{eff}^{QCM} = \dfrac{m^{QCM}}{\rho_{eff}}$ |

### 17.3.3.2 Solvation/Porosity

Through time-resolved correlation of $m_o$ and $m^{QCM}$, the amount of solvent that is coupled to the film can be determined. Because the time resolution of SE and QCM-D data is typically not identical, interpolation of one data set is required before time-resolved correlation can be pursued.

For laterally homogeneous films, the coupled solvent relates to film solvation or porosity. For the reader's convenience, Table 17.1 reviews the formulae for the calculation of quantities related to film solvation. For ultra-thin films, solvation/porosity can be accessed exclusively with combined SE/QCM-D data. For thicker films, the parameters listed in Table 17.1 can alternatively be determined from SE data alone, by taking advantage of the ellipsometric film thickness $d_{eff}^{SE}$. In this case, $m^{QCM}$ is replaced by $d_{eff}^{SE}\rho_{eff}$, with $\rho_{eff} = \rho_a + m_o/d_{eff}^{SE}(1 - \rho_a/\rho_o)$. We remind that thickness parameters are effective and instrument-specific. They may vary significantly, if the film exhibits a fuzzy boundary, or more generally, if the density within the film varies along the surface normal. Accordingly, solvation data determined through $m^{QCM}$ and $d_{eff}^{SE}$, respectively, may differ systematically.

For monolayers of discrete nanoscale objects, the interpretation of hydrodynamically coupled solvent as film solvation/porosity is only meaningful if the monolayer is dense. For sparsely covered surfaces, the simplifying approximation of a continuous film is not adequate for QCM-D. Here, the trapped-solvent coat model (Fig. 17.4) can be useful to extract information about the organization of particles (e.g., particle size and height-to-width ratio, or clustering) on the surface.

### 17.3.3.3 Optical and Mechanical Properties

Naturally, the optical and mechanical properties that are accessible with SE and QCM-D, respectively, remain accessible with the combined setup, and the reader is referred to specialized publications for a detailed treatment of how these properties are determined from the response of each technique. In principle, the combined setup enables a direct correlation of the mechanical and optical properties of a given organic film. To our knowledge, the potential of the SE/QCM-D combination in this regard remains unexplored and an interesting subject for further study.

The information about film thickness and porosity, determined as described in the preceding section, can be valuable for the quantification of optical and mechanical properties. In particular, knowledge of $\rho_{eff}$ is required to extract $G'$ and $G''$ from the quantities $\rho_{eff}G'$ and $\rho_{eff}G''$ that are provided by the viscoelastic model. Similarly, $n_{eff}$ of ultra-thin films can be estimated through (17.4), (17.5) or (17.8), if the volume fraction of adsorbate and solvent, or equivalently the film thickness, are known.

## 17.4  Applications

In this section, we shall illustrate current and emerging applications for the combined SE/QCM-D setup through selected examples. In some cases, data analysis approaches that are useful for the SE/QCM-D combination have first been developed based on data obtained through the combination of QCM-D with other optical mass-sensitive techniques such as surface plasmon resonance (SPR), single-wavelength ellipsometry or optical reflectometry. Where appropriate, we will therefore also refer to these techniques.

**Example 1: Formation of a Self-assembled Monolayer** Alkanethiols are hydrocarbons with a sulfur head group, a hydrocarbon chain body, and a functionalized tail group exhibiting a desired surface chemistry, and known to form self-assembled monolayers (SAMs) through the interaction of the head group with gold. SAMs are useful as uniform, cost-effective coatings for adjusting a substrate's surface properties [27]. Figure 17.10 shows data from a combined SE/QCM-D measurement for the formation of an 8-mercapto-1-octanol SAM in aqueous solution. SAM formation from ethanol solution can be monitored in a similar way [28]. From the data, it is evident that a fast initial growth step is followed by a slower second process, in agreement with the literature [27]. $f_o^V$ is uniform throughout the measurement, implying that the porosity is constant throughout the ultra-thin film growth. Note that the thickness measured by QCM-D, $d_{eff}^{QCM} = 6$ nm, exceeds the length of the molecules from which the SAM is made by about 3-fold. This indicates that the SAM is not perfect but that aggregates are likely to be present on the SAM.

**Example 2: Controlling the Preparation of Biofunctional Films for Analytical Applications** Self-assembled monolayers that incorporate single-stranded DNA molecules are widely used for analytical applications that include genotyping,

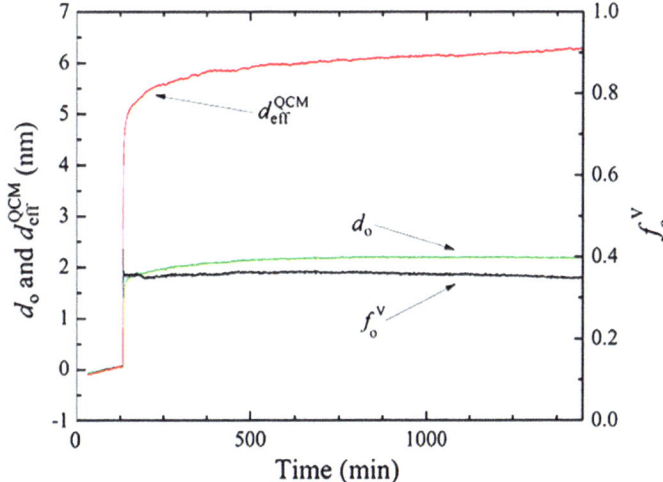

**Fig. 17.10** Thickness parameters $d_o$ and $d_{eff}^{QCM}$, and adsorbate fraction $f_o^V$ as a function of incubation time of 2 mM 8-mercapto-1-octanol solution. The measurement was performed with a closed fluid cell in purified water. At about 130 min (after a desired baseline had been reached), a 2 mM aqueous alkanethiol solution was injected. Flow was initially applied, at a rate of 0.1 ml/min, until about 130 min, and then stopped. $d_o$ was determined using the virtual separation approach (see Sect. 17.2.1.1) and $n_o = 1.484$; $d_{eff}^{QCM}$ was determined through (17.13) and $\rho_o = 0.93$ mg/ml ($n_o = 1.484$ and $\rho_o$ were provided by the manufacturer). From these data $f_o^V$ was determined from $\rho_a = 1.0$ mg/ml and equations in Table 17.1

protein and small molecule detection, and high-throughput affinity screening. Characterizing the formation and interrogation of DNA-based sensors using SE/QCM has the potential to elucidate factors that contribute to sensor response, such as surface conformation and hybridization efficiency.

Figure 17.11 demonstrates that SE/QCM is capable of characterizing sub-nanometer average thickness changes and the porosity of multiple-component, biological, ultra-thin films. The DNA probe, a stem-loop probe specific for a region surrounding codon 12 of the *K-ras* gene, mutations of which are often present in pancreatic cancer lesions [29], was conjugated to a six-carbon alkanethiol moiety for attachment to a gold substrate and the surface was subsequently backfilled with 6-mercapto-1-hexanol to enhance stability. Non-coding DNA (BRCA2) and *K-ras* target DNA (all DNA constructs from Biosearch Technologies, Novato, CA) were sequentially added. The resolution in $m_o$ and $d_{eff}^{QCM}$ in this measurement is 0.1 ng/cm$^2$ and 0.1 nm, respectively. It is notable that the adsorbate volume fraction $f_o^V$ after the first deposition step is close to 1, suggesting that the probe aptamer binds in a flat conformation, potentially driven by nonspecific interactions between DNA and gold.

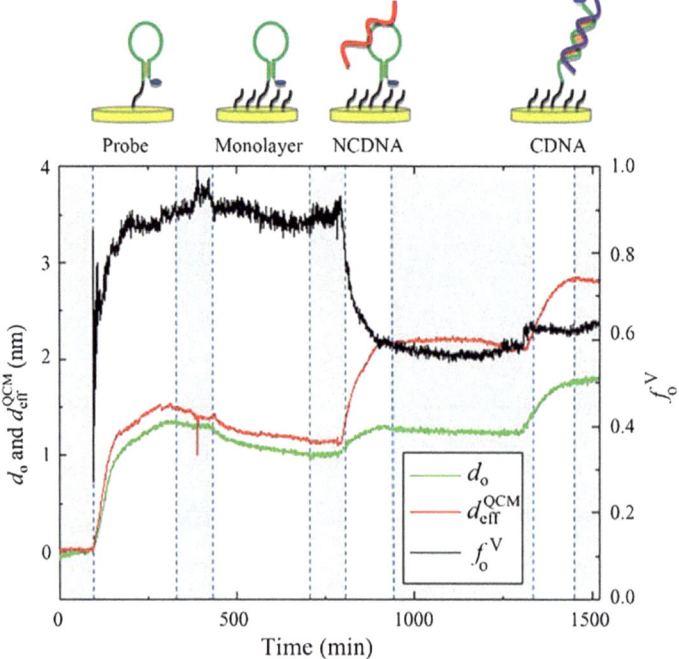

**Fig. 17.11** Evolution of thickness parameters $d_o$ and $d_{eff}^{QCM}$, and adsorbate fraction $f_o^V$ during the construction of a multiple-component organic film on gold. The measurement was performed with a closed fluid cell in buffer solution (20 mM Tris, 140 mM NaCl, 5 mM KCl, 1 mM CaCl$_2$, 1 mM MgCl$_2$ at pH 7.4) at a flow rate of 50 μl/min. $d_o$ was determined using the virtual separation approach (see Sect. 17.2.1.1) assuming $n_o = 1.50$; $d_{eff}^{QCM}$ was determined through (17.13) assuming $\rho_o = 1.0$ mg/ml. Immobilization of probe aptamer, backfilling with 6-mercapto-1-hexanol, exposure to non-complementary DNA, and complementary DNA are denoted by *Probe*, *Monolayer*, *NC DNA*, and *C DNA*, respectively. Note the decrease in thickness upon monolayer formation, the shifts in adsorbate fraction when non-complementary and complementary DNA are introduced, and the differences between SE and QCM-D thickness responses for non-complementary and complementary DNA interrogation. The graph is shaded when buffer solution rinses proceeded through the fluid cell

## 17.4.1 Solvation and Swelling of Polymer Films

Iturri Ramos et al. [30] applied the combined SE/QCM-D setup to quantify the hydration of polyelectrolyte multi-layers (PEMs), i.e. films that are formed through alternate adsorption of polycations and polyanions onto a charged substrate. Correlation of SE and QCM-D data revealed drastic variations in film hydration as a function of the polyelectrolyte species employed for multilayer construction, and the ionic strength of the ambient solution. In particular, films made of poly(diallyldimethylammonium chloride) and poly(sodium 4-styrenesulfonate) could be observed to swell and collapse in cycles, depending on which polyion was deposited last.

Stimuli-responsive polymer films are now popular for a wide range of applications. The ability to quantify both adsorption/desorption and swelling/collapse processes, and to distinguish between them, constitutes a particular strength of the combined setup for the study of such systems.

#### 17.4.1.1  Effective Thickness of Films with a Density Profile

In a pioneering study, Domack et al. [16] combined QCM and ellipsometry on the same substrate to study the swelling of polymer brushes, i.e. films made from polymer strands that are end-grafted at high density to a substrate. The polystyrene brushes under investigation had a thickness on the order of 100 nm, i.e. well above the ultra-thin film limit, allowing independent determination of $d_{\mathrm{eff}}^{\mathrm{SE}}$ and $d_{\mathrm{eff}}^{\mathrm{QCM}}$. Comparison of both quantities revealed that $d_{\mathrm{eff}}^{\mathrm{QCM}}$ was up to several times larger than $d_{\mathrm{eff}}^{\mathrm{SE}}$. In order to extract film thickness parameters, the authors assumed that the polymer density within the film is constant and that the film-ambient interface is sharp ("box profile"). The density in polymer brushes, however, is known to decrease smoothly with the distance from the substrate, and this effect is accentuated if the brush-forming polymers have a broad size distribution. By comparing theory and experiment, the authors demonstrated that the difference in the effective thickness parameters arises because the acoustic contrast is much higher than the optical contrast in such films [14].

More generally, $d_{\mathrm{eff}}^{\mathrm{QCM}}$ is expected to be larger than $d_{\mathrm{eff}}^{\mathrm{SE}}$ as soon as the interface between the organic film and the ambient solution becomes fuzzy. In many experimentally relevant cases, the density profile of the polymer film of interest is not known. Detailed theoretical analysis [14, 16, 31] shows that the data available from either QCM-D or SE alone are usually not sufficient to resolve the shape of the density profile for organic thin films. Here, the magnitude of the discrepancy between $d_{\mathrm{eff}}^{\mathrm{QCM}}$ and $d_{\mathrm{eff}}^{\mathrm{SE}}$ can provide insight about the "fuzzyness" of the film-ambient interface.

### 17.4.2  Calibration Curves

The requirements imposed by the combined setup make it difficult to miniaturize the fluid cell. As a consequence, the applicability of the SE/QCM-D combination may be limited, if the amount of available sample is restricted. In contrast, QCM-D and ellipsometry alone can operate with significantly smaller liquid volumes of 100 µl and less per sample incubation step. Moreover, parallelization remains a challenge with the combined setup while it is readily available for QCM-D (and feasible for ellipsometry [32]) alone. In systematic studies that involve repeated immobilization of a given molecule at different surface densities, it is often useful to quantitatively relate the QCM-D responses to surface coverage. To this end, a selected measurement with the combined setup can be performed to "calibrate" the QCM-D response.

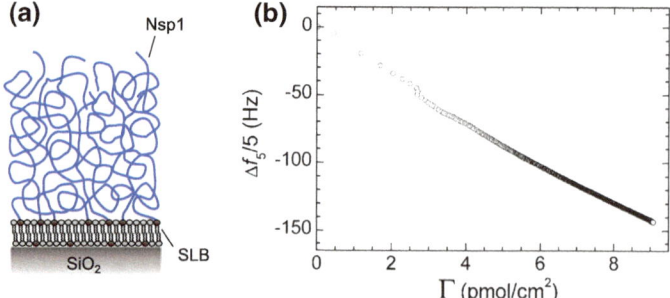

**Fig. 17.12** QCM-D frequency shifts $\Delta f$ *versus* grafting density $\Gamma$ for an ultra-thin film of FG domains of the nucleoporin Nsp1. Nucleoporin FG domains are intrinsically disordered protein domains rich in phenylalanine-glycine (FG) dipeptides. **a** In the present experiment, FG domains were end-grafted through N-terminal histidine tags to a $Ni^{2+}$-NTA functionalized supported lipid bilayer (SLB). The FG domains form a film of entangled and/or transiently cross-linked polymers of a few 10 nm in thickness. These FG domain films represent a well-defined model system of the so-called nuclear pore permeability barrier, which regulates transport of macromolecules into and out of the nucleus of eukaryotic cells [17, 35]. **b** The plot was obtained from a combined SE/QCM-D measurement with the open fluid cell, as described in [17]. Grafting densities were derived from $m_0$ and the FG domain's molecular mass. The plot represents a "calibration curve". With the aid of this curve, films of desired grafting density can be created in measurements with QCM-D alone, by interrupting the incubation with proteins at the equivalent frequency shift. The curve is approximately linear, with a slope of 16 Hz/(pmol/cm$^2$)

The calibration curve can then be used to correlate areal adsorbate densities with frequency (or dissipation) responses for a large set of measurements with QCM-D alone. Figure 17.12 provides an example of such a calibration curve, and [33, 34] illustrate how calibration curves can be put to practical use.

It is notable that the relationship between $\Delta f$ and $m_0$ in Fig. 17.12 is approximately linear. Such linear relationships have been found in a number of different organic films, such as monolayers of polymers that adsorb to the substrate through nonspecific [36] or biospecific [36] interactions, and polymer brushes [37]. The proportionality between $\Delta f$ and $m_0$ makes calibration conveniently simple. Further studies will be required to test how generalized this feature is and what the physical reasons for the linear response are.

## 17.4.3 Trapped Solvent in Monolayers of Discrete Nanoscale Objects

In the case of monolayers of discrete particles, the linearity seen in Fig. 17.12 is not typically observed. For sufficiently rigid films, i.e. in the Sauerbrey limit, the linearity would imply that $f_a^m$ does not depend on surface coverage.

Reimhult et al. [38] integrated two separate yet identically prepared sensing surfaces into one fluid cell for combined QCM-D/SPR measurements. The cell was designed such that mass-transport conditions on both surfaces were comparable, enabling time-resolved correlation of QCM-D and SPR data. A study on the formation of monolayers of the protein streptavidin revealed that $f_a^m$ depends sensitively on surface coverage. With a different setup, combining optical reflectometry and QCM-D on the same surface, Bingen et al. [22] later revealed that this phenomenon can be reproduced quantitatively through the empirical trapped-solvent coat model (Fig. 17.4). With such a model, it is now possible to make quantitative predictions about the QCM-D frequency response for monolayers of discrete nano-scale objects, as a function of surface coverage, particle size and weight, and lateral distribution and orientation of the particles on the surface [22, 23].

Frequently, the degree of solvation (or the fraction of trapped solvent) depends sensitively on the morphology of thin organic films. Consequently, quantification of solvation can inform about the organization of and morphological changes in such films. A well-known example is the formation of supported lipid bilayers by the method of vesicle spreading [38–40].

## 17.5 Conclusions and Perspectives

The combined SE and QCM-D setup is a powerful tool for the label-free characterization of organic films on surfaces. The method is versatile, because it can be applied on a large variety of substrates and over a large range of film thicknesses, from Angstroms to micrometers. Jointly, the two techniques can provide a wealth of qualitative and quantitative information, including film mass, thickness, solvation, morphology, optical and mechanical properties. The correlation of SE and QCM-D data can provide novel insight, that is not available with either technique alone. A central parameter is solvation (or porosity or trapped solvent) which becomes accessible for ultra-thin films (i.e. up to a few nm), or can be determined with greatly improved precision for films of intermediate thickness (i.e. up to a few 10 nm). Thanks to this parameter, additional quantities can be resolved, such as the film's effective refractive index, density and thickness, or the organization of surface-confined molecules. Data analysis and interpretation varies with the nature of the film under study, and we hope to have given the users the necessary tools for identifying and applying the methods that are appropriate for their films of interest.

With ongoing development and research, the future is likely to bring further improvements in SE or QCM-D instrumentation and in the qualitative and quantitative understanding of data from the respective methods. From these advances, additional and refined methods with combined SE/QCM-D setups should emerge, for the in-depth characterization of increasingly sophisticated and complex films. For example, optically anisotropic (birefringent) organic films or mixed organic/inorganic films that arise on topographically patterned/sculpted surfaces, or upon integration of inorganic inclusions (nanoparticles) into organic films, are interesting for a broad range of applications yet remain challenging to characterize. Here, generalized ellipsometry might provide access to an improved optical characterization.

Another promising perspective is the integration of additional techniques in the SE/QCM-D setup. For example, integration of SE/QCM-D with electrochemistry, should be of interest for a broad range of applications. Such a setup would provide access to optical, electrochemical and mechanical properties at the same time. Alternatively, electrochemical manipulation can be used to actuate stimuli-responsive films while SE/QCM-D provides direct access to the ensuing changes in optical, mechanical and morphological properties.

# References

1. G.E. Jellison, in *Handbook of Ellipsometry*, ed. by H.W. Tompkins, E.A. Irene (William Andrew, Norwich, 2004), p. 237
2. H. Fujiwara, *Spectroscopic Ellipsometry* (Wiley, New York, 2007)
3. G.E. Jellison, Thin Solid Films **313–314**, 33 (1998)
4. K.B. Rodenhausen, M. Schubert, Thin Solid Films **519**, 2772 (2011)
5. R.M. Azzam, N.M. Bashara, *Ellipsometry and Polarized Light* (North-Holland, Amsterdam, 1984)
6. H.A. Lorentz, *Theory of Electrons* (Teubner, Leipzig, 1906)
7. J.A. De Feijter, J. Benjamins, F.A. Veer, Biopolymers **17**, 1759 (1978)
8. H. Zhao, P.H. Brown, P. Schuck, Biophys. J. **100**, 2309 (2011)
9. P.A. Cuypers, J.W. Corsel, M.P. Janssen, J.M.M. Kop, W.T. Hermens, H.C. Hemker, J. Biol. Chem. **258**, 2426 (1983)
10. H.A. Lorentz, Ann. Phys. **9**, 641 (1880)
11. L. Lorenz, Ann. Phys. **11**, 70 (1880)
12. B. Borovsky, B.L. Mason, J. Krim, J. Appl. Phys. **88**, 4017 (2000)
13. I. Reviakine, D. Johannsmann, R.P. Richter, Anal. Chem. **83**, 8838 (2011)
14. D. Johannsmann, Phys. Chem. Chem. Phys. **10**, 4516 (2008)
15. G. Sauerbrey, Z. Phys. **155**, 206 (1959)
16. A. Domack, O. Prucker, J. Rühe, D. Johannsmann, Phys. Rev. E **56**, 680 (1997)
17. N.B. Eisele, F.I. Andersson, S. Frey, R.P. Richter, Biomacromolecules **13**, 2322 (2012)
18. D. Johannsmann, I. Reviakine, E. Rojas, M. Gallego, Anal. Chem. **80**, 8891 (2008)
19. D. Johannsmann, I. Reviakine, R.P. Richter, Anal. Chem. **81**, 8167 (2009)
20. L. Macakova, E. Blomberg, P.M. Claesson, Langmuir **23**, 12436 (2007)
21. R.P. Richter, A. Brisson, Langmuir **20**, 4609 (2004)
22. P. Bingen, G. Wang, N.F. Steinmetz, M. Rodahl, R.P. Richter, Anal. Chem. **80**, 8880 (2008)
23. I. Carton, A.R. Brisson, R.P. Richter, Anal. Chem. **82**, 9275 (2010)
24. K.B. Rodenhausen, T. Kasputis, A.K. Pannier, J.Y. Gerasimov, R.Y. Lai, M. Solinsky, T.E. Tiwald, H. Wang, A. Sarkar, T. Hofmann, N. Ianno, M. Schubert, Rev. Sci. Instrum. **82**, 103111 (2011)
25. J.W. Corsel, G.M. Willems, J.M.M. Kop, P.A. Cuypers, W.T.J. Hermens, J. Colloid Interface Sci. **111**, 544 (1986)
26. W.T. Hermens, M. Benes, R.P. Richter, H. Speijer, Biotechnol. Appl. Biochem. **39**, 277 (2004)
27. C.J. Love, L.A. Estroff, J.K. Kriebel, R.G. Nuzzo, G.M. Whitesides, Chem. Rev. **105**, 1103 (2005)
28. K.B. Rodenhausen, B.A. Duensing, T. Kasputis, A.K. Pannier, T. Hofmann, M. Schubert, T.E. Tiwald, M. Solinsky, M. Wagner, Thin Solid Films **519**, 2817 (2011)
29. W. Yang, J.Y. Gerasimov, R.Y. Lai, Chem. Commun. **2902** (2009)

30. J.J. Iturri Ramos, S. Stahl, R.P. Richter, S.E. Moya, Macromolecules **43**, 9063 (2010)
31. J.C. Charmet, P.G. de Gennes, J. Phys. (Paris) **44**(C10), 27 (1983)
32. C.W.N. Damen, H. Speijer, W.T. Hermens, J.H.M. Schellens, H. Rosing, J.H. Beijnen, Anal. Biochem. **393**, 73 (2009)
33. I. Carton, L. Malinina, R.P. Richter, Biophys. J. **99**, 2947 (2010)
34. P.M. Wolny, S. Banerji, C. Gounou, A.R. Brisson, A.J. Day, D.G. Jackson, R.P. Richter, J. Biol. Chem. **285**, 30170 (2010)
35. N.B. Eisele, S. Frey, J. Piehler, D. Görlich, R.P. Richter, EMBO Rep. **11**, 366 (2010)
36. M.A. Plunkett, Z. Wang, M.W. Rutland, D. Johannsmann, Langmuir **19**, 6837 (2003)
37. N.S. Baranova, E. Nilebäck, F.M. Haller, D.C. Briggs, S. Svedhem, A.J. Day, R.P. Richter, J. Biol. Chem. **286**, 25675 (2011)
38. E. Reimhult, C. Larsson, B. Kasemo, F. Höök, Anal. Chem. **76**, 7211 (2004)
39. C.A. Keller, K. Glasmästar, V.P. Zhdanov, B. Kasemo, Phys. Rev. Lett. **84**, 5443 (2000)
40. R.P. Richter, R. Bérat, A.R. Brisson, Langmuir **22**, 3497–3505 (2006)

# Chapter 18
# TIRE and SPR-Enhanced SE for Adsorption Processes

**Hans Arwin**

**Abstract** Ellipsometry configurations in internal reflection mode facilitate studies of adsorption processes without the light beam passing through the medium from which adsorption occurs. Monitoring of adsorption processes on surfaces in opaque media is thus possible. If the surface in addition has a thin semitransparent metal film in which surface plasmon polaritons can be excited, one can achieve very high sensitivity to small changes in surface mass density of an adsorbed biolayer. Thickness changes as small as one pm can be resolved. In this chapter the theory for Total Internal Reflection Ellipsometry (TIRE), also called surface plasmon resonance enhanced ellipsometry, will be described and instrumentation will be briefly discussed. TIRE applied in spectroscopic as well as in angle of incidence interrogation modes will be considered. Finally applications in the areas of bioadsorption processes, biosensing, gas adsorption and biolayer imaging will be reviewed.

## 18.1 Introduction

The surface plasmon-polariton phenomenon has been exploited extensively for biosensing and it is commonly referred to as the surface plasmon resonance technique or shortly SPR. The basic concept is that at a thin gold layer on a glass surface is exposed to biomolecules, whereby the irradiance of internally reflected light polarized parallel to the plane of incidence (*p*-polarized) strongly depends on the adsorbed mass of biomolecules on the gold surface. The physics/optics of the SPR sensor concept can be described briefly as follows. At a gold/liquid interface a surface electromagnetic mode can be excited by light incident through a prism onto which a thin gold layer has been deposited. If the angle of incidence, gold layer thickness and wavelength are optimized, the energy from the incident electric field component parallel to the plane of incidence is coupled to a surface plasmon wave and a quasi-

H. Arwin (✉)
Department of Physics, Chemistry and Biology,
Linköping University, 58183 Linköping, Sweden
e-mail: hans.arwin@liu.se

© Springer International Publishing AG, part of Springer Nature 2018     419
K. Hinrichs and K.-J. Eichhorn (eds.), *Ellipsometry of Functional Organic Surfaces and Films*, Springer Series in Surface Sciences 52,
https://doi.org/10.1007/978-3-319-75895-4_18

particle, a surface plasmon-polariton (SPP) is excited. A minimum in the reflectance curve, often called the SPR dip, is then observed and in principle all energy can be coupled from the $p$-component of the incident light to the SPP at resonance. The angular position of the minimum becomes very sensitive to the refractive index of the medium close to the gold surface. If biomolecules are adsorbing on the surface of the gold layer within the evanescent field depth, they will modify the refractive index and can therefore easily be detected and quantified.

In this chapter we will discuss the use of ellipsometry in internal mode with emphasis on SPP enhancement. Figure 18.1a shows an example of a basic configuration and

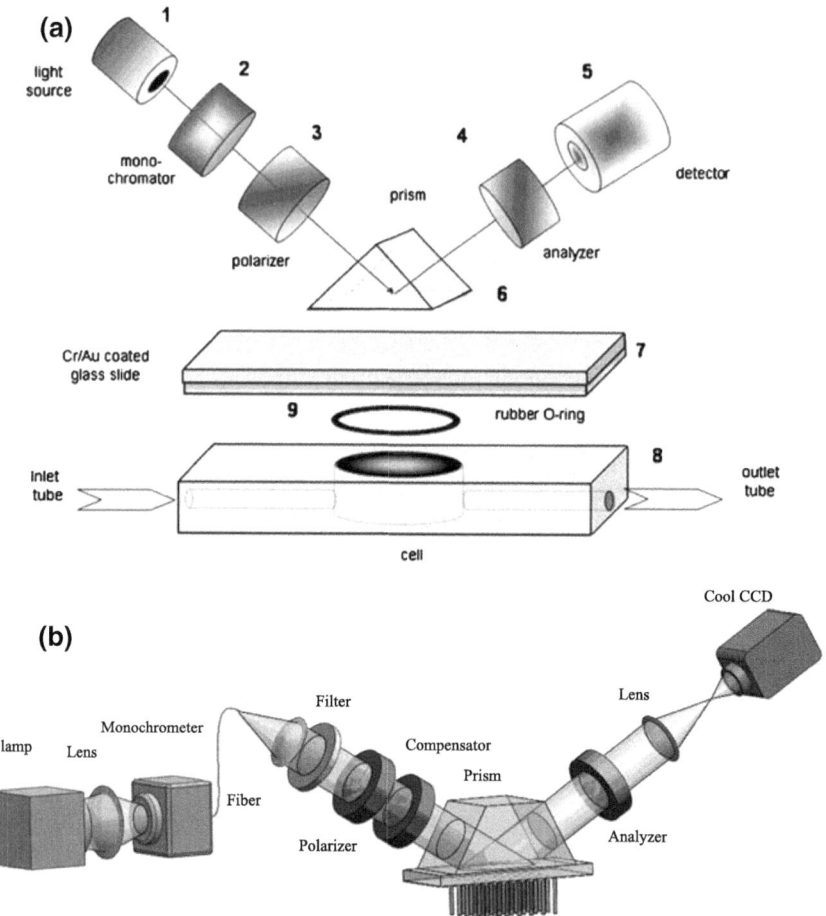

**Fig. 18.1  a** The Kretschmann configuration for SPP excitation exemplified with an ellipsometer in a rotating analyzer configuration. Reprinted from [1] with permission from Elsevier. **b** Internal reflection configuration based on a null ellipsometer in an imaging mode. Reprinted from [2] with permission from Elsevier

Fig. 18.1b an implementation in imaging mode for high throughput biosensor applications. Ellipsometry is established as a versatile tool to study surfaces and thin films. For thickness determinations the resolution depends on the sample studied but it is commonly agreed on that the detection limit is in the pm range. Aspnes [3] reports that in a fab-air environment, the standard deviation for determination of the effective thickness of a 1.57 nm thick native oxide layer on silicon can be as low as 130 fm. The high thickness sensitivity combined with the fact that measurements are non-destructive, does not require labeling of molecules and can be used in non-vacuum environments including liquids, have lead to that ellipsometry has been proposed as read-out principle in sensor systems [4, 5]. Both biosensors [6, 7] and chemical sensors [8] have been demonstrated.

Early work by Abelès [9] showed that ellipsometric monitoring of surface electromagnetic waves is very sensitive for superficial films. Bortchagovsky et al. [10] showed that ellipsometry with surface plasmon excitation provided enhanced sensitivity for studies of fullerene embedded in Langmuir–Blodgett films. Further developments to exploit the enhancement of sensitivity for in situ ellipsometry in a Kretschmann configuration, i.e. in internal reflection mode as shown in Fig. 18.1a and using a thin metal film on a glass prism similar to the configurations used in SPR techniques, were demonstrated later [11–13]. The concept is illustrated in Fig. 18.2, which shows internal reflection ellipsometric spectra recorded before and after adsorption of a monolayer of the protein human serum albumin (HSA) on a thin gold film [11]. The minima in the $\Psi$-curves resemble the minima observed in an SPR experiment. The shape of each minimum in $\Psi$ is close to the square root of the SPR-reflectance which implies that the $\Psi$ minima in fact are sharper than the corresponding SPR minima. Ellipsometry also provides access to the $\Delta$-spectra which gives additional and more sensitive readout about the adsorbed layer as we will discuss later. For adsorption of a protein monolayer on gold, changes in the ellipsometric parameter $\Delta$ of 90° or more are observed [13] in internal mode ellipsometry experiments compared to a few degrees in ordinary external mode ellipsometry [14].

The methodology has been called total internal reflection ellipsometry (TIRE) [11] and also surface plasmon resonance enhanced ellipsometry (SPREE) [12].

**Fig. 18.2**  $\Delta$ and $\Psi$ without and with a monolayer of the protein human serum albumin (HSA) on a 50 nm gold film on glass in phosphate buffered saline (PBS, pH 7.4) at an angle of incidence of 70°. Modified from [11]

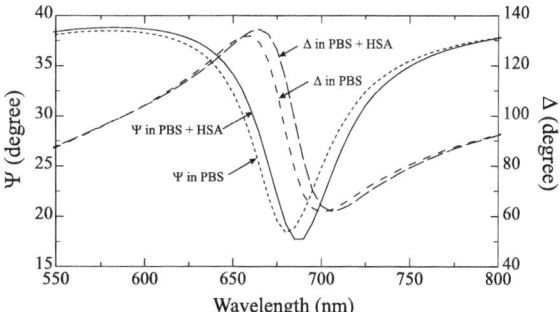

Applications without SPR-enhancement but in internal reflection mode are often refereed to as IRE [15]. The acronym TIRE will be used in this chapter. TIRE shows several similarities to SPR techniques for sensor applications. A major difference is, however, that in SPR only the intensity information for reflection of $p$-polarized light is utilized, whereas TIRE in addition utilizes the reflection properties in the direction perpendicular to the plane of incidence ($s$-polarization). Ellipsometry is technically more complex but has two major advantages over SPR techniques: (1) the $s$-polarization provides a reference for the overall irradiance transmittance and (2) not only the amplitude (irradiance) information in the reflected beam is utilized but also phase information in terms of the ellipsometric parameter $\Delta$. An approach related to ellipsometry has earlier been proposed by Kabashin et al. [16] who suggested to enhance the resolution of SPR-based bio- and chemical sensors by using phase-polarization contrast. Nelson et al. suggested a sensing configuration in an SPR heterodyne phase detection mode [17] and demonstrated that detection by means of phase modulation has three times higher resolution compared to a configuration based on angle of incidence or wavelength modulation.

The objective with this chapter is to review the development, recent applications and current status of TIRE. The outlined is as follows. In the next section the theory of ellipsometry is expanded to include SPP's. The objective is to discuss sample configuration requirements and how $\Psi$ and $\Delta$ depend on them. Then follows a very short and general presentation of experimental setups for TIRE. The applications section describes selected examples of bioadsorption/biosensing, gas adsorption and TIRE imaging. Finally a short outlook is given.

## 18.2 Theory

A brief theoretical background to TIRE is presented here. It should be noticed that the phenomena are fully included in the Fresnel formalism and are readily modeled in multilayer systems using e.g. scattering matrix formalisms [18]. However, such a treatment does not provide a detailed insight into the physics and dependence on system parameters. An analytic approach is therefore presented here to illustrate the phenomena even though some approximations will be necessary. The reader is assumed to be familiar with the basics of ellipsometry as described elsewhere in this book.

### 18.2.1 Surface Plasmon-Polaritons

**The Dispersion Relation of a Surface Plasmon-Polariton** For a transverse magnetic wave, corresponding to the $p$-polarization, the wave vector is $\mathbf{q} = (q_x, 0, q_z)$ using a coordinate system as defined in Fig. 18.3a with the $xz$-plane as the plane of incidence. By combining Maxwell's curl equations it can be shown that they have

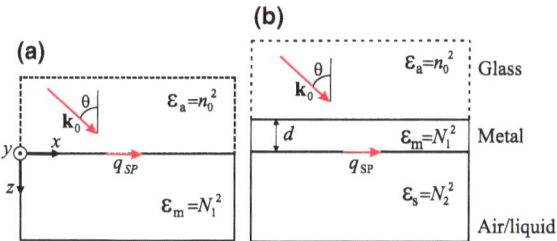

**Fig. 18.3 a** Two half-infinite media, an ambient and a metal, with dielectric functions $\epsilon_a$ and $\epsilon_m$ separated by an interface parallel to the $xy$-plane at $z = 0$. $\mathbf{k}_0$ is the wave vector of light incident at an angle $\theta$ and $q_{SP}$ the $x$-component of the wave vector of an SPP. $n_0$ is the refractive index of the ambient and is assumed real-valued whereas the second medium index $N_1 = n_1 + ik_1$ generally is complex-valued. **b** Two half-infinite dielectric media, an ambient and a substrate, with dielectric functions $\epsilon_a$ and $\epsilon_s$ separated by a metal layer with thickness $d$ and dielectric function $\epsilon_m$. The substrate refractive index is in a majority of cases real-valued but is generally denoted $N_2 = n_2 + ik_2$

a solution corresponding to a surface-bound wave propagating along an interface between two semi-infinite media as indicate by the $x$-component $q_{SP}$ of the wave vector $\mathbf{q}$ in Fig. 18.3a [19]. With the boundary conditions of continuity of the tangential fields across the interface, the allowed solutions must obey [19]

$$\frac{q_{z0}}{\epsilon_a} + \frac{q_{z1}}{\epsilon_m} = 0 \tag{18.1}$$

where $q_{z0}$ and $q_{z1}$ are $z$-components of $\mathbf{q}$ in the ambient and metal, respectively. $\epsilon_a$ and $\epsilon_m$ are the dielectric functions of the ambient and the metal, respectively, and are related to the corresponding refractive indices as $\epsilon_a = n_0^2$ and $\epsilon_m = N_1^2$. The dispersion relation for the surface wave is found to be [19]

$$q_{SP}^\infty = \frac{2\pi}{\lambda}\sqrt{\frac{\epsilon_a \epsilon_m}{\epsilon_a + \epsilon_m}} = q_{SP}'^\infty + i\Gamma^\infty. \tag{18.2}$$

Here $q_{SP}^\infty = q_{x0} = q_{x1}$ at the interface and $q_{SP}'^\infty$ and $\Gamma^\infty$ ($= q_{SP}''^\infty$) are the real and imaginary parts of $q_{SP}^\infty$ with $\infty$ indicating the semi-infinite geometry shown in Fig. 18.3a.

We are here only concerned with surface-bound waves and thus $q_{z0}$ and $q_{z1}$ ideally should be purely imaginary. Otherwise waves propagating in the $z$-direction would exist. If we therefore restrict to $q_{z0} = i|q_{z0}|$ and $q_{z1} = i|q_{z1}|$ we can rewrite (18.1) as

$$\epsilon_m = -\epsilon_a \frac{|q_{z0}|}{|q_{z1}|}. \tag{18.3}$$

For the configurations of interest for us, the ambient medium is dielectric with $\epsilon_a$ real-valued and positive, e.g. air, water or glass. Under these conditions it follows from (18.3) that the second medium should be a material with a real-valued and negative $\epsilon_m$. However, as we will discuss later, for the SPP-phenomenon to occur, $\epsilon_m = \epsilon_m' + i\epsilon_m''$ must have a small but non-zero imaginary part $\epsilon_m''$. Several metals including silver and gold fulfill the above requirements of $\epsilon_m' < 0$ and $\epsilon_m'' \ll |\epsilon_m'|$.

The solution to Maxwell's equations discussed here is called a surface plasmon-polariton (SPP). The field associated with an SPP has its maximum at the interface and decrease exponentially (evanescent fields) into both media as $q_{z0}$ and $q_{z1}$ are purely imaginary. The source of the field is a charge oscillation in the metal surface. The SPP wave propagates along the interface described by $q_x = q_{SP}^\infty$ in (18.2).

**Excitation of a Surface Plasmon-Polariton Using Light** In the geometry of Fig. 18.3a we can have an incident beam of light at an angle of incidence $\theta$. The wave vector for the light[1] is $\mathbf{k}_0 = (k_{x0}, 0, k_{z0})$ with $|\mathbf{k}_0| = 2\pi\sqrt{\epsilon_a}/\lambda$. The $x$-component could potentially be used to excite an SPP wave. However, for an ambient with a real-valued index of refraction $n_0 = \sqrt{\epsilon_a}$ we have

$$k_{x0} = \frac{2\pi}{\lambda}\sqrt{\epsilon_a}\sin\theta = \frac{2\pi}{\lambda}n_0\sin\theta \tag{18.4}$$

and for no $\theta$ we can find a match so that $k_{x0}$ becomes equal to $q_{SP}'^\infty$ in (18.2) as it always holds that $k_{x0} < q_{SP}'^\infty$. To resolve this problem, the trick is to use a thin semi-transparent metal film with thickness $d$ on glass as shown in Fig. 18.3b. As we will see later, the ambient in a TIRE setup is often in shape of a glass prism. The $x$-component of the wave vector of incident light will be the same as before, i.e. as described by (18.4). The $x$-components $k_{x0} = k_{x1} = k_{x2}$ are also invariant through the structure and can be tuned by $n_0$ or $\theta$ to match the $x$-component of the SPP at the metal/substrate interface. Notice that the SPP is now at the metal/substrate interface and not at the metal/ambient interface as in the two-phase model in Fig. 18.3a. $q_{SP}^\infty$ will thus be determined by $\epsilon_m$ and $\epsilon_s$ and $k_{x0}$ and $q_{SP}^\infty$ can be tuned independently to match each other.

A small complicating factor is that for a thin metal film, the SPP conditions at the metal/substrate interface in Fig. 18.3b will be slightly different compared to the case with a semi-infinite metal in Fig. 18.3a. Due to the presence of the prism, or rather that the prism is replacing metal except for the thin film region, a perturbation term $\Delta q_{SP}$ must be added to $q_{SP}^\infty$ and $\epsilon_a$ must be replaced with $\epsilon_s$. We now obtain a modified $q_{SP}$ according to

$$\begin{aligned} q_{SP} &= q_{SP}^\infty + \Delta q_{SP} = \frac{2\pi}{\lambda}\sqrt{\frac{\epsilon_a\epsilon_s}{\epsilon_a + \epsilon_s}} + \Delta q_{SP} \\ &= q_{SP}'^\infty + \Delta q_{SP}' + i\left(\Gamma^\infty + \Gamma^{rad}\right). \end{aligned} \tag{18.5}$$

Here the perturbation $\Delta q_{SP} = \Delta q_{SP}' + i\Gamma^{rad}$ depends on the metal layer thickness and can thus be controlled. For sufficiently large thickness, $\Delta q_{SP}$ goes to zero. The

---

[1]We use the notation $\mathbf{k}$ and $\mathbf{q}$ for the wave vector of light and SPP, respectively.

detailed derivation of $q_{SP}$ (or rather $\Delta q_{SP}$, as we know $q_{SP}^{\infty}$) for a double interface as in Fig. 18.3b is rather complicated and requires a strategy similar to the single interface case but is not done here [20].

To study the reflection of light when an SPP is excited we need an expression for $r_p$, the reflection coefficient for $p$-polarized light. The reflection of the $s$-component is of minor interest as its electric field has no $x$-component and can not couple to an SPP. For the $p$-component we have in the three-phase model

$$r_p = \frac{r_{01p} + r_{12p}e^{i2\beta}}{1 + r_{01p}r_{12p}e^{i2\beta}} = r_{01p}\frac{1 + r_{01p}^{-1}r_{12p}e^{i2\beta}}{1 + r_{01p}r_{12p}e^{i2\beta}} \tag{18.6}$$

where $\beta$ is the film phase thickness [18]. In the special situation with film properties and instrument settings selected so that an SPP resonance is excited, a series of approximations can be performed on (18.6) to derive a simplified expression for $r_p$ near the resonance. Following the derivation by Raether [20] gives

$$r_p = r_{01p}\frac{k_{x0} - (q_{SP}'^{\infty} + \Delta q_{SP}') - i(\Gamma^{\infty} - \Gamma^{rad})}{k_{x0} - (q_{SP}'^{\infty} + \Delta q_{SP}') - i(\Gamma^{\infty} + \Gamma^{rad})}. \tag{18.7}$$

Equation (18.7) provides the $p$-reflectance $R_p = |r_p|^2$ utilized in SPR-methods. The resonance condition is $k_{x0} = (q_{SP}'^{\infty} + \Delta q_{SP}')$ and from (18.7) we learn that $r_p = 0$ and $R_p = 0$ at resonance if $\Gamma^{\infty} = \Gamma^{rad}$. For a metal film with finite thickness it holds that $\Gamma^{rad} \neq 0$ and we require $\Gamma^{\infty} - \Gamma^{rad} = 0$, i.e. we must fulfill $\Gamma^{\infty} \neq 0$. A non-zero $\Gamma^{\infty}$ implies that there must be a non-zero imaginary part on $\epsilon_m$ as can be seen in (18.2).

## 18.2.2 Surface Plasmon-Polariton Resonances in Ellipsometric Mode

In ellipsometry the angle of incidence and/or wavelength dispersion in $\Delta$ and $\Psi$ are of interest when SPP's are monitored. Close to an SPP resonance the complex reflectance ratio $\rho = \tan\Psi \exp(i\Delta)$ is obtained by dividing (18.7) with $r_s$, the $s$-reflection coefficient, according to

$$\rho = \frac{r_p}{r_s} = \frac{r_{01p}}{r_s}\frac{k_{x0} - (q_{SP}'^{\infty} + \Delta q_{SP}') - i(\Gamma^{\infty} - \Gamma^{rad})}{k_{x0} - (q_{SP}'^{\infty} + \Delta q_{SP}') - i(\Gamma^{\infty} + \Gamma^{rad})}. \tag{18.8}$$

From (18.8) it follows

$$\tan \Psi = \frac{|r_{01p}|}{|r_s|} \sqrt{1 - \frac{4\Gamma^\infty \Gamma^{rad}}{Q^2 + [\Gamma^\infty + \Gamma^{rad}]^2}}, \tag{18.9a}$$

$$\Delta = \arg \frac{r_{01p}}{r_s} + \arctan \frac{2Q\Gamma^{rad}}{Q^2 + (\Gamma^\infty + \Gamma^{rad})(\Gamma^\infty - \Gamma^{rad})} \tag{18.9b}$$

where $Q = k_{x0} - (q_{SP}'^\infty + \Delta q_{SP}')$. Recall that $r_{01p}$ is the Fresnel reflection coefficient for the glass/metal interface. Both $r_{01p}$ and $r_s$ are slowly varying functions of $k_{x0}$ around the SPP resonance. The dispersion in $\tan \Psi$ is therefore dominated by the expression under the square root in (18.9a). The dispersion in $\Delta$ is dominated by the second term in (18.9b). If the metal layer thickness is matched for zero reflectance, i.e. $\Gamma^\infty = \Gamma^{rad}$, (18.9b) reduces to

$$\Delta = \arg \frac{r_{01p}}{r_s} + \arctan \frac{2\Gamma^{rad}}{k_{x0} - (q_{SP}'^\infty + \Delta q_{SP}')}. \tag{18.10}$$

Equation (18.10) shows that, at an SPP resonance, i.e. when $k_{x0} = q_{SP}'^\infty + \Delta q_{SP}'$, $\Delta$ is a step function with step $\pi$ if $k_{x0} = 2\pi \lambda^{-1} n_0 \sin \theta$ is varied by scanning either $\lambda$ or $\theta$ over the resonance. In a real measurement non-idealities like surface and interface roughness, not exactly matched metal layer thickness, etc., will come into play. If a small mismatch $\Delta\Gamma = \Gamma^\infty - \Gamma^{rad}$ is introduced, (18.9b) instead reduces to

$$\Delta = \arg \frac{r_{01p}}{r_s} + \arctan \frac{1}{\frac{k_{x0}-(q_{SP}'^\infty+\Delta q_{SP}')}{2\Gamma^\infty} + \frac{\Delta\Gamma}{k_{x0}-(q_{SP}'^\infty+\Delta q_{SP}')}}. \tag{18.11}$$

When either $\lambda$ or $\theta$ is scanned over the SPP resonance, the variation of $\Delta$ is dominated by the behavior of argument of the arctan function in the second term in (18.11). If $\Delta\Gamma$ is made very small by carefully controlling the metal film thickness, a sharp resonance will occur in a wavelength interrogation when $\lambda = \lambda^{res}$, where superscript res indicates the SPP resonance values.[2]

Let us finally discuss effects of adlayers. A sample with a gold layer thickness tuned to match the SPP condition becomes a useful tool in sensor applications as will be demonstrated in Sect. 18.4. This is also illustrated in Fig. 18.2 which shows the change in $\Delta$ and $\Psi$ induced by adsorption of a HSA monolayer from a buffer solution. Due to this layer, only a few nm thick, a shift in the wavelength position of the resonance is obtained but the shapes of the curves are preserved to a large extent.

---

[2]The angle of incidence is then assumed to be at $\theta^{res}$. In an angle of incidence interrogation the resonance occurs in an angle scan when $\theta = \theta^{res}$ if the wavelength is set at $\lambda^{res}$.

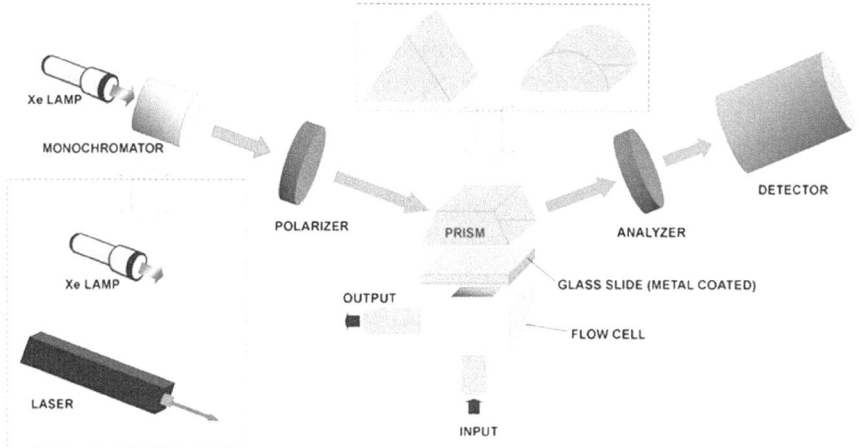

**Fig. 18.4** A schematic overview of a TIRE system. Reprinted from [25]

## 18.3 Ellipsometric Setups for TIRE

An experimental setup for TIRE is composed of an ellipsometer, a prism and a specially designed flow cell as schematically illustrated in Fig. 18.4. The flow cell is attached to the prism through which the light is incident. In a few applications, adsorption studies are performed directly on the glass surface utilizing the evanescent field at the interface [2]. However, in most applications a thin semitransparent metal film is included. It can be deposited directly on the prism [21] but repeated use requires regeneration of the metal surface which is not trivial. Therefore it is most common that a glass slide with a metal film is mounted with an index matching oil on the prism. In this way only the slide with metal film has to be replaced for each new experiment. The refractive index of the glass in the slide and the prism should be the same to minimize interface reflections.

As indicated the light source can be broad-band whereby a monochromator is required to select one wavelength or to scan the wavelength if a wavelength interrogation mode is used. Alternatively a fixed wavelength operation using a laser in an angle interrogation mode can be employed. Various prism geometries including semicylindrical, trapezoidal and triangular shapes are found. A rotating analyzer configuration is shown in Fig. 18.4 but also designs based on null ellipsometers [22], phase modulation ellipsometers [21, 23] and Stokes polarimetry [24] are found in the literature. The flow cell may be designed for gas or liquid flow depending on application. In addition one needs a flow control system, data acquisition and system control software. Further details of various TIRE configurations are discussed in [25] and in the references describing the specific applications presented in the next section.

The metal layer on the glass slide is in a majority of cases of gold due to its inertness and suitable optical properties for excitation of SPP's. The quality in terms of optical properties and the surface of the gold film is important and as TIRE is a resonance technique it is very important to optimize the system parameters like angle of incidence, wavelength and gold layer thickness [26]. Balevicius and coworkers have explored effects of gold surface roughness [27] and of gold nanoparticles [28]. Also, other metals like silver and copper can be used. Poksinksi et al. [29] studied copper corrosion in a TIRE configuration as an example on use of other metals than gold. Even metals with a larger absorption (larger $\epsilon_m''$) like chromium may be possible to use as discussed by Arwin et al. [30].

## 18.4  Applications

TIRE is applied for monitoring adsorption of proteins and other biomolecules with objectives to learn more about thin film surface dynamics and structure. In many cases biosensing concepts are suggested and often in high throughput mode employing imaging TIRE. These aspects will be discussed below. Also gas adsorption applications are included. In the literature one can find some early reports describing experiments involving phase detection of SPP's with similarities to determination of $\Delta$ in ellipsometry [9, 16, 17]. However, we will here limit the review to applications using traditional ellipsometric configurations.

### 18.4.1  TIRE for Bioadsorption Processes and Biosensing

Already when TIRE first was presented, the extremely high sensitivity for protein layer thicknesses was demonstrated [11]. In Fig. 18.5 it is seen that with TIRE it is possible to resolve in situ effective thickness variations smaller than 1 pm for a layer of the protein HSA in a buffer solution [25]. Similar sensitivities have been found also for ferritin [13] and fibrinogen [31].

These early but general demonstrations of very high sensitivity for bioadsorption have been followed by several studies addressing specific problems in biomedicine. One example is detection of $\beta$-amyloid peptide (1–16) and amyloid precursor protein (APP$_{770}$) with relevance for diagnostics of Alzheimer's disease [32]. The same investigators also studied receptor-protein specificity with TIRE. Chaperone receptors were bound to a gold surface and specific binding of different chaperones were monitored and quantified [33]. Balevicius and coworkers [34] studied immunosensor performance and compared antigen binding to fragmented and intact antibodies and concluded that fragmented antibodies could bind 2.5 times more antigen. Recently they extended these studies to include modeling of the time responses in $\Psi$ and $\Delta$ to analyze the dynamics of adsorption/interaction of antibody/antigen binding [35]. Model experiments for studies of antigen-antibody binding using bovine serum

**Fig. 18.5** Δ versus time in TIRE mode during adsorption of human serum albumin on a 45.8 nm gold layer in PBS buffer (pH 7.4). The measurements were done at an angle of incidence of 70° and λ = 650 nm. Reprinted from [25]

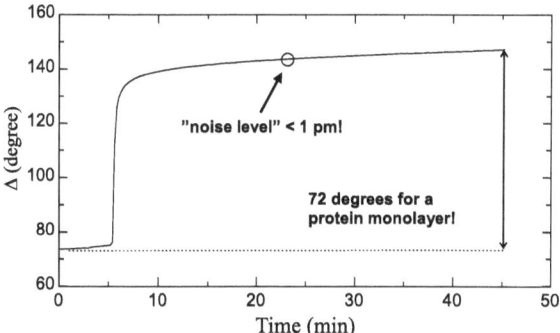

albumin (BSA) and anti-BSA were performed by Moirangthem et al. [36]. Both spectral shifts and adsorption dynamics were monitored and TIRE was compared with other SPR-related techniques. A more technical application is adsorption of proteins from milk suggested for in line monitoring of fouling in milk processing industry [37]. This is an example when the possibility of measurements in an opaque solution are taken advantage of.

Bioadsorption studies involving other biomolecules than proteins have also been performed. Nabok et al. [38] electrostatically adsorbed single strands (ss) of DNA from herring and salmon on a polyethylenimine coated gold film. With TIRE they then studied specific binding of ss-DNA from the same sources and found a thickness increase of around 20 nm for herring ss-DNA on herring ss-DNA and similar for salmon ss-DNA on salmon ss-DNA, whereas adsorption from alternate species, e.g. salmon ss-DNA on herring ss-DNA, only resulted in a layer a few nm thick. Also Le et al. [39] studied DNA hybridization as well as immunoassays with a functionalized polymer layer of zeonor. Growth of polymer brushes on gold was monitored with TIRE by Erber et al. [40]. The possibility to measure the non-linear polymerization kinetics permits control of polymer synthesis to optimize layer structure.

Low molecular weight toxins, such as herbicides simazine, atrazine, T2 mycotoxin and nonylphenol can be detected and quantified using immunoassays and TIRE [41, 42]. Concentrations of zearalanone and aflatoxin below 0.1 ng/ml can be detected [43–45]. A calibration curve for some toxins are shown in Fig. 18.6. An alternative to immunoassays to increase specificity is to use phthalocyanines layers in TIRE for detection of toxins [46]. In this way detection of simazine with concentration down to 1 ng/ml has been demonstrated. Several studies with TIRE for detection of the carcinogenic pesticide pentachlorophenol have also been performed [47–49].

### 18.4.2 Biolayer Imaging with TIRE

If a detector with spatial resolution is used in a TIRE or IRE setup, it is possible to measure high contrast images of biolayers. Various acronyms are used for

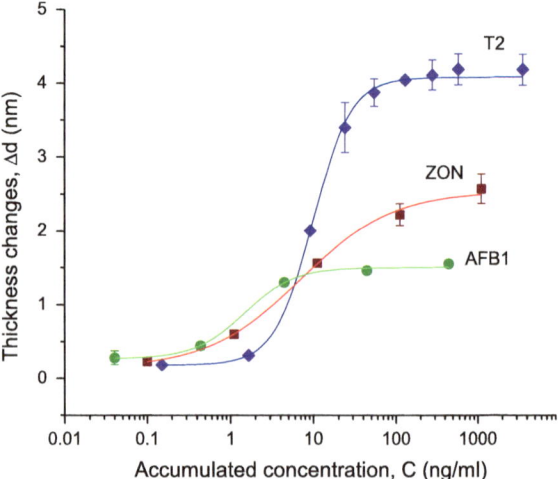

**Fig. 18.6** Change in thickness for adsorbed layers of T-2 mycotoxin (T-2), zearalanone (ZON) and aflatoxin (AFB1) based on direct immunosensing. Reprinted from [45]

this approach including TIRIE, Total Internal Reflection Imaging Ellipsometry [2], SPREE, Surface Plasmon Resonance Enhanced Ellipsometry [50] and SPRIE, Surface Plasmon Resonance Imaging Ellipsometry [22]. Jin and coworkers [51] have presented a review of TIRIE possibilities for biomedical applications including antibody screening, hepatitis B markers detection, cancer markers and virus recognition.

Jin and coworkers [2] have also developed imaging off-null ellipsometry [52] in TIRIE-mode for label-free optical biosensing using a micro-array for high throughput applications. By implementing TIRE in the system, they found an increase in sensitivity of one order of magnitude compared to ordinary imaging ellipsometry. Figure 18.7a shows an example of an ellipsometric image [53]. Applications of TIRIE in the area of cancer diagnostics were addressed by Zhang et al. [54]. Detection of serum tumor-associated antigen CA19-9 with TIRIE were compared with electrochemiluminence immunoassays (ECLIA) and a very good correlation was found as seen in Fig. 18.8. Compared to ECLIA, TIRIE has the additional advantage of allowing label-free and real-time measurements. Celen et al. [50] have addressed DNA detection and used TIRIE for detection of single-strand oligonucleotides with microarray platforms. TIRIE has also been used to monitor dynamics of living cells by Kim et al. [22]. They studied cell division, cell migration and cell-cell communication on an extracellular matrix on a gold surface on a prism in a TIRIE configuration.

The work by Otsuki et al. [15] demonstrates that increased sensitivity in internal reflection ellipsometry can be achieved without SPR enhancement. They used a high-index dielectric layer of $TiO_2$ on a silica prism to perform imaging of protein layers with a resolution of $\pm 0.2$ nm as shown in Fig. 18.7b.

**(a)**

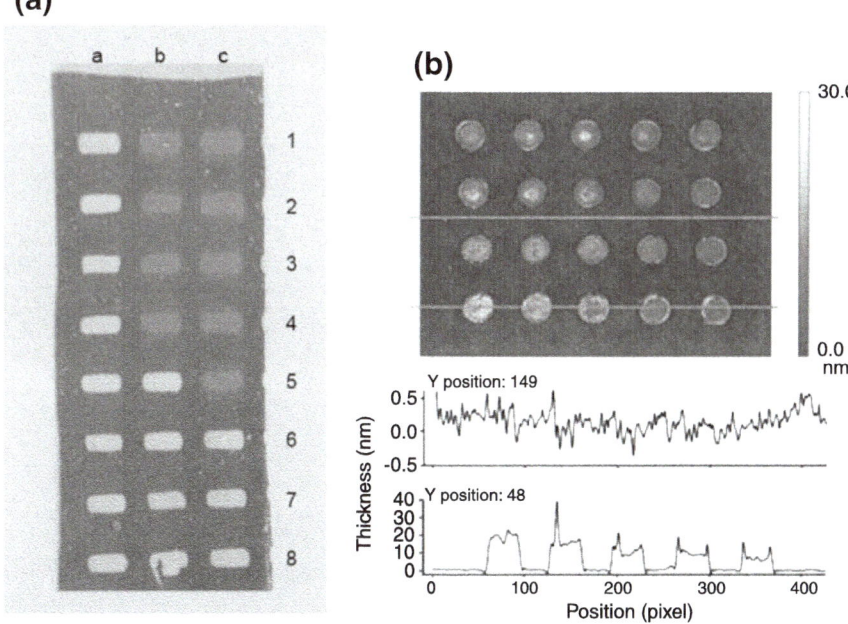

**Fig. 18.7 a** TIRIE image recorded on a biochip [53]. 9 out of 24 channels are included in this experiment. The 50 nm gold layers in each channel are chemically activated and bovine serum albumin (BSA) with concentration 10 mg/ml is delivered to channels b1 to b4 and c1 to c4 for 10 min and buffer to c5 as control. After rinsing with buffer, anti-BSA with concentration 5, 10, 20, 40 μg/ml are added to the unit c1 to c4, respectively. **b** Internal reflection ellipsometry image showing thickness variation of arrays of dried proteins which from top to bottom are lactoglobulin, albumin, ovalbumin and papain. *Below the image*, two line profiles along cross sections as indicated by lines in the image are shown. Reprinted from [15] with permission from Optical Society of America

**Fig. 18.8** Correlation between response (U/ml) to the serum tumor-associated antigen CA19-9 for an anti-CA19-9 modified surface in TIRIE and response (U/ml) as measured with ECLIA. Serum from 15 patients with three different types of cancer were tested. Reprinted from [54] with permission from Elsevier

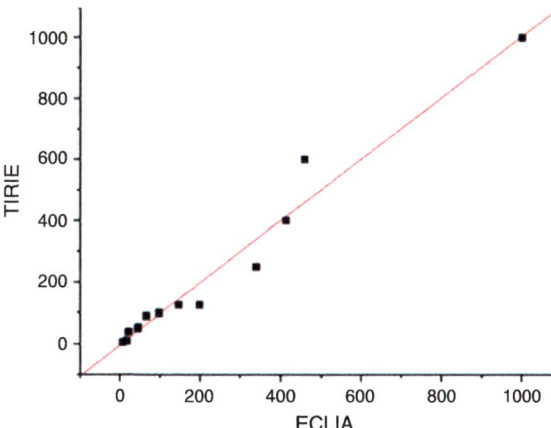

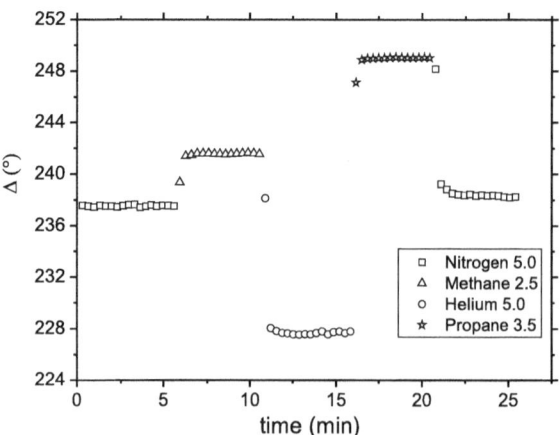

**Fig. 18.9** Change in Δ in a TIRE experiment with different gases at an angle of incidence of 44.3°. Reprinted from [55] with permission from Elsevier

## 18.4.3   TIRE for Gas Adsorption Studies

The sensitivity in TIRE is sufficient to detect changes in gas composition in a nitrogen flow as demonstrated by Nooke et al. [55]. They used angle of incidence interrogation with a null ellipsometer as well as with a rotating analyzer ellipsometer to monitor the change in refractive index in the evanescent field of an SPP. Ten different gases in nitrogen were studied and changes smaller than 0.00003 in index could be resolved for oxygen in nitrogen. The detection limit depends on the gas and could be as low as 0.1% for ammonia in nitrogen. In dynamic mode, as illustrated in Fig. 18.9, time constants of less than 1 s were possible to achieve. An interesting observation was that the change in Δ was much slower in polar gases compared to non-polar which was attributed to formation of a surface layer when exposed to polar gases whereas non-polar gases only affect the bulk refractive index.

Gas selectivity can be improved by depositing an appropriate thin film on the gold layer in a TIRE setup. For 5 nm films of copper phthalocyanine, Basova et al. [1] found a reversible response in Δ and Ψ to benzene and chloroform vapors whereas ethanol and butanol vapors did not affect the film. The changes in the TIRE data were attributed to film swelling in conjunction with a small change in refractive index due to interaction between gas and film. The nitrogen oxide sensitivity of copper phthalocyanine has earlier been studied with ellipsometry and in addition to thickness changes also changes in index due to shift in the Q-band were found upon gas exposure [56].

## 18.5    Concluding Remarks and Outlook

Main advantages of using TIRE as readout principle in sensor systems are that it is label-free, non-destructive, can be used in real time, in opaque media and has high sensitivity as it makes use of the phase information in the reflected light from a sample. Compared to other techniques it is technically more complex but in a sensor-type application, the system components can be reduced to a minimum. One drawback is that the selectivity is pour and one has to rely on chemical or biological discrimination by using sensing layers/surfaces with the required specificity.

Future technical developments include more advanced techniques like imaging TIRE spectroscopy. Metamaterials can be expected to play a role as tunable SPP layers. Gold nanoparticles can also be used to tune resonance conditions e.g. wavelength, to fit a suitable light source [57]. New types of applications can also be foreseen if TIRE in the near-infrared is developed. Patskovsky et al. [58] have shown that this indeed is possible using a silicon prism. With silicon as base material, the well developed silicon-based technology may be utilized. They also suggested the use of two SPP-modes simultaneously [59]. For biosensing an optimization of system and sample parameters can also be performed to make TIRE more competitive [26]. A real challenge is to optimize resolution to allow mass density depth profiling in thin organic layers similar to what can be done with neutron reflection.

**Acknowledgements**  Uwe Beck and Gang Jin are acknowledged for providing figures and unpublished material.

## References

1. T. Basova, A. Tsargorodskaya, A. Nabok, A.K. Hassan, A.G. Gürek, G. Gümüs, V. Ahsen, Mat. Sci. Eng. C **29**, 814 (2009)
2. L. Liu, Y.-Y. Chen, Y.-H. Meng, S. Chen, G. Jin, Thin Solid Films **519**, 2758 (2011)
3. D.E. Aspnes, Thin Solid Films **455–456**, 3 (2004)
4. H. Arwin, Sens. Actuators A **92**, 43 (2001)
5. H. Arwin, in *Encyclopedia of Sensors*, vol. 3, ed. by C.A. Grimes, E.C. Dickey, M.V. Pishko (American Scientific Publishers, 2006), p. 329
6. M. Ostroff, D. Maul, G.R. Bogart, S. Yang, J. Christian, D. Hopkins, D. Clark, B. Trotter, G. Moddel, Clin. Chem. **44**, 2031 (1998)
7. G. Jin, P. Tengvall, I. Lundström, H. Arwin, Anal. Biochem. **232**, 69 (1995)
8. G. Wang, H. Arwin, R. Jansson, IEEE Sens. J. **3**, 739 (2003)
9. F. Abelès, Thin Solid Films **34**, 291 (1976)
10. E. Bortchagovsky, I. Yurchenko, Z. Kazantseva, J. Humlicek, J. Hora, Thin Solid Films **313–314**, 795 (1998)
11. M. Poksinski, H. Dzuho, J.-O. Järrhed, H. Arwin, *Total Internal Reflection Ellipsometry, Proceedings Eurosensors XIV*, Copenhagen, Denmark, ed. by R. de Reuss, S. Bouwstra (MIC - Mikroelektronik Centret, Kgs. Lyngby, 2000), pp. 239–242, ISBN: 87-89935-50-0
12. P. Westphal, A. Bornmann, Sens. Actuators B **84**, 278 (2002)
13. M. Poksinski, H. Arwin, Protein monolayers monitored by internal reflection ellipsometry. Thin Solid Films **455–456**, 716 (2004)

14. J. Mårtensson, H. Arwin, H. Nygren, I. Lundström, J. Colloid Interface Sci. **174**, 79 (1995)
15. S. Otsuki, K. Tamada, S.-I. Wakida, Appl. Opt. **44**, 1410 (2005)
16. A.V. Kabashin, V.E. Kochergin, P.I. Nikitin, Sens. Actuators **54**, 51 (1999)
17. S.G. Nelson, K.S. Johnston, S.S. Yee, Sens. Actuators B **35–36**, 187 (1996)
18. R.M.A. Azzam, N.M. Bashara, *Ellipsometry and Polarized Light* (North-Holland Publishing Company, Amsterdam, 1986)
19. S.A. Maier, *Plasmonics: Fundamentals and Applications* (Springer, Berlin, 2007)
20. H. Raether, *Surface Plasmons*, Springer Tracts in Modern Physics (Springer, New York, 1988)
21. W. Yuan, H.P. Ho, S.Y. Wu, Y.K. Suen, S.K. Kong, Sens. Actuators A: Phys. **151**, 23 (2009)
22. S.-H. Kim, W. Chegal, J. Doh, H.M. Cho, D.W. Moon, Biophys. J. **100**, 1819 (2011)
23. W.-L. Hsu, S.-S. Lee, C.K. Lee, J. Biomed. Opt. **14**, 024036 (2009)
24. L.S. Maksimenko, I.E. Matyash, I.A. Minailova, O.N. Mishchuk, S.P. Rudenko, B.K. Serdega, Opt. Spectrosc. **109**, 808 (2010)
25. M. Poksinski, H. Arwin, Total Internal Reflection Ellipsometry: Monitoring of Proteins on Thin Metal Films, in *Proteins at Solid-Liquid Interfaces*, ed. by P. Déjardin (Springer, Berlin, 2006)
26. H. Arwin, M. Poksinski, K. Johansen, Phys. Stat. Solidi (a) **205**, 817 (2008)
27. Z. Balevicius, V. Vaicikauskas, G.-J. Babonas, Appl. Surf. Sci. **256**, 640 (2009)
28. Z. Balevicius, R. Drevinskas, M. Dapkus, G.J. Babonas, A. Ramanaviciene, A. Ramanavicius, Thin Solid Films **519**, 2959 (2011)
29. M. Poksinski, H. Dzuho, H. Arwin, J. Electrochem. Soc. **11**, B536 (2003)
30. H. Arwin, M. Poksinski, K. Johansen, Appl. Opt. **43**, 3028 (2004)
31. M. Poksinski, H. Arwin, Opt. Lett. **32**, 1308 (2007)
32. M.K. Mustafa, A. Nabok, D. Parkinson, I.E. Tothill, F. Salam, A. Tsargorodskaya, Biosens. Bioelectron. **26**, 1332 (2010)
33. V. Kriechbaumer, A. Tsargorodskaya, M.K. Mustafa, T. Vinogradova, J. Lacey, D.P. Smith, B.M. Abell, A. Nabok, Biophys. J. **101**, 504 (2011)
34. Z. Balevicius, A. Ramanaviciene, I. Baleviciute, A. Makaraviciute, L. Mikoliunaite, A. Ramanaviciusa, Sens. Actuators B: Chem. **160**, 555 (2011)
35. I. Baleviciute, Z. Balevicius, A. Makaraviciute, A. Ramanaviciene, A. Ramanavicius, Biosens. Bioelectron. **39**, 170 (2013)
36. R.S. Moirangthem, Y.-C. Chang, S.-H. Hsu, P.-K. Wei, Biosens. Bioelectron. **25**, 2633 (2010)
37. M. Poksinski, H. Arwin, Sens. Actuators B **94**, 247 (2003)
38. A. Nabok, A. Tsargorodskaya, F. Davis, S.P.J. Higson, Biosens. Bioeletron. **23**, 377 (2007)
39. N.C.H. Le, V. Gubala, R.P. Gandhiraman, C. Coyle, S. Daniels, D.E. Williams, Anal. Bioanal. Chem. **398**, 1927 (2010)
40. M. Erber, J. Stadermann, K.-J. Eichhorn, Macromol. Symp. **305**, 101 (2011)
41. A.V. Nabok, A. Tsargorodskaya, A.K. Hassan, N.F. Starodub, Appl. Surf. Sci. **246**, 381 (2005)
42. A. Nabok, A. Tsargorodskaya, A. Holloway, N.F. Starodub, A. Demchenko, Langmuir **23**, 8485 (2007)
43. A. Nabok, A. Tsargorodskaya, M.K. Mustafa, A. Székács, I. Székács, N.F. Starodub, Proced. Chem. **1**, 1491 (2009)
44. A.V. Nabok, A. Tsargorodskaya, A. Holloway, N.F. Starodub, O. Gojster, Biosens. Bioelectron. **22**, 885 (2007)
45. A.V. Nabok, M.K. Mustafa, A. Tsargorodskaya, N.F. Starodub, BioNanoScience **1**, 38 (2011)
46. A. Hassan, T. Basova, F. Yuksel, G. Gümüs, A.G. Gürek, V. Ahsen, Sens. Actuators B: Chem. **175**, 73 (2012)
47. T. Basova, V. Plyashkevich, A. Hassan, A.G. Gürek, G. Gümüs, V. Ahsen, Sens. Actuators. B Chem. **139**, 557 (2009)
48. T. Basova, A. Hassan, F. Yuksel, A.G. Gürek, V. Ahsen, Sens. Actuators B: Chem. **150**, 523 (2010)
49. A. Hassan, T. Basova, S. Tuncel, F. Yuksel, A.G. Gürek, V. Ahsen, Proced. Eng. **25**, 272 (2011)
50. B. Celėn, G. Demirel, E. Piskin, Nanotechnology **22**, 165501 (2011)

51. G. Jin, Y.H. Meng, L. Liu, Y. Niu, S. Chen, Q. Cai, T.J. Jiang, Thin Solid Films **519**, 2750 (2011)
52. H. Arwin, S. Welin-Klintström, R. Jansson, J. Colloid Interface Sci. **156**, 377 (1993)
53. G. Jin, Private communication. Also presented as Protein microarray biosensor based on imaging ellipsometry and biomedical applications, in *7th Asian congress for microcirculation and the 6th chinese national congress for microcirculation*, Tai'an, China, October 17–19 2008
54. Y. Zhang, Y. Chen, G. Jin, Sens. Actuators B: Chem. **159**, 121 (2011)
55. A. Nooke, U. Beck, A. Hertwig, A. Krause, H. Krüger, V. Lohse, D. Negendank, J. Steinbach, Sens. Actuators B: Chem. **149**, 194 (2010)
56. J. Mårtensson, H. Arwin, I. Lundström, Sens. Actuators B: Chem. **1**, 134 (1990)
57. R.S. Moirangthem, Y.-C. Chang, P.-K. Wei, Biomed. Opt. Express **1**, 2569 (2011)
58. S. Patskovsky, A.V. Kabashin, M. Meunier, Opt. Mater. **27**, 1093 (2005)
59. S. Patskovsky, A.V. Kabashin, M. Meunier, J.H.T. Luong, Appl. Opt. **34**, 6905 (2003)

# Chapter 19
# In-Line Quality Control of Organic Thin Film Fabrication on Rigid and Flexible Substrates

**Argiris Laskarakis and Stergios Logothetidis**

**Abstract** A major factor for the achievement of the required performance, efficiency and lifetime of organic electronic (OE) devices (Organic Photovoltaics—OPVs, Organic Light Emitting Diodes—OLEDs, etc.) is the quality control of the substrates, active layers, barrier materials and transparent electrode nanolayers that are used for the fabrication of these devices. The in-line optical characterization and modeling of the optical and electrical properties of the above nanolayers can give valuable information of the growth mechanisms and the structure-property relationships that can play a major role towards the optimization of the nanolayers performance. Also, the capability for in-line monitoring, at every single step, of the optical properties and the quality of the fabricated nanolayers, e.g. by roll-to-roll (r2r) process, will improve the process yield opening the way for the low cost fabrication of OE devices. In this chapter, we will discuss in detail the latest advances on the combination of optical sensing by Spectroscopic Ellipsometry with the processes (vacuum deposition, r2r printing) for fabrication of OE nanolayers and devices. These advances include the results of the determination of the optical constants, composition, refractive index, thickness with nm precision, stability and the uniformity of the OE nanolayers onto rigid (e.g. c-Si, glass) and flexible (as Polyethylene Terephthalate—PET) substrates.

## 19.1 Introduction

During the last years, there are significant advances in materials and processes for Organic Electronic (OE) devices [1–8]. The materials for organic electronics include organic semiconductors (small molecules, polymers) with sufficient electrical conductivity, transparent organic electrodes that can be deposited by printing processes,

A. Laskarakis (✉) · S. Logothetidis
Lab for Thin Films-Nanosystems & Nanometrology (LTFN),
Department of Physics, Aristotle University of Thessaloniki,
54124 Thessaloniki, Greece
e-mail: alask@physics.auth.gr

© Springer International Publishing AG, part of Springer Nature 2018
K. Hinrichs and K.-J. Eichhorn (eds.), *Ellipsometry of Functional Organic Surfaces and Films*, Springer Series in Surface Sciences 52,
https://doi.org/10.1007/978-3-319-75895-4_19

437

hybrid barriers for protection of the sensitive device materials from atmospheric gas permeation, and finally, flexible polymeric substrates, as PolyEthylene Terephthalate (PET), and PolyEthylene Naphthalate (PEN) to replace Si and glass substrates [9, 10]. The knowledge of their optical properties is of considerable importance and it can contribute to the understanding of the optical transparency, bonding and electronic structure and microstructure, as well as the morphology and growth mechanisms [6].

The capabilities of Spectroscopic Ellipsometry (SE) for in-situ and real-time monitoring has been proved by the several published works that report on its combination with vacuum deposition processes for the in-situ and real-time investigation of the optical properties, thickness and growth mechanisms of vacuum deposited thin films [11–18]. Although the implementation of SE for the in-line optical monitoring of r2r fabricated nanolayers has not been reported in detail yet, there are some works that provide promising results towards this direction [19–21]. The implementation of SE for the in-line monitoring of the optical properties of r2r printed nanolayers for organic electronics will lead to the optimization of the materials quality and increase of the production yield [22–24]. In this chapter, we will summarize on the latest advances of the implementation of in-line SE for the optical characterization of state-of-the-art nanomaterials (inorganic and organic barrier layers, transparent electrodes and bulk heterojunction active materials) by r2r gravure printing onto flexible polymer substrates. Gravure is a typical rotary printing method where ink is transferred from a physical printing form in a high-speed and high-throughput process and it is considered as one of the most reliable techniques for the low-cost formation of patterned nanolayers suitable for a successful high-volume manufacturing of flexible organic electronic devices, such as OPVs [25–28]. These will prove the potentiality of SE for implementation to: (a) lab scale processes as an important research tool that will contribute to the understanding and the optimization of the materials properties, and (b) industrial scale processes as a powerful quality control tool for the r2r fabrication of flexible organic electronic devices.

## 19.2   Adaptation of Spectroscopic Ellipsometry to r2r Fabrication Processes

The basic principles of SE have been described in detail in Chap. 1. The main requirement that has to be overcome for the successful adaptation of the SE technique on instrumentation for thin film (e.g. vacuum chambers or r2r printing systems) is the fulfillment of the necessary optical geometry between the optical elements before and after the light reflection on the sample surface and the sample itself. In the case of the adaptation of a SE unit onto a high vacuum deposition chamber, the optical elements (e.g. modulator and analyzer heads) are firmly attached by the use of suitable optical ports, including optical windows, such as $BaF_2$ [29]. This ensures

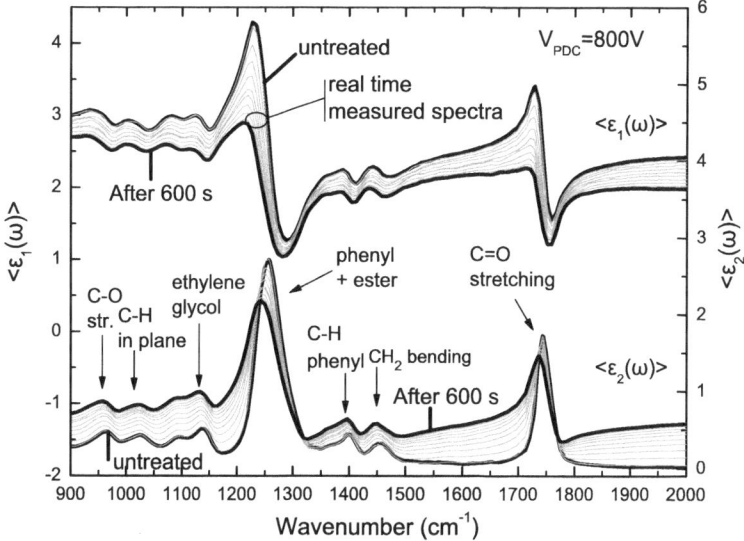

**Fig. 19.1** Measured $\langle \varepsilon(\omega) \rangle$ in the IR spectral region, before, during and after surface modification of PC using negative voltages of $V = 800$ V, and the characteristic bonding vibration bands of PC [35]

the minimization of possible vibrations of the optical elements allowing the accurate measurement of the optical properties from the sample surface. This has been reported in several works, where SE units working in a multiwavelength configuration, have been used for the in-situ and real-time investigation of the optical and electronic properties of growing thin films [12, 13, 17, 30–34]. In addition, these have been used for the investigation of the surface treatment of polymers, such as Polycarbonate (PC) [35]. For example, Fig. 19.1 shows the pseudo-dielectric function $\langle \varepsilon(\omega) \rangle = \langle \varepsilon_1(\omega) \rangle + i \langle \varepsilon_2(\omega) \rangle$ measured before, during and after surface modification of PC together with the characteristic bonding vibration of PC with a negative voltage of $V = 800$ V [35].

However, in the case of the adaptation of an SE unit onto a r2r fabrication system, there are several factors that influence the accuracy of the SE measurements, and the deduced results. These include the vibration of the system due to the various moving parts (e.g. rolls) and the moving sample under the point of measurement, which has to be planar (to allow the focusing of the light beam) and stable to allow the continuous measurement over the entire spectral region. For example, the flexible polymer rolls that are used as substrates usually they have a non-planar surface that wrinkles if it is not stretched appropriately with the result of defocusing the reflected light beam away from the detection head [24].

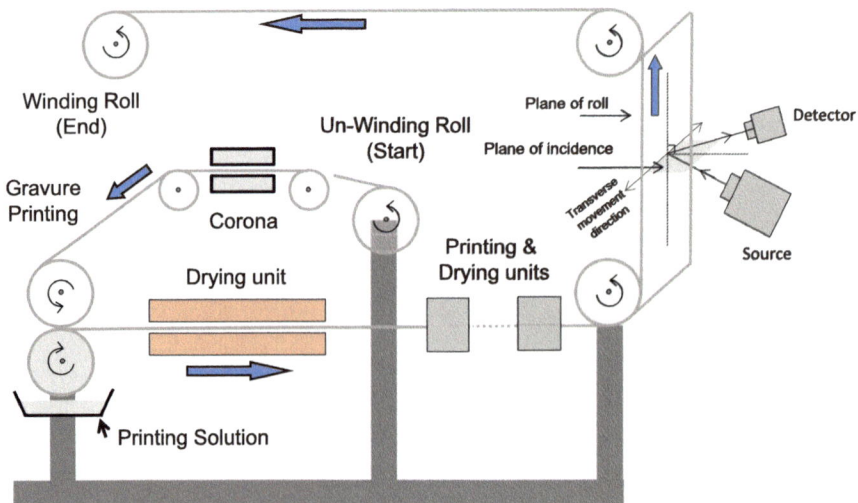

**Fig. 19.2** Schematic representation of the r2r printing unit onto which the in-line SE unit has been adapted. The plane of incidence of light at the measurement position is perpendicular to the plane of incidence of the polymer substrate roll. Modified after [24]

Figure 19.2 shows a schematic representation of a r2r system on which an in-line SE unit has been adapted. The measurement position is located on a vertical path of the roll direction, between two directional rolls. Between these rolls the polymer substrate stays absolutely flat and stable, without any wrinkles and tilting vibrations. The SE unit is equipped with an array detector that consists of 32 fiber optics for the simultaneous measurement of $\langle \varepsilon(\omega) \rangle$ at 32 specific photon energies at the spectral region 3.0–6.5 eV. The integration time (IT), which is the time for the completion of one multi-wavelength (MWE) spectrum, can be adjusted at 100 ms. The corresponding sampling time (ST), which is the total time for the recording of one spectrum after the integration of several measurements in order to calculate the final $\langle \varepsilon(\omega) \rangle$ spectra, should be higher than 100 ms. The stability of the foil, due to vibrations of the r2r system's mechanical parts has been taken care in order to minimize the artifacts and at the same time assure the quality of the measurements.

To assure that the roll surface remains flat, stable and without wrinkles during the rolling, a measurement of the optical and electronic properties of the plain PET roll over its entire length has been performed. In this way, the stability of the measured $\langle \varepsilon(\omega) \rangle$ as well as the values of the band gap energy and the characteristic electronic transitions of the substrate material will provide information on whether the conditions for accurate optical measurements are fulfilled. The SE measurements were performed with IT $= 150$ ms and ST $= 300$ ms, which accounts for the acquisition time for every step measuring, which embodied the 32-MWE. In this experiment, 200 spectra were recorded on the center of the PET roll during a length of 16 m.

Figure 19.3 shows the measured $\langle \varepsilon(\omega) \rangle$ of the PET roll at different positions along the direction of the rolling (Machine Direction-MD) [24]. Although more than 200

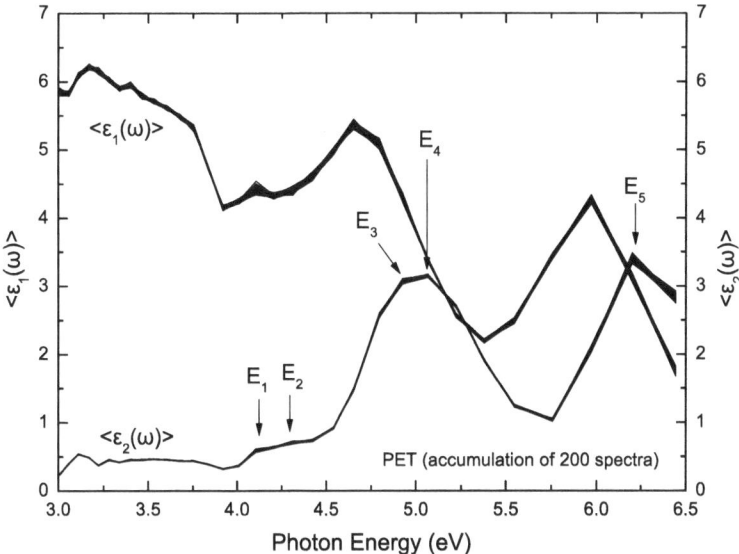

**Fig. 19.3** The time evolution of real and imaginary part of the measured pseudo-dielectric function $\langle \varepsilon(\omega) \rangle = \langle \varepsilon_1(\omega) \rangle + i \langle \varepsilon_2(\omega) \rangle$ from the PET substrate during its rolling from the unwinding to the winding roll. Modified after [24]

spectra have been recorded, we observe only a small deviation in $\langle \varepsilon(\omega) \rangle$, mostly at the energy region above 6 eV. The characteristic absorptions of PET are found at photon energies of 4.1, 4.2, 4.9, 5.2 and 6.3 eV in agreement to the literature [9]. The optical absorptions at 4.1 and 4.2 eV are attributed to the electronic transition of the non-bonded electron of the carbonyl O atom from the n state to the $\pi^*$ unoccupied valence state orbital ($n \rightarrow \pi^*$) transition [9]. The characteristic absorptions at 4.9 and 5.2 eV are attributed to the spin-allowed, orbitally forbidden $^1A_{1g} \rightarrow \, ^1B_{1u}$ transition, whereas the absorption at 6.3 eV is attributed to the $^1A_{1g} \rightarrow \, ^1B_{1u}$ electronic transition of the para-distributed benzene rings [9, 10, 36].

For the analysis of the measured $\langle \varepsilon(\omega) \rangle$, a theoretical model has been implemented that consists of the layer sequence air/PET/air. The optical properties of PET are described by the Tauc-Lorentz (TL) model (5 oscillators) in order to model the band gap energy and the characteristic electronic transitions of PET. The TL model is based on the combination of the classical Lorentz dispersion relation and the Tauc density of states in the proximity of the fundamental optical gap $\omega_g$. This results to an asymmetrical Lorentzian line-shape for $\langle \varepsilon_2(\omega) \rangle$. The TL dispersion model is described by the following relations in which the real part $\varepsilon_1(\omega)$ is determined by the imaginary part $\varepsilon_2(\omega)$ by the Kramers-Kronig integration [14, 37]:

**Fig. 19.4** Comparison
between the measured and
the calculated $\langle \varepsilon(\omega) \rangle$ spectra
of PET. Modified after [24]

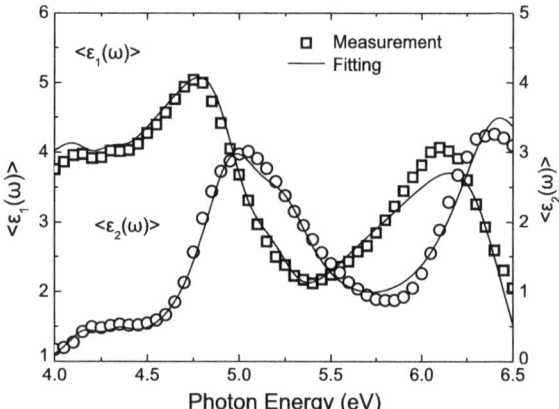

$$\varepsilon_2(\omega) = \frac{A\omega_o C(\omega - \omega_g)^2}{(\omega^2 - \omega_o^2)^2 + C\omega^2} \frac{1}{\omega}, \quad \omega > \omega_g \tag{19.1}$$

$$\varepsilon_2(\omega) = 0, \quad \omega \le \omega_g$$

$$\varepsilon_1(\omega) = \varepsilon_\infty + \frac{2}{\pi} P \int_{\omega_g}^{\infty} \frac{\xi \varepsilon_2(\xi)}{\xi^2 - \omega^2} d\xi \tag{19.2}$$

where $\varepsilon_\infty$ the non-dispersive term, $\omega_g$ the fundamental band gap energy, $A$ the amplitude, $\omega_0$ the Lorenz resonant energy, namely Penn gap, and $C$ the broadening term,
which is a measure of the materials disorder. The comparison between the mea-
sured and the calculated $\langle \varepsilon(\omega) \rangle$ by the above theoretical model is shown in Fig. 19.4.
The agreement between the measured and fitted $\langle \varepsilon(\omega) \rangle$ indicates the validity of the
followed approach.

The industrially supplied PET rolls are treated with mechanical biaxial stretching,
which induces a preferential orientation of the PET macromolecular chains towards
the stretching direction (or MD, as denoted before) [9, 10]. As a result, these materials
have anisotropic optical response, which require the realization of SE measurements
in multiple angles of incidence and at different angles between the MD and the plane
of incidence, in order to calculate the final bulk dielectric function [38]. However,
since the focus is the determination of the stability of the PET optical and electronic
response during its movement along the MD, which is a high symmetry orientation,
we can approximate its optical response with an isotropic optical model.

Figure 19.5 shows the evolution of the calculated values of the PET band-gap
as well as the electronic transition energies of PET that were calculated from the
analysis of $\langle \varepsilon(\omega) \rangle$. The characteristic electronic transition energies remain stable
over the whole length of the PET roll (16 m), leading to the conclusion that the
optical and structural properties of the PET roll are homogeneous over the entire
length and that the optical measurements are stable.

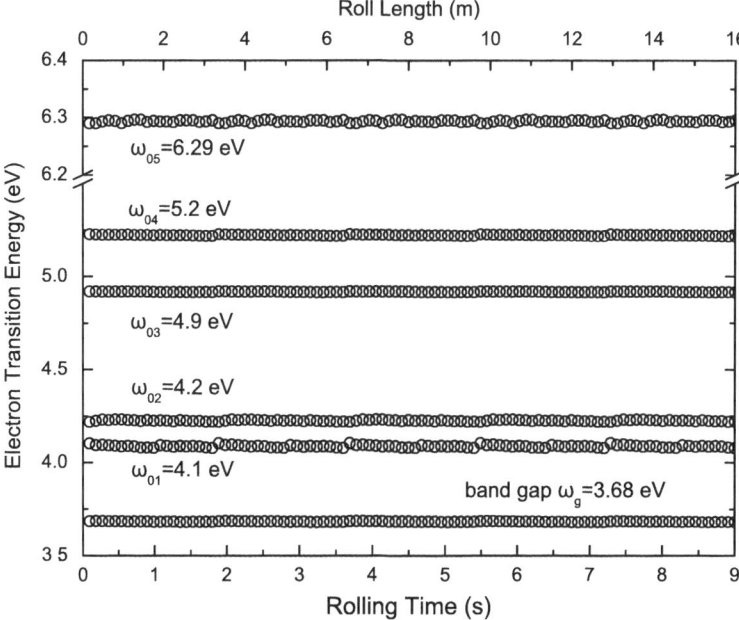

**Fig. 19.5** Evolution of the energy of band gap and the electronic transitions of PET substrate, calculated from the analysis of $\langle\varepsilon(\omega)\rangle$ measured during the rolling of the PET substrate [24]. Reprinted from S. Logothetidis, D. Georgiou, A. Laskarakis, C. Koidis, N. Kalfagiannis, In-line spectroscopic ellipsometry for the monitoring of the optical properties and quality of roll-to-roll printed nanolayers for organic photovoltaics, Sol. Energy Mater. Solar Cells **112**, 144. Copyright 2013 with permission from Elsevier

## 19.3 Optical Properties of OE Nanomaterials by In-Line SE

### 19.3.1 Inorganic Barrier Layers

Flexible organic electronic devices are fabricated on flexible polymer substrate rolls (e.g. PET, PEN), which have thickness in the range of 50–150 μm. These substrates can be incorporated to large-scale manufacturing processes, because they exhibit a combination of very important properties such as easy processing, flexibility, low cost, good mechanical properties that makes them ideal to be incorporated in large scale manufacturing processes [6, 9, 38–43]. However, one of the main drawbacks of these substrates is their relatively high permeability values for atmospheric $O_2$ and $H_2O$. These values are in the range of $10^{-1}$–$10^2$ cm$^3$/m$^2$ dbar (Oxygen Transmission Rate—OTR) and g/m$^2$ d (Water Vapor Transmission Rate—WVTR), which are sufficient only for food packaging applications [44, 45]. On the contrary, the requirements in OTR and WVTR for flexible OPVs and OLEDs are below than $10^{-5}$ cm$^3$/m$^2$ dbar (for OTR) and g/m$^2$ d (WVTR) [5, 46–48].

Initially, SiO$_x$ thin films have been used as encapsulation layers [49]. Their advantages include high optical transparency, recyclability and the possibility to deposited by several techniques such as metal–organic and plasma enhanced chemical vapor deposition and electron beam evaporation, which provides high deposition rate, thickness control and high production yield [50–52]. The barrier properties and the quality of SiO$_x$ depend on many factors, such as the process parameters, the deposition rate and the initial stages of the deposition as well as the stiffness and smoothness of the substrate [44, 53]. In this case, the molecule permeation is controlled by the defects of the flexible polymeric substrate and of the barrier nanolayers that induced by the intrinsic surface roughness, the surface and structure in-homogeneities and by cracks of the inorganic nano-layers created by the bending and/or tension of the flexible polymeric substrates [5, 46–48]. Therefore, it is crucial to understand the nucleation and growth of the oxide on polymer substrates, and in particular, the effect of the polymer surface characteristics on the processes.

In-situ and real-time SE has been extensively implemented for the monitoring of the growth mechanisms of vacuum deposited SiO$_x$ onto PET and PEN substrates, revealing significant information on the growth mechanisms. Figure 19.6a, b show the evolution of thickness and deposition rate of SiO$_x$ thin film deposited onto PET and PEN substrates. At the early stages of growth of SiO$_x$/PET, the deposition rate is increased up to 7.5 nm/s for $t = 6$ s, whereas at higher deposition times it follows an oscillating behavior. The oscillation of the deposition rate (Fig. 19.6b) is characteristic of the island-type growth and defines the distinct growth stages. At the initial stages of the SiO$_x$ deposition onto PET, separate clusters are formed with a size that increases until the first 6 s with incomplete film coverage. After $t = 6$ s, the reduction of deposition rate indicates the coalescence stage where the clusters merge. At $t > 12.5$ s, the deposition rate increase indicates the formation of SiO$_x$ clusters that merge again.

On the contrary, the deposition of SiO$_x$ onto PEN follows a layer-by-layer growth mechanism. The cluster size increases in the first 2.5 s and then merge until $t = 10$ s. For $t > 10$ s the SiO$_x$ film thickness increases linearly with time and the deposition rate is almost stable, resulting in a homogeneous layer deposition. The growth mechanism of SiO$_x$ is also affected by the substrate surface roughness, since PET and PEN have different roughness values (1.18 and 1.75 nm, respectively [54]).

In-line SE can be also monitor the thickness and optical properties of SiO$_x$ thin films deposited onto flexible substrates with the form of rolls. Commercially available SiO$_x$/PET can be purchased from specific suppliers in the form of roll with customized width, film thickness and structure, as well as roll length. Before its use as a base for organic electronic device fabrication, it is important to ensure that the SiO$_x$ film thickness is homogeneous over the whole length of the PET roll (in this case the length is 9.5 m). The in-line SE unit was utilized in order to monitor the thickness uniformity and optical and electronic properties of the SiO$_x$ films (deposited by electron beam evaporation onto the PET substrate) during the roll unwinding (speed = 4 m/min). The in-line SE measurements were performed on the middle of the SiO$_x$/PET roll. For the analysis of $\langle \varepsilon(\omega) \rangle$ we have used a theoretical model that

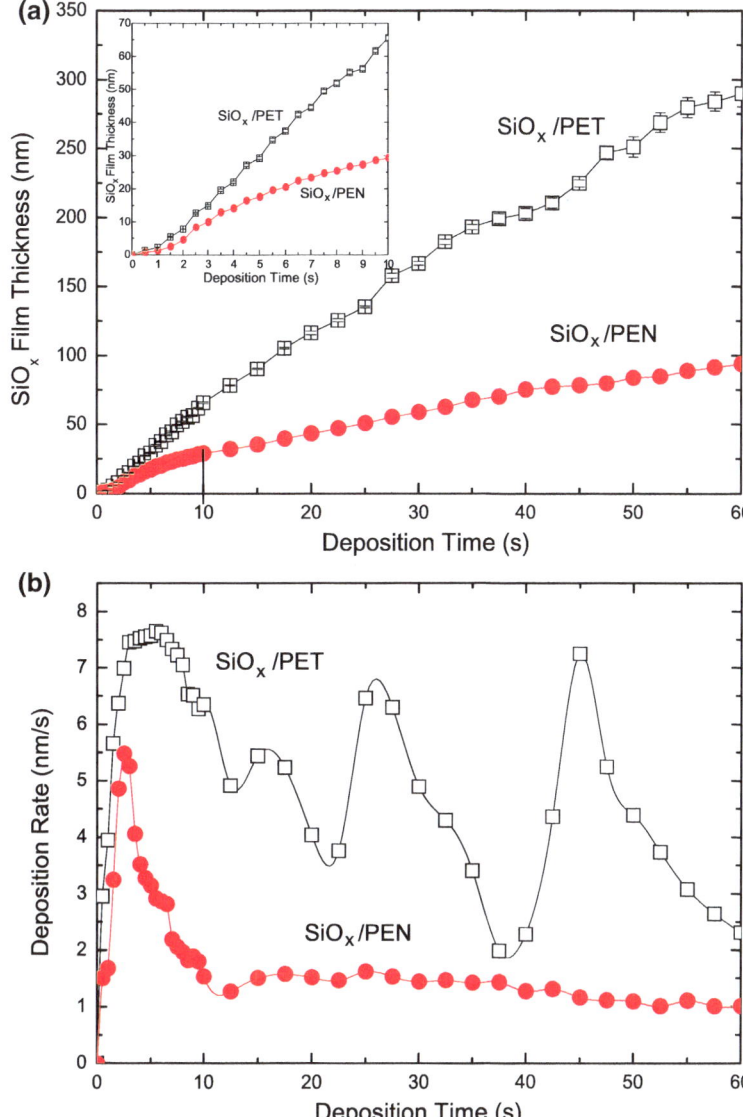

**Fig. 19.6** Time evolution of: **a** thickness and **b** deposition rate of the $SiO_x$ thin film deposited onto PET and PEN substrates, during the first 60 s of their deposition. The *inset* in Fig. 19.6a shows the thickness evolution in the first 10 s [54]. Reprinted from D. Georgiou, A. Laskarakis, C. Koidis, N. Goktsis, S. Logothetidis, Phys. Status Solidi C **5**, 3387. Copyright 2008 Wiley-VCH Verlag GmbH & Co. KGaA, Weinheim

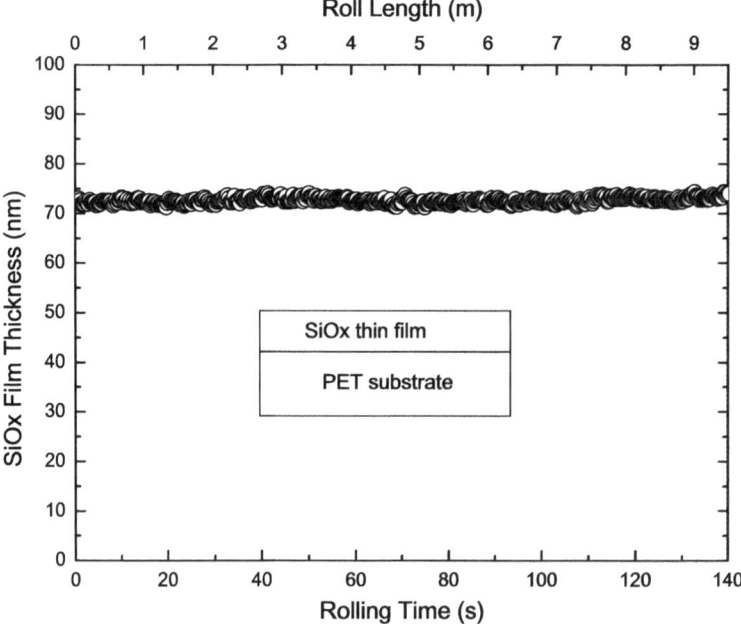

**Fig. 19.7** Thickness of $SiO_x$ film deposited onto PET as calculated by in-line SE

includes the layer sequence air/$SiO_x$/PET, whereas the optical properties of $SiO_x$ were described by 1 TL oscillator.

The analysis of $\langle \varepsilon(\omega) \rangle$ spectra revealed that the $SiO_x$ thickness was $73 \pm 1$ nm over the whole $SiO_x$/PET roll (Fig. 19.7). This indicates the good $SiO_x$ film uniformity along the roll. In addition, the optical parameters of the $SiO_x$ material during the PET rolling were quite stable, as it is shown in Fig. 19.8. The calculated band gap $\omega_g$ has been determined at 5.18 eV, whereas the maximum optical absorption $\omega_0$ has been calculated at 9.61 eV. The stability of the determined thickness and optical properties of $SiO_x$ demonstrates the stability of the evaporation process and the homogeneity of the vacuum deposited $SiO_x$ nanolayers. Finally, Fig. 19.9 shows the calculated bulk dielectric function $\varepsilon(\omega)$ of $SiO_x$ by the analysis of $\langle \varepsilon(\omega) \rangle$.

## 19.3.2  PEDOT:PSS Transparent Electrodes

Poly(3,4-ethylenedioxy-thiophene):poly(styrene-sulfonate) (PEDOT:PSS is one of the most promising candidates for use as transparent anode electrodes that combine metallic-like behavior with sufficient solubility for printing onto large areas [55]. PEDOT:PSS consists of a conducting part, PEDOT a low molecular weight polymer, which is insoluble and thus difficult to process and an insulating polymer PSS that is

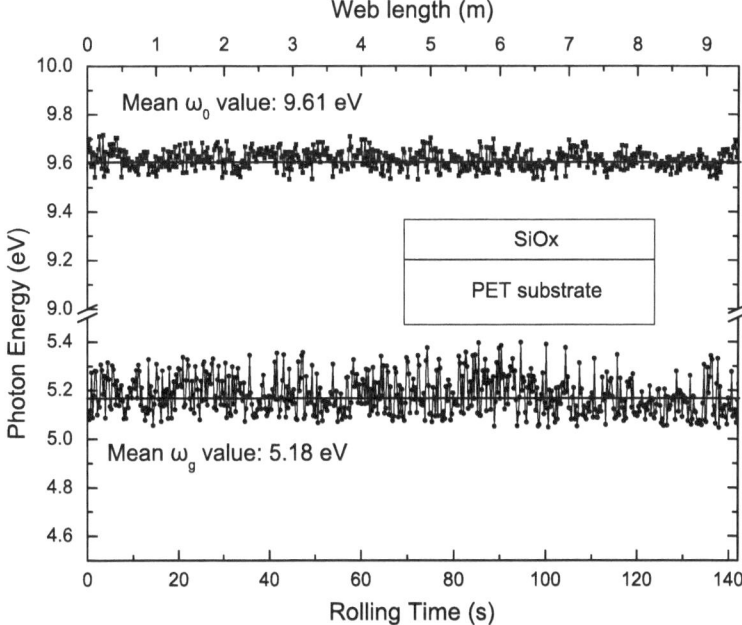

**Fig. 19.8** Determined values of fundamental energy band gap $\omega_g$ and maximum optical absorption $\omega_0$ of SiO$_x$ film deposited onto PET roll [24]. Reprinted from S. Logothetidis, D. Georgiou, A. Laskarakis, C. Koidis, N. Kalfagiannis, In-line spectroscopic ellipsometry for the monitoring of the optical properties and quality of roll-to-roll printed nanolayers for organic photovoltaics, Sol. Energy Mater. Solar Cells **112**, 144. Copyright 2013 with permission from Elsevier

**Fig. 19.9** Bulk dielectric function $\varepsilon(\omega)$ of SiO$_x$ calculated by the analysis of $\langle\varepsilon(\omega)\rangle$. Modified after [24]

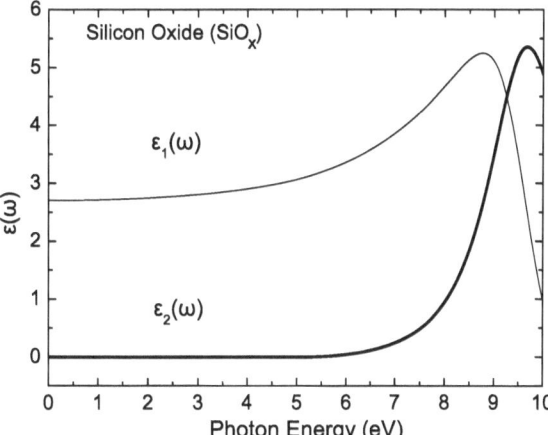

a high molecular weight polymer, which gives the desirable flexibility and solubility in water, making the polymer blend easy to process. The oligomer PEDOT segments are electrostatically attached on the PSS polymer chains [4–6, 56–63].

The conductivity of PEDOT:PSS depends strongly on their structure and morphology. Pristine PEDOT:PSS yields a conductivity of below 10 Scm$^{-1}$, much too low to be used as an electrode in an efficient OPV. However, the addition of solvents, that include ethylene glycol, sorbitol, glycerol, dimethyl-sulfoxide, significantly improves the conductivity of PEDOT:PSS [61, 64–70], whereas, post-deposition processing (e.g. annealing) has been also addressed [57, 71–73]. In this way, conductivities above 800 Scm$^{-1}$ have been reported [73, 74]. However, even if the required conductivity target of the material is achieved, it is crucial to ensure that the material structure, thickness and properties is homogeneous over the entire printed area and that the r2r fabrication process is stable. Thickness fluctuations and composition irregularities result to the degradation of the optical properties of the transparent electrode layer and to the non-optimum operation of the device.

The dielectric function of PEDOT:PSS consists of contributions from the conductive PEDOT and the non-conductive PSS part, at different photon energies. More specifically, it has been reported that there are three optical absorptions at 0.47, 5.37 and 6.38 eV [41, 55, 58, 61, 71, 75, 76]. The low energy absorption is attributed to PEDOT, whereas the two absorptions at the high energies are attributed to the $\pi$–$\pi^*$ electronic transition of PSS. In addition, a more intense absorption has been found at the infrared spectral region which can allow the determination of the metallic-like behavior of PEDOT:PSS [41, 55, 58, 61, 71, 75, 76].

Gravure printing has been widely used to print PEDOT:PSS onto flexible polymer substrates by r2r process. The most common printing patterns include stripes parallel to the rolling direction in various sizes and distances. For the determination of the capability to monitor in-line the thickness and optical and electronic properties of the gravure printed PEDOT:PSS, we have printed a PEDOT:PSS pattern onto the PET roll. This consists of a set of parallel stripes, each one having a width of 10 mm and a distance of 5 mm between each other. The power applied for the film curing was 100 W min/m$^2$ and the rolling speed was 4 m/min. For the analysis of the in-line measured $\langle \varepsilon(\omega) \rangle$ spectra, a theoretical model that includes air/PEDOT:PSS/PET/air has been used. The optical properties of the PEDOT:PSS layer were described by the use of 2 TL oscillators. These oscillators model the characteristic absorption peaks at $\sim$5.5 and $\sim$6.3 eV that originate from PSS [41, 55, 58, 61, 71, 75, 76]. The reason for the modeling of only these two electronic transitions is that the optical absorptions of the more conductive PEDOT part are reported to appear at lower energy values below 1.7 eV, which is outside of the measured energy region.

It has to be noted that spin coated PEDOT:PSS films are reported to show uniaxial optical anisotropy, with the ordinary complex refractive index in the plane of the film and the extraordinary complex refractive index normal to the film and parallel to the optical axis [41, 77, 78]. This is the result of the films structure that consists of PEDOT-rich flattened lamellas that are vertically segregated by PSS-rich layers. However, the printed PEDOT:PSS films that are describe in this chapter are characterized by non-directional structure and their optical response can be modeled by

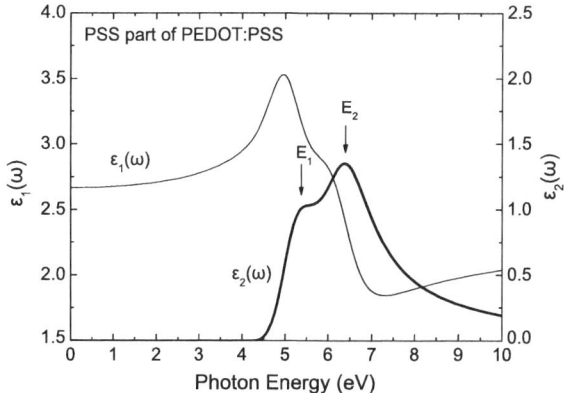

**Fig. 19.10** Bulk dielectric function $\varepsilon(\omega)$ of the PSS part of the PEDOT:PSS as calculated by the best-fit parameters from the analysis of $\langle \varepsilon(\omega) \rangle$. Modified after [24]

a isotropic optical model [24]. Figure 19.10 shows the bulk dielectric function of the PSS part of the PEDOT:PSS as calculated by the best-fit parameters from the analysis of the measured $\langle \varepsilon(\omega) \rangle$. The $\varepsilon(\omega)$ is characteristic of the optical properties of the PSS part and it is independent of the film thickness [24].

Figure 19.11 shows the evolution of the calculated electronic transition energies, thickness and the band gap energy of the insulating PSS part over the whole PET roll with a length of 5.5 m. The film thickness is calculated at $42 \pm 1$ nm and this value is stable over the whole length of the PET roll. This indicates a homogeneous printing process, whereas the stability of $\omega_g$ at $4.42 \pm 0.01$ eV leads to the conclusion that the optical and electronic properties of the material are stable in the entire printed area.

To test the capability of in-line SE to monitor sudden changes in the film thickness, and to verify the uniformity of the gravure printed PEDOT:PSS stripes, we have measured the $\langle \varepsilon(\omega) \rangle$ in 100 subsequent points in the transverse direction, keeping the roll stable. The film thickness has been modeled in real-time during the scanning of the PET roll with the SE unit. The thickness profile of the PEDOT:PSS pattern based on the calculated thickness values from the analysis of $\langle \varepsilon(\omega) \rangle$ is shown in Fig. 19.12. The film thickness is calculated at 30 nm for all seven stripes, whereas it can be seen that between the stripes (where there is no printed film) the film thickness drops to zero. The results from the in-line analysis procedure verify the uniformity of the PEDOT:PSS material laterally to the direction of roll.

For the investigation of the effect of the experimental parameters on the printed film thickness, we have modified the drying temperature from 40 to 160 °C. The drying temperature was increased from 40 °C at the starting of the rolling and it increased gradually until it will reach 160 °C, at $t = 600$ s. Figure 19.13 shows the $\langle \varepsilon_2(\omega) \rangle$ of PEDOT:PSS that has been subjected to drying at different temperatures from 40 °C to 160 °C. During the temperature increase, the PEDOT:PSS thickness decreases from $44 \pm 1$ nm at 40 °C to $35 \pm 1$ nm at 150 °C (Fig. 19.14) due to the

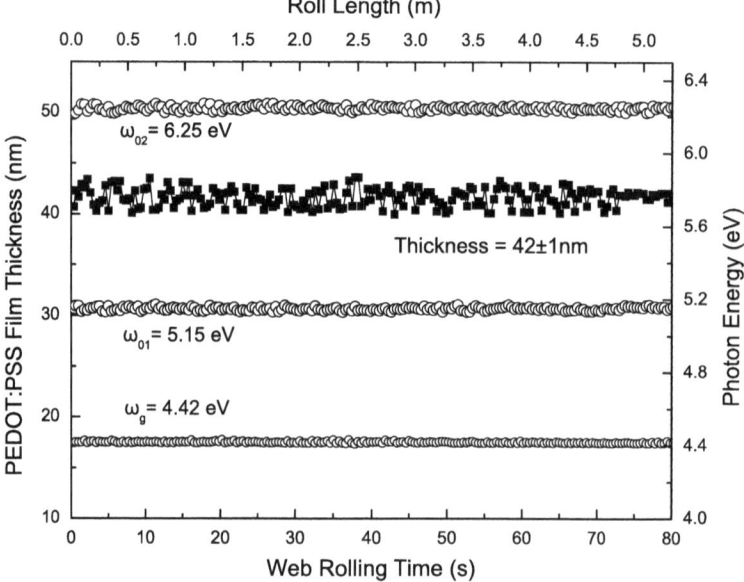

**Fig. 19.11** In-line calculated thickness, band gap energy and characteristic interband electronic transitions of PSS as function of the rolling time (*bottom* x-axis) and roll length (*top* x-axis). Modified after [24]

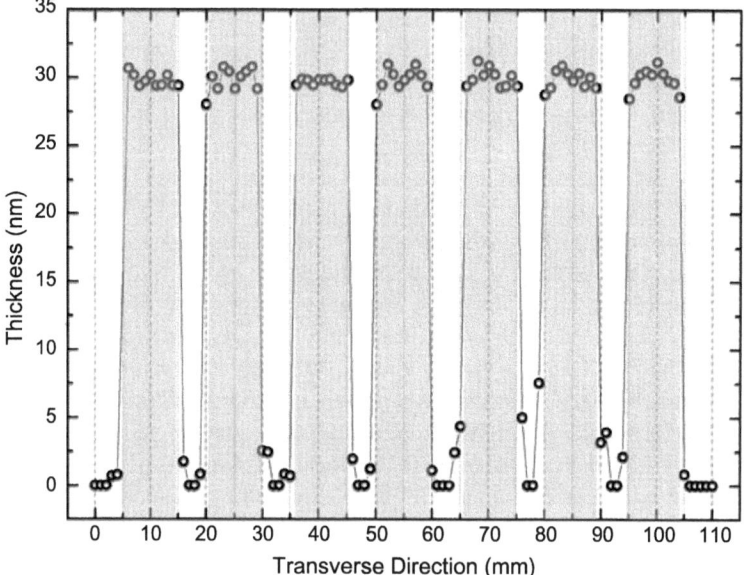

**Fig. 19.12** Pattern of gravure printed PEDOT:PSS and grid for the measurements and analysis of PEDOT:PSS in the lateral direction. Modified after [24]

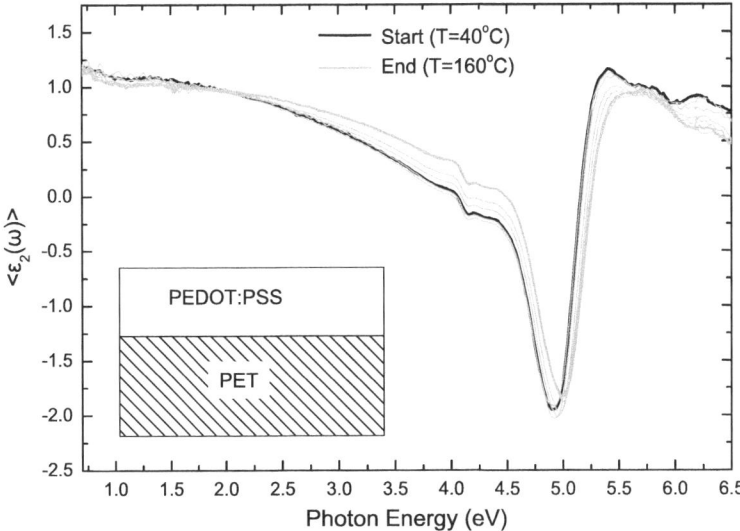

**Fig. 19.13** Time evolution of imaginary part of pseudo-dielectric function $\langle\varepsilon(\omega)\rangle = \langle\varepsilon_1(\omega)\rangle + i\langle\varepsilon_2(\omega)\rangle$ of PEDOT:PSS/PET layer structure. The thin gray lines represent the measurements at drying temperatures between 40 and 160 °C. Modified after [24]

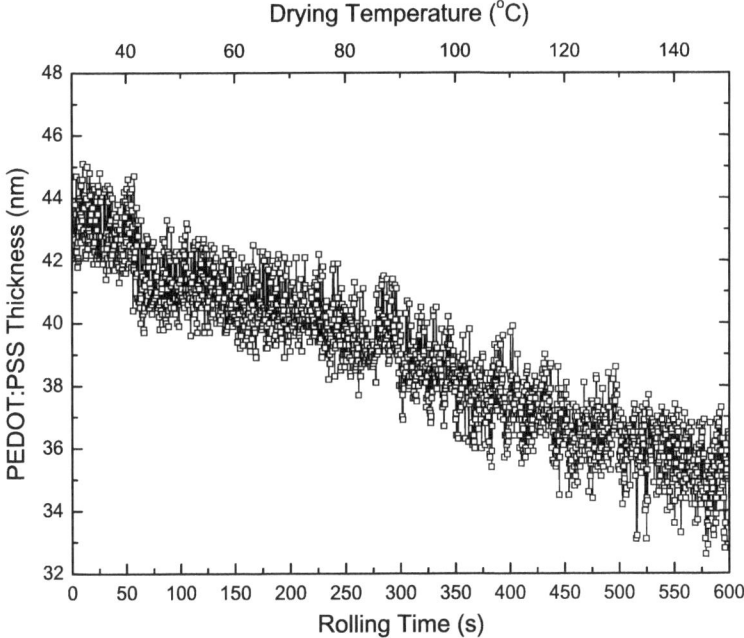

**Fig. 19.14** Evolution of PEDOT:PSS thickness with the drying temperature from 40 to 160 °C. Modified after [24]

evaporation of the solvent as the printed area passes below the drying unit. Also, the results verified the unit sensitivity to monitor fluctuation of ~1 nm during printing and drying.

### 19.3.3  Multilayer Structures onto PET Rolls

In order to verify the capability of in-line SE unit to measure and analyze multilayer stacks we gravure printed a hybrid polymer film onto $SiO_x$/PET and afterwards a PEDOT:PSS film onto the hybrid polymer/$SiO_x$/PET. The hybrid (inorganic-organic) polymer is a barrier nano-composite, synthesized via the sol-gel process. The physical properties of these materials have been discussed in detail elsewhere [44, 45, 79–81]. The optical properties of this multilayer structure have been measured and analyzed by in-line SE during the film printing.

Figure 19.15 shows the calculated thickness of $SiO_x$ film, hybrid polymer and PEDOT:PSS over a roll with a length of 7.5 m. The $SiO_x$ film thickness is 70 nm with a high uniformity across the roll. The thickness of PEDOT:PSS and hybrid polymer layers that were gravure printed show a relatively stable uniformity at 85 and 760 nm, respectively. This demonstrates the better thickness control of e-beam

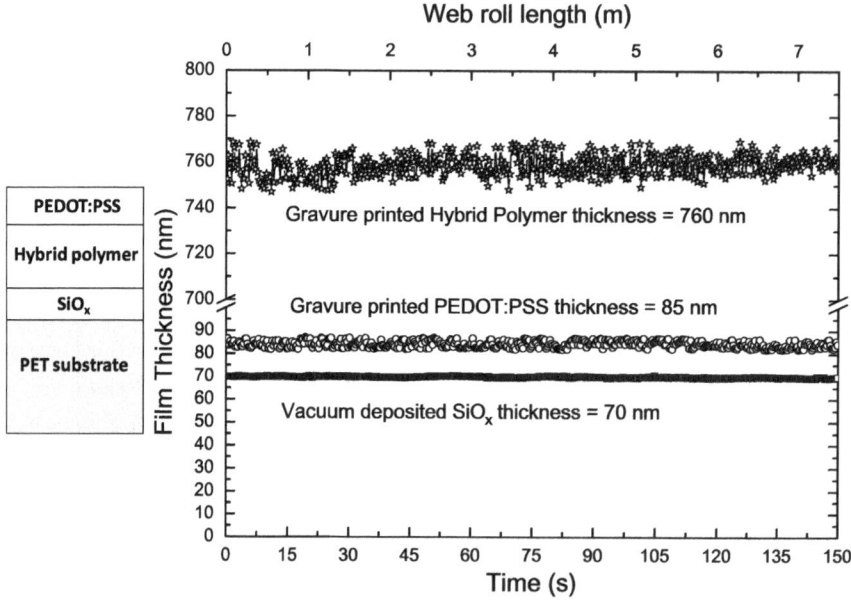

**Fig. 19.15** Thickness of individual layers of the multilayer structure (hybrid polymer)/PEDOT: PSS/$SiO_x$/PET as determined by in-line SE. Modified after [24]

evaporation than gravure printing. The above demonstrate the capability of in-line SE to determine the optical and electronic properties and thickness of multilayer stacks with complex optical response.

### 19.3.4  P3HT:PCBM Nanolayers

The currently most successful photoactive material for OPVs consists of a bulk heterojunction (BHJ) that is formed by a p-type semiconductor (electron donor), such as poly(3-hexylthiophene) (P3HT) with an n-type semiconductor (electron acceptor), such as methanofullerene derivatives (PCBM) [42, 62, 63]. In order to achieve maximum charge generation, a large interfacial area between these two organic semiconductors is required, which can be achieved by optimum nanoscale phase separation. The distribution of the constituents in the blend film plays an important role for efficient charge extraction toward the electrodes. Ideally more p-type material (polymer) should be located at the interface of the hole-collecting anode, and more n-type (fullerene) material should be at the electron collecting cathode facilitating collecting of charges from the photoactive layer. However, the morphology of a BHJ consisting of a binary blend cannot be easily controlled. The formation of the blend film structure is affected by several parameters, such as blend composition, viscos-

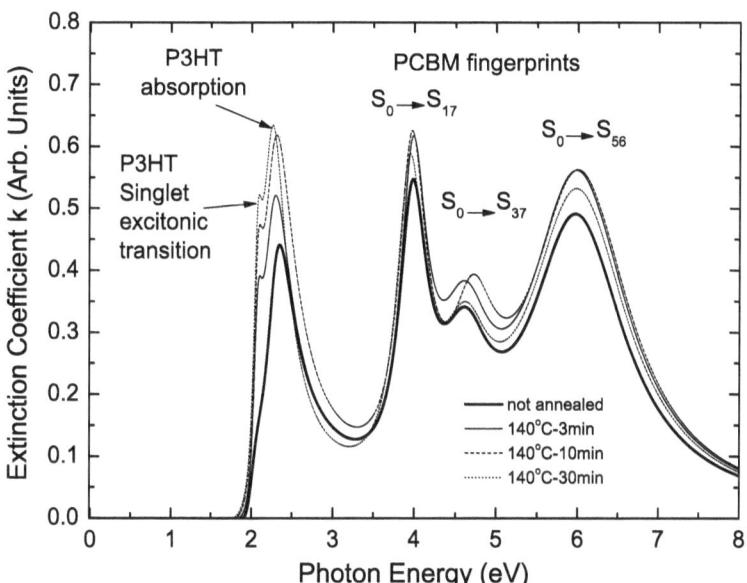

**Fig. 19.16** Extinction coefficient ($k$) of P3HT:PCBM after post-deposition thermal annealing at 140 °C for various periods of time based on the best-fit parameters from the analysis of $\langle \varepsilon(\omega) \rangle$. Modified after [84]

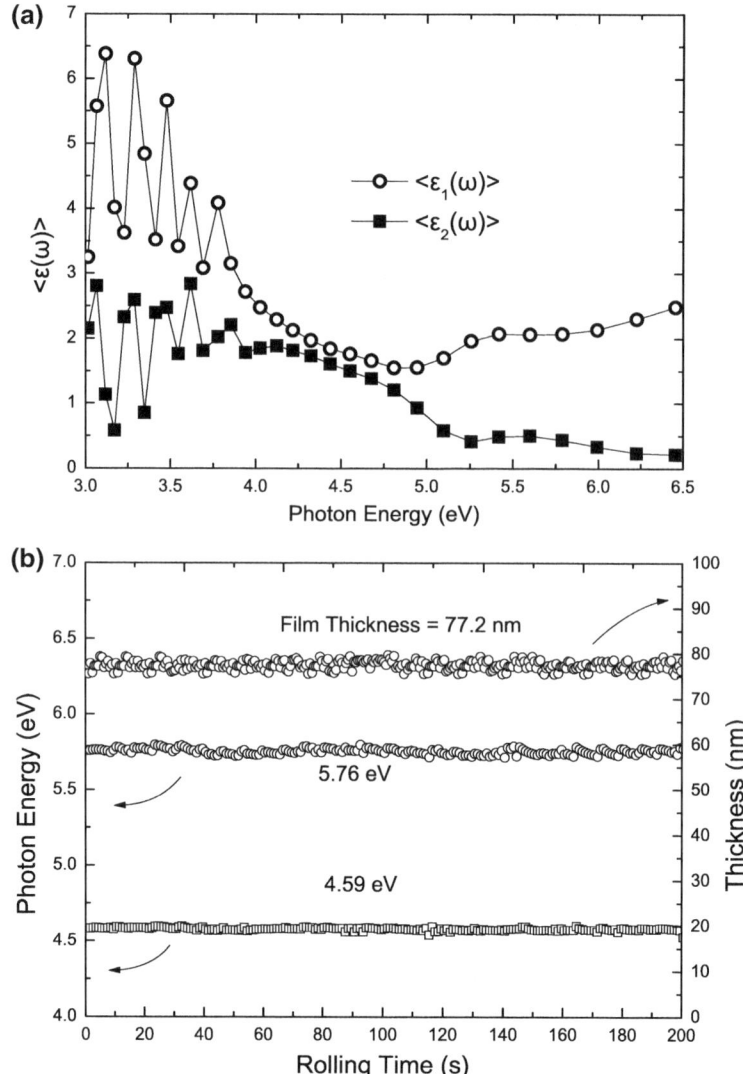

**Fig. 19.17** **a** Representative $\langle \varepsilon(\omega) \rangle$ spectrum of P3HT:PCBM/PET, **b** evolution of the blend thickness and the electronic transitions $S_0 \rightarrow S_{37}$ and $S_0 \rightarrow S_{56}$ at 4.59 and 5.76 eV, respectively of the PCBM, with the rolling time of the blend/PET film with length of 18 m, as determined by the in-line SE. Modified after [24]

ity, solvent evaporation rate or substrate surface energy, providing difficulties to the achievement of the desired blend morphology for maximum charge generation and transport [1, 4, 42, 62, 63, 82].

The optical properties of P3HT:PCBM blends have been investigated by ex-situ SE in a wide spectral region. These studies focused on the contribution of the optical response of the blend components as well as on their vertical distribution in the blend volume. For example, Karagiannidis et al., has reported on the investigation of the optical constants of P3HT:PCBM blends and on the effect of the post deposition thermal annealing on their optical and electronic properties [83, 84]. Figure 19.16 shows the calculated extinction coefficient ($k$) of the P3HT:PCBM blend films (as grown and annealed at 140 °C for various annealing times from 3 to 30 min). The optical and electronic response of the blends in the Vis–fUV spectral region includes five optical absorptions that are found at photon energies of 2.05, 2.24, 3.95, 4.65 and 5.89 eV. The first optical absorption at 2.05 eV is attributed to the singlet excitonic transition of the P3HT conjugated polymer whereas the transition at 2.24 eV corresponds to the formation of excitons with phonons. The other three electronic transitions at higher energies are originated from the PCBM and they can be assigned to the electronic transitions $S_0 \rightarrow S_{17}$, $S_0 \rightarrow S_{37}$ and $S_0 \rightarrow S_{56}$, respectively [84–86]. The increase of the annealing temperature leads to the excitonic enhancement of the P3HT, which has been correlated to the increase of the P3HT crystallization [84].

Figure 19.17a shows a representative $\langle \varepsilon(\omega) \rangle$ spectrum that has been recorded during the in-line measurements of the blend/PET rolling, whereas Fig. 19.17b shows the evolution of the P3HT:PCBM blend film thickness measured and analyzed during the passing of the roll under the in-line SE unit, and after the gravure printing of the blend onto the PET roll. The thickness of the P3HT:PCBM has been calculated at $77 \pm 2$ nm and it is stable over the whole length of the roll (above 18 m). This proves the stability of gravure printing and the thickness homogeneity of P3HT:PCBM at the different spots on the PET roll.

## 19.4 Summary and Outlook

In this chapter, we have focused on the implementation of in-line SE unit on r2r printing process for the real-time measurement and analysis of the thickness and optical properties of nanomaterials for the large scale fabrication of OE devices, such as OPVs. The in-line control of the optical and electronic properties, thickness and quality of the r2r gravure printed nanomaterials plays an important role on the optimization of their functionality as well as on the stability and property repeatability of the printed layers and OE devices.

We have described the application of in-line SE for the monitoring of a number of different material systems, from flexible substrates to multilayered film structures composed of barrier layers, transparent electrodes and organic semiconductors with thicknesses in the order of several nanometers. The measurement scans at the lateral and transverse directions of the rolls in combination with the real time analysis of the

$\langle \varepsilon(\omega) \rangle$ have provided the thickness profiles as well as the characteristic electronic transition energies. The above emphasize the significance of non-destructive in-line optical sensing tools for the quality control tools of r2r fabrication processes (in lab, pilot and large scale) of single and/or multilayer structures on polymeric substrates.

The challenges include the use of more complex devices structures with several layers, and the use of nano-materials with complex optical and electronic properties and possible optical anisotropy. Also, the subsequent printing of complicated film patterns (especially in the case of OLEDs, OTFTs and sensors) will provide difficulties on the robust quality control by in-line SE. Continuous advancements in the instrumentation that will enhance the speed and accuracy of the measured spectra, along with the rapid development of computational algorithms, strongly suggest that in-line SE will become an essential part of the r2r fabrication of flexible OE devices.

**Acknowledgements** The authors would like to thank Dr. Nikolaos Kalfagiannis, Dr. Despoina Georgiou, Dr. Christos Koidis, Dr. Panagiotis G. Karagiannidis and the other staff of the Lab for Thin Films, Nanosystems and Nanometrology (LTFN) for their contribution. The authors would also like to thank Amcor for supply of the SiO$_x$/PET rolls, Clevios for the supply of the PEDOT:PSS formulations and Fraunhofer-Institut für Silicatforschung for the supply of the hybrid polymer formulations. This work was partially supported by the EC STREP Project OLAtronics, Grand Agreement No. 216211, and by the EC REGPOT Project ROleMak No. 286022.

# References

1. D.M. de Leeuw, E. Cantatore, Mater. Sci. Semicond. Proces. **11**, 199 (2008)
2. F.C. Krebs, Org. Electron. **10**, 761 (2009)
3. M. Cavallini, M. Facchini, M. Massi, F. Biscarini, Synth. Met. **146**, 283 (2004)
4. P. Kopola, T. Aernouts, R. Sliz, S. Guillerez, M. Ylikunnari, D. Cheyns, M. Välimäki, M. Tuomikoski, J. Hast, G. Jabbour, R. Myllylä, A. Maaninen, Sol. Energy Mater. Sol. Cells **95**, 1344 (2011)
5. S. Logothetidis, Mater. Sci. Eng. B **152**, 96 (2008)
6. S. Logothetidis, A. Laskarakis, Eur. Phys. J. Appl. Phys. **46**, 12502 (2009)
7. White Paper OE—A Roadmap (2011)
8. P. Kumar, S. Chand, Prog. Photovolt. Res. Appl. **20**(4), 377 (2012)
9. A. Laskarakis, S. Logothetidis, J. Appl. Phys. **99**, 066101 (2006)
10. A. Laskarakis, S. Logothetidis, J. Appl. Phys. **101**, 053503 (2007)
11. S. Logothetidis, in *Thin Films Handbook*, ed. by H.S. Nalwa (Academic Press, Dublin, 2001)
12. V.G. Kechagias, M. Gioti, S. Logothetidis, R. Benferhat, D. Teer, Thin Solid Films **364**, 213 (2000)
13. S. Logothetidis, A. Laskarakis, A. Gika, P. Patsalas, Surf. Coat. Technol. **152**, 204 (2002)
14. G.E. Irene, H.G. Tompkins (eds.), *Handbook of Ellipsometry* (William Andrew Publishing, Norwich, 2005)
15. A. Laskarakis, S. Kassavetis, C. Gravalidis, S. Logothetidis, Nucl. Instrum. Methods Phys. Res. Sect. B: Beam Interact. Mater. At. **268**, 460 (2010)
16. M. Gioti, S. Logothetidis, C. Charitidis, Y. Panayiotatos, I. Varsano, Sens. Actuators **99**, 35 (2002)
17. R.W. Collins, J. Koh, H. Fujiwara, P.I. Rovira, A.S. Ferlauto, J.A. Zapien, C.R. Wronski, R. Messier, Appl. Surf. Sci. **154–155**, 217 (2000)
18. A. Laskarakis, S. Logothetidis, S. Kassavetis, E. Papaioannou, Thin Solid Films **516**, 1443 (2008)

19. L.R. Dahal, Z. Huang, D. Attygalle, M.N. Sestak, C. Salupo, S. Marsillac, R.W. Collins, *35th IEEE Photovoltaic Specialists Conference* (2010), p. 000631
20. K. Heymann, G. Mirschel, T. Scherzer, M. Buchmeiser, Vib. Spectrosc. **51**, 152 (2009)
21. B. Schmidt-Hansberg, M.F.G. Klein, K. Peters, F. Buss, J. Pfeifer, S. Walheim, A. Colsmann, U. Lemmer, P. Scharfer, W. Schabel, J. Appl. Phys. **106**, 124501 (2009)
22. S. Logothetidis, Method for the In-situ and Real-time Determination of the Thickness, Optical Properties and Quality of Transparent Coatings During their Growth onto Polymeric Substrates and Determination of the Modification, Activation and the Modification Depth of Polym, U.S. Patent 7,777,882 (2010)
23. S. Logothetidis, Method for In-line Determination of Film Thickness and Quality During Printing Processes for the Production of Organic Electronics, U.S. Patent PCT/GR2011/000018 (2011)
24. S. Logothetidis, D. Georgiou, A. Laskarakis, C. Koidis, N. Kalfagiannis, Sol. Energy Mater. Sol. Cells **112**, 144 (2013)
25. J. Noh, D. Yeom, C. Lim, H. Cha, J. Han, J. Kim, Y. Park, V. Subramanian, G. Cho, IEEE Trans. Electron. Packag. Manuf. **33**, 275 (2010)
26. Y.-J. Cheng, S.-H. Yang, C.-S. Hsu, Chem. Rev. **109**, 5868 (2009)
27. C.-Y. Lo, J. Hiitola-Keinänen, O.-H. Huttunen, J. Petäjä, J. Hast, A. Maaninen, H. Kopola, H. Fujita, H. Toshiyoshi, Microelectron. Eng. **86**, 979 (2009)
28. C. Koidis, S. Logothetidis, S. Kassavetis, C. Kapnopoulos, P.G. Karagiannidis, D. Georgiou, A. Laskarakis, Sol. Energy Mater. Sol. Cells **112**, 36 (2013)
29. A. Laskarakis, S. Logothetidis, M. Gioti, Phys. Rev. **64**, 1 (2001)
30. C. Chen, M.W. Horn, S. Pursel, C. Ross, R.W. Collins, Appl. Surf. Sci. **253**, 38 (2006)
31. S. Logothetidis, M. Gioti, P. Patsalas, Diam. Relat. Mater. **10**, 117 (2001)
32. C. Gravalidis, M. Gioti, A. Laskarakis, S. Logothetidis, Surf. Coat. Technol. **181**, 655 (2004)
33. M. Gioti, S. Logothetidis, P. Patsalas, A. Laskarakis, Y. Panayiotatos, V. Kechagias, Surf. Coat. Technol. **125**, 289 (2000)
34. C. Koidis, S. Logothetidis, D. Georgiou, A. Laskarakis, Phys. Status Solidi C **5**, 1366 (2008)
35. A. Laskarakis, S. Kassavetis, C. Gravalidis, S. Logothetidis, Nucl. Instrum. Methods Phys. Res. B **268**, 460 (2010)
36. A. Laskarakis, S. Logothetidis, Appl. Surf. Sci. **253**, 52 (2006)
37. G.E. Jellison, F.A. Modine, Appl. Phys. Lett. **69**, 371 (1996)
38. H.F. Dam, F.C. Krebs, Sol. Energy Mater. Sol. Cells **97**, 191 (2012)
39. A. Laskarakis, S. Logothetidis, E. Pavlopoulou, M. Gioti, Thin Solid Films **455–456**, 43 (2004)
40. T. Yoshioka, M. Tsuji, Y. Kawahara, S. Kohjiya, J. Appl. Polym. Sci. **44**, 7997 (2003)
41. L.A.A. Pettersson, S. Ghosh, O. Inganas, Org. Electron. **3**, 143 (2002)
42. F.C. Krebs, Sol. Energy Mater. Sol. Cells **93**, 394 (2009)
43. P.E. Burrows, G.L. Graff, M.E. Gross, P.M. Martin, M.K. Shi, M. Hall, E. Mast, C. Bonham, W. Bennett, M.B. Sullivan, Displays **22**, 65 (2001)
44. C. Charton, N. Schiller, M. Fahland, A. Hollander, A. Wedel, K. Noller, Thin Solid Films **502**, 99 (2006)
45. K. Haas, S. Amberg-Schwab, K. Rose, Thin Solid Films **351**, 198 (1999)
46. J. Fahlteich, M. Fahland, W. Schönberger, N. Schiller, Thin Solid Films **517**, 3075 (2009)
47. B.M. Hanika, H. Langowski, U. Moosheimer, W. Peukert, Chem. Eng. Technol. **26**, 605 (2003)
48. K. Haas, K. Rose, Rev. Adv. Mater. Sci. **5**, 47 (2003)
49. D. Georgiou, S. Logothetidis, C. Koidis, A. Laskarakis, Phys. Status Solidi C **5**, 1300 (2008)
50. D.G. Howells, B.M. Henry, J. Madocks, H.E. Assender, Thin Solid Films **516**, 3081 (2008)
51. A.P. Roberts, B.M. Henry, A.P. Sutton, C.R.M. Grovenor, G.A.D. Briggs, T. Miyamoto, M. Kano, Y. Tsukahara, M. Yanaka, J. Membr. Sci. **208**, 75 (2002)
52. A.S. da Silva Sobrinho, M. Latreche, G. Czeremuszkin, J.E. Klemberg-Sapieha, M.R. Wertheimer, J. Vac. Sci. Technol. A **16**(6), 3190 (1998)
53. M. Yanaka, B.M. Henry, A.P. Roberts, C.R.M. Grovenor, G.A.D. Briggs, A.P. Sutton, Thin Solid Films **397**, 176 (2001)

54. D. Georgiou, A. Laskarakis, C. Koidis, N. Goktsis, S. Logothetidis, Phys. Status Solidi C **5**, 3387 (2008)
55. Y. Chen, K.S. Kang, K.J. Han, K.H. Yoo, J. Kim, Synth. Met. **159**, 1701 (2009)
56. A.M. Nardes, R.A.J. Janssen, M. Kemerink, Adv. Funct. Mater. **18**, 865 (2008)
57. Z. Xiong, C. Liu, Org. Electron. **13**, 1532 (2012)
58. A. Nardes, M. Kemerink, R. Janssen, Phys. Rev. B **76**, 1 (2007)
59. G. Wang, S.-I. Na, T.-W. Kim, Y. Kim, S. Park, T. Lee, Org. Electron. **13**, 771 (2012)
60. T.P. Nguyen, P. Le Rendu, P.D. Long, S.A. De Vos, Surf. Coat. Technol. **180–181**, 646 (2004)
61. O.P. Dimitriev, D.A. Grinko, Y.V. Noskov, N.A. Ogurtsov, A.A. Pud, Synth. Met. **159**, 2237 (2009)
62. C.J. Brabec, Sol. Energy Mater. Sol. Cells **83**, 273 (2004)
63. H. Hoppe, N.S. Sariciftci, J. Mater. Res. **19**, 1924 (2011)
64. A. Onorato, M.A. Invernale, I.D. Berghorn, C. Pavlik, G.A. Sotzing, M.B. Smith, Synth. Met. **160**, 2284 (2010)
65. S.-I. Na, G. Wang, S.-S. Kim, T.-W. Kim, S.-H. Oh, B.-K. Yu, T. Lee, D.-Y. Kim, J. Mater. Chem. **19**, 9045 (2009)
66. N.G. Semaltianos, S. Logothetidis, N. Hastas, W. Perrie, S. Romani, R.J. Potter, G. Dearden, K.G. Watkins, P. French, M. Sharp, Chem. Phys. Lett. **484**, 283 (2010)
67. A. Nardes, M. Kemerink, M. Dekok, E. Vinken, K. Maturova, R. Janssen, Org. Electron. **9**, 727 (2008)
68. S. Jonsson, J. Birgerson, X. Crispin, G. Greczynski, W. Osikowicz, A.W.D. van der Gon, W.R. Salaneck, M. Fahlman, Synth. Met. **139**, 1 (2003)
69. H. Yan, H. Okuzaki, Synth. Met. **159**, 2225 (2009)
70. M. Fabretto, C. Hall, T. Vaithianathan, P.C. Innis, J. Mazurkiewicz, G.G. Wallace, P. Murphy, Thin Solid Films **516**, 7828 (2008)
71. A. Laskarakis, P.G. Karagiannidis, D. Georgiou, D.M. Nikolaidou, S. Logothetidis, Thin Solid Films (2013), https://doi.org/10.1016/j.tsf.2013.03.138
72. B. Friedel, P.E. Keivanidis, T.J.K. Brenner, A. Abrusci, C.R. McNeill, R.H. Friend, N.C. Greenham, Macromolecules **42**, 6741 (2009)
73. Y.H. Kim, C. Sachse, M.L. Machala, C. May, L. Müller-Meskamp, K. Leo, Adv. Funct. Mater. **21**, 1076 (2011)
74. S.-I. Na, S.-S. Kim, J. Jo, D.-Y. Kim, Adv. Mater. **20**, 4061 (2008)
75. M.V. Madsen, K.O. Sylvester-hvid, B. Dastmalchi, K. Hingerl, K. Norrman, T. Tromholt, M. Manceau, D. Angmo, F.C. Krebs, J. Phys. Chem. C **115**, 10817 (2011)
76. L.A.A. Pettersson, T. Johansson, F. Carlsson, H. Arwin, O. Inganäs, Synth. Met. **101**, 198 (1999)
77. K. Yim, R. Friend, J. Kim, J. Chem. Phys. **124**, 184706 (2006)
78. S. Logothetidis, A. Laskarakis, Eur. Phys. J. Appl. Phys. **46**, 12502 (2009)
79. D. Georgiou, A. Laskarakis, S. Logothetidis, S. Amberg-Schwab, U. Weber, M. Schmidt, K. Noller, Appl. Surf. Sci. **255**, 8023 (2009)
80. A. Laskarakis, S. Logothetidis, D. Georgiou, S. Amberg-Schwab, U. Weber, Thin Solid Films **517**, 6275 (2009)
81. Y. Leterrier, Prog. Mater. Sci. **48**, 1 (2003)
82. N. Koch, Chem. Phys. Chem. **8**, 1438 (2007)
83. P.G. Karagiannidis, N. Kalfagiannis, D. Georgiou, A. Laskarakis, N.A. Hastas, C. Pitsalidis, S. Logothetidis, J. Mater. Chem. **22**, 14624 (2012)
84. P.G. Karagiannidis, D. Georgiou, C. Pitsalidis, A. Laskarakis, S. Logothetidis, Mater. Chem. Phys. **129**, 1207 (2011)
85. M.M. Voigt, R.C.I. Mackenzie, C.P. Yau, P. Atienzar, J. Dane, P.E. Keivanidis, D.D.C. Bradley, J. Nelson, Sol. Energy Mater. Sol. Cells **95**, 731 (2011)
86. D.C. Harris, M.D. Bertolucci, *Symmetry and Spectroscopy* (Oxford University Press, New York, 1978)

# Chapter 20
# Application of In-Situ IR-Ellipsometry in Silicon Electrochemistry to Study Ultrathin Films

**Jörg Rappich, Karsten Hinrichs, Guoguang Sun and Xin Zhang**

**Abstract** This chapter provides an overview of in-situ application of infrared spectroscopic ellipsometry (IRSE) for the characterization of thin films on silicon (Si) prepared and modified by use of electrochemical surface treatments. In-situ IRSE investigations of Si surfaces during electrochemical grafting of ultra-thin layers via diazonium compounds (cathodic process), and thin polymeric layers of polypyrrole and polyaniline (both via anodic processes) are presented and discussed in detail. The film growth was monitored by an increase in layer specific vibrational signatures of molecules or monomers present on the surface or in the layer. Additionally, species postformed after electrochemical processing (i.e. after drying) and over-oxidation of the polymeric layer have been identified as well as the incorporation of dopants in the polymeric layer by their specific vibrational modes. The obtained results are discussed in frame of layer thickness, structure, and are compared to calculated IRSE spectra in case of very thin layers formed via cathodic reduction of diazonium compounds (nitrobenzene and maleimidobenzene) on Si.

J. Rappich (✉) · X. Zhang
Institut für Silizium Photovoltaik, Helmholtz-Zentrum Berlin für
Materialien und Energie GmbH, Kekuléstrasse 5, 12489 Berlin, Germany
e-mail: rappich@helmholtz-berlin.de

X. Zhang
e-mail: xin.zhang@helmholtz-berlin.de

K. Hinrichs · G. Sun
Leibniz-Institut für Analytische Wissenschaften – ISAS – e.V.,
Schwarzschildstr. 8, 12489 Berlin, Germany
e-mail: karsten.hinrichs@isas.de

G. Sun
e-mail: guoguang.sun@isas.de

© Springer International Publishing AG, part of Springer Nature 2018
K. Hinrichs and K.-J. Eichhorn (eds.), *Ellipsometry of Functional
Organic Surfaces and Films*, Springer Series in Surface Sciences 52,
https://doi.org/10.1007/978-3-319-75895-4_20

459

## 20.1   Introduction

In-situ investigations of Si surfaces during (electro)chemical modification and functionalization are of high interest to understand and develop reaction schemes and processing steps. Beside the characterization of the grafted material itself such studies provide information on intermediate surface species or side reactions [1] during processing as well as species formed after the treatment in solution [2] or during aging. The properties of organic/inorganic interfaces, as for example work function and band bending [3–5], conductivity or adsorption and binding characteristics [6–8], depend sensitively on their composition and structure. Ellipsometry in the visible spectral range has been used to investigate the adsorption of bacterial films at metal-electrolyte interfaces [3], to inspect the corrosion behaviour of metals in different media [4, 5, 9, 10] and to monitor changes in the Si surface oxide coverage [11, 12]. Additionally grafting of organic molecules and formation of ultrathin polymeric layers are widely used processes to modify Si surfaces with respect to nanopatterning [13, 14] and passivation [15, 16]. The electrochemical grafting/polymerization process, which is induced by changing the applied potential or the passed charge during deposition, is monitored in the infrared spectrum by the specific vibrational absorption signature of molecules or monomers present on the surface or in the layer [17–21]. The discussed sample geometry for in-situ IRSE measurements permits sensitive measurements of the surface species by a single reflection of the IR beam at incidence angles below total reflection regime. Mainly for two different electrochemical polymerization techniques (cathodic and anodic) results are discussed in frame of layer thickness and structure properties as determined from in-situ IRSE experiments compared to other techniques like X-ray photoelectron spectroscopy, electrochemical quartz crystal microbalance measurements, FT-IR and Raman spectroscopy.

## 20.2   Geometry for In-Situ Electrochemical and IR-Ellipsometric Measurements on Si Surfaces

Figure 20.1 shows the combined setup of the electrochemical and in-situ IR-ellipsometric experiment. The path of the IR beam is presented in Fig. 20.1a: the IR light of a Bruker FTIR (Bruker ifs 55) spectrometer is passing an analyzer, is reflected from the Si backside of the wedge, passes the polarizer, and is detected by a mercury cadmium telluride (MCT) detector. The electrochemical cell is sketched in more detail in Fig. 20.1b.

The electrode consists of an infrared transparent p-type silicon wedge with a polished section of 1.5° and the (111) surface was faced toward the solution (Fig. 20.1b, top). A wedge was used to suppress interferences which would arise from multiple reflections in a plane parallel substrate. The wedge had a size of $52 \times 20$ mm and was adjusted in the in-situ electrochemical cell (Fig. 20.1b, bottom). General details

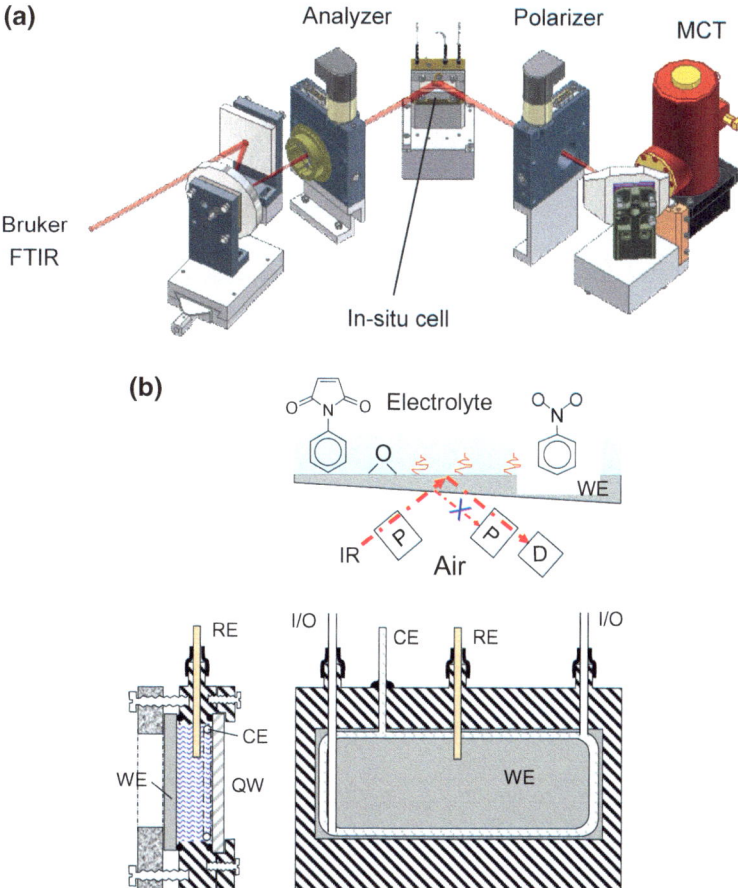

**Fig. 20.1** **a** Schematic drawing of the polarization dependent in-situ IRSE set-up; **b** optical paths of the beams reflected from the front and back sides and drawing of the in-situ cell with the Si wedge as working electrode, Pt counter and Au reference electrodes, respectively. The in-situ cell was placed in the ellipsometer with an incidence angle of 50° and afterwards adjusted by rotation on the reflex of the back-side (oxide/solution interface) leading to an incidence angle of about 59° (adapted from [12])

about IRSE set-ups can be found in Chap. 1 in this book. The reflected IR-radiation is described by tan $\Psi$ and $\Delta$, the absolute amplitude ratio and the phase difference between p- and s-polarized components of the reflected waves. For more details see also Chap. 1 in this book. For IR-SE measurements a Bruker ifs 55, Vertex 70 or Tensor 27 Fourier Transform spectrometer was used. The time to record a tan $\Psi$ spectrum was about 20 s, using a mercury-cadmium-telluride (MCT) detector [17].

The float-zone Si wedge (p-doped, 2–4 $\Omega$ cm) was fixed in the electrochemical cell by a steel frame on the backside where the IR beam is incident. The Si wedge was used as the working electrode (WE). A Pt ring and Au wire served as counter- (CE) and pseudo reference electrode (RE), respectively. The electrolyte was pumped through or was exchanged by a sliding pump and silicone hose via two in/out Teflon tubes (I/O) during the measurements. The cell has a volume of about 5 ml, a quartz window (QW) permits illumination with light if needed, and the electrode potential was controlled by a potentiostat (Bank PGS 88 or iviumstat). The Si wedge was H-terminated by use of standard Piranha-oxide/NH$_4$F (40%) treatment [22]. The used diazonium salts (i.e. 4-nitrobenzene diazonium tetrafluoroborate, 4-NBDT and 4-(N-Maleimido)benzene diazonium tetrafluoroborate, 4-MBDT) were cleaned by several re-crystallization procedures whereas pyrrole and aniline were freshly distilled prior to use.

## 20.3   Preconditioning of the Si Surface by H-Termination

To start all measurements with similar preconditioned Si surfaces, we use the standard H-termination procedure where the Si(111) sample was cleaned and oxidized in Piranha solution (H$_2$SO$_4$:H$_2$O$_2$ = 1:1) followed by a treatment in 40% NH$_4$F (pH 7.8) [22] to obtain flat and H-terminated (111) oriented terraces (see Fig. 20.2). The Si(111) surface treated in 5% HF shows a microscopical roughness of about 3 nm without any terraces visible.

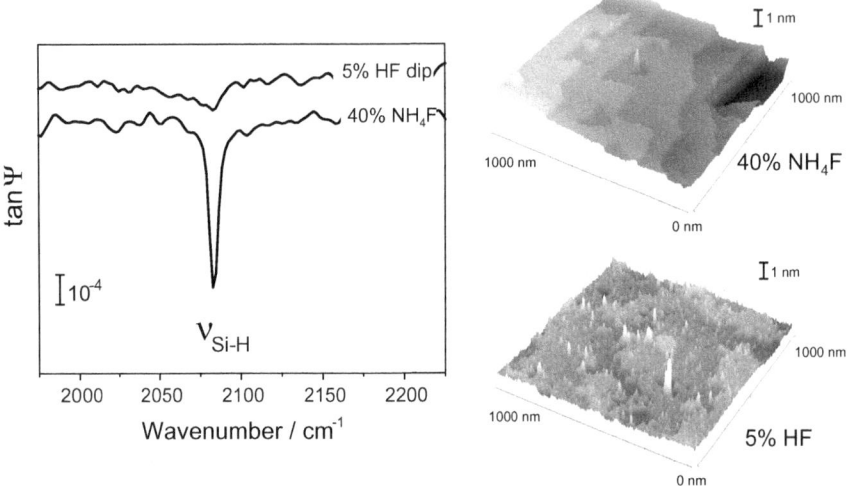

**Fig. 20.2** Left: ex-situ tan$\Psi$ spectra in the Si–H vibration regime of H-terminated Si(111) prepared by 5% HF or 40% NH$_4$F and right: respective AFM images of the 40% NH$_4$F and 5% HF treated Si(111) surfaces

## 20.4   Grafting from Diazonium Compound

Modification of surfaces by small molecules is of great interest for surface passiva-tion, functionalization for optical applications and biosensing, using surfaces as work benches and so on. The following paragraph addresses the in-situ characterization of surface modification by small molecules using diazonium compounds as respective species. Hereby the diazonium compound is cathodically reduced to form reactive radical intermediates which are able to bind covalently to surfaces [23, 24].

For silicon, the overall reaction of the grafting process is induced by the electro-chemically radical formation by the reduction of the diazonium group at the phenyl ring in solution leading to $N_2$ formation and phenyl radicals with the functional group X (radical-formation (a) in Scheme (20.1)).

$$(20.1)$$

These radicals activate the Si surface by abstracting H-atoms (b) and the inter-mediate Si dangling bonds at the surface react with the phenyl radicals by forming Si—C bonds (c) [25–27].

The insets in Fig. 20.3 show potential scans of p-Si(111) in 5 mM 4-NBDT solved in 0.01 M $H_2SO_4$ (a) and in 2 mM 4-MBDT solved in ACN + 0.1 M $Bu_4NBF_4$ as conducting salt (b). A broad peak can be seen during the first scan (solid lines) whereas the second scan (dashed lines) shows no pronounced structure due to passivation of the Si(111) surface by grafted nitrobenzene from 4-NBDT at about −1.1 V and maleimido-benzene from 4-MBDT around −0.5 V, respectively. The more negative potential of the 4-NBDT in comparison to the 4-MBDT molecules is mainly due to the –M/–I effects of the nitro group that stabilizes the intermediate radical. The increase in current at more cathodic potentials is due to the overlaying reaction with the solvent at such potentials. A more controllable technique is the injection of the diazonium salt into the solution during constant cathodic polarization. In both cases the potential was set to about −1 to −1.1 V as indicated by the arrow in the insets of Fig. 20.3. For 4-NBDT two different concentrations of 1 and 10 mM in 0.01 M $H_2SO_4$ have been used (Fig. 20.3a) whereas 2 mM was used for 4-MBDT in ACN

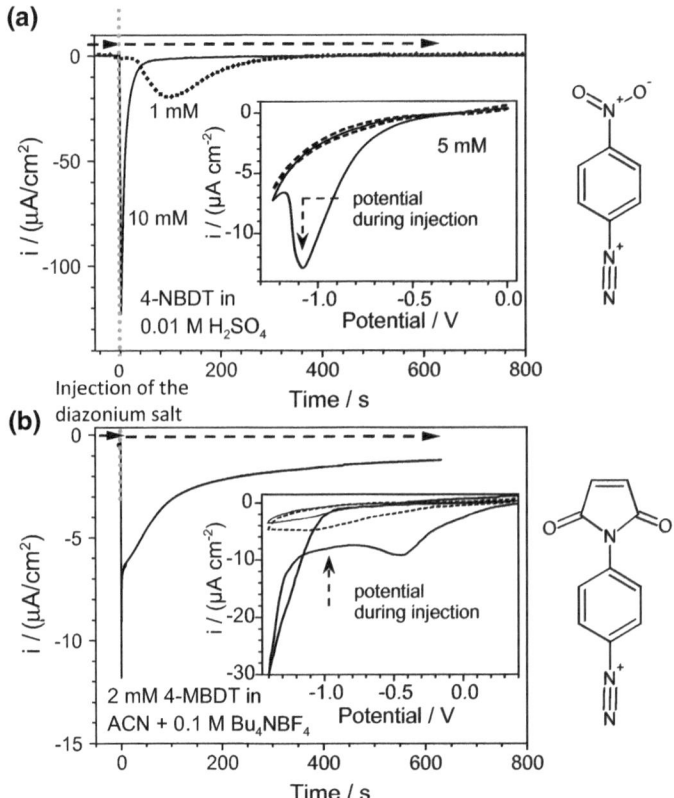

**Fig. 20.3** Chronoamperometry of p-Si(111) after injection of **a** 1 and 10 mM solution of 4-NBDT in 0.01 M $H_2SO_4$, polarization at $-1.1$ V, flown electrical charge: about 640 (1 mM) and 405 (10 mM) $\mu C/cm^2$; inset: potential scans of p-Si(111) in 5 mM 4-NBDT / 0.01 M $H_2SO_4$(⎯⎯ 1st , ⎯⎯ 2nd) and **b** 2 mM solution of 4-MBDT in acetonitrile with 0.1 M $Bu_4NBF_4$ as conducting salt, polarization at $-1$ V; inset: three potential scans in the same solution (⎯⎯ 1st, ⎯⎯ 2nd, ⎯⎯ 3rd). The molecules are sketched on the right

(Fig. 20.3b). The amount of flown electrical charge was about 640 and 405 $\mu C/cm^2$ for 4-NBDT and 1400 $\mu C/cm^2$ for 4-MBDT, respectively.

To get information about the layer thickness we did electrochemical quartz crystal microbalance (EQCM) and X-ray photoelectron spectroscopy (XPS) analysis, from which we get a layer thickness of about 2–3 nm (about 3–5 monolayers). What is in good agreement with the slightly higher amount of 4–5 nm when calculating the layer thickness from the flown electrical charge at constant potential by taking a two electron transfer process into account. This higher amount is a result of side reactions like dimerization, or reaction with the solvent so that electrochemically formed radicals are lost without any deposition [1, 26, 28]. There are also some hints that Azo-groups exist in such layers as sketched in the middle part of scheme (20.2) [24]. Therefore, the following surface structure can be assumed where X denotes the nitro or maleimido group.

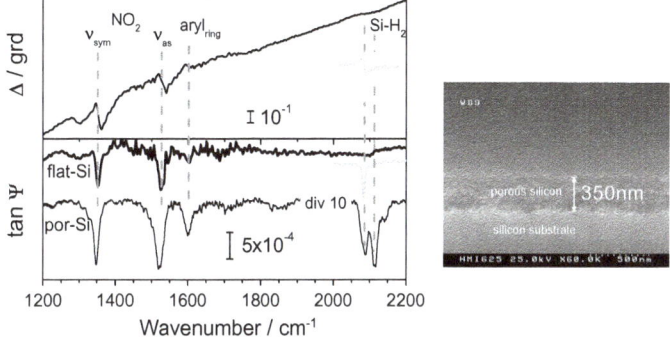

(20.2)

**IRSE Investigation of the Nitrobenzene Modified Si Surface**

**Fig. 20.4**  Left top: $\Delta$ spectra of the NB and H-terminated flat Si surface; left bottom: $\tan\Psi$ spectra of H-terminated flat (grey), electrochemically modified flat (thick line) and porous Si surfaces (thin line) using 4-NBDT. The vibrational species are indicated (Note: the spectrum of the porous Si layer modified by NB is divided by 10). Right: SEM image (30° tilted) of the 350 nm thick porous layer before grafting of NB;  adapted from [29]

IRSE was used to identify the surface species by their specific vibrational signature. Figure 20.4 shows ex-situ IRSE spectra of electrochemically nitrobenzene (NB) modified flat and porous Si surfaces. The porous structure of Si was used to enhance the amount of grafted NB molecules to clarify the absorption peak positions of the NB molecules on Si surfaces, which are the symmetric and asymmetric $NO_2$ stretching vibrations, and the aryl ring vibration at about 1348, 1522, and 1600 cm$^{-1}$, respectively [29].

The same absorption peaks are visible for NB grafted on flat surfaces (thick line in Fig. 20.4b), however with much smaller amplitudes because the amount of grafted

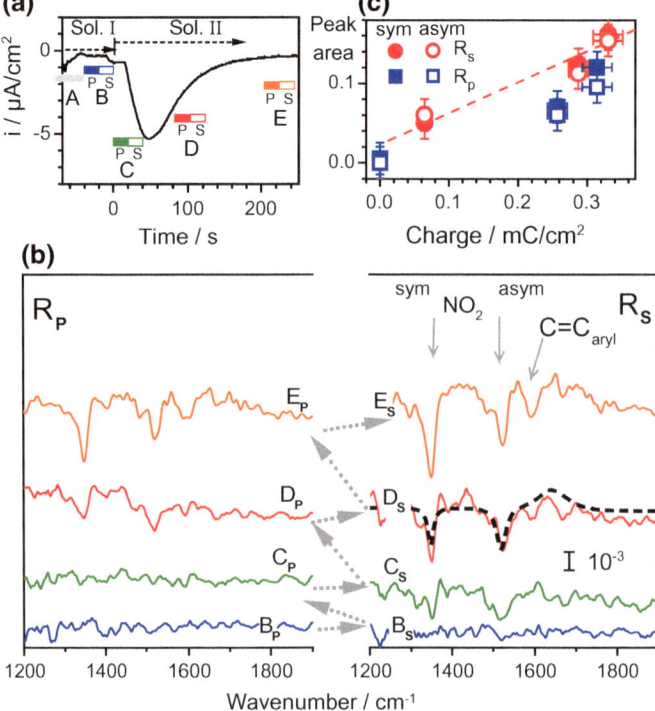

**Fig. 20.5** **a** Current time dependence during injection of the 4-NBDT (2mM) compound at $-1.1$ V, positions and times needed for the IRSE measurement (about 20 s each) are marked. **b** $r_P$ and $r_S$ spectra as measured during grafting of NB from 0.01 M $H_2SO_4$. The grey arrows denote the chronology of the measurements; the dashed line represents the calculated $\tan\Psi$ spectrum (2 nm thick layer of NB on Si). **c** IR peak intensities for the symmetric and asymmetric $NO_2$ stretching vibrations as a function of the flown charge (adapted from [2])

NB molecules is much smaller for the flat surface. Please note that the spectrum of the porous Si layer modified by NB is divided by 10. The absorption peaks at 2085 and 2113 cm$^{-1}$ are due to Si–H and Si–H$_2$ surface species of non-grafted porous Si which is H-terminated after preparation in HF solution [30–32].

The vibrational spectra in Fig. 20.4 prove the passivation and successful preparation of nitrobenzene films. The ellipsometric spectra of the NB-modified and H-terminated flat Si surface exhibit Kramers–Kronig consistent features of characteristic absorption bands: derivative-like structures in the $\Delta$ spectra and absorption like bands in the $\tan\Psi$ spectra.

The image on the right side of Fig. 20.4 shows the SEM image (30° tilted) of the 350 nm thick porous layer before grafting of NB.

For a better understanding of the processes during electrochemical grafting of NB, in-situ IRSE measurements as outlined in the experimental section were performed to investigate the surface species formation during the grafting process [2].

Figure 20.5a shows the well-known current time behavior (see Fig. 20.3) during injection of the diazonium compound at $-1.1$ V with the time positions and the times needed for the measurement of the p- and s-polarized reflection spectra (about 20 s each), which are plotted in Fig. 20.5b. The grey arrows denote the chronology of polarized reflection measurements. The IR absorption due to $NO_2$ surface groups sets on when the current peak maximum is reached and saturates at longer times of grafting. However, the peak intensities for the symmetric and asymmetric $NO_2$ stretching vibrations increase with increasing charge flow, pointing to a constant growth rate with the flown charge (see Fig. 20.5c). The dashed line in Fig. 20.5b reflects the calculated Rs spectrum using a 2 nm thick layer of NB on Si. This calculated spectrum fits very well with the measured one. From that fit, we can conclude that the layer has a thickness of about 2.5–3 nm at the end of the grafting process, what is in good agreement with the data obtained from EQCM and XPS measurements [33] as outlined before.

## 20.5   Oxidation as Consequence of Side Reactions

In order to understand the stability of prepared films a Si sample was dried under argon for further investigations and/or processing after the NB layer was formed on Si. The tan$\Psi$ spectra measured in solution and after drying reveal an IR absorption due to $SiO_2$ surface species at about $1180 \, \mathrm{cm}^{-1}$. The only option to explain this observation is the presence of neighbored Si-OH species which are able to condense and form at least $SiO_2$ [34]. The formation of these Si-OH surface species must occur via a side reaction of the intermediate Si dangling bonds (Si• in (20.1)) where Si• reacts with the solvent ($H_2O$) as suggested recently [1]. Additionally, Fig. 20.6 shows that no Si–H could be detected in solution since no peak up at about $2080 \, \mathrm{cm}^{-1}$ is observed as for the ex-situ measurement. This behavior is due to a broadening of

**Fig. 20.6** Tan$\Psi$ spectra of a Si/NB sample measured in-situ and ex-situ. The spectra were normalized to the spectra recorded for the H-terminated Si in the same environment (adapted from [2])

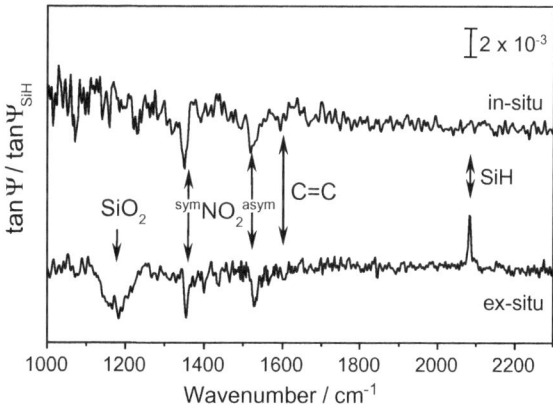

**Fig. 20.7** In-situ measured (dotted line) and calculated tanΨ spectra for different surface orientations (out-of-plane, in-plane, and isotropic as sketched in the figure) of a 2.5 nm thick NB layer ($n_\infty = 1.41$) in the regime of the $NO_2$ stretching vibrations (adapted from [2])

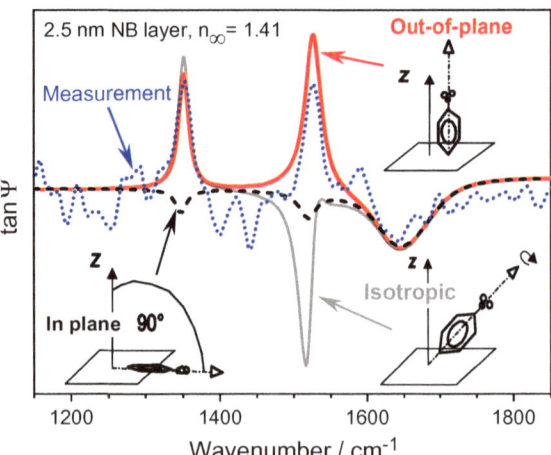

## 20.6 Molecular Orientation

Figure 20.7 shows the in-situ measured tanΨ spectrum (dotted line) and spectra calculated for different surface orientation of the NB molecules (out-of-plane, in-plane, and isotropic as sketched in the figure) using a 2.5 nm thick NB layer with a refractive index of $n_\infty = 1.41$. These calculations suggest that the NB molecules mostly have an out-of-plane orientation on the p-Si(111) surface, rather than in-plane or isotropic.

## 20.7 Thickness Determination of the NB Film

A thickness of 2.5 nm of the nitrobenzene film could be determined from the simulation of the in-situ spectra by the use of the known optical constants of nitrobenzene by use of an isotropic 3-phase model (solution/nitrobenzene film/silicon) [2]. Additionally the water band at about $3400 \, cm^{-1}$, which occurs due to the change of the effective dielectric function at the interface due to the growth of the nitrobenzene film, can be used for estimation of the film thickness, d.

Calculations of tanΨ spectra for some characteristic values of $n_\infty$ and d of the grown NB-film are shown in the Fig. 20.8 (dashed/dotted lines). However, the calculated line shapes do not match perfectly the measured one. The deviation can be

**Fig. 20.8** In-situ measured
tanΨ spectra of the NB
modified Si surface
referenced to the Si–H
covered surface in the same
environment (water or He
atmosphere). The
dashed/dotted lines represent
calculated tanΨ spectra for
different $n_\infty$ and d of the
nitrobenzene layer (adapted
from [2])

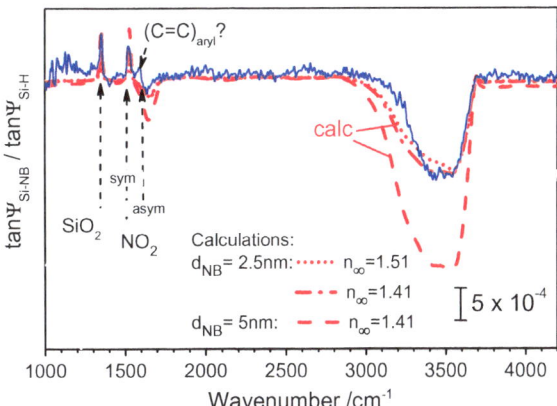

explained by contributions from overlapping Si–OH groups or those water molecules
aligned at the charged surface.

## 20.8  IRSE Investigation of the Maleimidobenzene Modified Si Surface

Maleimido functionalization of surfaces is of great interest due to the possibility
of further reaction with thiol containing biomolecules by Michael-addition reaction for biosensing purposes. This section addresses the functionalization of Si by
maleimido groups via electrochemical reduction of the respective diazonium compound, 4-MBDT. The corresponding CV and CA are shown in Fig. 20.3b. The tanΨ
spectra measured in solution and after drying of the grafted MB layer are presented in
Fig. 20.9 and reveal the maleimide related vibrational modes of the C=O, aryl ring,
and =C–H at about 1726, 1520–1540, and $\approx 1400\,cm^{-1}$, respectively [38]. Surprisingly and different to grafting of NB (see Fig. 20.6), both spectra in-situ and ex-situ
show the presence of $SiO_2$ associated vibrational absorption around $1200\,cm^{-1}$ with
a higher intensity for the ex-situ measurement when comparing with the appropriate C–H vibrational intensities. This behavior is all the more remarkable because
the acetonitrile solution has nominally very small amounts of water and there is no
direct $SiO_2$ formation in aqueous solution as used for NB grafting (see Fig. 20.6).

Therefore it can be concluded that acetonitrile supports oxide formation via reaction of Si dangling bonds with $H_2O$ even at low $H_2O$ concentrations since the wetting
of the Si–H surface is much better by the methyl groups of the acetonitrile than for
the water molecules since Si–H surfaces are highly hydrophobic and water molecules
are repelled.

**Fig. 20.9** In-situ and ex-situ
measured tanΨ spectra of the
MB modified Si surface
referenced to the Si–H
covered surface in the same
environment (acetonitrile or
He atmosphere), adapted
from [39]

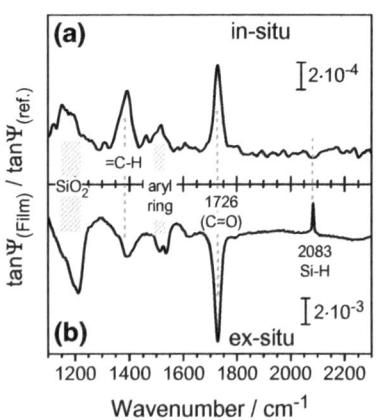

## 20.9 Thickness Determination of the MB Film

Figure 20.10a shows the optical constants of a dry MB film as derived from the ex-situ IRSE measurements. These n and k values have been applied to calculate the in-situ recorded spectra of MB film grafted on the Si(111) surface. The IRSE spectra have been calculated for different thicknesses and were normalized to the initial spectrum just before the deposition, $tanΨ_{(ref)}$. The calculated spectra and the in-situ measured spectrum (all are referenced to the spectrum without a layer) are plotted in Fig. 20.10b for comparison. From the C=O band amplitudes a thickness of $3.5 \pm 0.5$ nm, which is in qualitative agreement with the thickness as determined from ex-situ experiments, can be extracted [39]. However, there is a little uncertainty since the optical constants of the MB layer may not be the same in the liquid and dry environment and the assumed homogeneous thickness in the probed spot area of a few 10 mm² may not be fulfilled. However, this thickness agrees well with that obtained for the NB grafted layer on Si(111) of about 2.5–3 nm and the MB layer thickness as obtained via IR-AFM of transferred MB functionalized graphene layer [40].

## 20.10 Summary of the IRSE Investigations of Si Surfaces Modified by Diazonium Cations

This previous paragraph has presented IRSE results obtained for nitrobenzene and maleimido-benzene grafted on Si(111) surfaces. Characteristic IR absorptions due to $NO_2$ and C=O stretching modes are visible under in-situ conditions. Thereby different steps of the grafting process of nitrobenzene (NB) from 4-nitrobenzene diazonium tetrafluoroborate (4-NBDT) were monitored by in-situ IRSE during the electrochemically induced reaction at the surface but no 4-NBDT or intermediate NB radicals in solution contribute to the IR absorption in the $NO_2$ stretching mode region.

**Fig. 20.10**  **a** Optical
constants n and k of the MB
film as determined from
optical-layer simulations
applied to ex-situ measured
tanΨ spectrum of the MB
layer on Si(111); **b**
Comparison of calculated
referenced spectra for 3.0,
3.5 and 4.0 nm MB layer on
Si(111) with the in-situ
measured and normalized
tanΨ spectrum in the regime
of the C=O vibrational mode
at about $1726\,cm^{-1}$; adapted
from [39]

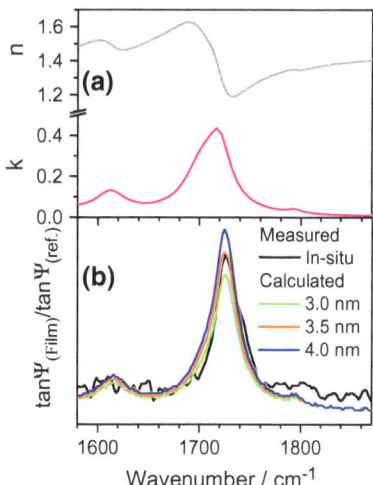

Si–H groups could not be detected in solution with a single reflection measurement
due to their weakness and strong interaction with water molecules of the electrolyte
that lead to a line broadening in the tanΨ spectra independent of the diazonium cation
used. Additionally, in aqueous solution $SiO_2$ formation during NB grafting seems
to be mainly a result of post-oxidation reaction of Si–OH surface species formed
during electrochemical reduction of 4-NBDT in 0.01 M $H_2SO_4$. On the other hand,
acetonitrile used as solvent for grafting of maleimido-benzene (MB) supports oxide
formation via reaction of intermediate Si dangling bonds with small amounts of water
present in solution due the much stronger wetting of the hydrophobic Si–H surface.

## 20.11   Deposition of Ultra-Thin Polymeric Layers from Pyrrole

This paragraph gives an overview of in-situ IRSE measurements and results on the
deposition of polypyrrole (PPy). Electrochemical oxidation of pyrrole is an example
for the anodic polymerization of monomers. The sketch in Fig. 20.11 presents the
main steps during polymerization, the formation of radical cations and the dimerization of them followed by further polymer formation [41–43].

Figure 20.12 shows the typical behavior of the current and the mass change as
obtained by electrochemical quartz crystal microbalance (EQCM) as a function of
the applied potential during PPy deposition on p-Si(111) in 0.1 M pyrrole + 0.1 M
$HNO_3$. The increase in anodic and cathodic current in a broad potential regime is due
to capacitive charging of the increasing PPy layer thickness induced by the potential
change during scanning [44]. Deposition of the PPy layer occurs only at potentials
above 0.6 V as can be seen from the EQCM measurement (Fig. 20.12b).

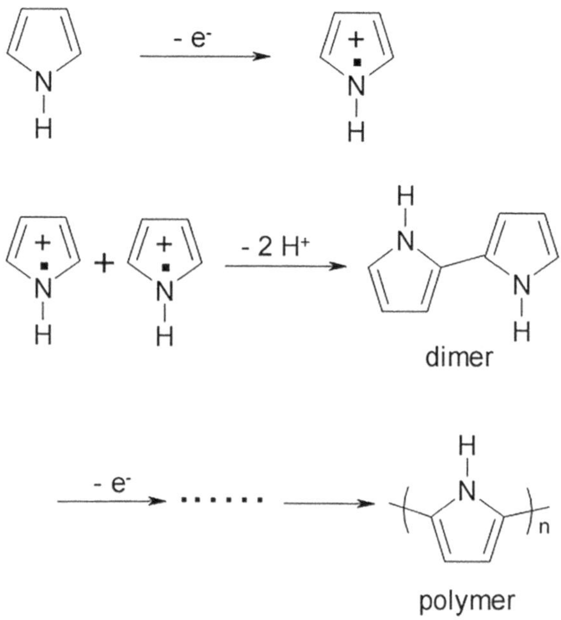

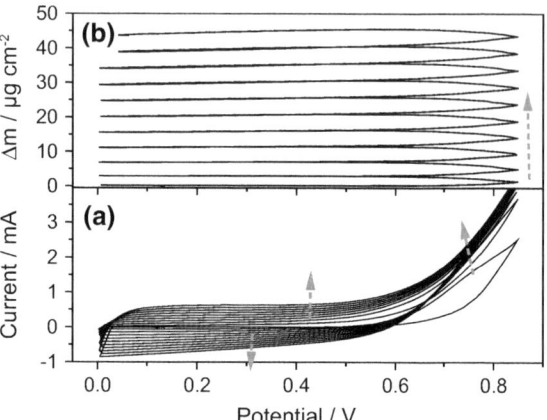

**Fig. 20.11** The main steps during polymerization of pyrrole: the formation of radical cations, dimerization and further polymer formation according to [41–43]

**Fig. 20.12** Current (**a**) and mass change (**b**) as a function of the applied potential during PPy deposition on p-Si(111) in 0.1 M pyrrole + 0.1 M HNO₃, scan rate 50 mV/s

To reduce this strong charging and discharging of the layer pulsed potential techniques can be used as presented in Fig. 20.13 what enhances the film quality and at least the Si/PPy interface properties [45] compared to scanning technique [46]. The potential of 0.7 V was chosen from the EQCM measurements since deposition occurs at this potential on a low level and over-oxidation of the PPy layer can be neglected.

Figure 20.14b reflects the potential pulse sequence as used for the in-situ tanΨ measurements during PPy deposition where 100 s of deposition was followed by

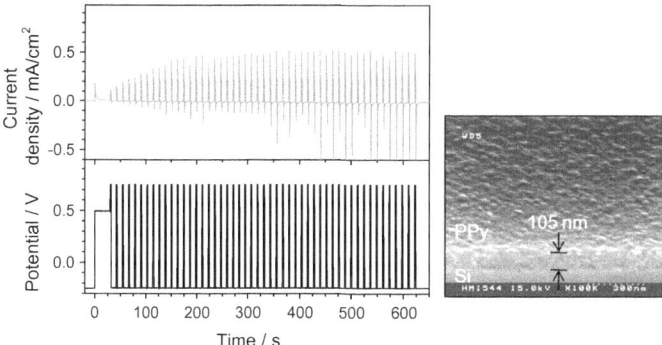

**Fig. 20.13** Left: potential pulse (bottom) and current (top) sequences during polypyrrole formation in 0.1 M Pyrrole + 0.1 M HNO$_3$ (pH 1.5); right: SEM image (tilt angle 30°) of a PPy layer on Si as prepared by such a pulse method. (modified according to [46])

about 25 s needed for the tan$\Psi$ measurement. These spectra are plotted in Fig. 20.14c normalized to the tan$\Psi$ spectrum of the uncovered Si(111)-H surface in solution in the beginning of the deposition sequence. IR absorption due to wagging modes of N–H ($\delta_{NH}$) and C–H ($\delta_{CH}$) develop with increasing amount of potential pulses which follow from the increase in the PPy layer thickness (see Fig. 20.14). Obviously, no or only very small amount of SiO$_2$ is formed during the anodic oxidation of pyrrole in 0.1 M HNO$_3$ since the IR absorption due to SiO$_2$ is broader than that of $\delta_{CH}$ as can be seen in Fig. 20.14 for an anodically oxidized Si surface (SiO$_2$).

Figure 20.15 compares IRSE measurements under in-situ and ex-situ conditions normalized to the Si–H covered surface in the same environment. The ring related vibrations are slightly enhanced in intensity after drying and measured ex-situ, whereas the wagging modes of C–H and N–H remain nearly constant.

Preparing the PPy layer on Si at higher anodic potentials (+1.2V) leads to totally different IRSE spectra as compared to +1.1 V as presented exemplarily in Fig. 20.16. A broad peak around $\sim$1719–1734 cm$^{-1}$ is visible in the ex-situ spectrum at +1.2 V whereas this IR absorption is very weak at 1.1 V and is identified around $\sim$1706 cm$^{-1}$. This absorption band can be related to the C=O stretching vibration as a result of the over-oxidation of the PPy at +1.2 V [47, 48]. This behavior was also confirmed by in-situ Raman measurements [48].

This paragraph summarizes IRSE results obtained for polypyrrole formation on Si surfaces. The thin PPy layer prepared by anodic oxidation is uniform and compact. The growth of the PPy layer was monitored by means of in-situ IRSE measurements observing the changes in the IR absorption of the C–H and N–H wagging modes. The pulsed deposition technique leads to PPy layers on Si without any SiO$_2$ interface species even in aqueous solution pointing to a fast charge transfer to the pyrrole molecules. IRSE measurements reveal a small potential range from which the over-oxidation of the PPy layer occurs (between +1.1 and 1.2 V using 0.1M

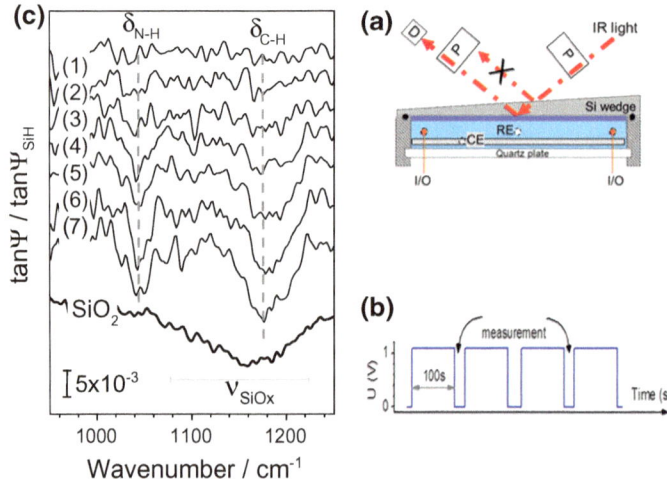

**Fig. 20.14** **a** In-situ pump through cell (see also Fig. 20.1) for combined IRSE and electrochemical measurements, **b** potential pulse sequence (100s of deposition in 0.1 M Pyrrole + 0.1 M HNO₃ followed by 25 s for the tanΨ measurement), **c** in-situ tanΨ spectra recorded during PPy deposition (2 to 7, measured after 5 potential pulses each step) normalized to the tanΨ spectrum of the uncovered Si(111)-H surface (1) in solution (respective vibrational modes are marked). A spectrum of a thin SiO₂ layer on the Si wedge is shown for comparison (adapted from [45])

**Fig. 20.15** IRSE spectra of a thin PPy layer on Si under in-situ and ex-situ conditions normalized to the Si–H covered surface in the same environment (respective vibrational modes are marked), 0.1M pyrrole in 0.01 M H₂SO₄ (adapted from [19])

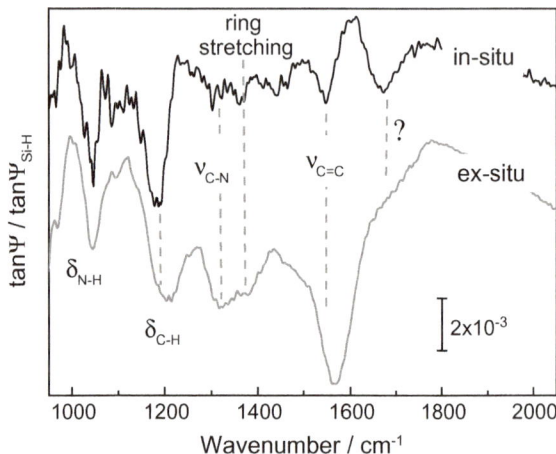

pyrrole in 0.01 M H₂SO₄) as reflected by the C=O related IR absorption band around 1730 cm⁻¹.

**Table 20.1** Assignments of IR bands of PPy prepared at different oxidation potentials (Data from [19])

| Wavenumber (cm$^{-1}$) | | Assignments |
|---|---|---|
| 1.2 V | 1.1 V | |
| 1734, 1719 | 1706 | C=O stretching |
| | 1667 | H$_2$O |
| 1600,1580 | 1577 | C=C stretching |
| 1525 | | Ring stretching |
| 1465, 1458 | | C–H in N–CH$_3$ |
| 1380 | | Ring stretching |
| 1290, 1282 | 1266 | C-N? |
| | 1229 | Si-O? |
| 1128 | 1113 | Ring breathing or in plane N–H def. |
| 1074 | 1065 | C–H in plane def. |
| 1045 | 1043, 1051 | C–H out-of-plane def. or N–H |
| 1014 | 1014 | C–H in plane def. Ring vibration |

**Fig. 20.16** Ex-situ tanΨ spectra of PPy layers on Si prepared by different applied potentials (1.1 and 1.2 V, 0.1M pyrrole in 0.01 M H$_2$SO$_4$). The spectra have been normalized to the tanΨ spectrum of the H-terminated Si(110) surface, characteristic bands are marked. The assignments are summarized in Table 20.1 (adapted from [19])

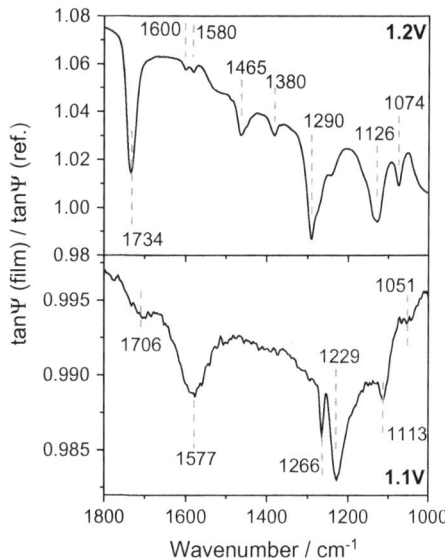

## 20.12   Doping of PANI Films with PSS

Poly (4-styrenesulfonate) (PSS) doped polyaniline (PANI) films were prepared by electrochemical deposition and characterized by in-situ IR spectroscopy. Figure 20.17 shows the molecular structure of both compounds. The aniline concentration was 0.3 M in diluted aqueous $H_2SO_4$ [0.1 M]. 0.1 g poly (sodium 4-styrenesulfonate) [-$CH_2CH$ ($C_6H_4SO_3Na$)-, Sigma-Aldrich, Mw~70.000] was added into 150 ml aniline solution prior the deposition.

The electrochemical growth was performed with following procedure: The oxidation of aniline monomer started at +2.3 V for 100 s and ended at −0.75 V for the other 100 s (1 loop). The whole process lasted for 50 loops. In all steps of the preparation process the solution was pumped in a cycle from a reservoir through the flow-cell. The polarized in-situ IR measurements in an ellipsometric set-up were made after 10, 20, 30, 40 and 50 loops, respectively.

Figure 20.18 shows the in-situ spectra of the as-deposited doped PANI film in the spectral range of PSS dopant. Beside the typical bands of PANI (not shown) two new bands appear at around 1032 and 1007 cm$^{-1}$, which can be attributed to S=O stretching of the sulfonate group from the dopant PSS [49], and the in-plane bending of C–H in the benzene ring [50], respectively. Obviously the band amplitudes of PSS related bands increase with the amount of deposited polymer material. However, unclear at present stage is the influence of the dopant on the structural properties of the PANI films which is indicated by deposition dependent shifts of PANI related absorption bands (not shown).

The possibility of characterization of the film properties by IR ellipsometry as well as identification of the dopant itself could be used for optimization of preparation schemes of doped polymer films.

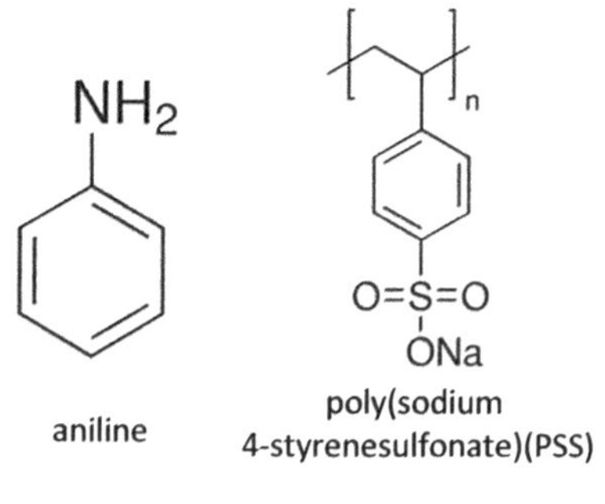

**Fig. 20.17** Molecular structure of used aniline monomer and PSS dopant

aniline

poly(sodium 4-styrenesulfonate)(PSS)

**Fig. 20.18** In-situ tanΨ
spectra of growth of PSS
doped PANI film; adapted
from [19]

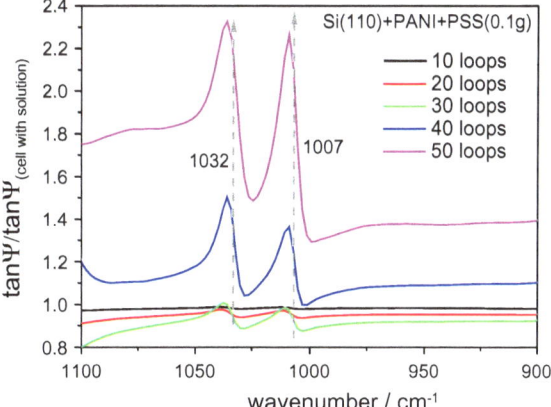

## 20.13   Summary

This chapter reviews the results obtained by in-situ infrared spectroscopic ellipsometry during electrochemical surface modification of gold and silicon. It shows that electrochemical treatment using cathodic reduction (diazonium compound, e.g. 4-Nitrobenzene diazonium tetrafluoroborate) or anodic oxidation (pyrrole) lead to the formation of thin polymeric surface layers. The increase in layer thickness is reflected by an increase in the absorption of compound specific vibrational modes. In case of cathodic reduction of the diazonium compound, a formation of $SiO_2$ layer was observed after drying obviously due to post-condensation of SiOH intermediate surface species present after the grafting process. The anodic oxidation of pyrrole shows the over-oxidation of the polymeric layer by the occurrence of C=O related IR absorption band at about $+1.2\,V$.

Overall, IRSE is a sensitive and contactless tool to investigate processing steps and intermediate species under in-situ conditions with respect to their specific vibrational signature.

**Acknowledgements** The financial support of the European Union through the EFRE program (ProFIT grant, contract no. 10131870/1 and no. 10144388), the Senatsverwaltung für Wissenschaft, Forschung und Kultur des Landes Berlin and the Bundesministerium für Bildung, Wissenschaft, Forschung und Technologie are gratefully acknowledged. The author thanks Dr. M. Gensch, Dr. K. Roodenko, Dr. V. Syritski, and Dr. M.C. Intelmann for valuable discussions.

## References

1. J. Rappich, A. Merson, K. Roodenko, T. Dittrich, M. Gensch, K. Hinrichs, Y. Shapira, J. Phys. Chem. B **110**, 1332 (2006)
2. J. Rappich, K. Hinrichs, Electrochem. Commun. **11**, 2316 (2009)

3. J.P. Busalmen, S.R.d Sánchez, D.J. Schiffrin, Appl. Environ. Microbiol. **64**, 3690 (1998)
4. J.J. Ritter, M.J. Rodriguez, Corrosion **38**, 223 (1982)
5. J.A. Petit, F. Dabosi, Corros. Sci. **20**, 745 (1980)
6. J.C. Harper, R. Polsky, D.R. Wheeler, S.M. Brozik, Langmuir **24**, 2206 (2008)
7. J.C. Harper, R. Polsky, D.R. Wheeler, D.M. Lopez, D.C. Arango, S.M. Brozik, Langmuir **25**, 3282 (2009)
8. X. Zhang, A. Tretjakov, M. Hovestadt, G.G. Sun, V. Syritski, J. Reut, R. Volkmer, K. Hinrichs, J. Rappich, Acta Biomater. **9**, 5838 (2013)
9. P.C.S. Hayfield, Surf. Sci. **56**, 488 (1976)
10. M. Poksinski, H. Dzuho, H. Arwin, J. Electrochem. Soc. **150**, B536 (2003)
11. S. Böhm, L.M. Peter, G. Schlichthörl, R. Greef, J. Electroanal. Chem. **500**, 178 (2001)
12. K. Hinrichs, K. Roodenko, J. Rappich, Electrochem. Commun. **10**, 315 (2008)
13. P.T. Hurley, A.E. Ribbe, J.M. Buriak, J. Am. Chem. Soc. **125**, 11334 (2003)
14. P. Wagner, S. Nock, J.A. Spudich, W.D. Volkmuth, S. Chu, R.L. Cicero, C.P. Wade, M.R. Linford, C.E.D. Chidsey, J. Struct. Biol. **119**, 189 (1997)
15. J. Rappich, P. Hartig, N.H. Nickel, I. Sieber, S. Schulze, T. Dittrich, Microelectron. Eng. **80**, 62 (2005)
16. J. Rappich, X. Zhang, S. Chapel, G. Sun, K. Hinrichs, Phys. Stat. Sol. C **7**, 210 (2010)
17. M. Gensch, K. Roodenko, K. Hinrichs, R. Hunger, A.G. Güell, A. Merson, U. Schade, Y. Shapira, Th Dittrich, J. Rappich, N. Esser, J. Vac. Sci. Technol. B **23**, 1838 (2005)
18. G. Sun, M. Hovestaedt, X. Zhang, K. Hinrichs, D.M. Rosu, I. Lauermann, C. Zielke, A. Vollmer, H. Löchel, B. Ay, H.-G. Holzhütter, U. Schade, N. Esser, R. Volkmer, J. Rappich, Surf. Interface Anal. **43**, 1203 (2010)
19. G. Sun, X. Zhang, C. Kaspari, K. Haberland, J. Rappich, K. Hinrichs, J. Electrochem. Soc. **159**, H811 (2012)
20. K. Roodenko, J. Rappich, M. Gensch, N. Esser, K. Hinrichs, Appl. Phys. A **90**, 175 (2008)
21. K. Hinrichs, M. Gensch, N. Esser, Appl. spectrosc. **59**, 272A (2005)
22. Y.J. Chabal, P. Dumas, P. Guyot-Sionnest, G.S. Higashi, Surf. Sci. **242**, 524 (1991)
23. J. Pinson, F. Podvorica, Chem. Soc. Rev. **34**, 429 (2005)
24. M.M. Chehimi (ed.), *Aryl Diazonium Salts: New Coupling Agents in Polymer and Surface Science* (Wiley, Weinheim, 2012)
25. P. Allongue, M. Delamar, B. Desbat, O. Fagebaume, R. Hitmi, J. Pinson, J.-M. Saveant, J. Am. Chem. Soc. **119**, 201 (1997)
26. P. Allongue, CHd Villeneuve, J. Pinson, F. Ozanam, J.N. Chazalviel, X. Wallart, Electrochim. Acta **43**, 2791 (1998)
27. P. Hartig, J. Rappich, T. Dittrich, Appl. Phys. Lett. **80**, 67 (2002)
28. P. Allongue, CHd Villeneuve, G. Cherouvrier, R. Corte's, M.-C. Bernard, J. Electroanal. Chem. **550/551**, 161 (2003)
29. K. Roodenko, J. Rappich, F. Yang, X. Zhang, N. Esser, K. Hinrichs, Langmuir **25**, 1445 (2009)
30. M. Estes, G. Moddel, Appl. Phys. Lett. **68**, 1814 (1996)
31. J. Rappich, Phys. Stat. Sol. C **1**, 1169 (2004)
32. W. Theiß, Surf. Sci. Rep. **29**, 91 (1997)
33. R. Hunger, W. Jaegermann, A. Merson, Y. Shapira, C. Pettenkofer, J. Rappich, J. Phys. Chem. B **110**, 15432 (2006)
34. R. Tomita, S. Urano, S. Kohiki, Chem. Lett. **30**, 684 (2001)
35. J. Rappich, H.J. Lewerenz, J. Electrochem. Soc. **142**, 1233 (1995)
36. J. Rappich, H.J. Lewerenz, Electrochim. Acta **41**, 675 (1996)
37. A. Belaïdi, J.-N. Chazalviel, F. Ozanam, O. Gorochov, A. Chari, B. Fotouhi, M. Etman, J. Electroanal. Chem. **444**, 55 (1998)
38. S.F. Parker, S.M. Mason, K.P.J. Williams, Spectrochim. Acta Part A Mol. Spectrosc. **46**, 315 (1990)
39. P. Kanyong, G. Sun, F. Rösicke, V. Syritski, U. Panne, K. Hinrichs, J. Rappich, Electrochem. Commun. **51**, 103 (2015)

40. F. Rösicke, M.A. Gluba, T. Shaykhutdinov, G. Sun, C. Kratz, J. Rappich, K. Hinrichs, N.H. Nickel, Chem. Commun. **53**, 9308 (2017)
41. S. Asavapiriyanont, G.K. Chandler, G.A. Gunawardena, D. Pletcher, J. Electroanal. Chem. **177**, 229 (1984)
42. G.P. Gardini, Adv. Heterocycl. Chem. **15**, 67 (1973)
43. P. Rapta, A. Neudeck, A. Petr, L. Dunsch, J. Chem. Soc., Faraday Trans. **94**, 3625 (1998)
44. U. Rammelt, S. Bischo, M. El-Dessouki, R. Schulze, W. Plieth, L. Dunsch, J. Solid State Electrochem. **3**, 406 (1999)
45. C.M. Intelmann, K. Hinrichs, V. Syritski, F. Yang, J. Rappich, Jpn. J. Appl. Phys. Part I **47**, 554 (2008)
46. C.M. Intelmann, V. Syritski, D. Tsankov, K. Hinrichs, J. Rappich, Electrochim. Acta **53**, 4046 (2008)
47. J.P. Wang, Y. Xua, J. Wang, X. Dua, F. Xiao, J. Li, Synth. Metals **160**, 1826 (2010)
48. Y.-C. Liu, B.-J. Hwang, W.-J. Jian, R. Santhanam, Thin Solid Films **374**, 85 (2000)
49. P.P. Sengupta, P. Kar, B. Adhikari, Thin Solid Films **517**, 3774 (2009)
50. S. Logothetidis, A. Laskarakis, Thin Solid Films **518**, 1248 (2009)

# Part VI
# Infared Spectroscopic Methods for Characterization of Thin Organic Films

# Chapter 21
# Characterization of Thin Organic Films with Surface-Sensitive FTIR Spectroscopy

**Katy Roodenko, Damien Aureau, Florent Yang, Peter Thissen and Jörg Rappich**

**Abstract** This chapter reviews the role of infrared spectroscopy in characterization of surfaces and interfaces of thin organic films. FTIR spectroscopy is widely utilized in studies of chemical bonds addressing questions concerning organization and orientation of the molecules in those films. In-situ FTIR spectroscopy frequently aids in studies of chemical reactions under a variety of experimental conditions, from high vacuum to aqueous solutions. FTIR spectroscopy can be realized in a multitude of setup geometries sensitive to a small amount of surface adsorbates. Anisotropic film properties can be studied by incorporating polarizing optics in an FTIR setup. FTIR modes of operation discussed in this chapter are Attenuated Total Reflection (ATR), transmission and reflection of the IR radiation through (or from) the sample, Polarization Modulation Infrared Reflection Absorption Spectroscopy (PM-IRRAS) and

K. Roodenko (✉)
Department of Materials Science and Engineering, Laboratory for Surface
and Nanostructure Modification, University of Texas at Dallas, Richardson, TX 75080, USA
e-mail: katy.roodenko@utdallas.edu

D. Aureau
Inst. Lavoisier, CNRS, UMR 8180, University of Versailles St Quentin Yvelines,
45 Av Etats Unis, 78035 Versailles, France
e-mail: damien.aureau@chimie.uvsq.fr

F. Yang
Institute for Heterogeneous Material Systems, Helmholtz-Zentrum Berlin für Materialien
und Energie GmbH, 14109 Berlin, Germany
e-mail: florent.yang@helmholtz-berlin.de

P. Thissen
Institute of Functional Interfaces (IFG), Karlsruhe Institute of Technology (KIT),
Hermann-von-Helmholtz-Platz 1, 76344 Eggenstein-Leopoldshafen, Germany
e-mail: peter.thissen@kit.edu

J. Rappich
Inst. für Si-Photovoltaik, Helmholtz-Zentrum Berlin für Materialien und Energie GmbH,
12489 Berlin, Germany
e-mail: rappich@helmholtz-berlin.de

© Springer International Publishing AG, part of Springer Nature 2018
K. Hinrichs and K.-J. Eichhorn (eds.), *Ellipsometry of Functional
Organic Surfaces and Films*, Springer Series in Surface Sciences 52,
https://doi.org/10.1007/978-3-319-75895-4_21

Infrared Spectroscopic Ellipsometry (IRSE). Practical considerations related to the sample properties (such as doping or roughness) and to the measurement conditions are discussed.

## 21.1 Introduction

Infrared spectroscopy is a widely used technique for studies of chemical composition of functional organic films as well as of the interfaces that are formed between the deposited layers and the surfaces in contact with them. Interaction of the IR radiation with the adsorbed organic molecules is characterized by infrared absorption lines positioned at the frequencies specific to the internal molecular vibrations of certain chemical groups. Identification of these absorption bands allows to verify molecular composition of the deposited organic films, the type of the intra- or intermolecular interactions, their structure and the nature of the interfacial bonding between the substrate and the organic layer. Several characteristic absorption bands, typical for surface adsorbates on most studied substrates such as Si or Ge, are listed in Table 21.1.

The position of the absorption bands depends on a variety of factors, such as temperature, material density, and interactions with the neighboring molecules. The frequency shifts of the absorption-bands are used for interpretation of the molecular structure and bonding on the surfaces. For instance, the quality of the surface functionalization with alkane molecules can be assessed through the position of the asymmetric methylene stretching vibration absorption band, the $v_{as}(CH_2)$ [2, 3, 13]. For hydrocarbon chains with the length of above 14 $CH_2$ units, a typical indication of a dense, well-ordered alkane layer is the appearance of the $v_{as}(CH_2)$ band close to $2920\,cm^{-1}$. Frequencies higher than $2920\,cm^{-1}$ indicate lower order of the alkane chains on the surfaces [1, 3].

Experimentally, FTIR spectroscopy can be realized in a multitude of setup geometries that can provide high sensitivity for a small amount of surface adsorbates. Each geometry is optimized for implementation under certain experimental conditions, such as in solutions or under vacuum. The dielectric properties of the investigated materials dictate the way that the radiation is reflected from, transmitted through and absorbed within the sample of interest. The signal-to-noise ratio (SNR) is highly dependent on the choice of the mode of measurement. Optimization of the optical setup is especially important for the in-situ experiments, where the fast spectral acquisition is critical for the detection of the spectral changes due to the dynamic processes taking place on the investigated surfaces.

This chapter describes the principles, experimental setup and application of most widely utilized surface-sensitive FTIR methods for studies of functional organic films, namely the ATR, IRSE, PM-IRRAS, as well as the transmission and reflection modes of operation.

**Table 21.1** Several characteristic infrared absorption bands of typical surface functional groups. $\delta$: bending modes; $\nu$: stretching modes

| Chemical group | Characteristic wavenumbers $cm^{-1}$ | Surface functionalization example |
|---|---|---|
| $\nu\,(CH_2)$ | 2960–2850 | alkyl chains on Si [1–3] |
| $\nu\,(C{=}O)$ | 1700–1735 | Aldehydes, ketones, carboxylic acids, ester [4] |
| $\nu\,(NO_2)_{sym}$ | 1380–1345 | Functional nitro-groups [5–7] |
| $\nu\,(NO_2)_{as}$ | 1570–1525 | Functional nitro-groups [5–7] |
| $\nu\,(Si{-}H_x)$ | 2070–2150 | H-terminated Si (mono-, di- and trihydrides) [8, 9] |
| $\delta\,(Si{-}H_x)$ | 656 | $Si{-}H_2$ in H-terminated Si [8, 9] |
| $\nu\,(Ge{-}H_x)$ | 2062–1960 | H-terminated Ge (mono-, di- and trihydrides) [10–12] |
| $\delta\,(Ge{-}H_2)$ | 830 | $Ge{-}H_2$ in H-terminated Ge [10–12] |

## 21.2 Attenuated Total Reflection (ATR)

Attenuated total internal reflection (ATR) technique was developed simultaneously and independently by Harrick [14] and Fahrenfort [15]. The detailed reviews of ATR-FTIR can be found in several review manuscripts [16–19] and books [20, 21]. ATR is based on the total internal reflection of the incident IR beam at the interface between two media (the film and the substrate). Figure 21.1 schematically presents the principles of the ATR approach. The light is reflected on the surface of the internal reflection element (IRE) and only molecules located in a (few) micrometer-range distance over the surface of the IRE are interacting with the probing infrared light. Since the refractive index of the IRE changes with wavelength, the internal angle $\theta$ is also wavelength-dependent. Normal incidence of the beam at the entrance and the exit from the IRE allows to maintain the same angle of incidence over the entire spectral range. Only the portion of the beam that is confined within the aperture $A$ at the beveled area of the IRE will be guided within the IRE at the desired internal reflection angle $\theta$ (Fig. 21.1). The total internal reflection occurs at angles of incidence above the critical angle $\theta_c$:

$$\theta_c = \arcsin\left(\frac{n_2}{n_1}\right) \tag{21.1}$$

where $n_1$ is the refractive index of the IRE and $n_2$ is the refractive index of the medium surrounding the IRE (air or solution for an isotropic two-layer system).

ATR measurements can be performed in single internal reflection (SIR) or multiple internal reflection (MIR) configurations. Figure 21.2 shows several possible configurations along with the proposed setups for in-situ implementation in electrochemical processes.

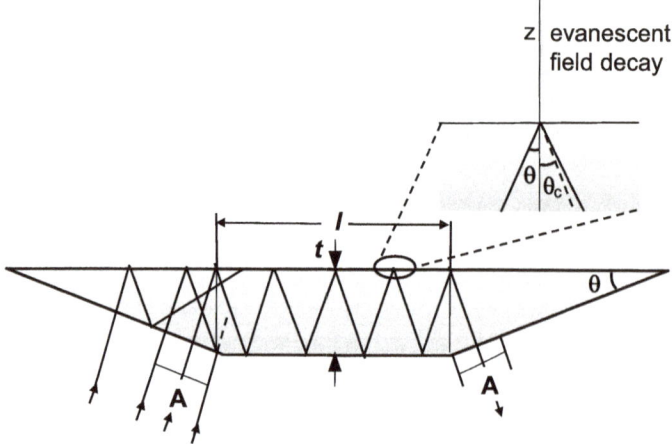

**Fig. 21.1** Schematical view of the beam path inside the internal reflection element with the length $l$ and thickness $t$. The aperture $A$ defines the area on the wedged side of the IRE that allows the beam guidance at the desired internal angle of incidence $\theta$. For the normal angle of incidence on the aperture, the angle of the internal reflection is the same as the wedge-angle $\theta$. On the *inset*: schematic representation of the evanescent field decay above the IRE

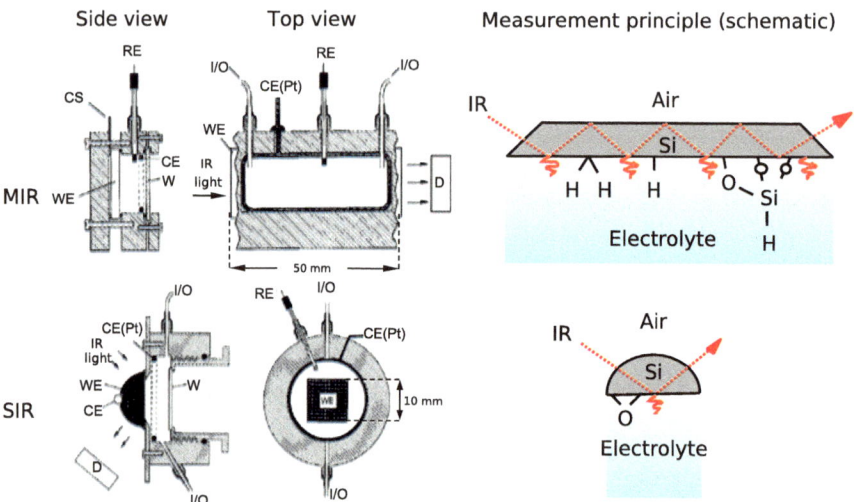

**Fig. 21.2** *Top*: multiple internal reflections (MIR); *bottom*: single internal reflections (SIR) ATR configurations. *Left*: electrochemical configurations. *Right*: schematic representation of the beam path in various optical elements. WE: working electrode; CE: counter electrode; RE: reference electrode; W: window; I/O: input/output; D: detector

In MIR configuration, the number of the internal reflections $N$ depends on the plate length $l$ and the plate thickness $t$ accordingly to the following formula:

$$N = (l/t)\cot(\theta). \tag{21.2}$$

The evanescent radiation that probes the film structure is confined to the interfacial region between the substrate and the outer probed medium. This feature of the ATR technique is frequently exploited in in-situ studies of solid/liquid interfaces, where the adsorption of molecules can be studied in presence of solvents that are strongly absorbent in the IR spectral range, such as for example water. The penetration depth $d_p$ is defined as the distance where the amplitude of the electric field falls to $e^{-1}$ of its value at the surface and is calculated accordingly to (21.3):

$$d_p = \frac{\lambda/n_1}{2\pi\sqrt{\sin^2\theta - (\frac{n_2}{n_1})^2}} \tag{21.3}$$

where $\lambda$ is the wavelength, $n_1$ is the refractive index of the IRE and $n_2$ is the refractive index of the medium surrounding the IRE. The choice of the plate length depends on the application. Although multiple interactions with the adsorbates should increase the SNR, yet, if the IRE itself absorbs in the IR spectral range of interest, the beam path in the IRE should be minimized. For instance, when silicon is used as an IRE in a multiple internal reflection configuration the spectral range below $1500\,\text{cm}^{-1}$ may become unaccessible if the beam path inside the IRE is too long.

In surface-modification experiments, ATR spectroscopy can be utilized both in-situ, where the film is casted on the IRE surface itself, or ex-situ, where the ex-situ modified surfaces are tightly pressed towards the surface of IRE and the absorption spectrum is recorded [20, 21]. In the latter case, inadequate contact to the IRE due to the irregularities of the sample surface can lead to the variations in the IR absorbance intensities in dependence on the clamping pressure applied to the sample. In-situ experiments are frequently carried out in flow-cells that allow to monitor molecular adsorption as a function of time [2]. Both in-situ and ex-situ ATR experiments have been reported in homemade as well as in commercial cells [22, 23]. Discussions regarding design of the in-situ ATR cells and the related technical aspects can be found in recent review articles [17, 18].

The choice of the IRE element depends on the spectral range of interest and the IRE stability to the required experimental conditions, such as the pH of the solvents for studies in solutions. The most commonly used materials are Si, Ge, ZnSe, ZnS, KRS-5, diamond and several others. The optical and mechanical properties of these materials and other IRE materials are summarized in [16, 18]. The IRE surfaces can be further modified to suit the experimental requirements. For instance, studies of catalytic reactions typically require deposition of a catalyst as a thin film or in a form of powdered layers [18, 24, 25]. The thickness of the film must be thinner than the penetration depth of the evanescent wave at any wavelength of interest. For powdered layers, quantitative analysis of the spectra is especially challenging due to the structural complexity of such films, where the factors such as layer porosity and the size of the nanoparticles may play an important role in the spectral lineshape and intensity. For both thin films and powdered layers a possibility of signal enhancement due to the surface-enhanced infrared absorption (SEIRA), an effect observed on rough metallic surfaces or upon addition of metallic nanoparticles to an organic layer, should be considered [24, 25].

In-situ applications of the ATR technique include monitoring of the thin film formation [2], catalysis [26], biomolecular sensing [27, 28], and many others.

ATR technique is frequently implemented for in-situ studies of reaction mechanisms at the solid/liquid interfaces. Studies of dynamic processes occurring in *aqueous solutions* present many challenges due to the strong absorption of water in the mid-IR spectral range, as discussed in [22]. Additional examples of the implementation of the ATR technique in studies of catalysis can be found, for example, in [17–19, 26, 29].

One of the examples of application of internal reflection techniques is in-situ electrochemical studies, as demonstrated in Fig. 21.3. This figure shows IR absorption spectra in the range of $SiO_2$ stretching modes (spectra obtained in SIR mode) and Si–H stretching modes (obtained in MIR operational mode) during anodic oxidation at different potentials for n-Si(111) in 0.2 M $Na_2SO_4$ (pH 3). The spectra are referenced to the spectrum of a H-terminated Si surface recorded in the same solution polarized at about −0.7 V. Applying a potential positive to −0.5 V versus Ag/AgCl leads firstly to the oxidation of Si–Si back bonds so that oxygen back bonded Si–H species develop on the former Si–H covered surface. This behavior is reflected by a

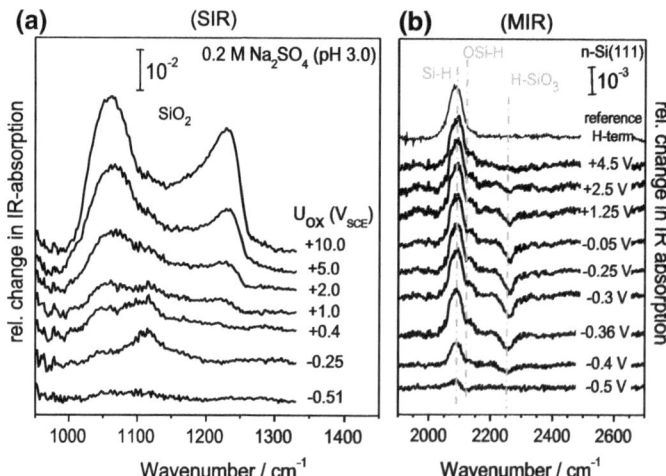

**Fig. 21.3** IR absorption spectra **a** in the range of $SiO_2$ stretching modes and **b** in the range of Si–H stretching modes during anodic oxidation at different potentials for n-Si(111) substrates in 0.2 M $Na_2SO_4$ (pH 3). The spectra are referenced to the hydrogen-passivated surface. The spectra are shifted for clarity. SIR: single internal reflection; MIR: multiple internal reflections

decrease in the IR absorption in the Si–H region (peak up at around 2090 cm$^{-1}$) and by an increase in Si–H species with an increasing amount of oxygen back bonds (O–Si–H 2118 cm$^{-1}$, $O_2$Si=$H_2$ 2200 cm$^{-1}$, and $O_3$Si–H 2255 cm$^{-1}$ [30–33], respectively). At about +1.25 V, the $O_3$Si–H related IR absorption starts to vanish. Simultaneously, the IR absorption at 1120 cm$^{-1}$ due to SiO$_x$ species splits into 2 peaks at 1070 and 1240 cm$^{-1}$, respectively, pointing to an SiO$_x$ island formation in the beginning and development of a continuous SiO$_2$ layer positive to +1 V [34].

ATR technique is also useful for in-situ studies of formation of self-assembled monolayers at the solid/liquid interfaces. Vallant et al. [2] have studied the formation of alkylsiloxane monolayers $O_x$Si–$(CH_2)_n$–$Y$ with different hydrocarbon chain lengths ($n = 10, 16, 17$) and different terminal substituents ($Y = CH_3$, $COOCH_3$, CN, Br) on native silicon (Si/SiO$_2$) by means of in-situ ATR in the precursor solution. Figure 21.4 presents the results of monitoring the formation of a monolayer film of octadecylsiloxane (ODS) on SiO$_2$ surface. Two groups of absorptions can be distinguished in Fig. 21.4: the CH$_x$ stretching vibrations of the monolayer between 2800 and 3000 cm$^{-1}$ that grow in the positive direction, and the CH stretching bands of the solvent (benzene) above 3000 cm$^{-1}$, growing in the negative direction with increasing time of adsorption. These spectral changes are characteristic for a successive replacement of solvent molecules with film molecules at the substrate/solution interface. After a certain time, the band intensities remain constant and indicate the complete formation of a monolayer film. The two major absorption bands in Fig. 21.4 can be assigned to $\nu_{as}$(CH$_2$) around 2920 cm$^{-1}$ and to $\nu_s$(CH$_2$) at 2850 cm$^{-1}$, respectively. As the surface coverage increases, the $\nu_{as}$(CH$_2$) absorption band shifts to the

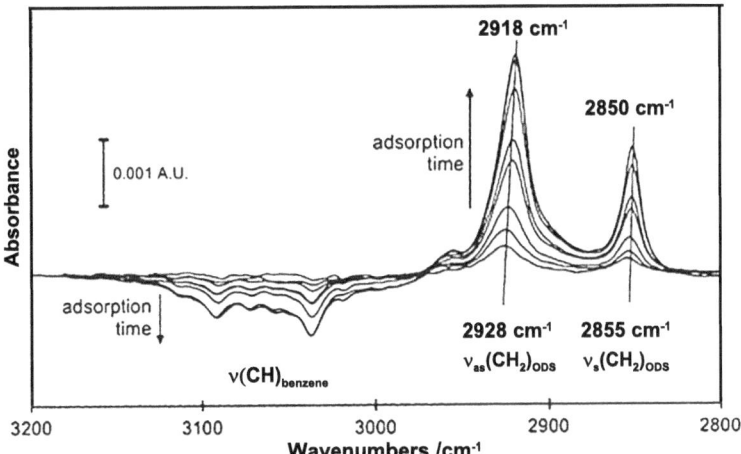

**Fig. 21.4** In situ ATR spectra (s polarization, 35° incidence) monitoring the formation of a monolayer film of octadecylsiloxane (ODS) at the interface between a silicon ATR crystal and the adsorbate solution (1 mmol/L octadecyltrichlorosilane in benzene). Reprinted with permission from [2]: T. Vallant, J. Kattner, H. Brunner, U. Mayer, and H. Hoffmann, Langmuir **15**, 5339 (1999). Copyright 1999 American Chemical Society

lower frequency. The frequency of the asymmetric methylene stretching vibration close to $2920\,cm^{-1}$ is a typical indication for crystalline alkanes, while the frequencies around $2928\,cm^{-1}$ are typical for liquid alkanes [1, 3]. This indicates the transition to highly packed, ordered molecular structure at higher coverages.

## 21.3  Infrared Spectroscopy in Transmission and Reflection Modes

### 21.3.1  Transmission Mode

Infrared spectroscopy in a transmission mode is especially well-adapted for transparent substrates. Since the infrared beam is transmitted through the sample, it is important that the substrate is transparent and homogeneous. This enables comparison between the surfaces at different steps of modification. At each measurement step, the sample has to be positioned at the exact same position in order to observe the influences of the modification steps performed on the upper monolayers. Unfortunately, for any configuration of the IR spectroscopy the presence of the substrate itself can sometimes distort the infrared spectrum (such as for example due to the frequency shifts of the bulk phonon modes upon changes of the ambient temperature) that may complicate the interpretation.

Various types of glasses based on silicon oxides exhibit good transparency in the visible region but unfortunately, they have low transparency in the infrared spectral range. Thus, it is not possible to use them for wavelengths longer than $5\,\mu m$, which corresponds to wavenumbers below $2000\,cm^{-1}$. Indeed, the strong absorption of the Si–O bond may hinder a proper analysis of the phenomena occurring at the surfaces. Another example is alkali halide crystals that exhibit a very good transparency from ultraviolet to infrared but due to their highly hygroscopic properties, a careful handling is required.

Finally, the good transparency of low-doped semiconductors such as silicon, germanium, indium phosphate or gallium arsenide make them the favorable materials for transmission infrared spectroscopy. Traditionally, silicon has been the material of choice since it is cheaper and easier to obtain with a *sufficient area* for being used in a transmission mode with a sufficient sensitivity to thin films. It is important to point out that among the large numbers of available silicon substrates, high-quality semiconductors are required for this technique. For instance, substrates obtained from float zone (FZ) method are typically preferred over the ones obtained by Czochralski (CZ) growth. The latter exhibit a strong absorption band around $1100\,cm^{-1}$ associated with interstitial oxide. In addition, FZ silicon has lower levels of carbon contamination than CZ silicon.

The surface roughness is another important aspect to be considered. One of the most common ways to avoid roughness-related problems is to use double-side polished material. It turns out that the rough side of a silicon wafer can scatter infrared light significantly, preventing a part of it from reaching the detector. Sometimes the rough side can be chemically polished, for instance in a mixture of hydrofluoric acid, nitric acid and acetic acid. It must be noted that for transmission spectroscopy, it is essential to use lightly doped substrates in order to reduce the effect of free carrier absorption. The use of polycrystalline substrates is generally not possible, due to the scattering of the infrared beam. Qualitative analysis can be performed by IR transmission on amorphous [35] or porous silicon [36]. The density of vibrators is naturally higher in comparison to crystalline substrates but not easy to quantify. The case of the porous substrates is also delicate. In its electrochemical preparation, the metallic back-contact must be controlled to avoid disturbances to the infrared beam. A contact on the corner of the sample can be used if the substrate is conductive enough, the deposition of a gold grid can be used as well. Monocrystalline silicon is generally the most adapted substrate for transmission infrared especially if quantification is required. Knowing the surface crystallographic orientation allows a good estimation of the surface coverage. For instance, a surface density of $8.3 \times 10^{14}$ Si–H bonds per $cm^2$ is obtained on flat H-terminated Si(111). Once the H-terminated Si surfaces are modified, the integration over the absorption band area due to the remaining Si–H bonds allows a quantitative analysis of the coverage. Yet care must be taken since the remaining Si–H bonds may interact with the adsorbed surface molecules, thus the spectral lines due to the Si–H absorption may broaden, change the amplitude and exhibit frequency shifts.

Special deposition compartments, such as vacuum chambers, can be especially adapted for the transmission infrared spectroscopy. The controlled atmosphere and the fact that the sample does not move between different measurements ensure that the features observed on the IR spectra directly relate to the surface modification and not to the factors associated with, for instance, changes in humidity or to the substrate displacement [37]. The exposure of substrates to gases and gas-phase reactions can be studied by this method. For instance, the reaction of H-terminated Si surfaces with $NH_3$ have been performed by transmission infrared in an atomic-layer deposition (ALD) reactor [38–40]. The baseline quality facilitated the analysis of specific small contributions, such as the Si–NH–Si at the steps. For gas-phase reactions, it is important to control the temperature of the sample. Each measurement (before the modification and thereafter, at various stages of the modification) must be conducted at the precisely same temperature, in order to avoid misinterpretation of absorption bands due to the substrate-related phonon features [41]. The main advantage of the transmission infrared spectroscopy is it's ease of use. The main disadvantage and the principal reason for the development of others geometries is the low sensitivity of this technique in comparison, for instance, with the MIR.

Nonetheless, recent examples showed that transmission infrared spectroscopy could be sufficient in terms of sensitivity to monolayers and to multistep–modifications of monolayers [4, 42, 43]. Among the thinnest monolayers, methoxy groups ($OCH_3$) in Si–$OCH_3$ was characterized in a transmission mode [44]. As confirmed by the DFT calculations, all the vibrations associated with the C–H groups were clearly assigned. In this work, the high quality of the IR spectra was achieved due to the fact that the experiments were performed inside a glove-box, ensuring the absence of bands associated with water vapor in the spectra. The integration over the absorption bands due to the remaining H–Si(111) bonds pointed out on a replacement of the third of the initial Si–H bonds by the methoxy groups (under an assumption that the strength and the width of the Si–H absorption bands were not affected by the interactions with the nearby molecules), demonstrating the high sensitivity of transmission infrared to ultrathin layers. Furthermore, this example demonstrated the applicability of the IR spectroscopy to nanopatterned surfaces. Indeed, the methoxy groups may be replaced by fluorine atoms, while the surrounding Si–H bonds stay unreacted. The band position of absorption bands due to the Si–H vibrations is then shifted depending on the number of halogen atoms in the nearest-neighbor sites [45]. It is rather challenging to obtain spectra of methyl groups on surfaces due to the presence of contaminants for all the methods of preparation. A solution to problematic IR characterization of the Si–CH$_3$ surfaces was suggested by the preparation of a porous silicon surface that allows to increase the density of vibrators [46]. Of course, the possibility to quantify the experimental yield is partially lost in this case. Recently, Lewis et al. also showed infrared characterization of ultrathin monolayers on germanium surfaces by transmission infrared spectroscopy [47].

On transparent substrates MIR technique is often advantageous over the transmission IR spectroscopy due to the enhanced SNR and the intensity of the infrared bands. Indeed, the infrared beam interacts with the modified surfaces several times (in comparison to two interactions using transmission IR). Another advantage of the internal reflection geometry for transparent substrates is the possibility for in-situ studies in liquids, where for instance, part of the sample can be in contact with liquid phase [48]. However, using MIR could be difficult for two main experimental reasons: first, it is never easy to prepare the sample at prism geometry with well-polished bevels. The polishing of the sides that couples the light in and out of the MIR element can break the fragile substrates that are less solid than silicon. Second, the necessity to place the sample at the exact same position for each measurement to avoid spectral artifacts is even more crucial in MIR than in transmission geometry.

## 21.3.2   Reflection Mode

Infrared reflection absorption spectroscopy (IRRAS), also named Reflection Absorption Infrared Spectroscopy (RAIRS), External Reflectance Infrared Spectroscopy (ERIS) or even Grazing Angle Infrared Spectroscopy (GIR) is a technique based on a single specular reflection of light from a reflective surface. Consequently, it is especially well-suited for studies of adsorption of small molecules on metal surfaces [49]. Based on the surface-selection rules on metallic substrates [50], IRRAS can help to determine the adsorption geometry and the adsorption site of a given molecule [51]. Historically, this method has been developed after the first studies of CO adsorption on supported metals by single transmission IR [52]. The need to use single-crystal as model surfaces to address issues related to specific site adsorption or reaction mechanisms motivated using a reflection mode geometry [53, 54]. The first studies were carried out to investigate CO adsorption on copper surfaces [55]. Due to the intensity of their dipole moments, CO and NO adsorption on a large variety of metals have been carried out at the beginning [56]. It should be noted that CO adsorption is still used to calibrate a system or to test the reactivity of a given surface.

The reflection of electromagnetic wave from a surface can be derived from the Maxwell equations, as discussed in the Introduction chapter of this book. The interaction of the electromagnetic wave with the adsorbate molecules is a dipole interaction between the electromagnetic field and the dipole moment of the molecule. The dipole selection rules derived from Fresnel equations dictate that on metallic surfaces, the only absorption bands that can be observed in the IR spectra are due to the molecular dipole moment perpendicular to the sample surface. These absorption bands would be typically enhanced due to the image-effect at the metallic interface [50]. The intensities of the electrical fields at the surface are functions of the angle of incidence and the refractive indices of substrate and film. These parameters influence the reflectivity $R$ of the system.

IRRAS spectrum is typically presented as $(R_0 - R)/R_0$, where $R_0$ is the reflectivity without any adsorbate and $R$ is the reflectivity of the surface with adsorbed material system. Alternatively, the measured absorbance of the thin film can be presented as $-\log(R/R_0)$.

On metallic surfaces, the experiment is most effective at a high angle of incidence relatively to the surface normal (grazing angle spectroscopy, GIR). Typically, the angle of incidence is chosen between 80° and 88°, depending on the metal. IRRAS can be applied not only on metals but also on transparent substrates and weakly absorbing substrates, such as semiconductors and dielectrics (including liquids) at appropriate angles of incidence [50]. For example external reflection infrared spectroscopy was used to study monolayers of octadecylsiloxane formed on native silicon (Si/SiO$_2$) and glass surfaces [57]. Spectral simulations based on classical electromagnetic theory allowed to calculate an average tilt angle of the hydrocarbon chains on both silicon and glass surfaces that was estimated to be 10° with respect to the surface normal.

### 21.3.3 Polarization Modulation Infrared Reflection Absorption Spectroscopy

The overlapping infrared bands of water present in the ambient atmosphere can considerably complicate the evaluation of the infrared spectrum. In the early 1990s [58], a differential reflectivity technique, Polarization Modulation-Infrared Reflection-Adsorption Spectroscopy (PM-IRRAS) has been developed. The combination of FT-IRRAS and fast polarization modulation of the incident beam (ideally between p- and s- linear states) gives a faster way to obtain infrared bands avoiding those that appear in both polarizations. For the molecular characterization of surfaces of silicon, gold or any reflective material, the use of PM-IRRAS brings the great advantage

**Fig. 21.5** Optical setup for
PM-IRRAS measurements

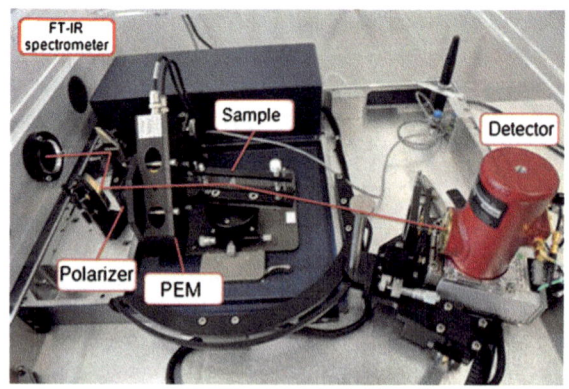

of easy and rapid analyses, without requiring a perfect positioning of the sample or
control of the surrounding atmosphere [59]. Figure 21.5 shows a photograph of a
PM-IRRAS system. A continuous light source emits infrared radiation that is
absorbed by the thin film. Radiation reflected from the surface is registered in a detector and electronically converted into an IR-spectrum. Following the earlier-mentioned
surface-selection rules on metals, enhanced absorbance of the reflected p-polarized
light is obtained. In contrast, in thin films nearly no absorbance is observed with
s-polarized light. This disparity in absorbance intensities offers the possibility to
obtain the differential reflectance spectrum of the surface species $\Delta R/R$ by polarization modulation.

It is important to use only polarization–insensitive elements behind the sample
(see Fig. 21.5), since any instrumental polarization differences will appear in the
PM-IRRAS spectrum. To obtain the differential spectrum, the detector signal is
processed by specialized electronics that generate the average and the difference
interferogram required for the PM-IRRAS spectrum. These two signals are digitized
simultaneously with the A/D converter of the FTIR spectrometer.

Another component of the setup is a photoelastic modulator (PEM), which is
based on a birefringent material such as ZnSe. The stress amplitude that dictates
the birefringent response of the PEM can be selected so that the PEM acts like a
'half-wave' plate. This means that the plane of linear polarized light is rotated by
$90°$, after passing through the PEM crystal. The stress, applied to the PEM crystal,
is sinusoidally modulated. Thus the state of polarization is modulated as well. The
effective polarization modulation frequency of the light is twice the mechanical
oscillation frequency of the PEM crystal. The PEM modulates the infrared beam by
rotating the polarization of the light sinusoidally at the frequency of the birefringence
modulation introduced mechanically in a ZnSe crystal [60, 61]. If a monochromatic
incident infrared beam that is linearly polarized at $45°$ to the strain axis is passed
through the modulator, the intensity of the output light is given by [61]:

$$I(t) = \left[I_p + I_s + (I_p - I_s) \cdot \cos\big(\Phi_0 \cdot \cos(\gamma \cdot t)\big)\right]/2 \qquad (21.4)$$

where

- $I_p$: polarization of the light beam prior to the PEM, experimentally set to p-polarized light, i.e. polarized so that the electric field is parallel to the plane of incidence.
- $I_s$: s-polarized light, i.e. polarized so that the electric field is perpendicular to the plane of incidence.
- $\cos(\gamma \cdot t)$: modulation frequency of the PEM.
- $\Phi_0$: constant that depends linearly on the amplitude of mechanical modulation of the PEM.

PEM provides a high frequency modulation on the classical interferogram, detected at the detector element. These two frequencies, interferometer and PEM modulation, are separated by the dedicated electronics.

Choosing the condition of the maximum retardation between the polarized light components (that occurs at one-half of the light wavelength), the PEM signal is defined as:

$$\Delta R / R = J_2 \cdot (I_p - I_s)/(I_p + I_s) \qquad (21.5)$$

where $J_2$ are the Bessel functions of the first kind. The amplitude of the mechanical excitation of the PEM crystal defines the points of the zero crossings of the Bessel function. The appropriate setting for the amplitude depends on the experiment, i.e. on the wavelength region of interest.

Figure 21.6 shows the effects of the background and the spectrum after the removal of the background on the example of adsorbed phosphonic acid on oxidized Al surface. The spectrum consists of the (P=O) stretching mode at $1227\,cm^{-1}$, the asymmetric and symmetric stretching modes of the (P–O) at 1078 and $1004\,cm^{-1}$, and the deformation mode of the (P–O–H) at $956\,cm^{-1}$. In the aliphatic region, the

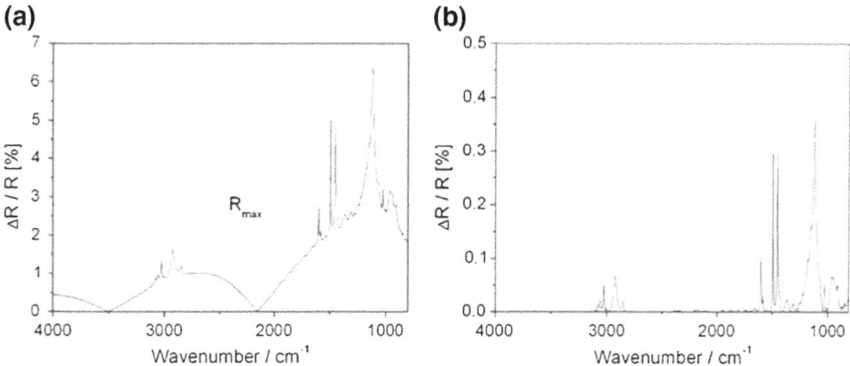

**Fig. 21.6  a** Differential reflectance spectrum of [12-(4-Benzophenone)-dodecyl] phosphonic acid monolayer on aluminum oxide surfaces. This spectrum exhibits IR absorption features of the sample and the Bessel function background. **b** Spectral features of the sample after removal of the background

spectral positions of the strong asymmetric and symmetric stretching modes of the $(CH_2)$ at 2918 and 2850 cm$^{-1}$, respectively, indicate an ordered arrangement on the surface.

After the measurement of the raw data $(\Delta R/R)$, the refinement of the spectra typically requires the correction of the lock-in amplifier amplification factor and the removal of the background that can be fitted by a polynomial algorithm or obtained by the measurement of a clean metal substrate. In comparison to the infrared transmission spectroscopy, PM-IRRAS spectra may exhibit spectral shifts and changes of the full width at half maximum of the peaks.

IR ellipsometry (discussed in the next section) can as well benefit from the incorporation of the photoelastic modulator element, which allows for faster acquisition times. The description of several PEM-based ellipsometric setup configurations can be found, for example, in [62–64]. The mathematical description of the PEM-based ellipsometers can be found in [65].

## 21.4   IR Spectroscopic Ellipsometry: IRSE

The major advantage of IRSE over other IR techniques is that in principle, it is reference-free technique. Since IRSE analyzes the changes in the polarization state of the beam upon transmission or reflection from the surface, the measurement of the ellipsometric parameters tan $\Psi$ and $\Delta$, or alternatively the real and imaginary parts of the complex (pseudo)-dielectric function, defined in the Introduction chapter, does not require a reference sample, as in all other IR techniques. In practice, however, the absorption-like features specific to the investigated *ultrathin* films are better observed on tan $\Psi$ and $\Delta$ spectra that are referenced to those of the bare unmodified substrate. Similar to the ellipsometry in visible and UV spectral ranges, IRSE is an indirect technique in a sense that it requires application of optical models in order to extract the thicknesses and the dielectric functions of the measured materials. While the IR bands do provide an immediate possibility to identify the molecular composition of thin films, the thickness and the dielectric function of the thin films must be fitted using optical models described in Chap. 13.

Similar to transmission and reflection spectroscopies, IRSE is sensitive to a monolayer coverage of the surface adsorbates [5].

IRSE can be performed in transmission, reflection and ATR modes. This opens up a possibility for characterization of different thin films deposited on both transparent and metallic substrates. IRSE can be applied to flat and porous surfaces. As had been discussed earlier in this chapter, porous surfaces are frequently employed in many IR modes of operation for SNR enhancement. A comparative IRSE study of adsorption of nitrobenzene and methyl groups on porous and flat Si surfaces can be found, for instance, in [9].

An instructive example of application of IRSE for the characterization of ultrathin organic films on flat Si surfaces is the study of the one-step electrochemical process by anodic decomposition of Grignard compounds that results in a direct grafting of molecules to Si surfaces [66]. In this process, the anodization is stimulated by the creation of radicals near the surface, which are controlled by the charge flow. The radicals R· are created (arising from Grignard compounds) when an anodic current is applied to the Si surface. The radicals react with the H-terminated Si(111) surface to create a Si dangling bond and, subsequently, another radical R· available in solution is able to react with this dangling bond to form a covalent Si–C bond. Interestingly, the halogen in the Grignard reagent has been found to strongly influence the grafting procedure, which affects the electronic properties (passivation) as observed for the methyl- and the ethynyl-terminated Si(111) surfaces, respectively [66–68].

One of the noticeable examples that demonstrates high sensitivity of IRSE to ultrathin films is the characterization of Si surfaces modified by methyl groups. IRSE studies performed on $CH_3$ and $CD_3$ terminated Si surfaces (see Fig. 21.7) clearly show a prominent vibrational band shift due to the isotopic substitution. IR bands are observed in the tan $\Psi$ spectra at 980 and 1253 $cm^{-1}$ and are assigned to the symmetric "umbrella" deformation (bending) mode of the methyl groups, i.e.,

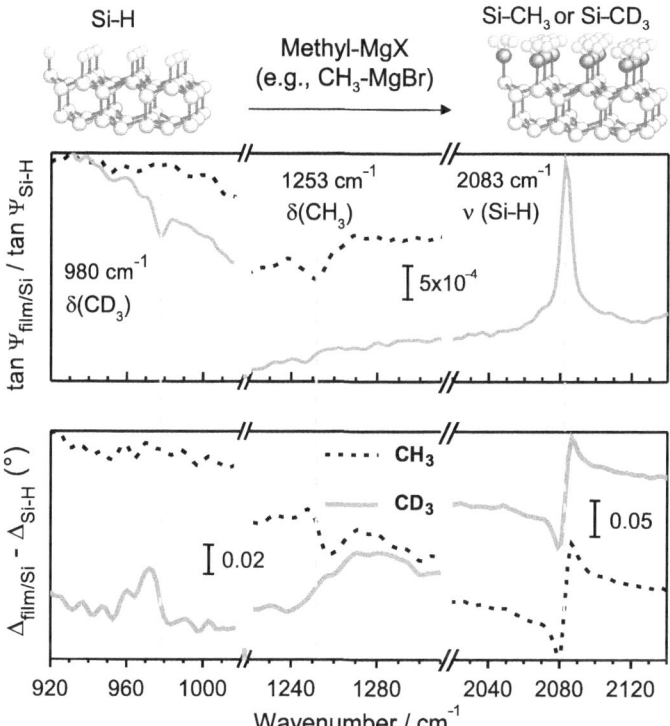

**Fig. 21.7** IRSE spectra (*top*: tan $\Psi$, *bottom*: $\Delta$) of methylated Si(111) surfaces after anodic treatment in $CH_3$–MgBr (*dotted line*) and $CD_3$–MgI (*solid grey line*) Grignard solutions. The corresponding IRSE spectra obtained from the H-terminated Si(111) surface served as the reference data [66]

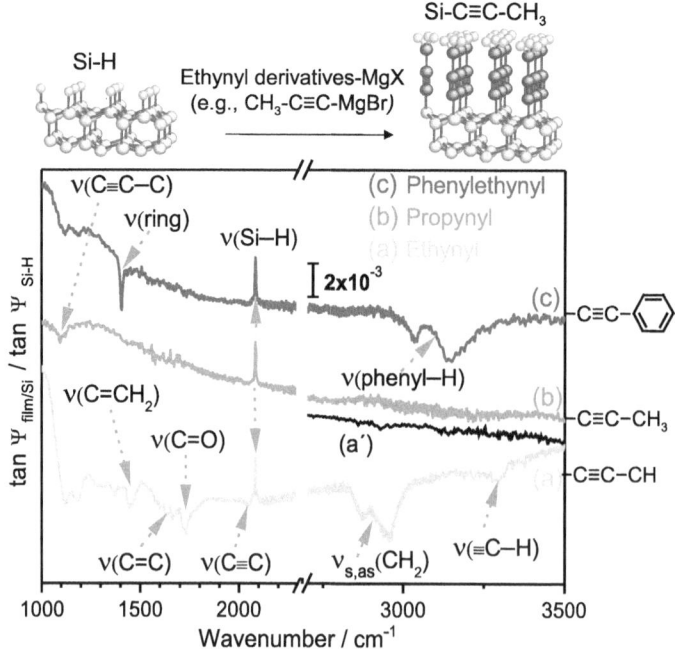

**Fig. 21.8** tan $\Psi$ spectra after anodic treatment in **a** H–C≡C–MgCl, **b** CH₃–C≡C–MgBr, and **c** C₆H₅–C≡C–MgBr referenced to the tan $\Psi$ spectrum of the H-terminated Si(111) surface. Please note: (**a**′) anodic treatment in H–C≡C–MgBr has been added to the figure for comparison [67]

$\delta_s(CD_3)$ and $\delta_s(CH_3)$ [69, 70]. The spectra were referenced to the H-terminated Si(111) surface in order to highlight the changes after the electrochemical grafting. The positive upward–pointing band in the tan $\Psi$ spectra at 2083 cm$^{-1}$ indicates the *loss* of the symmetric vibrational mode of Si–H surfaces species ($\nu_s$(Si–H)) after the grafting process. The infrared dynamic dipoles of the methyl groups are essentially perpendicular to the surface since only the symmetric "umbrella" mode can be observed in the spectrum (Fig. 21.7). Following Kramers–Kronig relations, the $\Delta$ spectrum exhibits the respective bands due to the $\delta_s(CD_3)$ and $\delta_s(CH_3)$ vibrational modes at the matching wavenumbers. The ability to access both parameters related through the Kramers–Kronig relations is particularly critical in evaluation of the optical constants of thin films [71]. The intensities of the C–H stretching modes of methyl bonded to silicon are very weak, yet they are detectable with IRSE and other IR-related spectroscopic techniques at a monolayer coverage [46, 69, 72, 73].

Although the methylated Si(111) surfaces strongly protect the surface against Si oxidation and other undesired chemical reactions, they are not suitable for further functionalization. Among potential candidates for this task that also provide a complete coverage of Si(111) surfaces are the unsaturated organic species, such as linear ethynyl or propynyl derivatives (–C≡C–R, with R=H, CH₃). Functionalization of Si surfaces by C≡C groups opens new reaction pathways for further organic modification of these surfaces by the introduction of functional groups through the "click" chemistry [74]. Figure 21.8 shows IRSE spectra obtained after the electrochemical grafting of ethynyl derivatives onto Si surfaces. Grafting of propynyl onto Si(111) surfaces is characterized by a small IR absorption band at about $1100\,cm^{-1}$ that can be assigned to the $-C{\equiv}C-C$ stretching vibrational mode [50]. The strong and broad IR-absorption bands, from contaminations and the layer itself, are attributed to symmetric and asymmetric stretching vibrational modes of $v(CH_2)$ bands around 2870 and $2958\,cm^{-1}$, respectively [75]. The weak and broad absorption peak around $3300\,cm^{-1}$ that is assigned to the acetylenic C–H stretching modes in the $-C{\equiv}C-H$ unit [75, 76] appears only on the spectrum obtained from the Si(111) surfaces modified in $H-C{\equiv}C-MgCl$. Moreover, the stretching vibration of the ethynyl groups $v(C{\equiv}C)$ at approximately $2046\,cm^{-1}$ is more distinguishable for the Si(111) modified by $H-C{\equiv}C-MgCl$ [75, 77]. A weak band also appears at $1725\,cm^{-1}$ after grafting of $H-C{\equiv}C-$. This band was described as related to the radical formation in the THF (the Grignard solvent) in [75]. Two other bands that appear at 1450 and $1650\,cm^{-1}$ are assigned to the symmetric bending and stretching vibrational modes of $C{=}CH_2$ and $C{=}C$ groups, respectively [75, 76]. All these features indicate the formation of a polymeric layer for ethynyl-modified Si surfaces in $H-C{\equiv}C-MgCl$ solution. An attack of intermediate radicals on the grafted $-C{\equiv}C-H$ species leads to C–C, $C{=}C$ and $C{\equiv}C$ bonds in a polymeric layer [75]. However, exchanging the halogen Cl atoms in the Grignard compound by Br atoms leads to a very thin polymeric layer under the identical experimental conditions [78]. Therefore, the halogen atoms present in the Grignard reagents play an important role in the grafting process.

Proper optical modeling and fitting of the IRSE data provides the information on the thickness and the complex dielectric function of the investigated thin films. This in turn enables one to address the anisotropy of the grafted organic material, such as for instance the anisotropy related to the orientation of self-assembled monolayers on various surfaces [79, 80]. IRSE investigations of thin films of benzenediazonium derivatives electrochemically grafted on various surfaces is one example where the determination of order and orientation in organic monolayers was successfully attempted. These studies have unequivocally shown that benzenediazonium molecules are isotropically distributed on the surfaces, with no particular order due to the tendency of the radicals to polymerize during the electrochemical process [6, 9]. On the other hand, examples of IRSE studies of molecular tilt in anisotropic films include, among the others, investigations of the molecular orientation in octanedithiol and hexadecanethiol (HDT) monolayers on GaAs and Au surfaces [81], or the changes in orientation of 2-[4-(N-dodecanoylamino)phenyl]-5-(4-nitrophenyl)-1,3,4-oxadiazole Langmuir–Blodgett films on gold surfaces upon annealing [7]. Figure 21.9a shows the results of the fitted IRSE spectrum obtained from HDT monolayer on gold surface that suggested molecular tilt of 22.5° to the surface normal. Figure 21.9b shows the simulated $\tan \Psi$ spectra for molecular tilt ranging from 12° to 42°. The molecular geometry along with the molecular dipole moments, $v_{as}$ and $v_{ss}$, the asymmetric and symmetric stretching vibrations, are shown in Fig. 21.9c.

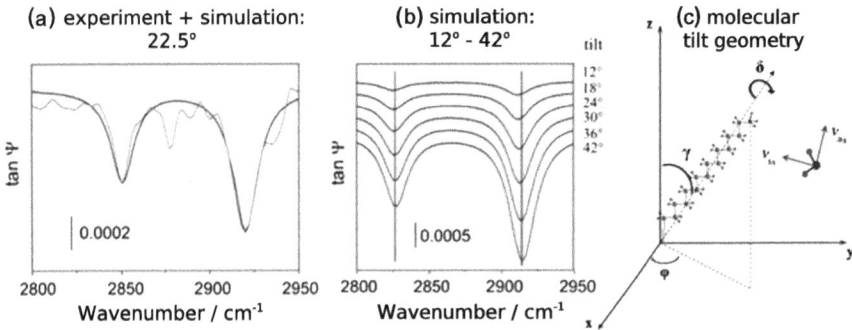

**Fig. 21.9** **a** Simulated (*black*) and measured (*gray*) tan $\Psi$ spectra of a HDT monolayer on Au surface. The incidence angle was set to 65°. **b** Simulations for tilt angles from 12° to 42°. **c** Schematic representation of the molecular geometry. Reprinted with permission from [81]: D.M. Rosu, J.C. Jones, J.W.P. Hsu, K.L. Kavanagh, D. Tsankov, U. Schade, N. Esser, and K. Hinrichs, Langmuir **25**, 919 (2009). Copyright 2009 American Chemical Society

IRSE examples provided in this section highlight the high sensitivity of this method to thin organic films. IRSE is a useful technique for addressing molecular composition and orientation of organic layers. Application of optical models and spectral simulations frequently help to verify the optical constants in the IR spectral range along with the film thickness and molecular composition [71, 82].

## 21.5  Summary

IR spectroscopy is an essential tool for characterization of thin organic layers. This chapter highlighted the versatility of modes of operation that can be employed to address molecular composition and structure of modified surfaces. Careful observation of spectral shifts due to a certain surface modification can help to address the nano-composition of thin films, such as in the case of nanopatterned surfaces where absorption bands due to Si–H bonds shift in response to the nature of the nearest-neighbor bonds [45]. IR spectroscopy is a useful tool in characterization of multi-step surface functionalization, where a careful control over the surface reactions is critical for the final device quality [4]. Proper choice of the IR mode of operation and experimental conditions is necessary in order to optimize the SNR. In specific cases, application of optical models and density-functional theory (DFT) calculations [83] is helpful for understanding the spectra obtained from novel materials.

# References

1. H. Shpaisman, O. Seitz, O. Yaffe, K. Roodenko, L. Scheres, H. Zuilhof, Y.J. Chabal, T. Sueyoshi, S. Kera, N. Ueno, A. Vilan, D. Cahen, Chem. Sci. **3**(3), 851 (2012)
2. T. Vallant, J. Kattner, H. Brunner, U. Mayer, H. Hoffmann, Langmuir **15**(16), 5339 (1999)
3. O. Seitz, T. Böcking, A. Salomon, J.J. Gooding, D. Cahen, Langmuir **22**(16), 6915 (2006)
4. O. Seitz, P.G. Fernandes, R. Tian, N. Karnik, H.-C. Wen, H. Stiegler, R.A. Chapman, E.M. Vogel, Y.J. Chabal, J. Mater. Chem. **21**, 4384 (2011)
5. K. Roodenko, F. Yang, R. Hunger, N. Esser, K. Hinrichs, J. Rappich, Surf. Sci. **604**, 1623 (2010)
6. M. Gensch, K. Roodenko, K. Hinrichs, R. Hunger, A.G. Guell, A. Merson, U. Schade, Y. Shapira, T. Dittrich, J. Rappich, N. Esser, J. Vac. Sci. Technol. B **23**(4), 1838 (2005)
7. D. Tsankov, K. Hinrichs, E.H. Korte, R. Dietel, A. Röseler, Langmuir **18**(17), 6559 (2002)
8. Y.J. Chabal, Phys. B **170**, 447 (1991)
9. K. Roodenko, J. Rappich, F. Yang, X. Zhang, N. Esser, K. Hinrichs, Langmuir **25**, 1445 (2009)
10. G. Lu, J.E. Crowell, J. Chem. Phys. **98**, 3415 (1993)
11. Y.J. Chabal, Surf. Sci. **168**, 594 (1986)
12. K. Choi, J.M. Buriak, Langmuir **16**, 7737 (2000)
13. K. Roodenko, O. Seitz, Y. Gogte, J.-F. Veyan, X.-M. Yan, Y.J. Chabal, J. Phys. Chem. C **114**, 22566 (2010)
14. N.J. Harrick, Phys. Rev. Lett. **4**, 224 (1960)
15. J. Fahrenfort, Spectrochim. Acta **17**, 698 (1961)
16. A.R. Hind, S.K. Bhargava, A. McKinnon, Adv. Colloid Interface Sci. **93**, 91 (2001)
17. T. Buergi, A. Baiker, Adv. Catal. **50**, 227 (2006)
18. J.-M. Andanson, A. Baiker, Chem. Soc. Rev. **39**, 4571 (2010)
19. A.V. Cheruvathur, E.H.G. Langner, J.W. Niemantsverdriet, P.C. Thüne, Langmuir **28**, 2643 (2012)
20. N.J. Harrick, *Internal Reflection Spectroscopy*, 1st edn. (Interscience, New York, 1967)
21. F.M. Mirabella Jr. (ed.), *Internal Reflection Spectroscopy*, 1st edn. (Dekker, New York, 1993)
22. B.L. Mojet, S.D. Ebbesen, L. Lefferts, Chem. Soc. Rev. **39**, 4643 (2010)
23. H. Schumacher, U. Künzelmann, B. Vasilev, K.-J. Eichhorn, J.W. Bartha, Thin Solid Films **525**, 97 (2012)
24. M. Watanabe, Y. Zhu, H. Uchida, J. Phys. Chem. B **104**, 1762 (2000)
25. H. Miyake, T. Okada, G. Samjeske, M. Osawa, Phys. Chem. Chem. Phys. **10**, 3662 (2008)
26. Z. Liu, A. Rittermeier, M. Becker, K. Kähler, E. Löffler, M. Muhler, Langmuir **27**, 4728 (2011)
27. S. Devouge, J. Conti, A. Goldsztein, E. Gosselin, A. Brans, M. Voue, J. De Coninck, F. Homble, E. Goormaghtigh, J. Marchand-Brynaert, J. Colloid Interface Sci. **332**, 408 (2009)
28. C.M. Pradier, Y.J. Chabal (eds.), *Biointerface Characterization by Advanced IR Spectroscopy* (Elsevier, Amsterdam, 2011)
29. J.T. Carneiro, J.A. Moulijn, G. Mul, J. Catal. **273**, 199 (2010)
30. M. Niwano, J. Kageyama, K. Kurita, K. Kinashi, I. Takahashi, N. Miyamoto, J. Appl. Phys. **76**, 2157 (1994)
31. J.C. Knights, R.A. Street, G. Lucovsky, J. Non-Cryst, Solids **35–36**, 279 (1980)
32. G. Lucovsky, Sol. Energy Mater. **8**, 165 (1982)
33. G. Lucovsky, J. Yang, S.S. Chao, J.E. Tyler, W. Czubatyj, Phys. Rev. B **28**, 3225 (1983)
34. F. Ozanam, A. Djebri, J.-N. Chazalviel, Electrochim. Acta **41**, 687 (1996)
35. L. Touahir, P. Allongue, D. Aureau, R. Boukherroub, J.-N. Chazalviel, E. Galopin, A.C. Gouget-Laemmel, C. Henry de Villeneuve, A. Moraillon, J. Niedziółka-Jönsson, F. Ozanam, J. Salvador Andresa, S. Sam, I. Solomon, S. Szunerits, Bioelectrochemistry **80**, 17 (2010)
36. S. Sam, L. Touahir, J. Salvador Andresa, P. Allongue, J.-N. Chazalviel, A.C. Gouget-Laemmel, C. Henry de Villeneuve, A. Moraillon, F. Ozanam, N. Gabouze, S. Djebbar, Langmuir **26**, 809 (2010)
37. J.-F. Veyan, D. Aureau, Y. Gogte, P. Campbell, X.-M. Yan, Y.J. Chabal, J. Appl. Phys. **108**, 114913 (2010)

38. Y. Wang, M.-T. Ho, L.V. Goncharova, L.S. Wielunski, S. Rivillon-Amy, Y.J. Chabal, T. Gustafsson, N. Moumen, M. Boleslawski, Chem. Mater. **19**, 3127 (2007)
39. M. Li, M. Dai, Y.J. Chabal, Langmuir **25**, 1911 (2009)
40. M. Dai, Y. Wang, J. Kwon, M.D. Halls, Y.J. Chabal, Nat. Mater. **8**, 825 (2009)
41. J. Kwon, Y.J. Chabal, J. Appl. Phys. **107**, 123505 (2010)
42. D. Aureau, Y. Varin, K. Roodenko, O. Seitz, O. Pluchery, Y.J. Chabal, J. Phys. Chem. C **114**, 14180 (2010)
43. N.A. Lapin, Y.J. Chabal, J. Phys. Chem. B **113**, 8776 (2009)
44. D.J. Michalak, S. Rivillon Amy, D. Aureau, M. Dai, A. Esteve, Y.J. Chabal, Nat. Mater. **9**, 266 (2010)
45. G.A. Ferguson, D. Aureau, Y. Chabal, K. Raghavachari, J. Phys. Chem. C **114**, 17644 (2010)
46. A. Fidélis, F. Ozanam, J.-N. Chazalviel, Surf. Sci. **444**, 7 (2000)
47. D. Knapp, B.S. Brunschwig, N.S. Lewis, J. Phys. Chem. C **115**, 16389 (2011)
48. D. Aureau, F. Ozanam, P. Allongue, J.-N. Chazalviel, Langmuir **24**, 9440 (2008)
49. V. Humblot, C. Methivier, C.-M. Pradier, Langmuir **22**, 3089 (2006)
50. V.P. Tolstoy, I.V. Chernyshova, V.A. Skryshevsky, *Handbook of Infrared Spectroscopy of Ultra-thin Films* (Wiley-VCH, New York, 2003)
51. S. Haq, N. Liu, V. Humblot, A.P.J. Hansen, R. Raval, Nat. Chem. **1**, 409 (2009)
52. R.P. Eischens, S.A. Francis, W.A. Pliskin, J. Phys. Chem. **60**, 194 (1956)
53. R.G. Greenler, J. Chem. Phys. **44**, 310 (1966)
54. R.G. Greenler, J. Chem. Phys. **50**, 1963 (1969)
55. J. Pritchard, M.L. Sims, Trans. Faraday Soc. **66**, 427 (1970)
56. R. Raval, Surf. Sci. **331–333**, 1 (1995)
57. H. Brunner, U. Mayer, H. Hoffmann, Appl. Spectrosc. **51**, 209 (1997)
58. D.T. Blaudez, T. Buffeteau, J.C. Cornut, N. Desbat, B. Escafre, M. Pezolet, J.M. Turlet, Appl. Spectrosc. **47**, 869 (1993)
59. M.A. Ramin, G. Le Bourdon, N. Daugey, B. Bennetau, L. Vellutini, T. Buffeteau, Langmuir **27**, 6076 (2011)
60. B.L. Wang, Spectroscopy **12**, 30 (1997)
61. B.J. Barner, M.J. Green, E.I. Saez, R.M. Corn, Anal. Chem. **63**, 55 (1991)
62. A. Cannilas, E. Pascual, B. Drevillon, Rev. Sci. Instrum. **64**, 2153 (1993)
63. L.J.K. Cross, D.K. Hore, Appl. Opt. **51**, 5100 (2012)
64. R.T. Graf, F. Eng, J.L. Koenig, H. Ishida, Appl. Spectrosc. **40**, 498 (1986)
65. H.G. Tompkins, E.A. Irene, *Handbook of Ellipsometry* (Springer, Heidelberg, 2005)
66. F. Yang, K. Roodenko, R. Hunger, K. Hinrichs, K. Rademann, J. Rappich, J. Phys. Chem. C **116**, 18684 (2012)
67. F. Yang, R. Hunger, K. Roodenko, K. Hinrichs, K. Rademann, J. Rappich, Langmuir **25**, 9313 (2009)
68. F. Yang, R. Hunger, K. Rademann, J. Rappich, Phys. Status Solidi C **7**, 161 (2010)
69. C.A. Canaria, I.N. Lees, A.W. Wun, G.M. Miskelly, M.J. Sailor, Inorg. Chem. Commun. **5**, 560 (2002)
70. T. Dubois, F. Ozanam, J.-N. Chazalviel, Proc. Electrochem. Soc. **97**, 296 (1997)
71. K. Hinrichs, S.D. Silaghi, C. Cobet, N. Esser, D.R.T. Zahn, Phys. Status Solidi B **242**, 2681 (2005)
72. S. Rivillon, Y.J. Chabal, J. Phys. IV **132**, 195 (2006)
73. M.J. Kong, S.S. Lee, J. Lyubovitsky, S.F. Bent, Chem. Phys. Lett. **263**, 1 (1996)
74. S. Ciampi, T. Bocking, K.A. Kilian, J.B. Harper, J.J. Gooding, Langmuir **24**, 5888 (2008)
75. S. Fellah, A. Amiar, F. Ozanam, J.-N. Chazalviel, J. Vigneron, A. Etcheberry, M. Stchakovsky, J. Phys. Chem. B **111**, 1310 (2007)
76. F. Tao, Z.H. Wang, Y.H. Lai, G.Q. Xu, J. Am. Chem. Soc. **125**, 6687 (2003)
77. S. Fleischmann, K. Hinrichs, U. Oertel, S. Reichelt, K.-J. Eichhorn, B. Voit, Macromol. Rapid Commun. **29**, 1177 (2008)
78. A. Teyssot, A. Fidélis, S. Fellah, F. Ozanam, J.-N. Chazalviel, Electrochim. Acta **47**, 2565 (2002)

79. K. Hinrichs, M. Levichkova, D. Wynands, K. Walzer, K.-J. Eichhorn, P. Bäuerle, K. Leo, M. Riede, Appl. Spectrosc. **64**, 1022 (2010)
80. R. Lovrincic, J. Trollmann, C. Pölking, J. Schöneboom, C. Lennartz, A. Pucci, J. Phys. Chem. C **116**, 5757 (2012)
81. D.M. Rosu, J.C. Jones, J.W.P. Hsu, K.L. Kavanagh, D. Tsankov, U. Schade, N. Esser, K. Hinrichs, Langmuir **25**, 919 (2009)
82. K. Hinrichs, A. Röseler, K. Roodenko, J. Rappich, Appl. Spectrosc. **62**, 121 (2008)
83. P. Thissen, T. Peixoto, R.C. Longo, W. Peng, W.G. Schmidt, K. Cho, Y.J. Chabal, J. Am. Chem. Soc. **134**, 8869 (2012)

# Chapter 22
# Brilliant Infrared Light Sources for Micro-ellipsometric Studies of Organic Thin Films

**Michael Gensch**

**Abstract** Micro-ellipsometric studies in the infrared spectral range are of increasing interest in particular for the determination of the optical constants of organic films and multilayers as in these cases the composition, thickness or roughness often vary on micro- and mesoscopic length scales. In cases where the aforementioned properties change across the probed spot, the degree of polarization of the reflected beam is deteriorated and sophisticated models have to be employed to derive the optical constants or other parameters from the determined ellipsometric angles. The achievable spot size in an ellipsometric set-up is now limited by the necessity of performing a specular reflectance measurement with a reasonably defined angle. In the optimal case the infrared radiation can be focused to near diffraction limited spot sizes with opening angles in the incoming beam of less than 7°. In other words such an experiment turns out to be limited by a source property that is typically called *brilliance* or *brightness* and makes the technique particularly suited for the use of accelerator based infrared sources such as 3rd generation synchrotron storage rings. The current status of such activities will be reviewed on the example of different pilot experiments. An outlook on future developments will also be given.

## 22.1  Introduction

Infrared spectroscopic ellipsometry (IRSE) has seen enormous progress in terms of sensitivity and data acquisition times over the past 30 years. While measurements of bulk properties [1] could in the early days easily require data acquisition times of one hour or more, today ultra-thin organic films on the few Angstrom scale can be measured within less than 30 min. Modern day infrared ellipsometers thereby enable routine measurements of optical constants on timescales that even allow

M. Gensch (✉)
Institut für Strahlenphysik/Institut für Ionenstrahlphysik
und Materialforschung, Helmholtz-Zentrum Dresden-Rossendorf,
Bautzner Landstr. 400, 01328 Dresden, Germany
e-mail: m.gensch@hzdr.de

© Springer International Publishing AG, part of Springer Nature 2018
K. Hinrichs and K.-J. Eichhorn (eds.), *Ellipsometry of Functional
Organic Surfaces and Films*, Springer Series in Surface Sciences 52,
https://doi.org/10.1007/978-3-319-75895-4_22

implementation of IRSE as a monitor for surface and interface dynamics e.g. during the deposition or growth of monomolecular films. These advances have been achieved by utilizing the vast progress in detector technology, infrared optical components and FTIR spectrometers over the past 10–15 years. Prototype instruments have been continuously improved and meanwhile even commercialized.

One additional important development was the combination of IRSE with broad band brilliant light sources based on synchrotron radiation from storage ring facilities, which started in the mid 1990s. These accelerator based sources provide broad band infrared radiation generated by relativistic electrons with more than 3 orders of magnitude larger brilliance than available from the conventionally used globar or Hg arc lamps (see [2, 3] and references therein). This exceptionally high brilliance allows combining the sensitivity and spectral resolution of modern day IRSE with a spatial resolution down to length scales of only 10 times the wavelength [4–7]. The Microfocus-mapping ellipsometer at BESSYII thereby extends the range of imaging ellipsometry techniques which previously where only available in the visible spectral range (see [8] and references therein). In the following, a brief overview of the developments of synchrotron based infrared micro-ellipsometry set-ups is given, then different successful applications of the techniques are discussed and finally an outlook on future developments is presented with a special emphasis on the potential of alternative sources of yet even more brilliant broad band infrared radiation.

## 22.2   Infrared Synchrotron Beamlines for Spectroscopic Ellipsometry Applications

One crucial parameter in a micro-focus ellipsometric experiment is the so called brilliance or brightness of the infrared source. The term brilliance or brightness essentially defines the property of a source to have its emitted flux $F$ concentrated by beam optics onto a small area [9]. Fundamentally, it can be shown that the product of the angular divergence of a beam $\Delta\theta$ and the beam size $\Delta x$ is a constant for a source with a given Brilliance and cannot be improved by optics. The exact definition of brilliance $B = (const\, F)/((\Delta\theta_x \Delta x)(\Delta\theta_y \Delta y))$ thereby defines the number of photons that can reach a unit sample area in a unit time and should ideally be as large as possible in a micro-ellipsometric experiment. In infrared spectroscopic ellipsometry the common consensus is that the divergence of the beam focused onto the sample should not exceed a value of $\Delta\theta < 3.5°$ (see e.g. [10]). This fundamental constraint to only moderately focusing optics together with the poor brilliance of conventional table top sources of mid infrared radiation makes infrared ellipsometric measurements of samples/sample areas of only a few $mm^2$ a so called "brilliance-limited" experiment. When using standard broad band infrared sources such as globars or Hg arc lamps ellipsometric measurements require unacceptably long acquisition times or become even unfeasible for small samples/sample areas.

As early as 1982, W. Duncan and G.P. Williams discussed infrared synchrotron radiation from synchrotron storage rings for the first time as broad band infrared sources of exceptional brilliance [11]. Early work focused on the use of the high brilliance for employing infrared synchrotron radiation for RAIRS measurements in gracing incidence of surface layers on metallic surfaces and IR microscopy. But in 1997 the group of Manual Cardona [12] eventually implemented a first real ellipsometric set-up at an infrared synchrotron beamline at the storage ring at Brookhaven National laboratory. This instrument was not aiming at micro-ellipsometric measurements in the mid-infrared but at the investigation of bulk solid samples in the far-infrared or low THz frequency range. It has been extremely successfully used to investigate the far infrared optical properties mainly of various correlated solids and meanwhile a further improved set-up of a similar kind has been taken into operation at the Angstroem Quelle Karlsruhe (ANKA) in Karlsruhe/Germany. A different approach was taken at the Berlin Synchrotron Storage Ring (BESSYII) in Berlin/Storage ring, where a collaboration of institutions from the analytical and bioanalytical chemistry community initiated the development of a micro-focus mapping ellipsometer working in the mid-infrared spectral range and targeting investigations of thin and ultra-thin organic films [4–7]. This set-up shown in Fig. 22.1a allows scanning an area of 50 by 50 mm with lateral resolution of down to 100 μm and a spectral range between 2.5–30 μm (4000–333 cm$^{-1}$/133–10 THz).

Although the brilliance advantage of infrared synchrotron radiation of more than two orders of magnitude is partially diminished by the increased noise due to source instabilities intrinsic to all accelerator based sources [13] the infrared ellipsometric set up at BESSY II provides an enormous improvement of the sensitivity of

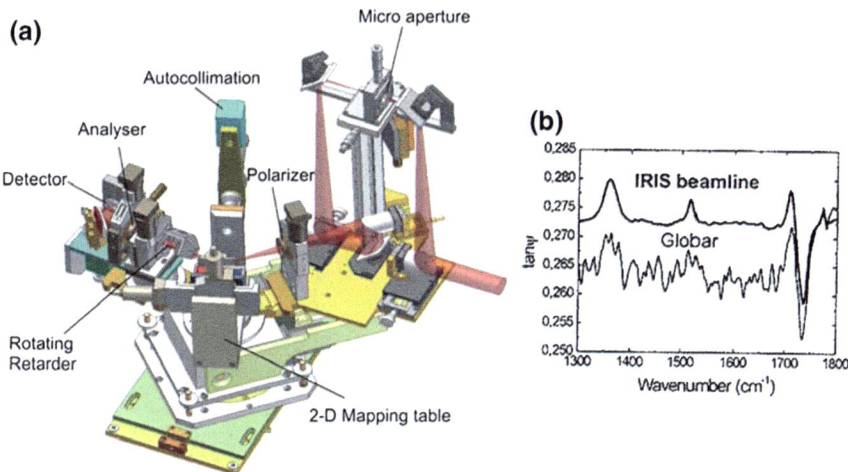

**Fig. 22.1** **a** Schematic of the microfocus mapping ellipsometer at the infrared beamline IRIS at BESSY II. **b** Comparison of the spectra of a 1 mm$^2$ area of a 32 nm polyimide film on silicon taken by a standard globar source and the radiation from the IRIS beamline

ellipsometry for small ultrathin organic film samples/sample areas (see Fig. 22.1b). Since its commissioning in 2003 it has been used in the investigation of various materials but mainly for the investigation of organic thin films. A selection of these investigations is given below.

## 22.3 Applications

In the following, examples of thin organic film samples investigated at the microfocus-mapping ellipsometer at BESSY II over the past 10 years are presented and discussed. It should be noted that the strength of the set up lies in its ability to investigate small samples or small samples areas with monolayer sensitivity in a broad spectral range in the mid-infrared. This allows investigating lateral changes/differences in composition, anisotropy or thickness of the organic film or the organic to inorganic interface.

### 22.3.1 Mesoscopic Samples: Biosensors and Single Flake Graphene

As mentioned above, the micro-ellipsometric set up at BESSY II allows scanning of the infrared optical constants of few Angstroem thick layers for a sample area of several $100 \, \text{mm}^2$ with a spatial resolution in the $100 \, \mu\text{m}$ regime. This turned out to be a good opportunity to investigate the architecture of different sensor designs for on-chip biosensors that consist of a sandwich of ultra-thin organic films on silicon chip. One such mapping measurement is shown in Fig. 22.2.

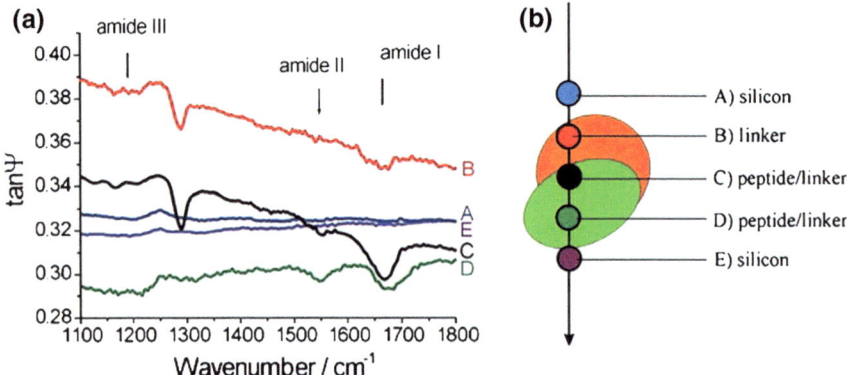

**Fig. 22.2** **a** tan $\Psi$ spectra at different spots of a biosensor and **b** schematic of the sensor geometry. Adapted from [14]

The eventual functionality of the sensor would require that specific target molecules would bind covalently to the top "anchorlayer" and that bonding of target molecules to specific sites could be read out electronically. The "anchorlayer" to the silicon substrate is thereby done via a monolayer of linker molecules. The overall process of the preparation of the sensor involves several complex chemical reactions involving a photo-electrochemical process to attach the linker molecules to the substrate (see [14] for details). As can be seen in Fig. 22.2, microfocus-mapping ellipsometry could readily establish the overall sensor structure and specific fingerprints for the linker molecule and the peptide "anchorlayer" could be identified. This could be used to map out the distribution of peptide and linker molecules across the sensor. The result, shown in Fig. 22.3, let to the conclusion that the film thicknesses within the sensor architecture, in particular that of the linker molecules varied.

This is a clear sign that the photo-electrochemical reaction of the linker molecules with the surface coincides with competing side reactions (e.g. polymerization of the linker-molecules with one another) and that the process needed to be further optimized.

Another example where the microfocus-mapping proved to be very valuable has been recent measurements of the infrared optical constants of single flake graphene. These measurements are an excellent example for the technological importance of micro-focus ellipsometry. Graphene has in the past 5 years turned into one of the most studied materials for its special electronic and optical properties. It can be nowadays prepared in many different ways and lateral dimensions of several 10 cm have been achieved. However, the actual properties and the quality of the graphene sheets vary and the most pristine and pure graphene samples to date are still derived from the exfoliation technique. Typical dimensions of the by these technique prepared small pieces, that are coined "flakes", are of the order of few 100 μm or less. A sample size too small for conventional infrared spectroscopic ellipsometry but ideally suited for microfocus-mapping ellipsometry with the set-up at BESSY II. The graphene flakes investigated in this study were deposited onto a silicon wafer with a thermal oxide layer of 98 nm on top. As a result of the preparation procedure the different single layer graphene flakes are scattered alongside graphite and multi-layer graphene pieces across the surface. In a first step the graphene flakes need to therefore be identified. A standard procedure here is imaging of the surface in the visible spectral range where the graphene sheets give a sufficient contrast due to their extreme optical properties. Afterwards the substrate was mounted into the micro-ellipsometer and the surface was mapped in order to identify the exact location of the graphene flake (see Fig. 22.4).

Finally the focus of the microfocus-mapping ellipsometer was optimized to the size of the graphene flake and the conductivity of the graphene flakes and of the graphite pieces were determined (see Fig. 22.5).

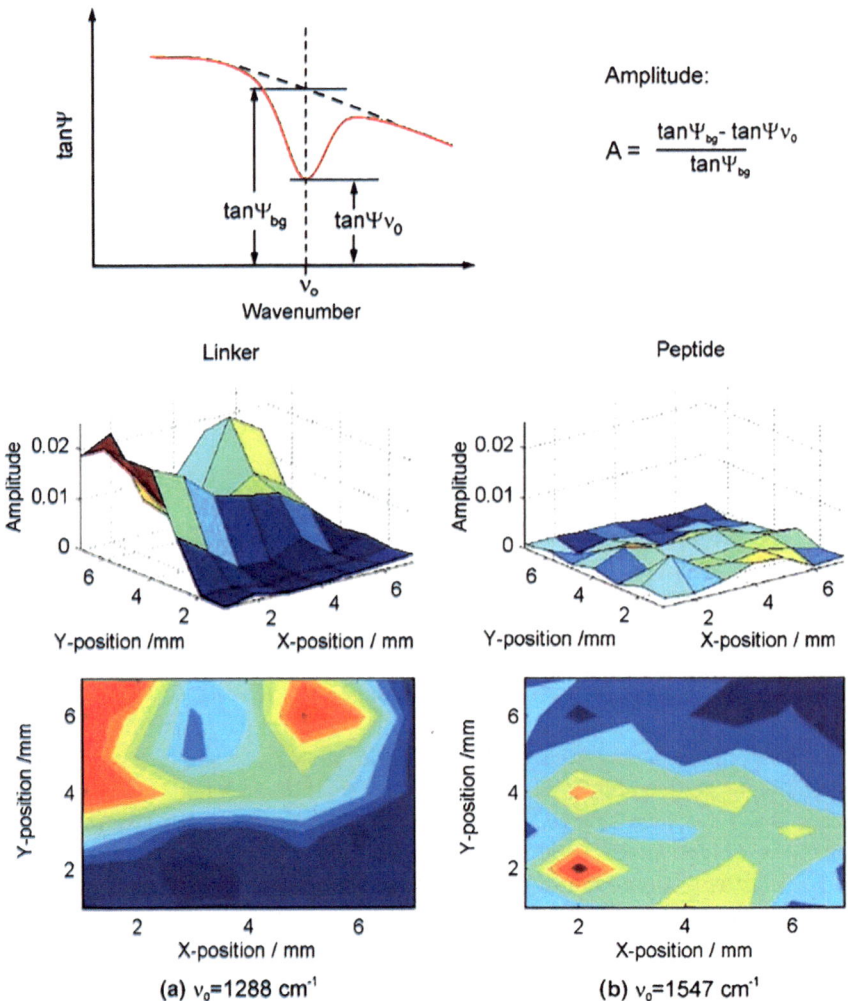

**Fig. 22.3  a** Three-dimensional representation and contour plots of 6 mm × 6 mm maps of the amplitude of the ellipsometric parameter tan $\Psi$ of **b** a linker band at $1288\,cm^{-1}$ and a peptide band at $1547\,cm^{-1}$ of a biosensor sample after it had been incubated partially in a peptide solution. The band amplitudes were determined for every spot as shown in the schematic (*top*). A resolution of $4\,cm^{-1}$ was used and the angle of incidence was $65°$. Adapted from [14]

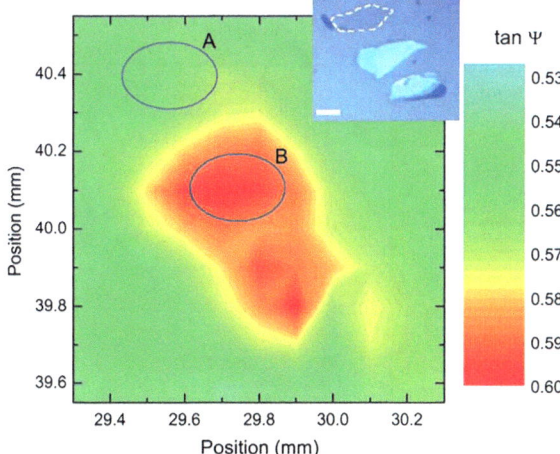

**Fig. 22.4** Map of tan $\Psi$ at 2992 cm$^{-1}$ showing the two graphite flakes and the spot size (80/20 knife edge) and locations of the scans A and B. The *inset* shows a digital photograph of the graphite flakes (*middle* and *bottom*) and the graphene flake (*white outline*) which is visible by eye. Scale bar approximately 200 μm [15]. Reprinted with permission from J.W. Weber, K. Hinrichs, M. Gensch, M.C.M. van de Sanden, T.W.H. Oates, Appl. Phys. Lett. (2011), American Institute of Physics

The optical conductivity of the investigated graphene flake turned out to be very similar to that of bulk gold. A comparison with model calculation allowed postulating a Fermi energy of 2 eV due to chemical doping and accumulated charge due to atmospheric exposure [15].

## 22.3.2   Nano Patterned Films: Polymer Brushes

As a second important application of microfocus-mapping ellipsometry is the determination of film inhomogeneities/lateral changes in ultra-thin organic layers. This shall be discussed in the following on the example of a special class of stimuli responsive polymer films, so called polymer brushes (for details see also Chap. 6). The name "brush" thereby stems from the fact that in this case the polymer chains are grafted to the solid substrates with one end at a sufficiently high density so that the polymer chains adopt a stretched conformation. Polymer brushes can now be prepared as uniform, patterned or gradient films and furthermore can be designed such that they respond to an external stimuli. In the presented example a polyelectrolyte brush made up from polyacrylic acid (PAA) chains of a length of roughly 3 nm was investigated that exhibits a response to the pH of a solution. In brief the PAA brushes show a swelling behavior when exposed to higher pH values. The underlying process

**Fig. 22.5** Real (*top*) and imaginary (*bottom*) parts of the dielectric functions of the graphene flakes (*thick solid black lines*) and graphite flakes (*thin solid red lines*) measured at positions A and B, respectively. Also shown are the measured curves for gold (*green line*). The inset shows a zoomed region comparing the thin graphite and bulk HOPG (*dashed blue line*) curves. Reprinted with permission from J.W. Weber, K. Hinrichs, M. Gensch, M.C.M. van de Sanden, T.W.H. Oates, Appl. Phys. Lett. (2011), American Institute of Physics

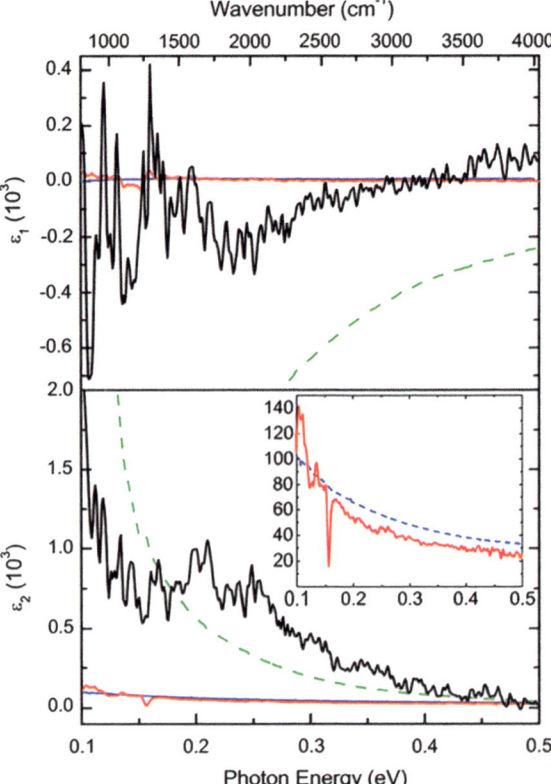

here is a dissociation of the carboxylic groups that leads to conformational changes due to osmotic pressure and repulsive forces between the fully ionized carboxylic groups (for details see [7] and references therein). In the presented example one part of the PAA brush was immersed in a KOH solution with a pH of 10. Thereafter micro-ellipsometric measurements were taken at different points across the generated chemical phase boundary.

It can clearly be seen in Fig. 22.6 that the ellipsometric measurement is sensitive enough to detect the chemically different states in the ultrathin films. Meanwhile, the ellipsometric set up that allows investigating such minute changes in the infrared optical properties of ultra-thin films on surfaces also allows investigating liquid-solid interfaces as can be seen in Fig. 22.7 [16]. This now opens up the opportunity to study surface chemical reactions in solution with a spatial resolution of few hundred micrometers.

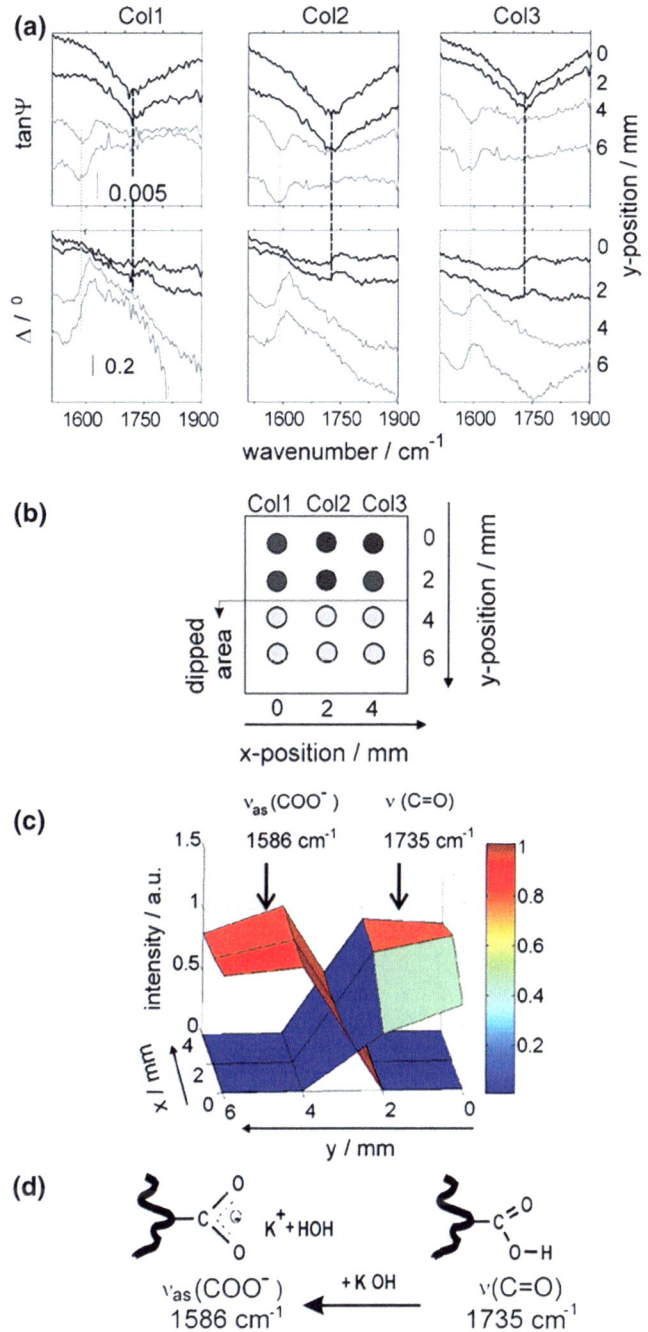

**Fig. 22.6** (Continued)

◄ **Fig. 22.6** **a** Ellipsometric parameters tan Ψ and Δ of a 3 nm thin polyacrylic acid brush film during the switching experiment obtained with 2 mm steps at 69° angle of incidence. *Black*: spectra from the nondipped area; *grey*: spectra from the dipped area of the sample. **b** A schematic drawing (top view) of the sample and the performed scan. **c** A mapping chart of the integrated tan Ψ IR band intensities from panel (**a**) normalized to the highest signal of each band. **d** Chemical reaction and peak assignments upon the PAA brush treatment with KOH solution (pH 10). Abbreviation "Col" refers to the scanned column [7]. Reprinted with permission from K. Roodenko, Y. Mikhaylova, L. Ionov, M. Gensch, S. Minko, U. Schade, K.J. Eichhorn, N. Esser, K. Hinrichs, Appl. Phys. Lett. **92**, 103102 (2008). American Institute of Physics

**Fig. 22.7** Four point "map" of a PAA-b-PS/PEG brush sample. Spectra show the ratio of spectra at pH 10 and pH 2. Increasing amplitudes of COO– bands and the decreasing amplitude of the COOH band are highlighted [16]. Reprinted with permission from D. Aulich et al., Phys. Status Solidi C **7**, 197 (2010). Copyright 2010, Wiley-VCH Verlag GmbH & Co. KGaA, Weinheim

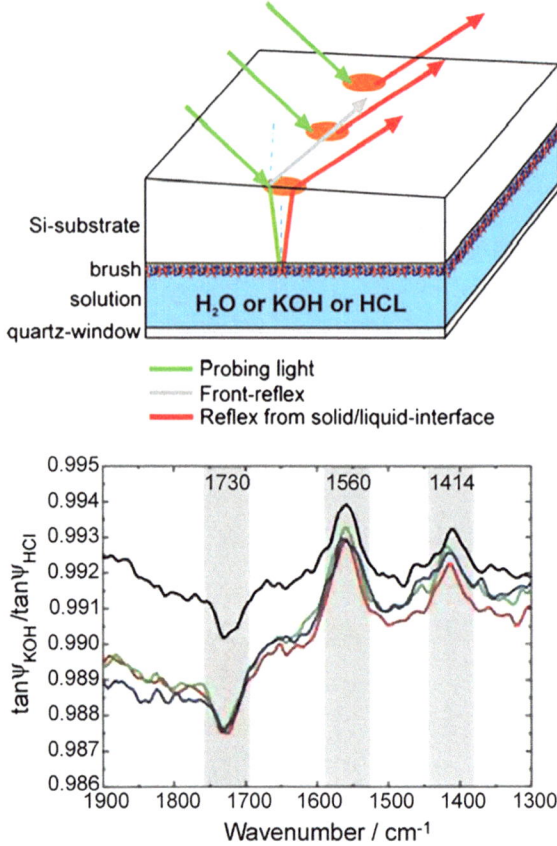

## 22.4 Summary and Outlook

Infrared spectroscopic ellipsometry has become an established technique at infrared synchrotron beamlines. The opportunities for the investigations of ultrathin organic films lie in the capability to determine minute changes in the infrared optical properties of ultra-thin films simultaneously in a wide spectral range and in sophisticated sample geometries. Since ellipsometric set-ups at multi user facilities require a relatively high degree of expertise for recalibrating and maintaining the instruments, permanent installations are typically initiated and run by expert user groups rather than the support groups of the storage ring facilities itself. Presently three instruments are installed permanently at infrared synchrotron beamlines at Brookhaven National Lab [17], ANKA [18] and BESSYII [2]. Of these only one focuses primarily on organic thin film samples [6]. It should be noted that infrared synchrotron beam-lines are not the primary focus of 3rd Generation synchrotron storage rings. These large scale facilities are specially developed for X-ray applications, and many of the current technological developments and upgrades to improve the X-ray beam qual-ity at these storage rings are not favorable for infrared spectroscopic applications. Without going into details here, it can in general be said the infrared synchrotron beamlines and with them the infrared ellipsometric set-ups are best suited for lower energy, higher beam current storage ring facilities. The most recently opened facili-ties aim for a further enhanced brilliance in the X-ray spectral range and are hence high energy/lower current storage ring facilities. From this point of view it is arguable whether further ellipsometric set-ups at 3rd generation storage ring facilities will be installed in the near future. At the existing instruments it would be of interest to make use of the intrinsic time structure of the infrared synchrotron radiation. Recent work shows that this can in principle be done [13]. At the microfocus-ellipsometer at BESSY II one could achieve a time resolution of roughly 50–70 ps (given by the duration of the emitted infrared pulses). Furthermore there are a number of new unique types of infrared radiation sources, which have been established over the past 10 years and that may open new avenues for micro-ellipsometric measurements in the mid infrared. A selection of those is briefly discussed below. Note that the focus lies on broad band infrared sources that are suited for spectroscopic ellipsometry, while narrow bandwidth sources such as infrared gas lasers or free electron lasers are omitted.

### 22.4.1   Femtosecond-Laser Based Broad Band Sources

Recently so called time-domain THz/IR spectroscopy approaches have become avail-able as a consequence of the development of robustly working femtosecond laser systems (for details see e.g. [19]). In principle such a time-domain measurement provides phase as well as amplitude information. Combined with polarization sen-sitive detection several proof of principle measurements of the ellipsometric angles from the lower THz to mid-infrared have recently been shown [20–22]. The main application for these time-domain ellipsometric approaches should be seen in the possibility of determining changes in the THz/infrared spectral range on femtosec-ond timescales. The presently achievable signal-to-noise ratios are not competitive with that of static FTIR based measurements and do not presently allow the inves-tigation of ultrathin films. However, the generation process by difference frequency mixing has an inherently better brilliance than globar sources and the future will show whether improvement in laser technology will make these sources suitably

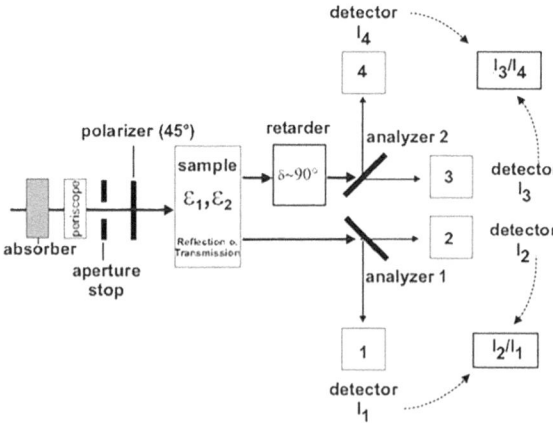

**Fig. 22.8** Sketch of the division of amplitude polarimeter (DOAP) approach developed at the THz FEL FELBE that is insensitive to power fluctuations due to calculation of the ellipsometric parameters tan $\Psi$ and $\Delta$ from 2 simultaneously measured intensity ratios (for details see [29]). Reprinted with permission from M. Gensch et al., *Joint 31st International Conference on Infrared and Millimeter Waves and 14th International Conference on Terahertz Electronics 2006, IRMMW-THz 2006*, 18–22 Sept. 2006. Copyright 2006, IEEE

stable and robust for micro-ellipsometric investigations of ultra-thin organic films. It should be noted that a time-domain micro-ellipsometry measurement intrinsically allows achieving a Fourier limited time resolution given by the IR pulse duration in the few 10 fs range.

### 22.4.2   Fourth Generation Broad Band IR/THz Light Sources

The next generation of accelerator-based IR/THz sources, which is currently under development, benefits from recent progress in accelerator technology that allows compressing electron bunches into the few 100 to even few femtosecond regime [23–27]. Thereby these electron bunches emit synchrotron radiation in the THz and even mid-infrared spectral range coherently, broad band and at unprecedented peak and average power. Given the superior brilliance of these sources and coupled with FT-IR spectrometers, microfocus-ellipsometric measurements may benefit tremendously. Similar to the case of the femtosecond-laser based approach, ellipsometry at a Fourth Generation IR/THz lightsource is intrinsically suited for time resolved measurements in the femtosecond—picosecond range (e.g. [28]). However as for the case of the femto-second laser based approaches the robustness and stability of the fourth generation IR/THz lightsources has yet to be proven and/or the detection schemes may have to be adapted.

Single-shot ellipsometry approaches (see also Fig. 22.8), making use of polarizing beamsplitters that have recently been tested with Quantum Cascade Laser (QCL) lasers and at THz free electron lasers could turn out to be crucial developments (see e.g. [29, 30]).

**Acknowledgements** Critical reading and fruitful discussions by and with K. Hinrichs (ISAS), T. Kampfrath (FHI) and G.P. Williams (Thomas Jefferson Laboratory) is gratefully acknowledged. B. Green (HZDR) is acknowledged for proofreading the manuscript.

# References

1. A. Roeseler, *Infrared Spectroscopic Ellipsometry* (Akademie Verlag, Berlin, 1990)
2. U. Schade, A. Roeseler, E.H. Korte, F. Bartl, K.P. Hofmann, T. Noll, W.B. Peatman, Rev- Sci. Instr. **73**, 1568 (2002)
3. E.J. Singley, M. Abo-Bakr, D.N. Basov, J. Feikes, P. Guptasarma, K. Holldack, H.W. Hubers, P. Kuske, M.C. Martin, W.B. Peatmann, U. Schade, G. Wustefeld, Phys. Rev. B **69**, 092512 (2004)
4. M. Gensch, K. Hinrichs, A. Roeseler, E.H. Korte, U. Schade, Anal. Bioanal. Chem. **376**, 621 (2003)
5. K. Hinrichs, M. Gensch, A. Roeseler, E.H. Korte, K. Sahre, K.J. Eichhorn, N. Esser, U. Schade, Appl. Spectrosc. **57**, 1200 (2003)
6. M. Gensch, E.H. Korte, N. Esser, U. Schade, K. Hinrichs, Infrared Phys. Technol. **49**, 74 (2006)
7. K. Roodenko, Y. Mikhaylova, L. Ionov, M. Gensch, S. Minko, U. Schade, K.J. Eichhorn, N. Esser, K. Hinrichs, Appl. Phys. Lett. **92**, 103102 (2008)
8. G. Jin, R. Jansson, H. Arwin, Rev. Sci. Instr. **67**, 2930 (1996)
9. K.J. Kim, Nucl. Instr. Meth. A **246**, 71 (1986)
10. K. Hinrichs, M. Gensch, N. Esser, Appl. Spectrosc. **59**, 272A (2005)
11. W.D. Duncan, G.P. William, Appl. Opt. **22**, 2914 (1983)
12. J. Kirchner, R. Henn, M. Cardona, P.L. Richards, G.P. Williams, J. Opt. Soc. Am. B **14**, 705 (1997)
13. L. Carroll, P. Friedli, Ph Lerch, J. Schneider, D. Treyer, S. Hunziker, S. Stutz, H. Sigg, Rev. Sci. Instr. **82**, 063101 (2011)
14. K. Hinrichs, M. Gensch, N. Esser, U. Schade, J. Rappich, S. Kröning, M. Portwich, R. Volkmer, Anal. Bioanal. Chem. **387**, 1823 (2007)
15. J.W. Weber, K. Hinrichs, M. Gensch, M.C.M. van de Sanden, T.W.H. Oates, Appl. Phys. Lett. **99**, 061909 (2011)
16. D. Aulich, O. Hoy, I. Luzinov, K.-J. Eichhorn, M. Stamm, M. Gensch, U. Schade, N. Esser, K. Hinrichs, Phys. Stat. Sol. C **7**, 197 (2010)
17. G.P. Williams, C.J. Hirschmugl, E.M. Kneedler, Rev. Sci. Instr. **60**, 2176 (1989)
18. Y.L. Mathis, B. Gasharova, D. Moss, J. Biol. Phys. **29**, 313 (2003)
19. U. Jepsen, D. Cooke, M. Koch, Laser Photonics Rev. **5**, 124 (2011)
20. T. Nagashima, M. Hangyo, Appl. Phys. Lett. **79**, 3917 (2001)
21. N. Matsumoto, T. Hosokura, T. Nagashima, M. Hangyo, Opt. Lett. **36**, 265 (2011)
22. A. Rubaro, L. Braun, M. Wolf, T. Kampfrath, Appl. Phys. Lett. **101**, 081103 (2012)
23. G.L. Carr et al., Nature **420**, 153 (2002)
24. M. Gensch et al., Infrared Phys. Technol. **51**, 423 (2008)
25. A.-S. Mueller, T. Baumbach, S. Casalbuoni, M. Hagelstein, E. Huttel, Y.-L. Mathis, D.A. Moss, A. Plech, R. Rossmanith, E. Bruendermann, M. Havenith, K.G. Sonnad, PAC Proceedings (2009)
26. F. Tavella, N. Stojanovic, G. Geloni, M. Gensch, Nat. Photon. **6**, 162 (2011)
27. G.P. Williams, Rep. Prog. Phys. **69**, 301 (2006)

28. M. Foerst, M.C. Hoffmann, S. Kaiser, A. Dienst, M. Rini, R.I. Tobey, M. Gensch, C. Manzoni, A. Cavalleri, THz control in correlated electron solids: sources and applications, in *Terahertz Spectroscopy and Imaging*, ed. by K.-E. Peiponen et al. Springer Series in Optical Sciences, vol. 171 (Springer, Berlin, 2012). https://doi.org/10.1007/978-3-642-29564-5_23

29. M. Gensch, J.S. Lee, K. Hinrichs, N. Esser, W. Seidel, A. Roeseler, U. Schade, in *Joint 31st International Conference on Infrared and Millimeter Waves and 14th International Conference on Terahertz Electronics 2006, IRMMW-THz 2006*, 18–22 Sept. 2006, p. 416

30. A. Furchner, C. Kratz, D. Gkogkou, H. Ketelsen, K. Hinrichs, Appl. Surf. Sci. **421**, 440 (2017)

# Part VII
# Optical Constants

# Chapter 23
# Common Polymers and Proteins

**Andreas Furchner and Dennis Aulich**

**Abstract** The optical constants $n$ and $k$ are important material properties necessary for interpreting ellipsometric spectra of novel, complex, microstructured, or mixed-material systems. This appendix provides thin-film optical constants of commonly used polymers and proteins.

Interpretation of ellipsometric measurements on novel materials requires a strong correlation between an optical model and the sample's material properties. When new materials, such as microstructured organic films or mixtures made from different materials, are analyzed, it is helpful—and often necessary—to have at hand ellipsometric thin-film data of the initial components. This appendix is intended to support the reader with the optical constants $n$ and $k$ of some widespread organic materials in thin films between several nm and µm. The focus of this chapter lies on materials used for functional films like responsive polymer brushes, functional coatings, etc. Additionally, optical constants of some thin protein films are given.

The following list comprises the publications in which the original graphs, data, and pictures presented in this chapter can be found:

1. L. Ionov, A. Sidorenko, K.-J. Eichhorn, M. Stamm, S. Minko, K. Hinrichs, Langmuir **21**, 8711 (2005).
2. A. Furchner, E. Bittrich, P. Uhlmann, K.-J. Eichhorn, K. Hinrichs, Thin Solid Films **541**, 41 (2013).
3. K. Hinrichs, K.-J. Eichhorn, Spectrosc. Eur. **19**(6), 11 (2007).
4. B.D. Vogt, S. Kang, V.M. Prabhu, E.K. Lin, S.K. Satija, K. Turnquest, W. Wu, Macromolecules **39**, 8311 (2006)
5. S. Kang, V.M. Prabhu, C.L. Soles, E.K. Lin, and W. Wu, Macromolecules **42**, 5296 (2009).

A. Furchner (✉) · D. Aulich
Leibniz-Institut für Analytische Wissenschaften – ISAS – e.V.,
Schwarzschildstraße 8, 12489 Berlin, Germany
e-mail: andreas.furchner@isas.de

© Springer International Publishing AG, part of Springer Nature 2018
K. Hinrichs and K.-J. Eichhorn (eds.), *Ellipsometry of Functional Organic Surfaces and Films*, Springer Series in Surface Sciences 52,
https://doi.org/10.1007/978-3-319-75895-4_23

6. H.G. Tompkins, T. Tiwald, C. Bungay, A.E. Hooper, J. Phys. Chem. B **108**, 3777 (2004), ©2004, American Chemical Society.

7. C.L. Bungay, T.E. Tiwald, D.W. Thompson, M.J. DeVries, J.A. Woollam, J.F. Elman, Thin Solid Films **313–314**, 713 (1998), ©1998, Elsevier.

8. L. Yan, X. Gao, C. Bungay, J.A. Woollam, J. Vac. Sci. Technol. A **19**(2), 447 (2001), ©2001, American Vacuum Society.

9. D. Blaudez, F. Boucher, T. Buffeteau, B. Desbat, M. Grandbois, C. Salesse, Appl. Spectrosc. **53**(10), 1299 (1999), ©1999, Society for Applied Spectroscopy.

10. H. Arwin, Thin Solid Films **519**, 2589 (2011), ©2011, Elsevier.

## PBA, PS, and P2VP [poly(*tert*-butyl acrylate), poly styrene, and poly(2-vinyl-pyridine)]

**Fig. 23.1** Optical constants *n* (*top*) and *k* (*bottom*) of several 10 nm thick spincoated PS, PBA, and P2VP films on gold-coated glass substrates. Data from L. Ionov, A. Sidorenko, K.-J. Eichhorn, M. Stamm, S. Minko, K. Hinrichs, Langmuir **21**, 8711 (2005)

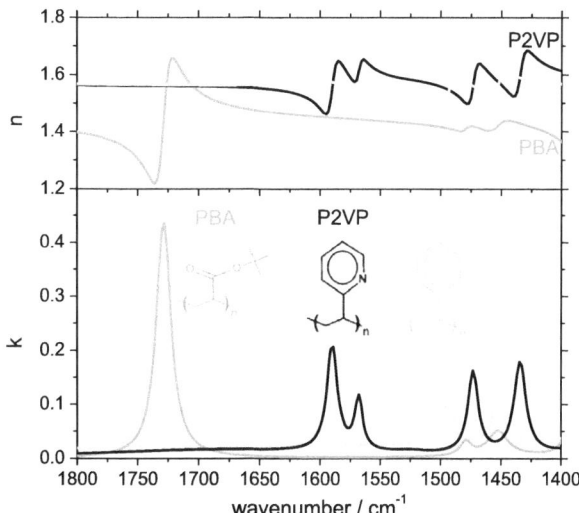

## PGMA [poly(glycidyl methacrylate)]

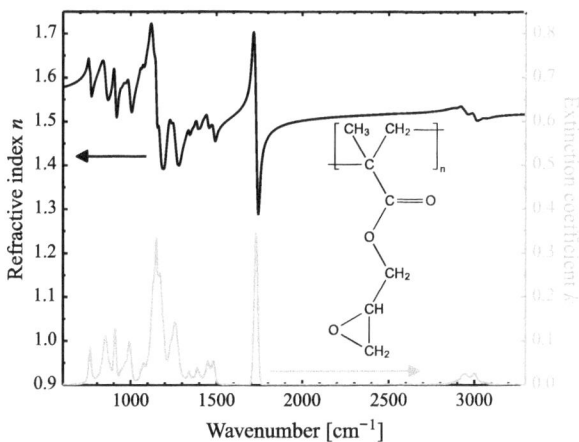

**Fig. 23.2** Optical constants *n* (*black*) and *k* (*gray*) of a 105 nm thick PGMA film on gold-coated glass substrate. PGMA (molecular weight $\overline{M}_n = 17\,500$ g/mol, $\overline{M}_w/\overline{M}_n = 1.70$) was deposited by spincoating 0.02 wt% PGMA in CHCl$_3$ solution with subsequent annealing at 100 °C in a vacuum oven for 20 min. Data from A. Furchner, E. Bittrich, P. Uhlmann, K.-J. Eichhorn, K. Hinrichs, Thin Solid Films **541**, 41 (2013)

## PNIPAAm [poly($N$-isopropyl acrylamide)]

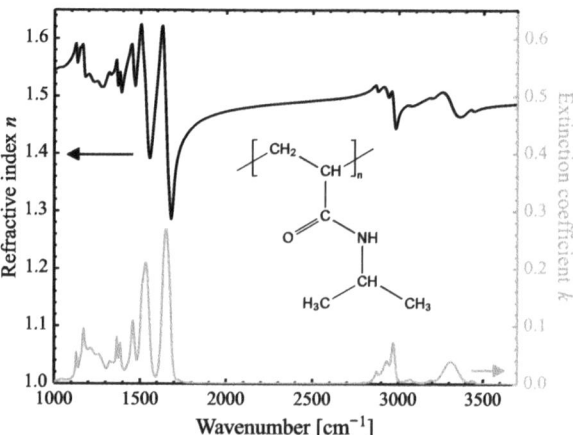

**Fig. 23.3** Optical constants $n$ (*black*) and $k$ (*gray*) of a 132 nm thick PNIPAAm film on gold-coated glass substrate. PNIPAAm (molecular weight $\overline{M}_n = 94\,000$ g/mol, $\overline{M}_w/\overline{M}_n = 1.27$) was deposited by spincoating a PNIPAAm solution in tetrahydrofuran (1 wt%) with subsequent rinsing with ethanol. Data from A. Furchner, E. Bittrich, P. Uhlmann, K.-J. Eichhorn, K. Hinrichs, Thin Solid Films **541**, 41 (2013)

## PnBMA and PVC [poly($n$-butyl methacrylate) and polyvinyl chloride]

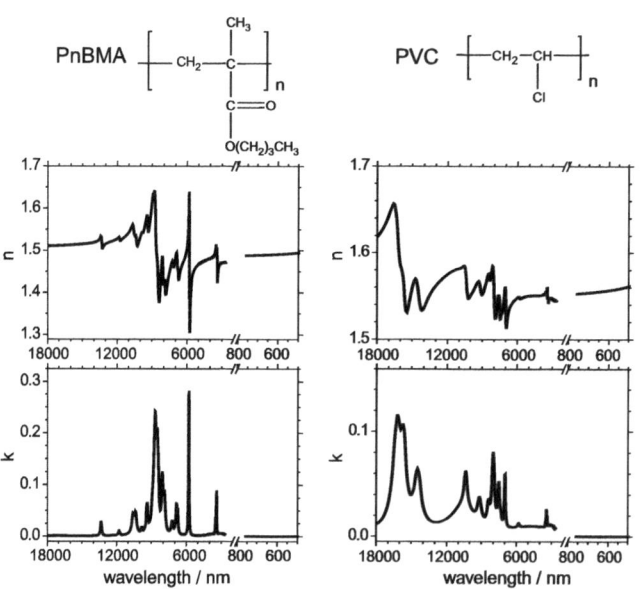

**Fig. 23.4** Optical constants $n$ (*top*) and $k$ (*bottom*) of 100 nm thick PnBMA and PVC layers on gold-coated glass substrates. Data from K. Hinrichs, K.-J. Eichhorn, Spectrosc. Eur. **19**(6), 11 (2007)

## PHOSt and PMAdMA [Poly(4-hydroxystyrene) and poly(methyladamantyl methacrylate)]

**Fig. 23.5** Chemical structures of PHOSt and PMAdMA. These polymers were investigated for their optical properties by Kang et al. (see Fig. 23.6). Partly adapted with kind permission from B.D. Vogt, S. Kang, V.M. Prabhu, E.K. Lin, S.K. Satija, K. Turnquest, W. Wu, Macromolecules **39**, 8311 (2006), ©2006 American Chemical Society

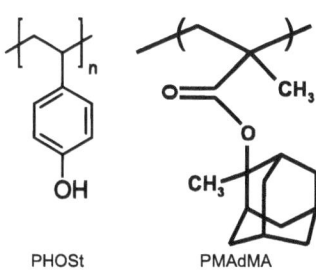

PHOSt          PMAdMA

**Fig. 23.6** Refractive indices *n* and extinction coefficients *k* of PHOSt (*dashed lines*) and PMAdMA (*solid lines*), determined from multiple spincoated sub-100 nm thin films on silicon and gold substrates. Reprinted with kind permission from S. Kang, V.M. Prabhu, C.L. Soles, E.K. Lin, and W. Wu, Macromolecules **42**, 5296 (2009), ©2006 American Chemical Society

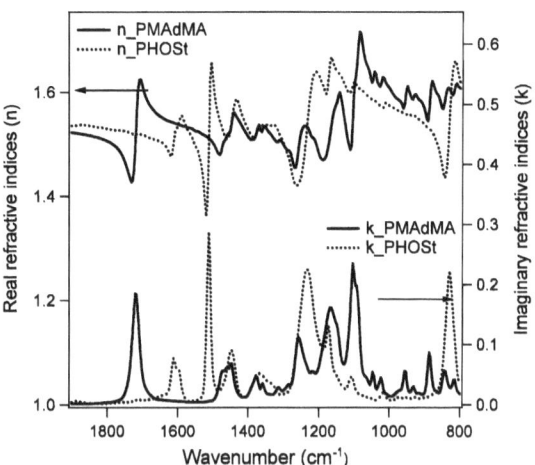

## Nylon 6 [polyamide 6]

**Fig. 23.7** Optical constants *n* (*solid line*) and *k* (*dashed line*) for nylon 6 on glass. The values were determined by the analysis of films with thicknesses ranging from 5 to 100 nm. Adapted with kind permission from H.G. Tompkins, T. Tiwald, C. Bungay, A.E. Hooper, J. Phys. Chem. B **108**, 3777 (2004), ©2004, American Chemical Society

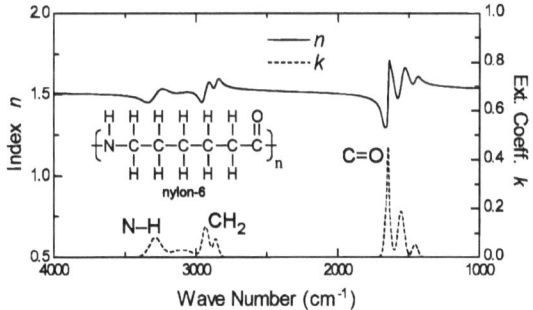

## PI (Polyimide): PI2611 [poly(biphenyl dianhydride-p-phenylenediamine)]

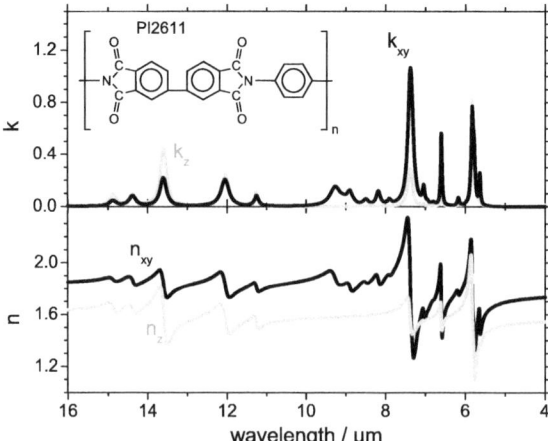

**Fig. 23.8** Anisotropic optical constants $k_{xy}$, $n_{xy}$ (in-plane) and $k_z$, $n_z$ (out-of-plane) of a 1.81 µm thick spincoated polyimide layer (PI2611 [BPDA-PPD, poly(biphenyl dianhydride-p-phenylenediamine)]) on silicon substrate. Data from K. Hinrichs, K.-J. Eichhorn, Spectrosc. Eur. **19**(6), 11 (2007)

## Silicone CV-1144-O [poly(dimethyl-co-diphenyl siloxane)]

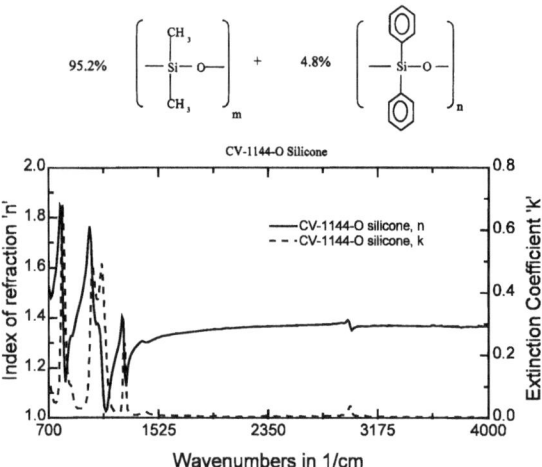

**Fig. 23.9** Infrared optical constants of CV-1144-O silicone from spincoated 100 nm thick silicone films on optically thick gold films on silicon substrates after 7-days curing. With kind permission from C.L. Bungay, T.E. Tiwald, D.W. Thompson, M.J. DeVries, J.A. Woollam, J.F. Elman, Thin Solid Films **313–314**, 713–717 (1998), ©1998, Elsevier, and from L. Yan, X. Gao, C. Bungay, J.A. Woollam, J. Vac. Sci. Technol. A **19**(2), 447–454 (2001), ©2001, American Vacuum Society

## Bacteriorhodopsin

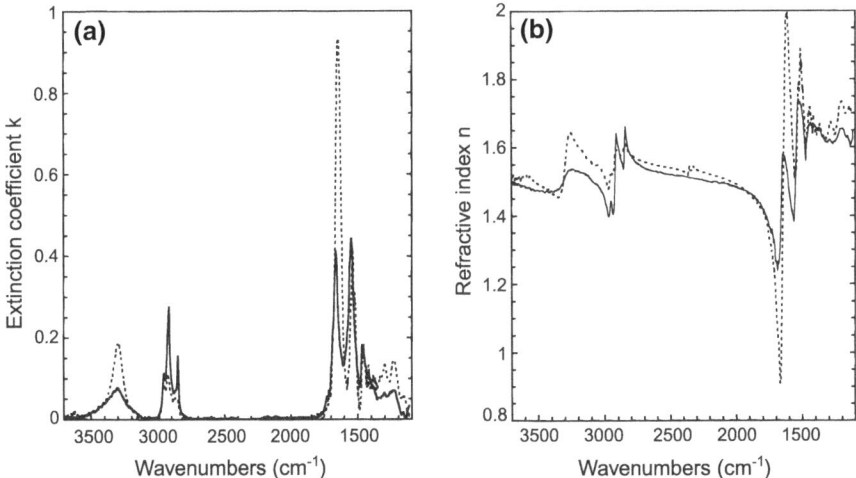

**Fig. 23.10**  In-plane (*solid lines*) and out-of-plane (*dashed lines*) extinction coefficients *k* (*left*) and refractive indices *n* (*right*) of a bacteriorhodopsin monolayer deposited by Langmuir–Blodgett film preparation on gold-coated mirrors and on CaF$_2$ plates.  With kind permission from D. Blaudez, F. Boucher, T. Buffeteau, B. Desbat, M. Grandbois, C. Salesse, Appl. Spectrosc. **53**(10), 1299 (1999), ©1999, Society for Applied Spectroscopy

## Fibrinogen

**Fig. 23.11**  Refractive index *n* (*solid line*) and extinction coefficient *k* (*dashed line*) of a 4.54 nm thick Fibrinogen layer on gold substrate.  With kind permission from H. Arwin, Thin Solid Films **519**, 2589–2592 (2011), ©2011, Elsevier

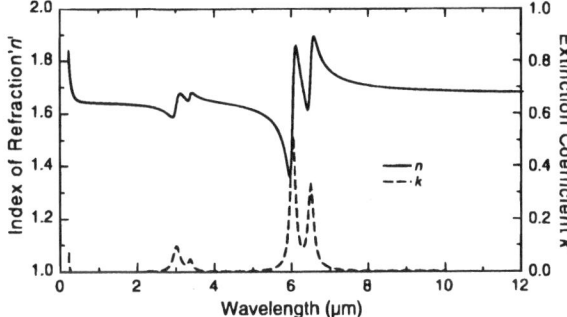

# Chapter 24
# Organic Materials for Optoelectronic Applications

**Andreas Furchner and Dennis Aulich**

**Abstract** Material characterization in the wide field of optoelectronics often involves ellipsometric measurements. A deep knowledge of the sample properties, including anisotropy, is required for successful material analysis via optical modeling. This appendix provides thin-film optical constants $n$ and $k$ for commonly used organic materials used for transparent electrodes, solar cells, etc.

This chapter gives an overview on the optical constants $n$ and $k$ of materials with relevance for technical applications, for example, transparent electrodes, solar cells, and optoelectronics in general. The overview focuses on basic, mostly mono-component materials with thickness-*independent* optical constants. For the optical properties of blend materials, that is, mixtures of two or more components at different volume fractions, and for materials with thickness-*dependent* optical properties, the reader is referred to the wide literature on these subjects. An example for blend films are the two organic-photovoltaics materials PCBM and P3HT studied by S. Engmann et al., Adv. Energy Mater. **1**, 684 (2011). Polyaniline is an example for strongly thickness-dependent optical constants, see H.A. Al-Attar et al., Thin Solid Films **429**, 286 (2003). It becomes apparent that the ellipsometric determination of organic-thin-film optical constants can benefit from input parameters obtained from theoretical calculations by means of density functional theory, see, for example, R. Lovrinčić et al., J. Phys. Chem. C **116**, 5757 (2012).

The following list comprises the publications in which the original graphs, data, and pictures presented in this chapter can be found:

1. D. Wynands, M. Erber, R. Rentenberger, M. Levichkova, K. Walzer, K.-J. Eichhorn, M. Stamm, Org. Electron. **13**, 885 (2012), ©2012, Elsevier.

A. Furchner (✉) · D. Aulich
Leibniz-Institut für Analytische Wissenschaften – ISAS – e.V.,
Schwarzschildstraße 8, 12489 Berlin, Germany
e-mail: andreas.furchner@isas.de

© Springer International Publishing AG, part of Springer Nature 2018
K. Hinrichs and K.-J. Eichhorn (eds.), *Ellipsometry of Functional Organic Surfaces and Films*, Springer Series in Surface Sciences 52,
https://doi.org/10.1007/978-3-319-75895-4_24

2. M. Losurdo, M.M. Giangregorio, P. Capezzuto, G. Bruno, F. Babudri, D. Colangiuli, G.M. Farinola, F. Naso, Macromolecules **36**, 4492 (2003), ©2003, American Chemical Society.
3. L.A.A. Petterson, S. Ghosh, O. Inganäs, Org. Electron. **3**, 143 (2002), ©2002, with permission from Elsevier.
4. M. Campoy-Quiles, P.G. Etchegoin, D.D.C. Bradley, Synth. Met. **155**, 279 (2005), ©2005, Elsevier B.V.
5. C.M. Ramsdale, N.C. Greenham, J. Phys. D, Appl. Phys. **36**, L29 (2003), ©2003 IOP Publishing Ltd.
6. H.-W. Lin, C.-L. Lin, H.-H. Chuang, Y.-T. Lin, C.-C. Wu et al., J. Appl. Phys. **95**(3), 881 (2004), ©2004, American Institute of Physics.
7. S.D. Silaghi, T. Spehr, C. Cobet, T.P.I. Saragi, C. Werner, J. Salbeck, N. Esser, J. Appl. Phys. **103**, 043503 (2008).
8. S. Pop, P. Kate, N. Esser, K. Hinrichs, Tunable optical constants of thermally grown thin porphyrin films on silicon for photovoltaic applications, in preparation (2013).
9. K. Hinrichs, S.D. Silaghi, C. Cobet, N. Esser, D.R.T. Zahn, Phys. Status Solidi B **242**(13), 2681 (2005).
10. NIM_NIL project, funded by the European Community's 7th Framework Programme under grant agreement no 228637, in cooperation between Profactor GmbH, SENTECH Instruments GmbH, and Micro Resist Technology. To be published.

## DCV6T [α,ω-bis-dicyanovinylene-sexithiophene derivative DCV6T-Bu(1,2,5,6)]

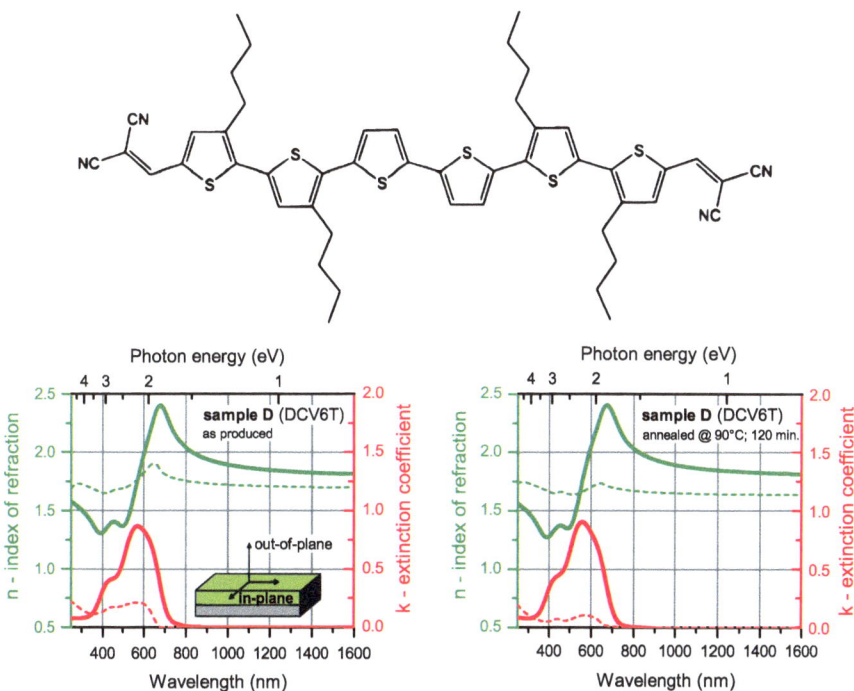

**Fig. 24.1** Anisotropic optical constants $n$ (*green*) and $k$ (*red*) of a 116 nm thick DCV6T layer deposited on an interference-enhanced silicon substrate with 964 nm SiO$_x$, before annealing (*left*) and after annealing (*right*) at 90 °C for 120 min. The DCV6T thin film was prepared by thermal vacuum-deposition in a vacuum chamber with a base pressure of about $10^{-8}$ mbar. Reprinted, with permission from Elsevier, from D. Wynands, M. Erber, R. Rentenberger, M. Levichkova, K. Walzer, K.-J. Eichhorn, M. Stamm, Spectroscopic ellipsometry characterization of vacuum-deposited organic films for the application in organic solar cells, Org. Electron. **13**, 885–893 (2012). ©2012, Elsevier

## PArPs [Poly(arylenephenylene) polymers]

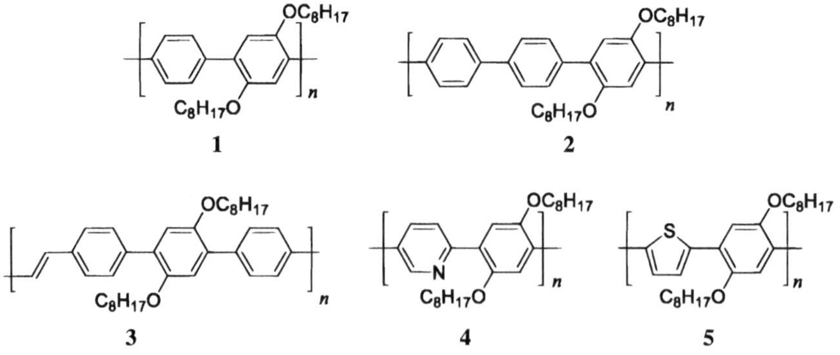

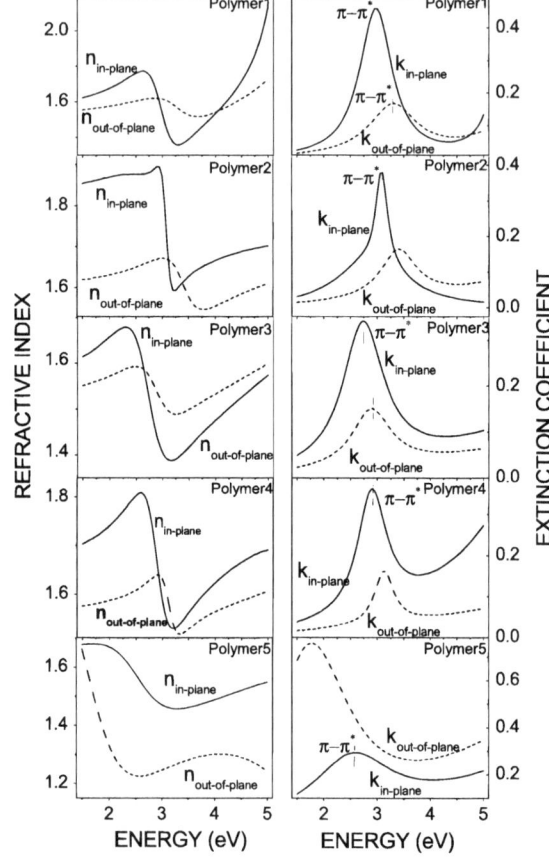

**Fig. 24.2** Poly(arylenephenylene) polymers investigated for their anisotropic optical properties by Losurdo et al. (see Fig. 24.3). Reprinted with permission from M. Losurdo, M.M. Giangregorio, P. Capezzuto, G. Bruno, F. Babudri, D. Colangiuli, G.M. Farinola, F. Naso, Macromolecules **36**, 4492 (2003). ©2003, American Chemical Society

**Fig. 24.3** In-plane and out-of-plane refractive indices $n$ and extinction coefficients $k$ for thin films of poly(arylenephenylene) polymers 1–5 (see Fig. 24.2) spin coated on glass substrates. Reprinted with permission from M. Losurdo, M.M. Giangregorio, P. Capezzuto, G. Bruno, F. Babudri, D. Colangiuli, G.M. Farinola, F. Naso, Macromolecules **36**, 4492 (2003). ©2003, American Chemical Society

## PEDOT:PSS [poly(3,4-ethylenedioxythiophene)–poly(4-styrenesulfonate)]

**Fig. 24.4** Uniaxial anisotropic refractive indices $n_j$ and extinction coefficients $k_j$ parallel (ordinary, $j = \parallel$) and perpendicular (extraordinary, $j = \perp$) to the surface plane of thin PEDOT-PSS films, determined from multiple samples. Reprinted from L.A.A. Petterson, S. Ghosh, O. Inganäs, Optical anisotropy in thin films of poly(3,4-ethylenedioxythiophene)—poly(4-styrenesulfonate), Org. Electron. **3**, 143–148 (2002). ©2002, with permission from Elsevier

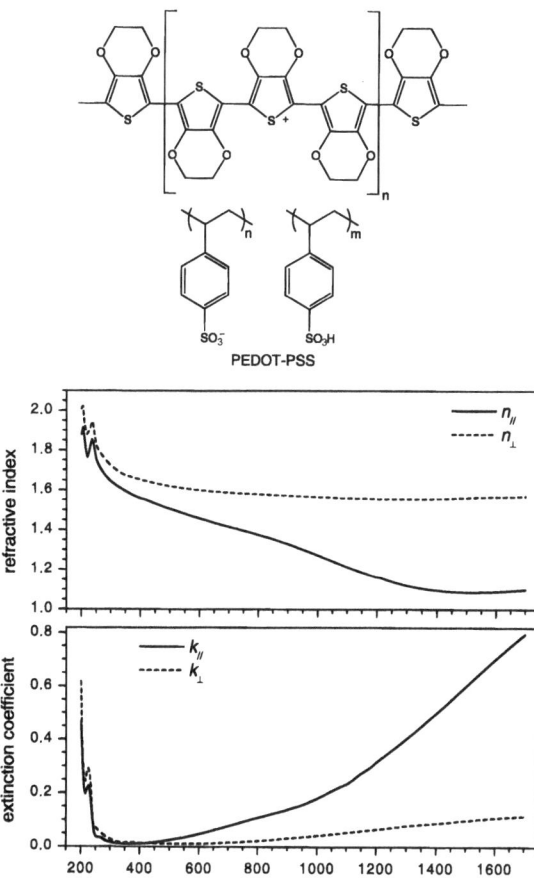

## PFO [poly(9,9-dioctylfluorene)]

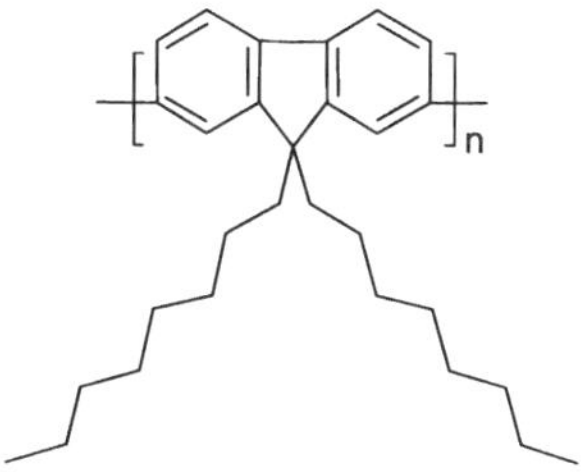

**Fig. 24.5** Chemical structure of PFO [poly(9,9-di-octyl-fluorene)]

**Fig. 24.6** Ordinary (*solid line*) and extraordinary (*dashed line*) complex refractive indices $N = n + i\kappa$ for spin coated PFO, determined with interference-enhancement variable angle spectroscopic ellipsometry. With kind permission from M. Campoy-Quiles, P.G. Etchegoin, D.D.C. Bradley, Synth. Met. **155**, 279–282 (2005). ©2005, Elsevier B.V

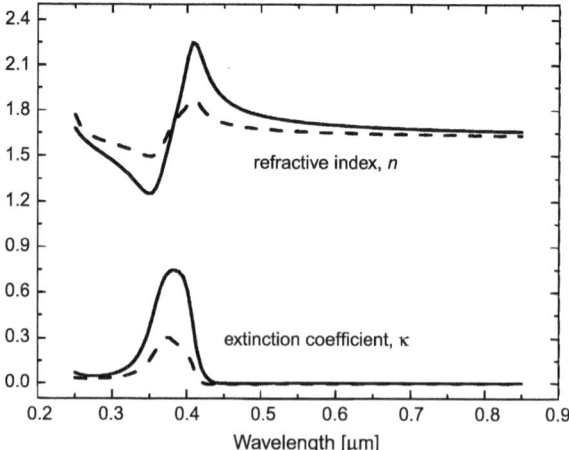

**Polyfluorenes F8BT, PFB, and TFB [poly(9,9′-dioctylfluorene-*co*-benzothiadiazole), poly(9,9′-dioctylfluorene-*co*-bis-N,N′-(4-butylphenyl)-bis-N,N′-phenyl-1,4-phenylenediamine), and poly(9,9′-dioctylfluorene-*co*-bis-N,N′-(4-butylphenyl)diphenylamine)]**

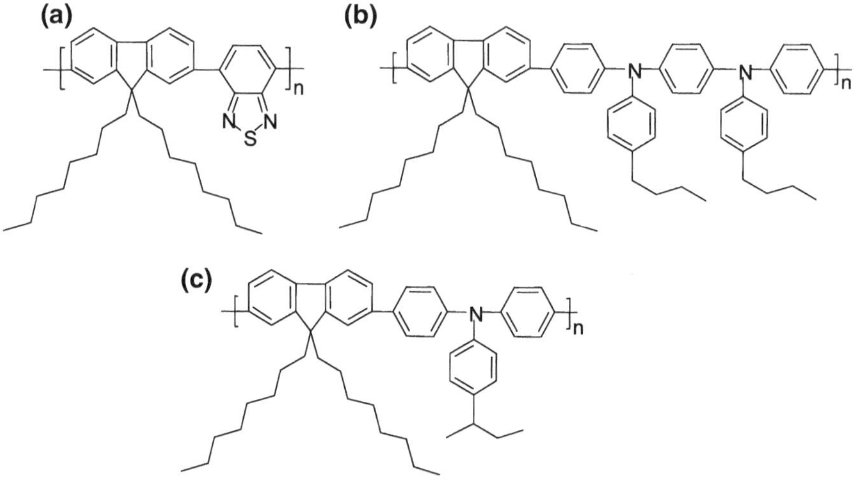

**Fig. 24.7** Chemical structures of (**a**) F8BT, (**b**) PFB, and (**c**) TFB. These polymers were investigated for their anisotropic optical properties by Ramsdale et al. (see Fig. 24.8). With kind permission from C.M. Ramsdale, N.C. Greenham, J. Phys. D, Appl. Phys. **36**, L29 (2003). ©2003 IOP Publishing Ltd.

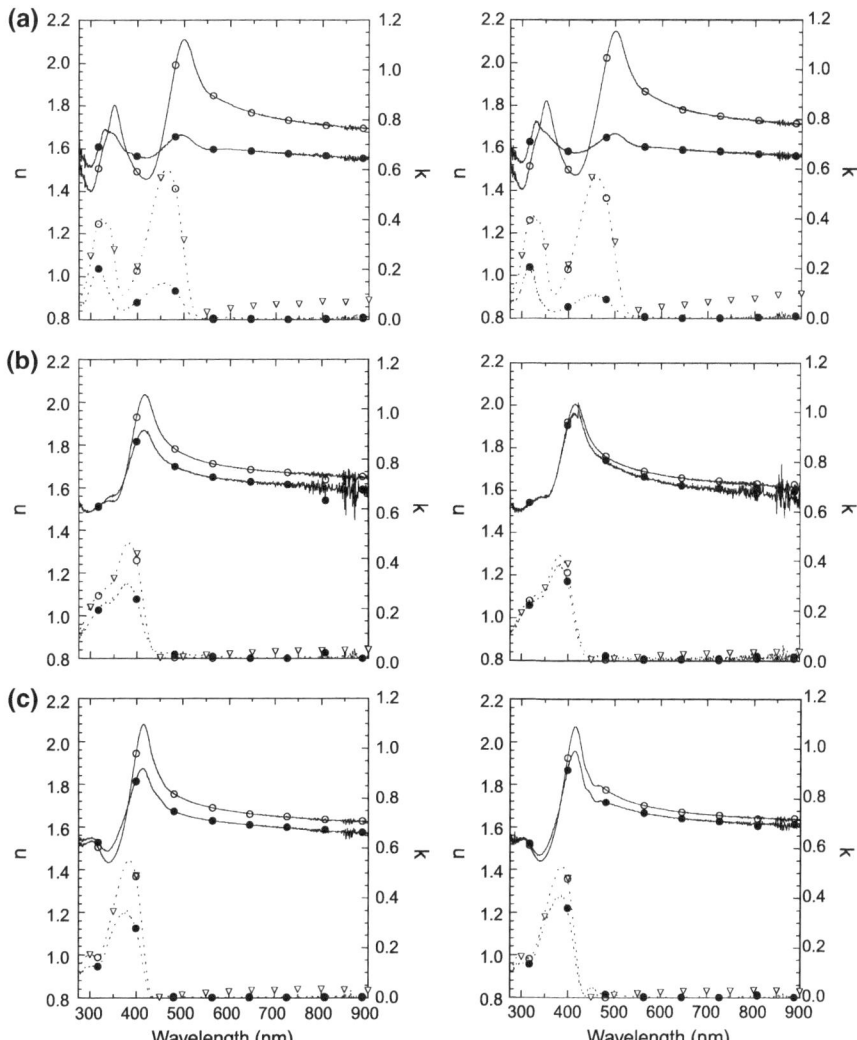

**Fig. 24.8** The ordinary (○) and extraordinary (●) values for the refractive indices *n* (*solid lines*) and extinction coefficients *k* (*dashed lines*) for unannealed (*left*) and annealed (*right*) spincoated films of (**a**) F8BT, (**b**) PFB, and (**c**) TFB (see Fig. 24.7) on quartz substrates. The ordinary extinction coefficients are extracted from UV—visible transmission measurements are also plotted (∇). With kind permission from C.M. Ramsdale, N.C. Greenham, J. Phys. D, Appl. Phys. **36**, L29 (2003). ©2003 IOP Publishing Ltd.

## TDAF [ter(9,9-diarylfluorene)]

**Fig. 24.9** *Top*: Chemical structures of the ter(9,9-diarylfluorene)s TDAF 1 and TDAF 2. *Bottom*: Ordinary and extraordinary refractive indices *n* and extinction coefficients *k* of TDAF 1 determined from multiple samples with film thicknesses between 30 and 200 nm. Reprinted with permission from H.-W. Lin, C.-L. Lin, H.-H. Chuang, Y.-T. Lin, C.-C. Wu et al., J. Appl. Phys. **95**(3), 881–886 (2004). ©2004, American Institute of Physics

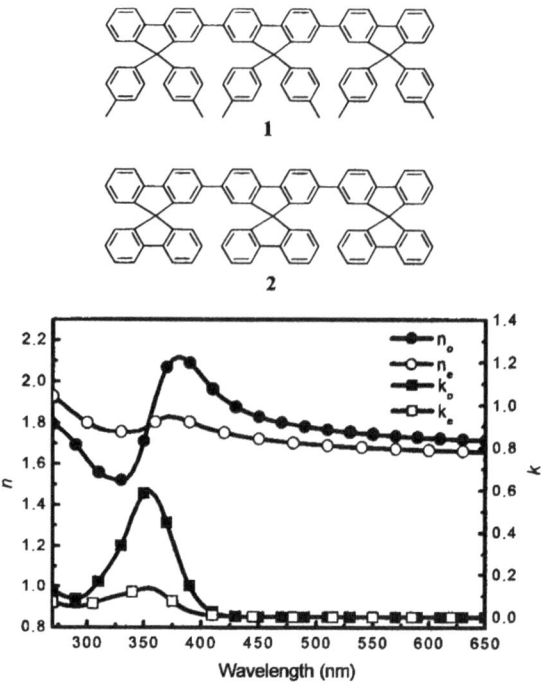

## Spiro-octo-1 and spiro-octo-2 [2,2′,4,4′,7,7′-hexaphenyl-9,9′-spirobifluorene and 2,2′,4,4′,7,7′-hexakis(biphenyl-4-yl)-9,9 ′-spirobifluorene]

**Fig. 24.10** Optical constants *n* (*top*) and *k* (*bottom*) of 133 nm thick spiro-octo-1 (*black*) and spiro-octo-2 (*gray*) films on silicon substrate. The films were deposited by organic molecular beam deposition. Data from S.D. Silaghi, T. Spehr, C. Cobet, T.P.I. Saragi, C. Werner, J. Salbeck, N. Esser, J. Appl. Phys. **103**, 043503 (2008)

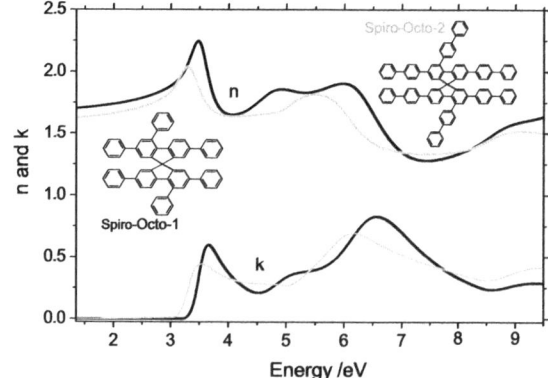

## 5,10,15,20-tetraphenyl porphyrins H$_2$TPP, CoTPP, NiTPP

**Fig. 24.11**  Optical constants *n* (*black*) and *k* (*gray*) of H$_2$TPP (33 nm), CoTPP (38 nm), and NiTPP (30 nm) films on silicon substrate. The films were deposited by organic molecular beam deposition in a vacuum chamber with a base pressure of about $10^{-8}$ mbar and evaporation temperatures between 200 and 225 °C. Data from Pop et al. (S. Pop, P. Kate, N. Esser, K. Hinrichs, Tunable optical constants of thermally grown thin porphyrin films on silicon for photovoltaic applications, in preparation (2013))

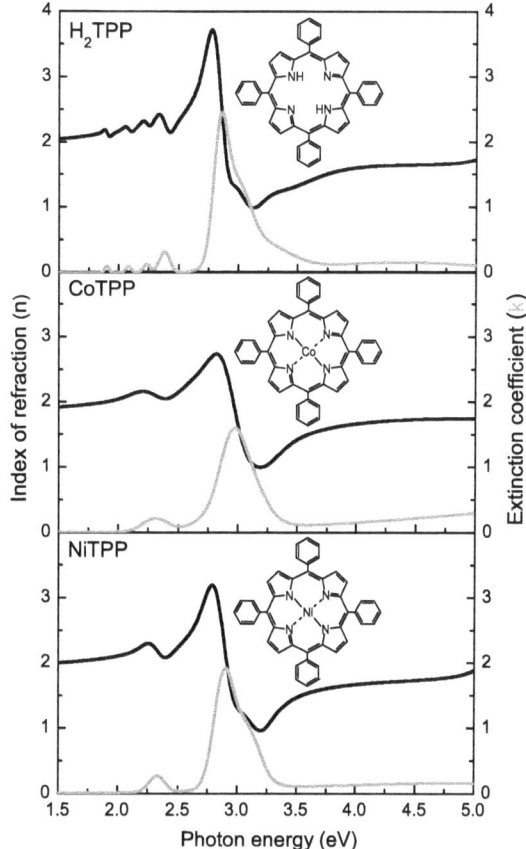

## Guanine [2-amino-1*H*-purin-6(9*H*)-one]

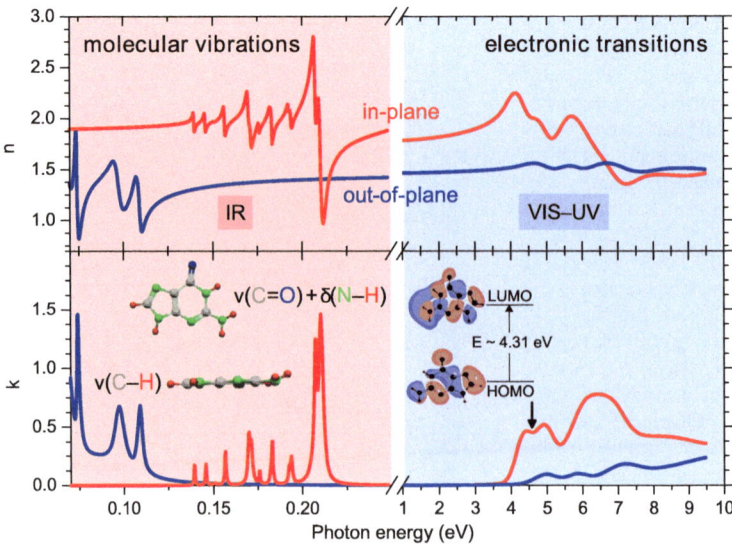

**Fig. 24.12** Mid-infrared and VIS-UV in-plane and out-of-plane optical constants *n* and *k* of guanine films prepared by organic molecular beam deposition. Data from K. Hinrichs, S.D. Silaghi, C. Cobet, N. Esser, D.R.T. Zahn, Phys. Status Solidi B **242**(13), 2681 (2005)

## Ormostamp

**Fig. 24.13** Optical constants of 155 nm Ormostamp on silicon substrate determined in the framework of NIM_NIL, funded by the European Community's 7th Framework Programme under grant agreement no. 228637, in cooperation between Profactor GmbH, SENTECH Instruments GmbH, and Micro Resist Technology

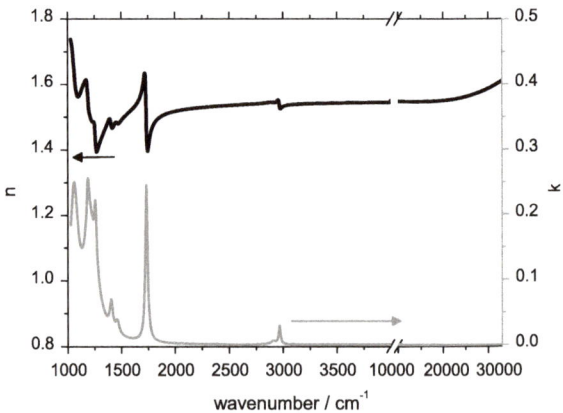

# Index

Printed by Printforce, the Netherlands